AF520701

Springer Proceedings in Mathematics & Statistics

Volume 143

Springer Proceedings in Mathematics & Statistics

This book series features volumes composed of selected contributions from workshops and conferences in all areas of current research in mathematics and statistics, including operation research and optimization. In addition to an overall evaluation of the interest, scientific quality, and timeliness of each proposal at the hands of the publisher, individual contributions are all refereed to the high quality standards of leading journals in the field. Thus, this series provides the research community with well-edited, authoritative reports on developments in the most exciting areas of mathematical and statistical research today.

More information about this series at http://www.springer.com/series/10533

P.N. Agrawal · R.N. Mohapatra
Uaday Singh · H.M. Srivastava
Editors

Mathematical Analysis and its Applications

Roorkee, India, December 2014

Editors
P.N. Agrawal
Department of Mathematics
Indian Institute of Technology Roorkee
Roorkee, Uttarakhand
India

Uaday Singh
Department of Mathematics
Indian Institute of Technology Roorkee
Roorkee, Uttarakhand
India

R.N. Mohapatra
Department of Mathematics
University of Central Florida
Orlando, FL
USA

H.M. Srivastava
Department of Mathematics and Statistics
University of Victoria
Victoria, BC
Canada

ISSN 2194-1009 ISSN 2194-1017 (electronic)
Springer Proceedings in Mathematics & Statistics
ISBN 978-81-322-2484-6 ISBN 978-81-322-2485-3 (eBook)
DOI 10.1007/978-81-322-2485-3

Library of Congress Control Number: 2015942480

Mathematics Subject Classification: 34H05, 41A25, 41A36, 65M12, 68T10, 70J25, 76M10

Springer New Delhi Heidelberg New York Dordrecht London

Printed on acid-free paper

Springer (India) Pvt. Ltd. is part of Springer Science+Business Media
(www.springer.com)

Preface

Welcome to the Proceedings of the International Conference on Recent Trends in Mathematical Analysis and its Applications 2014 (ICRTMAA-2014). With a view to bringing together experts and young researchers working on various areas of Mathematical Analysis and its Applications, the Department of Mathematics at the Indian Institute of Technology, Roorkee, decided to organize this international conference from December 21 to 23, 2014. We invited many mathematicians of repute to deliver invited lectures on topics of their interest and also young researchers to present their research during the conference. The conference provided an opportunity to exchange ideas and share cutting-edge research.

The conference had invited speakers and participants from United States of America, Germany, South Africa, the Sultanate of Oman, and reputed institutions from India. When we called for papers, we received 94 papers. Each of these was sent to two carefully chosen referees, who went through the papers and gave their recommendations on whether the paper should be included in the proceedings of the conference. Based on the recommendations, we selected 60 papers for inclusion in the proceedings.

Mathematical Analysis is an interesting subject which has applications in many different areas of pure and applied mathematics. The papers included in this volume demonstrate the versatility and inherent beauty of Analysis. The areas included are Compressive sensing, Approximation Theory, Solitons and nonlinear waves, Galerkin methods, singularly perturbed differential and difference equations, fractal interpolations and surface fitting, equilibrium problems, optimization using various techniques, different methods to approach pricing European options, Fractional, functional and other types of differential equations, stochastic integro-differential equations, mathematical models from Biology, wavelet frames, Frames in semi-inner product spaces, and Frames in Hilbert C* module.

The main theme in each of these papers is to employ analytical techniques to solve the problems at hand and wherever possible to find their numerical solutions. The large collection of papers in Approximation theory shows the use of operators, q-integers, and splines. The authors of these papers have carefully described the problem and showed appropriate methods to obtain the solution. For a problem in mathematical biology one may need to validate the model by showing that the model is a realistic representation of the problem, and its solution may yield valuable insights. In these proceedings, the papers have been classified into two parts namely, "Mathematical Analysis" and "Applications."

The trend in research is constantly changing due to developments of new tools, and a conference of this type is a valuable means for knowledge transfer. This volume which contains the research papers and a few carefully selected survey papers will provide the readers with an opportunity to see how mathematical analysis can be applied in various contexts to solve problems.

As the Conference Chairs of ICRTMAA 2014, we are thankful to all the funding agencies for their grant support for the successful completion of this international conference. The conference was supported partially by the following funding agencies:

1. Department of Science and Technology (DST), Government of India, New Delhi
2. Uttarakhand State Council for Science and Technology (UCOST), India
3. Quality Improvement Programme Centre, Indian Institute of Technology (IIT) Roorkee, India
4. International Society for Analysis, Applications and Computation (ISAAC)

We would like to express our thanks to Prof. Pradipta Banerji, Director IIT Roorkee, Roorkee India, for his constant encouragement, motivation, and support. We also extend our profound thanks to all the authorities of IIT Roorkee as well as the faculty members and research scholars of the Department of Mathematics, IIT Roorkee, Roorkee.

We are grateful to the Chairs and Members of the Screening Committee, Registration Committee, Publication Committee, Academic Programme Committee, Catering Committee, Cultural Committee, Finance Committee, and Advisory Committee which worked as a team by investing their invaluable time and hard work to make this event a success.

We extend our hearty thanks to the keynote speakers who kindly accepted our invitation. Especially, we would like to thank the following experts:

1. Prof. R.A. Zalik, Auburn University, USA
2. Prof. R.N. Mohapatra, University of Central Florida, USA
3. Prof. Zhiseng Shuai, University of Central Florida, USA
4. Prof. Margareta Heilmann, Germany
5. Prof. Elena Berdysheva, Muskat, Oman
6. Prof. M.C. Joshi, IIT Gandhinagar, India
7. Prof. R.K. Mohanty, South Asian University, India

A total of 71 subject experts from around the world contributed to the peer-review process. We express our sincere gratitude to the reviewers for spending their valuable time to review the papers and for sorting out the papers for presentation at the conference.

Our goal will be accomplished if the readers find this volume useful and informative for their future research. We are thankful to Springer for publishing the proceedings of the conference.

February 2015

P.N. Agrawal
R.N. Mohapatra
Uaday Singh
H.M. Srivastava

International Program Committee

Prof. R.P. Agarwal, Texas A&M University, USA
Prof. T. Amarnath, Central University of Hyderabad, India
Prof. D. Bahuguna, IIT Kanpur, India
Prof. Ismail Naci Cangul, Uludag University, Turkey
Prof. B.K. Dass, University of Delhi, India
Prof. Octay Duman, TOBB University, Turkey
Prof. Zoltan Finta, Babes-Bolyai University, Romania
Prof. Raju K. George, IIST, Thiruvananthapuram, India
Prof. N.K. Govil, Auburn University, USA
Prof. Vijay Gupta, NSIT New Delhi, India
Prof. Nurhayat Ispir, Gazi University, Turkey
Prof. G.P. Kapoor, IIT Kanpur, India
Prof. Harun Karsli, Abant Izzet Baysal University, Turkey
Prof. Imdad Khan, AMU, India
Prof. N.I. Mahmudov, Eastern Mediterranean University, Turkey
Prof. S.K. Mishra, BHU, Varanasi, India
Prof. R.N. Mohapatra, University of Central Florida, USA
Prof. A.J. Lopez Moreno, Universidad de Jaen, Spain
Prof. M. Mursaleen, AMU, India
Prof. M.T. Nair, IIT Madras, India
Prof. A.K. Nandakumaran, IISc Bangalore, India
Prof. A.K. Pani, IIT Bombay, India
Prof. R.K.S. Rathore, IIT Kanpur, India
Prof. H.M. Srivastava, University of Victoria, Canada
Prof. P.D. Srivastava, IIT Kharagpur, India
Prof. Richard A. Zalik, Auburn University, USA

Local Organizing Committee

Patron: Prof. Pradipta Banerji, Director IIT Roorkee
President: Prof. R.C. Mittal, Head, Department of Mathematics, IIT Roorkee
Conveners: Prof. P.N. Agrawal, Department of Mathematics, IIT Roorkee
Dr. Uaday Singh, Department of Mathematics, IIT Roorkee

Members:

Prof. S.P. Sharma, Prof. Rama Bhargava, Prof. V.K. Katiyar, Prof. N. Sukavanam, Prof. Kusum Deep, Prof. Tanuja Srivastava, Dr. A. Swaminathan, Dr. Sanjeev Kumar, Dr. D.N. Pandey

Sub-committees:

Academic Program Committee 1. Prof. Tanuja Srivastava 2. Dr. D.N. Pandey 3. Dr. Ankik Giri	Catering Committee 1. Prof. Rama Bhargava 2. Prof. V.K. Katiyar 3. Dr. D.N. Pandey
Publication Committee 1. Prof. R.C. Mittal 2. Prof. Rama Bhargava 3. Prof. N. Sukavanam 4. Dr. A. Swaminathan 5. Dr. P. Bera	Purchase Committee 1. Prof. Roshan Lal 2. Prof. P.N. Agrawal 3. Prof. Sunita Gakkhar 4. Dr. Sanjeev Kumar
Transport Committee 1. Dr. Sandip Banerjee 2. Dr. Uaday Singh 3. Dr. S.K. Gupta 4. Dr. Ram Jiwari	Accommodation Committee 1. Prof. V.K. Katiyar 2. Prof. M.L. Mittal 3. Prof. Kusum Deep 4. Dr. R.K. Pandey
Cultural Program Committee 1. Prof. Kusum Deep 2. Dr. Madhu Jain 3. Dr. Pratibha 4. Dr. A.K. Nayak	Registration Committee 1. Prof. S.P. Sharma 2. Prof. S.P. Yadav 3. Dr. Maheshanand 4. Dr. Aditi Gangopadhyay
Finance Committee 1. Prof. S.P. Sharma 2. Prof. P.N. Agrawal	

Contents

Part II Applications

Editors and Contributors

About the Editors

P.N. Agrawal is Professor at the Department of Mathematics, Indian Institute of Technology Roorkee, India. He obtained his Ph.D. degree from Indian Institute of Technology Kanpur, India, in 1980. Professor Agrawal has 110 research papers to his credit published in several journals of repute as well as guided many post-graduate and Ph.D. students. He has presented research papers at several international conferences in India and abroad and also delivered invited lectures.

Ram N. Mohapatra is professor of Mathematics, University of Central Florida, Orlando, USA. He received his Ph.D. degree from the University of Jabalpur, India, in 1968. Earlier, he taught at Sambalpur University in India, American University in Beirut, Lebanon, University of Alberta, and York University, Canada, prior to coming to Orlando. His area of research is Mathematical Analysis and he is the author of two books, two edited monographs, and over 120 research papers to his credit. He referees articles for professional journals and serves as a member of the editorial board of a number of journals.

Uaday Singh is assistant professor at the Department of Mathematics, Indian Institute of Technology Roorkee, India. Earlier, Dr. Singh worked as lecturer at Banaras Hindu University, Varanasi. His areas of research are fourier approximation and summability theory. He has more than 20 publications to his credit that include research papers, monograph and book chapters. He is actively involved in teaching undergraduate and postgraduate students as well as Ph.D. students. He is a member of many professional societies such as American Mathematical Society (AMS), International Society for Augmentative and Alternative Communication (ISAAC), International Association of Engineers (IAENG), and Indian Mathematical Society (IMS).

H.M. Srivastava is professor emeritus at the Department of Mathematics and Statistics, University of Victoria, Canada. He joined the University of Victoria in 1969, first as associate professor (1969–1974), and then as full professor (1974–2006).

Professor Srivastava has held numerous visiting positions including (for example) those at West Virginia University in the USA (1967–1969), Universite Laval in Canada (1975), and the University of Glasgow in the UK (1975–1976), and indeed also at many other universities and research institutes in different parts of the world. Professor Srivastava has published 21 books (monographs and edited volumes), 30 book chapters, 45 papers in international conference proceedings, and over 1000 scientific research journal articles on various topics of mathematical analysis and applicable mathematics. In addition, he has written forewords to several books by other authors and to several special issues of scientific journals. He has also edited (and contributed to) many volumes, which are dedicated to the memories of famous mathematical scientists. Citations of his research contributions can be found in many books and monographs, Ph.D. and D.Sc. theses, and scientific journal articles, much too numerous to be recorded here. Currently, he is actively associated editorially (that is, as editor, honorary editor, senior editor, associate editor or editorial board member) with over 200 international scientific research journals. His biographical sketches (many of which are illustrated with his photograph) have appeared in various issues of more than 50 international biographies, directories, and Who's Who's.

Contributors

Syed Abbas School of Basic Sciences, Indian Institute of Technology Mandi, Mandi, Himachal Pradesh, India

P.N. Agrawal Department of Mathematics, Indian Institute of Technology Roorkee, Roorkee, India

Neha Ahlawat Department of Mathematics, Indian Institute of Technology Roorkee, Roorkee, India

Urvashi Arora Department of Mathematics, Indian Institute of Technology Roorkee, Roorkee, India

Ashish Awasthi Department of Mathematics, National Institute of Technology Calicut, Kozhikode, India

D. Bahuguna Department of Mathematics and Statistics, Indian Institute of Technology Kanpur, Kanpur, India

P. Balasubramaniam Gandhigram Rural Institute-Deemed University, Gandhigram, Tamil Nadu, India

Komal Bansal Department of Mathematics, Panjab University, Chandigarh, India

Ramandeep Behl School of Mathematics, Statistics and Computer Science, University of KwaZulu-Natal, Pietermaritzburg, South Africa

Elena E. Berdysheva Department of Mathematics and Science, German University of Technology in Oman, Muscat, Sultanate of Oman

Rama Bhargava Department of Mathematics, Indian Institute of Technology Roorkee, Roorkee, India

Alka Chadha Department of Mathematics, Indian Institute of Technology Roorkee, Roorkee, India

A.K.B. Chand Department of Mathematics, Indian Institute of Technology Madras, Chennai, India

S. Chandra Sekhara Rao Department of Mathematics, Indian Institute of Technology Delhi, Hauz Khas, New Delhi, India

Renu Chaudhary Department of Mathematics, Indian Institute of Technology Roorkee, Roorkee, Uttarakhand, India

Sheetal Chawla Department of Mathematics, Indian Institute of Technology Delhi, Hauz Khas, New Delhi, India

Jaydev Dabas Department of Applied Science and Engineering, IIT Roorkee, Saharanpur Campus, Saharanpur, India

Sumita Dahiya Department of Mathematics, IIT Roorkee, Roorkee, Uttarakhand, India

Amiya Das IIT Kharagpur, Kharagpur, India

Payel Das Mathematics Department, IIT Kharagpur, Kharagpur, India

Sanjukta Das Department of Mathematics, IIT Roorkee, Roorkee, India

P. Devaraj Department of Mathematics, College of Engineering Guindy Campus, Anna University, Chennai, India

Girish Dobhal University of Petroleum and Energy Studies, Dehradun, India

L. Jones Tarcius Doss College of Engineering Guindy, Anna University, Chennai, India

Ramu Dubey Department of Mathematics, Indian Institute of Technology Roorkee, Roorkee, India

Shruti Dubey Department of Mathematics, Indian Institute of Technology Madras, Chennai, India

Sharad Dwivedi Department of Mathematics, Indian Institute of Technology Madras, Chennai, India

Asha Ram Gairola University of Petroleum and Energy Studies, Dehradun, India

Asish Ganguly IIT Kharagpur, Kharagpur, India

Ganga Ram Gautam Department of Applied Science and Engineering, IIT Roorkee, Saharanpur Campus, Saharanpur, India

Mania Goyal Department of Mathematics, Indian Institute of Technology Roorkee, Roorkee, India

Meenu Goyal Department of Mathematics, Indian Institute of Technology Roorkee, Roorkee, India

Rangoli Goyal Department of Mathematics, Indian Institute of Technology Roorkee, Roorkee, India

S.K. Gupta Department of Mathematics, Indian Institute of Technology Roorkee, Roorkee, India

Vidushi Gupta Department of Applied Science and Engineering, IIT Roorkee, Saharanpur, India

Vijay Gupta Department of Mathematics, Netaji Subhas Institute of Technology, New Delhi, India

Margareta Heilmann Faculty of Mathematics and Natural Sciences, University of Wuppertal, Wuppertal, Germany

J.P. Jaiswal Department of Mathematics, Maulana Azad National Institute of Technology, Bhopal, Madhya Pradesh, India

S. Jayaraj Department of Mechanical Engineering, National Institute of Technology Calicut, Kozhikode, India

Arun Kajla Department of Mathematics, Indian Institute of Technology Roorkee, Roorkee, India

Munish Kansal University Institute of Engineering and Technology, Panjab University, Chandigarh, India

Shashi Kant Department of Applied Mathematics, Delhi Technological University, Delhi, India

V. Kanwar University Institute of Engineering and Technology, Panjab University, Chandigarh, India

S.K. Katiyar Department of Mathematics, Indian Institute of Technology Madras, Chennai, India

S.D. Kendre Department of Mathematics, Savitribai Phule Pune University, Pune (MS), India

Pradeep Kumar Department of Mathematics and Statistics, Indian Institute of Technology Kanpur, Kanpur, India

Vivek Kumar Department of Applied Mathematics, Delhi Technological University, Delhi, India

Roshan Lal Department of Mathematics, Indian Institute of Technology Roorkee, Roorkee, India

S.G. Latpate Department of Mathematics, Nowrosjee Wadia College of Arts and Science, Pune, India

N.K. Mahato Indian Institute of Information Technology, Design and Manufacturing Jabalpur, Jabalpur, India

R.A. Malekar National Defence Academy, Pune, India

Neha Mathur Department of Mathematics, Career Convent College, Lucknow, India

Pankaj Mathur Department of Mathematics and Astronomy, Lucknow University, Lucknow, India

Mani Mehra Indian Institute of Technology, Delhi, India

M.L. Mittal Department of Mathematics, Indian Institute of Technology Roorkee, Roorkee, India

R.C. Mittal Department of Mathematics, IIT Roorkee, Roorkee, Uttarakhand, India

R.N. Mohapatra Department of Mathematics, University of Central Florida, Orlando, FL, USA

S.S. Motsa School of Mathematics, Statistics and Computer Science, University of KwaZulu-Natal, Pietermaritzburg, South Africa

Mohd Nadeem Department of Applied Science and Engineering, IIT Roorkee, Saharanpur, India

R. Ramu Naidu Department of Mathematics, IIT Hyderabad, Hyderabad, India

Nageswara Rao Narni Rajiv Gandhi University of Knowledge Technologies, Hyderabad, Andhra Pradesh, India

Ameeya K. Nayak Department of Mathematics, Indian Institute of Technology Roorkee, Roorkee, India

Gnaneshwar Nelakanti Mathematics Department, IIT Kharagpur, Kharagpur, India

Edgard Ngounda Department of Mathematics and Applied Mathematics, University of the Western Cape, Bellville, South Africa

Dwijendra N. Pandey Department of Mathematics, Indian Institute of Technology Roorkee, Roorkee, India

Sabyasachi Pani IIT Bhubaneswar, Odisha, India

Gayatri Pany IIT Bhubaneswar, Odisha, India

Mamata Parida Department of Mathematics, Utkal University, Bhubaneswar, India

Kuldip Singh Patel Indian Institute of Technology Delhi, Delhi, India

Kailash C. Patidar Department of Mathematics and Applied Mathematics, University of the Western Cape, Bellville, South Africa

Donald Porchia Department of Mathematics, University of Central Florida, Orlando, FL, USA

S. Prabhakaran College of Engineering Guindy, Anna University, Chennai, India

B. Mayur Prakash Department of Mechanical Engineering, National Institute of Technology Calicut, Kozhikode, India

Pratibha Department of Mathematics, Indian Institute of Technology Roorkee, Roorkee, India

Pratima Rai Department of Mathematics, University of Delhi, Delhi, India

R. Rakkiyappan Department of Mathematics, Bharathiar University, Coimbatore, Tamil Nadu, India

Rashmi Rani Indian Institute of Technology Roorkee, Roorkee, India

Rajendra K. Ray School of Basic Sciences, Indian Institute of Technology, Mandi, Himachal Pradesh, India

Ioan Raşa Department of Mathematics, Technical University, Cluj-Napoca, Romania

K.M. Reddy Department of Mathematics, Indian Institute of Technology Madras, Chennai, India

N.K. Sahu Dhirubhai Ambani Institute of Information and Communication Technology, Gandhinagar, India

Renu Saini Department of Mathematics, Indian Institute of Technology Roorkee, Roorkee, India

Soshal Saini Department of Mathematics, Indian Institute of Technology Roorkee, Roorkee, India

T. Sathiyaraj Gandhigram Rural Institute-Deemed University, Gandhigram, Tamil Nadu, India

Kapil K. Sharma Department of Mathematics, South Asian University (SAU), Chanakyapuri, New Delhi, India

Zhisheng Shuai Department of Mathematics, University of Central Florida, Orlando, FL, USA

Anurag Shukla Department of Mathematics, Indian Institute of Technology Roorkee, Roorkee, India

Abdelmgid O.M. Sidahmed Department of Mathematics and Applied Mathematics, University of the Western Cape, Bellville, South Africa

Manjari Sidharth Department of Mathematics, Indian Institute of Technology Roorkee, Roorkee, India

Anuradha Singh Department of Mathematics, Maulana Azad National Institute of Technology, Bhopal, Madhya Pradesh, India

Karunesh Kumar Singh Government Polytechnic, Rampur, India

Mradul Veer Singh Department of Mathematics, Indian Institute of Technology Roorkee, Roorkee, India

Uaday Singh Department of Mathematics, Indian Institute of Technology Roorkee, Roorkee, India

Smita Sonker Department of Mathematics, National Institute of Technology Kurukshetra, Kurukshetra, India

M. Srinivas Department of Computer Science and Engineering, IIT Hyderabad, Hyderabad, India

N. Sukavanam Department of Mathematics, Indian Institute of Technology Roorkee, Roorkee, India

P. Tamilalagan Department of Mathematics, Gandhigram Rural Institute—Deemed University, Gandhigram, Tamil Nadu, India

Jai Prakash Tripathi School of Basic Sciences, Indian Institute of Technology Mandi, Mandi, Himachal Pradesh, India

K.R. Tyada Department of Mathematics, Indian Institute of Technology Madras, Chennai, India

Swati Tyagi School of Basic Sciences, Indian Institute of Technology, Mandi, Himachal Pradesh, India

G. Velmurugan Department of Mathematics, Bharathiar University, Coimbatore, Tamil Nadu, India

P. Viswanathan Department of Mathematics, Indian Institute of Technology Madras, Chennai, India

S. Yugesh Department of Mathematics, College of Engineering Guindy Campus, Anna University, Chennai, India

Part I
Mathematical Analysis

Convergence Analysis of Legendre Spectral Galerkin Method for Volterra-Fredholm-Hammerstein Integral Equations

Payel Das and Gnaneshwar Nelakanti

Abstract In this paper, we analyze the Legendre spectral Galerkin method for a class of nonlinear Volterra-Fredholm mixed-type integral equations. Existence and convergence of the approximate and iterated approximate solutions to the exact solution are discussed and the rates of convergence are obtained. We prove that the iterated approximate solution improves over the approximate solution for Volterra-Fredholm-Hammerstein integral equations with smooth kernels. Also, we obtain optimal order of convergence for the iterated Legendre Galerkin method.

Keywords Volterra-Fredholm-Hammerstein integral equation · Spectral method · Legendre Galerkin · Convergence rates

1 Introduction

In this section, we consider the following Volterra-Fredholm-Hammerstein integral equations

$$x(t) - \int_{-1}^{t} k_1(t,s)\psi_1(s,x(s))ds - \int_{-1}^{1} k_2(t,s)\psi_2(s,x(s))ds = f(t), \quad -1 \le t \le 1, \tag{1}$$

where k_1, k_2, f, ψ_1 and ψ_2 are known functions and x is the unknown solution to be found in a Banach space $\mathbb{X}$. The existence and uniqueness of the solution of Eq. (1) are discussed in [1, 2].

The Volterra-Fredholm integral equation of type (1) arises from parabolic boundary value problems and in various other physical and biological models (see [3–6]). The essential features of these models are of wide applicability.

Several numerical methods for approximating the solution of nonlinear integral equations of type (1) are available in the literature. Classical projection methods

P. Das (✉) · G. Nelakanti
Mathematics Department, IIT Kharagpur, Kharagpur, India
e-mail: dasPayel2@gmail.com

G. Nelakanti
e-mail: gnanesh@maths.iitkgp.ernet.in

P.N. Agrawal et al. (eds.), *Mathematical Analysis and its Applications*,
Springer Proceedings in Mathematics & Statistics 143,
DOI 10.1007/978-81-322-2485-3_1

such as spline-based collocation and Galerkin methods are applied to investigate the approximate solutions of nonlinear Volterra-Fredholm integral equations of type (1) (see [2, 7–9]). However, it is necessary to increase the number of partitioning points, to obtain more accuracy in spline-based projection methods. This leads to solve a large system of nonlinear equations, which is computationally very expensive. Since use of global polynomials implies smaller nonlinear system, we will use orthogonal projection method using global polynomial basis functions to overcome the difficulties in terms of computational work encountered in the existing techniques.

In this paper, we apply Galerkin method to the Eq. (1) using Legendre polynomial basis functions, which can be generated recursively with ease and possess nice property of orthogonality. We prove that the approximated solution of the Legendre Galerkin method converges to the exact solution with the order $\mathcal{O}(n^{\frac{1}{2}-r})$ in infinity norm, and the iterated Legendre Galerkin solution converges with the order $\mathcal{O}(n^{-r})$ in infinity norm, n being the highest degree of Legendre polynomial employed in the approximation and r being the smoothness of the kernels.

We organize this paper as follows. In Sect. 2, we discuss the Legendre spectral Galerkin method for the equation of type (1). In Sect. 3, we obtain the existence and convergence results for the approximate and iterated approximate solutions. Throughout this paper, we assume that c is a generic constant.

2 Legendre Spectral Galerkin Method: Volterra-Fredholm-Hammerstein Integral Equations

In this section, we describe the Galerkin method for approximating the solution of Volterra-Fredholm mixed type Hammerstein integral equations using Legendre polynomial basis functions.

Let $\mathbb{X} = C[-1, 1]$ and consider the following Volterra-Fredholm-Hammerstein integral equation:

$$x(t) - \int_{-1}^{t} k_1(t, s)\psi_1(s, x(s))\, ds - \int_{-1}^{1} k_2(t, s)\psi_2(s, x(s))\, ds = f(t), \quad -1 \le t \le 1, \tag{2}$$

where k_1, k_2, f, ψ_1 and ψ_2 are known functions and x is the unknown function to be determined.

Throughout the paper, the following assumptions are made on $f, k_1(.,.), k_2(.,.)$, $\psi_1(., x(.))$ and $\psi_2(., x(.))$:

(i) $f \in C[-1, 1]$.
(ii) $\lim_{t \to t'} \|k_i(t, .) - k_i(t', .)\|_\infty = 0, \quad t, t' \in [-1, 1], \; i = 1, 2.$

(iii) For $i = 1, 2$, $k_i(.,.) \in \mathcal{C}^r([-1,1] \times [-1,1])$, $r \geq 1$. Let $M_i = \|k_i\|_\infty = \sup_{t,s\in[-1,1]} |k_i(t,s)| < \infty$, $i = 1, 2$, and

$$\|k_l\|_{r,\infty} = \max_{\substack{0\leq i,j\leq r \\ t,s\in[-1,1]}} \left| \frac{\partial^{i+j}}{\partial t^i \partial s^j} k_l(t,s) \right|, \; l = 1, 2.$$

(iv) The nonlinear functions $\psi_i(., x(.))$ are continuous in $s \in [-1, 1]$ and are Lipschitz continuous in x, i.e., for any $x_1, x_2 \in \mathbb{R}$, there exist constants $l_i > 0$, $i = 1, 2$ such that

$$|\psi_i(s, x_1) - \psi_i(s, x_2)| \leq l_i |x_1 - x_2|, \; \forall s \in [-1, 1].$$

(v) The partial derivatives $\psi_i^{(0,1)}(s, x)$ of $\psi_i(s, x)$ with respect to the second variable exist and $\psi_i^{(0,1)}(., x(.)) \in \mathcal{C}([-1, 1] \times \mathbb{R})$. The functions $\psi_i^{(0,1)}(., x(.))$ are Lipschitz continuous in x, i.e., for any $x_1, x_2 \in \mathbb{R}$, there exist constants $c_i > 0$, $i = 1, 2$ such that

$$|\psi_i^{(0,1)}(s, x_1) - \psi_i^{(0,1)}(s, x_2)| \leq c_i |x_1 - x_2|, \; \forall s \in [-1, 1].$$

For convenience, we define the operators $\mathcal{K}_1, \mathcal{K}_2$ on $\mathbb{X}$ as

$$(\mathcal{K}_1\psi_1)(x)(t) = \int_{-1}^{t} k_1(t,s)\psi_1(s, x(s))\, ds \; ; \; (\mathcal{K}_2\psi_2)(x)(t) = \int_{-1}^{1} k_2(t,s)\psi_2(s, x(s))\, ds.$$

Using $\mathcal{K}_1$ and $\mathcal{K}_2$, Eq. (2) can be written as

$$x - \mathcal{K}_1\psi_1(x) - \mathcal{K}_2\psi_2(x) = f. \tag{3}$$

Letting $\mathcal{T}(x) := f + \mathcal{K}_1\psi_1(x) + \mathcal{K}_2\psi_2(x)$, $x \in \mathbb{X}$, Eq. (3) takes the form

$$x = \mathcal{T}x. \tag{4}$$

Using similar technique given in Theorem 2.4 of [10], it can be easily proved that $\mathcal{T}$ has a unique fixed point in $\mathbb{X}$. We assume x_0 to be a isolated solution of Eq. (4) in $\mathbb{X}$. We denote $d_i = \sup_{s\in[-1,1]} |\psi_i^{(0,1)}(s, x_0(s))|$.

Using Leibniz rule, we have

$$\|[(\mathcal{K}_1\psi_1)'(x_0)x]^{(1)}\|_\infty$$

$$= \sup_{t\in[-1,1]} \left| \frac{\partial}{\partial t} \int_{-1}^{t} k_1(t,s)\psi_1^{(0,1)}(s, x_0(s))x(s)ds \right|$$

$$
\begin{aligned}
&\leq \sup_{t\in[-1,1]}\left[|k_1(t,t)||\psi^{(0,1)}(t,x_0(t))||x(t)| + \left|\int_{-1}^{t}\left\{\frac{\partial}{\partial t}k_1(t,s)\right\}\psi_1^{(0,1)}(s,x_0(s))x(s)ds\right|\right]\\
&\leq \sup_{t\in[-1,1]}\left[|k_1(t,t)||\psi_1^{(0,1)}(t,x_0(t))||x(t)|\right]\\
&+ \sup_{t,s\in[-1,1]}\left|\frac{\partial}{\partial t}k_1(t,s)\right| \sup_{s\in[-1,1]}\left[|\psi_1^{(0,1)}(s,x_0(s))||x(s)|\right]\int_{-1}^{t}ds\\
&\leq M_1 d_1\|x\|_\infty + 2\|k_1\|_{1,\infty}d_1\|x\|_\infty < \infty.
\end{aligned} \tag{5}
$$

And for $j = 0, 1, 2, ..., r$, we have

$$
\begin{aligned}
\|[(\mathcal{K}_2\psi_2)'(x_0)x]^{(j)}\|_\infty &= \sup_{t\in[-1,1]}\left|\frac{\partial^j}{\partial t^j}\int_{-1}^{1}k_2(t,s)\psi_2^{(0,1)}(s,x_0(s))x(s)ds\right|\\
&\leq 2\sup_{t,s\in[-1,1]}\left|\frac{\partial^j}{\partial t^j}k_2(t,s)\right| \sup_{s\in[-1,1]}\left[|\psi_2^{(0,1)}(s,x_0(s))||x(s)|\right]\\
&\leq 2\|k_2\|_{j,\infty}d_2\|x\|_\infty < \infty.
\end{aligned} \tag{6}
$$

Next we will apply Legendre Galerkin method to the Eq. (2). To do this, we let $\mathbb{X}_n = \text{span}\{\phi_0, \phi_1, \phi_2, \ldots, \phi_n\}$ be the sequence of Legendre polynomial subspaces of $\mathbb{X}$ of degree $\leq n$, where $\{\phi_0, \phi_1, \phi_2, \ldots, \phi_n\}$ forms an orthonormal set and ϕ_i's are given by

$$
\phi_i(s) = \sqrt{\frac{2i+1}{2}}L_i(s), \quad i = 0, 1, ..., n, \tag{7}
$$

where L_i's are the Legendre polynomials of degree $\leq i$. These Legendre polynomials can be generated by the following three-term recurrence relation:

$$
L_0(s) = 1, L_1(s) = s, \quad s \in [-1, 1], \tag{8}
$$

and for $i = 1, 2, \ldots, n-1$

$$
(i+1)L_{i+1}(s) = (2i+1)sL_i(s) - iL_{i-1}(s), \quad s \in [-1, 1]. \tag{9}
$$

Orthogonal Projection Operator: Let $\mathbb{X} = C[-1, 1]$ and let the operator $\mathcal{P}_n : \mathbb{X} \to \mathbb{X}_n$ be the orthogonal projection defined as

$$
\mathcal{P}_n x = \sum_{j=0}^{n}\langle x, \phi_j\rangle\phi_j, \quad x \in \mathbb{X}, \tag{10}
$$

where $\langle x, \phi_j\rangle = \int_{-1}^{1} x(t)\phi_j(t)dt$.

We quote some crucial properties of $\mathcal{P}_n$ from Canuto et al. [11] (pp. 283–287).

Lemma 1 *Let $\mathcal{P}_n : \mathbb{X} \to \mathbb{X}_n$ denote the orthogonal projection defined by (10). Then the projection $\mathcal{P}_n$ satisfies the following properties:*

(i) $\{\mathcal{P}_n : n \in \mathbb{N}\}$ is uniformly bounded in L^2-norm.

(ii) There exists a constant $c > 0$ such that for any $n \in \mathbb{N}$ and $u \in \mathbb{X}$,

$$\|\mathcal{P}_n u - u\|_{L^2} \le c \inf_{\phi \in \mathbb{X}_n} \|u - \phi\|_{L^2} \to 0, \text{ as } n \to \infty. \tag{11}$$

Hence there exists a constant $p_1 \ge 0$, independent of n, such that

$$\|\mathcal{P}_n u\|_{L^2} \le p_1 \|u\|_\infty, \; u \in \mathbb{X}. \tag{12}$$

Lemma 2 *Let $\mathcal{P}_n$ be the orthogonal projection defined as (10). Then for any $u \in C^r[-1, 1]$, there hold*

$$\|u - \mathcal{P}_n u\|_{L^2} \le cn^{-r}\|u^{(r)}\|_{L^2}, \tag{13}$$

$$\|u - \mathcal{P}_n u\|_\infty \le cn^{\frac{3}{4}-r}\|u^{(r)}\|_{L^2}, \tag{14}$$

$$\|u - \mathcal{P}_n u\|_\infty \le cn^{\frac{1}{2}-r} V(u^{(r)}), \tag{15}$$

where c is a constant independent of n and $V(u^{(r)})$ denotes the total variation of $u^{(r)}$.

The Legendre spectral Galerkin method for Eq. (3) is seeking an approximate solution $x_n \in \mathbb{X}_n$, such that

$$x_n - \mathcal{P}_n \mathcal{K}_1 \psi_1(x_n) - \mathcal{P}_n \mathcal{K}_2 \psi_2(x_n) = \mathcal{P}_n f, \tag{16}$$

where $\mathcal{P}_n : \mathbb{X} \to \mathbb{X}_n$ is the orthogonal projection operator.

Let $\mathcal{T}_n$ be the operator defined as

$$\mathcal{T}_n(u) := \mathcal{P}_n f + \mathcal{P}_n \mathcal{K}_1 \psi_1(u) + \mathcal{P}_n \mathcal{K}_2 \psi_2(u), \; u \in \mathbb{X}. \tag{17}$$

Then the Eq. (16) can be written as $x_n = \mathcal{T}_n x_n$.

To obtain greater accuracy in approximate solution, we further consider the iterated Legendre Galerkin method for (3). To this end, we define the iterated solution as

$$\tilde{x}_n = f + \mathcal{K}_1 \psi_1(x_n) + \mathcal{K}_2 \psi_2(x_n). \tag{18}$$

Since $\mathcal{P}_n \tilde{x}_n = x_n$, it follows that the iterated approximate solution $\tilde{x}_n$ satisfies

$$\tilde{x}_n = f + \mathcal{K}_1 \psi_1(\mathcal{P}_n \tilde{x}_n) + \mathcal{K}_2 \psi_2(\mathcal{P}_n \tilde{x}_n). \tag{19}$$

Letting $\widetilde{\mathcal{T}}_n(u) := f + \mathcal{K}_1 \psi_1(\mathcal{P}_n u) + \mathcal{K}_2 \psi_2(\mathcal{P}_n u), u \in \mathbb{X}$, the Eq. (19) can be written as $\tilde{x}_n = \widetilde{\mathcal{T}}_n \tilde{x}_n$.

3 Convergence Results

In this section, we analyze the existence and convergence of the approximate solutions in the Legendre Spectral Galerkin method for the Eq. (2). To do this, we will use the well-known Vainikko's Theorem (Theorem 2 in [12]). We first give the following lemma, which we need for our analysis.

Lemma 3 *For any $x, y \in \mathbb{X}$, the following hold for $i = 1, 2$,*

$$\|(\mathcal{K}_i\psi_i)(x_0) - (\mathcal{K}_i\psi_i)(x)\|_\infty \leq \sqrt{2}M_i l_i \|x_0 - x\|_{L^2}, \tag{20}$$

$$\|[(\mathcal{K}_i\psi_i)'(x_0) - (\mathcal{K}_i\psi_i)'(x)]y\|_\infty \leq M_i c_i \|x_0 - x\|_{L^2}\|y\|_{L^2}. \tag{21}$$

Proof Using Lipschitz continuity of $\psi_i(., x(.))$, $\psi_i^{(0,1)}(., x(.))$ and Cauchy-Schwarz inequality, the proof of the above lemma follows.

In the following theorem, we give the error bounds for the approximate solution x_n to x_0.

Theorem 1 *Let $x_0 \in C^r[-1, 1]$, $r \geq 1$, be an isolated solution of the Eq. (3). Assume that 1 is not an eigenvalue of the linear operator $\mathcal{T}'(x_0)$, where $\mathcal{T}'(x_0)$ denotes the Frechet derivative of $\mathcal{T}(x)$ at x_0. Then the Eq. (16) has a unique solution $x_n \in B(x_0, \delta) = \{x : \|x - x_0\|_\infty < \delta\}$ for some $\delta > 0$ and for sufficiently large n. Moreover, there exists a constant $0 < q < 1$, independent of n such that*

$$\frac{\alpha_n}{1+q} \leq \|x_n - x_0\|_\infty \leq \frac{\alpha_n}{1-q},$$

where $\alpha_n = \|(\mathcal{I} - \mathcal{T}_n{}'(x_0))^{-1}(\mathcal{T}_n(x_0) - \mathcal{T}(x_0))\|_\infty$. Further, we obtain

$$\|x_n - x_0\|_\infty = \mathcal{O}\left(n^{\frac{1}{2}-r}\right).$$

Proof Using estimates (5), (6) and (14), we have
$\|[\mathcal{T}_n{}'(x_0) - \mathcal{T}'(x_0)]x\|_\infty$

$$\begin{aligned} &= \|[\mathcal{P}_n(\mathcal{K}_1\psi_1)'(x_0) + \mathcal{P}_n(\mathcal{K}_2\psi_2)'(x_0) - (\mathcal{K}_1\psi_1)'(x_0) - (\mathcal{K}_2\psi_2)'(x_0)]x\|_\infty \\ &\leq \|(\mathcal{P}_n - \mathcal{I})((\mathcal{K}_1\psi_1)'(x_0)x)\|_\infty + \|(\mathcal{P}_n - \mathcal{I})((\mathcal{K}_2\psi_2)'(x_0)x)\|_\infty \\ &\leq cn^{-\frac{1}{4}}\|[(\mathcal{K}_1\psi_1)'(x_0)x]^{(1)}\|_\infty + cn^{\frac{3}{4}-r}\|[(\mathcal{K}_2\psi_2)'(x_0)x]^{(r)}\|_\infty \\ &\leq c\left[n^{-\frac{1}{4}}(M_1 d_1 + 2\|k_1\|_{1,\infty}d_1) + 2n^{\frac{3}{4}-r}\|k_2\|_{r,\infty}d_2\right]\|x\|_\infty. \end{aligned} \tag{22}$$

Since $r \geq 1$, it follows that

$$\|\mathcal{T}_n{}'(x_0) - \mathcal{T}'(x_0)\|_\infty \to 0, \text{ as } n \to \infty.$$

This shows that $\mathcal{T}_n'(x_0)$ is norm convergent to $\mathcal{T}'(x_0)$. Since 1 is not an eigenvalue of the linear operator $\mathcal{T}'(x_0)$, we have $(\mathcal{I} - \mathcal{T}'(x_0))^{-1}$ is invertible on $\mathbb{X}$. Hence by a result from Ahues et al. [13], we have, for some sufficiently large n, $(\mathcal{I} - \mathcal{T}_n'(x_0))^{-1}$ exists and uniformly bounded on $\mathbb{X}$, i.e., there exists some $A_1 > 0$, such that $\|(\mathcal{I} - \mathcal{T}_n'(x_0))^{-1}\|_\infty \leq A_1 < \infty$.

For any $x \in B(x_0, \delta)$, we have

$$\begin{aligned}
&\|\mathcal{T}_n{}'(x_0) - \mathcal{T}_n{}'(x)\|_\infty \\
&= \|\mathcal{P}_n(\mathcal{K}_1\psi_1)'(x_0) + \mathcal{P}_n(\mathcal{K}_2\psi_2)'(x_0) - \mathcal{P}_n(\mathcal{K}_1\psi_1)'(x) - \mathcal{P}_n(\mathcal{K}_2\psi_2)'(x)\|_\infty \\
&\quad \le \|\mathcal{P}_n(\mathcal{K}_1\psi_1)'(x_0) - \mathcal{P}_n(\mathcal{K}_1\psi_1)'(x)\|_\infty + \|\mathcal{P}_n(\mathcal{K}_2\psi_2)'(x_0) - \mathcal{P}_n(\mathcal{K}_2\psi_2)'(x)\|_\infty.
\end{aligned} \tag{23}$$

Using Lipschitz continuity of $\psi_1^{(0,1)}(., x(.))$ and Leibniz rule, we obtain

$$\begin{aligned}
&\|[((\mathcal{K}_1\psi_1)'(x_0) - (\mathcal{K}_1\psi_1)'(x))y]^{(1)}\|_\infty \\
&= \sup_{t\in[-1,1]} \left|\frac{\partial}{\partial t}\int_{-1}^{t} k_1(t,s)\left[\psi_1^{(0,1)}(s, x_0(s)) - \psi_1^{(0,1)}(s, x(s))\right]y(s)ds\right| \\
&\le c_1 \sup_{t\in[-1,1]} \left[|k_1(t,t)||(x_0 - x)(t)||y(t)| + \left|\int_{-1}^{t}\left\{\frac{\partial}{\partial t}k_1(t,s)\right\}(x_0 - x)(s)y(s)ds\right|\right] \\
&\le c_1 \sup_{t\in[-1,1]} [|k_1(t,t)||(x_0 - x)(t)||y(t)|] \\
&\quad + c_1 \sup_{t,s\in[-1,1]} \left|\frac{\partial}{\partial t}k_1(t,s)\right| \sup_{s\in[-1,1]} [|(x_0 - x)(s)|y(s)|] \int_{-1}^{t} ds \\
&\le c_1 M_1\|x_0 - x\|_\infty\|y\|_\infty + 2c_1\|k_1\|_{1,\infty}\|x_0 - x\|_\infty\|y\|_\infty.
\end{aligned} \tag{24}$$

From estimates (14), (21), (24), for the first term of (23), we have

$$\begin{aligned}
&\|[\mathcal{P}_n(\mathcal{K}_1\psi_1)'(x_0) - \mathcal{P}_n(\mathcal{K}_1\psi_1)'(x)]y\|_\infty \\
&\le \|(\mathcal{P}_n - \mathcal{I})[(\mathcal{K}_1\psi_1)'(x_0) - (\mathcal{K}_1\psi_1)'(x)]y\|_\infty + \|[(\mathcal{K}_1\psi_1)'(x_0) - (\mathcal{K}_1\psi_1)'(x)]y\|_\infty \\
&\le cn^{-\frac{1}{4}}\|[((\mathcal{K}_1\psi_1)'(x_0) - (\mathcal{K}_1\psi_1)'(x))y]^{(1)}\|_\infty + \|[(\mathcal{K}_1\psi_1)'(x_0) - (\mathcal{K}_1\psi_1)'(x)]y\|_\infty \\
&\le cn^{-\frac{1}{4}}(c_1M_1 + 2c_1\|k_1\|_{1,\infty})\|x_0 - x\|_\infty\|y\|_\infty + 2M_1c_1\|x - x_0\|_\infty\|y\|_\infty \\
&\le \left[cn^{-\frac{1}{4}}(c_1M_1 + 2c_1\|k_1\|_{1,\infty}) + 2M_1c_1\right]\delta\|y\|_\infty.
\end{aligned} \tag{25}$$

Using Lipschitz continuity of $\psi_2^{(0,1)}(., x(.))$, we get

$$\begin{aligned}
&\|[((\mathcal{K}_2\psi_2)'(x_0) - (\mathcal{K}_2\psi_2)'(x_0))y]^{(r)}\|_\infty \\
&= \sup_{t\in[-1,1]} \left|\frac{\partial^r}{\partial t^r}\int_{-1}^{1} k_2(t,s)\left[\psi_2^{(0,1)}(s, x_0(s)) - \psi_2^{(0,1)}(s, x(s))\right]y(s)ds\right| \\
&\le c_2 \sup_{t,s\in[-1,1]} \left|\frac{\partial^r}{\partial t^r}k_2(t,s)\right| \sup_{s\in[-1,1]} |(x_0 - x)(s)| \int_{-1}^{1} |y(s)|ds \\
&\le 2c_2\|k_2\|_{r,\infty}\|x_0 - x\|_\infty\|y\|_\infty \le 2c_2\|k_2\|_{r,\infty}\delta\|y\|_\infty.
\end{aligned} \tag{26}$$

Now using estimates (14), (21), (26), for the second term of (23), we obtain

$$\|[\mathcal{P}_n(\mathcal{K}_2\psi_2)'(x_0) - \mathcal{P}_n(\mathcal{K}_2\psi_2)'(x)]y\|_\infty$$

$$\leq \|(\mathcal{P}_n - \mathcal{I})[(\mathcal{K}_2\psi_2)'(x_0) - (\mathcal{K}_2\psi_2)'(x)]y\|_\infty + \|[(\mathcal{K}_2\psi_2)'(x_0) - (\mathcal{K}_2\psi_2)'(x)]y\|_\infty$$
$$\leq cn^{\frac{3}{4}-r}\|[((\mathcal{K}_2\psi_2)'(x_0) - (\mathcal{K}_2\psi_2)'(x))y]^r\|_\infty + \|[(\mathcal{K}_2\psi_2)'(x_0) - (\mathcal{K}_2\psi_2)'(x)]y\|_\infty$$
$$\leq [2c_2cn^{\frac{3}{4}-r}\|k_2\|_{r,\infty} + 2M_2c_2]\delta\|y\|_\infty. \quad (27)$$

Combining estimates (23), (25) and (27), we have

$$\|\mathcal{T}_n'(x_0) - \mathcal{T}_n'(x)\|_\infty \leq \left(cn^{-\frac{1}{4}}(c_1M_1 + 2c_1\|k_1\|_{1,\infty}) + 2M_1c_1 + 2c_2cn^{\frac{3}{4}-r}\|k_2\|_{r,\infty} + 2M_2c_2\right)\delta.$$

Hence, we have

$$\sup_{\|x-x_0\|_\infty \leq \delta} \|(\mathcal{I} - \mathcal{T}_n'(x_0))^{-1}(\mathcal{T}_n'(x_0) - \mathcal{T}_n'(x))\|_\infty$$
$$\leq A_1\left(cn^{-\frac{1}{4}}(c_1M_1 + 2c_1\|k_1\|_{1,\infty}) + 2M_1c_1 + 2c_2cn^{\frac{3}{4}-r}\|k_2\|_{r,\infty} + 2M_2c_2\right)\delta \leq q \ (say).$$

We choose δ in such a way that $q \in (0, 1)$. This proves the estimate (4.4) of Theorem 2 in [12].
Since $r \geq 1$, taking use of estimates (3) and (15), we have

$$\alpha_n = \|(\mathcal{I} - \mathcal{T}_n'(x_0))^{-1}(\mathcal{T}_n(x_0) - \mathcal{T}(x_0))\|_\infty$$
$$\leq A_1\|\mathcal{T}_n(x_0) - \mathcal{T}(x_0)\|_\infty$$
$$= A_1\|\mathcal{P}_n\mathcal{K}_1\psi_1(x_0) + \mathcal{P}_n\mathcal{K}_2\psi_2(x_0) + \mathcal{P}_nf - \mathcal{K}_1\psi_1(x_0) - \mathcal{K}_2\psi_2(x_0) - f\|_\infty$$
$$= A_1\|(\mathcal{P}_n - \mathcal{I})(\mathcal{K}_1\psi_1(x_0) + \mathcal{K}_2\psi_2(x_0) + f)\|_\infty$$
$$= A_1\|(\mathcal{P}_n - \mathcal{I})x_0\|_\infty \leq A_1cn^{\frac{1}{2}-r}V\left(x_0^{(r)}\right) \to 0, \ as\ n \to \infty. \quad (28)$$

By choosing n large enough such that $\alpha_n \leq \delta(1-q)$, the Eq. (4.5) of Theorem 2 in [12] is satisfied. Hence applying Theorem 2 of [12], we obtain

$$\frac{\alpha_n}{1+q} \leq \|x_n - x_0\|_\infty \leq \frac{\alpha_n}{1-q},$$

where $\alpha_n = \|(\mathcal{I} - \mathcal{T}_n'(x_0))^{-1}(\mathcal{T}_n(x_0) - \mathcal{T}(x_0))\|_\infty$.
And the estimate (28) implies

$$\|x_n - x_0\|_\infty = \mathcal{O}\left(n^{\frac{1}{2}-r}\right).$$

This completes the proof. □

Next, we discuss the existence and convergence of the iterated approximate solution $\tilde{x}_n$ to x_0.

Theorem 2 *Let* $x_0 \in \mathcal{C}[-1, 1]$, *be an isolated solution of the Eq. (3). Assume that* 1 *is not an eigenvalue of* $\mathcal{T}'(x_0)$. *Then for sufficiently large n, the operator* $\mathcal{I} - \widetilde{\mathcal{T}}_n'(x_0)$

is invertible on $\mathbb{X}$ *and there exist constants* $A_2 > 0$ *independent of* n *such that* $\|(\mathcal{I} - \widetilde{\mathcal{T}}_n'(x_0))^{-1}\|_\infty \leq A_2$.

Proof Consider

$$\begin{aligned}|\widetilde{\mathcal{T}}_n'(x_0)x(t)| &= |[(\mathcal{K}_1\psi_1)'(\mathcal{P}_n x_0) + (\mathcal{K}_2\psi_2)'(\mathcal{P}_n x_0)]\mathcal{P}_n x(t)| \\ &\leq |(\mathcal{K}_1\psi_1)'(\mathcal{P}_n x_0)\mathcal{P}_n x(t)| + |(\mathcal{K}_2\psi_2)'(\mathcal{P}_n x_0)\mathcal{P}_n x(t)|. \end{aligned} \tag{29}$$

We have

$$\begin{aligned}\|(\mathcal{K}_1\psi_1)'(x_0)\mathcal{P}_n x\|_\infty &= \sup_{t\in[-1,1]} \left| \int_{-1}^{t} k_1(t,s)\psi_1^{(0,1)}(s, x_0(s))\mathcal{P}_n x(s)ds \right| \\ &\leq \sup_{t,s\in[-1,1]} |k_1(t,s)| \sup_{s\in[-1,1]} |\psi_1^{(0,1)}(s, x_0(s))| \int_{-1}^{1} |\mathcal{P}_n x(s)|ds \\ &\leq \sqrt{2}M_1 d_1 \|\mathcal{P}_n x\|_{L^2}. \end{aligned} \tag{30}$$

Using estimates (12), (21) and (30), we have

$$\begin{aligned}&\|(\mathcal{K}_1\psi_1)'(\mathcal{P}_n x_0)\mathcal{P}_n x\|_\infty \\ &\quad\leq \|[(\mathcal{K}_1\psi_1)'(\mathcal{P}_n x_0) - (\mathcal{K}_1\psi_1)'(x_0)](\mathcal{P}_n x)\|_\infty + \|(\mathcal{K}_1\psi_1)'(x_0)(\mathcal{P}_n x)\|_\infty \\ &\quad\leq M_1 c_1 \|\mathcal{P}_n x_0 - x_0\|_{L^2} \|\mathcal{P}_n x\|_{L^2} + \sqrt{2}M_1 d_1 \|\mathcal{P}_n x\|_{L^2} \\ &\quad\leq (M_1 c_1 p_1 \|\mathcal{P}_n x_0 - x_0\|_{L^2} + \sqrt{2}M_1 p_1 d_1)\|x\|_\infty. \end{aligned} \tag{31}$$

In a similar manner, we obtain

$$\|(\mathcal{K}_2\psi_2)'(\mathcal{P}_n x_0)\mathcal{P}_n x\|_\infty \leq (M_2 c_2 p_1 \|\mathcal{P}_n x_0 - x_0\|_{L^2} + \sqrt{2}M_2 p_1 d_2)\|x\|_\infty. \tag{32}$$

Combining estimates (11), (29), (31) and (32), we get

$$\|\widetilde{\mathcal{T}}_n'(x_0)\|_\infty \leq (M_1 c_1 + M_2 c_2) p_1 \|\mathcal{P}_n x_0 - x_0\|_{L^2} + (M_1 d_1 + M_2 d_2)\sqrt{2} p_1 < \infty. \tag{33}$$

This shows that $\|\widetilde{\mathcal{T}}_n'(x_0)\|_\infty$ is uniformly bounded.

Next we consider

$$\begin{aligned}|\widetilde{\mathcal{T}}_n'(x_0)x(t) - \widetilde{\mathcal{T}}_n'(x_0)x(t')| &\leq |(\mathcal{K}_1\psi_1)'(\mathcal{P}_n x_0)\mathcal{P}_n x(t) - (\mathcal{K}_1\psi_1)'(\mathcal{P}_n x_0)\mathcal{P}_n x(t')| \\ &\quad + |(\mathcal{K}_2\psi_2)'(\mathcal{P}_n x_0)\mathcal{P}_n x(t) - (\mathcal{K}_2\psi_2)'(\mathcal{P}_n x_0)\mathcal{P}_n x(t')|. \end{aligned} \tag{34}$$

We can write

$$|(\mathcal{K}_1\psi_1)'(\mathcal{P}_n x_0)\mathcal{P}_n x(t) - (\mathcal{K}_1\psi_1)'(\mathcal{P}_n x_0)\mathcal{P}_n x(t')| \leq T_1 + T_2 + T_3, \tag{35}$$

where

$$T_1 = |(\mathcal{K}_1\psi_1)'(\mathcal{P}_n x_0)\mathcal{P}_n x(t) - (\mathcal{K}_1\psi_1)'(x_0)\mathcal{P}_n x(t)|,$$
$$T_2 = |(\mathcal{K}_1\psi_1)'(x_0)\mathcal{P}_n x(t) - (\mathcal{K}_1\psi_1)'(x_0)\mathcal{P}_n x(t')|,$$
$$T_3 = |(\mathcal{K}_1\psi_1)'(x_0)\mathcal{P}_n x(t') - (\mathcal{K}_1\psi_1)'(\mathcal{P}_n x_0)\mathcal{P}_n x(t')|.$$

Now using the estimates (11), (12), and (21), we have

$$\begin{aligned} T_1 &= |[(\mathcal{K}_1\psi_1)'(\mathcal{P}_n x_0) - (\mathcal{K}_1\psi_1)'(x_0)]\mathcal{P}_n x(t)| \\ &\leq \|[(\mathcal{K}_1\psi_1)'(\mathcal{P}_n x_0) - (\mathcal{K}_1\psi_1)'(x_0)]\mathcal{P}_n x\|_\infty \\ &\leq M_1 c_1 \|(I - \mathcal{P}_n)x_0\|_{L^2} \|\mathcal{P}_n x\|_{L^2} \\ &\leq M_1 c_1 p_1 \|\mathcal{P}_n x_0 - x_0\|_{L^2} \|x\|_\infty \to 0, \ as\ n \to \infty, \end{aligned} \tag{36}$$

and similarly

$$\begin{aligned} T_3 &= |[(\mathcal{K}_1\psi_1)'(\mathcal{P}_n x_0) - (\mathcal{K}_1\psi_1)'(x_0)]\mathcal{P}_n x(t')| \\ &\leq M_1 c_1 p_1 \|\mathcal{P}_n x_0 - x_0\|_{L^2} \|x\|_\infty \to 0, \ as\ n \to \infty. \end{aligned} \tag{37}$$

Since $k_1(t, s) \in \mathcal{C}([-1, 1]\times[-1, 1]), k_1(t, s)$ is uniformly continuous in first variable t. Hence for any $\epsilon > 0$, however small, there exists some number $\delta > 0$ such that

$$|k_1(t, s) - k_1(t', s)| < \epsilon, \text{ whenever } |t - t'| < \delta.$$

Using estimate (12), we get

$$\begin{aligned} T_2 &= \left| \int_{-1}^{t} [k_1(t, s) - k_1(t', s)]\psi_1^{(0,1)}(s, x_0(s))\mathcal{P}_n x(s) ds \right| \\ &\leq \sup_{-1\leq s\leq 1} |k_1(t, s) - k_1(t', s)| \sup_{s\in[-1,1]} |\psi_1^{(0,1)}(s, x_0(s))| \int_{-1}^{1} |\mathcal{P}_n x(s)| ds \\ &\leq \epsilon d_1 \sqrt{2} \|\mathcal{P}_n x\|_{L^2} \leq \epsilon\sqrt{2} d_1 p_1 \|x\|_\infty \to 0, \ as\ t \to t'. \end{aligned} \tag{38}$$

Hence from estimates (35), (36), (37) and (38), we have

$$|(\mathcal{K}_1\psi_1)'(\mathcal{P}_n x_0)\mathcal{P}_n x(t) - (\mathcal{K}_1\psi_1)'(\mathcal{P}_n x_0)\mathcal{P}_n x(t')| \to 0, \ as\ t \to t' \text{ and } n \to \infty. \tag{39}$$

On similar lines, it can be proved that

$$|(\mathcal{K}_2\psi_2)'(\mathcal{P}_n x_0)\mathcal{P}_n x(t) - (\mathcal{K}_2\psi_2)'(\mathcal{P}_n x_0)\mathcal{P}_n x(t')| \to 0, \ as\ t \to t' \text{ and } n \to \infty. \tag{40}$$

Hence combining estimates (34), (39) and (40), we have

$$|\widetilde{\mathcal{T}}_n'(x_0)x(t) - \widetilde{\mathcal{T}}_n'(x_0)x(t')| \to 0, \ as\ t \to t' \text{ and } n \to \infty. \tag{41}$$

This implies $\{\widetilde{\mathcal{T}}_n'(x_0)\}_{n=1}^{\infty}$ is a family of collectively compact operators. Since $\mathcal{T}'(x_0)$ is compact and $(\mathcal{I} - \mathcal{T}'(x_0))^{-1}$ exists, it follows from the theory of collectively compact operators (see Anselone [14]) that $(\mathcal{I} - \widetilde{\mathcal{I}}_n'(x_0))^{-1}$ exists and is uniformly bounded for sufficiently large n, i.e., $\exists$ some $A_2 > 0$, independent of n such that $\|(\mathcal{I} - \widetilde{\mathcal{T}}_n'(x_0))^{-1}\|_\infty \leq A_2$.
This completes the proof. □

Theorem 3 *Let $x_0 \in \mathcal{C}^r[-1, 1]$, $r \geq 1$ be an isolated solution of the Eq. (3). Assume that 1 is not an eigenvalue of $\mathcal{T}'(x_0)$, then for sufficiently large n, the iterated solution $\tilde{x}_n$ defined by (19) is the unique solution in the sphere $B(x_0, \delta) = \{x : \|x - x_0\|_\infty < \delta\}$. Moreover, there exists a constant $0 < q < 1$, independent of n such that*

$$\frac{\beta_n}{1+q} \leq \|\tilde{x}_n - x_0\|_\infty \leq \frac{\beta_n}{1-q},$$

where $\beta_n = \|(\mathcal{I} - \widetilde{\mathcal{T}}_n'(x_0))^{-1}(\widetilde{\mathcal{T}}_n(x_0) - \mathcal{T}(x_0))\|_\infty$. Also, we obtain

$$\|\tilde{x}_n - x_0\|_\infty = \mathcal{O}(n^{-r}).$$

Proof From Theorem 2, we have $(\mathcal{I} - \widetilde{\mathcal{T}}_n'(x_0))^{-1}$ exists and it is uniformly bounded on $\mathbb{X}$ for sufficiently large n, i.e., there exists a constant $A_2 > 0$, such that $\|(\mathcal{I} - \widetilde{\mathcal{T}}_n'(x_0))^{-1}\|_\infty \leq A_2$.

For any $x \in B(x_0, \delta)$, consider

$$\begin{aligned} &\|[\widetilde{\mathcal{T}}_n'(x) - \widetilde{\mathcal{T}}_n'(x_0)]y\|_\infty \\ &= \|[(\mathcal{K}_1\psi_1)'(\mathcal{P}_n x) + (\mathcal{K}_2\psi_2)'(\mathcal{P}_n x) - (\mathcal{K}_1\psi_1)'(\mathcal{P}_n x_0) - (\mathcal{K}_2\psi_2)'(\mathcal{P}_n x_0)]\mathcal{P}_n y\|_\infty \\ &\leq \|[(\mathcal{K}_1\psi_1)'(\mathcal{P}_n x_0) - (\mathcal{K}_1\psi_1)'(\mathcal{P}_n x)]\mathcal{P}_n y\|_\infty + \|[(\mathcal{K}_2\psi_2)'(\mathcal{P}_n x) - (\mathcal{K}_2\psi_2)'(\mathcal{P}_n x_0)]\mathcal{P}_n y\|_\infty. \end{aligned} \tag{42}$$

Using estimates (12) and (21), we get

$$\begin{aligned} \|[(\mathcal{K}_1\psi_1)'(\mathcal{P}_n x_0) - (\mathcal{K}_1\psi_1)'(\mathcal{P}_n x)]\mathcal{P}_n y\|_\infty &\leq M_1 c_1 \|\mathcal{P}_n(x_0 - x)\|_{L^2}\|\mathcal{P}_n y\|_{L^2} \\ &\leq M_1 c_1 p_1^2 \|x - x_0\|_\infty \|y\|_\infty \\ &\leq M_1 c_1 p_1^2 \delta \|y\|_\infty. \end{aligned} \tag{43}$$

Again following similar lines, we can deduce that

$$\|[(\mathcal{K}_2\psi_2)'(\mathcal{P}_n x_0) - (\mathcal{K}_2\psi_2)'(\mathcal{P}_n x)]\mathcal{P}_n y\|_\infty \leq M_2 c_2 p_1^2 \delta \|y\|_\infty. \tag{44}$$

Thus we obtain

$$\sup_{\|x - x_0\|_\infty \leq \delta} \|(\mathcal{I} - \widetilde{\mathcal{T}}_n'(x_0))^{-1}(\widetilde{\mathcal{T}}_n'(x) - \widetilde{\mathcal{T}}_n'(x_0))\|_\infty \leq A_2(M_1c_1 + M_2c_2)p_1^2\delta \leq q \ \ (say).$$

We choose δ in such a way that $0 < q < 1$. Hence the estimate (4.4) of Theorem 2 in [12] is proved.

Next using estimates (13) and (20), we have

$$\begin{aligned}\|\widetilde{\mathcal{T}}_n(x_0) - \mathcal{T}(x_0)\|_\infty &= \|\mathcal{K}_1\psi_1(\mathcal{P}_n x_0) + \mathcal{K}_2\psi_2(\mathcal{P}_n x_0) - \mathcal{K}_1\psi_1(x_0) - \mathcal{K}_2\psi_2(x_0)\|_\infty \\ &\le \|\mathcal{K}_1\psi_1(\mathcal{P}_n x_0) - \mathcal{K}_1\psi_1(x_0)\|_\infty + \|\mathcal{K}_2\psi_2(\mathcal{P}_n x_0) - \mathcal{K}_2\psi_2(x_0)\|_\infty \\ &\le \sqrt{2}(M_1 l_1 + M_2 l_2)\|\mathcal{P}_n x_0 - x_0\|_{L^2} \\ &\le \sqrt{2}(M_1 l_1 + M_2 l_2) c n^{-r}\|x_0^{(r)}\|_\infty. \end{aligned} \tag{45}$$

This implies

$$\begin{aligned}\beta_n &= \|(\mathcal{I} - \widetilde{\mathcal{T}}_n'(x_0))^{-1}(\widetilde{\mathcal{T}}_n(x_0) - \mathcal{T}(x_0))\|_\infty \\ &\le \|(\mathcal{I} - \widetilde{\mathcal{T}}_n'(x_0))^{-1}\|_\infty \|\widetilde{\mathcal{T}}_n(x_0) - \mathcal{T}(x_0)\|_\infty \\ &\le A_2\sqrt{2}(M_1 l_1 + M_2 l_2) c n^{-r}\|x_0^{(r)}\|_\infty \to 0, \ as\ n \to \infty. \end{aligned} \tag{46}$$

Choose n large enough such that $\beta_n \le \delta(1-q)$. Thus Eq. (4.5) of Theorem 2 in [12] is satisfied. Hence by Theorem 2 of [12],

$$\frac{\beta_n}{1+q} \le \|\widetilde{x}_n - x_0\|_\infty \le \frac{\beta_n}{1-q},$$

where

$$\beta_n = \|(\mathcal{I} - \widetilde{\mathcal{T}}_n'(x_0))^{-1}(\widetilde{\mathcal{T}}_n(x_0) - \mathcal{T}(x_0))\|_\infty.$$

Hence from estimate (46), we have

$$\|\tilde{x}_n - x_0\|_\infty - \mathcal{O}(n^{-r}).$$

This completes the proof. □

Remark 1 From Theorems 1 and 3, we observe that the Legendre Galerkin solution and the iterated Legendre Galerkin solution converges with the orders $\mathcal{O}\left(n^{\frac{1}{2}-r}\right)$ and $\mathcal{O}(n^{-r})$ in the infinity norm, respectively. This shows that iterated Legendre Galerkin approximation improves over the Legndre Galerkin approxiamtion and optimal convergence rate is obtained in case of the iterated Legendre Galerkin approximate solution.

References

1. Pachpatte, B.G.: On a nonlinear Volterra-Fredholm integral equation. Sarajevo J. Math. **16**, 61–71 (2008)
2. Brunner, H.: On the numerical solution of nonlinear Volterra-Fredholm integral equation by collocation methods. SIAM J. Numer. Anal. **27**, 987–1000 (1990)

3. Wazwaz, A.M.: A reliable treatment for mixed Volterra-Fredholm integral equations. Appl. Math. Comput. **127**, 405–414 (2002)
4. Bakodah, H.O., Darwish, M.A.: On discrete Adomian decomposition method with Chebyshev abscissa for nonlinear integral equations of Hammerstein Type. Adv. Pure Math. **2**, 310–313 (2012)
5. Darwish, M.A.: Fredholm-Volterra integral equation with singular kernel. Korean J. Comput. Appl. Math. **6**, 163–174 (1999)
6. Darwish, M.A.: Note on stability theorems for nonlinear mixed integral equations. J. Appl. Math. Comput. **6**, 633–637 (1999)
7. Maleknejad, K., Karami, M., Rabbani, M.: Using the Petrov-Galerkin elements for solving Hammerstein integral equations. Appl. Math. Comput. **172**, 831–845 (2006)
8. Maleknejad, K., Derili, H.: The collocation method for Hammerstein equations by Daubechies wavelets. Appl. Math. Comput. **172**, 846–864 (2006)
9. Delves, L.M., Mohamed, J.L.: Computational Methods for Integral Equations. Cambridge University Press, Cambridge, UK (1985)
10. Kaneko, H., Noren, R.D., Xu, Y.: Regularity of the solution of Hammerstein equations with weakly singular kernels. Integr. Eqn. Oper. Theory **13**, 660–670 (1990)
11. Canuto, C., Hussaini, M.Y., Quarteroni, A., Zang, T.A.: Spectral Methods, Fundamentals in Single Domains. Springer-Verlag, Berlin (2006)
12. Vainikko, G.M.: A perturbed Galerkin method and the general theory of approximate methods for nonlinear equations. USSR Comput. Math. Math. Phys. **7**, 1–41 (1967)
13. Ahues, M., Largillier, A., Limaye, B.: Spectral Computations for Bounded Operators. CRC Press, Boca Raton (2010)
14. Anselone, P.M.: Collectively Compact Operator Approximation Theory and Application to Integral equations. Prentice Hall, Englewood cliffs, NJ (1971)

Analysis of an Eco-Epidemiological Model with Migrating and Refuging Prey

Shashi Kant and Vivek Kumar

Abstract This paper concerns a predator–prey system with migrating and refuging prey with disease infection. Analysis of the model regarding stability has been performed. The effect of time delay on the above system is also studied. By assuming the time delay a bifurcation parameter, the stability of the positive equilibrium, and Hopf-bifurcation is studied. Further, the directions of Hopf-bifurcation and the stability of bifurcated periodic solutions are calculated using the famous normal form theory, Riesz representation theorem and central manifold theorem. This is not a case study, hence real data is not available. However, to verify our theoretical predictions, some numerical simulations are also included.

Keywords Predator–prey model · Stability · Hopf-bifurcation · Migration · Refuge · Delay

1 Introduction

The dynamic relation between prey and predator has been studied extensively in the literature. At first sight, prey–predator dynamics may seem very simple mathematically, but they are, in fact very difficult and challenging. The classical Lotka-Volterra model is a first stepping stone in the study of prey–predator dynamics and interactions [1, 23]. In mathematical ecology, this model is extensively used and cited and proved a milestone in the progress of mathematical ecology. On the other hand, the famous work of Kermack-Mckendric [25] in epidemiological studies received much attention among applied mathematicians, scientists, and ecologists. After the work of [1, 23, 25], many mathematical models have been published for reference

S. Kant (✉) · V. Kumar
Department of Applied Mathematics, Delhi Technological University,
Delhi 110042, India
e-mail: onlineskmishra@gmail.com

V. Kumar
e-mail: vivekkumar.ag@gmail.com

P.N. Agrawal et al. (eds.), *Mathematical Analysis and its Applications*,
Springer Proceedings in Mathematics & Statistics 143,
DOI 10.1007/978-81-322-2485-3_2

(see [2, 6, 7, 13–17, 21, 24, 27, 29], etc., and references therein). Combined and/or overlapping study of ecology and epidemiology is termed as Eco-epidemiology. Eco-epidemiological models are gaining popularity day-by-day. The present study also falls under the purview of Eco-epidemiology.

To study the environmental impact on prey–predator models, the 'time delay' has been investigated by the researchers. A good number of papers are available in the literature for instance (see [4, 8, 18]). In these papers, most of the authors investigate the 'time delay' as a game changing. Time delay may cause changes in stability, occurrence for limit cycle, bifurcation, etc.

Further, migration of species especially prey is also evolved and few references are available in the literature, for reference we can refer [11, 19, 20]. Prey-refuge in prey-predator models also play an important role in dynamical nature. Further if prey-refuge is more than outbreak of the prey population occurs. To understand a role of prey-refuge in mathematical ecology few publications are available. At this juncture we may refer readers to ([9, 22, 24, 26, 29] and references therein).

Pal and Samanta [3] proposed the following mathematical model by incorporating prey-refuge in the model proposed of Xiao and Chen [28]:

$$\begin{cases} \frac{dS}{dt} = r_1 S(1 - \frac{S+I}{k}) - SI\beta, \\ \frac{dI}{dt} = SI\beta - cI - \frac{bIY}{aY+I}, \\ \frac{dY}{dt} = -dY + \frac{pbIY}{aY+I}. \end{cases} \tag{1}$$

Motivated by the model of Samanta [18] and model in [12], Hu and Li [10] proposed the following model:

$$\begin{cases} \frac{dS}{dt} = rS\left(1 - \frac{S+I}{k}\right) - SI\beta - p_1 SY, \\ \frac{dI}{dt} = -cI + SI\beta - p_2 IY, \\ \frac{dY}{dt} = -dY + qp_1 S(t-\tau)Y(t-\tau) + qp_2 I(t-\tau)Y(t-\tau). \end{cases} \tag{2}$$

In order to study the influence of prey-refuge, migration, and disease on the Prey-predator system, in this paper, we concentrate on an eco-epidemiological prey-predator system consisting of three species as in [10]. Motivated by the models in [10] and [3], we propose a mathematical model in which prey is migrating and refuging with disease in both species. We present stability and Hopf-bifurcation analysis of the mathematical model. Detailed assumptions for model formulation are listed in the next section.

The rest of the paper is structured as follows: In the next section, we formulate our main mathematical model with the help of biological and ecological assumptions. In Sect. 3, we consider the model without delay. In Sect. 4, we discuss the stability of mathematical model with delay. In Sect. 5, we discuss the direction and stability of Hopf-bifurcation using the normal form theory, Riesz representation theorem and central manifold theorem as in [10]. Numerical simulations have been done in Sect. 6 followed by discussion in the last Sect. 7.

2 The Model

In this paper, we propose to study a prey-predator system by means of mathematical modeling. To formulate the model and in view of simplicity we make the following assumptions:

- In the absence of disease and predation the healthy (susceptible) prey population has logistic growth with growth rate r and carrying capacity k, i.e.,

$$\frac{dN}{dt} = rN\left(1 - \frac{N}{k}\right). \tag{3}$$

- Disease is spreading in both populations. After disease prey population is divided into two parts susceptible prey(S) and infected prey (I). Thus total biomass of prey population is $S(t) + I(t)$.
- Due to mathematical complexity, the bifurcation of predator population and the detailed dynamics of the disease infection in the predator population is omitted. Further, it is also assumed that disease infection in predator occurs due to eating of the infected prey and not due to outside infection. In other words, it is easy to understand that the disease infection starts from prey and then carries forward to predator. For example H1N1, H5N1, etc., may be pointed out here to understand the physical phenomenon better. Thus total biomass of predator population is Y.
- Infected prey population does not become immune as well as they have no reproduction rate. However, infected prey population contributes the carrying capacity k.
- Predator population consumes both susceptible as well as infected prey population.
- Due to environmental and fear factors, we consider out migration in prey population, i.e., once prey migrated they will not return. Let m_1 and m_2 be the migration rates of susceptible and infected prey respectively. Further, healthy prey population is more active compared to infected one. Hence, healthy prey can migrate more easily than infected prey before their predation. Hence, by using this ecological information, we can impose the mathematical condition $m_1 > m_2$.
- Let d_2 and d_3 be natural death rates for infected prey and predator population respectively.
- Death (mortality) rate due to disease for infected prey population and predator population are denoted by c and d_4 respectively.
- The coefficient for S-prey and I-prey to predator are denoted by q_1 and q_2 respectively. The relationship between q_1 and q_2 is established later.
- Let a refuge protecting m_3S of healthy prey and m_4I that of infected prey, where $m_3, m_4 \in [0, 1)$. Hence, $(1 - m_3)S$ and $(1 - m_4)I$ of healthy and infected prey, respectively, are available to the predator for predation.

Based on these assumptions, model takes the following form:

$$\begin{cases} \frac{dS}{dt} = rS\left(1 - \frac{S+I}{k}\right) - SI\beta - p_1(1-m_3)SY - m_1S, \\ \frac{dI}{dt} = SI\beta - p_2(1-m_4)IY - d_2I - m_2I - cI, \\ \frac{dY}{dt} = q_1p_1(1-m_3)S(t-\tau)Y(t-\tau) + q_2p_2(1-m_4)I(t-\tau)Y(t-\tau) - d_3Y - d_4Y. \end{cases} \tag{4}$$

The initial conditions are

$$\begin{cases} S(t) = \phi_1(t) > 0, \\ I(t) = \phi_2(t) > 0, \\ Y(t) = \phi_3(t) > 0, \\ (\phi_1(t), \phi_2(t), \phi_3(t)) \in C = C\left([-\tau, 0], R^3_+\right), \\ R^3_+ = \{(x, y, z) | x \geq 0, y \geq 0, z \geq 0\}, \end{cases} \tag{5}$$

where,

β : Disease Contact Rate
p_1, p_2 : Predation Coefficients of Susceptible (S) and Infected (I) Prey
τ : Gestation period (delay).

Ecological and biological assumptions suggests the following relationship between q_1 and q_2:

$$q_2 \neq q_1 \text{ and } 0 < q_1 \leq 1,$$

$$q_2 > q_1 \text{ and } 0 < q_2 \leq 1.$$

3 Analysis of the Model Without Delay

In this section, model (4) is investigated under the condition $\tau = 0$. Before going to main analysis, we state two lemmas for our model without proof.

Lemma 1 *Each solution of the system (4) without delay with the initial conditions (5) are strictly positive for all* $t \geq 0$.

Lemma 2 *Solutions of the system (4) without delay with the initial conditions (5) are eventually bounded, i.e., uniformity bounded in* R^3_+.

3.1 Equilibrium Points

System of ODEs under consideration has the following equilibrium points:

(i) $E_1(0,0,0) = (0,0,0)$.
(ii) $E_2(\widehat{S},0,0)$. where $\widehat{S} = (r - m_1)\frac{k}{r}$.
(iii) $E_3(S^*,0,Y^*)$, where

$$\begin{cases} S^* = & \dfrac{d_3 + d_4}{q_1 p_1 (1 - m_3)}, \\ Y^* = & \dfrac{k p_1 (1 - m_3) q_1 (r - m_1) - r(d_3 + d_4)}{k q_1 (1 - m_3) p_1^2}. \end{cases}$$

(iv) $E_4(\overline{S},\overline{I},0)$, where

$$\begin{cases} \overline{S} = & \dfrac{c + d_2 + m_2}{\beta}, \\ \overline{I} = & \dfrac{\{(r - m_1)k\beta - r(c + d_2 + m_2)\}}{(\beta(r + k\beta))}. \end{cases}$$

(v) $E_5(\widetilde{S},\widetilde{I},\widetilde{Y})$, where

$$\begin{cases} \widetilde{S} = & \dfrac{(r - m_1)q_2(1 - m_4)p_2 k + (c + d_2 + m_2)q_2 p_1 (1 - m_3)(1 - m_4)p_2 k - (r + k\beta)(d_3 + d_4)}{r q_2 p_2 (1 - m_4) - q_1 p_1 (1 - m_3)(r + k\beta) + q_2 (1 - m_4) p_2 \beta k}, \\ \widetilde{I} = & \dfrac{(d_3 + d_4) - q_1 p_1 (1 - m_3)\widetilde{S}}{q_2 p_2 (1 - m_4)}, \\ \widetilde{Y} = & \dfrac{\beta\widetilde{S} - (c + d_2 + m_2)}{p_2 (1 - m_4)}. \end{cases}$$

3.1.1 Existence Conditions

We have the following existence conditions:

(i) Trivial equilibrium E_1 always exists.
(ii) E_2 exists provided $(r - m_1) > 0$ or $r > m_1$. Physical meaning implies that existence of E_2 is independent of other parameters and depends only on growth rate and migration of S, viz., r and m_1. E_2 exists if growth rate of S is greater than migration of itself.
(iii) E_3 exists provided $\frac{r - m_1}{d_3 + d_4} > \frac{r}{k q_1 p_1 (1 - m_3)}$. This is the case when no disease infection occurs in the prey population.
(iv) E_4 exists provided $\frac{r - m_1}{c + d_2 + m_2} > \frac{r}{k\beta}$. This is the case when predator does not survive.
(v) E_5 exists provided the following conditions are satisfied:

$$\begin{cases} d_3 + d_4 > q_1 p_1 (1 - m_3)\widetilde{S}, \\ \beta\widetilde{S} > (c + d_2 + m_2), \\ (r - m_1)q_2 p_2 (1 - m_4)k + (c + d_2 + m_2)q_2 p_2 (1 - m_4) p_1 (1 - m_3)k > (r + k\beta)(d_3 + d_4), \\ (r q_2 p_2 (1 - m_4) + q_2 p_1 (1 - m_3)\beta k) > q_1 p_1 (1 - m_3)(r + k\beta). \end{cases}$$

This equilibrium point is very important, since it provides the coexistence of all the three populations.

3.2 Stability Analysis

Jacobian matrix of the system is given by

$$J = \begin{pmatrix} (r - m_1 - \frac{2rS}{K} - \frac{rI}{k} - \beta I - p_1(1-m_3)Y) & \left(-\frac{rS}{k} - \beta S\right) & (-p_1(1-m_3)S) \\ (\beta I) & (\beta S - p_2(1-m_4)Y - c - d_2 - m_2) & (-p_2(1-m_4)I) \\ (q_1(1-m_3)p_1 Y) & (q_2 p_2(1-m_4)Y) & (q_1 p_1(1-m_3)S + q_2 p_2(1-m_4)I - d_3 - d_4) \end{pmatrix}, \tag{6}$$

with this matrix stability analysis is carried out. We will focus on the non zero equilibrium point.

After a little calculation we see that trivial equilibrium point is locally stable if $r < (m_1)$. Equilibrium (E_2) is locally asymptotically stable provided the following conditions are satisfied:

$$\begin{cases} (\beta\widehat{S} - c - d_2 - m_2) = \frac{\beta k(r-m_1)}{r} < 0, \\ (q_1 p_1(1-m_3)\widehat{S} - d_3 - d_4) = q_1 p_1(1-m_3)\frac{k(r-m_1)}{r} - d_3 - d_4 < 0. \end{cases}$$

Equilibrium (E_3) is locally asymptotically stable if

$$\begin{cases} \left[\frac{\beta(d_3+d_4)}{q_1(1-m_3)p_1} - \frac{p_2(1-m_4)(r-m_1)}{p_1(1-m_3)} + \frac{p_2(1-m_3)r(d_3+d_4)}{kq_1(1-m_3)p_1^2} - (c + d_2 + m_2)\right] < 0; \\ \text{Quadratic equation } (\lambda^2 - \xi\lambda + \zeta) \text{ have roots with negative real parts, where} \\ \xi = \frac{-(d_3+d_4)[kq_1 p_1(1-m_3)+r]}{kq_1(1-m_3)p_1}, \\ \zeta = \frac{(d_3+d_4)[kq_1 p_1(1-m_3)(r-m_1)-(d_3+d_4)r]}{kq_1 p_1(1-m_3)}. \end{cases}$$

Equilibrium (E_4) is locally asymptotically stable if

$$\begin{cases} (q_1 p_1\overline{S} + q_2 p_2\overline{I} - d_3 - d_4) = -(d_3 + d_4) + \frac{q_1 p_1(c+d_2+m_2)}{\beta} + \frac{q_2 p_2[(r-m_1)k\beta - r(c+d_2+m_2)]}{\beta(r+k\beta)} < 0; \\ \text{Equation } (\lambda^2 - B\lambda + C) \text{ have roots with negative real parts, where} \\ B = \frac{-3r(c+d_2+m_2)}{k\beta}, \\ C = (c + d_2 + m_2)\left[(r - m_1) - \frac{r(c+d_2+m_2)}{k\beta}\right]. \end{cases}$$

Remark 1 In this case no infection occurs in the system, hence ecologically mortality due to infection in predator population may be omitted. Similarly, parameter c may also be deleted. Jacobian matrix at E_3 is now reduced to

$$J = \begin{bmatrix} \left(r - m_1 - \frac{2rS^*}{k} - p_1(1-m_3)Y^*\right) & \left(-\frac{rS^*}{k} - \beta S\right) & (-p_1(1-m_3)S^*) \\ 0 & (\beta S^* - p_2(1-m_4)Y^* - d_2 - m_2) & 0 \\ (q_1 p_1(1-m_3)Y^*) & (q_2 p_2(1-m_4)Y^*) & (q_1 p_1(1-m_3)S^* - d_3) \end{bmatrix}, \tag{7}$$

where $S^* = \frac{d_3}{q_1 p_1(1-m_3)}$ and $Y^* = \frac{kq_1 p_1(1-m_3)(r-m_1)-r(d_3)}{kq_1(1-m_3)p_1^2}$.

Characteristic equation is given by $(\lambda - \lambda_1)(\lambda^2 - \xi\lambda + \zeta) = 0$, where $\lambda_1 = (\beta S^* - p_2(1-m_4)Y^* - d_2 - m_2) = \frac{\beta(d_3)}{q_1 p_1(1-m_3)} - \frac{p_2(1-m_4)(r-m_1)}{p_1(1-m_3)} + \frac{p_2(1-m_4)r(d_3)}{kq_1(1-m_3)p_1^2} - (d_2+m_2)$,

$\xi = \frac{-(d_3)[kq_1p_1(1-m_3)+r]}{kq_1(1-m_3)p_1}$, $\zeta = \frac{(d_3)[kq_1p_1(1-m_3)(r-m_1)-(d_3)r]}{kq_{1(1-m_3)}p_1}$.

Thus E_3 is locally asymptotically stable if the following conditions are satisfied:

$$\begin{cases} \left[\frac{\beta(d_3)}{q_1(1-m_3)p_1} - \frac{p_2(1-m_4)(r-m_1)}{p_1(1-m_3)} + \frac{p_2(1-m_4)r(d_3)}{kq_1(1-m_3)p_1^2} - (d_2+m_2)\right] < 0, \\ \text{Equation } (\lambda^2 - \xi\lambda + \zeta) \text{ have roots with negative real parts.} \end{cases}$$

3.2.1 Positive Equilibrium

In this case, populations of all the three species exists simultaneously. As promised, we will furnish the detail of the stability of the positive equilibrium point. For the stability of the positive equilibrium E_5, we state the following theorem:

Theorem 1 *System (4) with $\tau = 0$ is locally asymptotically stable at E_5 if the following conditions are satisfied:*

(i) $\Gamma\widetilde{S} + \Delta\widetilde{I} + \Theta\widetilde{Y} + \Lambda < 0$.

(ii) $A_1A_2 + A_3 > 0$,

where $\Gamma = (q_1p_1(1-m_3) + \beta - \frac{2r}{k})$,
$\Delta = (-(\frac{r}{k}+\beta) + q_2(1-m_4)p_2)$,
$\Theta = -(p_1(1-m_3) + p_2(1-m_4))$,
$\Lambda = (r - m_1 - m_2 - d_2 - d_3 - d_4 - c)$,
$A_1 = \Gamma\widetilde{S} + \Delta\widetilde{I} + \Theta\widetilde{Y} + \Lambda$,
$A_2 = \widetilde{S}^2[\beta q_1p_1(1-m_3) - \frac{2rq_1(1-m_3)p_1}{k} - \frac{2r\beta}{k}] + \widetilde{Y}^2[p_1(1-m_3)(1-m_4)p_2] + \widetilde{I}^2[-(\frac{r}{k}+\beta)q_2p_2(1-m_4)] + \widetilde{S}\widetilde{I}[-(\frac{r}{k}+\beta)q_1p_1(1-m_3) + \beta q_2p_2(1-m_4) - \frac{2rq_1p_1(1-m_3)}{k}] + \widetilde{Y}\widetilde{I}[-q_2p_2(1-m_4)(1-m_3)p_1 + p_2(1-m_4)(\frac{r}{k}+\beta)] + \widetilde{S}\widetilde{Y}[-q_1p_2p_1(1-m_4)(1-m_3) + \frac{2rp_2(1-m_4)}{k} - p_1(1-m_3)\beta] + \widetilde{S}[-\beta(d_3+d_4) - (c+m_2+d_2)q_1p_1(1-m_3) + (r-m_1)q_1p_1(1-m_3) + \frac{2r(d_3+d_4)}{k} + \beta(r-m_1) + \frac{2r(c+d_2+m_2)}{k}] + \widetilde{I}[-(c+d_2+m_2)(1-m_4)q_2p_2 + (r-m_1)q_2(1-m_4)p_2 + (\frac{r}{k}+\beta)(d_3+d_4) + (\frac{r}{k}+\beta)(c+d_2+m_2)] + \widetilde{Y}[p_1(1-m_3)(d_3+d_4) - p_2(r-m_1)(1-m_4) + p_1(1-m_3)(c+m_2+d_2)] + [p_2(1-m_4)(d_3+d_4) + (c+m_2+d_2)(d_3+d_4) - (d_3+d_4)(r-m_1) - (r-m_1)(c+m_2+d_2)]$,
$A_3 = \widetilde{S}^3[-\frac{2r\beta q_1p_1(1-m_3)}{k}] + \widetilde{S}^2\widetilde{Y}[\frac{2rq_1p_1p_2(1-m_4)(1-m_3)}{k}] + \widetilde{I}^2\widetilde{S}[q_2p_2(1-m_4)\beta(\frac{r}{k}+\beta)] + \widetilde{S}^2\widetilde{I}[\frac{2r\beta q_2p_2(1-m_4)}{k}] + \widetilde{S}^2[(r-m_1)\beta q_1(1-m_3)p_1 + \frac{2r\beta(d_3+d_4)}{k} + \frac{2rq_1p_1(1-m_3)(c+m_2+d_2)}{k}] + \widetilde{I}2[(\frac{r}{k}+\beta)q_2(1-m_4)p_2(c+d_2+m_2)] + \widetilde{S}\widetilde{I}\widetilde{Y}[-2\beta q_2p_1p_2(1-m_4)(1-m_3) + q_1p_1p_2(1-m_4)(1-m_3)(\frac{r}{k}+\beta)] + \widetilde{S}\widetilde{I}[(r-m_1)\beta q_2p_2(1-m_4) + (\frac{r}{k}+\beta)(c+m_2+d_2)q_1p_1(1-m_3) + \frac{2r}{k}(c+m_2+d_2)q_2p_2(1-m_4)] + \widetilde{S}\widetilde{Y}[\beta p_1(1-m_3)(d_3+d_4)] + \widetilde{I}\widetilde{Y}[(c+m_2+d_2)q_2p_1p_2(1-m_4)(1-m_3)] +$

$\widetilde{S}[(r - m_1)\{-\beta(d_3 + d_4) - (c + m_2 + d_2)q_1 p_1(1 - m_3)\} - \frac{2rp_2(1-m_4)(d_3+d_4)}{k} - \frac{2r(c+m_2+d_2)(d_3+d_4)}{k}] + \widetilde{I}[-(r - m_1)(c + m_2 + d_2)q_2 p_2(1 - m_4) - (\frac{r}{k} + \beta)(c + m_2 + d_2)(d_3 + d_4)] + [p_2(1 - m_4)(d_3 + d_4)(r - m_1) + (r - m_1)(c + m_2 + d_2)(d_3 + d_4)]$.

Proof 1 Jacobian matrix at E_5 is given by

$$J = \begin{pmatrix} (r - \frac{2r\widetilde{S}}{k} - fracr\widetilde{I}k - \beta\widetilde{I} - p_1(1-m_3)\widetilde{Y}) & (-\frac{r\widetilde{S}}{k} - \beta\widetilde{S}) & (-p_1(1-m_3)\widetilde{S}) \\ (\beta\widetilde{I}) & (\beta\widetilde{S} - p_2(1-m_4)\widetilde{Y} - c - d_2 - m_2) & (-p_2(1-m_4)\widetilde{I}) \\ (q_1 p_1(1-m_3)\widetilde{Y}) & (q_2 p_2(1-m_4)\widetilde{Y}) & (q_1 p_1(1-m_3)\widetilde{S} + q_2 p_2(1-m_4)\widetilde{I} - d_3 - d_4) \end{pmatrix}. \tag{8}$$

The characteristic equation is given by $\lambda^3 + A_1\lambda^2 + A_2\lambda + A_3 = 0$, where A_1, A_2 and A_3 are the same as defined in statement of the theorem. By Routh-Hurwitz criteria the theorem follows.

4 Stability Analysis of the Model with Time Delay

In this section, model (4) with $\tau \neq 0$ is considered. It is also important to mention that we will consider the positive equilibrium (E_*) only. At any point, jacobian matrix of system (4) is given by

$$J = \begin{bmatrix} \left(r\left(1 - \frac{S+I}{k}\right) - \frac{rS}{k} - \beta I - p_1(1-m_3)Y - m_1\right) & \left(-\frac{rS}{k} - \beta S\right) & (-p_1(1-m_3)S) \\ (\beta I) & (\beta S - p_2(1-m_4)Y - c - d_2 - m_2) & (-p_2(1-m_4)I) \\ 0 & 0 & (-d_3 - d_4) \end{bmatrix}$$
$$+ \begin{bmatrix} 0 & 0 & 0 \\ 0 & 0 & 0 \\ (q_1 p_1(1-m_3)Y) & (q_2 p_2(1-m_4)Y) & (q_1 p_1(1-m_3)S + q_2 p_2(1-m_4)I) \end{bmatrix}(e^{-\lambda\tau}),$$

here λ being a complex number. Now we state two more lemmas from [10, 27];

Lemma 3 *Let* $\lambda = (A + iB)$ $A > 0$, $B > 0$ *then,*

- *if* $A < B$, *all roots of the equation* $\lambda + A - Be^{-\lambda\tau} = 0$ *have positive real parts for* $\tau < \frac{1}{\sqrt{B^2 - A^2}} cos^{-1}\left(\frac{A}{B}\right)$.
- *if* $A > B$, *all roots of the equation* $\lambda + A - Be^{-\lambda\tau} = 0$ *have negative real parts for any* τ.

Lemma 4 *Let the polynomial,* $h(z) = z^3 + p_0 z^2 + q_0 z + r_0 = 0$

(i) if $r_0 < 0$, *then this equation has at least one positive root;*
(ii) if $r_0 \geq 0$ *and* $\Delta = (p_0^2 - 3q_0) \leq 0$ *then this equation has no positive roots;*
(iii) if $r_0 \geq 0$ *and* $\Delta = (p_0^2 - 3q_0) > 0$, *then this equation has positive roots if and only if* $z_1^* = \frac{-p_0 + \sqrt{\Delta}}{3}$ *and* $h(z_1^*) \leq 0$.

Let us study nonzero equilibrium $E_* = (S_*, I_*, Y_*)$. The jacobian matrix at $E_* = (S_*, I_*, Y_*)$ is given as

$$J(E_*) = \begin{pmatrix} \left(r\left(1-\frac{S_*+I_*}{k}\right)-\frac{rS_*}{k}-p_1(1-m_3)Y_*-\beta I_*-m_1\right) & \left(-\frac{rS_*}{k}-\beta S_*\right) & (-p_1(1-m_3)S_*) \\ \beta I_* & (\beta S_*-p_2(1-m_4)Y_*-c-d_2-m_2) & p_2(1-,m_4)I_* \\ (q_1p_1(1-m_3)Y_*e^{-\lambda\tau}) & (q_2p_2(1-m_4)Y_*e^{-\lambda\tau}) & \begin{matrix}(q_1p_1(1-m_3)S_* \\ +q_2p_2(1-m_4)I_*)e^{-\lambda\tau}-d_3-d_4)\end{matrix} \end{pmatrix}.$$

The characteristics equation is given as $(\lambda^3+M_2\lambda^2+M_1\lambda+M_0)+(n_2\lambda^2+n_1\lambda+n_0)e^{-\lambda\tau}=0$, where,

$$M_2 = \left(\frac{2r}{k}+\frac{p_1(1-m_3)\beta}{p_2(1-m_4)}-\frac{q_1p_1(1-m_3)\beta}{q_2p_2}(1-m_4)S_*\right) + \left(\frac{r(d_3+d_4)}{q_2p_2(1-m_4)K}+d_3+d_4-\frac{p_1(1-m_3)(c+d_2+m_2)}{p_2(1-m_4)}-r\right),$$

$M_1 = (\{(d_3+d_4)(c+d_2+m_2)-(r-m_1)(c+d_2+m_2)\}+S_*\{-(d_3+d_4)\beta+\frac{2r}{K}(d_3+d_4)+(r-m_1)\beta+\frac{2r}{k}(c+d_2+m_2)\}+I_*\{(\frac{r}{k}+\beta)(d_3+d_4+c+d_2+m_2)\}+Y_*\{(d_3+d_4)p_2(1-m_4)+(d_3+d_4)p_1(1-m_3)+(c+d_2+m_2)p_1(1-m_3)+(r-m_1)p_2(1-m_4)\}+S_*^2(-\frac{2r\beta}{k})+Y_*^2(p_1p_2(1-m_3)(1-m_4))+S_*Y_*(-\beta p_1(1-m_3)+\frac{2r}{k}p_2(1-m_4))+I_*Y_*(\frac{r}{k}+\beta)p_2(1-m_4))$,

$M_0 = (\{-(r-m_1)(c+d_2+m_2)(d_3+d_4)+S_*\{(d_3+d_4)\beta(r-m_1)+\frac{2r}{k}(d_3+d_4)\}(c+d_2+m_2)\}+I_*\{(d_3+d_4)(c+d_2+m_2)(\frac{r}{k}+\beta)\}+Y_*\{(d_3+d_4)(c+d_2+m_2)p_1(1-m_3)-(r-m_1)(d_3+d_4)p_2\}+S_*^2(-\frac{2r}{k}(d_3+d_4)\beta)+Y_*^2(d_3+d_4)p_2p_1(1-m_3)(1-m_4))+S_*Y_*\{-(d_3+d_4)\beta p_1(1-m_3)-\frac{2r}{k}(d_3+d_4)p_2(1-m_4)\}+I_*Y_*\{(\frac{r}{k}+\beta)p_2(1-m_4)(d_3+d_4)\})$,

$n_2 = -(d_3+d_4)$,

$n_1 = (S_*^2\{q_1p_1(1-m_3)\beta-\frac{-2r}{k}q_1p_1(1-m_3)\}+I_*^2\{-(\frac{r}{k}+\beta)q_2p_2(1-m_4)\}+S_*Y_*\{-q_1p_2p_1(1-m_3)(1-m_4)\}+S_*I_*\{q_2p_2(1-m_4)\beta+\frac{2r}{k}q_2p_2(1-m_4)\}+Y_*I_*\{-q_2p_2p_1(1-m_3)(1-m_4)\}+S_*\{\{-(c+d_2+m_2)-(r-m_1)\}q_1p_1(1-m_3)\}+I_*\{-(c+d_2+m_2)q_2p_2(1-m_4)+(r-m_1)q_2p_2(1-m_4)\})$,

$n_0 = -(S_*^2\{q_1p_1(1-m_3)\beta(r-m_1-\frac{2r}{k})+\frac{2r}{K}(c+d_2+m_1)q_1p_1(1-m_3)-q_1\{p_1(1-m_3)\}^2\beta\}+I_*^2\{(\frac{r}{k}+\beta)(c+d_2+m_2)q_2p_2(1-m_4)\}+S_*^2Y_*\{\frac{2r}{k}q_1p_1p_2(1-m_3)(1-m_3)+q_1\{p_1(1-m_3)\}^2\beta\}+S_*^2I_*\{-\frac{2r}{k}q_2p_2(1-m_4)\beta\}+S_*I_*Y_*\{(\frac{r}{k}+\beta)q_1p_1p_2(1-m_3)(1-m_4)-2\beta q_2p_1p_2(1-m_3)(1-m_4)\}+S_*Y_*\{\{-(r-m_1)+(\frac{r}{k}+\beta)\}q_1p_1p_2(1-m_3)(1-m_4)\}+S_*I_*\{\beta(r-m_1)q_2p_2(1-m_4)+\frac{2r}{k}(c+d_2+m_2)q_2p_2(1-m_4)+(c+d_2+m_2)(\frac{r}{k}+\beta)q_1p_1(1-m_3)\}+Y_*I_*\{(c+d_2+m_2)q_2p_2p_1(1-m_3)(1-m_4)\}+S_*\{(c+d_2+m_2)(r-m_1)q_1p_1(1-m_3)\}+I_*\{-(c+d_2+m_2)(r-m_1)q_2p_2(1-m_4)\})$.

Now we put $\lambda = i\omega$ $(\omega > 0)$ we get

Real Part: $\{n_2\omega^2+n_0\}\cos\omega\tau+\{n_1\omega\sin\omega\tau-M_2\omega^2+M_0\}$, **Imaginary Part**: $n_1\omega\cos\omega\tau-(-n_2\omega^2+n_0)\sin\omega\tau+M_1\omega-\omega^3$ (**Real Part**)^2+(**Imaginary Part**)^2 $=\omega^6+p_0\omega^4+q_0\omega^2+r_0$. Hence, we have $\omega^6+p_0\omega^4+q_0\omega^2+r_0=0$, where

$p_0=(M_2^2-2M_1-n_2^2)$ $q_0=(M_1^2-2M_2M_0+2n_2n_0-n_1^2)$ $r_0=(M_0^2-n_0^2)$. If we put $z=\omega^2$, then we have the equation $z^3+p_0z^2+q_0z+r_0=0$. If $M_0^2\geq n_0^2$, then we will have $r_0 \geq 0$, we have two situations for Δ (i)$\Delta = (p_0^2 - 3q_0) \leq 0$. (ii)$\Delta = (p_0^2 - 3q_0) > 0$.

In situation (i) we have to say that E_* is absolutely stable if $r_0 \geq 0$ and $\Delta = (p_0^2 - 3q_0) \leq 0$. Also, if we have and $r_0 \geq 0$ $\Delta = (p_0^2 - 3q_0) > 0$ then equation has negative roots if and only if $h(z_1^*) > 0$ where $z_1^* = \frac{-p_0+\sqrt{\Delta}}{3}$ thus we have the following theorem for the stability of E_*.

Theorem 2 *$E_*(S_*, I_*, Y_*)$ is absolutely stable if one of the following conditions holds:*

(i) $\Delta = (p_0^2 - 3q_0) \leq 0$.
(ii) $\Delta = (p_0^2 - 3q_0) > 0$ *and* $z_1^* = \frac{-p_0+\sqrt{\Delta}}{3} < 0$.
(iii) $\Delta = (p_0^2 - 3q_0) > 0$, $z_1^* = \frac{-p_0+\sqrt{\Delta}}{3} > 0$ *and* $h(z_1^*) > 0$ *provided* $r_0 \geq 0$.

Next, if we consider the case when $r_0 < 0$ or $\{r_0 \geq 0, \Delta = (p_0^2 - 3q_0) > 0, z_1^* > 0, h(z_1^*) < 0\}$. Then, according to lemma , equation will have one positive root say ω_0 that is the characteristic equation has a pair of purely imaginary roots say $\pm i\omega_0$. Now assume that $i\omega_0, \omega_0 > 0$ is a root of $h(z)$, then we have real and imaginary parts as under.

Real Part$= \{n_2\omega^2 + n_0\}\cos\omega\tau + \{n_1\omega\sin\omega\tau - M_2\omega^2 + M_0\} = 0$.

Imaginary Part $= n_1\omega\cos\omega\tau - (-n_2\omega^2 + n_0)\sin\omega\tau + M_1\omega - \omega^3 = 0$.

Solving the above equation for τ, we have (by eliminating $\sin\omega\tau$ between these equations)

$$\tau = \frac{1}{\omega_0}\cos^{-1}\left(\frac{n_1\omega_0^2\{\omega_0 - M_1\} - \{M_2\omega_0^2 - M_0\}\{n_2\omega_0^2 - n_0\}}{n_1^2\omega_0^2 + n_2\omega_0^2 - n_0}\right) + \frac{2k\pi}{\omega_0}, (k = 0, 1, 2, \ldots)$$

We call it as a 'critical value' and may be denoted as $\tau_k = \frac{1}{\omega_0}\cos^{-1}\left(\frac{n_1\omega_0^2\{\omega_0 - M_1\} - \{M_2\omega_0^2 - M_0\}\{n_2\omega_0^2 - n_0\}}{n_1^2\omega_0^2 + n_2\omega_0^2 - n_0}\right) + \frac{2k\pi}{\omega_0}$, $(k = 0, 1, 2, \ldots)$. This is corresponding to the characteristic equation as it has purely imaginary roots $\pm i\omega$, which is a result similar to that of Hu et al. 2012 [10]. Differentiating the characteristics equation w.r.t. τ, we get $(\frac{d\lambda}{d\tau})^{-1} = \frac{(3\lambda^2+2M_2\lambda+M_1)e^{\lambda\tau}}{(\lambda^2 n_2+\lambda n_1+n_0)\lambda} + \frac{2n_2\lambda+n_1}{(\lambda^2 n_2+\lambda n_1+n_0)\lambda} - \frac{\tau}{\lambda}$ or $(\frac{d\lambda}{d\tau})^{-1} = \frac{(3\lambda^2+2M_2\lambda+M_1)e^{\lambda\tau} + (2n_2\lambda+n_1) - \tau(2n_2\lambda+n_1)}{(\lambda^2 n_2+\lambda n_1+n_0)\lambda}$. As proved in [10], it is easy to prove the transversality condition at τ_k e.g. $\frac{d(Re\lambda)}{d\tau} \neq 0$. τ_k is used as a point for direction of Hopf Bifurcation as in the next section.

Remark 2 The equilibrium points E_5 of model (4) with $\tau = 0$ and E_* of model (4) with $\tau \neq 0$ are ecologically similar. Both convey the message that all the species exist simultaneously.

5 Direction and Stability of the Hopf-Bifurcation

With the symbols used in [10] and procedure explained in [5], we have the following system of functional differential equation, $\dot{u}(t) = L_\mu(\mu_t) + F(\mu, u_t)$, where $u_t(\theta) = u(t+\theta) \in R^3$ and $L_\mu : \mathbb{R} \times \mathbb{C} \to \mathbb{R}^3$ and $F : \mathbb{R} \times \mathbb{C} \to \mathbb{R}^3$ are given as

$$L_\mu\phi = (\tau_k + \mu)\begin{bmatrix} -\frac{rS_*}{k} & -\left(\frac{r}{k} + \beta\right)S_* & (-p_1(1-m_3)S_*) \\ \beta I_* & (\beta S_* - p_2(1-m_4)Y_* - c - d_2 - m_2) & -p_2(1-m_4)I_* \\ 0 & 0 & -d_3 - d_4) \end{bmatrix} \times \phi(0)$$

$$+ (\tau_k + \mu)\begin{bmatrix} 0 & 0 & 0 \\ 0 & 0 & 0 \\ q_1p_1(1-m_3)Y_* & q_2p_2(1-m_4)Y_* & q_1p_1(1-m_3)S_* + q_2p_{2(1-m_4)}I_* \end{bmatrix} \times \phi(-1),$$

and

$$F(\mu, \theta) = \begin{pmatrix} -\frac{r}{k}\phi_1^2(0) - \left(\frac{r}{k} + \beta\right)\phi_1(0)\phi_2(0) - p_1(1-m_3)\phi_1(0)\phi_3(0) \\ \beta\phi_1(0)\phi_2(0) - p_2(1-m_4)\phi_2(0)\phi_3(0) \\ q_1p_1(1-m_3)\phi_1(-1)\phi_3(-1) + q_2p_2(1-m_4)\phi_1(-1)\phi_2(-1) \end{pmatrix},$$

$\phi(0) \equiv (\phi_1(0), \phi_1(0), \phi_1(0))^T \in \mathbb{C}$ i.e.

$$L_\mu\phi = (\tau_k + \mu)\begin{bmatrix} -\frac{rS_*}{k} & -\left(\frac{r}{k} + \beta\right)S_* & (-p_1(1-m_3)S_*) \\ \beta I_* & (\beta S_* - p_2(1-m_4)Y_* - c - d_2 - m_2) & -p_2(1-m_4)I_* \\ 0 & 0 & -d_3 - d_4 \end{bmatrix} \times \begin{pmatrix} \phi_1(0) \\ \phi_2(0) \\ \phi_3(0) \end{pmatrix}$$

$$+ (\tau_k + \mu)\begin{bmatrix} 0 & 0 & 0 \\ 0 & 0 & 0 \\ q_1p_1(1-m_3)Y_* & q_2(1-m_4)p_2Y_* & q_1p_1(1-m_3)S_* + q_2(1-m_4)p_2I_* \end{bmatrix} \times \begin{pmatrix} \phi_1(-1) \\ \phi_2(-1) \\ \phi_3(-1) \end{pmatrix},$$

we have considered, $\tau = (\tau_k + \mu)$, $\mu = 0$ gives the hopf bifurcation value for the mathematical model with delay as promised in the previous section. Normalizing delay τ by timescaling $t \to \frac{t}{\tau}$ the model is written in the Banach Space $\mathbb{C} \equiv \mathbb{C}([-1, 0], \mathbb{R}^3)$. By the Riesz representation theorem, we found that there exists a matrix function whose components are bounded variation function $\eta(\theta, \mu)$ in $\theta \in [-1, 0]$ such that

$L_\mu\phi = \int_\Omega d\eta(\theta, \mu)\phi(\theta)$, $\phi \in \mathbb{C}$, $\Omega \in [-1, 0)$.

We can choose

$$\eta(\theta, \mu) = (\tau_k + \mu)\begin{bmatrix} -\frac{rS_*}{k} & -\left(\frac{r}{k} + \beta\right)S_* & (-p_1(1-m_3)S_*) \\ \beta I_* & (\beta S_* - p_2(1-m_4)Y_* - c - d_2 - m_2) & -p_2(1-m_4)I_* \\ 0 & 0 & -d_3 - d_4) \end{bmatrix} \times \delta(\theta)$$

$$- (\tau_k + \mu)\begin{bmatrix} 0 & 0 & 0 \\ 0 & 0 & 0 \\ q_1p_1(1-m_3)Y_* & q_2p_2(1-m_4)Y_* & q_1(1-m_3)p_1S_* + q_2(1-m_4)p_2I_* \end{bmatrix} \times \delta(\theta + 1).$$

where $\delta(\theta)$ denotes the dirac delta function, viz.,

$$\delta(\theta) = \begin{cases} 0, & \theta \neq 0 \\ 1, & \theta \doteq 0, \end{cases}$$

for $\phi \in \mathbb{C}^1([-1,0], \mathbb{R}^3)$, define

$$A(\mu)\phi(\theta) = \begin{cases} \frac{d\phi(\theta)}{d\theta} & \theta \in [-1,0) \\ \int_{-1}^{0} d\eta(\theta,\mu)\phi(\theta) & \theta \doteq 0 \end{cases}$$

$$\text{or } A(\mu)\phi(\theta) = \begin{cases} \frac{d\phi(\theta)}{d\theta}, & -1 \leq \theta < 0 \\ \int_{-1}^{0} d\eta(\theta,\mu)\phi(\theta), & \theta = 0 \end{cases}$$

and

$$\mathbb{R}(\mu)\phi(\theta) = \begin{cases} 0, & \theta \in [-1,0) \\ F(\mu,\phi), & \theta \doteq 0 \end{cases} = \begin{cases} 0, & -1 \leq \theta < 0 \\ F(\mu,\phi), & \theta \doteq 0 \end{cases}$$

with these symbols, $u\dot{(t)} = L_\mu(\mu_t) + F(\mu, \mu_t)$ may be written as

$$u\dot{(t)} = A\mu(\mu_t) + \mathbb{R}(\mu)\mu_t \tag{9}$$

which is an abstract differential equation. Where $u_t(\theta) = u(t+\theta), -1 \leq \theta < 0$. Now we come to operator theory, for $\psi \in \mathbb{C}^1\big([0,1], (\mathbb{R}^3)^*\big)$ we define A^*, adjoint operator of A,

$$A^*\psi(S) = \begin{cases} -\frac{d\psi(S)}{dS} & S \in (0,1] \\ \int_{-1}^{0} d\eta^T(S,\mu)\psi(-S) & S = 0. \end{cases}$$

And a bilinear product $< \psi(S), \phi(\theta) >= \overline{\psi}(0)\phi(0) - \int_1^0 \int_{\xi=0}^{\theta} \overline{\psi}^T(\xi - \theta)d\eta(\theta)$ $\phi(\xi)d\xi$ where $\eta(\theta) = \eta(\theta, 0)$. Then $A(0)$ and A_* are adjoint operators. Now $\pm i\omega_0\tau_k$ are eigen values of $A(0)$. Hence they are eigenvalues of A^* also. To determine the poincare normal form of the operator A, we first need to evaluate the eigenvectors of $A(0)$ and A^* corresponding to $i\omega_0\tau_k$ and $-i\omega_0\tau_k$ respectively. Suppose that $q(\theta) = (1, \alpha_1, \alpha_2)^I \exp(i\omega_0\tau_k\theta)$ is the eigen vector of $A(0)$ corresponding to $i\omega_0\tau_k$, then we have $A(0)q(\theta) = i\omega_0 q(\theta)$ from the definition of $A(0)$,we have

$$\left[\begin{bmatrix} -\frac{rS_*}{k} & -\left(\frac{r}{k}+\beta\right)S_* & (-p_1(1-m_3)S_*) \\ \beta I_* & (\beta S_* - p_2(1-m_4)Y_* - c - d_2 - m_2) & -p_2(1-m_4)I_* \\ 0 & 0 & -d_3 - d_4 \end{bmatrix} + \begin{bmatrix} 0 & 0 & 0 \\ 0 & 0 & 0 \\ q_1p_1(1-m_3)Y_* & q_2p_2(1-m_4)Y_* & q_1p_1(1-m_3)S_* + q_2(1-m_4)p_2I_* \end{bmatrix} \times \exp(i\omega_0\tau_k)\right] \begin{pmatrix} 1 \\ \alpha_1 \\ \alpha_2 \end{pmatrix} = i\omega_0 \begin{pmatrix} 1 \\ \alpha_1 \\ \alpha_2 \end{pmatrix},$$

$$\left[\begin{bmatrix} -\frac{rS_*}{k} & -\left(\frac{r}{k}+\beta\right)S_* & (-p_1(1-m_3)S_*) \\ \beta I_* & (\beta S_* - p_2(1-m_4)Y_* - c - d_2 - m_2) & -p_2(1-m_4)I_* \\ 0 & 0 & -d_3 - d_4 \end{bmatrix} \times \begin{pmatrix} 1 \\ \alpha_1 \\ \alpha_2 \end{pmatrix} + \right.$$

$$\left[\begin{array}{ccc} 0 & 0 & 0 \\ 0 & 0 & 0 \\ q_1p_1(1-m_3)Y_* & q_2p_2(1-m_4)Y_* & q_1p_1(1-m_3)S_*+q_2p_2(1-m_4)I_* \end{array}\right] \times \exp(i\omega_0\tau_k)\Bigg] \begin{pmatrix} \exp(-i\omega_0\tau_k) \\ \alpha_1\exp(-i\omega_0\tau_k) \\ \alpha_2\exp(-i\omega_0\tau_k) \end{pmatrix}$$

$$= i\omega_0 \begin{pmatrix} 1 \\ \alpha_1 \\ \alpha_2 \end{pmatrix}.$$

We obtain, $\alpha_1 = \frac{-p_2(1-m_4)I_*(i\omega_0+\frac{rS_*}{k})-p_1(1-m_3)\beta S_* I_*}{p_2(1-m_4)(\frac{r}{k}+\beta)S_*I_*-p_1(1-m_3)S_*(i\omega_0-\beta S_*+c+d_2+m_2+p_2(1-m_4)Y_*)}$,

$\alpha_2 = \frac{q_1p_1(1-m_3)Y_*\exp(-i\omega_0\tau_k+q_2p_2(1-m_4)Y_*\exp(-i\omega_0\tau_k)}{i\omega_0+d_3+d_4-q_1p_1(1-m_3)S_*+q_2p_2(1-m_4)I_*}$.

Next, suppose that $q_*(s) = B(1,\alpha_1^*,\alpha_2^*)\exp(i\omega_0\tau_k s)$ is the eigen vector of A^* corresponding to $-i\omega_0\tau k$ similarly, we have,

$\alpha_1^* = \frac{-p_1(1-m_3)(\frac{r}{k}+\beta)S_*-p_2(1-m_4)(i\omega_0-\frac{rS_*}{k})}{p_2(1-m_4)\beta I_*-p_1(1-m_3)(i\omega_0+\beta S_*-c-d_2-m_2-p_2(1-m_4)Y_*)}$,

$\alpha_2^* = \frac{-p_1(1-m_3)S_*-p_2(1-m_4)I_*\alpha_1^*}{-i\omega_0+d_3+d_4-(q_1p_1(1-m_3)S_*+q_2p_2(1-m_4)I_*)\exp(-i\omega_0\tau_k)}$,

where B has to be calculated. We have the conditions

$\left.\begin{array}{l} < q^*, q(\theta) >= 1 \\ < q^*, \overline{q(\theta)} >= 0 \end{array}\right\}$ which may be verified.

$< q^*, q(\theta) >= \overline{q^*}(0)q(0) - \int_{-1}^{0}\int_{\xi=0}^{\infty}\overline{q^*}^T(\xi-\theta)d\eta(\theta)q(\xi)d\xi$

$= \overline{B}(1,\overline{\alpha_1}^*,\overline{\alpha_2}^*)(1,\alpha_1,\alpha_2)^T - \int_{-1}^{0}\int_{\xi=0}^{\infty}\overline{B}(1,\overline{\alpha_1}^*,\overline{\alpha_2}^*)\exp(-i\omega_0\tau_k(\xi-\theta))$ $d\eta(\theta)\times(1,\alpha_1,\alpha_2)^T\exp(i\omega_0\tau_k\xi)d\xi = \overline{B}\{1+\alpha_1\overline{\alpha_1}^*+\alpha_2\overline{\alpha_2}^* - \int_{-1}^{0}(1,\overline{\alpha_1}^*,\overline{\alpha_2}^*)$ $\exp(i\omega_0\tau_k)d\eta(\theta)(1,\alpha_1,\alpha_2)^T\} = \overline{B}\{1+\alpha_1\overline{\alpha_1}^*+\alpha_2\overline{\alpha_2}^*+\tau_k[q_2p_2(1-m_4)\overline{\alpha_2}^*Y_* +$ $q_2p_2(1-m_4)\alpha_1\overline{\alpha_2}^*Y_*+(q_1p_1(1-m_3)S_*+q_2p_2(1-m_4)I_*)\alpha_2\overline{\alpha_2}^*]\exp(-i\omega_0\tau_k)\}$

which gives:

$$\overline{B} = \frac{1}{\{1+\alpha_1\overline{\alpha_1}^*+\alpha_2\overline{\alpha_2}^*+\tau_k[q_2p_2(1-m_4)\overline{\alpha_2}^*Y_*+q_2p_2(1-m_4)\alpha_1\overline{\alpha_2}^*Y_* + (q_1p_1(1-m_3)S_*+q_2p_2(1-m_4)I_*)\alpha_2\overline{\alpha_2}^*]\exp(-i\omega_0\tau_k)\}}.$$

5.1 Stability of Bifurcated Periodic Solutions

We first compute the coordinates to describe the Center Manifold $\mathbb{C}_0$ at $\mu = 0$. Let u_t be the solution of $\dot{u}(t) = L_\mu(u_t) + F(\mu,\mu_t)$ and define, $z(t) =< q^*, u_t >$, q^* being the eigenvalue of A^*. And $W(t,\theta) = u_t(\theta) - 2Re\{z(t)q(\theta)\}$ on the Center Manifold $\mathbb{C}_0$, we have, $W(t,\theta) = W(z(t),\overline{z(t)},\theta)$, where,

$W(z,\overline{z},\theta) = W_{20}(\theta)\frac{z^2}{2} + W_{02}(\theta)\frac{\overline{z}^2}{2} + W_{11}(\theta)z\overline{z} + W_{30}\frac{z^3}{\lfloor 3} + \cdots$

In fact, z and $\overline{z}$ are local coordinates for the Center Manifold $\mathbb{C}_0$ in the direction of q^* and $\overline{q}^*$ respectively. The existence of $\mathbb{C}_0$ will provide an opportunity to reduce the system $\dot{u}(t) = L_\mu(u_t) + F(\mu,\mu_t)$ into an Ordinary Differential Equation ODE(in a single complex variable z) on $\mathbb{C}_0$ which is very interesting. u_t is the solution of system under consideration. $u_t \in \mathbb{C}_0$, we have

$\dot{z}(t) =< q^*, \dot{u_t} >$

$=< q^*, A(u_t) + R(u_t) >$

$= < q^*, A(u_t) > + < q^*, R(u_t) >$
$= < A^*q^*, (u_t) > + < q^*, R(u_t) >$
$= i\omega_0 \overline{\tau} z + \overline{q}^* \cdot F(0, W(t, 0) + 2Re[z(t)q(\theta)])$

Rewrite it as $\dot{z}(t) = i\omega_0 \overline{\tau} z + g(z, \overline{z})$, where $g(z, \overline{z}) = g_{20}(\theta)\frac{z^2}{2} + g_{02}(\theta)\frac{\overline{z}^2}{2} + g_{11}(\theta)z\overline{z} + g_{21}\frac{\overline{z}z^2}{\lfloor 3} + \cdots$

The above two equations give us $g(z, \overline{z}) = (\overline{q}^*)^T F(z, \overline{z})$

$$= \tau_k \overline{B}(1, \alpha_1^*, \alpha_2^*) \begin{pmatrix} -\frac{r}{k}u_1^2(t) - (\frac{r}{k} + \beta)u_1(t)u_2(t) - p_1(1 - m_3)u_1(t)u_3(t) \\ \beta u_1(t)u_2(t) - p_2(1 - m_4)u_2(t)u_3(t) \\ p_1 q_1(1 - m_3)u_1(t - 1)(t)u_3(t - 1) + p_2 q_2(1 - m_4)u_1(t - 1)u_2(t - 1) \end{pmatrix}$$

Further,
$u(t + \theta) = W(t, \theta) + z(t)q(\theta) + \overline{z}(t)\overline{q}(\theta)$,
$u_1(t) = z + \overline{z} + W^{(1)}(t, 0)$,
$u_2(t) = \alpha_1 z + \overline{\alpha_1 z} + W^{(2)}(t, 0)$,
$u_3(t) = \alpha_2 z + \overline{\alpha_2 z} + W^{(3)}(t, 0)$,
$u_1(t - 1) = z\exp(-i\omega_0\tau_k) + \overline{z}\exp(i\omega_0\tau_k) + W^{(1)}(t, -1)$,
$u_2(t - 1) = \alpha_1 z\exp(-i\omega_0\tau_k) + \overline{\alpha_1 z}\exp(i\omega_0\tau_k) + W^{(2)}(t, -1)$,
$u_3(t - 1) = \alpha_2 z\exp(-i\omega_0\tau_k) + \overline{\alpha_2 z}\exp(i\omega_0\tau_k) + W^{(3)}(t, -1)$. Hence, $g(z, \overline{z}) = \tau_k\overline{B}[-\frac{r}{k}u_1^2(t) - (\frac{r}{k} + \beta)u_1(t)u_2(t) - p_1(1 - m_3)u_1(t)u_3(t) + \overline{\alpha_1}^*\{\beta u_1(t)u_2(t) - p_2(1 - m_4)u_2(t)u_3(t)\} + \overline{\alpha_2}^*\{p_1(1 - m_3)q_1 u_1(t - 1)(t)u_3(t - 1) + p_2(1 - m_4)q_2 u_1 (t - 1)u_2(t - 1)\}]$.

Putting the values of $u_1, u_2, u_3, u_1(t - 1), u_2(t - 1), u_3(t - 1)$ etc. in $g(z, \overline{z})$, we get

$g(z, \overline{z}) = \tau_k\overline{B}\Big(-\frac{r}{k}[z + \overline{z} + W^{(1)}(t, 0)]^2 - (\frac{r}{k} + \beta)[z + \overline{z} + W^{(1)}(t, 0)][\alpha_1 z + \overline{\alpha_1 z} + W^{(2)}(t, 0)] - p_1[z + \overline{z} + W^{(1)}(t, 0)][\alpha_2 z + \overline{\alpha_2 z} + W^{(3)}(t, 0)] + \overline{\alpha_1}^*(\beta[z + \overline{z} + W^{(1)}(t, 0)][\alpha_1 z + \overline{\alpha_1 z} + W^{(2)}(t, 0)] - p_2(1 - m_4)[\alpha_1 z + \overline{\alpha_1 z} + W^{(2)}(t, 0)][\alpha_2 z + \overline{\alpha_2 z} + W^{(3)}(t, 0)]) + \overline{\alpha_2}^*(p_1(1 - m_3)q_1[z\exp(-i\omega_0\tau_k) + \overline{z}\exp(i\omega_0\tau_k) + W^{(1)}(t, -1)][\alpha_2 z \exp(-i\omega_0\tau_k) + \overline{\alpha_2 z}\exp(i\omega_0\tau_k) + W^{(3)}(t, -1)] + p_2(1 - m_4)q_2[z\exp(-i\omega_0\tau_k) + \overline{z}\exp(i\omega_0\tau_k) + W^{(1)}(t, -1)][\alpha_1 z\exp(-i\omega_0\tau_k) + \overline{\alpha_1 z}\exp(i\omega_0\tau_k) + W^{(2)}(t, -1)])\Big)$.

From this equation we can find the values of the coefficients $g_{20}(\theta), g_{02}(\theta), g_{11}(\theta), g_{21}(\theta)$, etc., by comparing the same powers of z, we have

$g_{20} = 2\tau_k\overline{B}\{-\frac{r}{k} - (\frac{r}{k} + \beta)\alpha_1 - p_1(1 - m_3)\alpha_2 + \beta\overline{\alpha_1}^*\alpha_1 - \overline{\alpha_1}^*\alpha_1\alpha_2 p_2 (1 - m_4) + \overline{\alpha_2}^*(p_1(1 - m_3)q_1\alpha_2 + p_2(1 - m_4)q_2\alpha_1)\exp(-2i\omega_0\tau_k)\}$,

$g_{11} = \tau_k\overline{B}(-\frac{2r}{k} + (\overline{\alpha_1}^*)\beta + \overline{\alpha_2}^* p_2(1 - m_4)q_2 - \frac{r}{k} + \beta)(\overline{\alpha_1} + \alpha_1) + (\overline{\alpha_2}^* p_1 (1 - m_3)q_1 - p_1(1 - m_3))(\overline{\alpha_2} + \alpha_2) - \overline{\alpha_1}^* p_2(\overline{\alpha_2}\alpha_1 + \overline{\alpha_1}\alpha_2))$,

$g_{02} = 2\tau_k\overline{B}\{-\frac{r}{k} - (\frac{r}{k} + \beta)\overline{\alpha_1} - p_1(1 - m_3)\overline{\alpha_2} + \beta\overline{\alpha_1}^*\overline{\alpha_1} - \overline{\alpha_1}^*\overline{\alpha_1}\overline{\alpha_2}p_2(1 - m_4) + \overline{\alpha_2}^*(p_1(1 - m_3)q_1\overline{\alpha_2} + p_2 q_2(1 - m_4)\overline{\alpha_1})\exp(2i\omega_0\tau_k)\}$,

$g_{21} = 2\tau_k\overline{B}(-\frac{r}{k}(2W_{11}^{(1)}(0) + W_{20}^{(1)}(0)) - (\frac{r}{k} + \beta)[W_{11}^{(2)}(0) + \alpha_1 W_{11}^{(1)}(0) + \frac{1}{2}\overline{\alpha_1}W_{20}^{(1)}(0) + \frac{1}{2}W_{20}^{(1)}(0)] - p_1(1 - m_3)[W_{11}^{(3)}(0) + \frac{1}{2}W_{20}^{(3)}(0) + \alpha_2 W_{11}^{(1)}(0) + \frac{1}{2}W_{20}^{(1)}(0)] + \overline{\alpha_1}^*\beta[W_{11}^{(2)}(0) + \frac{1}{2}W_{20}^{(2)}(0) + \alpha_1 W_{11}^{(1)}(0) + \frac{1}{2}\overline{\alpha_1}W_{20}^{(1)}(0)]$

These coefficients are used in calculating $\mathbb{C}_0$, etc. Now we need to calculate $W_{20}(\theta)$ and $W_{11}(\theta)$. Now $\dot{u}_t = A(\mu)u_t + R(\mu)u_t$ and $z(t) =< q^*, u_t >, W(t,\theta) = u_t(\theta) - 2Re\{z(t)q(\theta)\}$ gives us

$\dot{W} = \dot{u_t} - zq - \dot{\overline{zq}}$

$$= \begin{cases} AW - 2Re\overline{q}^*(0)F_0 q(\theta), & \text{for} -1 \le \theta < 0 \\ AW - 2Re\overline{q}^*(0)F_0 q(\theta) + F_0, & \text{for} \theta = 0. \end{cases}$$

Rewrite the above equation as, $\dot{W} = AW + H(z,\overline{z},\theta)$, where,

$H(z,\overline{z},\theta) = H_{20}(\theta)\frac{z^2}{2} + H_{11}(\theta)z\overline{z} + H_{02}(\theta)\frac{\overline{z^2}}{2} + H_{21}(\theta)\frac{z^2\overline{z}}{2} + \cdots$ Near to the origin on $\mathbb{C}_0$, $\dot{W} = W_z\dot{z} + W_{\overline{z}}\overline{z}$ $(A - 2i\omega_0\tau_k)W_{20}(\theta) = -H_{20}(\theta)$ and $AW_{11}(\theta) = -H_{11}(\theta)$ hence for $-1 \le \theta < 0$ we have, $H(z,\overline{z},\theta) = -2Re(\overline{q}^*(0)F_0 q(\theta)) = -g(z,\overline{z})q(\theta) - \overline{g}(z,\overline{z})\overline{q}(\theta)$, by comparing the coefficients of z, we have $H_{20}(\theta) = -g_{20}q(\theta) - \overline{g}_{02}\overline{q}(\theta)$ and $H_{11}(\theta) = -g_{11}q(\theta) - \overline{g}_{11}\overline{q}(\theta)$,

$\dot{W}_{20}(\theta) = 2i\omega_0\tau_k W_{20}(\theta) + g_{20}q(\theta) + \overline{g}_{02}\overline{q}(\theta)$,

$\dot{W}_{11}(\theta) = g_{11}q(\theta) + \overline{g}_{11}\overline{q}(\theta)$.

Integrating, we have

$W_{20}(\theta) = \frac{ig_{20}}{\omega_0\tau_k}q(0)\exp(i\omega_0\tau_k\theta) + \frac{i\overline{g}_{20}\overline{q}(0)}{3\omega_0\tau_k}\exp(-i\omega_0\tau_k\theta) + E_1\exp(2i\omega_0\tau_k\theta)$,

$W_{11}(\theta) = \frac{g_{21}}{i\omega_0\tau_k}q(0)\exp(i\omega_0\tau_k\theta) + \frac{i\overline{g}_{11}\overline{q}(0)}{\omega_0\tau_k}\exp(-i\omega_0\tau_k\theta) + E_2$.

where E_1 and E_2 are to be determined. From definitions of A and $(A - 2i\omega_0\tau_k)$ $W_{20}(\theta) = -H_{20}(\theta)$

$(A - 2i\omega_0\tau_k)W_{20}(\theta) = -H_{20}(\theta)$ gives us $\int_{-1}^{0} d\eta(\theta)W_{20}(\theta) = 2i\omega_0\tau_k W_{20}(0) - H_{20}(0)$ which gives us $H_{20}(0) = -g_{20}q(0) - \overline{g}_{02}\overline{q}(0)$

$$+2\tau_k\begin{pmatrix} -\frac{r}{k} - (\frac{r}{k} + \beta)\alpha_1 - p_1(1 - m_3)\alpha_2 \\ \beta\alpha_1 - p_2(1 - m_4)\alpha_1\alpha_2 \\ (q_1p_1(1 - m_3)\alpha_2 + q_2(1 - m_4)p_2\alpha_1)\exp(-2i\omega_0\tau_k) \end{pmatrix}.$$

Now, $(i\omega_0\tau_k I - \int_{-1}^{0}\exp(i\omega_0\tau_k\theta)d\eta(\theta))q(0) = 0$

$$\left(-i\omega_0\tau_k I - \int_{-1}^{0}\exp(-i\omega_0\tau_k\theta)d\eta(\theta)\right)\overline{q(0)} = 0$$

And we have $\left(2i\omega_0\tau_k I - \int_{-1}^{0}\exp(i\omega_0\tau_k\theta)d\eta(\theta)\right)$

$$E_1 = 2\tau_k\begin{pmatrix} -\frac{r}{k} - (\frac{r}{k} + \beta)\alpha_1 - p_1(1 - m_3)\alpha_2 \\ \beta\alpha_1 - p_2(1 - m_4)\alpha_1\alpha_2 \\ (q_1p_1(1 - m_3)\alpha_2 + q_2p_2(1 - m_4)\alpha_1)\exp(-2i\omega_0\tau_k) \end{pmatrix},$$ which leads to

$$\begin{pmatrix} 2i\omega_0 + \frac{rS_*}{k} & S_*(\frac{r}{k} + \beta) & p_1(1-m_3)S_* \\ -\beta I_* & 2i\omega_0 - \beta S_* + c + d_2 + m_2 + p_2(1-m_4)Y_* & p_2(1-m_4)I_* \\ -q_1(1-m_3)p_1Y_*\exp(-2i\omega_0\tau_k) & -q_2p_2(1-m_4)Y_*\exp(-2i\omega_0\tau_k) & 2i\omega_0 + d_3 + d_4 - (q_1p_1(1-m_3)S_* + q_2p_2(1-m_4)I_*)\exp(-2i\omega_0\tau_k) \end{pmatrix}\times E_1$$

$$= 2\begin{pmatrix} -\frac{r}{k} - (\frac{r}{k} + \beta)\alpha_1 - p_1(1 - m_3)\alpha_2 \\ \beta\alpha_1 - p_2(1 - m_4)\alpha_1\alpha_2 \\ (q_1p_1(1 - m_3)\alpha_2 + q_2p_2(1 - m_4)\alpha_1)\exp(-2i\omega_0\tau_k) \end{pmatrix}.$$

E_1 can be calculated from this equation. Now, $\int_{-1}^{0} d\eta(\theta) W_{11}(\theta) = -H_{11}(0)$

$$H_{11}(0) = -g_{11}q(0) - \overline{g}_{11}\overline{q}(0) + 2\tau_k \begin{pmatrix} -\frac{r}{k} - (\frac{r}{k} + \beta)Re(\alpha_1) - p_1(1 - m_3)Re(\alpha_2) \\ \beta Re(\alpha_1) - p_2(1 - m_4)Re(\alpha_1\alpha_2) \\ (q_1 p_1(1 - m_3)Re(\alpha_2) + q_2 p_2(1 - m_3)Re(\alpha_1) \end{pmatrix},$$

$$\begin{pmatrix} \frac{rS_*}{k} & S_*(\frac{r}{k} + \beta) & p_1(1 - m_3)S_* \\ -\beta I_* & -\beta S_* + c + d_2 + m_2 + p_2(1 - m_4)Y_* & p_2(1 - m_4)I_* \\ -q_1 p_1(1 - m_3)Y_* & -q_2 p_2(1 - m_4)Y_* & d_3 + d_4 - (q_1 p_1(1 - m_3)S_* + q_2 p_2(1 - m_4)I_*) \end{pmatrix}$$

$$\times E_2 = 2 \begin{pmatrix} -\frac{r}{k} - (\frac{r}{k} + \beta)Re(\alpha_1) - p_1(1 - m_3)Re(\alpha_2) \\ \beta Re(\alpha_1) - p_2(1 - m_4)Re(\alpha_1\alpha_2) \\ (q_1 p_1(1 - m_3)Re(\alpha_2) + q_2 p_2(1 - m_4)Re(\alpha_1) \end{pmatrix}.$$

E_2 can be obtained from this equation. By putting values of E_1 and E_2 we can obtain $W_{20}(\theta)$ and $W_{11}(\theta)$ and hence g_{20},g_{11}, g_{02}, g_{21} etc. Hence as stated in [5, 10], we can obtain the following values;

$$\begin{cases} c_1(0) = & \frac{i}{2\omega_0\tau_k}(g_{11}g_{20} - 2 \mid g_{11} \mid^2 - \frac{|g_{02}|^2}{3}) + \frac{g_{21}}{2}, \\ \mu_2 = & -\frac{Re(c_1(0))}{Re(\lambda'(\tau_k))}, \\ \beta_2 = & 2Re(c_1(0)), \\ T_2 = & -\frac{1}{\omega_0\tau_k}[Im(c_1(0)) + \mu_2 Im(\lambda'(\tau_k))], \end{cases} \tag{10}$$

which determine the direction and stability of the model with delay at the critical value τ_k. Now, we state the following theorem due to [5, 10, 21], which is the main result of this section:

Theorem 3 *(i) The sign of μ_2 determined the direction of Hopf bifurcation: if $\mu_2 > 0(\mu_2 < 0)$, then the Hopf bifurcation is supercritical (subcritical).*

(ii)The stability of bifurcated periodic solutions is determined by β_2: the periodic solutions are stable if $\beta_2 < 0$ and unstable if $\beta_2 > 0$.

(iii)The period of bifurcated periodic solutions is determined by T_2: the period increases if $T_2 > 0$ and decreases if $T_2 < 0$.

From part (i) of this theorem, it is clear that Hopf bifurcation is supercritical if either $Re(c_1(0)) < 0$ or $Re(\lambda'(\tau_k)) < 0$. Similarly, Hopf bifurcation is subcritical if $Re(\lambda'(\tau_k)) > 0$ and $Re(c_1(0)) > 0$.

6 Numerical Simulation

In this section, we consider a hypothetical set of parameters $P_1 = \{r = 0.8, k = 1, \beta = 1, p_1 = 0.12, p_2 = 6, m_1 = 0.02, m_2 = 0.06, m_3 = 0.5, m_4 = 0.2, d_2 = 0.05, d_3 = 0.6, d_4 = 0.5, c = 0.025, q_1 = 0.75, q_2 = 0.75\}$. We will focus on positive equilibrium. Calculation shows that $\widetilde{S} = .2339, \widetilde{I} = .2749, \widetilde{Y} = .0487$, thus model has the positive equilibrium $E_5(.2339, .2749, 0.0487)$. Also, $\Gamma = -0.5550, \Delta = 1.8, \Theta = -4.860, \Lambda = -0.380, A_1 = .3698, A_2 = .1647, A_3 = .0217$, therefore $\Gamma\widetilde{S} + \Delta\widetilde{I} + \Theta\widetilde{Y} + \Lambda = -0.0335 < 0$ and $A_1A_2 + A_3 = .0826 > 0$,

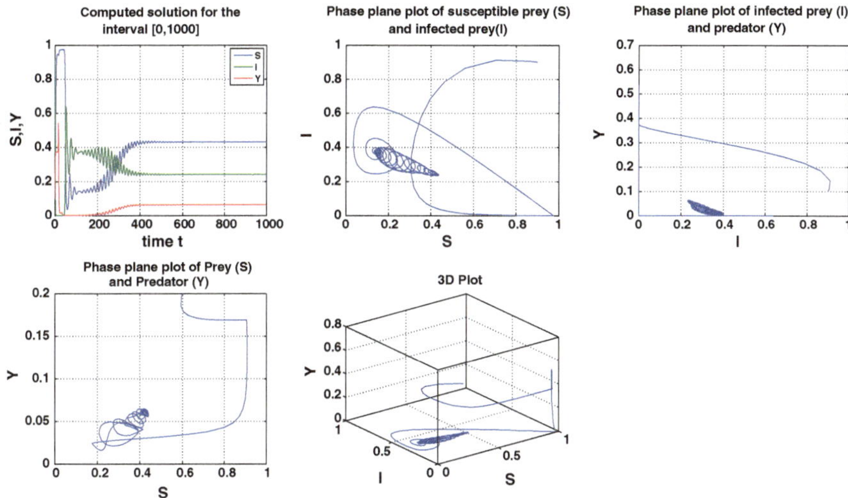

Fig. 1 Solution of system (4) for initial function $S(0) = 0.6, I(0) = 0.2, Y(0) = 0.2$ with parameter set P_1, $\tau = 15.14 < \tau_0$, the positive equilibrium point is stable

hence $E_5(.2339, .2749, 0.0487)$ is stable. Indeed, we also have the jacobian matrix at E_5;

$$\begin{pmatrix} -0.1349 & -0.4210 & -0.0140 \\ 0.2749 & -0.1349 & -1.3195 \\ 0.0022 & 0.0029 & -0.1 \end{pmatrix},$$

this has the characteristics equation $\lambda^3 + 0.3698\lambda^2 + 0.1647\lambda + 0.0217$. It has three roots, viz.,

$$\begin{cases} -0.1020 + 0.3471i, \\ -0.1020 - 0.3471i, \\ -0.1658, \end{cases}$$

hence $E_5(0.2339, 0.2749, 0.0487)$ is stable. It is also calculated that $n_0 = -0.1941$, $n_2 = -1.1$, $n_1 = 0.6895$, $m_0 = 0.2158$, $m_1 = -0.5248$, $m_2 = 1.4698$. Therefore, $p_0 = 3.1633$, $q_0 = -0.4073$, $r_0 = 0.0089$, $h(z) = z^3 + 3.1633z^2 - 0.4073z + 0.0089$, $p_0^2 - 3q_0 = 11.2284 > 0$ and $z_1^* = 0.063$. From this $h(z_1^*) = 11.20998528$, hence E_* is stable. Further $\omega_0 = 0.6382$ and $\tau_0 = 33.14$. Thus, Hopf bifurcation occurs as the τ passes through τ_0 which is depicted by numerical simulation in Figs. 1 and 2.

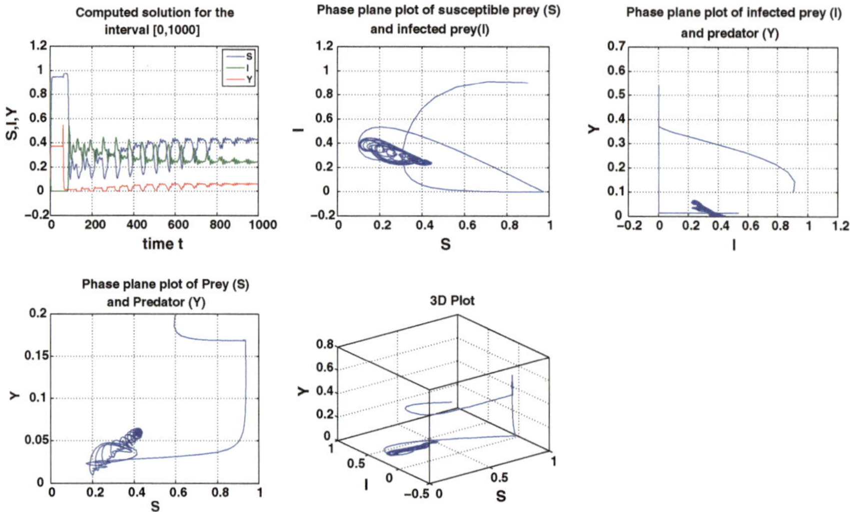

Fig. 2 Solution of system (4) for initial function $S(0) = 0.6, I(0) = 0.2, Y(0) = 0.2$ with parameter set P_1, $\tau = 60.30 > \tau_0$, the positive equilibrium point is unstable

7 Discussion

In this paper, we have considered a delayed prey–predator system with infection. Migration has been allowed among prey population only. It is also considered that prey population has self-defence in the form of prey refuge. This decreases the availability of prey population for predation to predators. For instance, only $(1-m_3)S$ of sound prey are available for predation. Similarly, $(1-m_4)I$ of infected prey are available for predation. Stability results have been investigated.

Similar to the study of [10], in this paper the time delay τ is the gestation period of predator. In our analysis this is found to be the bifurcation parameter. It is proved that beyond some specific value of τ, Hopf-bifurcation occurs. The direction of Hopf-bifurcation and stability of bifurcated periodic solutions have been derived using the central manifold reduction technique and normal form theory.

In this paper, bifurcation of predator into two parts, viz., healthy predator and infected predator has been ignored. The same may be done in the future. Further, for simplification, parameters are taken as time independent. In real-life the parameters are time dependent, this may also considered in the future.

The main issue in applied mathematical modeling is to identify the real parameters. The present study is not a case study, hence real parameters are not available. Hence, the main scope of this study is to study a real eco-system and to identify the real/experimental parameters.

References

1. Lotka, A.: Elements of Physical Biology. Williams and Wilkins, Baltimore (1925)
2. Bate, A.M., Hilker, F.M.: Predator-prey oscillations can shift when disease become endemic. J. Theor. Biol. **316**, 1–8 (2013)
3. Pal, A.K., Samanta, G.P.: Stability analysis of an eco-epidimiolgical model incorporating a prey refuge. Nonlinear Anal.: Model. Control **15**, 473–491 (2010)
4. Mukhopadhyaya, B., Bhattacharyya, B.: Dynamics of a delay-diffusion prey-predator model with disease in the prey. J. Appl. Math Comput. **17**(12), 361–377 (2005)
5. Celik Karaaslanli, C.: Bifurcation analysis and its applications, Chapter 1. In: Andriychuk, M. (ed.) Numerical Simulation: From Theory to Industry. INTECH, Vienna (2012)
6. Greenhalgh, D., Haque, M.: A predator-prey model with disease in the prey species only. Math. Methods Appl. Sci. **30**, 911–929 (2007)
7. Hilker, F.M., Schmitz, K.: Disease-induced stabilization of predator prey oscillations. J. Theor. Biol. **255**, 299–306 (2008)
8. Wang, F., et al.: Stability and bifurcation of a stage-structured predator-prey model with both discete and distributed delays. Chaos Solitons Fractals **46**, 19–27 (2013)
9. Faina, S., et al.: Role of prey dispersal and refuges on predator-prey dynamics. SIAM J. Appl. Math. **70**(6), 1821–1839 (2010)
10. Hu, G.-P., Li, X.-L.: Stability and Hopf bifurcation for a delayed predator-prey model with disease in the prey. Chaos Solitons Fractals **45**, 229–237 (2012)
11. Sun, G.-Q., et al.: Dynamical complexity of a spatial predator-prey model with migration. Ecol. Modell. **219**, 248–255 (2008)
12. Chattopadhyay, J., Arino, O.: A predator-prey model with disease in prey. Non-Linear Anal. **36**, 747–766 (1999)
13. Berthier, K., et al.: Dynamics of a feline virus with two transmission Medels within exponentially growing host populations. Proc. R. Soc. Lond. **267**, 2049–2056 (2000)
14. Maoxing, L., et al.: An Impulsive predator-prey model with communicable disease in the prey species only. Non-Linear Anal.: Real World Appl. **10**, 3098–3111 (2009)
15. Haque, M., et al.: An eco-epidemiological predator-prey model with standard disease incidence. Math. Methods Appl. Sci. (2008). doi:10.1002/mma.1071
16. Haque, M., Venturino, E.: Increase of the prey may decrease the healthy prtedator population in presence of a disease in predator. HERMISS **7**, 38–59 (2006)
17. Naji, R.K.: The dynamics of prey-predator model with disease in the prey. J. Math. Comput. Sci. **2**(4), 1052–1072 (2012)
18. Samanta, G.P.: Analysis of a delay nonautonomous predator-prey system with disease in the prey. Non-Linear Analy.: Model. Control **15**(1), 97–108 (2010)
19. Sinha, S., et al.: Modelling a predator-prey system with infected prey in polluted environment. Appl. Math. Model. **34**, 1861–1872 (2010)
20. Bhattacharyya, R., Mukhopadhyay, B.: Spatial dynamics of nonlinear prey-predator models with prey migration and predator switching. Ecol. Complex. **3**, 160–169 (2006)
21. Jana, S., Kar, T.K.: Modeling and analysis of a prey-predator system with disease in the prey. Chaos Solitons Fractals **47**, 42–53 (2013)
22. Kar, T.K.: Stability Analysis of a prey-predator model incorporating a prey refuge. Commun. Nonlinear Sci. Numer. Simul. **10**, 681–691 (2005)
23. Volterra, V.: Variazioni e fluttauazionidel numero d individui in species animals conviventii. Mem Acd. Linciei **2**, 31–33 (1926)
24. Krivan, V.: On the Gauss predator-prey model with a refuge: a fresh look at the history. J. Theor. Biol. **274**, 67–73 (2011)
25. Kermack, W.O., Mckendrick, A.G.: Contributons to the mathematical theory of epidimics, Part 1. Proc. R. Soc. A **115**(5), 700–721 (1927)
26. Yu, X., Sun, F.: Global dynamics of a predator-prey model incorporating a constant prey-refuge. Electron. J. Differ. Equ. **4**, 1–8 (2013)

27. Xiao, Y., Chen, L.: Modeling and analysis of a predator-prey model with disease in the prey. Math. Biosci. **171**, 59–82 (2001)
28. Xiao, Y., Chen, L.: A ratio-dependent predator-prey model with disease in the prey. Appl. Math. Comput. **131**, 397–414 (2002)
29. Ma, Z., et al.: Effect of prey refuges on a predator-prey model with a class of functional responses: the role of refuges. Math. Biosci. **218**, 73–79 (2009)

Stability Analysis of an Integro Differential Equation Model of Ring Neural Network with Delay

Swati Tyagi, Syed Abbas and Rajendra K. Ray

Abstract In this paper we present and study a ring neural network model with delays. We study existence and uniqueness of equilibrium point and global stability of the model system. At the end few examples have been given to illustrate the analytical findings.

Keywords Ring network · Neural network · Time delay · Equilibrium point · Existence and uniqueness · Global asymptotic stability

AMS Subject Classification: 92B20 · 34D23 · 34K20 · 35A35

The term neural networks denote the collection of interconnected, interacting neurons, which can be biological/artificial. A system of connected nodes constitutes an artificial neural network. Arranging the nodes in different configurations yields distinct artificial neural networks with characteristic properties. For performing parallel computation the model of neural networks are also very much promising. Several authors studied the dynamics of neural network [1–7] theoretically and numerically. The integration and communication delays are present everywhere in all physical systems, due to which several questions arise in our mind about their effects on the dynamic and various other properties of neural networks. While implementing artificial neural systems, due to the finite switching speed of neurons and amplifiers, time delays are ineluctable and so we cannot avoid them. Over some recent years, these effects of delays on almost all physical dynamical networks and the stability analysis of time delayed system have received a remarkable attention. Recurrent neural networks have been extensively used to study dynamical varying data and have been used in many practical applications [8]. In 1994, Baldi and Atiya studied the effects of delays on neural dynamics by using additive neural network model given as [9]:

S. Tyagi (✉) · S. Abbas · R.K. Ray
School of Basic Sciences, Indian Institute of Technology, Mandi 175001, H.P., India
e-mail: styagi99@gmail.com

S. Abbas
e-mail: sabbas.iitk@gmail.com

P.N. Agrawal et al. (eds.), *Mathematical Analysis and its Applications*,
Springer Proceedings in Mathematics & Statistics 143,
DOI 10.1007/978-81-322-2485-3_3

$$\frac{du_i}{dt} = -\frac{u_i}{\tau_i} + \sum_j T_{ij} f_j(u_j) + I_i \quad (i = 1, 2, ..., n),$$

T_{ij} are synaptic connection strengths, τ_i are time constants, I_i are external inputs, f_i are input/output transfer functions. Recently Feng and Plamondon [10] studied the following recurrent neural network. They considered a ring of neurons connected cyclically with delayed interactions, which was modelled as:

$$\frac{du_i(t)}{dt} = -\frac{1}{r_i}u_i(t) + w_{i,i-1} f_{i-1}\Big(u_{i-1}(t - \tau_{i,i-1})\Big)(i \ mod \ n), \tag{1}$$

They discussed the conditions under which a ring network could be exploited as an oscillatory pattern generator. Using Chafee's theory, they guaranteed the existence of permanent oscillations in a time delayed neural network model. Motivated by their work, in the present paper, we will be particularly studying stability of the following delayed neural ring network model,

$$\frac{du_i(t)}{dt} = -\frac{1}{r_i}u_i(t) + w_{i,i-1} f_{i-1}\Big(t, u_{i-1}(t - \tau_{i,i-1}), \int_{-\infty}^{t} k_{i-1}(t-s)u_{i-1}(s)ds\Big)$$
$$(i \ mod \ n), \tag{2}$$

with the passive decay rates given by $-\frac{1}{r_i} > 0$, $\tau_{i,i-1} \geq 0$ are the time delays present in the system, $w_{i,i-1} \neq 0 \quad (i = 1, 2, ..., n)$ are the weights. By taking the transformation $u = (u_1, u_2, ..., u_n)^T$ the system (2) can be rewritten as:

$$\frac{du(t)}{dt} = Ru(t) + Wf\Big(t, u(t-\tau), \int_{-\infty}^{t} K(t-s)u(s)ds\Big), \tag{3}$$

where R is a diagonal matrix given by, $R = diag(-\frac{1}{r_1}, -\frac{1}{r_2}, ..., -\frac{1}{r_n})$, and the weight matrix W is given by,

$$W = \begin{pmatrix} 0 & 0 & \cdots & \cdots & \cdots & w_{1,n} \\ w_{2,1} & 0 & \cdots & \cdots & 0 & 0 \\ \cdots & \cdots\cdots & & \cdots & \cdots & \cdots \\ 0 & 0 & \cdots & w_{n-1,n-2} & 0 & 0 \\ 0 & 0 & \cdots & \cdots & w_{n,n-1} & 0 \end{pmatrix},$$

The kernel matrix K is given by,

$$K(t-s) = \begin{pmatrix} 0 & 0 & \cdots & \cdots & 0 & k_n(t-s) \\ k_1(t-s) & 0 & \cdots & \cdots & 0 & 0 \\ \cdots & \cdots\cdots & & \cdots & \cdots & \cdots \\ 0 & 0 & \cdots & k_{n-2}(t-s) & 0 & 0 \\ 0 & 0 & \cdots & \cdots & k_{n-1}(t-s) & 0 \end{pmatrix},$$

and the activation functions are given by,

$$
\begin{aligned}
f(t,u(t-\tau), & \int_{-\infty}^{t} K(t-s)u(s)ds) \\
& = (f_1(t,u_1(t-\tau_{2,1}), \int_{-\infty}^{t} k_1(t-s)u_1(s)ds), \\
& f_2(t,u_2(t-\tau_{3,2}), \int_{-\infty}^{t} k_2(t-s)u_2(s)ds), \\
& ...., f_{n-1}(t,u_{n-1}(t-\tau_{n,n-1}), \int_{-\infty}^{t} k_{n-1}(t-s)u_{n-1}(s)ds), \\
& f_n(t,u_n(t-\tau_{1,n}), \int_{-\infty}^{t} k_n(t-s)u_n(s)ds))^T.
\end{aligned}
$$

For any matrix $A=(a_{ij})_{n\times n}$, we denote $|A|=(|a_{ij}|)_{n\times n}$ and $A>0$ denotes A to be a positive definite matrix, $\|A\|=\max\{\lambda\}$: where λ is an eigen value of $\sqrt{A^TA}$ and for $x=(x_1,x_2,...,x_n)^T$, $\|x\|=(\sum_{i=1}^n x_i^2)^{\frac{1}{2}}$. The remaining of this paper is organized as follows. In Sect. 1, some definitions and lemmas are given, which help us in deriving the existence and uniqueness of the equilibrium point, which is derived in Sect. 3. Global stability of the solution of the system (2) is studied in Sect. 4, and in Sect. 5, some examples have been given to validate the theoretical findings.

Remark 1 In this paper, the model that we are studying is a generalization of [10, 11], and after using suitable transformations and assumptions, it becomes similar to [10, 11].

1 Preliminaries

Assumptions

1. The activation functions $f_i(u_i)(i=1,2,...,n)$ are bounded, continuous and non linear Lipschitz functions:

$$f: I\times X\times C\rightarrow X$$

$$
\begin{aligned}
& \|f(t,x,\phi)-f(t,y,\psi)\| \le c_1\|x-y\|_a + c_2\|\phi-\psi\|_b, \\
or \quad & \|f(t,x,\phi)-f(t,y,\psi)\| \le c_i'\|x-y\|,
\end{aligned} \tag{4}
$$

where $x=u(t-\tau), y=v(t-\tau), \phi=\int_{-\infty}^{t} k_{i-1}(t-s)u_{i-1}(s)ds, \psi=\int_{-\infty}^{t} k_{i-1}(t-s)v_{i-1}(s)ds$, $x,y\in R^n$, $c_1,c_2>0$, $c_i'=\max c_i\ (i=1,2,...,n)$

and $f_i(0) = 0 \quad (i = 1, 2, ..., n)$, where a norm is defined on $C([-r, 0], X)$ and b norm is defined on $((-\infty, 0], X)$.

Remark 2 To prove result (4), we have,

$$\begin{aligned}
\|\phi - \psi\| &= \| \int_{-\infty}^{t} k_{i-1}(t-s)u_{i-1}(s)ds - \int_{-\infty}^{t} k_{i-1}(t-s)v_{i-1}(s)ds \|, \\
&= \| \int_{-\infty}^{t} k_{i-1}(t-s)[u_{i-1}(s) - v_{i-1}(s)]ds \|, \\
&\leq \| \int_{-\infty}^{t} k_{i-1}(t-s)ds \| \|u_{i-1} - v_{i-1}\|, \\
&= \| \int_{0}^{\infty} k_{i-1}(s)ds \| \|u_{i-1} - v_{i-1}\|, \quad k \in L^1(0, \infty), \\
&\leq c_2 \|u_{i-1} - v_{i-1}\|.
\end{aligned}$$

Since we have assumed that $k \in L^1(0, \infty)$, so it is bounded by some constant, say c_2, and hence we get the above result.

2. In the neighbourhood of zero point, the function $f_i(u_i)$ $(i = 1, 2, ..., n)$ are differentiable.

Definition 1 [10] A continuous mapping $H : R^n \to R^n$ is called a homeomorphism, if it satisfies the following properties:

1. H is bijection (one-one and onto).
2. The inverse map H^{-1} is also continuous.

Lemma 1 *[10]: A continuous mapping $H(x) : R^n \to R^n$ is a homeomorphism if it satisfies the following properties:*

1. H(x) is injective.

2. $\lim_{\|x\| \to \infty} \|H(x)\| \to \infty$, where $\|x\| = \left(\sum_{i=1}^{n} x_i^2\right)^{1/2}$.

2 Existence and Uniqueness

Lemma 2 *Suppose that there is a positive diagonal matrix $D = diag(d_1, d_2, ..., d_n)$, $d_i > 0 \quad (i = 1, 2, ..., n)$ such that,*

$$r_1 d_1 |w_{1,n}| l_n d_n^{-1} < 1 \text{ and } r_i d_i |w_{i,i-1}| l_{i-1} d_{i-1}^{-1} < 1, \tag{5}$$

where $l_i^{'}s$ are constants with $0 < l_i \leq c_i^{'}(i = 2, 3, ..., n)$, then the system (2) has a unique equilibrium point v^*, where $v^* = [v_1^*, v_2^*, ..., v_n^*]^T$ and is equal to $(0, 0, ..., 0)^T$.

Proof Let we define a map corresponding to (2):

$$H(u) = (H_1(u), H_2(u), ..., H_n(u))^T : R^n \to R^n, \tag{6}$$

where,

$$\begin{aligned} H_1(u) &= -\frac{1}{r_1}u_1 + w_{1,n} f_n\Big(t, u_n, \int_{-\infty}^{t} k_n(t-s)u_n(s)ds\Big), \\ H_i(u) &= -\frac{1}{r_i}u_i + w_{i,i-1} f_{i-1}\Big(t, u_{i-1}, \int_{-\infty}^{t} k_{i-1}(t-s)u_{i-1}(s)ds\Big), \\ &\quad (i = 2, 3, ..., n). \end{aligned} \tag{7}$$

Let if v^* be an equilibrium point of (2), then it must satisfy $H(v^*) = 0$. Now we show the existence and uniqueness of the equilibrium point v^*. For this, we need to show that $H(v)$ is an homeomorphism. First we show that $H(v)$ is injective on R^n. Assume on the contrary that, there exist two vectors $u, v \in R^n$ where $u = [u_1, u_2, ..., u_n]^T$ and $v = [v_1, v_2, ..., v_n]^T$ with $u \neq v$ such that $H(u) = H(v)$, then we get,

$$\begin{aligned} \frac{1}{r_1}(v_1 - u_1) &= w_{1,n}[f_n(t, v_n, \psi_n) - f_n(t, u_n, \phi_n)], \\ \frac{1}{r_i}(v_i - u_i) &= w_{i,i-1}[f_{i-1}(t, v_{i-1}, \psi_{i-1}) - f_{i-1}(t, u_{i-1}, \phi_{i-1})], \\ &\qquad (i = 2, 3, ..., n) \end{aligned} \tag{8}$$

where $\phi_i = \int_{-\infty}^{t} k_i(t-s)u_i(s)ds$ and $\psi_i = \int_{-\infty}^{t} k_i(t-s)v_i(s)ds$, $(i = 1, 2, ..., n)$. Since we have $r_i > 0 \quad (i = 1, 2, ..., n)$, we get,

$$\begin{aligned} \frac{1}{r_1}|v_1 - u_1| &= w_{1,n}|f_n(t, v_n, \psi_n) - f_n(t, u_n, \phi_n)|, \\ \frac{1}{r_i}|v_i - u_i| &= w_{i,i-1}|f_{i-1}(t, v_{i-1}, \psi_{i-1}) - f_{i-1}(t, u_{i-1}, \phi_{i-1})| \\ &\qquad (i = 2, 3, ..., n). \end{aligned} \tag{9}$$

From (4), there exist $0 \le l_i' \le c_i'$ such that

$$\begin{aligned} \|f_n(t, v_n, \phi) - f_n(t, u_n, \psi)\| &= l_n'\Big[\|v - u\| + \|\phi - \psi\|\Big], \\ \|f_{i-1}(t, v_{i-1}, \phi) - f_{i-1}(t, u_{i-1}, \psi)\| &= l_{i-1}'\Big[\|v_{i-1} - u_{i-1}\| + \|\phi - \psi\|\Big] \\ &\qquad (i = 2, 3, ..., n), \end{aligned} \tag{10}$$

Using (10) in (9), we get,

$$
\begin{aligned}
\|v_1 - u_1\| &= r_1|w_{1,n}|l_n'\Big[\|v_n - u_n\| + \|\phi - \psi\|\Big],\\
&= r_1|w_{1,n}|l_n'\Big[\|v_n - u_n\| + M_n\|v_n - u_n\|\Big],\\
&= r_1|w_{1,n}|l_n'(1 + M_n)\|v_n - u_n\|,\\
&= r_1|w_{1,n}|l_n\|v_n - u_n\|, \text{(where } l_n = l_n'(1 + M_n)),\\
&= r_1|w_{1,n}|l_n \cdot r_n|w_{n,n-1}|l_{n-1}\|v_{n-1} - u_{n-1}\|,\\
&= \cdots = r_1|w_{1,n}|l_n \cdot r_n|w_{n,n-1}|l_{n-1} \cdot r_{n-1}|w_{n-1,n-2}|l_{n-2}\\
&\quad \cdots r_2|w_{2,1}|l_1\|v_1 - u_1\|,\\
&= \frac{r_1d_1|w_{1,n}|l_n}{d_n} \cdot \frac{r_nd_n|w_{n,n-1}|l_{n-1}}{d_{n-1}}\\
&\quad \cdot \frac{r_{n-1}d_{n-1}|w_{n-1,n-2}|l_{n-2}}{d_{n-2}} \cdots \frac{r_2d_2|w_{2,1}|l_1}{d_1}\|v_1 - u_1\|\\
&< \|v_1 - u_1\|,
\end{aligned}
$$

which is a contradiction, Thus we have $v_1 = u_1$. Similarly for $i = 2, 3, ..., n$, using similar above calculations, we have

$$\|v_i - u_i\| < \|v_i - u_i\|,$$

which is again a contradiction. From this we can say that $v_i = u_i$ $(i = 2, 3, ...n)$. This implies that $v = u$, which is a contradiction. Hence map H is injective. Also $f(u)$ is bounded from our assumption, which clearly implies that as $\|u\| \to \infty$, each $\|H_i(u)\| \to \infty (i = 1, 2, ..., n)$, So $\|H(u)\| \to \infty$ and H is a homeomorphism, and thus system (2) has only one equilibrium point. As we assumed $f_i(0) = 0$, $(i = 1, 2, ..., n)$, we can say that $(0, 0, ..., 0)^T$ is the unique equilibrium point.

Note 1 If the activation function is monotone increasing or a monotone decreasing bounded continuous function, then (4) changes to

$$0 < f_i(t, u, \phi) - f_i(t, v, \psi) \le l_i(u - v) \quad (u > v), \quad i = 1, 2, ..., n, \tag{11}$$

or

$$-l_i(u - v) \le f_i(t, u, \phi) - f_i(t, v, \psi) < 0, \quad (u > v), \quad i = 1, 2, ..., n. \tag{12}$$

Note 2 In justification of the above remark, for example, functions such as $tanh(u)$, $arctan(u)$ satisfy condition (11), while for functions like $\frac{1}{1+exp(u)}$, condition (12) is satisfied.

Now, we have the following lemma:

Lemma 3 *Assume that condition (11) holds, and*

$$r_1 w_{1,n} l_n \cdot r_2 w_{2,1} l_1 \cdot r_3 w_{3,2} l_2 \cdots r_n w_{n,n-1} l_{n-1} < 1, \tag{13}$$

or that the condition (12) holds, and

$$|r_1 w_{1,n}(-l_n) \cdot r_2 w_{2,1}(-l_1) \cdot r_3 w_{3,2}(-l_2) \cdots r_n w_{n,n-1}(-l_{n-1})| > 1. \tag{14}$$

Then system (2) has unique equilibrium point.

Proof Doing the similar calculations as done in Lemma (1), if u, v be any two equilibrium points and condition (11) holds, then we have,

$$\begin{aligned} &\frac{1}{r_1}(v_1 - u_1) \leq w_{1,n} l_n (v_n - u_n), \\ &\frac{1}{r_i}(v_i - u_i) \leq w_{i,i-1} l_{i-1} (v_{i-1} - u_{i-1}), \quad (i = 2, 3, ..., n). \end{aligned} \tag{15}$$

Thus if $v_1 \neq u_1$, then we have

$$\begin{aligned} v_1 - u_1 &\leq r_1 w_{1,n} l_n (v_n - u_n), \\ &\leq r_1 w_{1,n} l_n \cdot r_n w_{n,n-1} l_{n-1} (v_{n-1} - u_{n-1}), \\ &\leq \cdots \leq r_1 w_{1,n} l_n \cdot r_n w_{n,n-1} l_{n-1} \\ &\quad \cdot r_{n-1} w_{n-1,n-2} l_{n-2} \cdots r_2 w_{2,1} l_1 (v_1 - u_1), \\ &= r_1 w_{1,n} l_n \cdot r_2 w_{2,1} l_1 \\ &\quad \cdots r_{n-1} w_{n-1,n-2} l_{n-2} \cdot r_n w_{n,n-1} l_{n-1} (v_1 - u_1), \\ &< v_1 - u_1. \end{aligned}$$

This is a contradiction. So we have $v_1 = u_1$. Similarly for $i = 2, 3, ..., n$, we get,

$$v_i - u_i < v_i - u_i,$$

which is also a contradiction. So $v_i = u_i (i = 2, 3, ..., n)$. This implies that $v = u$, from which we can conclude that the system (2) has a unique equilibrium point. Similarly, If condition (12) holds, and if $v_1 \neq u_1$, then we have,

$$\begin{aligned} v_1 - u_1 &\geq r_1 w_{1,n}(-l_n)(v_n - u_n), \\ &\geq r_1 w_{1,n}(-l_n) \cdot r_n w_{n,n-1}(-l_{n-1})(v_{n-1} - u_{n-1}), \\ &\geq \cdots \geq r_1 w_{1,n}(-l_n) \cdot r_n w_{n,n-1}(-l_{n-1}) \\ &\quad \cdot r_{n-1} w_{n-1,n-2}(-l_{n-2}) \cdots r_2 w_{2,1}(-l_1)(v_1 - u_1). \end{aligned}$$

From (12), it implies that $|v_1 - u_1| > |v_1 - u_1|$, which is a contradiction. Hence $v_1 = u_1$ and in the similar way, we can conclude that $v_i = u_i (i = 2, 3, ..., n)$. Thus the system (2) has a unique equilibrium point.

Lemma 4 *Suppose that the condition (11) (or condition (12)) holds. If there exists some constants* $0 < d_i \leq c_i^{'} (or - c_i^{'} \leq d_i < 0 \ (i = 1, 2, ..., n))$ *for the system (3) such that the matrix* $R + WD$ *is a non-singular matrix, where* $D = diag(d_1, d_2, ..., d_n)^T$, *then the given system (3) has a unique equilibrium point.*

Proof Let we define a map associated with system (3) given by:

$$H(u) = Ru + Wf(t, u, \phi). \tag{16}$$

Assume on the contrary that there exist two vectors $v, u \in R^n$ with $v \neq u$, where $v = [v_1, v_2, ..., v_n]^T$ and $u = [u_1, u_2, ..., u_n]^T$ such that $H(v) = H(u)$, then we get,

$$R(v - u) + W(f(t, v, \psi) - f(t, u, \phi)) = 0. \tag{17}$$

Since (11) (or (12)) holds, so there exists $0 < d_i \leq k_i \ and - k_i \leq d_i < 0$ $(i = 1, 2, ..., n)$ such that

$$f_i(t, v_i, \psi_i) - f_i(t, u_i, \phi_i) = d_i(v_i - u_i), \quad (i = 1, 2, ..., n). \tag{18}$$

Using (18) in (17), we obtain

$$(R + WD)(v - u) = 0, \tag{19}$$

where

$$R + WD = \begin{pmatrix} -\frac{1}{r_1} & 0 & 0 & 0 & \cdots & \cdots & d_n w_{1,n} \\ w_{2,1}d_1 & -\frac{1}{r_2} & 0 & \cdots & \cdots & 0 & 0 \\ \cdots & \cdots & \cdots & \cdots & \cdots & \cdots & \cdots \\ \cdots & \cdots & \cdots & \cdots & \cdots & \cdots & \cdots \\ 0 & 0 & 0 & \cdots & w_{n-1,n-2}d_{n-2} & -\frac{1}{r_{n-1}} & 0 \\ 0 & 0 & 0 & \cdots & \cdots & w_{n,n-1}d_{n-1} & -\frac{1}{r_n} \end{pmatrix}.$$

Since it is given that $(R + WD)$ is a non-singular matrix, so $\det(R + WD) \neq 0$. Hence from (19), $v = u$, which is a contradiction. So $H(u)$ is injective on R^n. Also since each $f_i(u_i)$ is a bounded continuous function, so $f(u)$ is bounded, and it is clear from the defined mapping that $\|H(u)\| \rightarrow \infty$, as $\|u\| \rightarrow \infty$. Thus H is a homeomorphism and hence system (3) has a unique equilibrium point.

3 Global Stability

Theorem 1 *Under the Assumptions (1)–(2) , there exist a unique equilibrium point for the system (3). Furthermore, system (3) is globally asymptotically stable.*

Proof For the given neural ring network model, existence and uniqueness of the equilibrium point has already been proved in Sect. 1. In this section, We prove the global stability of system (3) by constructing suitable Lyapunov function. Throughout the calculations, for simplicity, we denote $f(u, u(t-\tau), \int_{-\infty}^{t} K(t-s)u(s)ds) = f$. Consider the Lyapunov function given as:

$$V = V_1 + V_2 + V_3 + V_4, \tag{20}$$

where

$$\begin{aligned}
V_1 &= \Big(u(t) - R\int_{t-\tau}^{t} u(s)ds\Big)^2,\\
V_2 &= 2\Big(\int_{t-\tau}^{t} u(s)\Big)Wfds,\\
V_3 &= -2RWf\int_{t-\tau}^{t}(\theta - t + \tau)u(\theta)d\theta - 2R^2\Big(\int_{t-\tau}^{t}(\theta - t + \tau)u(\theta)d\theta\Big)u(t-\tau),\\
V_4 &= \tau R^2\int_{t-\tau}^{t} u^2(s)ds - R\int_{t-\tau}^{t} u^2(s)ds - 2\Big(\int_{t-\tau}^{t} u(s)ds\Big)WM.
\end{aligned} \tag{21}$$

Taking derivative with respect to t, we get

$$\dot{V} = \dot{V}_1 + \dot{V}_2 + \dot{V}_3 + \dot{V}_4. \tag{22}$$

Differentiating each V_i with respect to t, we get,

$$\begin{aligned}
\dot{V}_1 &= 2\Big(u(t) - R\int_{t-\tau}^{t} u(s)ds\Big)\Big(Wf + Ru(t-\tau)\Big),\\
&= 2\Big[u(t)Wf + u(t)Ru(t-\tau) - R\Big(\int_{t-\tau}^{t} u(s)ds\Big)Wf\\
&\quad - R^2\Big(\int_{t-\tau}^{t} u(s)ds\Big)u(t-\tau)\Big].
\end{aligned} \tag{23}$$

For V_2, taking M as bound of f (because we have assumed earlier that the activation functions are monotonic, bounded and Lipschitz continuous), we have

$$\dot{V}_2 \leq 2\Big(u(t) - u(t-\tau)\Big)WM, \tag{24}$$

$$\dot{V}_3 = -2\tau RWfu(t) + 2RWf\Big(\int_{t-\tau}^{t} u(s)ds\Big) - 2\tau R^2 u(t-\tau)u(t)$$
$$+2R^2\Big(\int_{t-\tau}^{t} u(s)ds\Big)u(t-\tau), \tag{25}$$

$$\dot{V}_4 = \tau R^2 u^2(t) - \tau R^2 u^2(t-\tau) - Ru^2(t) + Ru^2(t-\tau) - 2u(t)WM + 2u(t-\tau)WM. \tag{26}$$

Using (24), (25) in (26) in (22) and taking modulus with use of inequality $-2ab \leq (a^2 + b^2)$, we obtain

$$\begin{aligned}
\dot{V} &\leq -|R|u^2(t) + 2u(t)|W|M - 2\tau|R||W||f|u(t) - 2\tau|R^2||u(t-\tau)||u(t)| \\
&\quad + \tau|R^2|u^2(t) - \tau|R^2|u^2(t-\tau) - |R|u^2(t), \\
&\leq -2|R|u^2(t) + 2u(t)|W|M - 2\tau|R||W||f|u(t) - 2\tau|R^2||u(t-\tau)||u(t)| \\
&\quad + \tau|R^2|u^2(t) - \tau|R^2|u^2(t-\tau), \\
&\leq -2|R|u^2(t) + 2Mu(t)|W| - 2\tau M|R||W|u(t) + \tau|R^2|u^2(t) \\
&\quad + \tau|R^2|u^2(t-\tau) + \tau|R^2|u^2(t) - \tau|R^2|u^2(t-\tau), \\
&= -2|R|u^2(t) + 2Mu(t)|W| - 2\tau M|R||W|u(t) + 2\tau|R^2|u^2(t), \\
&\leq -2|R|u^T(t)u(t), \\
&\leq 0.
\end{aligned} \tag{27}$$

Hence the given neural ring network model (3) is globally asymptotically stable.

Remark 3 In proof of global stability, throughout the calculations, $u^2(t) = u^T(t)u(t)$.

4 Examples

Example 1 Consider the following system, in which the activation function is taken to be $tanh(u)$ with value of delay kernel term to be 0, so our system takes the following form as:

$$\begin{aligned}
u_1'(t) &= -0.3u_1(t) + w_{1,2} \cdot tanh(u_2(t-\tau_{1,2})), \\
u_2'(t) &= -0.5u_2(t) + w_{2,1} \cdot tanh(u_1(t-\tau_{2,1})),
\end{aligned} \tag{28}$$

where $R = diag(-0.3, -0.5)$. Since the activation function we considered here $tanh(u)$ is a monotone increasing function. Selecting $l_i's$ to be equal to 1, and taking

$w_{1,2} = 1.4$ and $w_{2,1} = -1.2$, we get that $r_1 w_{1,2} l_2 \cdot r_2 w_{2,1} l_1 = -0.252 < 1$ or $|r_1 w_{1,2} l_2 \cdot r_2 w_{2,1} l_1| > 1$. Thus by Lemma (2), we get that the system (2) has a unique equilibrium point.

Example 2 Consider the following three-node system, with $tanh(u)$ as the activation function.

$$
\begin{aligned}
u_1'(t) &= -0.3u_1(t) + w_{1,3} \cdot f_3\Big(t, u_3(t-\tau_{1,3}), \int_{-\infty}^{t} k_3(t-s)u_3(s)ds\Big), \\
u_2'(t) &= -0.4u_2(t) + w_{2,1} \cdot f_1\Big(t, u_1(t-\tau_{2,1}), \int_{-\infty}^{t} k_1(t-s)u_1(s)ds\Big), \\
u_3'(t) &= -0.2u_2(t) + w_{3,2} \cdot f_2\Big(t, u_2(t-\tau_{3,2}), \int_{-\infty}^{t} k_2(t-s)u_2(s)ds\Big).
\end{aligned} \quad (29)
$$

Selecting the integral term in activation function in the following manner,

$$
\begin{aligned}
& f_i\Big(t, u_i(t-\tau_{i,i-1}), \int_{-\infty}^{t} k_{i-1}(t-s)u_{i-1}(s)ds\Big) \\
&= tanh(u_i(t-\tau_{i,i-1})) + \int_{-\infty}^{t} k_{i-1}(t-s)u_{i-1}(s)ds, \\
&= tanh(u_i(t-\tau_{i,i-1})) + \int_{0}^{\infty} k_{i-1}(s)u_{i-1}(t-s)ds.
\end{aligned}
$$

Approximating the integral term with summation term, we have

$$
= tanh(u_i(t-\tau_{i,i-1})) + \sum_{0}^{\infty} k_{i-1}(s)u_{i-1}(t-s).
$$

Now to simplify the term $\sum_0^\infty k_{i-1}(s)u_{i-1}(t-s)$, let we choose kernel term as $k_{i-1}(s) = e^s$, which can be solved as:

$$
\begin{aligned}
\sum_{s=0}^{\infty} k_{i-1}(s)u_{i-1}(t-s) &= \sum_{s=0}^{\infty} e^s u_{i-1}(t-s), \\
&= u_{i-1}(t) + e^1 u_{i-1}(t-1) + e^2 u_{i-1}(t-2) \\
&\quad + e^3 u_{i-1}(t-3) + \cdots, \\
&\le u_{i-1}(t) + e^1 u_{i-1}(t) + e^2 u_{i-1}(t) + e^3 u_{i-1}(t) + \cdots,
\end{aligned}
$$

(because we have assumed activation functions to be monotonic increasing in this example)

$$= u_{i-1}(t)(1 + e^1 + e^2 + e^3 + \cdots),$$
$$= \frac{u_{i-1}(t)}{1 - e(1)}.$$

Substituting all these values in our example, our system reduces to following form:

$$\begin{aligned}
u_1'(t) &\approx -0.3u_1(t) + w_{1,3} \cdot \Big(tanh(u_3(t - \tau_{1,3})) + \frac{u_3(t)}{1 - e^1}\Big),\\
u_2'(t) &\approx -0.4u_2(t) + w_{2,1} \cdot \Big(tanh(u_1(t - \tau_{2,1})) + \frac{u_1(t)}{1 - e^1}\Big),\\
u_3'(t) &\approx -0.2u_2(t) + w_{3,2} \cdot \Big(tanh(u_2(t - \tau_{3,2})) + \frac{u_2(t)}{1 - e^1}\Big),
\end{aligned} \tag{30}$$

where $R = diag(-0.3, -0.4, -0.2), w_{1,3} = 1.2, w_{2,1} = -3.5, w_{3,2} = -2.6$. Taking $d_1 = 1, d_2 = 1/2, d_3 = 1$, we get $det(R + WD) = 5.4630 \neq 0$, so by Lemma (3), the system (2) has a unique equilibrium point.

5 Discussion

In this paper, a ring neural network model (2) with delay has been studied and various results for the existence and uniqueness of the equilibrium point are obtained. Global stability of the equilibrium point is also studied with the help of suitable Lyapunov function. At the end, some examples are also given to validate our results.

References

1. Pan, L., Cao, J.: Anti-periodic solution for delayed cellular neural networks with impulsive effects. Nonlinear Anal. Real World Appl. **12**(6), 3014–3027 (2011)
2. Xiang, H., Cao, J.: Almost periodic solutions of recurrent neural networks with continuously distributed delays. Nonlinear Anal. **71**(12), 6097–6108 (2009)
3. Yang, Y., Cao, J.: Stability and periodicity in delayed cellular neural networks with impulsive effects. Nonlinear Anal. Real World Appl. **8**(1), 362–374 (2007)
4. Cao, J., Liang, J.: Boundedness and stability for Cohen-Grossberg neural network with time-varying delays. J. Math. Anal. Appl. **296**(2), 665–685 (2004)
5. Abbas, S., Xia, Y.: Existence and attractivity of k-almost automorphic sequence solution of a model of cellular neural networks with delay. Acta Math. Sci. Ser. B Engl. Ed. **33**(1), 290–302 (2013)
6. Abbas, S.: Existence and attractivity of k-pseudo almost automorphic sequence solution of a model of bidirectional neural networks. Acta Appl. Math. **119**, 57–74 (2012)
7. Abbas, S.: Pseudo almost periodic sequence solutions of discrete time cellular neural networks. Nonlinear Anal. Model. Control **14**(3), 283–301 (2009)
8. Gao, B., Zhang, W.: Equilibria and their bifurcations in a recurrent network involving iterates of a transcendental function. IEEE Trans. Neural Netw. **19**(5), 782–794 (2008)

9. Baldi, P., Atiya, A.F.: How delays affect neural dynamics and learning. IEEE Trans. Neural Netw. **5**(4), 612–621 (1994)
10. Feng, C., Plamondon, R.: An oscillatory criterion for a delayed neural ring network model. Neural Netw. **29–30**, 70–79 (2012)
11. Wei, J., Zhang, C.: Bifurcation analysis of a class of neural networks with delays. Nonlinear Anal: Real World Appl. **9**, 2234–2252 (2008)

Approximation of Solutions of a Stochastic Differential Equation

Sanjukta Das, D.N. Pandey and N. Sukavanam

Abstract The existence, uniqueness, and convergence of approximate solutions of a stochastic differential equation with deviated argument is studied using analytic semigroup theory and fixed point method. Then we consider Faedo-Galerkin approximation of solution and prove some convergence results. We also study an example to illustrate our result.

Keywords Analytic semigroup · Deviated argument · Stochastic Fractional differential equation

Mathematics Subject Classification (2010): 34G10, 34G20, 34K30, 35K90, 47D20

1 Introduction

Fractional differential equations appear abundantly in the theory of fractals, visco-elasticity, seismology, polymers, etc. Stochastic evolution equations are natural generalizations of ordinary differential equations incorporating the random noise which causes fluctuations in deterministic models. For details refer [1]. In certain real-world problems, delay depends not only on the time but also on the unknown quantity as we can see in [2]. Das et al. [3, 4] can be referred for related work with deviated argument. Bahuguna et. al. [5] discussed the Faedo-Galerkin approximation of solution.

S. Das (✉) · D.N. Pandey · N. Sukavanam
Department of Mathematics, IIT Roorkee, Roorkee, India
e-mail: sanjukta_das44@yahoo.com

D.N. Pandey
e-mail: nsukvfma@iitr.ernet.in

N. Sukavanam
e-mail: nsukvfma@iitr.ernet.in

P.N. Agrawal et al. (eds.), *Mathematical Analysis and its Applications*,
Springer Proceedings in Mathematics & Statistics 143,
DOI 10.1007/978-81-322-2485-3_4

So far the Faedo-Galerkin approximation of solution stochastic fractional differential equation with deviated argument is neglected in the literature. In an attempt to fill this gap we study the following stochastic fractional differential equation with deviated argument in a separable Hilbert space $(H, (.,.))$.

$$
\begin{aligned}
{}^cD_t^\beta u(t) + Au(t) &= f(u(t), u(h(u(t))))\frac{dw(t)}{dt}, \quad t \in [0, T] \\
u(0) &= u_0 \in H
\end{aligned} \tag{1}
$$

where $0 < \beta < 1$ and $0 < T < \infty$. ${}^cD_t^\beta$ denotes the Caputo fractional derivative of order β and $A : D(A) \subset X \to H$ is a linear operator. A and the functions f, h are defined in the hypotheses $(H1) - (H3)$ of Sect. 2.

2 Preliminaries

Here we deal with two separable Hilbert spaces H and K.

(H1) A is a closed, densely defined, self-adjoint operator with pure point spectrum $0 \le \lambda_0 \le \lambda_1 \le \cdots \le \lambda_m \le \cdots$ with $\lambda_m \to \infty$ and $m \to \infty$ and corresponding complete orthonormal system of eigenfunctions ϕ_j such that

$$A\phi_j = \lambda_j \phi_j \; and \; < \phi_i, \phi_j >= \delta_{i,j}$$

(H2) The function $f : [O, T] \times H_\alpha \times H_{\alpha-1} \to L(K, H)$ is continuous and $\exists$ constant L_f such that

$$\|f(u, u_1) - f(v, v_1)\|_Q^2 \le L_f[+\|u - v\|_\alpha + \|u_1 - v_1\|_{\alpha-1}]$$

(H3) The map $h : H_\alpha \times \mathcal{R}_+ \to \mathcal{R}_+$ satisfies $\|h(u,) - h(v,)\| \le L_h(\|u - v\|_\alpha)$

If $(H1)$ is satisfied then $-A$ is the infinitesimal generator of an analytic semigroup $\{e^{-tA} : t \ge 0\}$ in H. We also note that $\exists$ constant C such that $\|S(t)\| \le Ce^{\omega t}$ and constants C_i 's such that $\|\frac{d^i}{dt^i}S(t)\| \le C_i$, $t > 0$, $i = 1, 2$. Also $\|AS(t)\| \le Ct^{-1}$ and $\|A^\alpha S(t)\| \le C_\alpha t^{-\alpha}$.

We define the space H_α as $D(A^\alpha)$ endowed with the norm $\|.\|_\alpha$. Let $(\Omega, \mathfrak{F}, P)$ be a complete probability space endowed with complete family of right continuous increasing sub σ—algebras $\{\mathfrak{F}_t, t \in J\}$ such that $\mathfrak{F}_t \subset \mathfrak{F}$. A H—valued random variable is a $\mathcal{F}$—measurable process. We also assume that W is a Wiener process on K with covariance operator Q. Suppose Q is symmetric, positive, linear and bounded operator with $TrQ < \infty$. Let $K_0 = Q^{\frac{1}{2}}(K)$. The space $L_2^0 = L_2(K_0, H_\alpha)$ is a separable Hilbert space with norm $\|\psi\|_{L_2^0} = \|\psi Q^{\frac{1}{2}}\|_{L_2(K, H_\alpha)}$. Let $L_2(\Omega, \mathfrak{F}, P; H_\alpha) \equiv L_2(\Omega; H_\alpha)$ be the Banach space of all strongly measurable, square integrable, H_α—valued random variables equipped with the norm

$\|u(.)\|^2_{L_2} = E\|u(.;w)\|^2_{H_\alpha}$. C^α_T denotes the Banach space of all continuous maps from $J = (0, T]$ into $L_2(\Omega; H_\alpha)$ which satisfy $sup_{t\in J} E\|u(t)\|^2_{C^\alpha} < \infty$. $L^0_2(\Omega, H_\alpha) = \{f \in L_2(\Omega, H_\alpha) : f\ is\ \mathcal{F}_0 - measurable\}$ denotes an important subspace. For $0 \le \alpha < 1$ define

$$C^{\alpha-1}_T = \{u \in C^\alpha_T : \|u(t) - u(s)\|_{\alpha-1} \le L|t-s|, \forall t, s \in [0, T]\}.$$

Now let us define mild solution of (1):

Definition 1 The mild solution of (1) is a continuous $\mathfrak{F}_t$ adapted stochastic process $u \in C^\alpha_T \cap C^{\alpha-1}_T$ which satisfies the following:

1. $u(t) \in H_\alpha$ has *Càdlàg* paths on $t \in [0, T]$.
2. $\forall t \in [0, T]$, $u(t)$ is the solution of the integral equation

$$u(t) = T_\beta(t)u_0 + \int_0^t (t-s)^{\beta-1} S_\beta(t-s) f(u(s), u(h(u(s), s)))dw(s),\ t \in [0, T] \tag{2}$$

where $S_\beta(t) = \int_0^\infty \zeta_\beta(\theta) S(t^\beta\theta)d\theta$; and $T_\beta(t) = q\int_0^\infty \theta\zeta_\beta(\theta)S(t^\beta\theta)d\theta$; ζ_β is a probability density function defined on $(0, \infty)$, i.e. $\zeta_\beta(\theta) \ge 0,\ \theta \in (0, \infty)$ and $\int_0^\infty \zeta_\beta(\theta)d\theta = 1$. Also $\|T_\beta(t)u\| \le C\|u\|$, $\|S_\beta(t)u\| \le \frac{\beta C}{\Gamma(1+\beta)}\|u\|$, $\|A^\alpha S_\beta(t)u\| \le \frac{\beta C_\alpha \Gamma(2-\alpha)}{\Gamma(1+\beta(1-\alpha))} t^{-\alpha\beta}\|u\|$.

Lemma 1 *Let $f : J \times \Omega \times \Omega \to L^0_2$ be a strongly measurable mapping with $\int_0^T E\|f(t)\|^p_{L^0_2}dt < \infty$. Then*

$$E\|\int_0^t f(s)dw(s)\|^p \le l_s \int_0^t E\|f(s)\|^p_{L^0_2}ds$$

$\forall t \in [0, T]$ and $p \ge 2$ where l_s is a constant containing p and T.

l_s is incorporated into the constants in the following sections.

3 Existence and Uniqueness of Approximate Solutions

In this section we consider a sequence of approximate integrals and establish the existence and uniqueness of solution for each of the approximate integral equations. For $0 \le \alpha < 1$ and $u \in C^\alpha_{T_0}$, the hypotheses $(H2) - (H3)$, imply that $f(u(s), u(h(u(s), s)))$ is continuous on $[0, T_0]$. Therefore, $\exists$ a positive constant

$$N = 2L_f[T_0^{\theta_1} + 2R(1 + LL_h) + LL_h T_0^{\theta_2}] + 2N_0, \quad N_0 = E\|f(u_0, u_0)\|^2$$

such that $\|f(s, u(s), u(h(u(s), s)))\| \le N,\ t \in [0, T]$. Choose T_0, $0 < T_0 \le T$ such that

$$\left(\frac{\beta C_\alpha \Gamma(2-\alpha)}{\Gamma(1+\beta(1-\alpha))}\right)^2 N \frac{T_0^{\beta(1-\alpha)-1}}{\beta(1-\alpha)-1} \leq \frac{R}{4},$$

$$D = \left(\frac{\beta C_\alpha \Gamma(2-\alpha)}{\Gamma(1+\beta(1-\alpha))}\right)^2 2L_f \frac{T_0^{\beta(1-\alpha)-1}}{2\beta(1-\alpha)-1} \leq 1 \tag{3}$$

Let

$$B_R = \{u \in \mathcal{C}_{T_0}^{\alpha} \cap \mathcal{C}_{T_0}^{\alpha-1} : u(0) = u_0, \quad \|u - u_0\|_{T_0,\alpha} \leq R\}$$

It is easy to see that B_R is a closed and bounded subset of $\mathcal{C}_{T_0}^{\alpha-1}$ and complete. Let us define the operator $\mathcal{F}_n : B_R :\to B_R$ by

$$(\mathcal{F}_n u)(t) = T_\beta(t)u_0 + \int_0^t (t-s)^{\beta-1} S_\beta(t-s) f_n(u(s), u(h(u(s), s)))dw(s). \tag{4}$$

Theorem 1 *If the hypotheses* $(H1)$, $(H2)$ *and* $(H3)$ *are satisfied and* $u_0 \in L_2^0(\Omega, X_\alpha)$, $0 \leq \alpha < 1$, *then* $\exists$ *a unique* $u_n \in B_R$ *such that* $\mathcal{F}_n u_n = u_n$, $\forall\, n = 0, 1, 2, \cdots$, *i.e.,* u_n *satisfies the approximate integral equation*

$$u_n(t) = T_\beta(t)u_0 + \int_0^t (t-s)^{\beta-1} S_\beta(t-s) f_n(s, u_n(s), u_n(h(u_n(s), s)))dw(s),$$
$$t \in [0, T] \tag{5}$$

Proof Step1 : We need to show that $\mathcal{F}_n u \in \mathcal{C}_{T_0}^{\alpha-1}$, $\forall u \in \mathcal{C}_{T_0}^{\alpha-1}$. It is easy to check that $\mathcal{F}_n : \mathcal{C}_T^{\alpha} \to \mathcal{C}_T^{\alpha}$. If $u \in \mathcal{C}_{T_0}^{\alpha-1}$, $0 < t_1 < t_2 < T_0$ and $0 \leq \alpha < 1$ then

$$\begin{aligned}
&E\|\mathcal{F}_n u(t_2) - \mathcal{F}_n u(t_1)\|_{\alpha-1}^2 \\
&\leq 3E\|[T_\beta(t_2) - T_\beta(t_1)]u_0\|_{\alpha-1}^2 \\
&\quad + 3E\|\int_{t_1}^{t_2} (t_2-s)^{\beta-1} A^{\alpha-1} S_\beta(t_2-s) f_n(u(s), u(h(u(s)))dw(s)\|_Q^2 \\
&\quad + 3E\|\int_0^{t_1} A[(t_2-s)^{\beta-1} S_\beta(t_2-s) - (t_1-s)^{\beta-1} S_\beta(t_1-s)] \\
&\quad A^{\alpha-2} \times f_n(u(s), u(h(u(s))))dw(s)\|_Q \\
&\leq 3E\|[T_\beta(t_2) - T_\beta(t_1)]u_0\|_{\alpha-1}^2 + 3\frac{\beta^2 C_\alpha^2 \Gamma^2(2-\alpha)}{\Gamma^2(1+\beta(1-\alpha))} \int_{t_1}^{t_2} \|(t_2-s)^{2\beta(1-\alpha)-2}\| \\
&\quad \times \|A^{-1}\|^2 E\|f_n(u(s), u(h(u(s),)))\|^2 ds \\
&\quad + 3\int_0^{t_1} \|A[(t_2-s)^{\beta-1} S_\beta(t_2-s) - (t_1-s)^{\beta-1} S_\beta(t_1-s)] \\
&\quad \times \|A^{\alpha-2}\|^2 E\|f_n(u(s), u(h(u(s))))\|^2 ds
\end{aligned} \tag{6}$$

$\forall u \in H$, we can write

$$[S(t_2^\beta\theta) - S(t_1^\beta\theta)]u = \int_{t_1}^{t_2} \frac{d}{dt} S(t^\beta\theta)u dt = \int_{t_1}^{t_2} \theta\beta t^{\beta-1} AS(t^\beta\theta)dt.$$

The first term of (6) can be estimated as follows:

$$\begin{aligned}
\|[T_\beta(t_2) - T_\beta(t_1)]u_0\|_{\alpha-1}^2 &\leq \left(\int_0^\infty \zeta_\beta(\theta)\|S(t_2^\beta\theta) - S(t_1^\beta\theta)\|\|A^{\alpha-1}u_0\|d\theta\right)^2 \\
&\leq \left(\int_0^\infty \zeta_\beta(\theta)[\int_{t_1}^{t_2} \|\frac{d}{dt}S(t^\beta\theta)\|dt]\|u_0\|_\alpha d\theta\right)^2 \\
&\leq C_1^2\|u_0\|_{\alpha-1}^2(t_2 - t_1)^2 \qquad (7)
\end{aligned}$$

For the second term of (6) we get the following estimate

$$\begin{aligned}
&\int_{t_1}^{t_2} (t_2 - s)^{2\beta(1-\alpha)-2} E\|f_n(u(s), u(h(u(s))))\|^2 ds \\
&\quad\leq \frac{N(t_2 - t_1)^{2\beta(1-\alpha)-1}}{2\beta(1-\alpha) - 1} \qquad (8)
\end{aligned}$$

For the third term we will use the following estimate

$$\begin{aligned}
&\int_0^{t_1} \|A[(t_2 - s)^{\beta-1}S_\beta(t_2 - s) - (t_1 - s)^{\beta-1}S_\beta(t_1 - s)]\|^2 \\
&\quad \times \|A^{\alpha-2}\|^2 E\|f_n(u(s), u(h(u(s))))\|^2 ds \\
&\leq \int_0^{t_1} \left(\int_0^\infty \zeta_\beta(\theta)\|[\frac{d}{dt}S((t-s)^\beta\theta)|_{t=t_2} - \frac{d}{dt}S((t-s)^\beta\theta)|_{t=t_1}]\|d\theta\right)^2 \\
&\quad \times E\|f(u(s), u(h(u(s))))\|^2 ds \\
&\leq \int_0^{t_1} \left(\int_0^\infty \zeta_\beta(\theta)[\int_{t_1}^{t_2} \|A^{\alpha-2}\frac{d^2}{dt^2}S((t-s)^\beta\theta)\|dt]d\theta\right)^2 N ds \\
&\leq C_2^2\|A^{\alpha-2}\|^2(t_2 - t_1)^2 N T_0 \qquad (9)
\end{aligned}$$

Hence from inequalities (7)–(9) we see that the map $\mathcal{F}_n : \mathcal{C}_{T_0}^{\alpha-1} \to \mathcal{C}_{T_0}^{\alpha-1}$ is well-defined. Now we prove that $\mathcal{F}_n : B_R \to B_R$. So for $t \in [0, T_0]$ and $u \in B_R$.

$$\begin{aligned}
&E\|(\mathcal{F}_n u)(t) - u_0\|_\alpha^2 \\
&\quad \leq 2E\|(T_\beta(t) - I)u_0\|_\alpha^2 \\
&\quad + 2E\|\int_0^t (t-s)^{\beta-1}S_\beta(t-s)f(u(s), u(h(u(s))))dw(s)\|_Q^2
\end{aligned}$$

$$\leq 2E\|(T_\beta(t)-I)u_0\|_\alpha^2 + 2\left(\frac{\beta C_\alpha \Gamma(2-\alpha)}{\Gamma(1+\beta(1-\alpha))}\right)^2 \int_0^t \|(t_2-s)^{2\beta(1-\alpha)-2}\|^2$$
$$\times\ E\|f_n(u(s),u(h(u(s))))\|^2 ds$$
$$\leq \frac{R}{2} + 2\left(\frac{\beta C_\alpha \Gamma(2-\alpha)}{\Gamma(1+\beta(1-\alpha))}\right)^2 N \frac{T_0^{\beta(1-\alpha)-1}}{\beta(1-\alpha)-1} \leq \frac{R}{2}+\frac{R}{2} = R$$

Now we show that $\mathcal{F}_n$ is a contraction map by using (3) in last but one inequality. $\forall u, v \in B_R$

$$E\|(\mathcal{F}_n u)(t)-(\mathcal{F}_n v)(t)\|_\alpha^2 = E\|\int_0^t (t-s)^{\beta-1}A^\alpha S_\beta(t-s)$$
$$\times\ [f(u(s),u(h(u(s)))) - f(s,v(s),v(h(v(s),s)))dw(s)]\|_Q^2$$
$$\leq \left(\frac{\beta C_\alpha \Gamma(2-\alpha)}{\Gamma(1+\beta(1-\alpha))}\right)^2 \int_0^t (t_2-s)^{2\beta(1-\alpha)-2}$$
$$\times\ E\|f(u(s),u(h(u(s)))) - f(v(s),v(h(v(s))))\|^2 ds$$
$$\leq \left(\frac{\beta C_\alpha \Gamma(2-\alpha)}{\Gamma(1+\beta(1-\alpha))}\right)^2 2L_f(1+2LLh)\|u-v\|_\alpha^2 \frac{T^2\beta(1-\alpha)-1}{2\beta(1-\alpha)-1}$$
$$\leq \|u-v\|_\alpha^2.$$

This implies that $\exists$ a unique fixed point u_n of $\mathcal{F}_n$. Thus there a unique mild approximate solution of (1)

Lemma 2 *Let $(H1)-(H3)$ hold. If $u_0 \in L_2^0(\Omega, D(A^\alpha))$, $\forall 0<\alpha<\eta<1$, then $u_n(t) \in D(A^\gamma)$ for all $t \in [0, T_0]$ with $0<\gamma<\eta<1$. Also if $u_0 \in D(A)$, then $u_n(t) \in D(A^\gamma)$ $\forall t \subset [0, T_0]$, where $0<\gamma<\eta<1$.*

Proof By Theorem (1) we get the existence of a unique $u_n \in B_R$, satisfying (5). Theorem 2.6.13 of [6] implies for $t>0$, $0\leq\gamma<1$, $S(t): H \to D(A^\gamma)$ and for $0\leq\gamma<\eta<1$, $D(A^\eta)\subset D(A^\gamma)$. It is easy to see that Holder continuity of u_n can be proved using the similar arguments from (6) to (9). Also from Theorem 1.2.4 in [6], we have $S(t)u \in D(A)$ if $u \in D(A)$. The result follows from these facts and that $D(A)\subset D(A^\gamma)$ for $0\leq\gamma<1$.

Lemma 3 *Let $(H1)-(H3)$ hold and $u_0 \in L_2^0(\Omega, X_\alpha)$. Then for any $t_0 \in (0, T_0]$ $\exists$ a constant U_{t_0}, independent of n such that $E\|u_n(t)\|_\gamma^2 \leq U_{t_0}$ $\forall t \in [t_0, T_0]$, $n = 1, 2, \cdots$. Also if $u_0 \in L_2^0(\Omega, D(A))$ then $\exists$ constant U_0 independent of n such that $E\|u_n(t)\|_\gamma^2 \leq U_0$ $\forall t \in [t_0, T_0]$, $n = 1, 2, \cdots,$ $\forall 0<\gamma\leq 1$.*

Proof Let $u_0 \in L_2^0(\Omega, H_\alpha)$. Applying A^γ on both sides of (4)

$$
\begin{aligned}
&E\|u_n(t)\|_\gamma^2 \\
&\leq 2E\|T_\beta(t)u_0\|_\gamma^2 + 2\|\int_0^t (t-s)^{\beta-1}S_\beta(t-s)f_n(u(s),u(h(u(s))))dw(s)\|_Q^2 \\
&\leq 2C_\gamma^2 t_0^{-2\gamma\beta}\|u_0\|^2 + \left(\frac{\beta C_\gamma \Gamma(2-\gamma)}{\Gamma(1+\beta(1-\gamma))}\right)^2 \frac{N(T_0)^{2\beta(1-\gamma)-1}}{2\beta(1-\gamma)-1} = U_{t_0}.
\end{aligned}
$$

Also if $u_0 \in L_2^0(\Omega, D(A))$, then we have that $u_0 \in L_2^0(\Omega, D(A^\gamma))$ for $0 \leq \gamma < 1$. Hence,

$$
\begin{aligned}
&E\|u_n(t)\|_\gamma^2 \\
&\leq 2E\|T_\beta(t)u_0\|_\gamma^2 + 2\|\int_0^t (t-s)^{\beta-1}S_\beta(t-s)f_n(u(s),u(h(u(s))))dw(s)\|_Q^2 \\
&\leq 2C^2\|u_0\|^2 + \left(\frac{\beta C_\gamma \Gamma(2-\gamma)}{\Gamma(1+\beta(1-\gamma))}\right)^2 \frac{N(T_0)^{2\beta(1-\gamma)-1}}{2\beta(1-\gamma)-1} = U_0.
\end{aligned}
$$

Hence proved.

4 Convergence of Solutions

In this section the convergence of the solution $u_n \in H_\alpha$ of the approximate integral equation (5) to a unique solution u of (2), is discussed.

Theorem 2 *Let the hypotheses $(H1) - (H3)$ hold and if $u_0 \in L_2^0(\Omega, H_\alpha)$ then $\forall t_0 \in (0, T]$,*

$$\lim_{m\to\infty} \sup_{\{n\geq M, t_0\leq t\leq T_0\}} \|u_n(t) - u_m(t)\|_\alpha = 0.$$

Proof Let $0 < \alpha < \gamma < \eta$. For $t_0 \in (0, T_0]$

$$
\begin{aligned}
&E\|f_n(u_n(t),u_n(h(u_n(t)))) - f_m(t,u_m(t),u_m(h(u_m(t))))\|^2 \\
&\quad\leq 2E\|f_n(u_n(t),u_n(h(u_n(t)))) - f_n(t,u_m(t),u_m(h(u_m(t))))\|^2 \\
&\quad\leq 2E\|f_n(u_m(t),u_m(h(u_m(t)))) - f_m(t,u_m(t),u_m(h(u_m(t))))\|^2 \\
&\quad\leq 2(2L_f(1+2LL_h)[E\|u_n - u_m\|_\alpha^2 + E\|(P^n - P^m)u_m(t)\|_\alpha^2]) \qquad (10)
\end{aligned}
$$

Now,

$$E\|(P^n - P^m)u_m(t)\|^2 \leq E\|A^{\alpha-\gamma}(P^n - P^m)A^\gamma u_m(t)\|^2 \leq \frac{1}{\lambda_m^{2(\gamma-\alpha)}} E\|A^\gamma u_m(t)\|^2$$

Then we have

$$
\begin{aligned}
&E\|f_n(t,u_n(t),u_n(h(u_n(t)))) - f_m(t,u_m(t),u_m(h(u_m(t))))\|^2 \\
&\le 2\left(2L_f(1+2LL_h)\left[E\|u_n - u_m\|_\alpha^2 + \frac{1}{\lambda_m^{2(\gamma-\alpha)}}E\|A^\gamma u_m(t)\|^2\right]\right)
\end{aligned}
$$

For $0 < t_0' < t_0$

$$
\begin{aligned}
E\|u_n(t) - u_m(t)\|_\alpha^2 &\le 2\left(\int_0^{t_0'} + \int_{t_0'}^t\right)\|(t-s)^{\beta-1}A^\alpha S_\beta(t-s)\|^2 \\
&\quad \times E\|f_n(u_n(t),u_n(h(u_n(t)))) - f_m(u_m(t),u_m(h(u_m(t))))\|^2 ds \qquad (11)
\end{aligned}
$$

The estimate of first integral of the above inequality is

$$
\begin{aligned}
&E\|u_n(t) - u_m(t)\|_\alpha^2 \\
&\quad \le \int_0^{t_0'} \|(t-s)^{\beta-1}A^\alpha S_\beta(t-s)\|^2 \\
&\quad \times E\|f_n(u_n(t),u_n(h(u_n(t)))) - f_m(u_m(t),u_m(h(u_m(t))))\|^2 ds \\
&\quad \le \left(\frac{\beta C_\gamma \Gamma(2-\gamma)}{\Gamma(1+\beta(1-\gamma))}\right)^2 \frac{2N(t_0-\delta_1 t_0')^{2\beta(1-\gamma)-2}}{2\beta(1-\gamma)-1} t_0', \quad 0 < \delta < 1 \qquad (12)
\end{aligned}
$$

The estimate of second integral is

$$
\begin{aligned}
E\|u_n(t) - u_m(t)\|_\alpha^2 &\le \int_{t_0'}^t \|(t-s)^{\beta-1}A^\alpha S_\beta(t-s)\|^2 \\
&\quad \times E\|f_n(u_n(t),u_n(h(u_n(t)))) - f_m(u_m(t),u_m(h(u_m(t))))\|^2 ds \\
&\le \left(\frac{\beta C_\gamma \Gamma(2-\gamma)}{\Gamma(1+\beta(1-\gamma))}\right)^2 \int_{t_0'}^t (t-s)^{2\beta(\alpha-1)-2} \\
&\quad \times 4L_f(1+2LL_h)\left[E\|u_n - u_m\|_\alpha^2 + \frac{E\|A^\gamma u_m(s)\|^2}{\lambda^2(\gamma-\alpha)}\right] ds \\
&\le 4L_f(1+2LL_h)\left(\frac{\beta C_\gamma \Gamma(2-\gamma)}{\Gamma(1+\beta(1-\gamma))}\right)^2 [\int_{t_0'}^t (t-s)^{2\beta(\alpha-1)-2} \\
&\quad \times E\|u_n - u_m\|_\alpha^2 ds + \frac{U_{t_0}}{\lambda_m^{2(\gamma-\alpha)}}\frac{T_0^{2\beta(1-\alpha)-1}}{2\beta(1-\alpha)-1}] \qquad (13)
\end{aligned}
$$

Substituting inequalities (12), (13) into (11) we get

$$\begin{aligned}
E\|u_n(t) - u_m(t)\|_\alpha^2 &\\
\leq \left(\frac{\beta C_\gamma \Gamma(2-\gamma)}{\Gamma(1+\beta(1-\gamma))}\right)^2 & \frac{4N(t_0 - \delta_1 t_0')^{2\beta(1-\gamma)-2}}{2\beta(1-\gamma)-1} t_0' \\
+ 8L_f(1+2LL_h) & \left(\frac{\beta C_\gamma \Gamma(2-\gamma)}{\Gamma(1+\beta(1-\gamma))}\right)^2 [\int_{t_0'}^t (t-s)^{2\beta(\alpha-1)-2} \\
\times E\|u_n - u_m\|_\alpha^2 ds & + \frac{U_{t_0}}{\lambda_m^{2(\gamma-\alpha)}} \frac{T_0^{2\beta(1-\alpha)-1}}{2\beta(1-\alpha)-1}]
\end{aligned}$$

By using Gronwall's inequality, $\exists$ a constant D such that

$$\begin{aligned}
E\|u_n(t) - u_m(t)\|_\alpha^2 \leq & \left[\left(\frac{\beta C_\gamma \Gamma(2-\gamma)}{\Gamma(1+\beta(1-\gamma))}\right)^2 \frac{4N(t_0 - \delta_1 t_0')^{2\beta(1-\gamma)-2}}{2\beta(1-\gamma)-1} t_0' \right. \\
& \left. + 8L_f(1+2LL_h)\left(\frac{\beta C_\gamma \Gamma(2-\gamma)}{\Gamma(1+\beta(1-\gamma))}\right)^2 \frac{U_{t_0}}{\lambda_m^{2(\gamma-\alpha)}} \frac{T_0^{2\beta(1-\alpha)-1}}{2\beta(1-\alpha)-1}\right] \times D
\end{aligned}$$

Let $m \to \infty$. Taking supremum over $[t_0, T_0]$ we get the following inequality:

$$E\|u_n(t) - u_m(t)\|_\alpha^2 \leq \left[\left(\frac{\beta C_\gamma \Gamma(2-\gamma)}{\Gamma(1+\beta(1-\gamma))}\right)^2 \frac{4N(t_0 - \delta_1 t_0')^{2\beta(1-\gamma)-2}}{2\beta(1-\gamma)-1} t_0'\right] \times D$$

Since t_0' is arbitrary, the right-hand side can be made infinitesimally small by choosing t_0' sufficiently small. Thus the lemma is proved.

Corollary 1 *If $u_0 \in D(A)$, then* $\lim_{m\to\infty} \sup_{\{n\geq m,\ 0\leq t\leq T_0\}} E\|u_n(t) - u_m(t)\|_\alpha^2 = 0$

Proof By using Lemmas (2) and (3) we can take $t_0 = 0$ in the proof of Theorem (2) and hence the corollary follows.

Theorem 3 *Let us assume that $(H1) - (H3)$ are satisfied and suppose $u_0 \in L_2^0(\Omega, X_\alpha)$. Then for $t \in [0, T_0]$, $\exists$ a unique function $u_n \in B_R$ where $u_n(t) = T_\beta u_0 + \int_0^t (t-s)^{\beta-1} S_\beta(t-s) f_n(u_n(s), u_n(h_n(u_n(s))))dw(s)$, and $u(t) \in B_R$, where $u(t) = T_\beta u_0 + \int_0^t (t-s)^{\beta-1} S_\beta(t-s) f(u(s), u(h(u(s))))dw(s), t \in [0, T_0]$, such that $u_n \to u$ as $n \to \infty$ in B_R and u satisfies (2) on $[0, T_0]$.*

Proof By using the above Corollary, Theorems 1 and 2 it is to see that $\exists\ u(t) \in B_R$ such that
$\lim_{n\to\infty} E\|u_n(t) - u(t)\|_\alpha^2 = 0$ on $[0, T_0]$. Now

$$
\begin{aligned}
E\|u_n(t) - T_\beta u_0 + \int_{t_0}^{t}(t-s)^{\beta-1}S_\beta(t-s)f_n(u_n(s),u_n(h_n(u_n(s))))dw(s)\|^2 \\
\leq E\|\int_{0}^{t_0}(t-s)^{\beta-1}S_\beta(t-s)f_n(u_n(s),u_n(h_n(u_n(s))))dw(s)\|^2 \\
\leq \left(\frac{\beta C}{\Gamma(1+\beta)}\right)^2 N\frac{T_0^{2\beta-2}}{2\beta-2}t_0 \qquad (14)
\end{aligned}
$$

Let $n \to \infty$ then
$E\|u_n(t) - T_\beta u_0 + \int_{t_0}^{t}(t-s)^{\beta-1}S_\beta(t-s)f_n(u_n(s),u_n(h_n(u_n(s))))dw(s)\|^2$
$\leq \left(\frac{\beta C}{\Gamma(1+\beta)}\right)^2 N\frac{T_0^{2\beta-2}}{2\beta-2}t_0$ and since t_0 is arbitrary we conclude $u(t)$ satisfies (2). Uniqueness follows easily from Theorems 1, 2 and Gronwall's inequality.

4.1 Faedo-Galerkin Approximations

We know from the previous sections that for any $0 \leq T_0 \leq T$, we have a unique $u \in C^\alpha_{T_0}$ satisfying the integral equation
$u(t) = T_\beta u_0 + \int_0^t(t-s)^{\beta-1}S_\beta(t-s)f(u(s),u(h(u(s))))dw(s)$, $t \in [0, T_0]$ Also, $\exists$ a unique solution $u_n \in C^\alpha_{T_0}$ of the approximate integral equation
$u_n(t) = T_\beta u_0 + \int_0^t(t-s)^{\beta-1}S_\beta(t-s)f_n(u_n(s),u_n(h(u_n(s))))dw(s)$, $t \in [0, T_0]$.
Faedo-Galerkin approximation $\bar{u}_n = P^n u_n$ is given by

$P^n u_n(t) = \bar{u}_n(t) = T_\beta(t)P^n u_0$
$+ \int_0^t(t-s)^{\beta-1}S_\beta(t-s)P^n f(u_n(s),u_n(h(u_n(s))))dw(s)$, $t \in [0, T_0]$. If the solution $u(t)$ to (2) exists on $[0, T_0]$ then it has the representation

$u(t) = \sum_{i=0}^{\infty}\alpha_i(t)\phi_i$, where $\alpha_i(t) = (u(t), \phi_i)$ for $i = 0, 1, 2, 3, \cdots$ and

$\bar{u}_n(t) = \sum_{i=0}^{n}\alpha_i^n(t)\phi_i$, where $\alpha_i^n(t) = (\bar{u}_n(t), \phi_i)$ for $i = 0, 1, 2, 3, \cdots$.

As a consequence of Theorems 1 and 2, we have the following result.

Theorem 4 *Let us assume that $(H1) - (H3)$ are satisfied and suppose $u_0 \in L_2^0(\Omega, X_\alpha)$. Then for $t \in [0, T_0]$, $\exists$ a unique function $u_n \in B_R$ where*
$u_n(t) = T_\beta P^n u_0 + \int_0^t(t-s)^{\beta-1}S_\beta(t-s)P^n f_n(u_n(s),u_n(h(u_n(s))))dw(s)$, and $u(t) \in B_R$, where
$u(t) = T_\beta u_0 + \int_0^t(t-s)^{\beta-1}S_\beta(t-s)f(u(s),u(h(u(s))))dw(s)$, $t \in [0, T_0]$, such that $u_n \to u$ as $n \to \infty$ in B_R and u satisfies (2) on $[0, T_0]$.

Now the convergence of $\alpha_i^n(t) \to \alpha_i(t)$ is shown. It is easily seen that

$$A^\alpha[u(t) - \bar{u}_n(t)] = A^\alpha\Big[\sum_{i=0}^{n}\{\alpha_i(t) - \alpha_i^n(t)\}\phi_i\Big] + A^\alpha\sum_{i=n+1}^{\infty}\alpha_i(t)\phi_i$$

$$= \sum_{i=0}^{n} \lambda_i^{\alpha}\{\alpha_i(t) - \alpha_i^n(t)\}\phi_i + \sum_{i=n+1}^{\infty} \lambda_i^{\alpha}\alpha_i(t)\phi_i.$$ Thus we have $E\|A^{\alpha}[u(t) - \bar{u}_n(t)\|^2 \geq \sum_{i=0}^{n} \lambda_i^{2\alpha} E|\alpha_i(t) - \alpha_i^n(t)|^2$.

Theorem 5 *Let us assume* $(H1) - (H3)$ *hold.*

(*i*) *If* $u_0 \in L_2^0(\Omega, X_\alpha)$ *then* $\lim_{n\to\infty} \sup_{t\in[t_0,T_0]} \left[\sum_{i=0}^{n} \lambda_i(t)^{2\alpha} E\|\alpha_i(t) - \alpha_i^n(t)\|^2\right] = 0$

(*ii*) *If* $u_0 \in L_2^0(\Omega, D(A))$ *then* $\lim_{n\to\infty} \sup_{t\in[0,T_0]} \left[\sum_{i=0}^{n} \lambda_i(t)^{2\alpha} E\|\alpha_i(t) - \alpha_i^n(t)\|^2\right] = 0$

Theorem 5 follows from the facts mentioned above the theorem.

Corollary 2 *Let us assume* $(H1) - (H3)$ *hold.*
(*i*) *If* $u_0 \in L_2^0(\Omega, X_\alpha)$ *then* $\lim_{n\to\infty} \sup_{t\in[t_0,T_0],n\geq m} E\|A^{\alpha}[\bar{u}_n(t) - \bar{u}_m(t)]\|^2 = 0$
(*ii*) *If* $u_0 \in L_2^0(\Omega, D(A))$ *then* $\lim_{n\to\infty} \sup_{t\in[0,T_0],n\geq m} E\|A^{\alpha}[\bar{u}_n(t) - \bar{u}_m(t)]\|^2 = 0$

Proof

$$\begin{aligned} E\|A^{\alpha}[\bar{u}_n(t) - \bar{u}_m(t)]\|^2 &= E\|P^n u_n(t) - P^m u_m(t)\|_\alpha^2 \\ &\leq 2E\|P^n[u_n(t) - u_m(t)]\|_\alpha^2 + 2E\|(P^n - P^m)y_m(t)\|_\alpha^2 \\ &\leq 2E\|[u_n(t) - u_m(t)]\|_\alpha^2 + 2\frac{1}{\lambda_m^{\gamma-\alpha}} E\|A^{\gamma} u_m(t)\|^2 \end{aligned}$$

Then the result (i) follows from Theorem 2 and result (ii) follows from Corollary 1.

5 Example

Suppose for $t \geq 0,\ x \in (0, 1), 0 < \beta \leq 1$

$$\begin{aligned} {}^cD^{\beta} v_t(t, x) &= v_{xx}(t, x) + F(v(t, x), v(h(t, v(x))))\frac{dw(t)}{dt}, \\ v(t, x) &= v_0,\ t = 0,\ x \in (0, 1) \quad and \quad v(t, 0) = v(t, 1) = 0,\ t \geq 0 \end{aligned} \tag{15}$$

Let F be an appropriate Holder continuous function satisfying $(H2)$ in $L_2^0(K, (0, 1))$. w is a standard $L_2(0, 1)$ valued Weiner process. Let us define $A = -\frac{d^2}{dx^2}$, $f := F$, $v(x) = u(t)$ and let $D(A) = H_0^1(0, 1) \cap H^2(0, 1)$, $D(A^{1/2}) = H_0^1(0, 1)$. Then (15) can be reformulated into (1). Now from Theorems (1), (2) we can similarly prove the existence, uniqueness, and approximation of the mild solution of (15).

References

1. Xua, D., Yanga, Z., Huang, Y.: Existence uniqueness and continuation theorems for stochastic functional differential equations. J. Differ. Equ. **245**(6), 1681–1703 (2008)
2. Das, S., Pandey, D.N., Sukavanam, N.: Approximate controllability of a second-order neutral differential equation with state dependent delay. Differ. Equa. Dyn. Syst. **218**, (2014). doi:10.1007/s12591-014-0218-6
3. Das, S., Pandey, D.N., Sukavanam, N.: Approximate controllability of a functional differential equation with deviated argument. Nonlinear Dyn. Syst. Theory **14**(3), 265–277 (2014)
4. Das, S., Pandey, D.N., Sukavanam, N.: Exact controllability of an impulsive semilinear system with deviated argument in a banach space. J. Differ. Equ. **461086**, 6 (2014). doi:10.1155/2014/461086
5. Bahuguna, D., Muslim, M.: Approximation of solutions to retarded differential equations with applications to population dynamics. J. Appl. Math. Stochast. Anal. **01**, 1–11 (2005)
6. Pazy, A.: Semigroups of Linear Operators and Applications to Partial Differential Equations. Springer (1983)

Reconstruction of Multiply Generated Splines from Local Average Samples

P. Devaraj and S. Yugesh

Abstract We analyze the following average sampling problem: Let h be a non-negative measurable function supported in $\left[-\frac{1}{2}, \frac{1}{2}\right]$. Given a sequence of samples $\{y_n\}_{n\in\mathbb{Z}} \in \mathbb{R}^{\mathbb{Z}}$ of polynomial growth, find a multiply generated spline f of polynomial growth such that $\int_{-\frac{1}{2}}^{\frac{1}{2}} f(n-t)h(t)dt = y_n$, $n \in \mathbb{Z}$. It is shown that the solution to this problem is unique over certain subspaces of the multiply generated spline space of polynomial growth.

Keywords Interpolation · Multiply generated splines · Average sampling

1 Introduction

The sampling theorem is one of the widely used results in the signal processing field. The well-known Shannon sampling theorem states that, any bandlimited signal f is completely determined by its samples [4, 8]. Although the Shannon sampling theorem is very useful, it has a number of problems when using it for practical applications. The bandlimited functions have analytic continuation to the entire complex plane and hence they are of infinite duration which is not always realistic. On the other hand, the sinc function has a very slow decay. Further, the measured samples are not exact in practical problems and they are the average of the signal around the sampling point and the averaging function depends on the aperture device used for capturing the samples. For these reasons, sampling and local average sampling have been investigated in several other classes of signals. In general, spline spaces yield many advantages in their generation and numerical treatment so that there are

P. Devaraj · S. Yugesh (✉)
Department of Mathematics, College of Engineering Guindy Campus,
Anna University, Chennai 600 025, India
e-mail: devaraj@annauniv.edu

S. Yugesh
e-mail: mathsyugesh@gmail.com

P.N. Agrawal et al. (eds.), *Mathematical Analysis and its Applications*,
Springer Proceedings in Mathematics & Statistics 143,
DOI 10.1007/978-81-322-2485-3_5

many practical applications in signal, image processing, and communication theory. In the literature [1–8] many authors have investigated the generalized sampling technique for multiply generated shift-invariant spaces and spline subspaces. The multiply generated spline space is defined in [5, 6] as

$$\mathscr{S} = \left\{ f : f = \sum_{i=1}^{r} \sum_{n \in \mathbb{Z}} a_i(n) \beta_{d_i}(t-n) \right\}$$

with suitable coefficients $a_i(n)$, where β_{d_i} is the cardinal central B-spline of degree d_i and is defined by,

$$\beta_{d_i} = \chi_{[-\frac{1}{2},\frac{1}{2}]} \star \chi_{[-\frac{1}{2},\frac{1}{2}]} \star \cdots \star \chi_{[-\frac{1}{2},\frac{1}{2}]} (d_i + 1 \text{ terms}),$$

where $\star$ represents the convolution (The convolution of two functions f and g is defined as $f \star g(n) = \int f(t)g(n-t)dt$). We consider the following subspace of the multiply generated spline space:

$$\mathscr{S}_N := \left\{ f : f = \sum_{n \in \mathbb{Z}} a_n \sum_{i=1}^{r} \beta_{d_i}(t-n) \right\}$$

If $M = max\{d_1, d_2, \ldots, d_r\}$ and $m = \min\{d_1, d_2, \ldots, d_r\}$, then $f \in \mathscr{S}_N$ provided that $f(x) \in C^{m-1}(\mathbb{R})$ and that the restriction of $f(x)$ to any interval between consecutive knots is identical with a polynomial of degree not exceeding M. If d_i's are distinct, then $\sum_{i=1}^{r} \beta_{d_i}(. - n)$, $n \in \mathbb{Z}$ are globally linearly independent.

We consider the following local average sampling problem:

Problem: Let $\{y_n\}_{n\in\mathbb{Z}}$ be a given sequence of real numbers. Find a spline $f \in \mathscr{S}_N$ such that $f \star h(n) = y_n, n \in \mathbb{Z}$, where $h \in L^1(\mathbb{R})$ and $\left(h \star \sum_{i=1}^{r} \beta_{d_i}\right)(n) \neq 0$, for some $n \in \mathbb{Z}$ and

$$supp(h) \subseteq \left[-\frac{1}{2}, \frac{1}{2}\right], h(t) \geq 0, t \in \mathbb{R}, 0 < \int_{-\frac{1}{2}}^{0} h(t)dt < \infty, 0 < \int_{0}^{\frac{1}{2}} h(t)dt < \infty. \tag{1}$$

We show that this problem has infinitely many solutions. The uniqueness of solution is obtained by imposing the following growth conditions on the samples and the splines as that of Schoenberg [9]:

$$\mathscr{S}_{\mathscr{N},\gamma} = \left\{ f(t) \in \mathscr{S}_N : f(t) = O(|t|^\gamma) \; as \; t \to \pm\infty \right\}$$

and

$$\mathscr{D}_\gamma = \left\{ \{y_n\} : y_n = O(|n|^\gamma) \; as \; n \to \pm\infty \right\}.$$

This problem over the singly generated spline space is analyzed in [10]. It is shown in [10] that the local average sampling problem has a unique solution for $d \leq 4$ when the spline space is generated by a single central B-spline. For $d > 4$ the same problem has been posed as an open problem. The same authors have analyzed the problem for $d \geq 5$ by reducing the support of h. They have shown in [11] that the local average sampling problem for singly generated spline has a unique solution when h is supported in $\left[-\frac{l}{2}, \frac{l}{2}\right]$, $l < 1$.

Lemma 1 *Let* $\psi(x) = \sum_{i=1}^{r} \beta_{d_i}(x)$ *and let* A *be the greatest integer such that* $h \star \psi(n) = 0, \forall n < A$, *and let* N_1 *be the smallest nonnegative integer such that* $h \star \psi(n) = 0, \forall n > A + N_1$. *Then the solutions of the problem form a linear manifold in* $\mathscr{S}_N$ *of dimension* N_1. *Moreover,* $N_1 = M + 1$, *if* M *is odd and* $N_1 = M$, *if* M *is even.*

Proof When $N_1 = 0$ this problem has a unique solution. For $N_1 > 0$, we consider the linear map from $\mathbb{C}^{\mathbb{Z}}$ to $\mathscr{S}_N$ defined by

$$\{a_n\}_{n \in \mathbb{Z}} \longmapsto \sum_{n \in \mathbb{Z}} a_n \psi(t - n).$$

Since the integer translates of ψ are globally linearly independent, this map is an isomorphism from $\mathbb{C}^{\mathbb{Z}}$ onto $\mathscr{S}_N$. Therefore $h \star f(n) = y_n$ in $\mathbb{C}^{\mathbb{Z}}$ if and only if,

$$\sum_{k=0}^{N_1} h \star \psi(A + k) a_{n-A-k} = y_n, \forall n \in \mathbb{Z}.$$

This forms a linear difference equation of order N_1 with constant coefficients. Hence the solution space is an N_1 dimensional manifold in $\mathscr{S}_N$. □

2 Local Average Sampling Theorems

Theorem 1 (Main Theorem) *Let* $d_i \leq 4$ *and let* $h(t)$ *be an integrable function satisfying condition* (1). *Then for a given sequence of numbers* $\{y_n\}_{n \in \mathbb{Z}} \in \mathscr{D}_\gamma$, *there exists a unique* $f \in \mathscr{S}_{N,\gamma}$ *such that*

$$f \star h(n) = y_n, n \in \mathbb{Z}. \tag{2}$$

We define the function

$$G(z) := \sum_{i=1}^{r} G_i(z)$$

where

$$G_i(z) := \sum_{n\in\mathbb{Z}} h \star \beta_{d_i}(n) z^n.$$

The exponential Euler spline is defined as

$$\Upsilon_{z,d_i}(t) = \sum_{n\in\mathbb{Z}} z^n \beta_{d_i}(n-t).$$

In terms of the exponential Euler spline we can write $G_i(z) = \int_{-\frac{1}{2}}^{\frac{1}{2}} h(t)\Upsilon_{z,d_i}(t)dt$. Hence

$$G(z) = \int_{-\frac{1}{2}}^{\frac{1}{2}} h(t)\Upsilon_z(t)dt,$$

where $\Upsilon_z(t) = \sum_{i=1}^{r} \Upsilon_{z,d_i}(t) = \sum_{n\in\mathbb{Z}} z^n \sum_{i=1}^{r} \beta_{d_i}(n-t)$.

We need some properties of $\Upsilon_z(t)$.

Lemma 2 *For $d \in \mathbb{N}$, $n \in \mathbb{Z}$ and $z \in \mathbb{C} \setminus \{0\}$, we have:*

(i) $\Upsilon_{z^{-1}}(-t) = \Upsilon_z(t)$,
(ii) $\Upsilon_z(t+n) = (z)^n \Upsilon_z(t)$,
(iii) $\frac{d}{dt}(\Upsilon_{z,d_i+1}(t)) = \left(1 - \frac{1}{z}\right)\Upsilon_{z,d_i}\left(t + \frac{1}{2}\right)$,
(iv) $\Upsilon_{-1,d_i}\left(\frac{1}{2}\right) = 0$ *and* $\Upsilon_{-1,d_i}(t) > 0$ *for* $t \in \left(-\frac{1}{2}, \frac{1}{2}\right)$.

Proof (i)

$$\begin{aligned}\Upsilon_{z^{-1}}(\ t) &= \sum_{n\in\mathbb{Z}} z^{-n} \sum_{i=1}^{r} \beta_{d_i}(n+t)\\ &= \sum_{n\in\mathbb{Z}} z^{-n} \sum_{i=1}^{r} \beta_{d_i}(-n-t) = \sum_{n\in\mathbb{Z}} z^{n} \sum_{i=1}^{r} \beta_{d_i}(n-t) = \Upsilon_z(t).\end{aligned}$$

(ii)

$$\Upsilon_z(t+n) = \sum_{k\in\mathbb{Z}} z^k \sum_{i=1}^{r} \beta_{d_i}(k-t-n) = z^n \sum_{k\in\mathbb{Z}} z^k \sum_{i=1}^{r} \beta_{d_i}(k-t) = z^n \Upsilon_z(t).$$

(iii)

$$\begin{aligned}\frac{d}{dt}(\Upsilon_{z,d_i+1}(t)) &= \sum_{n\in\mathbb{Z}} z^n \frac{d}{dt}(\beta_{d_i+1}(n-t))\\ &= \sum_{n\in\mathbb{Z}} z^n \left[\beta_{d_i}\left(n - \left(t + \frac{1}{2}\right)\right) - \beta_{d_i}\left(n - \left(t - \frac{1}{2}\right)\right)\right]\end{aligned}$$

$$
\begin{aligned}
&= \Upsilon_{z,d_i}\left(t+\frac{1}{2}\right) - \sum_{n\in\mathbb{Z}} z^{n-1}\beta_{d_i}\left(n-1-t+\frac{1}{2}\right)\\
&= \Upsilon_{z,d_i}\left(t+\frac{1}{2}\right) - \frac{1}{z}\sum_{n\in\mathbb{Z}} z^{n}\beta_{d_i}\left(n-\left(t+\frac{1}{2}\right)\right)\\
&= \Upsilon_{z,d_i}\left(t+\frac{1}{2}\right) - \frac{1}{z}\Upsilon_{z,d_i}\left(t+\frac{1}{2}\right)\\
&= \left(1-\frac{1}{z}\right)\Upsilon_{z,d_i}\left(t+\frac{1}{2}\right)
\end{aligned}
$$

(iv)

$$
\Upsilon_{-1,d_i}\left(\frac{1}{2}\right) = \sum_{n\in\mathbb{Z}}(-1)^n\beta_{d_i}\left(n-\frac{1}{2}\right) = 0.
$$

We shall show that $\Upsilon_{-1,d_i}(t) > 0$ for $t \in \left(-\frac{1}{2}, \frac{1}{2}\right)$, by using induction on d_i. For $d_i = 1$ by simple manipulation we get $\Upsilon_{-1,1}(t) > 0$ for $t \in (-\frac{1}{2}, \frac{1}{2})$. Assume that it is true for d_i and using (iii) we get

$$
\frac{d}{dt}(\Upsilon_{-1,d_i+1}(t)) = 2\Upsilon_{-1,d_i}\left(t+\frac{1}{2}\right) > 0, t \in \left(-\frac{1}{2}, 0\right).
$$

Using $\Upsilon_{-1,d_i+1}\left(-\frac{1}{2}\right) = 0$ and Υ_{-1,d_i+1} and being an even function, we obtain that $\Upsilon_{-1,d_i}(t) > 0$ for $t \in \left(-\frac{1}{2}, \frac{1}{2}\right)$. □

2.1 Uniqueness Theorem

Theorem 2 *Let $\Lambda = \{f \in \mathscr{S}_N : f \star h(n) = 0, n \in \mathbb{Z}\}$ and $z_1, z_2, \ldots, z_l$ be the roots of $G(z)$. If the roots of $G(z)$ are simple, then the set of functions $\Upsilon_{z_j^{-1}}$, where $j = 1, 2, \ldots, l$ form a basis of Λ.*

Proof By Lemma 1, Λ is a $l = N_1$ dimensional subspace of $\mathscr{S}_N$.
Using Lemma 2, we get

$$
\begin{aligned}
h \star \Upsilon_{z_j^{-1}}(n) &= \int_{-\frac{1}{2}}^{\frac{1}{2}} h(t)\Upsilon_{z_j^{-1}}(n-t)dt\\
&= z_j^n \int_{-\frac{1}{2}}^{\frac{1}{2}} h(t)\Upsilon_{z_j^{-1}}(-t)dt\\
&= z_j^n \int_{-\frac{1}{2}}^{\frac{1}{2}} h(t)\Upsilon_{z_j}(t)dt\\
&= z_j^n G(z_j)
\end{aligned}
$$

Therefore, $\Upsilon_{z_j^{-1}} \in \Lambda$ for $j = 1, 2, \ldots, l$.

Next, we have to prove that the elements of Λ are linearly independent.

$$\sum_{j=1}^{l} c_j \Upsilon_{z_j^{-1}}(t) = 0 \Leftrightarrow \sum_{j=l}^{l} c_j \left[\sum_{n\in\mathbb{Z}} z_j^{-n} \sum_{i=1}^{r} \beta_{d_i}(n-t) \right] = 0$$

$$\Leftrightarrow \sum_{n\in\mathbb{Z}} \left[\sum_{j=1}^{l} c_j z_j^{-n} \right] \left\{ \sum_{i=1}^{r} \beta_{d_i}(n-t) \right\} = 0.$$

As $\left\{\sum_{i=1}^{r} \beta_{d_i}(n-t)\right\}$ are linearly independent, we obtain

$$\sum_{j=1}^{l} c_j z_j^{-n} = 0.$$

This is a linear system of equation in the variable $c_1, c_2, \ldots, c_l$ with coefficient matrix, the Vandermonde's determinant. Therefore $c_j = 0$.
Hence, the functions $\Upsilon_{z_j^{-1}}(t),\ j = 1, 2, \ldots, l$ form a basis of Λ. □

Theorem 3 *Let $d_i \in \mathbb{N}$ and $h(t)$ be an integrable function satisfying condition* (1)*. If the roots of $G(z)$ are simple and no roots on the unit circle $|z| = 1$, then for a given sequence of numbers $\{y_n\}_{n\in\mathbb{Z}} \in \mathscr{D}_\gamma$, there exists a unique $f \in \mathscr{S}_{N,\gamma}$, such that*

$$f \star h(n) = y_n, n \in \mathbb{Z}. \tag{3}$$

Moreover, the solution can be written as

$$f(t) = \sum_{n\in\mathbb{Z}} y_n L(t-n),$$

where the reconstruction function L is given by $L(t) := \sum_{i=1}^{r} L_i(t) := \sum_{i=1}^{r} \sum_{n\in\mathbb{Z}} c_n \beta_{d_i}(t-n)$ and c_n are the coefficients of the Laurent series expansion of $G(z)^{-1}$. Further the reconstruction function L is of exponential decay.

Proof Let $C(z) = G^{-1}(z) = \sum_{n\in\mathbb{Z}} c_n z^n$. Then there exist $\mu \in (0, 1)$ such that $c_n = O\left(\mu^{|n|}\right)$. As β_{d_i} has compact support, we obtain that $O(L) = O(\mu^{|t|})$. Now for $|t| > 2$, we have

$$\begin{aligned}
\frac{\sum_{n\in\mathbb{Z}} |n|^\gamma \mu^{|t-n|}}{(|t|+1)^\gamma} &\le \frac{\sum_{n\in\mathbb{Z}} |n|^\gamma \mu^{|[t]-n|-1}}{(|[t]|)^\gamma} \\
&= \frac{\sum_{n\in\mathbb{Z}} (|[t]-n|)^\gamma \mu^{|n|-1}}{(|[t]|)^\gamma} \\
&\le \sum_{n\in\mathbb{Z}} (1+|n|)^\gamma \mu^{|n|-1} \\
&< \infty.
\end{aligned}$$

As a consequence of the above inequality we obtain that

$$f(t) = \sum_{n \in \mathbb{Z}} y_n L(t-n) = O(|t|^{\gamma}),$$

as $t \to \pm\infty$. Since $y_n L(t-n) = O(|n|^{\gamma} \mu^{|t-n|})$, it is easy to see that the series

$$\sum_{n \in \mathbb{Z}} y_n L(t-n)$$

converges uniformly and absolutely in every finite interval. Also,

$$\begin{aligned} f(t) &= \sum_{n \in \mathbb{Z}} y_n L(t-n) \\ &= \sum_{n \in \mathbb{Z}} y_n \sum_{i=1}^{r} \sum_{k \in \mathbb{Z}} c_k \beta_{d_i}(t-n-k) \\ &= \sum_{k \in \mathbb{Z}} \left(\sum_{n \in \mathbb{Z}} y_n c_{k-n} \right) \sum_{i=1}^{r} \beta_{d_i}(t-k). \end{aligned}$$

Therefore $f \in \mathscr{S}_{N,\gamma}$.
Using $C(z)G(z) = 1$, we obtain that

$$(h \star L)(n) = \sum_{i=1}^{r} \sum_{k \in \mathbb{Z}} c_k h \star \beta_{d_i}(n-k) = \delta_0(n).$$

Hence $f(t) = \sum_{n \in \mathbb{Z}} y_n L(t-n)$ satisfies

$$(h \star f)(n) = y_n, n \in \mathbb{Z}. \tag{4}$$

Now we shall show the uniqueness. Assume that $f, g \in \mathscr{S}_{N,\gamma}$ satisfy (4). Then $f - g \in \Lambda$. Using Theorem 2, there exist a constant c_j such that

$$f(t) - g(t) = \sum_{j=1}^{l} c_j \left(\Upsilon_{z_j^{-1}} \right).$$

As $f, g \in \mathscr{S}_{N,\gamma}$, we get $f(t) - g(t) = O(|t|^{\gamma})$.

Using Lemma 2 and the behavior of $\left(\Upsilon_{z_j^{-1}} \right)(t)$ at $\pm\infty$, we get $c_j = 0$ and hence $f = g$.

□

For $d_i = 1, 2, 3, 4$ we shall show that the roots of $G(z)$ are simple and not on the unit circle $|z| = 1$.

Proof (*Main Theorem*) As a consequence of Theorem 1 it is sufficient to prove that, all the roots of $G(z)$ are simple and not on the unit circle $|z| = 1$ for distinct $d_i = 1, 2, 3, 4$.

We have $G(z) = \sum_{i=1}^{r} G_i(z)$. We can write

$$G(z) = \sum_{i=1}^{r} z^{\frac{-l_i}{2}} P_i(z)$$

where $l_i := \begin{cases} d_i + 1 & \text{if } d_i \text{ is odd} \\ d_i & \text{if } d_i \text{ is even} \end{cases}$ and $P_i(z)$ is a polynomial of degree l_i. Therefore,

$$G(z) = z^{\frac{-m}{2}} \sum_{i=1}^{r} z^{\frac{m-l_i}{2}} P_i(z) = z^{\frac{-m}{2}} P(z),$$

where $P(z)$ is a polynomial of degree $m = max(l_1, l_2, \ldots, l_r)$.

As d_i's are distinct, we can take $d_1 = 1$, $d_2 = 2$, $d_3 = 3$, and $d_4 = 4$. Therefore $m = 4$ and we obtain

$$\begin{aligned} P(z) &= z^2 G(z) \\ &= z^4 \left\{ h \star \beta_{d_4}(2) + h \star \beta_{d_3}(2) \right\} + z^3 \left\{ h \star \beta_{d_4}(1) + h \star \beta_{d_3}(1) + h \star \beta_{d_2}(1) \right. \\ &\quad \left. + h \star \beta_{d_1}(1) \right\} + z^2 \left\{ h \star \beta_{d_4}(0) + h \star \beta_{d_3}(0) + h \star \beta_{d_2}(0) + h \star \beta_{d_1}(0) \right\} \\ &\quad + z \left\{ h \star \beta_{d_4}(-1) + h \star \beta_{d_3}(-1) + h \star \beta_{d_2}(-1) + h \star \beta_{d_1}(-1) \right\} + \left\{ h \star \beta_{d_4}(-2) \right. \\ &\quad \left. + h \star \beta_{d_3}(-2) \right\}. \end{aligned}$$

Hence $P(0) > 0$ and $P(1) > 0$.

We can write

$$P(z) = z^2 \sum_{i=1}^{4} \int_{-\frac{1}{2}}^{\frac{1}{2}} h(t) \Upsilon_{z,d_i}(t) dt. \tag{5}$$

Using Lemma 2 and Eq. (5) we get

$$P(-1) = \sum_{i=1}^{4} \int_{-\frac{1}{2}}^{\frac{1}{2}} h(t) \Upsilon_{-1,d_i}(t) dt > 0.$$

Since $\lim_{z \longrightarrow \infty} P(z) = \infty$, It is suffices to find $z_0 \in (-1, 0)$ such that

$$\sum_{i=1}^{4} \Upsilon_{z_0,d_i}(t) < 0, \text{ for all } t \in \left(-\frac{1}{2}, \frac{1}{2} \right), \tag{6}$$

since for such a z_0, we have

$$P(z_0) = z_0^2 \sum_{i=1}^{4} \int_{-\frac{1}{2}}^{\frac{1}{2}} h(t) \Upsilon_{z_0,d_i}(t)dt < 0, z_0 \in (-1, 0)$$

and $P\left(\frac{1}{z_0}\right) = \frac{1}{z_0^2} \sum_{i=1}^{4} \int_{-\frac{1}{2}}^{\frac{1}{2}} h(t) \Upsilon_{z_0^{-1},d_i}(t)dt = \frac{1}{z_0^2} \sum_{i=1}^{4} \int_{-\frac{1}{2}}^{\frac{1}{2}} h(t) \Upsilon_{z_0,d_i}(-t)dt < 0$ and $z_0^{-1} \in (-\infty, -1)$. By solving $\sum_{i=1}^{4} \Upsilon_{z_0,d_i}\left(\frac{1}{2}\right) = 0$, we get a unique $z_0 \in (-1, 0)$.

Now $\sum_{i=1}^{4} \Upsilon_{z_0,d_i}\left(\frac{1}{2}\right) = 0 \Leftrightarrow \Upsilon_{z_0,1}\left(\frac{1}{2}\right) + \Upsilon_{z_0,2}\left(\frac{1}{2}\right) + \Upsilon_{z_0,3}\left(\frac{1}{2}\right) + \Upsilon_{z_0,4}\left(\frac{1}{2}\right) = 0$

$$\begin{aligned}
\Leftrightarrow z_0^3 &\left\{\beta_4\left(\frac{3}{2}\right) + \beta_3\left(\frac{3}{2}\right)\right\} + z_0^2\left\{\beta_4\left(\frac{1}{2}\right) + \beta_3\left(\frac{1}{2}\right) + \beta_2\left(\frac{1}{2}\right) + \beta_1\left(\frac{1}{2}\right)\right\} \\
&+ z_0\left\{\beta_4\left(-\frac{1}{2}\right) + \beta_3\left(-\frac{1}{2}\right) + \beta_2\left(-\frac{1}{2}\right) + \beta_1\left(-\frac{1}{2}\right)\right\} \\
&+ \left\{\beta_4\left(-\frac{3}{2}\right) + \beta_3\left(-\frac{3}{2}\right)\right\} = 0 \\
\Leftrightarrow z_0^3 &\frac{3}{48} + z_0^2\frac{93}{48} + z_0\frac{93}{48} + \frac{3}{48} = 0.
\end{aligned}$$

The possible solutions of z_0 are -1, $-15 - 4\sqrt{14}$, $-15 + 4\sqrt{14}$. The unique solution $z_0 \in (-1, 0)$ is $-15 + 4\sqrt{14}$. For this z_0 value

$$\sum_{i=1}^{4} \Upsilon_{z_0,d_i}(t) < 0, \text{ for all } t \in \left(-\frac{1}{2}, \frac{1}{2}\right).$$

Thus we can conclude that all the roots of $G(z)$ are simple and not on the unit circle $|z| = 1$ for $d_i = 1, 2, 3, 4$. □

Remark 1 The condition that the zeros of the Laurent polynomial $G(z)$ are simple and do not lie on the unit circle $|z| = 1$ is a sufficient condition for uniqueness of solution for the local average sampling problem.

3 Conclusion

We proved local average sampling theorem over certain subspaces of the multiply generated spline spaces of polynomial growth. Let $h(t)$ be an integrable function satisfying condition (1). We have shown that if the roots of $G(z)$ are simple and no roots on the unit circle $|z| = 1$, then for a given sequence of numbers $\{y_n\}_{n\in\mathbb{Z}} \in \mathscr{D}_\gamma$, there exists a unique $f \in \mathscr{S}_{N,\gamma}$ such that $f \star h(n) = y_n$, $n \in \mathbb{Z}$, for the distinct $d_i \leq 4$. Also, we have shown that the roots of $G(z)$ are simple and not on the unit circle $|z| = 1$, for $d_i \leq 4$. We could not find a proof for $d_i \geq 5$.

Acknowledgments The authors thank Anna University, Chennai-25, India for providing the Anna Centenary Research fellowship to the second author to carry out this research work.

References

1. Aldroubi, A., Sun, Q., Wai-Shing, T.: Nonuniform average sampling and reconstruction in multiply generated shift-invariant spaces. Constr. Approx. **20**, 173–189 (2004)
2. Garcia, A.G., Hernandez-Medina, M.A., Perez-Villalon, G.: Generalized sampling in shift-invariant spaces with multiple stable generators. J. Math. Anal. Appl. **337**(1), 69–84 (2008)
3. de Boor, C., Devore, R., Ron, A.: The structure of finitely generated shift invariant spaces in $L^2(R^d)$. J. Funct. Anal. **119**, 37–78 (1994)
4. Gröcheni: Reconstruction algorithms in irregular samplings. Math. Comput. **59**, 181–194 (1992)
5. Xian, J., Zhu, J., Wang, J.: Local sampling and reconstruction in spline signal spaces and multiply generated spline signal spaces. International Joint Conference on Computational Sciences and Optimization, 2000
6. Xian, J., Sun, W.: Local sampling and reconstruction in shift-invariant spaces and their applications in spline subspaces. Numer. Funct. Anal. Optim. **31**(3), 366–386 (2010)
7. Xian, J., Qiang, X.: Non-uniform sampling and reconstruction in weighted finitely generated shift-invariant spaces. Far East. J. Math. Sci. **8**(3), 281–293 (2003)
8. Butzer, P.L., Engels, W., Ries, S., Stens, R.L.: The shannon sampling series and the reconstruction of signals in terms of linear, quadratic and cubic splines. SIAM J. Appl. Math. **46**(2), 299–323 (1986)
9. Schoenberg, I.J.: Cardinal spline interpolation. SIAM Regional Conference Series in Applied Mathematics, 1973
10. Perez-Villalon, G., Portal, A.: Reconstruction of splines from local average samples. Appl. Math. Lett. **25**, 1315–1319 (2012)
11. Perez-Villalon, G., Portal, A.: Sampling in shift-invariant spaces of functions with polynomial growth. Appl. Anal. **92**(12), 2536–2546 (2013)

Approximation of Periodic Functions Belonging to $W(L^r, \xi(t), (\beta \geq 0))$-Class By $(C^1 \cdot T)$ Means of Fourier Series

Smita Sonker

Abstract Various investigators such as Khan [3], Qureshi [8–10], Qureshi and Nema [11], Leindler [6] and Chandra [1] have determined the degree of approximation of functions belonging to the classes $W(L^r, \xi(t))$, $Lip(\xi(t), r)$, $Lip(\alpha, r)$ and $Lip\alpha$ using different summability methods with monotonocity conditions. Recently, Lal [5] has determined the degree of approximation of the functions belonging to $Lip\alpha$ and $W(L^r, \xi(t))$ classes by using Cesàro-Nörlund $(C^1 \cdot N_p)$—summability with non-increasing weights $\{p_n\}$. In this paper, we shall determine the degree of approximation of 2π-periodic function f belonging to the function classes $Lip\alpha$ and $W(L^r, \xi(t))$ by $C^1 \cdot T$—means of Fourier series of f. Our theorems generalize the results of Lal [5], and we also improve these results in the light of [7, 12, 13]. From our results, we derive some corollaries also.

Keywords Trigonometric fourier series · $W(L^r, \xi(t), (\beta \geq 0))$-class · Fourier series · Matrix means · Signals · Trigonometric polynomials

1 Introduction

For a given signal $f \in L^r := L^r[0, 2\pi], r \geq 1$, let

$$s_n(f) := s_n(f; x) = \frac{a_0}{2} + \sum_{k=1}^{n}(a_k \cos kx + b_k \sin kx) = \sum_{k=0}^{n} u_k(f; x), \tag{1}$$

denote the partial sums, called trigonometric polynomial of degree (or order) n, of the first $(n + 1)$ terms of the Fourier series of f. The matrix means of (1) are defined by

S. Sonker (✉)
Department of Mathematics, National Institute of Technology Kurukshetra,
Kurukshetra 136119, India
e-mail: smita.sonker@gmail.com

P.N. Agrawal et al. (eds.), *Mathematical Analysis and its Applications*,
Springer Proceedings in Mathematics & Statistics 143,
DOI 10.1007/978-81-322-2485-3_6

$$t_n(f) := t_n(f; x) = \sum_{k=0}^{n} a_{n,k} s_k, \quad n = 0, 1, 2, ...,$$

where $T \equiv (a_{n,k})$ is a lower triangular matrix with non-negative entries such that $a_{n,-1} = 0$, $A_{n,k} = \sum_{r=k}^{n} a_{n,r}$ so that $A_{n,0} = 1, \forall n \geq 0$. The Fourier series of f is said to be T-summable to s, if $t_n(f) \to s$ as $n \to \infty$.

By superimposing C^1 summability upon T summability, we get the $C^1 \cdot T$ summability. Thus the $C^1 \cdot T$ means of $\{s_n(f)\}$ denoted by $t_n^{C^1 \cdot T}(f)$ are given by

$$t_n^{C^1 \cdot T}(f) := (n+1)^{-1} \sum_{r=0}^{n} \left(\sum_{k=0}^{r} a_{r,k} s_k(f) \right).$$

If $t_n^{C^1 \cdot T} \to s_1$ as $n \to \infty$, then the Fourier series of f is said to be $C^1 \cdot T$—summable to the sum s_1. The regularity of methods C^1 and T implies regularity of method $C^1 \cdot T$. A function $f \in Lip\alpha$ if $|f(x+t) - f(x)| = O(|t|^\alpha)$, for $0 < \alpha \leq 1$, $f \in Lip(\alpha, r)$ if $\left(\int_0^{2\pi} |f(x+t) - f(x)|^r dx \right)^{1/r} = O(|t|^\alpha), 0 < \alpha \leq 1, r \geq 1$,
$f \in Lip(\xi(t), r)$ if $\left(\int_0^{2\pi} |f(x+t) - f(x)|^r dx \right)^{1/r} = O(\xi(t))$ and
$f \in W(L^r, \xi(t))$ if $\left(\int_0^{2\pi} |(f(x+t) - f(x)) \sin^\beta(x/2)|^r dx \right)^{1/r} = O(\xi(t))$,
$\beta \geq 0, r \geq 1$, where $\xi(t)$ is a positive increasing function of t.
If $\beta = 0$, $W(L^r, \xi(t),) \equiv Lip(\xi(t), r)$ and for $\xi(t) = t^\alpha (\alpha > 0)$, $Lip(\xi(t), r) \equiv Lip(\alpha, r)$. $Lip(\alpha, r) \to Lip\alpha$ as $r \to \infty$. Thus

$$Lip\alpha \subseteq Lip(\alpha, r) \subseteq Lip(\xi(t), r) \subseteq W(L^r, \xi(t)).$$

The L^r-norm of $f \in L^r[0, 2\pi]$ is defined by
$\|f\|_r = \left\{ \frac{1}{2\pi} \int_0^{2\pi} |f(x)|^r dx \right\}^{1/r} (1 \leq r < \infty)$ and $\|f\|_\infty = \sup_{x \in [0,2\pi]} |f(x)|$.
The degree of approximation of $f \in L^r$ denoted by $E_n(f)$ is given by

$$E_n(f) = \min_{T_n} \| f(x) - T_n(x) \|_r,$$

in terms of n , where $T_n(x)$ is a trigonometric polynomial of degree n.

This method of approximation is called trigonometric Fourier approximation. We also write

$$\phi(t) = f(x+t) + f(x-t) - 2f(x),$$

$$(C^1 \cdot T)_n(t) = \frac{1}{2\pi(n+1)} \sum_{r=0}^{n} \sum_{k=0}^{r} a_{r,r-k} \frac{\sin(r-k+1/2)t}{\sin(t/2)},$$

and $\tau = [1/t]$, the integral part of $1/t$.

2 Known Results

Various investigators such as Khan [3], Qureshi [8–10], Qureshi and Nema [11], Leindler [6] and Chandra [1] have determined the degree of approximation of functions belonging to the classes $W(L^r, \xi(t))$, $Lip(\xi(t), r)$, $Lip(\alpha, r)$ and $Lip\alpha$ with $r \geq 1$ and $0 < \alpha \leq 1$ using different summability methods with monotonocity conditions on the rows of summability matrices. Recently, Lal [5] has determined the degree of approximation of the functions belonging to $Lip\alpha$ and $W(L^r, \xi(t))$ classes by using Cesáro-Nörlund $(C^1 \cdot N_p)$—summability with non-increasing weights $\{p_n\}$. He proved:

Theorem 1 *Let N_p be a regular Nörlund method defined by a sequence $\{p_n\}$ such that*

$$P_\tau \sum_{v=\tau}^{n} P_v^{-1} = O(n+1). \tag{2}$$

Let $f \in L^1[0, 2\pi]$ be a 2π-periodic function belonging to $Lip\,\alpha$ $(0 < \alpha \leq 1)$, then the degree of approximation of f by $C^1 \cdot N_p$ means of its Fourier series is given by

$$\sup_{0\leq x\leq 2\pi} |t_n^{CN}(x) - f(x)| = \|t_n^{CN} - f\|_\infty = \begin{cases} O((n+1)^{-\alpha}), & 0 < \alpha < 1, \\ O\left(\log(n+1)\pi e/(n+1)\right), & \alpha = 1. \end{cases}$$

Theorem 2 *If f is a 2π-periodic function and Lebesgue integrable on $[0, 2\pi]$ and is belonging to $W(L^r, \xi(t))$ class then its degree of approximation by $C^1 \cdot N_p$ means of its Fourier series is given by*

$$\|t_n^{CN} - f\|_r = O\left((n+1)^{\beta+1/r}\xi\left(1/(n+1)\right)\right),$$

provided $\xi(t)$ satisfies the following conditions:

$$\{\xi(t)/t\} \text{ be a decreasing sequence}, \tag{3}$$

$$\left(\int_0^{\pi/(n+1)} \left(t|\phi(t)|\sin^\beta(t)/\xi(t)\right)^r dt\right)^{1/r} = O((n+1)^{-1}), \tag{4}$$

$$\left(\int_{\pi/(n+1)}^{\pi} \left(t^{-\delta}|\phi(t)|/\xi(t)\right)^r dt\right)^{1/r} = O((n+1)^{\delta}), \tag{5}$$

where δ is an arbitrary number such that $s(1-\delta) - 1 > 0$, $r^{-1} + s^{-1} = 1$, $r \geq 1$, conditions (4) and (5) hold uniformly in x.

The improved version of above theorems with their generalization to non-monotone weights $\{p_n\}$ can be seen in [13].

3 Main Results

In this paper, we generalize Theorems 1 and 2 by replacing matrix N_p with matrix T in the light of Remarks 2.3 and 2.4 of [13, pp. 3–4]. More precisely, we prove:

Theorem 3 *If $T \equiv (a_{n,k})$ is a lower triangular regular matrix with non-negative and non-decreasing (with respect to k) entries which satisfy*

$$\sum_{r=\tau}^{n} A_{r,r-\tau} = O(n+1), \tag{6}$$

hold uniformly in $\tau = [1/t]$, then the degree of approximation of a 2π-periodic function $f \in Lip\alpha$ $(0 < \alpha \leq 1) \subset L^1[0, 2\pi]$ by $C^1 \cdot T$ means of its Fourier series is given by

$$\|t_n^{C^1.T}(f) - f(x)\|_\infty = \begin{cases} O((n+1)^{-\alpha}), & 0 < \alpha < 1, \\ O\left((\log(n+1))/(n+1)\right), & \alpha = 1. \end{cases} \tag{7}$$

Theorem 4 *If $T \equiv (a_{n,k})$ be a lower triangular regular matrix with non-negative and non-decreasing (with respect to k) entries which satisfy condition (6), then the degree of approximation of a 2π-periodic function with $r > 1$ and $0 < \beta s < 1$ by $C^1 \cdot T$ means of its Fourier series is given by*

$$\|t_n^{C^1.T}(f) - f(x)\|_r = O\left((n+1)^{\beta+1/r}\xi\left(1/(n+1)\right)\right), \tag{8}$$

provided positive increasing function $\xi(t)$ satisfies the conditions:

$$\xi(t)/t \text{ be a decreasing function}, \tag{9}$$

$$\left(\int_0^{\pi/(n+1)} \left(|\phi(t)| \sin^\beta(t/2)/\xi(t)\right)^r dt\right)^{1/r} = O((n+1)^{-1/r}), \tag{10}$$

$$\left(\int_{\pi/(n+1)}^{\pi} \left(t^{-\delta}|\phi(t)| \sin^\beta(t/2)/\xi(t)\right)^r dt\right)^{1/r} = O((n+1)^{\delta-1/r}), \tag{11}$$

where δ is a real number such that $\beta + 1/r > \delta > r^{-1}, r^{-1} + s^{-1} = 1, r > 1$. Also, conditions (10) and (11) hold uniformly in x.

Remark 1 If we take $a_{n,k} = p_{n-k}/P_n$ for $k \leq n$ and $a_{n,k} = 0$ for $k > n$ such that $P_n (= \sum_{k=0}^{n} p_k \neq 0) \to \infty$ as $n \to \infty$ and $P_{-1} = 0 = p_{-1}$, then $C^1 \cdot T$ means reduce to $C^1 \cdot N_p$ means and

$$\sum_{r=\tau}^{n} A_{r,r-\tau} = \sum_{r=\tau}^{n}\sum_{k=r-\tau}^{r} a_{r,k} = \sum_{r=\tau}^{n}\sum_{k=r-\tau}^{r} (p_{r-k}/P_r) = \sum_{r=\tau}^{n}(P_\tau/P_r) = P_\tau \sum_{r=\tau}^{n} P_r^{-1}.$$

Therefore, condition (6) reduces to condition (2) and $t_n^{C^1 \cdot T}$ means reduce to t_n^{CN} *means*. Hence our Theorems 3 and 4 are generalization of Theorems 1 and 2, respectively.

4 Lemmas

We need the following lemmas for the proof of our theorems.

Lemma 1 *Let $\{a_{r,k}\}$ be a non-negative sequence of real numbers, then*

$$(C^1 \cdot T)_n(t) = O(n+1), \text{ for } 0 < t \le \pi/(n+1).$$

Proof Using $\sin nt \le nt$ and $\sin(t/2) \ge t/\pi$ for $0 < t \le \pi/(n+1)$, we have

$$\begin{aligned}
\left|(C^1 \cdot T)_n(t)\right| &= (2\pi(n+1))^{-1} \left|\sum_{r=0}^{n}\sum_{k=0}^{r} a_{r,r-k} \sin((r-k+1/2)t)/\sin(t/2)\right| \\
&= (2\pi(n+1))^{-1} \sum_{r=0}^{n}\sum_{k=0}^{r} a_{r,r-k} \left|\sin((r-k+1/2)t)/\sin(t/2)\right| \\
&\le (2\pi(n+1))^{-1} \sum_{r=0}^{n}\sum_{k=0}^{r} a_{r,r-k}(r-k+1/2)t/(t/\pi) \\
&\le (4(n+1))^{-1} \sum_{r=0}^{n}\sum_{k=0}^{r} a_{r,r-k}(2r-2k+1) \\
&\le (4(n+1))^{-1} \sum_{r=0}^{n}(2r+1)\sum_{k=0}^{r} a_{r,r-k} \\
&= (4(n+1))^{-1} \sum_{r=0}^{n}(2r+1)A_{r,0} = O(n+1).
\end{aligned}$$

Lemma 2 *[4] If $\{a_{r,k}\}$ is a non-negative and non-decreasing (with respect to k) sequence, then for $0 \le a < b \le \infty$, $0 < t \le \pi$ and for every r*

$$\left|\sum_{k=a}^{b} a_{r,r-k} e^{i(r-k)t}\right| = O(A_{r,r-\tau}).$$

Lemma 3 *If* $\{a_{r,k}\}$ *is non-negative and non-decreasing (with respect to k) sequence, then for* $0 < t \leq \pi$

$$\left|\sum_{r=0}^{n}\sum_{k=0}^{r} a_{r,r-k}e^{i(r-k)t}\right| = O(t^{-1}) + O\left(\sum_{r=\tau}^{n} A_{r,r-\tau}\right),$$

holds uniformly in $\tau = [1/t]$.

Proof For $0 < t \leq \pi$, we have

$$\begin{aligned}\left|\sum_{r=0}^{n}\sum_{k=0}^{r} a_{r,r-k}e^{i(r-k)t}\right| &\leq \left|\sum_{r=0}^{\tau-1}\sum_{k=0}^{r} a_{r,r-k}e^{i(r-k)t} + \sum_{r=\tau}^{n}\sum_{k=0}^{r} a_{r,r-k}e^{i(r-k)t}\right| \\ &\leq \sum_{r=0}^{\tau-1}\sum_{k=0}^{r} a_{r,r-k}|e^{i(r-k)t}| + \left|\sum_{r=\tau}^{n}\sum_{k=0}^{r} a_{r,r-k}e^{i(r-k)t}\right| \\ &\leq \sum_{r=0}^{\tau-1}\sum_{k=0}^{r} a_{r,r-k} + \sum_{r=\tau}^{n}\left|\sum_{k=0}^{r} a_{r,r-k}e^{i(r-k)t}\right| \\ &\leq \sum_{r=0}^{\tau-1} 1 + \sum_{r=\tau}^{n} O(A_{r,r-\tau}) = (\tau - 1 + 1) + O\left(\sum_{r=\tau}^{n} A_{r,r-\tau}\right) \\ &= O(t^{-1}) + O\left(\sum_{r=\tau}^{n} A_{r,r-\tau}\right),\end{aligned}$$

in view of Lemma 2.

Lemma 4 *If* $\{a_{r,k}\}$ *is non-negative and non-decreasing (with respect to k) sequence and satisfies the condition (6), then*

$$|(C^1 \cdot T)_n(t)| = O\left(t^{-2}/(n+1)\right) + O(t^{-1}),\ \textit{for}\ \pi/(n+1) < t \leq \pi.$$

Proof Using $\sin(t/2) \geq t/\pi$, for $\pi/(n+1) < t \leq \pi$ and Lemma 3, we have

$$\begin{aligned}|(C^1 \cdot T)_n(t)| &= (2\pi(n+1))^{-1}\left|\sum_{r=0}^{n}\sum_{k=0}^{r} a_{r,r-k}\sin((r-k+1/2)t)/\sin(t/2)\right| \\ &\leq (2\pi(n+1))^{-1}\left|\sum_{r=0}^{n}\sum_{k=0}^{r} a_{r,r-k}\sin((r-k+1/2)t)/(t/\pi)\right| \\ &= (2t(n+1))^{-1}\left|\sum_{r=0}^{n}\sum_{k=0}^{r} a_{r,r-k}\sin(r-k+1/2)t\right| \\ &\leq (2t(n+1))^{-1}\left|\sum_{r=0}^{n}\sum_{k=0}^{r} a_{r,r-k}e^{i(r-k+1/2)t}\right|\end{aligned}$$

$$= (2t(n+1))^{-1}\left|e^{it/2}\sum_{r=0}^{n}\sum_{k=0}^{r}a_{r,r-k}e^{i(r-k)t}\right|$$

$$= (2t(n+1))^{-1}\left|\sum_{r=0}^{n}\sum_{k=0}^{r}a_{r,r-k}e^{i(r-k)t}\right|$$

$$= (2t(n+1))^{-1}\left|O(t^{-1}) + O\left(\sum_{r=\tau}^{n}A_{r,r-\tau}\right)\right| = O\left(t^{-2}/(n+1)\right) + O(t^{-1}),$$

in view of condition (6).

5 Proof of Theorem 3

Following Titchmarsh [14], we have

$$s_n(f;x) - f(x) = \frac{1}{2\pi}\int_0^{\pi}\phi(t)(\sin(n+1/2)t/\sin(t/2))dt$$

Denoting $C^1 \cdot T$ means of $\{s_n(f;x)\}$ by $t_n^{C^1\cdot T}(f)$, we write

$$\begin{aligned} t_n^{C^1\cdot T}(f) - f(x) &= \int_0^{\pi}\phi(t)(2\pi(n+1))^{-1}\sum_{r=0}^{n}\sum_{k=0}^{r}a_{r,r-k}\sin((r-k+1/2)t)/\sin(t/2)dt \\ &= \int_0^{\pi/(n+1)}\phi(t)(C^1\cdot T)_n(t)dt + \int_{\pi/(n+1)}^{\pi}\phi(t)(C^1\cdot T)_n(t)dt \\ &= I_1 + I_2, \text{ say.} \end{aligned} \quad (12)$$

Using Lemma 1 and the fact that $f \in Lip\,\alpha \Rightarrow \phi \in Lip\,\alpha$ {[2], Lemma 5.27}, we have

$$\begin{aligned} |I_1| &\le \int_0^{\pi/(n+1)}|\phi(t)||(C^1\cdot T)_n(t)|dt = O(n+1)\int_0^{\pi/(n+1)}t^{\alpha}dt \\ &= O(n+1)((n+1)^{-\alpha-1}) = O((n+1)^{-\alpha}). \end{aligned} \quad (13)$$

Now, using Lemma 4 and the fact that $f \in Lip\,\alpha \Rightarrow \phi \in Lip\,\alpha$,

$$\begin{aligned} |I_2| &\le \int_{\pi/(n+1)}^{\pi}|\phi(t)|\;\left|(C^1\cdot T)_n(t)\right|dt \le \int_{\pi/(n+1)}^{\pi}|\phi(t)|O\left[(t^{-2}/(n+1)) + t^{-1}\right]dt \\ &= O(I_{21}) + O(I_{22}),\;\text{ say,} \end{aligned} \quad (14)$$

where

$$I_{21} = (n+1)^{-1}\int_{\pi/(n+1)}^{\pi}t^{\alpha-2}dt = \begin{cases} O((n+1)^{-\alpha}), & 0<\alpha<1, \\ O\left(\log(n+1)/(n+1)\right), & \alpha = 1. \end{cases} \quad (15)$$

and

$$I_{22} = O\left(\int_{\pi/(n+1)}^{\pi} t^{\alpha-1}dt\right) = O((n+1)^{-\alpha}). \tag{16}$$

Collecting (12)–(16), we get

$$t_n^{C^1 \cdot T}(f) - f(x) = \begin{cases} O((n+1)^{-\alpha}), & 0 < \alpha < 1, \\ O(\log(n+1)/(n+1)), & \alpha = 1. \end{cases}$$

Thus

$$\|t_n^{C^1 \cdot T}(f) - f\|_\infty = \sup_{0 \le x \le 2\pi} \{|t_n^{C^1 \cdot T}(x) - f(x)|\} = \begin{cases} O((n+1)^{-\alpha}), & 0 < \alpha < 1, \\ O((\log(n+1))/(n+1)), & \alpha = 1. \end{cases}$$

6 Proof of Theorem 4

Following the proof of Theorem 3,

$$\begin{aligned} t_n^{C^1 \cdot T}(f) - f(x) &= \int_0^{\pi/(n+1)} \phi(t)(C^1 \cdot T)_n(t)dt + \int_{\pi/(n+1)}^{\pi} \phi(t)(C^1 \cdot T)_n(t)dt \\ &= I_1^{'} + I_2^{'}, \text{ say.} \end{aligned} \tag{17}$$

Using Hölder's inequality, $\phi(t) \in W(L^r, \xi(t))$, condition (10), Lemma 1 and mean value theorem for integrals, we have

$$\begin{aligned} |I_1^{'}| &= \left|\lim_{\varepsilon \to 0} \int_{\varepsilon}^{\pi/(n+1)} \left[(\phi(t)\sin^{\beta}(t/2)/\xi(t)) \cdot (\xi(t)(C^1 \cdot T)_n(t))/(\sin^{\beta}(t/2))\right] dt\right| \\ &\le \left[\int_0^{\pi/(n+1)} \left(|\phi(t)| \sin^{\beta}(t/2)/\xi(t)\right)^r dt\right]^{1/r} \\ &\cdot \left[\lim_{\varepsilon \to 0} \int_{\varepsilon}^{\pi/(n+1)} \left(\xi(t)|(C^1 \cdot T)_n(t)|/(\sin^{\beta}(t/2)\right)^s dt\right]^{1/s} \\ &= O((n+1)^{-1/r}) \left[\lim_{\varepsilon \to 0} \int_{\varepsilon}^{\pi/(n+1)} \left|\xi(t)(n+1)/(\sin^{\beta}(t/2))\right|^s dt\right]^{1/s} \\ &= O(n+1)^{1-1/r}(\xi(\pi/(n+1)) \left[\lim_{\varepsilon \to 0} \int_{\varepsilon}^{\pi/(n+1)} t^{-\beta s} dt\right]^{1/s} \\ &= O(\xi(1/(n+1)(n+1)^{\beta+1-1/r-1/s}) = O((n+1)^{\beta}\xi(1/(n+1)), \end{aligned} \tag{18}$$

in view of condition (9), i.e. $(\xi(\pi/(n+1))/(\pi/(n+1))) \leq (\xi(1/(n+1))/(1/(n+1)))$.

Using Lemma 4, we have

$$
\begin{aligned}
|I_2'| &= \left[\int_{\pi/(n+1)}^{\pi} |\phi(t)| \left[O\left(t^{-2}/(n+1)\right) + O\left(t^{-1}\right)\right] dt\right] \\
&= O\left[\int_{\pi/(n+1)}^{\pi} t^{-2}|\phi(t)|/(n+1)dt\right] + O\left[\int_{\pi/(n+1)}^{\pi} t^{-1}|\phi(t)|dt\right] \\
&= O(I_{21}') + O(I_{22}'). \qquad (19)
\end{aligned}
$$

Using Hölder's inequality, $|\sin t| \leq 1$, $\sin(t/2) \geq (t/\pi)$ and condition (11), we have

$$
\begin{aligned}
|I_{21}'| &= (n+1)^{-1}\left[\int_{\pi/(n+1)}^{\pi} \left\{(t^{-\delta}|\phi(t)|\sin^{\beta}(t/2)/\xi(t)) \cdot (\xi(t)/(t^{-\delta+2}\sin^{\beta}(t/2)))\right\} dt\right] \\
&\leq ((n+1)^{-1})\left[\int_{\pi/(n+1)}^{\pi} \left|t^{-\delta}|\phi(t)|\sin^{\beta}(t/2)/\xi(t)\right|^{r} dt\right]^{1/r} \\
&\quad \cdot \left[\int_{\pi/(n+1)}^{\pi} \left|\xi(t)/\left(t^{-\delta+2}\sin^{\beta}(t/2)\right)\right|^{s} dt\right]^{1/s} \\
&= O((n+1)^{-1})\left[\int_{\pi/(n+1)}^{\pi} \left|t^{-\delta}|\phi(t)|\sin^{\beta}(t/2)/\xi(t)\right|^{r} dt\right]^{1/r} \\
&\quad \cdot \left[\int_{\pi/(n+1)}^{\pi} \left|\xi(t)/\left(t^{-\delta+2}\sin^{\beta}(t/2)\right)\right|^{s} dt\right]^{1/s} \\
&= O((n+1)^{-1})O\left((n+1)^{\delta-1/r}\right)\left[\int_{\pi/(n+1)}^{\pi} \left|\xi(t)/\left(t^{-\delta+2}\sin^{\beta}(t/2)\right)\right|^{s} dt\right]^{1/s} \\
&= O((n+1)^{\delta-1-1/r})\left[\int_{\pi/(n+1)}^{\pi} \left(\xi(t)/t^{-\delta+2+\beta}\right)^{s} dt\right]^{1/s} \\
&= O((n+1)^{\delta-1/r})\xi(\pi/(n+1))\left[\int_{\pi/(n+1)}^{\pi} t^{-(-\delta+1+\beta)s} dt\right]^{1/s} \\
&= O((n+1)^{\delta-1/r})\xi(\pi/(n+1))\left[(n+1)^{(-\delta+1+\beta)-1/s} dt\right] \\
&= O(\xi(1/(n+1)(n+1)^{\beta}) \qquad (20)
\end{aligned}
$$

in view of decreasing nature of $\xi(t)/t$ and $r^{-1} + s^{-1} = 1$.

Similarly, as above, we have

$$
\begin{aligned}
|I_{22}'| &= \int_{\pi/(n+1)}^{\pi} t^{-1}|\phi(t)|dt = \int_{\pi/(n+1)}^{\pi} \left(t^{-\delta}|\phi(t)|\sin^{\beta}(t/2)/\xi(t)\right)\left(\xi(t)/(t^{1-\delta}\sin^{\beta}(t/2))\right) dt \\
&\leq \left[\int_{\pi/(n+1)}^{\pi} \left|t^{-\delta}|\phi(t)|\sin^{\beta}(t/2)/\xi(t)\right|^{r} dt\right]^{1/r}\left[\int_{\pi/(n+1)}^{\pi} \left|\xi(t)/\left(t^{1-\delta}\sin^{\beta}(t/2)\right)\right|^{s} dt\right]^{1/s} \\
&= O\left((n+1)^{\delta-1/r}\right)\left[\int_{\pi/(n+1)}^{\pi} \left(\xi(t)/t^{1-\delta+\beta}\right)^{s} dt\right]^{1/s}
\end{aligned}
$$

$$
\begin{aligned}
&= O\left((n+1)^{\delta+1-1/r}\right)\xi(1/(n+1))\left[\int_{\pi/(n+1)}^{\pi} t^{(\delta-\beta)s}dt\right]^{1/s}\\
&= O\left((n+1)^{\delta+1-1/r}\right)\xi(1/(n+1))(n+1)^{(-\delta+\beta)-1/s}\\
&= O(\xi(1/(n+1))(n+1)^{\beta+1-1/r-1/s}\\
&= O(\xi(1/(n+1))(n+1)^{\beta}.
\end{aligned}
\tag{21}
$$

Collecting (17)–(21), we have

$$|t_n^{C^1.T}(f)-f(x)| = O\left((n+1)^{\beta}\xi(1/(n+1))\right).$$

Hence,

$$\|t_n^{C^1.T}(f)-f(x)\|_r = \left(1/2\pi\int_0^{2\pi}|t_n^{C^1.T}(f)-f(x)|^r dx\right)^{1/r} = O\left((n+1)^{\beta}\xi\left(1/(n+1)\right)\right).$$

Remark 2 The proof of Theorem 3, for $r=1$, *i.e.* $s=\infty$ can be written by using sup norm while using Hölder's inequality.

7 Corollaries

The following corollaries can be derived from Theorem 4
1. If $\beta=0$, then for $f\in Lip(\xi(t),r)$, $\|t_n^{C^1.T}(f)-f(x)\|_r = O\left(\xi(1/n)\right)$.
2. If $\beta=0$, $\xi(t)=t^{\alpha}(0<\alpha\le 1)$, then for $f\in Lip(\alpha,r)(\alpha>1/r)$,

$$\|t_n^{C^1.T}(f)-f(x)\|_r = O\left(n^{-\alpha}\right). \tag{22}$$

3. If $r\to\infty$ in Corollary 2, then for $f\in Lip\alpha(0<\alpha<1)$, (22) gives

$$\|t_n^{C^1.T}(f)-f(x)\|_{\infty} = O(n^{-\alpha}).$$

Remark 3 In view of Remark 2, corollaries of Lal [5, p. 350] are particular cases of our Corollaries 2 and 3, respectively.

References

1. Chandra, P.: Trigonometric approximation of functions in L_p-norm. J. Math. Anal. Appl. **275**(1), 13–26 (2002)
2. Faddeen, L.M.: Absolute Nörlund summability. Duke Math. J. **9**, 168–207 (1942)
3. Khan, H.: On the degree of approximation of functions belonging to the class $Lip(\alpha, p)$, Indian. J. Pure Appl. Math. **5**(2), 132–136 (1974)

4. Kishore, N., Hotta, G.C.: On absolute matrix summability of Fourier series Indian. J. Math. **13**(2), 99–110 (1971)
5. Lal, S.: Approximation of functions belonging to the generalized Lipschitz Class by $C^1.Np$ summability method of Fourier series. Appl. Math. Comput. **209**, 346–350 (2009)
6. Leindler, L.: Trigonometric approximation in L_p-norm. J. Math. Anal. Appl. **302**(1), 129–136 (2005)
7. Lenski, W., Szal, B.: Approximation of functions belonging to the class $L^p(w)_\beta$ by linear operators. Acta ET Commentationes Universitatis Tartuensis De Mathematica **13**, 11–24 (2009)
8. Qureshi, K.: On the degree of approximation of a periodic function f by almost Nörlund means. Tamkang J. Math. **12**(1), 35–38 (1981)
9. Qureshi, K.: On the degree of approximation of a function belonging to the Class $Lip\alpha$, Indian. J. Pure Appl. Math. **13**(8), 560–563 (1982)
10. Qureshi, K.: On the degree of approximation of a function belonging to weighted $W(L^p, \xi(t))$ - class. Indian J. Pure Appl. Math. **13**(4), 471–475 (1982)
11. Qureshi, K., Nema, H.K.: On the degree of approximation of functions belonging to weighted class. Ganita **41**(1–2), 17–22 (1990)
12. Rhoades, B.E., Ozkoklu, K., Albayrak, I.: On the degree of approximation of functions belonging to a Lipschitz class by Hausdorff means of its Fourier series. Appl. Math. Comput. **217**, 6868–6871 (2011)
13. Singh, U., Mittal, M.L., Sonker, S.: Trigonometric approximation of signals (functions) belonging to $W(L^p, \xi(t))$–class by matirx $(C^1.N_p)$ operator. Int. J. Math. Math. Sci. **2012**, 1–11 (2012)
14. Titchmarsh, E.C.: The Theory of Functions, pp. 402–403. Oxford University Press, Oxford (1939)

Modified Baskakov-Szász Operators Based on q-Integers

P.N. Agrawal and Arun Kajla

Abstract In the present paper we introduce the Stancu variant of certain q-modified Baskakov *Szász* operators. We estimate the moments of the operators and obtain some direct results in terms of the modulus of continuity. Then, we study the Voronovskaja type theorem and the rate of convergence of these operators in terms of the weighted modulus of continuity. Further, we discuss the point-wise estimation using the Lipschitz type maximal function. Finally, we investigate the rate of statistical convergence of these operators using weighted modulus of continuity.

Keywords q-Baskakov-Szasz operators · q-integers · Modulus of smoothness · Point-wise estimates · Statistical convergence

Mathematics Subject Classification (2010): 26A15 · 40A35

1 Introduction

In recent years, the most interesting area of research in approximation theory is the application of q-calculus. In 1997, Phillips [20] first considered a modification of Bernstein polynomials based on q-integers. He studied the rate of convergence and Voronovskaja-type asymptotic formula for these operators. Very recently, Gupta and Kim [14] considered q-Baskakov operators and they obtained some direct local results and the degree of approximation in terms of modulus of continuity. Subsequently, several researchers have considered the different types of operators in this direction and studied their approximation properties.

Let α and β be any two real numbers satisfying the condition that $0 \leq \alpha \leq \beta$, Stancu [21] defined in the following operators:

P.N. Agrawal · A. Kajla (✉)
Department of Mathematics, Indian Institute of Technology Roorkee, Roorkee 247667, India
e-mail: rachitkajla47@gmail.com

P.N. Agrawal
e-mail: pna_iitr@yahoo.co.in

P.N. Agrawal et al. (eds.), *Mathematical Analysis and its Applications*,
Springer Proceedings in Mathematics & Statistics 143,
DOI 10.1007/978-81-322-2485-3_7

$$S_n^{\alpha,\beta}(f,x)=\sum_{k=0}^{n}p_{n,k}(x)\left(\frac{k+\alpha}{n+\beta}\right),\quad 0\leq x\leq 1,$$

where $p_{n,k}(x)$ is the Bernstein basis function.

Recently, Büyükyazici [7] considered the Stancu–Chlodowsky polynomials and investigated their convergence. In 2012, Verma et al. [22] introduced a Stancu type generalization of certain q-Baskakov Durrmeyer operators and discussed some local direct results of these operators. For some other research papers where Stancu type operators have been considered, we refer to [1, 3, 4, 13, 15], etc.

Now, we give some basic definitions and concepts of q-calculus [6, 17]. For any real number $q>0$, the q-integer $[n]_q$ and q-factorial $[n]_q!$ are defined as

$$[n]_q=\begin{cases}\dfrac{(1-q^n)}{(1-q)}, & \text{if } q\neq 1\\ n, & \text{if } q=1\end{cases}$$

and

$$[n]_q!=\begin{cases}[n]_q[n-1]_q\ldots 1, & n=1,2,\ldots\\ 1, & n=0.\end{cases}$$

The q-Pochhammer symbol is defined as

$$(x;q)_n=\begin{cases}(1+x)(1+qx)\ldots(1+q^{n-1}x), & n=1,2,\ldots\\ 1, & n=0.\end{cases}$$

The q-binomial coefficients are given by

$$\binom{n}{k}_q=\frac{[n]_q!}{[k]_q![n-k]_q!},\quad 0\leq k\leq n.$$

The q-derivative D_q of a function f is given by

$$(D_qf)(x)=\frac{f(x)-f(qx)}{(1-q)x},\quad \text{if } x\neq 0.$$

The q-Jackson integrals and q-improper integrals are defined as

$$\int_0^a f(x)d_q(x)=(1-q)a\sum_{n=0}^{\infty}f(aq^n)q^n,\quad a>0,$$

and

$$\int_0^{\infty/A} f(x)d_q(x)=(1-q)\sum_{n=-\infty}^{\infty}f\left(\frac{q^n}{A}\right)\frac{q^n}{A},\quad A>0.$$

The q-Beta integral is defined by

$$\Gamma_q(t) = \int_0^{\frac{1}{1-q}} x^{t-1} E_q(-qx) d_q x, \quad t > 0 \tag{1}$$

which satisfies the following functional equation:

$$\Gamma_q(t+1) = [t]_q \Gamma_q(t), \quad \Gamma_q(1) = 1.$$

To approximate Lebesgue integrable functions on the interval $[0, \infty)$, Agrawal and Mohammad [2] introduced the following operators:

$$M_n(f(t); x) = n \sum_{v=1}^{\infty} p_{n,v}(x) \int_0^{\infty} q_{n,v-1}(t) f(t) dt + (1+x)^{-n} f(0). \tag{2}$$

where

$$p_{n,v}(x) = \binom{n+v-1}{v} x^v (1+x)^{-(n+v)}, \quad x \in [0, \infty)$$

and

$$q_{n,v}(t) = \frac{e^{-nt}(nt)^v}{v!}, \quad \forall\, t \in [0, \infty).$$

In [2], Agrawal et al. studied the asymptotic approximation and error estimates in terms of modulus of continuity in simultaneous approximation by (2).

In [16], Gupta and Srivastava considered a sequence of positive linear operators combining the Baskakov and Száz basis functions. Deo [8] studied the simultaneous approximation by Lupas operators with the weight functions of Száz operators.

Definition 1 For $f \in C_\gamma[0, \infty) := \{f \in C[0, \infty) : f(t) = O(e^{\gamma t})\ as\ t \to \infty$ *for some* $\gamma > 0\}$ and each positive integer n, the q-Baskakov operators [5] are defined as

$$\begin{aligned} V_{n,q}(f; x) &= \sum_{k=0}^{\infty} \binom{n+k-1}{k}_q q^{\frac{k(k-1)}{2}} \frac{x^k}{(1+x)_q^{n+k}} f\left(\frac{[k]_q}{q^{k-1}[n]_q}\right) \\ &= \sum_{k=0}^{\infty} p_{n,k}^q(x) f\left(\frac{[k]_q}{q^{k-1}[n]_q}\right). \end{aligned} \tag{3}$$

Remark 1 The first three moments of the q-Baskakov operators (see [5]) are given by

$$V_{n,q}(1;x)=1,\ \ V_{n,q}(t;x)=x,\ \ V_{n,q}(t^2;x)=x^2+\frac{x}{[n]_q}\left(1+\frac{x}{q}\right).$$

Definition 2 For $f\in C_\gamma[0,\infty), 0<q<1$ and each positive integer n, the q-Baskakov Szász operators defined as

$$B_{n,q}(f;x)=[n]_q\sum_{k=0}^{\infty}p_{n,k}^q(x)\int_0^{\frac{q}{(1-q^n)}}q^{-k-1}s_{n,k}^q(t)f\left(\frac{t}{q^k}\right)d_qt, \tag{4}$$

$$\text{where } p_{n,k}^q(x)=\binom{n+k-1}{k}q^{\frac{k(k-1)}{2}}\frac{x^k}{(1+x)_q^{(n+k)}}$$

$$\text{and } s_{n,k}^q(t)=E_q(-[n]_qt)\frac{([n]_qt)^k}{[k]_q!} \tag{5}$$

have been considered by Gupta [12].

2 Construction of Operators

For $f\in C_\gamma[0,\infty)$, $0<q<1$ and each positive integer n, the Stancu-type generalization of the operators (2) based on q-integers is defined as follows:

$$\begin{aligned}M_{n,q}^{(\alpha,\beta)}(f;x)&=[n]_q\sum_{k=1}^{\infty}p_{n,k}^q(x)\int_0^{\frac{q}{(1-q^n)}}q^{-k}s_{n,k-1}^q(t)f\left(\frac{[n]_qtq^{-k}+\alpha}{[n]_q+\beta}\right)d_qt\\&+f\left(\frac{\alpha}{[n]_q+\beta}\right)p_{n,0}^q(x),\end{aligned} \tag{6}$$

where $p_{n,k}^q(x)$ and $s_{n,k}^q(t)$ are as defined in (5).

If $\alpha=\beta=0$ and $q\to 1-$, the operators (6) reduce to the operators (2), which is a modification of the operator given by (4) where the value of the function at zero is considered explicitly. The aim of this paper is to study some direct results and asymptotic formula for the operators (6). We also discuss the rate of convergence and point-wise estimation. Lastly, we study the statistical approximation properties of these operators.

3 Basic Results

3.1 Moment Estimates

For $\alpha = \beta = 0$, we denote the operator $M_{n,q}^{(\alpha,\beta)}$ by $M_{n,q}$.

Lemma 1 *For the operators $M_{n,q}(f; x)$, the following equalities hold:*

(i) $M_{n,q}(1; x) = 1$;
(ii) $M_{n,q}(t; x) = x$;
(iii) $M_{n,q}(t^2; x) = x^2\left(1 + \frac{1}{q[n]_q}\right) + \frac{[2]_q x}{[n]_q}$.

Proof First, for $f(t) = 1$, we have

$$M_{n,q}(1; x) = [n]_q \sum_{k=1}^{\infty} p_{n,k}^q(x) \int_0^{\frac{q}{(1-q^n)}} q^{-k} s_{n,k-1}(t) d_q t + p_{n,0}^q(x).$$

Substituting $[n]_q t = qy$ and using (1)

$$\begin{aligned} M_{n,q}(1; x) &= [n]_q \sum_{k=1}^{\infty} p_{n,k}^q(x) \int_0^{\frac{1}{1-q}} q^{-k+1} \frac{(qy)^{k-1}}{[k-1]_q!} \frac{E_q[-qy]}{[n]_q} d_q y + p_{n,0}^q(x) \\ &= \sum_{k=1}^{\infty} p_{n,k}^q(x) \frac{\Gamma_q k}{[k-1]_q!} + p_{n,0}^q(x) \\ &= \sum_{k=0}^{\infty} p_{n,k}^q(x) \\ &= V_{n,q}(1; x) = 1, \quad \text{in view of Remark 1.} \end{aligned}$$

Next, let $f(t) = t$, we have

$$M_{n,q}(t; x) = [n]_q \sum_{k=1}^{\infty} p_{n,k}^q(x) \int_0^{\frac{q}{(1-q^n)}} q^{-2k} t^k E_q(-[n]_q t) \frac{([n]_q t)^{k-1}}{[k-1]_q!} d_q t.$$

Again, substituting $[n]_q t = qy$ and using (1)

$$\begin{aligned} M_{n,q}(t; x) &= [n]_q \sum_{k=1}^{\infty} p_{n,k}^q(x) \int_0^{\frac{1}{1-q}} q^{-2k+1} E_q(-qy) \frac{(qy)^k}{[k-1]_q!([n]_q)^2} d_q y \\ &= \sum_{k=1}^{\infty} p_{n,k}^q(x) \frac{1}{[n]_q [k-1]_q! q^{k-1}} \int_0^{\frac{1}{1-q}} E_q(-qy) y^k d_q y \\ &= \sum_{k=1}^{\infty} p_{n,k}^q(x) \frac{\Gamma_q(k+1)}{[n]_q [k-1]_q! q^{k-1}} \end{aligned}$$

$$= \sum_{k=1}^{\infty} p_{n,k}^{q}(x) \frac{[k]_q}{[n]_q q^{k-1}}$$

$$= \sum_{k=0}^{\infty} p_{n,k}^{q}(x) \frac{[k]_q}{[n]_q q^{k-1}} = V_{n,q}(t; x) = x, \quad \text{on applying Remark 1.}$$

Finally, we give the second moment as follows:

$$M_{n,q}(t^2; x) = [n]_q \sum_{k=1}^{\infty} p_{n,k}^{q}(x) \int_0^{\frac{q}{(1-q^n)}} \left(\frac{t}{q^k}\right)^2 q^{-k} \frac{([n]_q t)^{k-1}}{[k-1]_q!} E_q(-[n]_q t) d_q t.$$

Again, substituting $[n]_q t = qy$, using (1) and $[k+1]_q = [k]_q + q^k$, we have

$$\begin{aligned}
M_{n,q}(t^2; x) &= [n]_q \sum_{k=1}^{\infty} p_{n,k}^{q}(x) \int_0^{\frac{1}{1-q}} q^{-k} \frac{E_q(-qy)}{q^{2k}} \frac{(qy)^2}{([n]_q)^2} \frac{(qy)^{k-1} q}{[k-1]_q! [n]_q} d_q y \\
&= \sum_{k=1}^{\infty} p_{n,k}^{q}(x) \frac{1}{q^{2k-2}([n]_q)^2 [k-1]_q!} \int_0^{\frac{1}{1-q}} E_q(-qy) y^{k+1} d_q y \\
&= \sum_{k=1}^{\infty} p_{n,k}^{q}(x) \frac{1}{q^{2k-2}([n]_q)^2 [k-1]_q!} \Gamma(k+2)_q \\
&= \sum_{k=1}^{\infty} p_{n,k}^{q}(x) \frac{1}{q^{2k-2}([n]_q)^2} [k]_q([k]_q + q^k) \\
&= \sum_{k=1}^{\infty} p_{n,k}^{q}(x) \frac{([k]_q)^2}{([n]_q)^2 q^{2k-2}} + \frac{q}{[n]_q} \sum_{k=1}^{\infty} p_{n,k}^{q}(x) \frac{[k]_q}{q^{k-1}[n]_q} \\
&= V_{n,q}(t^2; x) + \frac{q}{[n]_q} V_{n,q}(t; x) \\
&= x^2 + \frac{x}{[n]_q}\left(1 + \frac{x}{q}\right) + \frac{qx}{[n]_q} \\
&= x^2\left(1 + \frac{1}{q[n]_q}\right) + \frac{[2]_q x}{[n]q}, \quad \text{on using Remark 1.}
\end{aligned}$$

□

Lemma 2 *For $M_{n,q}^{(\alpha,\beta)}(t^m; x)$, $m = 0, 1, 2$ we have*

(i) $M_{n,q}^{(\alpha,\beta)}(1; x) = 1;$

(ii) $M_{n,q}^{(\alpha,\beta)}(t; x) = \dfrac{[n]_q x + \alpha}{[n]_q + \beta};$

(iii) $M_{n,q}^{(\alpha,\beta)}(t^2; x) = \dfrac{[n]_q(1 + q[n]_q)x^2}{q([n]_q + \beta)^2} + \dfrac{[n]_q([2]_q + 2\alpha)x}{([n_q] + \beta)^2} + \dfrac{\alpha^2}{([n]_q + \beta)^2}.$

Proof Using Lemma 1, we estimate the moments as follows:

For $f(t) = 1$, we have

$$M_{n,q}^{(\alpha,\beta)}(1;x) = [n]_q \sum_{k=1}^{\infty} p_{n,k}^q(x) \int_0^{\frac{q}{1-q^n}} q^{-k} s_{n,k-1}^q(t) d_q t + p_{n,0}^q(x) = M_{n,q}(1;x) = 1.$$

Next, we obtain the first-order moment

$$\begin{aligned} M_{n,q}^{(\alpha,\beta)}(t;x) &= [n]_q \sum_{k=1}^{\infty} p_{n,k}^q(x) \int_0^{\frac{q}{1-q^n}} q^{-k} s_{n,k-1}^q(t) \left(\frac{[n]_q t q^{-k} + \alpha}{[n]_q + \beta} \right) d_q t + p_{n,0}^q(x) \left(\frac{\alpha}{[n]_q + \beta} \right) \\ &= \frac{[n]_q}{[n]_q + \beta} M_{n,q}(t;x) + \frac{\alpha}{[n]_q + \beta} M_{n,q}(1;x) \\ &= \frac{[n]_q}{[n]_q + \beta} x + \frac{\alpha}{([n]_q + \beta)} \\ &= \frac{[n]_q x + \alpha}{[n]_q + \beta}. \end{aligned}$$

Finally, for $f(t) = t^2$ we obtain

$$\begin{aligned} M_{n,q}^{(\alpha,\beta)}(t^2;x) &= [n]_q \sum_{k=1}^{\infty} p_{n,k}^q(x) \int_0^{\frac{q}{1-q^n}} q^{-k} s_{n,k-1}^q(t) \left(\frac{[n]_q t q^{-k} + \alpha}{[n]_q + \beta} \right)^2 d_q t + p_{n,0}^q(x) \left(\frac{\alpha}{[n]_q + \beta} \right)^2 \\ &= \frac{([n]_q)^2}{([n]_q + \beta)^2} M_{n,q}(t^2;x) + \frac{2[n]_q \alpha}{([n]_q + \beta)^2} M_{n,q}(t,x) + \frac{\alpha^2}{([n]_q + \beta)^2} M_{n,q}(1;x) \\ &= \frac{[n]_q^2}{([n]_q + \beta)^2} \left\{ x^2 \left(1 + \frac{1}{q[n]_q} \right) + \frac{x(1+q)}{[n]_q} \right\} + \frac{2[n]_q \alpha}{([n]_q + \beta)^2} x + \frac{\alpha^2}{([n]_q + \beta)^2} \\ &= \frac{[n]_q (1 + q[n]_q)}{q([n]_q + \beta)^2} x^2 + \frac{[n]_q ([2]_q + 2\alpha)}{([n]_q + \beta)^2} x + \frac{\alpha^2}{([n]_q + \beta)^2}. \end{aligned}$$

Hence, the proof is completed. □

Remark 2 By simple computation, we have

$$\begin{aligned} M_{n,q}^{(\alpha,\beta)}((t-x);x) &= \frac{\alpha - \beta x}{[n]_q + \beta}, \\ M_{n,q}^{(\alpha,\beta)}((t-x)^2;x) &= \frac{x^2([n]_q + q\beta^2)}{q([n]_q + \beta)^2} + \frac{x([2]_q[n]_q - 2\alpha\beta)}{([n]_q + \beta)^2} + \frac{\alpha^2}{([n]_q + \beta)^2}. \end{aligned}$$

Lemma 3 *For every $q \in (0, 1)$ we have*

$$M_{n,q}^{(\alpha,\beta)}((t-x)^2;x) \le \frac{[2]_q(1 + \beta^2)}{q([n]_q + \beta)} \left(\phi^2(x) + \frac{1}{([n]_q + \beta)} \right),$$

where $\phi(x) = \sqrt{x(1+x)}$, $x \in [0, \infty)$.

Proof

$$\begin{aligned}
M_{n,q}^{(\alpha,\beta)}((t-x)^2;x) &= \frac{x^2([n]_q+q\beta^2)}{q([n]_q+\beta)^2}+\frac{x([2]_q[n]_q-2\alpha\beta)}{([n]_q+\beta)^2}+\frac{\alpha^2}{([n]_q+\beta)^2}\\
&\leq \frac{([2]_q[n]_q+\beta^2)}{q([n]_q+\beta)^2}(x^2+x)+\frac{\alpha^2}{([n]_q+\beta)^2}\\
&\leq \frac{[2]_q([n]_q+\beta^2)}{q([n]_q+\beta)^2}(x^2+x)+\frac{\alpha^2}{([n]_q+\beta)^2}\\
&\leq \frac{[2]_q[n]_q(1+\beta^2)}{q([n]_q+\beta)^2}\phi^2(x)+\frac{\alpha^2}{([n]_q+\beta)^2}\\
&\leq \frac{[2]_q(1+\beta^2)}{q([n]_q+\beta)}\phi^2(x)+\frac{\alpha^2}{([n]_q+\beta)^2}\\
&\leq \frac{[2]_q(1+\beta^2)}{q([n]_q+\beta)}\left(\phi^2(x)+\frac{1}{[n]_q+\beta}\right).
\end{aligned}$$

This completes the proof. □

4 Main Results

If $q=\{q_n\}$ be a sequence in (0, 1) satisfying the following conditions:

$$\lim_{n\to\infty} q_n = 1 \text{ and } \lim_{n\to\infty} q_n^n = c,\ (0\leq c<1). \tag{7}$$

Our first result is a basic convergence theorem for the operators $M_{n,q_n}^{(\alpha,\beta)}$.

Theorem 1 *Let* $q_n \in (0,1)$ *and* $\lim_{n\to\infty} q_n^n = c,\ (0\leq c<1)$. *Then the sequence* $M_{n,q_n}^{(\alpha,\beta)}(f;x)$ *converges to* f *uniformly on* $[0,A]$, $A>0$, *for each* $f\in C_\gamma[0,\infty)$ *if and only if* $\lim_{n\to\infty} q_n = 1$.

Remark 3 If $\lim_{n\to\infty} q_n = 1$, then in view of Remark 2, $M_{n,q_n}^{(\alpha,\beta)}((t-x)^2;x)\to 0$ uniformly on $[0,A]$ as $n\to\infty$. Therefore, the well-known Korovkin theorem implies that $\{M_{n,q_n}^{(\alpha,\beta)}(f;x)\}$ converges to f uniformly on $[0,A]$ for each $f\in C_\gamma[0,\infty)$. The converse part follows on proceeding in a manner similar to the proof of [3], Theorem 1.

4.1 Direct Theorem

Let $C_B[0,\infty)$ be the space of all continuous and bounded functions f defined on the interval $[0,\infty)$, endowed with the norm $\|.\|$ on the space given by

$$\|f\| = \sup_{0\leq x<\infty} \mid f(x) \mid . \tag{8}$$

If $\delta > 0$ and $W^2 = \{g \in C_B[0,\infty) : g', g'' \in C_B[0,\infty)\}$, then the K-functional is defined as

$$K_2(f,\delta) = \inf\{\|f - g\| + \delta\|g''\| : g \in W^2\}. \tag{9}$$

By ([9], p. 177, Theorem 2, 4) there exists an absolute constant $C > 0$ such that $K_2(f,\delta) \leq C\omega_2(f,\sqrt{\delta})$,
where second order modulus of the smoothness of $f \in C_B[0,\infty)$ is defined as

$$\omega_2(f,\sqrt{\delta}) = \sup_{0<h\leq\sqrt{\delta}} \sup_{0\leq x<\infty} \mid f(x+2h) - 2f(x+h) + f(x) \mid .$$

The first-order modulus of continuity is defined as

$$\omega(f,\delta) = \sup_{0<h\leq\sqrt{\delta}} \sup_{0\leq x<\infty} \mid f(x+h) - f(x) \mid .$$

The next result is a direct local approximation theorem for the operators $M_{n,q}^{(\alpha,\beta)}$.

Theorem 2 *Let $f \in C_B[0,\infty)$ and let $\{q_n\}$ be sequence satisfying the conditions (7). Then, for every $x \in [0,\infty)$ we have*

$$\mid M_{n,q}^{(\alpha,\beta)}(f;x) - f(x) \mid \leq C\omega_2\left(f, \sqrt{\frac{4(1+\beta^2)}{q([n]_q+\beta)}\left\{\phi^2(x) + \frac{1}{([n]_q+\beta)}\right\}}\right) + \omega\left(f, \frac{\mid \alpha - \beta x \mid}{[n]_q+\beta}\right).$$

Proof We introduce auxiliary operator $L_{n,q}^{(\alpha,\beta)}$ as follows:

$$L_{n,q}^{(\alpha,\beta)}(f;x) = M_{n,q}^{(\alpha,\beta)}(f;x) - f\left(x + \frac{\alpha - \beta x}{([n]_q+\beta)}\right) + f(x). \tag{10}$$

These operators are linear and preserve the linear functions. Hence, we have

$$L_{n,q}^{(\alpha,\beta)}(t-x;x) = 0. \tag{11}$$

Let $g \in W^2$. From the Taylor's expansion of g, we get

$$g(t) = g(x) + g'(x)(t-x) + \int_x^t (t-u)g''(u)du,\ \ t \in [0,\infty).$$

In view of (10), we get

$$\begin{aligned}
L_{n,q}^{(\alpha,\beta)}(g;x) &= g(x) + L_{n,q}^{(\alpha,\beta)}\left(\int_x^t (t-u)g''(u)du; x\right) \\
\mid L_{n,q}^{(\alpha,\beta)}(g;x) - g(x) \mid &= \left| L_{n,q}^{(\alpha,\beta)}\left(\int_x^t (t-u)g''(u)du; x\right)\right| \\
&\le \left| M_{n,q}^{(\alpha,\beta)}\left(\int_x^t (t-u)g''(u)du; x\right)\right| \\
&+ \left| \int_x^{x+\frac{\alpha-\beta x}{([n]_q+\beta)}} \left(x + \frac{\alpha-\beta x}{[n]_q+\beta} - u\right) g''(u)du\right| \\
&\le M_{n,q}^{(\alpha,\beta)}\left(\left|\int_x^t (t-u)g''(u)du\right|; x\right) \\
&+ \left| \int_x^{x+\frac{\alpha-\beta x}{([n]_q+\beta)}} \left|(x + \frac{\alpha-\beta x}{([n]_q+\beta)} - u\right)\right| |g''(u)|du\right| \\
&\le \left\{ M_{n,q}^{(\alpha,\beta)}((t-x)^2;x) + \left(\frac{(\alpha-\beta x)}{([n]_q+\beta)}\right)^2\right\} \|g''\|.
\end{aligned} \tag{12}$$

$$\begin{aligned}
\left(\frac{\alpha-\beta x}{([n]_q+\beta)}\right)^2 &= \frac{(\alpha^2 - 2\alpha\beta x + \beta^2 x^2)}{([n]_q+\beta)^2} \le \frac{\alpha^2 + 2\alpha\beta x + \beta^2 x^2}{([n]_q+\beta)^2} \le \frac{\beta^2(1+2x+x^2)}{([n]_q+\beta)^2} \\
&\le \frac{2(1+\beta^2)}{q([n]_q+\beta)}\left\{x(1+x) + \frac{1}{([n]_q+\beta)}\right\} \\
&= \frac{2(1+\beta^2)}{q([n]_q+\beta)}\left\{\phi^2(x) + \frac{1}{([n]_q+\beta)}\right\}.
\end{aligned} \tag{13}$$

On the other hand, from (6), (10) and Lemma 2, we have

$$|L_{n,q}^{(\alpha,\beta)}(f;x)| \le |M_{n,q}^{(\alpha,\beta)}(f,x)| + 2\|f\| \le \|f\| M_{n,q}^{(\alpha,\beta)}(1;x) + 2\|f\| \le 3\|f\|.$$

From (12) and (13), we have

$$\mid L_{n,q}^{(\alpha,\beta)}(g;x) - g(x) \mid \ \ \le \frac{4(1+\beta^2)}{q([n]_q+\beta)}\left\{\phi^2(x) + \frac{1}{([n]_q+\beta)}\right\}\|g''\|.$$

Hence

$$
\begin{aligned}
| M_{n,q}^{(\alpha,\beta)}(f;x) - f(x) | &\leq | L_{n,q}^{(\alpha,\beta)}(f-g;x) - (f-g)(x) | + | L_{n,q}^{(\alpha,\beta)}(g;x) - g(x) | \\
&+ \left| f\left(x + \frac{\alpha - \beta x}{([n]_q + \beta)}\right) - f(x) \right| \\
&\leq 4\|f - g\| + \frac{4(1+\beta^2)}{q([n]_q + \beta)} \left\{ \phi^2(x) + \frac{1}{([n]_q + \beta)} \right\} \|g''\| \\
&+ \omega\left(f, \frac{| \alpha - \beta x |}{([n]_q + \beta)}\right).
\end{aligned}
$$

Now, taking infimum on the right-hand side over all $g \in W^2$, we get

$$
\begin{aligned}
| M_{n,q}^{(\alpha,\beta)}(f;x) - f(x) | &\leq C\omega_2\left(f, \sqrt{\frac{4(1+\beta^2)}{q([n]_q + \beta)} \left\{ \phi^2(x) + \frac{1}{([n]_q + \beta)} \right\}}\right) \\
&+ \omega\left(f, \frac{| \alpha - \beta x |}{[n]_q + \beta}\right).
\end{aligned}
$$

Hence, the proof is completed. □

4.2 Rate of Convergence

Let $B_{x^2}[0,\infty)$ be the space of all functions defined on $[0,\infty)$ and satisfying the condition $|f(x)| \leq M_f(1+x^2)$, where M_f is a constant depending on f. Let $C_{x^2}[0,\infty)$ be the subspace of all continuous functions belonging to $B_{x^2}[0,\infty)$. Also, $C^*_{x^2}[0,\infty)$ is the subspace of all functions $f \in C_{x^2}[0,\infty)$, for which $\lim_{x\to\infty} \frac{f(x)}{1+x^2}$ is finite. The norm on $C^*_{x^2}[0,\infty)$ is defined as $\|f\|_{x^2} := \sup_{x\in[0,\infty)} \frac{|f(x)|}{1+x^2}$. For any positive number a, the usual modulus of continuity is defined as

$$\omega_a(f,\delta) = \sup_{|t-x|\leq\delta,\ x,t\in[0,a]} |f(t) - f(x)|.$$

We observe that for a function $f \in C_{x^2}[0,\infty)$, the modulus of continuity $\omega_a(f,\delta)$ tends to zero as $\delta \to 0$. Now we give a rate of convergence theorem for the operator $M_{n,q_n}^{(\alpha,\beta)}$.

Theorem 3 *Let $f \in C_{x^2}[0,\infty)$, $q_n \in (0,1)$ such that $q_n \to 1$ as $n \to \infty$ and ω_{a+1} be its modulus of continuity on the finite interval $[0, a+1] \subset [0,\infty)$, where $a > 0$, then we have the following inequality:*

$$|M_{n,q_n}^{(\alpha,\beta)}(f;x)-f(x)| \leq \frac{K}{q_n([n]_{q_n}+\beta)}\left\{\phi^2(x)+\frac{1}{([n]_{q_n}+\beta)}\right\}$$
$$+ 2\omega_{a+1}\left(f,\sqrt{\frac{2(1+\beta^2)}{q_n([n]_{q_n}+\beta)}\left(\phi^2(x)+\frac{1}{[n]_{q_n}+\beta}\right)}\right),$$

where $K = 8M_f(1+a^2)(1+\beta^2)$.

Proof For $x \in [0, a]$ and $t > a+1$, since $t-x > 1$, we have

$$|f(t)-f(x)| \leq M_f(2+x^2+t^2) \leq M_f(2+3x^2+2(t-x)^2)$$
$$\leq M_f(t-x)^2(2+3x^2+2) \leq M_f(t-x)^2(4+3a^2)$$
$$|f(t)-f(x)| \leq 4M_f(1+a^2)(t-x)^2. \tag{14}$$

For $x \in [0, a]$ and $t \leq a+1$, we have

$$|f(t)-f(x)| \leq \omega_{a+1}(f,|t-x|) \leq \left(1+\frac{|t-x|}{\delta}\right)\omega_{a+1}(f,\delta), \quad \text{with } \delta > 0. \tag{15}$$

From (14) and (15), for all $t \in [0,\infty)$ and $x \in [0, a]$ we can write

$$|f(t)-f(x)| \leq 4M_f(1+a^2)(t-x)^2 + \left(1+\frac{|t-x|}{\delta}\right)\omega_{a+1}(f,\delta). \tag{16}$$

Hence, using Schwarz inequality,

$$|M_{n,q_n}^{(\alpha,\beta)}(f;x)-f(x)| \leq M_{n,q_n}^{(\alpha,\beta)}(|f(t)-f(x)|,x)$$
$$\leq 4M_f(1+a^2)M_{n,q_n}^{(\alpha,\beta)}((t-x)^2;x)$$
$$+ \omega_{a+1}(f,\delta)\left(1+\frac{1}{\delta}\{M_{n,q_n}^{(\alpha,\beta)}((t-x)^2;x)\}^{\frac{1}{2}}\right).$$

In view of Lemma 3, for $x \in [0, a]$

$$|M_{n,q_n}^{(\alpha,\beta)}(f;x)-f(x)| \leq \frac{8M_f(1+a^2)(1+\beta^2)}{q_n([n]_{q_n}+\beta)}\left\{\phi^2(x)+\frac{1}{([n]_{q_n}+\beta)}\right\}$$
$$+ \omega_{a+1}(f,\delta)\left\{1+\frac{1}{\delta}\left[\frac{2(1+\beta^2)}{q_n([n]_{q_n}+\beta)}\left(\phi^2(x)+\frac{1}{([n]_{q_n}+\beta)}\right)\right]^{\frac{1}{2}}\right\}.$$

Now, by choosing $\delta = \sqrt{\frac{2(1+\beta^2)}{q_n([n]_{q_n}+\beta)}\left(\phi^2(x)+\frac{1}{([n]_{q_n}+\beta)}\right)}$, we get the desired result. □

4.3 Voronovskaja Type Theorem

In this section we establish a Voronovskaja type asymptotic formula for the operators $M_{n,q}^{(\alpha,\beta)}$.

Lemma 4 *Assume that* $q_n \in (0, 1)$, $q_n \to 1$ *as* $n \to \infty$. *Then, for every* $x \in [0, \infty)$ *there hold*

$$\lim_{n\to\infty} [n]_{q_n} M_{n,q_n}^{(\alpha,\beta)}(t - x; x) = \alpha - \beta x$$

and

$$\lim_{n\to\infty} [n]_{q_n} M_{n,q_n}^{(\alpha,\beta)}((t - x)^2; x) = x^2 + 2x.$$

In view of Remark 2, the proof of this Lemma easily follows. Hence the details are omitted.

Theorem 4 *Let* $0 < q_n < 1$ *and* $q_n \to 1$ *as* $n \to \infty$. *Then, for all* $f \in C_{x^2}[0, \infty)$ *we have*

$$\lim_{n\to\infty} \|M_{n,q_n}^{(\alpha,\beta)}(f) - f\|_{x^2} = 0.$$

Proof Using [11], it is sufficient to verify the following conditions:

$$\lim_{n\to\infty} \|M_{n,q_n}^{(\alpha,\beta)}(t^m; x) - x^m\|_{x^2} = 0, \text{ for } m = 0, 1, 2. \tag{17}$$

Since $M_{n,q_n}^{(\alpha,\beta)}(1; x) = 1$, for $m = 0$, (17) holds. By Lemma 2, we have

$$\begin{aligned}
\|M_{n,q_n}^{(\alpha,\beta)}(t; x) - x\|_{x^2} &= \sup_{x\in[0,\infty)} \frac{|M_{n,q_n}^{(\alpha,\beta)}(t; x) - x|}{(1 + x^2)} \\
&\le \sup_{x\in[0,\infty)} \frac{\left|\frac{[n]_{q_n}x + \alpha}{([n]_{q_n} + \beta)} - x\right|}{1 + x^2} \\
&\le \frac{\beta}{([n]_{q_n} + \beta)} \sup_{x\in[0,\infty)} \frac{x}{(1 + x^2)} + \frac{\alpha}{([n]_{q_n} + \beta)} \sup_{x\in[0,\infty)} \frac{1}{1 + x^2} \\
&\le \frac{\alpha + \beta}{([n]_{q_n} + \beta)} = o(1) \quad as \quad n \to \infty.
\end{aligned}$$

Hence, the condition (17) holds for $m = 1$.

Again, by Lemma 2, we obtain

$$\begin{aligned}
\|M_{n,q_n}^{(\alpha,\beta)}(t^2;x)-x^2\|_{x^2} &= \sup_{x\in[0,\infty)} \frac{|M_{n,q_n}^{(\alpha,\beta)}(t^2;x)-x^2|}{(1+x^2)}\\
&= \sup_{x\in[0,\infty)} \frac{\left|\frac{[n]_{q_n}(1+q_n[n]_{q_n})x^2}{q_n([n]_{q_n}+\beta)^2}+\frac{[n]_{q_n}(1+q_n+2\alpha)x}{([n]_{q_n}+\beta)^2}+\frac{\alpha^2}{([n]_{q_n}+\beta)^2}-x^2\right|}{1+x^2}\\
&\le \frac{([n]_{q_n}(1+2q_n\beta)+\beta^2)}{q_n([n]_{q_n}+\beta)^2}\sup_{x\in[0,\infty)}\frac{x^2}{(1+x^2)}\\
&+\frac{[n]_{q_n}(1+q_n+2\alpha)}{([n]_{q_n}+\beta)^2}\sup_{x\in[0,\infty)}\frac{x}{(1+x^2)}\\
&+\frac{\alpha^2}{([n]_{q_n}+\beta)^2}\sup_{x\in[0,\infty)}\frac{1}{1+x^2}=o(1)\ \ as\ \ n\to\infty,
\end{aligned}$$

which implies that the condition (17) holds for $m=2$. This completes the proof. □

Theorem 5 *Assume that $q_n\in(0,1)$, $q_n\to 1$ as $n\to\infty$. Then, for any $f\in C_{x^2}^*[0,\infty)$ such that $f',f''\in C_{x^2}^*[0,\infty)$ we have*

$$\lim_{n\to\infty}[n]_{q_n}(M_{n,q_n}^{(\alpha,\beta)}(f;x)-f(x))=(\alpha-\beta x)f'(x)+\frac{1}{2}f''(x)(x^2+2x),$$

uniformly in $x\in[0,A]$, $A>0$.

Proof Let $f,f',f''\in C_{x^2}^*[0,\infty)$ and $x\in[0,A]$ be fixed. By Taylor's expansion, we may write

$$f(t)=f(x)+f'(x)(t-x)+\frac{1}{2}f''(x)(t-x)^2+r(t,x)(t-x)^2, \qquad (18)$$

where $r(t,x)$ is Peano form of the remainder, $r(.,x)\in C_{x^2}^*[0,\infty)$ and $\lim_{t\to x}r(t,x)=0$.

Applying $M_{n,q_n}^{(\alpha,\beta)}$ to the above Eq. (18) we obtain

$$\begin{aligned}
[n]_{q_n}(M_{n,q_n}^{(\alpha,\beta)}(f;x)-f(x)) &= f'(x)[n]_{q_n}M_{n,q}^{(\alpha,\beta)}(t-x;x)\\
&+\frac{1}{2}f''(x)[n]_{q_n}M_{n,q}^{(\alpha,\beta)}((t-x)^2;x)\\
&+[n]_{q_n}M_{n,q}^{(\alpha,\beta)}\left(r(t,x)(t-x)^2;x\right).
\end{aligned}$$

By Cauchy Schwarz inequality, we have

$$M_{n,q_n}^{(\alpha,\beta)}\Big(r(t,x)(t-x)^2;x\Big) \le \sqrt{M_{n,q_n}^{(\alpha,\beta)}\Big(r^2(t,x);x\Big)}\sqrt{M_{n,q_n}^{(\alpha,\beta)}\Big((t-x)^4;x\Big)}. \tag{19}$$

We observe that $r^2(x,x) = 0$ and $r^2(.,x) \in C_{x^2}^*[0,\infty))$. Then, it follows from Theorem 3 that

$$\lim_{n\to\infty}[n]_{q_n}(M_{n,q_n}^{(\alpha,\beta)}(r^2(t,x),x) = r^2(x,x) = 0, \tag{20}$$

uniformly with respect to $x \in [0,A]$. Now, from (19)–(20) and in view of the fact that

$$M_{n,q_n}^{(\alpha,\beta)}((t-x)^4;x) = O\left(\frac{1}{[n]_{q_n}}\right)^2$$

we obtain

$$\lim_{n\to\infty}[n]_{q_n}M_{n,q_n}^{(\alpha,\beta)}(r(t,x)(t-x)^2,x) = 0,$$

uniformly in $x \in [0,A]$. Thus, we obtain

$$\begin{aligned}\lim_{n\to\infty}[n]_{q_n}\left(M_{n,q_n}^{(\alpha,\beta)}(f,x) - f(x)\right) &= \lim_{n\to\infty}[n]_{q_n}\Big(f'(x)M_{n,q_n}^{(\alpha,\beta)}((t-x);x)\\ &\quad + \frac{1}{2}f''(x)M_{n,q_n}^{(\alpha,\beta)}((t-x)^2;x)\\ &\quad + M_{n,q_n}^{(\alpha,\beta)}(r(t,x)(t-x)^2,x)\Big)\\ &= (\alpha-\beta x)f'(x) + \frac{1}{2}f''(x)(x^2+2x),\end{aligned}$$

uniformly in $x \in [0,A]$. □

Corollary 1 *Let $q = q_n$ satisfy $0 < q_n < 1$ and let $q_n \to 1$ as $n \to \infty$. For each $f \in C_{x^2}[0,\infty)$ and $p > 0$, we have*

$$\sup_{x\in[0,\infty)} \frac{|M_{n,q_n}^{(\alpha,\beta)}(f;x) - f(x)|}{(1+x^2)^{1+p}} = 0.$$

Proof For any fixed $x_0 > 0$

$$\sup_{x\in[0,\infty)} \frac{|M_{n,q_n}^{\alpha,\beta}(f;x)-f(x)|}{(1+x^2)^{1+p}} \leq \sup_{x\leq x_0} \frac{|M_{n,q_n}^{(\alpha,\beta)}(f;x)-f(x)|}{(1+x^2)^{1+p}} + \sup_{x\geq x_0} \frac{|M_{n,q_n}^{(\alpha,\beta)}(f;x)-f(x)|}{(1+x^2)^{1+p}}$$

$$\leq \|M_{n,q_n}^{(\alpha,\beta)}(f)-f\|_{C[0,x_0]} + \|f\|_{x^2} \sup_{x\geq x_0} \frac{M_{n,q_n}^{(\alpha,\beta)}(1+t^2,x)}{(1+x^2)^{1+p}}$$

$$+ \sup_{x\geq x_0} \frac{|f(x)|}{(1+x^2)^{1+p}}. \tag{21}$$

Since $|f(x)| \leq M_f(1+x^2)$, we have

$$\sup_{x\geq x_0} \frac{|f(x)|}{(1+x^2)^{1+p}} \leq \sup_{x\geq x_0} \frac{M_f}{(1+x^2)^p} \leq \frac{M_f}{(1+x_0^2)^p}.$$

Let $\varepsilon > 0$ be arbitrary. Then, we can choose x_0 to be so large that

$$\frac{M_f}{(1+x_0^2)^p} < \frac{\varepsilon}{3} \tag{22}$$

and in view of Theorem 4, we obtain

$$\|f\|_{x^2} \lim_{n\to\infty} \frac{M_{n,q_n}^{(\alpha,\beta)}(1+t^2,x)}{(1+x^2)^{1+p}} = \frac{(1+x^2)\|f\|_{x^2}}{(1+x^2)^{1+p}} = \frac{\|f\|_{x^2}}{(1+x^2)^p} \leq \frac{\|f\|_{x^2}}{(1+x_0^2)^p} < \frac{\varepsilon}{3}. \tag{23}$$

Using Theorem 3, we see that the first term of inequality (21) implies that

$$\|M_{n,q_n}^{(\alpha,\beta)}(f)-f\|_{C[0,x_0]} < \frac{\varepsilon}{3} \quad \text{as} \quad n\to\infty. \tag{24}$$

Combining (22)–(24), we get the desired result. □

4.4 Point-Wise Estimates

Now, we establish some pointwise estimates of the rate of convergence of the operators (6). First, we give the relationship between the local smoothness of f and local approximation.

We know that a function $f \in C_B[0,\infty)$ is in $Lip_M\gamma$ on D, $\gamma \in (0,1]$, $D \subset [0,\infty)$ if it satisfies the condition

$$|f(t)-f(x)| \leq M|t-x|^\gamma, \quad t\in[0,\infty) \text{ and } x\in D,$$

where M is a constant depending only on γ and f.

Theorem 6 *Let* $f \in C_B[0,\infty) \bigcap Lip_M\gamma$, $\gamma \in (0,1]$, *and D be any bounded subset of the interval* $[0,\infty)$. *Then, for each* $x \in [0,\infty)$ *we have*

$$|M_{n,q}^{(\alpha,\beta)}(f;x)-f(x)| \le M\left(\left\{\frac{[2]_q(1+\beta^2)}{q([n]_q+\beta)}\left(\phi^2(x)+\frac{1}{([n]_q+\beta)}\right)\right\}^{\frac{\gamma}{2}}+2(d(x,D))^{\gamma}\right),$$

where $d(x,D)$ *represents the distance between* x *and* D.

Proof Let $\overline{D}$ be the closure of the set D in $[0,\infty)$. Then, there exists at least one point $x_0 \in \overline{D}$ such that

$$d(x,D)=|x-x_0|.$$

By the definition of $Lip_M\gamma$, we get

$$\begin{aligned}|M_{n,q}^{(\alpha,\beta)}(f;x)-f(x)| &\le M_{n,q}^{(\alpha,\beta)}(|f(t)-f(x_0)|;x)+M_{n,q}^{(\alpha,\beta)}(|f(x_0)-f(x)|;x))\\ &\le M\left\{M_{n,q}^{(\alpha,\beta)}(|t-x_0|^{\gamma};x)+|x_0-x|^{\gamma}\right\}\\ &\le M\left\{M_{n,q}^{(\alpha,\beta)}(|t-x|^{\gamma},x)+2|x-x_0|^{\gamma}\right\}.\end{aligned}$$

Now, by Holder's inequality with $p=\frac{2}{\gamma}$ and $\frac{1}{q}=1-\frac{1}{p}$, we have

$$\begin{aligned}|M_{n,q}^{(\alpha,\beta)}(f;x)-f(x)| &\le M\left\{\left[M_{n,q}^{(\alpha,\beta)}(|t-x|^{\gamma p};x)\right]^{\frac{1}{p}}\left[M_{n,q}^{(\alpha,\beta)}(1^q,x)\right]^{\frac{1}{q}}+2(d(x,D))^{\gamma}\right\}\\ &\le M\left\{\left[M_{n,q}^{(\alpha,\beta)}(|t-x|^2;x)\right]^{\frac{\gamma}{2}}+2(d(x,D))^{\gamma}\right\}\\ &\le M\left(\left\{\frac{[2]_q(1+\beta^2)}{q([n]_q+\beta)}\left(\phi^2(x)+\frac{1}{([n]_q+\beta)}\right)\right\}^{\frac{\gamma}{2}}+2(d(x,D))^{\gamma}\right).\end{aligned}$$

Hence, the proof is completed. □

Now, we give local direct estimate for the operators $M_{n,q}^{(\alpha,\beta)}$ using the Lipschitz type maximal function of order γ studied by Lenze [18]

$$\widetilde{\omega}_{\gamma}(f,x)=\sup_{t\neq x,t\in[0,\infty)}\frac{|f(t)-f(x)|}{|t-x|^{\gamma}},\quad x\in[0,\infty)\ \text{and}\ \gamma\in(0,1]. \tag{25}$$

Theorem 7 *Let* $\gamma \in (0,1]$ *and* $f \in C_B[0,\infty)$. *Then, for all* $x \in [0,\infty)$, *we have*

$$|M_{n,q}^{(\alpha,\beta)}(f;x)-f(x)| \le \widetilde{\omega}_{\gamma}(f,x)\left\{\frac{[2]_q(1+\beta^2)}{q([n]_q+\beta)}\left(\phi^2(x)+\frac{1}{([n]_q+\beta)}\right)\right\}^{\frac{\gamma}{2}}.$$

Proof From (25), we have

$$|f(t) - f(x)| \leq \widetilde{\omega}_\gamma(f, x)|t - x|^\gamma$$

and hence

$$|M_{n,q}^{(\alpha,\beta)}(f; x) - f(x)| \leq M_{n,q}^{(\alpha,\beta)}(|f(t) - f(x)|; x) \leq \widetilde{\omega}_\gamma(f, x)M_{n,q}^{(\alpha,\beta)}(|t - x|^\gamma; x).$$

Now, applying Holder's inequality with $p = \frac{2}{\gamma}$ and $\frac{1}{q} = 1 - \frac{1}{p}$, we have

$$|M_{n,q}^{(\alpha,\beta)}(f; x) - f(x)| \leq \widetilde{\omega}_\gamma(f, x)M_{n,q}^{(\alpha,\beta)}((t - x)^2; x)^{\frac{\gamma}{2}}.$$

On using Lemma 3, we have our assertion. □

4.5 Statistical Approximation

A sequence $(x_n)_n$ is said to be statistically convergent to a number L denoted by $st - \lim_n x_n = L$ if for every $\varepsilon > 0$,

$$\delta\{n \in \mathbb{N} : |x_n - L| \geq \varepsilon\} = 0,$$

where

$$\delta(K) = \lim_n \frac{1}{n} \sum_{j-1}^{n} \chi_K(j)$$

is the natural density of $K \subseteq \mathbb{N}$ and χ_K is the characteristic function of K. We note that every convergent sequence is statistically convergent, but the converse need not be true.

For example, let

$$x_n = \begin{cases} \log_{10} n, & n \in \{10^k, k \in \mathbb{N}\} \\ 1, & \text{otherwise.} \end{cases}$$

It follows that the sequence $\{x_n\}$ converges statistically to 1, but $\lim_n x_n$ does not exit.

Theorem 8 *For any $f \in C_{x^2}^*[0, \infty)$ and a sequence $(q_n)_n$ in $(0, 1)$ such that*

$$st - \lim_n q_n = 1, \; st - \lim_n (q_n)^n = a, \; (0 \leq a < 1), st - \lim_n \frac{1}{[n]_{q_n}} = 0, \tag{26}$$

the operator $M_{n,q}^{(\alpha,\beta)}(f;x)$ *statistically converges to* $f(x)$, *that is*

$$st-\lim_{n}\parallel M_{n,q}^{(\alpha,\beta)}(f)-f\parallel_{x^2}=0.$$

Proof Let us define $e_i(x) = x^i, i = 0,1,2$. It is sufficient to prove that $st-\lim_{n}\|M_{n,q_n}^{(\alpha,\beta)}(e_i)-e_i\|_{x^2}=0$, for $i=0,1,2$. It is clear that

$$st-\lim_{n}\|M_{n,q_n}^{(\alpha,\beta)}(e_0;.)-e_0\|_{x^2}=0.$$

From Lemma 2

$$\begin{aligned}
\|M_{n,q_n}^{(\alpha,\beta)}(e_1;.)-e_1\|_{x^2} &= \sup_{x\in[0,\infty)}\frac{|M_{n,q_n}^{(\alpha,\beta)}(e_1;x)-e_1(x)|}{(1+x^2)}\\
&\le \sup_{x\in[0,\infty)}\frac{\left|\frac{[n]_{q_n}x+\alpha}{([n]_{q_n}+\beta)}-x\right|}{1+x^2}\\
&\le \|e_0\|_{x^2}\frac{\alpha}{([n]_{q_n}+\beta)}+\frac{\beta}{([n]_{q_n}+\beta)}\|e_1\|_{x^2}\\
&\le \frac{\alpha}{([n]_{q_n}+\beta)}+\frac{\beta}{([n]_{q_n}+\beta)}.
\end{aligned}\tag{27}$$

Since, by the conditions (26), we get

$$st-\lim_{n}\frac{\alpha}{([n]_{q_n}+\beta)}=0$$

and

$$st-\lim_{n}\frac{\beta}{([n]_{q_n}+\beta)}=0.$$

For $\varepsilon>0$, let us define the following sets:

$$\begin{aligned}
E &:= \left\{n\in\mathbb{N}:\parallel M_{n,q_n}^{(\alpha,\beta)}(e_1;.)-e_1\parallel_{x^2}\ge\varepsilon\right\},\\
E_1 &:= \left\{n\in\mathbb{N}:\frac{\alpha}{([n]_{q_n}+\beta)}\ge\frac{\varepsilon}{2}\right\},\\
E_2 &:= \left\{n\in\mathbb{N}:\frac{\beta}{([n]_{q_n}+\beta)}\ge\frac{\varepsilon}{2}\right\}.
\end{aligned}$$

By (27), it is clear that $E\subseteq E_1\bigcup E_2$ which implies that $\delta(E)\le\delta(E_1)+\delta(E_2)=0$, and hence

$$st-\lim_{n}\|M_{n,q_n}^{(\alpha,\beta)}(e_1;.)-e_1\|_{x^2}=0.$$

Similarly, we can estimate

$$\begin{aligned}\|M_{n,q_n}^{(\alpha,\beta)}(e_2;.)-e_2\|_{x^2} &= \sup_{x\in[0,\infty)}\frac{|M_{n,q_n}^{(\alpha,\beta)}(e_2;x)-e_2(x)|}{(1+x^2)}\\ &= \sup_{x\in[0,\infty)}\frac{\left|\frac{[n]_{q_n}(1+q_n[n]_{q_n})x^2}{q_n([n]_{q_n}+\beta)^2}+\frac{[n]_{q_n}(1+q_n+2\alpha)x}{([n]_{q_n}+\beta)^2}+\frac{\alpha^2}{([n]_{q_n}+\beta)^2}-x^2\right|}{1+x^2}\\ &\le \frac{([n]_{q_n}(1+2q_n\beta)+\beta^2)}{q_n([n]_{q_n}+\beta)^2}\|e_2\|_{x^2}+\frac{[n]_{q_n}(1+q_n+2\alpha)}{([n]_{q_n}+\beta)^2}\|e_1\|_{x^2}\\ &+\frac{\alpha^2}{([n]_{q_n}+\beta)^2}\|e_0\|_{x^2}\\ &\le \frac{([n]_{q_n}(1+2q_n\beta)+\beta^2)}{q_n([n]_{q_n}+\beta)^2}+\frac{[n]_{q_n}(1+q_n+2\alpha)}{([n]_{q_n}+\beta)^2}+\frac{\alpha^2}{([n]_{q_n}+\beta)^2}.\end{aligned} \tag{28}$$

Again, using (26), we get

$$\begin{aligned} st-\lim_n\frac{([n]_{q_n}(1+2q_n\beta)+\beta^2)}{q_n([n]_{q_n}+\beta)^2}&=0,\\ st-\lim_n\frac{[n]_{q_n}(1+q_n+2\alpha)}{([n]_{q_n}+\beta)^2}&=0,\\ st-\lim_n\frac{\alpha^2}{([n]_{q_n}+\beta)^2}&=0.\end{aligned}$$

For a given $\varepsilon > 0$, we consider the following sets:

$$\begin{aligned} F&:=\left\{n\in\mathbb{N}:\|M_{n,q_n}^{(\alpha,\beta)}(e_2;.)-e_2\|_{x^2}\ge\varepsilon\right\},\\ F_1&:=\left\{n\in\mathbb{N}:\frac{([n]_{q_n}(1+2q_n\beta)+\beta^2)}{q_n([n]_{q_n}+\beta)^2}\ge\frac{\varepsilon}{3}\right\},\\ F_2&:=\left\{n\in\mathbb{N}:\frac{[n]_{q_n}(1+q_n+2\alpha)}{([n]_{q_n}+\beta)^2}\ge\frac{\varepsilon}{3}\right\},\\ F_3&:=\left\{n\in\mathbb{N}:\frac{\alpha^2}{([n]_{q_n}+\beta)^2}\ge\frac{\varepsilon}{3}\right\}.\end{aligned}$$

Consequently, by (28) we obtain $F\subseteq F_1\bigcup F_2\bigcup F_3$, which implies that $\delta(F)\le\delta(F_1)+\delta(F_2)+\delta(F_3)=0$. Hence, we get

$$st-\lim_n\|M_{n,q_n}^{(\alpha,\beta)}(e_2;.)-e_2\|_{x^2}=0.$$

This completes the proof of the theorem. □

4.5.1 Rate of Statistical Convergence

For $f \in C^*_{x^2}[0, \infty)$, following Freud [10], the weighted modulus of continuity of f is defined as

$$\Omega_2(f, \delta) = \sup_{x \geq 0, 0 < h \leq \delta} \frac{|f(x + h) - f(x)|}{1 + (x + h)^2}.$$

Lemma 5 *[19]. Let $f \in C^*_{x^2}[0, \infty)$. Then,*

(i) $\Omega_2(f, \delta)$ is a monotone increasing function of δ,
(ii) $\lim_{\delta \to 0} \Omega_2(f, \delta) = 0$,
(iii) For any $\lambda \in [0, \infty)$, $\Omega_2(f, \lambda\delta) \leq (1 + \lambda)\Omega_2(f, \delta)$.

Theorem 9 *Let $f \in C^*_{x^2}[0, \infty)$ and $(q_n)_n$ be a sequence satisfying (26). Then, for sufficiently large n.*

$$|M^{(\alpha,\beta)}_{n,q_n}(f; x) - f(x)| \leq K\Omega_2(f, \delta_n)(1 + x^{2+\lambda}), \quad x \in [0, \infty),$$

where $\lambda \geq 1$, $\delta_n = \sqrt{\frac{[2]_{q_n}(1+\beta^2)}{q_n([n]_{q_n}+\beta)}}$ and K is a positive constant independent f and n.

Proof

$$\begin{aligned}
|M^{(\alpha,\beta)}_{n,q_n}(f; x) - f(x)| &\leq M^{(\alpha,\beta)}_{n,q_n}(|f(t) - f(x)|; x) \\
&\leq M^{(\alpha,\beta)}_{n,q_n}\left\{(1 + (x + |t - x|)^2)\left(1 + \frac{|t - x|}{\delta}\right); x\right\}\Omega_2(f, \delta) \\
&\leq M^{(\alpha,\beta)}_{n,q_n}\left\{(1 + (t + 2x)^2)\left(1 + \frac{|t - x|}{\delta}\right); x\right\}\Omega_2(f, \delta) \\
&\leq \left(M^{(\alpha,\beta)}_{n,q_n}(\mu_x(t); x) + \frac{1}{\delta}M^{(\alpha,\beta)}_{n,q_n}(\mu_x(t)\psi_x(t); x)\right)\Omega_2(f, \delta),
\end{aligned}$$

where $\mu_x(t) = 1 + (t + 2x)^2$ and $\psi_x(t) = |t - x|$.

Now, using Cauchy–Schwarz inequality to the second term on the right-hand side, we obtain

$$|M^{(\alpha,\beta)}_{n,q_n}(f; x) - f(x)| \leq \left(M^{(\alpha,\beta)}_{n,q_n}(\mu_x; x) + \frac{1}{\delta}\sqrt{M^{(\alpha,\beta)}_{n,q_n}(\psi_x^2; x)}\sqrt{M^{(\alpha,\beta)}_{n,q_n}(\mu_x^2; x)}\right)\Omega_2(f, \delta). \tag{29}$$

From Lemma 2

$$M^{(\alpha,\beta)}_{n,q_n}(1 + t^2; x) = \left(1 + \frac{[n]_{q_n}(1 + q_n[n]_{q_n})}{q_n([n]_{q_n} + \beta)^2}x^2 + \frac{[n]_{q_n}([2]_q + 2\alpha)}{([n]_{q_n} + \beta)^2}x + \frac{\alpha^2}{([n]_{q_n} + \beta)^2}\right),$$

which implies that there exists a constant $C_1 > 0$ such that

$$\frac{1}{1+x^2}M_{n,q_n}^{(\alpha,\beta)}(1+t^2;x) = \frac{1}{1+x^2} + \frac{[n]_{q_n}(1+q_n[n]_{q_n})}{q_n([n]_{q_n}+\beta)^2}\frac{x^2}{1+x^2}$$
$$+\frac{[n]_{q_n}([2]_q+2\alpha)}{([n]_{q_n}+\beta)^2}\frac{x}{1+x^2}$$
$$+\frac{\alpha^2}{([n]_{q_n}+\beta)^2}\frac{1}{1+x^2},$$
$$\leq (1+C_1), \text{ for sufficiently large n .} \tag{30}$$

We have

$$\mu_x(t) = 1 + (2x+t)^2 \leq 1 + 2(4x^2+2t^2). \tag{31}$$

From (30) and (31), there is a positive constant K_1, such that

$$M_{n,q_n}^{(\alpha,\beta)}(\mu_x(t);x) \leq K_1(1+x^2), \quad \text{for sufficiently large n.}$$

Similarly, from Lemma 2

$$M_{n,q_n}^{(\alpha,\beta)}(\mu_x^2(t);x) = M_{n,q_n}^{(\alpha,\beta)}\Big((1+(2x+t)^2)^2;x\Big),$$
$$\leq M_{n,q_n}^{(\alpha,\beta)}\Big((1+2(4x^2+2t^2))^2;x\Big),$$
$$\leq 64\Big(M_{n,q_n}^{(\alpha,\beta)}(1+t^4;x) + (1+x^2)M_{n,q_n}^{(\alpha,\beta)}(1+t^2;x)$$
$$+(1+x^2)M_{n,q_n}^{(\alpha,\beta)}(1;x)\Big).$$

Since

$$\frac{1}{1+x^4}M_{n,q_n}^{(\alpha,\beta)}(1+t^4;x) \leq (1+C_2), \text{ for some constant } C_2>0 \text{ when n is sufficiently large ,}$$

there exists a positive constant K_2 such that

$$\sqrt{M_{n,q_n}^{(\alpha,\beta)}(\mu_x^2(t);x)} \leq K_2(1+x^2), \quad \text{for sufficiently large n.}$$

Also, from Lemma 3 we have

$$M_{n,q_n}^{(\alpha,\beta)}(\psi_x^2(t);x) \leq \frac{[2]_{q_n}(1+\beta^2)}{q_n([n]_{q_n}+\beta)}\phi^2(x) + \frac{[2]_{q_n}(1+\beta^2)}{q_n([n]_{q_n}+\beta)^2}.$$

Now from (29), we have

$$|M_{n,q_n}^{(\alpha,\beta)}(f;x)-f(x)| \leq \Omega_2(f,\delta) \left(K_1(1+x^2)+K_2(1+x^2)\frac{1}{\delta}\sqrt{\frac{[2]_{q_n}(1+\beta^2)}{q_n([n]_{q_n}+\beta)}\phi^2(x)+\frac{[2]_{q_n}(1+\beta^2)}{q_n([n]_{q_n}+\beta)^2}}\right).$$

Choosing $\delta = \sqrt{\frac{[2]_{q_n}(1+\beta^2)}{q_n([n]_{q_n}+\beta)}} = \delta_n$, we obtain

$$|M_{n,q_n}^{(\alpha,\beta)}(f;x)-f(x)| \leq \Omega_2(f,\delta_n)(1+x^2)(K_1+K_2\sqrt{1+\phi^2(x)}), \text{ for sufficiently large n.}$$

Hence, for sufficiently large n

$$|M_{n,q_n}^{(\alpha,\beta)}(f;x)-f(x)| \leq K\Omega_2(f,\delta_n)(1+x^{2+\lambda}), \; x \in [0,\infty),$$

where $\lambda \geq 1$ and K is a positive constant. This completes the proof of the theorem. □

Acknowledgments The authors are extremely grateful to the reviewers for careful reading of the manuscript and for making valuable suggestions leading to better presentation of the paper. The last author is thankful to the "University Grants Commission" India, for financial support to carry out the above research work.

References

1. Agrawal, P.N., Gupta, V., Kumar, A.S.: On q-analogue of Bernstein-Schurer-Stancu operators. Appl. Math. Comput. **219**, 7754–7764 (2013)
2. Agrawal, P.N., Mohammad, A.J.: On simultaneous approximation by a linear combination of a new sequence of linear positive operators. Turkish J. Math. **28**(4), 353–365 (2004)
3. Agrawal, P.N., Kumar, A.S., Sinha, T.A.K.: Stancu type generalization of modified Schurer operators based on q-intgers. Appl. Math. Comput. **226**, 765–776 (2014)
4. Aral, A., Gupta, V.: On the q-analogue of Stancu-Beta operators. Appl. Math. Lett. **25**(1), 67–71 (2012)
5. Aral, A., Gupta, V.: Generalized q-Baskakov operators. Mathematica Slovaca **61**(4), 619–634 (2011)
6. Aral, A., Gupta, V., Agrawal, R.P.: Applications of q-Calculus in Operator Theory. Springer, New York (2013)
7. Büyükyazici, I.: Approximation by Stancu-Chlodowsky polynomials. Comput. Math. Appl. **59**(1), 274–282 (2010)
8. Deo, N.: Simultaneous approximation by Lupas modified operators with weighted function of Sazsz operators, JIPAM. J. Inequal Pure Appl. Math. **5**(4) Article 113 (2004) 5pp
9. DeVore, R.A., Lorentz, G.G.: Constructive Approximation. Springer, Berlin (1993)
10. Freud, G.: Investigations on weighted approximation by polynomials. Studia Sci. Math. Hungar **8**, 285–305 (1973)
11. Gadzjiev, A.D.: Therems of the type of P. P. Korovkin type theorems, Math. Zametki. **20**(5), 781–786 (1976)
12. Gupta, V.: A note on q-Baskakov-Szasz operators. Lobachevskii J. Math. **31**(4), 359–366 (2010)
13. Gupta, V., Agarwal, R.P.: Convergence Estimates in Approximation Theory. Springer, Berlin (2014)

14. Gupta, V., Kim, T.: On a q-analogue of Baskakov basis functions. Russ. J. Math. Phys. **20**(3), 276–282 (2013)
15. Gupta, V., Karsli, H.: Some approximation properties by q-Szasz-Mirakyan-Baskakov-Stancu operators. Lobachevskii J. Math. **33**(2), 175–182 (2012)
16. Gupta, V., Srivastava, G.S.: Simultaneous approximation by Baskakov-Szász type operators, Bull. Math. Soc. Sci. Math. Roumanie (N.S.), **37**(85), 73–85 (1993)
17. Kac, V., Cheung, P.: Quantum Calculus (Universitext). Springer, New York (2002)
18. Lenze, B.: On Lipschitz-type maximal functions and their smoothness spaces. Nederl. Akad. Wetensch. Indag. Math. **50**(1), 53–63 (1988)
19. Lopez-Moreno, A.J.: Weighted simultaneous approximation with Baskakov type operators. Acta Math. Hungar. **104**(1–2), 143–151 (2004)
20. Phillips, G.M.: Bernstein polynomials based on the q-integers. Ann. Numer. Math. **4**, 511–518 (1997)
21. Stancu, D.D.: Approximation of functions by a new class of linear polynomial operators. Rev. Roumaine Math. Pures Appl. **13**, 1173–1194 (1968)
22. Verma, D.K., Agrawal, P.N.: Approximation by Baskakov-Durrmeyer-Stancu operators based on q-integers. Lobachevskii J. Math. **34**(2), 187–196 (2013)

Approximation of Functions of Class Lip(α, p) in L_p-Norm

M.L. Mittal and Mradul Veer Singh

Abstract Mittal and Rhoades (Int. J. Math. Game Theory Algebra **9**(4), 259–267, 1999 [9]; J. Comput. Anal. Appl. **2**(1) 1–10, 2000 [10]) and Mittal et al. (J. Math. Anal. Appl. **326**(1) 667–676, 2007 [7]; Appl. Math. Comput. **217**(9), 4483–4489, 2011 [8]) initiated the studies of error estimates $E_n(f)$ through trigonometric-Fourier approximation (tfa) for situations in which the summability matrix T does not have monotone rows. In this paper, we extend the results of Mittal et al. (Appl. Math. Comput. **217**(9), 4483–4489, 2011 [8]) to a more general C_λ-method in view of Armitage and Maddox (Analysis **9**, 195–204, 1989 [1]), which in turn generalizes the several previous known results due to Mittal and Singh (Int. J. Math. Math. Sci., Art. ID **267383**, 1–6, 2014 [11]), Değer et al. (Proc. Jangjeon Math. Soc. **15**(2), 203–213, 2012 [4]), Leindler (J. Math. Anal. Appl. **302**, 129–136, 2005 [6]), Chandra (J. Math. Anal. Appl. **275**, 13–26, 2002 [3]) and Quade (Duke Math. J. **3**(3), 529–543, 1937 [15]).

Keywords Trigonometric Fourier approximation · C_λ-method · L_p-norm · Class $Lip(\alpha, p)$

1 Introduction

For a given function $f \in L_p := L_p[0, 2\pi]$, $p \geq 1$, let

$$s_n(f) := s_n(f; x) = \frac{a_0}{2} + \sum_{k=1}^{n} (a_k \cos kx + b_k \sin kx) = \sum_{k=0}^{n} u_k(f; x) \quad (1)$$

M.L. Mittal · M.V. Singh (✉)
Department of Mathematics, Indian Institute of Technology Roorkee, Roorkee 247667, India
e-mail: mradul.singh@gmail.com

M.L. Mittal
e-mail: mlmittal_iit@yahoo.co.in

P.N. Agrawal et al. (eds.), *Mathematical Analysis and its Applications*, Springer Proceedings in Mathematics & Statistics 143,
DOI 10.1007/978-81-322-2485-3_8

denote the partial sums, called trigonometric polynomials of degree (or order) n, of the first $(n+1)$ terms of the Fourier series of f at a point x.

A positive sequence $\mathbf{c} := \{c_n\}$ is called almost monotone decreasing (increasing) if there exists a constant $K := K(\mathbf{c})$, depending on the sequence $\mathbf{c}$ only, such that for all $n \geq m$, $c_n \leq K c_m (K c_n \geq c_m)$. Such sequences will be denoted by $\mathbf{c} \in AMDS$ and $\mathbf{c} \in AMIS$ respectively. A sequence which is either *AMDS* or *AMIS* is called almost monotone sequence and will be denoted by $\mathbf{c} \in AMS$.

Let $\mathbb{F}$ be an infinite subset of $\mathbb{N}$ and $\mathbb{F}$ the range of strictly increasing sequence of positive integers, say $\mathbb{F} = \{\lambda(n)\}_{n=1}^{\infty}$. The Cesàro submethod C_λ is defined as

$$(C_\lambda x)_n = \frac{1}{\lambda(n)} \sum_{k=1}^{\lambda(n)} x_k, \ (n = 1, 2, 3, ...),$$

where $\{x_k\}$ is a sequence of real or complex numbers. Therefore, the C_λ-method yields a subsequence of the Cesàro method C_1, and hence it is regular for any λ. Matrix-C_λ is obtained by deleting a set of rows from Cesàro matrix. The basic properties of C_λ-method can be found in [1, 14].

Define

$$\tau_n^\lambda(f) = \tau_n^\lambda(f; x) = \sum_{k=0}^{\lambda(n)} a_{\lambda(n),k} s_k(f; x), \ \forall n \geq 0.$$

The trigonometric Fourier series of the signal f is said to be T^λ-summable to s if $\tau_n^\lambda(f) \to s$ as $n \to \infty$.

Throughout $T \equiv (a_{n,k})$, a linear operator, will denote an infinite lower triangular matrix with nonnegative entries and row sums 1. Such a matrix T is said to have monotone rows if , $\forall n$, $\{a_{n,k}\}$ is either nonincreasing or nondecreasing in k, $0 \leq k \leq n$. A linear operator T is said to be regular if it is limit-preserving over the space of convergent sequences.

We write

$$s_n(f; x) = \frac{1}{\pi} \int_0^{2\pi} f(x+t) D_n(t) \, \mathrm{d}t, \quad D_n(t) = (\sin(n+1/2)t)/2\sin(t/2),$$

$$A_{\lambda(n),k} = \sum_{r=k}^{\lambda(n)} a_{\lambda(n),r}, \quad A_{\lambda(n),0} \equiv 1, \forall n \geq 0.$$

The notation $[x]$ means the greatest integer contained in x.

2 Known Results

Chandra [3] proved three theorems on the trigonometric approximation using Nörlund and Riesz matrices. Some of them give sharper estimates than the results proved by Quade [15], Mohapatra and Russell [12] and himself earlier [2]. Similar results were proved by Khan [5] for generalized N_p-mean and Mohapatra et al. [13] for Taylor mean. Leindler [6] extended the results of Chandra [3] without the assumption of monotonicity on the generating sequence $\{p_n\}$. Leindler [6] proved the following:

Theorem 1 ([6]) *If $f \in Lip(\alpha, p)$ and $\{p_n\}$ be positive. If one of the conditions*
(i) $p > 1, 0 < \alpha < 1$ and $\{p_n\} \in AMDS$,
(ii) $p > 1, 0 < \alpha < 1$ and $\{p_n\} \in AMIS$ and

$$(n+1)p_n = O(P_n)\, holds, \tag{2}$$

(iii) $p > 1, \alpha = 1$ and $\sum_{k=1}^{n-1} k|\triangle p_k| = O(P_n)$,
(iv) $p > 1, \alpha = 1, \sum_{k=0}^{n-1} |\triangle p_k| = O(P_n/n)$ and (2) holds,
(v) $p = 1, 0 < \alpha < 1$ and $\sum_{k=-1}^{n-1} |\triangle p_k| = O(P_n/n)$,
maintains, then

$$||f - N_n(f)||_p = O(n^{-\alpha}). \tag{3}$$

Theorem 2 ([6]) *Let $f \in Lip(\alpha, 1), 0 < \alpha < 1$. If the positive $\{p_n\}$ satisfies conditions (2) and $\sum_{k=0}^{n-1} |\triangle p_k| = O(P_n/n)$ hold, then*

$$||f - R_n(f)||_1 = O(n^{-\alpha}).$$

Mittal et al. [7, 8] extended the work of Chandra to general matrices. Mittal et al. [8] proved the following:

Theorem 3 ([8]) *Let $f \in Lip(\alpha, p)$ and let $T = (a_{n,k})$ be an infinite regular triangular matrix.*
(i) If $p > 1, 0 < \alpha < 1$, $\{a_{n,k}\} \in AMS$ in k and satisfies

$$(n+1)\max\{a_{n,0}, a_{n,r}\} = O(1). \tag{4}$$

where $r := [n/2]$ then

$$||f - \tau_n(f)||_p = O(n^{-\alpha}). \tag{5}$$

(ii) If $p > 1, \alpha = 1$ and $\sum_{k=0}^{n-1}(n-k)|\triangle_k a_{n,k}| = O(1)$, or
(iii)If $p > 1, \alpha = 1$ and $\sum_{k=0}^{n} |\triangle_k a_{n,k}| = O(a_{n,0})$, or
(iv) If $p = 1, 0 < \alpha < 1$ and $\sum_{k=0}^{n} |\triangle_k a_{n,k}| = O(a_{n,0})$,
and also $(n+1)a_{n,0} = O(1)$, holds then (5) is satisfied.

Recently, Değer et al. [4] extended the results of Chandra [3] to more general C_λ-method in view of Armitage and Maddox [1]. Değer et al. [4] proved:

Theorem 4 ([4]) *Let $f \in Lip(\alpha, p)$ and $\{p_n\}$ be positive such that*

$$(\lambda(n)+1)p_{\lambda(n)} = O(P_{\lambda(n)}), \tag{6}$$

If either (i) $p > 1, 0 < \alpha \leq 1$ and $\{p_n\}$ is monotonic or (ii) $p = 1, 0 < \alpha < 1$ and $\{p_n\}$ is nondecreasing then

$$||f - N_n^\lambda(f)||_p = O(n^{-\alpha}).$$

Theorem 5 ([4]) *Let $f \in Lip(\alpha, 1), 0 < \alpha < 1$. If the positive $\{p_n\}$ satisfies condition (6) and nondecreasing, then $||f - R_n^\lambda(f)||_1 = O(n^{-\alpha})$.*

Very recently, in [11], the authors of this paper generalized two theorems of Değer et al. [4], by dropping the monotonicity on the elements of the matrix rows. These results also generalize the results of Leindler [6] to more general C_λ-method.

Theorem 6 ([11]) *If $f \in Lip(\alpha, p)$ and $\{p_n\}$ be positive. If one of the following conditions*
(i) $p > 1, 0 < \alpha < 1$ and $\{p_n\} \in AMDS$,
(ii) $p > 1, 0 < \alpha < 1$ and $\{p_n\} \in AMIS$ and (6) holds,
(iii) $p > 1, \alpha = 1$ and $\sum_{k=1}^{\lambda(n)-1} k|\Delta p_k| = O(P_{\lambda(n)})$,
(iv) $p > 1, \alpha = 1, \sum_{k=0}^{\lambda(n)-1} |\Delta p_k| = O\left(\frac{P_{\lambda(n)}}{\lambda(n)}\right)$ and (6) holds,
(v) $p = 1, 0 < \alpha < 1$ and $\sum_{k=-1}^{\lambda(n)-1} |\Delta p_k| = O\left(\frac{P_{\lambda(n)}}{\lambda(n)}\right)$,
maintains, then

$$||f - N_n^\lambda(f)||_p = O\left((\lambda(n))^{-\alpha}\right). \tag{7}$$

Theorem 7 ([11]) *Let $f \in Lip(\alpha, 1), 0 < \alpha < 1$. If the positive $\{p_n\}$ satisfies (6) and the condition $\sum_{k=0}^{\lambda(n)-1} |\Delta p_k| = O\left(\frac{P_{\lambda(n)}}{\lambda(n)}\right)$ holds, then*

$$||f - R_n^\lambda(f)||_1 = O\left((\lambda(n))^{-\alpha}\right). \tag{8}$$

3 Main Results

Mittal and Rhoades [9, 10] initiated the studies of error estimates through trigonometric-Fourier approximation (tfa) for situations in which the summability matrix T does not have monotone rows. In continuation of Mittal and Singh [11], in this paper, we generalize Theorem 3 of Mittal et al. [8] using more general C_λ-method. We prove the following:

Theorem 8 *Let $f \in Lip(\alpha, p)$ and let $T = (a_{n,k})$ be an infinite regular triangular matrix.*
(i) If $p > 1, 0 < \alpha < 1, \{a_{n,k}\} \in AMS$ in k and satisfies

$$(\lambda(n) + 1)\max\{a_{\lambda(n),0}, a_{\lambda(n),r}\} = O(1), \tag{9}$$

where $r := [\lambda(n)/2]$ then

$$||f - \tau_n^\lambda(f)||_p = O((\lambda(n))^{-\alpha}). \tag{10}$$

(ii) If $p > 1, \alpha = 1$ and

$$\sum_{k=0}^{\lambda(n)-1} (\lambda(n) - k)|\Delta_k a_{\lambda(n),k}| = O(1), or \tag{11}$$

(iii) If $p > 1, \alpha = 1$ *and*

$$\sum_{k=0}^{\lambda(n)} |\Delta_k a_{\lambda(n),k}| = O(a_{\lambda(n),0}), or \tag{12}$$

(iv) If $p = 1, 0 < \alpha < 1$ *and*

$$\sum_{k=0}^{\lambda(n)} |\Delta_k a_{\lambda(n),k}| = O(a_{\lambda(n),0}), \tag{13}$$

and also

$$(\lambda(n) + 1)a_{\lambda(n),0} = O(1), \tag{14}$$

holds then (10) is satisfied.

Remarks (1) If $\lambda(n) = n$, then our Theorem 8 generalizes Theorem 3.
(2) If $T \equiv (a_{n,k})$ is a Nörlund N_p (or weighted R_p) matrix then-
(a) If $\lambda(n) = n$, then condition (9) (or (14)) reduces to (2) while the conditions (11), (12), (13) reduce to conditions in (iii), (iv) and (v) of Theorem 1 respectively. Thus our Theorem 8 generalizes Theorems 1 and 2.
(b) Değer et al. [4] used the monotone sequences $\{p_n\}$ in Theorems 4 and 5 while our Theorem 8 claims less than the requirement of their theorems. For example, condition (11) of Theorem 8 is automatically satisfied if $\{p_n\}$ is nonincreasing sequence, i.e., L.H.S. of (11) gives

$$\begin{aligned}\sum_{k=0}^{\lambda(n)-1} (\lambda(n) - k)\left|\frac{\Delta_k p_{\lambda(n)-k}}{P_{\lambda(n)}}\right| &= \frac{1}{P_{\lambda(n)}} \sum_{k=0}^{\lambda(n)-1} (\lambda(n) - k)|p_{\lambda(n)-k} - p_{\lambda(n)-k-1}| \\ &= \frac{P_{\lambda(n)-1} - \lambda(n)p_{\lambda(n)}}{P_{\lambda(n)}} = O(1) = R.H.S.,\end{aligned}$$

while the condition (12) is always satisfied if $\{p_n\}$ is nondecreasing, i.e.,

$$\begin{aligned}\sum_{k=0}^{\lambda(n)} \left|\frac{\Delta_k p_{\lambda(n)-k}}{P_{\lambda(n)}}\right| &= \frac{1}{P_{\lambda(n)}} \sum_{k=0}^{\lambda(n)} |p_{\lambda(n)-k} - p_{\lambda(n)-k-1}| \\ &= \frac{1}{P_{\lambda(n)}}[p_{\lambda(n)} - p_{\lambda(n)-1} + p_{\lambda(n)-1} - p_{\lambda(n)-2} + \ldots + p_0 - p_{-1}] \\ &= O\left(\frac{p_{\lambda(n)}}{P_{\lambda(n)}}\right).\end{aligned}$$

Further, condition (9) (or (14)) of Theorem 8 reduces to (6) of Theorem 4. Thus our Theorem 8 generalizes the Theorems 4 and 5 of Değer et al. [4] under weaker assumptions and gives sharper estimate because all the estimates of Değer et al. [4] are in terms of n while our estimates are in terms of $\lambda(n)$ and $(\lambda(n))^{-\alpha} \leq n^{-\alpha}$ for $0 < \alpha \leq 1$.
(c) Also, Theorem 8 extends Theorems 6 and 7 of Mittal, Singh [11] where two theorems of Değer et al. [4] were generalized by dropping the monotonicity on the elements of matrix rows.

4 Lemmas

We shall use the following lemmas in the proof of our Theorem:

Lemma 1 ([15]) *If $f \in Lip(1, p)$, for $p > 1$ then*

$$||\sigma_n(f) - s_n(f)||_p = O(n^{-1}), \ \forall n > 0.$$

Lemma 2 ([15]) *If $f \in Lip(\alpha, p)$, for $0 < \alpha \leq 1$ and $p > 1$. Then*

$$||f - s_n(f)||_p = O(n^{-\alpha}), \ \forall n > 0.$$

Note: We are using sums upto $\lambda(n)$ in the nth partial sums s_n and σ_n and writing these sums s_n^{λ} and σ_n^{λ}, respectively, in the above lemmas for our purpose in this paper.

Lemma 3 *Let T have AMS rows and satisfy (4). Then, for $0 < \alpha < 1$,*

$$\sum_{k=0}^{\lambda(n)} a_{\lambda(n),k} \ (k+1)^{-\alpha} = O\left((\lambda(n)+1)^{-\alpha}\right).$$

Proof Suppose that the rows of T are *AMDS*. Then there exists a $K > 0$ such that

$$\begin{aligned}\sum_{k=0}^{\lambda(n)} a_{\lambda(n),k} \ (k+1)^{-\alpha} &= \sum_{k=0}^{\lambda(n)} K a_{\lambda(n),0}(k+1)^{-\alpha} = K a_{\lambda(n),0} \sum_{k=0}^{\lambda(n)} (k+1)^{-\alpha} \\ &= O(a_{\lambda(n),0}(\lambda(n)+1)^{1-\alpha}) = O((\lambda(n)+1)^{-\alpha}).\end{aligned}$$

A similar result can be proved if the rows of T are *AMIS*.

5 Proof of the Theorem 8

Case I. $p > 1, 0 < \alpha < 1$. We have

$$\tau_n^\lambda(f) - f = \sum_{k=0}^{\lambda(n)} a_{\lambda(n),k} s_k(f) - f = \sum_{k=0}^{\lambda(n)} a_{\lambda(n),k}(s_k(f) - f) \tag{15}$$

Thus in view of Lemmas 2 and 3 we have

$$\begin{aligned} ||\tau_n^\lambda(f) - f||_p \leq \sum_{k=0}^{\lambda(n)} a_{\lambda(n),k}||s_k(f) - f||_p &= \sum_{k=0}^{\lambda(n)} a_{\lambda(n),k} O((k+1)^{-\alpha}) \\ &= O\left((\lambda(n)+1)^{-\alpha}\right). \end{aligned}$$

Case III. $p > 1, \alpha = 1$. We have

$$||\tau_n^\lambda(f) - f||_p \leq ||\tau_n^\lambda(f) - s_n^\lambda(f)||_p + ||s_n^\lambda(f) - f||_p.$$

Again using the Lemma 2, we get

$$||\tau_n^\lambda(f) - f||_p \leq ||\tau_n^\lambda(f) - s_n^\lambda(f)||_p + O\left((\lambda(n))^{-1}\right). \tag{16}$$

So, it remains to show that

$$||\tau_n^\lambda(f) - s_n^\lambda(f)||_p = O\left((\lambda(n))^{-1}\right). \tag{17}$$

Since $A_{\lambda(n),0} = 1$, we have

$$\tau_n^\lambda(f) - s_n^\lambda(f) = \sum_{k=1}^{\lambda(n)} (A_{\lambda(n),k} - A_{\lambda(n),0}) u_k(f) = \sum_{k=1}^{\lambda(n)} \left(\frac{A_{\lambda(n),k} - A_{\lambda(n),0}}{k}\right)(k u_k(f)).$$

Thus using Abel's transformation, we get

$$\begin{aligned} ||\tau_n^\lambda(f;x) - s_n^\lambda(f;x)||_p \leq \sum_{k=1}^{\lambda(n)-1} \left|\Delta_k \left(\frac{A_{\lambda(n),k} - A_{\lambda(n),0}}{k}\right)\right| .||\sum_{j=1}^{k} j u_j(f)||_p \\ + \left|\frac{A_{\lambda(n),\lambda(n)} - A_{\lambda(n),0}}{\lambda(n)}\right| .||\sum_{j=1}^{\lambda(n)} j u_j(f)||_p. \end{aligned} \tag{18}$$

Let $\sigma_n(s)$ denote the nth term of the $(C, 1)$ transform of the sequence s, then

$$s_n^\lambda(f) - \sigma_n^\lambda(f) = \frac{1}{(\lambda(n)+1)} \sum_{j=1}^{\lambda(n)} j u_j(f).$$

Using Lemma 1, we get

$$\|\sum_{j=1}^{\lambda(n)} j u_j\|_p = (\lambda(n)+1)\|s_n^\lambda(f) - \sigma_n^\lambda(f)\|_p = (\lambda(n)+1) O\left((\lambda(n))^{-1}\right) = O(1). \tag{19}$$

Note that

$$\left|\frac{A_{\lambda(n),0} - A_{\lambda(n),\lambda(n)}}{\lambda(n)}\right| \leq (\lambda(n))^{-1} A_{\lambda(n),0} = O\left((\lambda(n))^{-1}\right).$$

Thus

$$\left|\frac{A_{\lambda(n),0} - A_{\lambda(n),\lambda(n)}}{\lambda(n)}\right| . \|\sum_{j=1}^{\lambda(n)} j u_j(f)\|_p = O\left((\lambda(n))^{-1}\right). \tag{20}$$

Now

$$\begin{aligned}\triangle_k\left(\frac{A_{\lambda(n),k} - A_{\lambda(n),0}}{k}\right) &= \frac{1}{k}\triangle_k(A_{\lambda(n),k} - A_{\lambda(n),0}) + \frac{A_{\lambda(n),k+1} - A_{\lambda(n),0}}{k(k+1)}\\ &= \frac{1}{k(k+1)}\left[(k+1)\triangle_k A_{\lambda(n),k} + \sum_{r=k+1}^{\lambda(n)} a_{\lambda(n),r} - \sum_{r=0}^{\lambda(n)} a_{\lambda(n),r}\right]\\ &= \frac{1}{k(k+1)}\left[(k+1)a_{\lambda(n),k} - \sum_{r=0}^{k} a_{\lambda(n),r}\right].\end{aligned} \tag{21}$$

Next we claim that $\forall k \in N$,

$$|\sum_{r=0}^{k} a_{\lambda(n),r} - (k+1)a_{\lambda(n),k}| \leq \sum_{r=0}^{k-1}(r+1)|a_{\lambda(n),r} - a_{\lambda(n),r+1}|, \tag{22}$$

If $k = 1$, then the inequality (22) reduces to

$$|\sum_{r=0}^{1} a_{\lambda(n),r} - 2a_{\lambda(n),1}| = |a_{\lambda(n),0} - a_{\lambda(n),1}|.$$

Thus (22) holds for $k = 1$. Now let us assume that (22) is true for $k = m$, i.e.,

$$|\sum_{r=0}^{m} a_{\lambda(n),r} - (k+1)a_{\lambda(n),m}| \leq \sum_{r=0}^{m-1} (r+1)|a_{\lambda(n),r} - a_{\lambda(n),r+1}|. \tag{23}$$

Let $k = m + 1$, using (23), we get

$$\begin{aligned}
&|\sum_{r=0}^{m+1} a_{\lambda(n),r} - (m+2)a_{\lambda(n),m+1}| \\
&= |\sum_{r=0}^{m} a_{\lambda(n),r} - (m+1)a_{\lambda(n),m} + (m+1)a_{\lambda(n),m} - (m+1)a_{\lambda(n),m+1}| \\
&\leq \sum_{r=0}^{m-1} (r+1)|a_{\lambda(n),r} - a_{\lambda(n),r+1}| + (m+1)|a_{\lambda(n),m} - a_{\lambda(n),m+1}| \\
&= \sum_{r=0}^{(m+1)-1} (r+1)|a_{\lambda(n),r} - a_{\lambda(n),r+1}|.
\end{aligned}$$

Thus (22) is true $\forall k$. Using (12), (14), (21), (22), we get

$$\begin{aligned}
\sum_{k=1}^{\lambda(n)} |\triangle_k \left(\frac{A_{\lambda(n),k} - A_{\lambda(n),0}}{k} \right)| &= \sum_{k=1}^{\lambda(n)} \frac{1}{k(k+1)} \left| (k+1)a_{\lambda(n),k} - \sum_{r=0}^{k} a_{\lambda(n),r} \right| \\
&\leq \sum_{k=1}^{\lambda(n)} \frac{1}{k(k+1)} \sum_{m=0}^{k-1} (m+1)|a_{\lambda(n),m} - a_{\lambda(n),m+1}| \\
&= \sum_{k=1}^{\lambda(n)} \frac{1}{k(k+1)} \sum_{m=1}^{k} m|a_{\lambda(n),m-1} - a_{\lambda(n),m}| \\
&\leq \sum_{m=1}^{\lambda(n)} m|\triangle_m a_{\lambda(n),m-1}| \sum_{k=m}^{\infty} \frac{1}{k(k+1)} \\
&= \sum_{k=0}^{\lambda(n)-1} |\triangle_k a_{\lambda(n),k}| = O(a_{\lambda(n),0}) = O\left((\lambda(n))^{-1}\right).
\end{aligned} \tag{24}$$

Combining (18), (19), (20) and (24) yields (17). From (17) and (16), we get

$$||\tau_n^{\lambda}(f) - f||_p = O\left((\lambda(n))^{-1}\right).$$

Case II. $p > 1, \alpha = 1$. For this we first prove that the condition $\sum_{k=0}^{\lambda(n)-1}(\lambda(n) - k)|\triangle_k a_{\lambda(n),k}| = O(1)$ implies that

$$\sum_{k=1}^{\lambda(n)}\left[\triangle_k\left(\frac{A_{\lambda(n),k} - A_{\lambda(n),0}}{k}\right)\right] = O\left((\lambda(n))^{-1}\right). \tag{25}$$

As in case (iii), using (22) and taking $r := [\lambda(n)/2]$ throughout the case, we have

$$\begin{aligned}
\sum_{k=1}^{\lambda(n)}\left|\triangle_k\left(\frac{A_{\lambda(n),k} - A_{\lambda(n),0}}{k}\right)\right| &= \sum_{k=1}^{\lambda(n)}\frac{1}{k(k+1)}\left|(k+1)a_{\lambda(n),k} - \sum_{m=0}^{k} a_{\lambda(n),m}\right| \\
&= \sum_{k=1}^{\lambda(n)}\frac{1}{k(k+1)}\sum_{m=0}^{k-1}(m+1)|a_{\lambda(n),m} - a_{\lambda(n),m+1}| \\
&= \left(\sum_{k=1}^{r} + \sum_{k=r+1}^{\lambda(n)}\right)k^{-1}(k+1)^{-1}\sum_{m=1}^{k} m|\triangle_m a_{\lambda(n),m-1}| \\
&:= B_1 + B_2, say.
\end{aligned}$$

Now interchanging the order of summation and using (11), we get

$$\begin{aligned}
B_1 &= \sum_{k=1}^{r} k^{-1}(k+1)^{-1}\sum_{m=1}^{k} m|\triangle_m a_{\lambda(n),m-1}| \le \sum_{m=1}^{r} m|\triangle_m a_{\lambda(n),m-1}|\sum_{k=m}^{\infty} k^{-1}(k+1)^{-1} \\
&= \sum_{m=1}^{r}|\triangle_m a_{\lambda(n),m-1}| = \sum_{m=\lambda(n)-r+1}^{\lambda(n)}|\triangle_{\lambda(n)-m} a_{\lambda(n),\lambda(n)-m}| \\
&= \sum_{m=r-1}^{\lambda(n)}|\triangle_{\lambda(n)-m} a_{\lambda(n),\lambda(n)-m}|\cdot\left(\frac{m}{r-1}\right) \\
&\le \frac{1}{r-1}\sum_{m=1}^{\lambda(n)} m|\triangle_{\lambda(n)-m} a_{\lambda(n),\lambda(n)-m}| = \frac{1}{r-1}\sum_{k=0}^{\lambda(n)-1}(\lambda(n)-k)|\triangle_k a_{\lambda(n),k}| \\
&= \frac{1}{r-1}O(1) = O\left((\lambda(n))^{-1}\right).
\end{aligned} \tag{26}$$

$$\begin{aligned}
\text{Now } B_2 &= \sum_{k=r}^{\lambda(n)} k^{-1}(k+1)^{-1}\sum_{m=1}^{k} m|\triangle_m a_{\lambda(n),m-1}| \\
&\le \sum_{k=r}^{\lambda(n)} k^{-1}(k+1)^{-1}\left[\left(\sum_{m=1}^{r} + \sum_{m=r}^{k}\right) m|\triangle_m a_{\lambda(n),m-1}|\right] := B_{21} + B_{22}, say.
\end{aligned}$$

Furthermore, using again our assumption, we get

$$\begin{aligned}
B_{21} &= \sum_{k=r}^{\lambda(n)} k^{-1}(k+1)^{-1} \sum_{m=1}^{r} m|\triangle_m a_{\lambda(n),m-1}| \\
&\leq r^{-1} \sum_{k=r}^{\lambda(n)} (k+1)^{-1} \sum_{m=1}^{\lambda(n)} m|\triangle_{\lambda(n)-m} a_{\lambda(n),\lambda(n)-m}| \\
&= r^{-1} \sum_{k=r}^{\lambda(n)} (k+1)^{-1} \sum_{k=0}^{\lambda(n)-1} (\lambda(n)-k)|\triangle_k a_{\lambda(n),k}| \\
&= O(r^{-1}) \sum_{k=r}^{\lambda(n)} (k+1)^{-1} = O\left((\lambda(n))^{-1}\right). \qquad (27)
\end{aligned}$$

Again interchanging the order of summation and using (11), we get

$$\begin{aligned}
B_{22} &= \sum_{k=r}^{\lambda(n)} k^{-1}(k+1)^{-1} \sum_{m=r}^{k} m|\triangle_m a_{\lambda(n),m-1}| \leq \sum_{k=r}^{\lambda(n)} (k+1)^{-1} \sum_{m=r}^{k} |\triangle_m a_{\lambda(n),m-1}| \\
&\leq \sum_{m=r}^{\lambda(n)} |\triangle_m a_{\lambda(n),m-1}| \sum_{k=m}^{\lambda(n)} (k+1)^{-1} \leq (r+1)^{-1} \sum_{m=r}^{\lambda(n)} |\triangle_m a_{\lambda(n),m-1}| \sum_{k=m}^{\lambda(n)} 1 \\
&= (r+1)^{-1} \sum_{m=r}^{\lambda(n)} (\lambda(n)-m+1)|\triangle_m a_{\lambda(n),m-1}| \\
&= (r+1)^{-1} \sum_{k=r-1}^{\lambda(n)-1} (\lambda(n)-k)|\triangle_k a_{\lambda(n),k}| \\
&= (r+1)^{-1} O(1) = O\left((\lambda(n))^{-1}\right). \qquad (28)
\end{aligned}$$

Summing up our partial results (26), (27), (28) we verified (25). Thus (16), (18), (19), (25) and Lemma 2, again yield

$$||f - \tau_n^{\lambda}(f)||_p = O\left((\lambda(n))^{-1}\right).$$

Case IV. $p = 1, 0 < \alpha < 1$.
Using Abel's transformation, conditions (13), (14), convention $a_{n,n+1} = 0$ and the result of Quade [15], we obtain

$$\|\tau_n^{\lambda}(f) - f\|_1 = \|\sum_{k=0}^{\lambda(n)} a_{\lambda(n),k} s_k(f) - f\|_1 = \|\sum_{k=0}^{\lambda(n)} a_{\lambda(n),k}(s_k(f) - f)\|_1$$

$$= \|\sum_{k=0}^{\lambda(n)-1} \left(\Delta_k a_{\lambda(n),k}\right) \sum_{r=0}^{k}(s_r(f) - f) + (a_{\lambda(n),\lambda(n)} - a_{\lambda(n),\lambda(n)+1}) \sum_{r=0}^{\lambda(n)}(s_r(f) - f)\|_1$$

$$= \|\sum_{k=0}^{\lambda(n)} \left(\Delta_k a_{\lambda(n),k}\right) \sum_{r=0}^{k}(s_r(f) - f)\|_1 = \|\sum_{k=0}^{\lambda(n)} \left(\Delta_k a_{\lambda(n),k}\right)(k+1)(\sigma_k(f) - f)\|_1$$

$$\leq \sum_{k=0}^{\lambda(n)} (k+1)|\Delta_k a_{\lambda(n),k}|.\|\sigma_k(f) - f)\|_1 = O\left(\sum_{k=0}^{\lambda(n)} (k+1)^{1-\alpha}|\Delta_k a_{\lambda(n),k}|\right)$$

$$= O\left(\lambda(n)^{1-\alpha}\right) \sum_{k=0}^{\lambda(n)} |\Delta_k a_{\lambda(n),k}| = O\left(\lambda(n)^{1-\alpha}\right) O\left(a_{\lambda(n),0}\right) = O\left((\lambda(n))^{-\alpha}\right).$$

This completes the proof of case (iv) and hence the proof of Theorem 8 is complete.

References

1. Armitage, D.H., Maddox, I.J.: A new type of Cesáro mean. Analysis **9**, 195–204 (1989)
2. Chandra, P.: A note on degree of approximation by Nörlund and Riesz operators. Mat. Vestnik **42**, 9–10 (1990)
3. Chandra, P.: Trigonometric approximation of functions in L_p-norm. J. Math. Anal. Appl. **275**, 13–26 (2002)
4. Değer, U., Dağadur, İ., Küçükaslan, M.: Approximation by trigonometric polynomials to functions in L_p-norm. Proc. Jangjeon Math. Soc. **15**(2), 203–213 (2012)
5. Khan, H.H.: On the degree of approximation of functions belonging to class Lip(α, p). Indian J. Pure Appl. Math. **5**(2), 132–136 (1974)
6. Leindler, L.: Trigonometric approximation in L_p-norm. J. Math. Anal. Appl. **302**, 129–136 (2005)
7. Mittal, M.L., Rhoades, B.E., Mishra, V.N., Singh, U.: Using infinite matrices to approximate functions of class Lip(α, p) using trigonometric polynomials. J. Math. Anal. Appl. **326**(1), 667–676 (2007)
8. Mittal, M.L., Rhoades, B.E., Sonker, S., Singh, U.: Approximation of signals of class Lip(α, p) by linear operators. Appl. Math. Comput. **217**(9), 4483–4489 (2011)
9. Mittal, M.L., Rhoades, B.E.: On the degree of approximation of continuous functions by using linear operators on their Fourier series. Int. J. Math. Game Theory Algebra **9**(4), 259–267 (1999)
10. Mittal, M.L., Rhoades, B.E.: Degree of approximation to functions in a normed space. J. Comput. Anal. Appl. **2**(1), 1–10 (2000)
11. Mittal, M.L., Singh, M.V.: Approximation of signals (functions) by trigonometric polynomials in L_p-norm. Int. J. Math. Math. Sci. Art. ID **267383**, 1–6 (2014)
12. Mohapatra, R.N., Russell, D.C.: Some direct and inverse theorems in approximation of functions. J. Aust. Math. Soc. Ser. A **34**(2), 143–154 (1983)
13. Mohapatra, R.N., Holland, A.S.B., Sahney, B.N.: Functions of class Lip(α, p) and their Taylor mean. J. Approx. Theory **45**, 363–374 (1985)
14. Osikiewicz, J.A.: Equivalence results for Cesáro submethods. Analysis **20**(1), 35–43 (2000)
15. Quade, E.S.: Trigonometric approximation in the mean. Duke Math. J. **3**(3), 529–543 (1937)

A New Genuine Durrmeyer Operator

Vijay Gupta

Abstract The generalization of the Bernstein polynomials based on certain parameter was considered by Stancu (Rew. Roum. Math. Pure. Appl. **13**, 1173–1194, 1968 [14]). Recently, Gupta and Rassias (J. Math. Anal. **8**(2), 146–155, 2014 [11]) proposed a Durrmeyer-type modification of the Lupaş operators and established some results. Actually, the genuine operators are important as far as the approximation is concerned. Here we propose genuine Durrmeyer-type operators, which preserve linear functions. We establish moments using generalized hypergeometric function and obtain an asymptotic formula and a direct result in terms of second-order modulus of continuity. In the end we propose an open problem for the readers.

Keywords Bernstein polynomials · Moments · Asymptotic formula · Genuine operators · Linear functions

1 Introduction

In the year 1967 Durrmeyer [4] introduced the integral modification of the Bernstein polynomials as

$$M_n(f,x) = (n+1)\sum_{k=0}^{n} p_{n,k}(x)\int_0^1 p_{n,k}(t)f(t)dt, x \in 0, 1, \tag{1}$$

where

$$p_{n,k}(t) = \binom{n}{k} t^k (1-t)^{n-k}.$$

Derriennic [3] first studied these operators in detail and she estimated some results in ordinary and simultaneous approximation. Later some direct estimates in simultaneous approximation for linear combinations were discussed by Agrawal and Gupta

V. Gupta (✉)
Department of Mathematics, Netaji Subhas Institute of Technology, Sector 3 Dwarka, New Delhi 110078, India
e-mail: vijaygupta2001@hotmail.com

P.N. Agrawal et al. (eds.), *Mathematical Analysis and its Applications*,
Springer Proceedings in Mathematics & Statistics 143,
DOI 10.1007/978-81-322-2485-3_9

[9]. Also Gupta in [7] and Finta and Gupta in [5] considered the q analogue of the Bernstein–Durrmeyer operators and established some direct results. The rate of convergence in simultaneous approximation in some other form has been discussed in [10].

For $P_n^{(\alpha)}: C[0,1] \to C[0,1]$, with a non-negative parameter α, Stancu in [14] considered a sequence of positive linear operators, which is defined as

$$P_n^{(\alpha)}(f,x) = \sum_{k=0}^{n} f\left(\frac{k}{n}\right) p_{n,k}^{(\alpha)}(x), \tag{2}$$

where $p_{n,k}^{(\alpha)}(x)$ is the Pólya distribution with density function given by

$$p_{n,k}^{(\alpha)}(x) = \binom{n}{k} \frac{\prod_{\nu=0}^{k-1}(x+\nu\alpha)\prod_{\mu=0}^{n-k-1}(1-x+\mu\alpha)}{\prod_{\lambda=0}^{n-1}(1+\lambda\alpha)}, x \in [0,1].$$

As a special case $p_{n,k}^{(0)}(x)$ is the density function of the binomial distribution and $P_n^{(0)}(f,x)$ reduces to the classical Bernstein polynomials. For convergence point of view Lupaş and Lupaş [12] considered the operators (2) by taking $\alpha = 1/n$, later in [13] some approximation properties have been discussed for this case. In an alternate form such operators can be represented as

$$P_n^{(1/n)}(f,x) = \frac{2(n!)}{(2n)!} \sum_{k=0}^{n} f\binom{n}{k}\left(\frac{k}{n}\right)(nx)_k(n-nx)_{n-k}, \tag{3}$$

where the Pochhammer symbol is given as $(x)_i = x(x+1)(x+2)\dots(x+i-1)$.

Very recently, Gupta and Rassias [11] introduced the Durrmeyer-type integral modification of the operators (3) and established some direct results, but their operators preserve only the constant functions. We propose here a genuine Durrmeyer-type modification of the operators (3), which also preserve linear functions and can be defined as

$$\begin{aligned} D_n^{(1/n)}(f,x) = (n-1)\sum_{k=1}^{n-1} p_{n,k}^{(1/n)}(x)\int_0^1 p_{n-2,k-1}(t)\, f(t)\, dt \\ + p_{n,0}^{(1/n)}(x) f(0) + p_{n,n}^{(1/n)}(x) f(1), \end{aligned} \tag{4}$$

where

$$p_{n,k}^{(1/n)}(x) = \frac{2(n!)}{(2n)!}\binom{n}{k}(nx)_k(n-nx)_{n-k}$$

and the Bernstein basis function is defined as in (1).

In the present paper, we establish the moments of the operators (4) using generalized hypergeometric series. We also establish an asymptotic formula and a direct result in terms of second-order modulus of continuity.

2 Auxiliary Results

We estimate the moments using the recurrence relation. Such relation for some other form was given by Greubel [6] in an open problem raised by the author [8]. To make the paper self-content, we provide the detailed proof below:

Lemma 1 *For $r \geq 1$, if we denote $T_{n,r}(x) = D_n^{(1/n)}(e_r, x)$, then we have*

$$T_{n,r+1}(x) = \frac{(r+1)(2r - nx + 3n - 3) + n(nx + x - 3) + 3}{(r+n)^2} T_{n,r}(x) + \frac{r(r-1)(nx + 1 - r - 2n)}{(n+r-1)(n+r)^2} T_{n,r-1}(x)$$

Proof By definition, we have

$$T_{n,r}(x) = (n-1) \sum_{k=1}^{n-1} p_{n,k}^{(1/n)}(x) \int_0^1 p_{n-2,k-1}(t)\, t^r\, dt + p_{n,n}^{(1/n)}(x)$$

Also, by simple computation, we have

$$(n-1) \int_0^1 p_{n-2,k-1}(t) t^r dt = \frac{(n-1)!(k+r-1)!}{(n+r-1)!(k-1)!}.$$

Thus

$$\begin{aligned} T_{n,r}(x) &= \sum_{k=1}^{n-1} p_{n,k}^{(1/n)}(x) \frac{(n-1)!(k+r-1)!}{(n+r-1)!(k-1)!} + p_{n,n}^{(1/n)}(x) \\ &= \sum_{k=1}^{n} p_{n,k}^{(1/n)}(x) \frac{(n-1)!(k+r-1)!}{(n+r-1)!(k-1)!}. \end{aligned}$$

Substituting the value of $p_{n,k}^{(1/n)}(x)$ and using

$$\binom{n}{k} = \frac{(-1)^k(-n)_k}{k!}, \ (a)_{n-k} = \frac{(-1)^k (a)_n}{(1-a-n)_k}, 0 \leq k \leq n$$

we can write, by using the identity $(k+r-1)! = (r)_k.(r-1)!$ in the next step

$$T_{n,r}(x) = \sum_{k=1}^{n} \frac{2.n!}{(2n)!} \cdot \frac{(-1)^k(-n)_k}{k!}(nx)_k \frac{(-1)^k(n-nx)_n}{(1-2n+nx)_k} \cdot \frac{(n-1)!(k+r-1)!}{(n+r-1)!(k-1)!}$$
$$= \frac{2.(r-1)!.(n-1)!n!(n-nx)_n}{(n+r-1)!(2n)!} \sum_{k=1}^{n} \frac{(-n)_k(nx)_k(r)_k}{(k-1)!(1-2n+nx)_k} \cdot \frac{1}{k!}$$
$$= \frac{2.(r-1)!.(n-1)!n!(n-nx)_n}{(n+r-1)!(2n)!} \sum_{k=0}^{n} \frac{(-n)_{k+1}(nx)_{k+1}(r)_{k+1}}{k!(1-2n+nx)_{k+1}} \cdot \frac{1}{(k+1)!}.$$

Next, using $(k+1)! = (2)_k$, $(a)_{k+1} = a(a+1)_k$, we have

$$T_{n,r}(x) = \frac{2.(r-1)!.(n-1)!n!(n-nx)_n}{(n+r-1)!(2n)!} \sum_{k=0}^{n} \frac{(-n)(-n+1)_k nx(nx+1)_k r(r+1)_k}{(2)_k(1-2n+nx)(2-2n+nx)_k} \cdot \frac{1}{k!}$$
$$= \frac{2.(r)!.nx(n-nx)_n}{(2n-nx-1)(n+r-1)!\binom{2n}{n}} \sum_{k=0}^{n} \frac{(-n+1)_k(nx+1)_k(r+1)_k}{(2)_k(2-2n+nx)_k} \cdot \frac{1}{k!}$$
$$= \frac{2nx.\Gamma(r+1).(n-nx)_n}{(2n-nx-1)\Gamma(n+r)\binom{2n}{n}} \cdot {}_3F_2(-n+1, nx+1, r+1; 2, 2-2n+nx; 1),$$

where ${}_3F_2$ is the Hypergeometric polynomial. Applying the recurrence relations between the parameters of a ${}_3F_2$ series

$$\begin{aligned} &(a-d)(a-e){}_3F_2(a-1, b, c; d, e; z) \\ =\, &a(a+1)(1-z){}_3F_2(a+2, b, c; d, e; z) \\ &+a[d+e-3a-2+z(2a-b-c+1)]{}_3F_2(a+1, b, c; d, e; z) \\ &+[(2a-d)(2a-e)-a(a-1)-z(a-b)(a-c)]{}_3F_2(a, b, c; d, e; z). \end{aligned}$$

Substituting $a = r+1$, $b = -n+1$, $c = nx+1$, $d = 2$, $e = nx-2n+2$ and $z = 1$, the above relation reduces to the following:

$$\begin{aligned} &(r-1)(r+2n-nx-1){}_3F_2(-n+1, nx+1, r; 2, nx-2n+2; 1) \\ =\, &-(r+1)(n+r){}_3F_2(-n+1, nx+1, r+2; 2, nx-2n+2; 1) \\ &+\phi\, {}_3F_2(-n+1, nx+1, r+1; 2, nx-2n+2; 1), \end{aligned}$$

where $\phi = (r+1)(2r-nx+3n-3) + n(nx+x-3) + 3$. Since,

$$\frac{(2n-nx-1)\binom{2n}{n}}{2nx(n-nx)_n} \cdot \frac{\Gamma(n+r)}{\Gamma(r+1)} T_{n,r}(x) = {}_3F_2(-n+1, nx+1, r+1; 2, nx-2n+2; 1),$$

then the result follows after simple computation, i.e.

$$T_{n,r+1}(x) = \frac{(r+1)(2r - nx + 3n - 3) + n(nx + x - 3) + 3}{(r+n)^2} T_{n,r}(x)$$
$$+ \frac{r(r-1)(nx + 1 - r - 2n)}{(n+r-1)(n+r)^2} T_{n,r-1}(x).$$

Remark 1 By definition of operator using $P_n^{(1/n)}(e_0, x) = 1$ we have $T_{n,0}(x) = 1$, by applying Lemma 1, we get

$$T_{n,1}(x) = x, T_{n,2}(x) = \frac{n(n-1)x^2 + (3n+1)x}{(n+1)^2}.$$

Remark 2 If we denote $\mu_{n,r}(x) = D_n^{(1/n)}((t-x)^r, x)$, then by Remark 1, we get

$$\mu_{n,1}(x) = 0, \mu_{n,2}(x) = \frac{(3n+1)x(1-x)}{(n+1)^2}.$$

Moreover, we have

$$\mu_{n,1}(x) = O(n^{-[(m+1)/2]}).$$

Lemma 2 *For* $f \in C[0, 1]$, *we have* $\left\| D_n^{(1/n)}(f, x) \right\| \leq \|f\|$, *where* $||.||$ *is the sup-norm on* $[0, 1]$.

Proof From the definition of operator and Remark 1, we get

$$\left| D_n^{(1/n)}(f, x) \right| \leq \|f\| D_n^{(1/n)}(1, x) = \|f\|.$$

Lemma 3 *For* $n \in N$, *we have*

$$D_n^{(1/n)}\left((t-x)^2, x\right) \leq \frac{3}{n+1} \delta_n^2(x),$$

where $\delta_n^2(x) = \varphi^2(x) + \frac{1}{n+1}$, *where* $\varphi^2(x) = x(1-x)$.

Proof By Remark 2, we have

$$D_n^{(1/n)}\left((t-x)^2, x\right) = \frac{(3n+1)x(1-x)}{(n+1)^2} \leq \frac{3}{n+1}\left[\varphi^2(x) + \frac{1}{n+1}\right],$$

which is desired.

3 Convergence Estimates

In this section, we present some convergence estimates of the operators $D_n^{(1/n)}(f,x)$.

Theorem 1 *Let $f \in C[0,1]$ and if f'' exists at a point $x \in [0,1]$, then*

$$\lim_{n\to\infty} n\left[D_n^{(1/n)}(f,x) - f(x)\right] = \frac{3x(1-x)}{2} f''(x).$$

Proof By Taylor's expansion of f, we have

$$f(t) = f(x) + (t-x)f'(x) + \frac{(t-x)^2}{2} f''(x) + \varepsilon(t,x)(t-x)^2,$$

where $\varepsilon(t,x) \to 0$ as $t \to x$. Applying $D_n^{(1/n)}$ on above Taylor's expansion and using Remark 2, we have

$$\begin{aligned} D_n^{(1/n)}(f,x) - f(x) &= f'(x) D_n^{(1/n)}((t-x),x) + \frac{1}{2} f''(x) D_n^{(1/n)}((t-x)^2,x) \\ &\quad + D_n^{(1/n)}(\varepsilon(t,x)(t-x)^2,x). \end{aligned}$$

Thus

$$\begin{aligned} &\lim_{n\to\infty} n\left[D_n^{(1/n)}(f,x) - f(x)\right] \\ &= \lim_{n\to\infty} n\frac{1}{2} f''(x) D_n^{(1/n)}((t-x)^2,x) + \lim_{n\to\infty} n D_n^{(1/n)}(\varepsilon(t,x)(t-x)^2,x) \\ &= \frac{3x(1-x)}{2} f''(x) + \lim_{n\to\infty} n D_n^{(1/n)}\left(\varepsilon(t,x)(t-x)^2, x\right) \\ &=: \frac{3x(1-x)}{2} f''(x) + F. \end{aligned}$$

In order to complete the proof, it is sufficient to show that $F = 0$. By Cauchy–Schwarz inequality, we have

$$F = \lim_{n\to\infty} n D_n^{(1/n)}\left(\varepsilon^2(t,x), x\right)^{1/2} D_n^{(1/n)}\left((t-x)^4, x\right)^{1/2}. \tag{5}$$

Furthermore, since $\varepsilon^2(x,x) = 0$ and $\varepsilon^2(.,x) \in C[0,1]$, it follows that

$$\lim_{n\to\infty} n D_n^{(1/n)}\left(\varepsilon^2(t,x), x\right) = 0, \tag{6}$$

uniformly with respect to $x \in [0,1]$. Thus from (5), (6) and application of Remark 2, we get

$$\lim_{n\to\infty} n D_n^{(1/n)}\left(\varepsilon^2(t,x),x\right)^{1/2} D_n^{(1/n)}\left((t-x)^4,x\right)^{1/2}=0.$$

Thus, we have

$$\lim_{n\to\infty} n\left[D_n^{(1/n)}(f,x)-f(x)\right]=\frac{3x(1-x)}{2}f''(x),$$

which completes the proof.

To prove the next direct result, we need the following auxiliary function viz. Peetre's K-functional which for $W^2=\left\{g\in C[0,1]: g', g''\in C[0,1]\right\}$ is defined as:

$$K_2(f,\delta)=\inf\left\{\|f-g\|+\delta\left\|g''\right\|: g\in W^2\right\}(\delta>0),$$

where $\|.\|$ is the uniform norm on $C[0,1]$.

Theorem 2 *For the operators $D_n^{(1/n)}$, there exists a constant $C>0$ such that*

$$\left|D_n^{(1/n)}(f,x)-f(x)\right|\le C\omega_2\left(f,(n+1)^{-1}\delta_n(x)\right),$$

where $f\in C[0,1]$, $\delta_n(x)=\left[\varphi^2(x)+\frac{1}{n+1}\right]^{1/2}$, $\varphi(x)=\sqrt{x(1-x)}$ and $x\in[0,1]$ and the second-order modulus of continuity is given by

$$\omega_2(f,\eta)=\sup_{0<h\le\eta}\ \sup_{x,x+2h\in[0,1]}|f(x+2h)-2f(x+h)+f(x)|.$$

Proof By Taylor's formula, we can write

$$g(t)=g(x)+(t-x)g'(x)+\int_x^t(t-u)g''(u)\,du.$$

Applying the above Taylor's formula, we have

$$D_n^{(1/n)}(g,x)=g(x)+D_n^{(1/n)}\left(\int_x^t(t-u)g''(u)\,du,x\right).$$

Hence

$$\begin{aligned}\left|D_n^{(1/n)}(g,x)-g(x)\right|&\le D_n^{(1/n)}\left(\int_x^t|t-u|\left|g''(u)\right|du,x\right)\\&\le D_n^{(1/n)}\left((t-x)^2,x\right)\|g''\|.\end{aligned}$$

For $f\in C[0,1]$ and $g\in W^2$, using Lemmas 2 and 3 we have

$$\begin{aligned}\left|D_n^{(1/n)}(f,x)-f(x)\right| &\le \left|D_n^{(1/n)}(f-g,x)-(f-g)(x)\right|+\left|D_n^{(1/n)}(g,x)-g(x)\right|\\ &\le 2\|f-g\|+\frac{3}{n+1}\delta_n^2(x)\left\|g''\right\|.\end{aligned}$$

Taking infimum over all $g \in W^2$, we obtain

$$\left|D_n^{(1/n)}(f,x)-f(x)\right| \le 3K_2\left(f,\frac{1}{n+1}\delta_n^2(x)\right)$$

Using the inequality due to DeVore and Lorentz [2], there exists a positive constant $C > 0$ such that

$$K_2(f,\delta) \le C\omega_2\left(f,\sqrt{\delta}\right),$$

we get at once

$$\left|D_n^{(1/n)}(f,x)-f(x)\right| \le C\omega_2\left(f,(n+1)^{-1}\delta_n(x)\right),$$

so the proof is completed.

Remark 3 It is easy to construct operators of summation-integral type, but under the integral sign in (4), the Pólya basis functions are not possible at this moment. Also the simultaneous approximation as done in [1] for usual Bernstein–Durrmeyer polynomials are not analogous for (4). We left it for readers, this can be considered as on open problem.

References

1. Agrawal, P.N., Gupta, V.: Simultaneous approximation by linear combination of modified Bernstein polynomials. Bull. Greek Math. Soc. **39**, 29–43 (1989)
2. DeVore, R.A., Lorentz, G.G.: Constructive Approximation. Springer, Berlin (1993)
3. Derriennic, M.M.: Sur l'approximation de functionsintegrable sur [0, 1] par des polynomes de Bernstein modifies. J. Approx. Theory **31**, 323–343 (1981)
4. Durrmeyer, J.L.: Une formule d' inversion de la Transformee de Laplace: applications a la Theorie des Moments, These de 3e Cycle, Faculte des Sciences de l' Universite de Paris (1967)
5. Finta, Z., Gupta, V.: Approximation by q-Durrmeyer operators. J. Appl. Math. Comput. **29**(1–2), 401–415 (2009)
6. Greubel, G.C.: Solution of the Problem 130 (Europeon Mathematical Socity Newsletter, March 2014). Eur. Math. Soc. Newsl, Sept (2014)
7. Gupta, V.: Some approximation properties on q-Durrmeyer operators. Appl. Math. Comput. **197**(1), 172–178 (2008)
8. Gupta, V.: Europeon mathematical socity newsletter. Problem **130**, 64 (2014)
9. Gupta, V., Agarwal R.P.: Convergence Estimates in Approximation Theory. Springer, New York (2014)
10. Gupta, V., Lopez-Moreno, A.J., Latorre-Palacios, J.-M.: On simultaneous approximation of the Bernstein Durrmeyer operators. Appl. Math. Comput. **213**(1), 112–120 (2009)

11. Gupta, V., Rassias, ThM: Lupaş-Durrmeyer operators based on Polya distribution. Banach J. Math. Anal. **8**(2), 146–155 (2014)
12. Lupaş, L., Lupaş, A.: Polynomials of binomial type and approximation operators, Studia Univ. Babes-Bolyai. Mathematica **32**(4), 61–69 (1987)
13. Miclăuş, D.: The revision of some results for Bernstein Stancu type operators. Carpathian J. Math. **28**(2), 289–300 (2012)
14. Stancu, D.D.: Approximation of functions by a new class of linear polynomial operators. Rew. Roum. Math. Pure. Appl. **13**, 1173–1194 (1968)

Degree of Approximation by Certain Genuine Hybrid Operators

Meenu Goyal and P.N. Agrawal

Abstract This paper is in continuation of our work on certain genuine hybrid operators in (Positivity (Under review)) [3]. First, we discuss some direct results in simultaneous approximation by these operators, e.g. pointwise convergence theorem, Voronovskaja-type theorem and an error estimate in terms of the modulus of continuity. Next, we estimate the rate of convergence for functions having a derivative that coincides a.e. with a function of bounded variation.

Keywords Rate of convergence · Modulus of continuity · Simultaneous approximation · Bounded variation

Mathematics Subject Classication (2010): 41A25 · 26A15 · 26A45

1 Introduction

Recently, Gupta and Rassias [5] introduced the Lupaş-Durrmeyer operators based on Polya distribution and discussed some local and global direct results. Also, Gupta [2] studied some other hybrid operators of Durrmeyer type. Păltănea [11] (see also [10]) considered a Durrmeyer-type modification of the genuine Szász-Mirakjan operators based on two parameters $\alpha, \rho > 0$. Inspired by his work, in [3] Gupta et al. introduced certain genuine hybrid operators as follows:

For $c \in \{0, 1\}$ and $f \in C_\gamma[0, \infty) := \{f \in C[0, \infty) : |f(t)| \leq M_f e^{\gamma t}$, for some $\gamma > 0, M_f > 0\}$, we define

M. Goyal (✉) · P.N. Agrawal
Department of Mathematics, Indian Institute of Technology Roorkee, Roorkee 247667, India
e-mail: meenu.goyaliitr@yahoo.com

P.N. Agrawal
e-mail: pna_iitr@yahoo.co.in

P.N. Agrawal et al. (eds.), *Mathematical Analysis and its Applications*, Springer Proceedings in Mathematics & Statistics 143,
DOI 10.1007/978-81-322-2485-3_10

$$B_\alpha^\rho(f,x) = \sum_{k=1}^{\infty} p_{\alpha,k}(x,c)\int_0^\infty \theta_{\alpha,k}^\rho(t)f(t)dt + p_{\alpha,0}(x,c)f(0), \tag{1}$$

$$= \int_0^\infty K_\alpha^\rho(x,t)f(t)dt, \tag{2}$$

where

$$p_{\alpha,k}(x,c) = \frac{(-x)^k}{k!}\phi_{\alpha,c}^{(k)}(x), \theta_{\alpha,k}^\rho(t) = \frac{\alpha\rho}{\Gamma(k\rho)}e^{-\alpha\rho t}(\alpha\rho t)^{k\rho-1}$$

$$\text{and} \quad K_\alpha^\rho(x,t) = \sum_{k=1}^{\infty} p_{\alpha,k}(x,c)\theta_{\alpha,k}^\rho(t) + p_{\alpha,0}(x,c)\delta(t);\ x \in (0,\infty).$$

It is observed that the operators $B_\alpha^\rho(f,x)$ are well-defined for $\alpha\rho > \gamma$. We assume that

$$\phi_{\alpha,c}(x) = \begin{cases} e^{-\alpha x}, & \text{for } c = 0, \\ (1+x)^{-\alpha}, & \text{for } c = 1. \end{cases}$$

As shown in paper [3], the operators (1) include several linear positive operators as special cases. Further, we note that the operators (1) preserve the linear functions. In [3], we studied some direct results, e.g. Voronovskaja-type theorems in ordinary and simultaneous approximation for first-order derivatives as well as results in local and weighted approximation. In this paper, we continue this work by discussing simultaneous approximation for $f^{(r)}(x), r \in \mathbb{N}$ and the rate of convergence of the operators (1) for the functions with derivatives of bounded variation on each finite subinterval of $(0,\infty)$. The paper is organized as follows:

In Sect. 2, we discuss some auxiliary results and then in Sect. 3, we obtain the main results of this paper.

2 Auxiliary Results

For $f : [0,\infty) \to R$, we define

$$S_\alpha(f;x) = \sum_{k=0}^{\infty} p_{\alpha,k}(x,c)f\left(\frac{k}{\alpha}\right) \tag{3}$$

such that (3) makes sense for all $x \geq 0$.

For $m \in \mathbb{N}^0 = \mathbb{N} \cup \{0\}$, the mth order central moment of the operators S_α is given by

$$\upsilon_{\alpha,m}(x) := S_\alpha((t-x)^m; x) = \sum_{k=0}^{\infty} p_{\alpha,k}(x,c)\left(\frac{k}{\alpha} - x\right)^m.$$

Lemma 1 *For the function $\upsilon_{\alpha,m}(x)$, we have*

$$\upsilon_{\alpha,0}(x) = 1, \;\; \upsilon_{\alpha,1}(x) = 0$$

and

$$x(1+cx)[\upsilon'_{\alpha,m}(x) + m\upsilon_{\alpha,m-1}(x)] = \alpha\upsilon_{\alpha,m+1}(x).$$

Thus,

(i) $\upsilon_{\alpha,m}(x)$ *is a polynomial in x of degree $[m/2]$;*
(ii) *for each $x \in [0,\infty)$, $\upsilon_{\alpha,m}(x) = O(\alpha^{-[(m+1)/2]})$, where $[\beta]$ denotes the integral part of β.*

Proof For the cases $c = 0$ and 1, the proof of this lemma can be found in [8, 12] respectively.

Lemma 2 *For the mth order ($m \in \mathbb{N}^0$) moment of the operators (1) defined as $u_{\alpha,m}(x) := B_\alpha^\rho(t^m; x)$, we have*

$$u_{\alpha,0}(x) = 1, \;\; u_{\alpha,1}(x) = x, \;\; u_{\alpha,2}(x) = x^2 + \frac{x}{\alpha}\left(\frac{1}{\rho} + (1+cx)\right)$$

and

$$x(1+cx)u'_{\alpha,m}(x) = \alpha u_{\alpha,m+1}(x) - \left(\tfrac{m}{\rho} + \alpha x\right)u_{\alpha,m}(x), \;\; m \in \mathbb{N}.$$

Consequently, for each $x \in (0,\infty)$ and $m \in \mathbb{N}$, $u_{\alpha,m}(x) = x^m + \alpha^{-1}(p_m(x,c) + o(1))$,
where $p_m(x,c)$ is a rational function of x depending on the parameters m and c.

Lemma 3 *[3] For $m \in \mathbb{N}^0$, if the mth order central moment $\mu_{\alpha,m}(x)$ for the operators B_α^ρ is defined as*

$$\mu_{\alpha,m}(x) := B_\alpha^\rho((t-x)^m, x) = \sum_{k=1}^{\infty} p_{\alpha,k}(x,c)\int_0^\infty \theta_{\alpha,k}^\rho(t)(t-x)^m dt + p_{\alpha,0}(x,c)(-x)^m,$$

then we have the following recurrence relation:

$$\alpha\mu_{\alpha,m+1}(x) = x(1+cx)\mu'_{\alpha,m}(x) + mx\left[\frac{1}{\rho} + (1+cx)\right]\mu_{\alpha,m-1}(x) + \frac{m}{\rho}\mu_{\alpha,m}(x).$$

Consequently,

(i) $\mu_{\alpha,0}(x) = 1,\ \mu_{\alpha,1}(x) = 0,\ \mu_{\alpha,2}(x) = \dfrac{\{1 + \rho(1 + cx)\}x}{\alpha\rho}$;

(ii) $\mu_{\alpha,m}(x)$ *is a polynomial in* x *of degree atmost* m;

(iii) *for every* $x \in (0, \infty)$, $\mu_{\alpha,m}(x) = O\left(\alpha^{-[(m+1)/2]}\right)$;

(iv) *the coefficients of* α^{-m} *in* $\mu_{\alpha,2m}(x)$ *and* $\mu_{\alpha,2m-1}(x)$ *are* $(2m-1)!!\left\{x\left(\dfrac{1}{\rho} + (1 + cx)\right)\right\}^m$ *and* $\dfrac{(2m-1)!!(m-1)}{3}x^{m-1}\left(\dfrac{1}{\rho} + (1 + cx)\right)^{m-2}\left\{(1 + cx)\left(\dfrac{1}{\rho} + (1 + 2cx)\right) + \dfrac{2}{\rho}\left(\dfrac{1}{\rho} + (1 + cx)\right)\right\}$ *respectively.*

Corollary 1 *For* $x \in [0, \infty)$ *and* $\alpha > 0$, *it is observed that*

$$\mu_{\alpha,2}(x) \leq \frac{\lambda x(1 + cx)}{\alpha}, \quad \text{where } \lambda = 1 + \frac{1}{\rho} > 1.$$

Corollary 2 *[3] Let* γ *and* δ *be any two positive real numbers and* $[a, b] \subset (0, \infty)$ *be any bounded interval. Then, for any* $m > 0$ *there exists a constant* M' *independent of* α *such that*

$$\left\| \sum_{k=1}^{\infty} p_{\alpha,k}(x, c) \int_{|t-x|\geq\delta} \theta^{\rho}_{\alpha,k}(t) e^{\gamma t} dt \right\| \leq M'\alpha^{-m},$$

where $\|.\|$ *is the sup-norm over* $[a, b]$.

Lemma 4 *For every* $x \in (0, \infty)$ *and* $r \in \mathbb{N}^0$, *there exist polynomials* $q_{i,j,r}(x)$ *in* x *independent of* α *and* k *such that*

$$\frac{d^r}{dx^r} p_{\alpha,k}(x, c) = p_{\alpha,k}(x, c) \sum_{\substack{2i+j\leq r \\ i,j\geq 0}} \alpha^i (k - \alpha x)^j \frac{(q_{i,j,r}(x, c))}{(p(x, c))^r},$$

where $p(x, c) = x(1 + cx)$.

Proof For the cases $c = 0, 1$, the proof of this lemma can be seen in [8, 12] respectively.

3 Main Results

3.1 Simultaneous Approximation

Throughout this section, we assume that $0 < a < b < \infty$.

In the following theorem, we show that the derivative $B_\alpha^{\rho(r)}(f;.)$ is also an approximation process for $f^{(r)}$.

Theorem 1 (Basic convergence theorem) *Let $f \in C_\gamma[0,\infty)$. If $f^{(r)}$ exists at a point $x \in (0,\infty)$, then we have*

$$\lim_{\alpha\to\infty}\left(\frac{d^r}{d\omega^r}B_\alpha^\rho(f;\omega)\right)_{\omega=x} = f^{(r)}(x). \tag{4}$$

Further, if $f^{(r)}$ is continuous on $(a-\eta, b+\eta)$, $\eta > 0$, then the limit in (4) holds uniformly in $[a,b]$.

Proof By our hypothesis, we have

$$f(t) = \sum_{\nu=0}^{r}\frac{f^{(\nu)}(x)}{\nu!}(t-x)^\nu + \psi(t,x)(t-x)^r,\ \ t\in[0,\infty),$$

where the function $\psi(t,x)\to 0$ as $t\to x$. From the above equation, we may write

$$\begin{aligned}\left(\frac{d^r}{d\omega^r}B_\alpha^\rho(f(t);\omega)\right)_{\omega=x} &= \sum_{\nu=0}^{r}\frac{f^{(\nu)}(x)}{\nu!}\left(\frac{d^r}{d\omega^r}B_\alpha^\rho((t-x)^\nu;\omega)\right)_{\omega=x}\\ &\quad+\left(\frac{d^r}{d\omega^r}B_\alpha^\rho(\psi(t,x)(t-x)^r;\omega)\right)_{\omega=x}\\ &=: I_1 + I_2, \text{ say.}\end{aligned}$$

First, we estimate I_1.

$$\begin{aligned}I_1 &= \sum_{\nu=0}^{r}\frac{f^{(\nu)}(x)}{\nu!}\left\{\frac{d^r}{d\omega^r}\left(\sum_{j=0}^{\nu}\binom{\nu}{j}(-x)^{\nu-j}B_\alpha^\rho(t^j;\omega)\right)_{\omega=x}\right\}\\ &= \sum_{\nu=0}^{r}\frac{f^{(\nu)}(x)}{\nu!}\sum_{j=0}^{\nu}\binom{\nu}{j}(-x)^{\nu-j}\left(\frac{d^r}{d\omega^r}B_\alpha^\rho(t^j;\omega)\right)_{\omega=x}\\ &= \sum_{\nu=0}^{r-1}\frac{f^{(\nu)}(x)}{\nu!}\sum_{j=0}^{\nu}\binom{\nu}{j}(-x)^{\nu-j}\left(\frac{d^r}{d\omega^r}B_\alpha^\rho(t^j;\omega)\right)_{\omega=x}\end{aligned}$$

$$+\frac{f^{(r)}(x)}{r!}\sum_{j=0}^{r}\binom{r}{j}(-x)^{r-j}\left(\frac{d^r}{d\omega^r}B_\alpha^\rho(t^j;\omega)\right)_{\omega=x}$$
$$:= I_3 + I_4, \text{ say.}$$

First, we estimate I_4.

$$I_4 = \frac{f^{(r)}(x)}{r!}\sum_{j=0}^{r-1}\binom{r}{j}(-x)^{r-j}\left(\frac{d^r}{d\omega^r}B_\alpha^\rho(t^j;\omega)\right)_{\omega=x} + \frac{f^{(r)}(x)}{r!}\left(\frac{d^r}{d\omega^r}B_\alpha^\rho(t^r;\omega)\right)_{\omega=x}$$
$$:= I_5 + I_6, \text{ say.}$$

Using Lemma 2, we get
$I_6 = f^{(r)}(x) + O\left(\frac{1}{\alpha}\right)$, $I_3 = O\left(\frac{1}{\alpha}\right)$ and $I_5 = O\left(\frac{1}{\alpha}\right)$, as $\alpha \to \infty$.
Combining the above estimates, for each $x \in (0, \infty)$ we obtain $I_1 \to f^{(r)}(x)$ as $\alpha \to \infty$.

Next, we estimate I_2. By making use of Lemma 4, we have

$$|I_2| \le \sum_{k=1}^{\infty}\frac{p_{\alpha,k}(x,c)}{(p(x,c))^r}\sum_{\substack{2i+j\le r\\ i,j\ge 0}}\alpha^i|k-\alpha x|^j|q_{i,j,r}(x,c)|\int_0^\infty \theta_{\alpha,k}^\rho(t)|\psi(t,x)||(t-x)^r|dt$$
$$+\left|\left(\frac{d^r}{d\omega^r}p_{\alpha,0}(\omega,c)\right)_{\omega=x}\right||\psi(0,x)(-x)^r|$$
$$:= I_7 + I_8, \text{ say.}$$

Since $\psi(t,x) \to 0$ as $t \to x$, for a given $\varepsilon > 0$ there exists a $\delta > 0$ such that $|\psi(t,x)| < \varepsilon$ whenever $|t-x| < \delta$. For $|t-x| \ge \delta$, $|(t-x)^r\psi(t,x)| \le Me^{\gamma t}$, for some constant $M > 0$.

Again, using Lemma 4, we have

$$|I_7| \le \sum_{k=1}^{\infty}\sum_{\substack{2i+j\le r\\ i,j\ge 0}}\alpha^i|k-\alpha x|^j\frac{|q_{i,j,r}(x,c)|}{(p(x,c))^r}p_{\alpha,k}(x,c)\left(\varepsilon\int_{|t-x|<\delta}\theta_{\alpha,k}^\rho(t)|t-x|^r dt\right.$$
$$\left.+M\int_{|t-x|\ge\delta}\theta_{\alpha,k}^\rho(t)e^{\gamma t}dt\right) := I_9 + I_{10}, \text{ say.}$$

Let $K = \sup_{\substack{2i+j\le r\\ i,j\ge 0}}\frac{|q_{i,j,r}(x,c)|}{(p(x,c))^r}$. By applying the Schwarz inequality, Lemmas 1 and 3, we get

$$\begin{aligned}
|I_9| &\leq \varepsilon K \sum_{k=1}^{\infty} \sum_{\substack{2i+j\leq r\\ i,j\geq 0}} \alpha^i |k-\alpha x|^j p_{\alpha,k}(x,c)\left(\int_0^{\infty} \theta_{\alpha,k}^{\rho}(t)(t-x)^{2r}dt\right)^{\frac{1}{2}}\\
&\leq \varepsilon K \sum_{\substack{2i+j\leq r\\ i,j\geq 0}} \alpha^{i+j}\left(\sum_{k=1}^{\infty}\left(\frac{k}{\alpha}-x\right)^{2j} p_{\alpha,k}(x,c)\right)^{\frac{1}{2}}\\
&\quad \left(\sum_{k=1}^{\infty} p_{\alpha,k}(x,c)\int_0^{\infty} \theta_{\alpha,k}^{\rho}(t)(t-x)^{2r}dt\right)^{\frac{1}{2}}\\
&\leq \varepsilon K \sum_{\substack{2i+j\leq r\\ i,j\geq 0}} \alpha^{i+j}\left(\upsilon_{\alpha,2j}(x)-x^{2j}\phi_{\alpha,c}(x)\right)^{\frac{1}{2}}\\
&\quad \left(B_{\alpha}^{\rho}((t-x)^{2r};x))-x^{2r}\phi_{\alpha,c}(x)\right)^{\frac{1}{2}}\\
&= \varepsilon \sum_{\substack{2i+j\leq r\\ i,j\geq 0}} \alpha^{i+j}\{O(\alpha^{-j})+O(\alpha^{-s_1})\}^{1/2}\\
&\quad \times\{O(\alpha^{-r})+O(\alpha^{-s_2})\}^{1/2},\ \text{for any } s_1, s_2>0.
\end{aligned}$$

Choosing s_1, s_2 such that $s_1 > j$ and $s_2 > r$, we have $|I_9| = \varepsilon \sum_{\substack{2i+j\leq r\\ i,j\geq 0}} \alpha^{i+j} O(\alpha^{-j/2})O(\alpha^{-r/2}) = \varepsilon.O(1)$.

Since $\varepsilon > 0$ is arbitrary, $I_9 \to 0$ as $\alpha \to \infty$.
Now, we estimate I_{10}. By applying Cauchy–Schwarz inequality, Lemma 1 and Corollary 2, we obtain

$$\begin{aligned}
|I_{10}| &\leq MK \sum_{k=1}^{\infty} \sum_{\substack{2i+j\leq r\\ i,j\geq 0}} \alpha^i |k-\alpha x|^j p_{\alpha,k}(x,c) \int_{|t-x|\geq\delta} \theta_{\alpha,k}^{\rho}(t)e^{\gamma t}dt\\
&\leq M_1 \sum_{\substack{2i+j\leq r\\ i,j\geq 0}} \alpha^{i+j}\left(\sum_{k=1}^{\infty}\left(\frac{k}{\alpha}-x\right)^{2j} p_{\alpha,k}(x,c)\right)^{1/2}\\
&\quad \times\left(\sum_{k=1}^{\infty} p_{\alpha,k}(x,c)\int_{|t-x|\geq\delta} \theta_{\alpha,k}^{\rho}(t)e^{2\gamma t}dt\right)^{1/2},\ \text{where } M_1 = MK\\
&\leq M_1 \sum_{\substack{2i+j\leq r\\ i,j\geq 0}} \alpha^{i+j}\left(\upsilon_{\alpha,2j}(x)-x^{2j}\phi_{\alpha,c}(x)\right)^{1/2}
\end{aligned}$$

$$\times\left(\sum_{k=1}^{\infty} p_{\alpha,k}(x,c)\int\limits_{|t-x|\geq\delta}\theta_{\alpha,k}^{\rho}(t)e^{2\gamma t}dt\right)^{1/2}$$

$$=\sum_{\substack{2i+j\leq r\\ i,j\geq 0}}\alpha^{i+j}\{O(\alpha^{-j})+O(\alpha^{-m_1})\}^{1/2}$$

$$\times\{O(\alpha^{-m_2})\}^{1/2},\ \text{for any } m_1, m_2>0.$$

Choosing $m_1 > j$, we get

$$|I_{10}|=\sum_{\substack{2i+j\leq r\\ i,j\geq 0}}\alpha^{i+j}O(\alpha^{-j/2})O(\alpha^{-m_2/2})=O(\alpha^{(r-m_2)/2}),$$

which implies that $I_{10}=o(1)$, as $\alpha\to\infty$, on choosing $m_2>r$. Next, we estimate I_8. We may write

$$|I_8|=\left|\left(\frac{d^r}{d\omega^r}p_{\alpha,0}(\omega,c)\right)_{\omega=x}\right||\psi(0,x)|x^r$$

$$=|\phi_{\alpha,c}^{(r)}(x)|\ |\psi(0,x)|x^r.$$

Now, we observe that $\phi_{\alpha,0}^{(r)}(x)=e^{-\alpha x}(-\alpha)^r$ and $\phi_{\alpha,1}^{(r)}(x)=\dfrac{(-1)^r(\alpha)_r}{(1+x)^{\alpha+r}}$, which implies that $I_8=O(\alpha^{-p})$ for any $p>0$, in view of the fact that $|\psi(0,x)x^r|\leq N_1$, for some $N_1>0$.

By combining the estimates I_7-I_{10}, we obtain $I_2\to 0$ as $\alpha\to\infty$.

To prove the uniformity assertion, it is sufficient to remark that $\delta(\varepsilon)$ in the above proof can be chosen to be independent of $x\in[a,b]$ and also that the other estimates hold uniformity in $x\in[a,b]$. This completes the proof of the theorem.

Next, we establish an asymptotic formula.

Theorem 2 (Voronovskaja type result) *Let $f\in C_\gamma[0,\infty)$. If f admits a derivative of order $(r+2)$ at a fixed point $x\in(0,\infty)$, then we have*

$$\lim_{\alpha\to\infty}\alpha\left(\left(\frac{d^r}{d\omega^r}B_\alpha^\rho(f;\omega)\right)_{\omega=x}-f^{(r)}(x)\right)=\sum_{\nu=1}^{r+2}Q(\nu,r,c,a,x)f^{(\nu)}(x),\qquad(5)$$

where $Q(\nu,r,c,a,x)$ are certain rational functions of x independent of α. Further, if $f^{(r+2)}$ is continuous on $(a-\eta,b+\eta)$, $\eta>0$, then the limit in (5) holds uniformly in $[a,b]$.

Proof From the Taylor's theorem, for $t \in [0, \infty)$ we may write

$$f(t) = \sum_{\nu=0}^{r+2} \frac{f^{(\nu)}(x)}{\nu!}(t-x)^{\nu} + \psi(t,x)(t-x)^{r+2}, \tag{6}$$

where the function $\psi(t,x) \to 0$ as $t \to x$.
Now, from Eq. (6), we have

$$\begin{aligned}
\left(\frac{d^r}{d\omega^r} B_\alpha^\rho(f(t);\omega)\right)_{\omega=x} &= \sum_{\nu=0}^{r+2} \frac{f^{(\nu)}(x)}{\nu!} \left(\frac{d^r}{dw^r}(B_\alpha^\rho((t-x)^\nu;\omega)\right)_{\omega=x} \\
&\quad + \left(\frac{d^r}{d\omega^r} B_\alpha^\rho(\psi(t,x)(t-x)^{r+2};\omega)\right)_{\omega=x} \\
&= \sum_{\nu=0}^{r+2} \frac{f^{(\nu)}(x)}{\nu!} \sum_{j=0}^{\nu} \binom{\nu}{j} (-x)^{\nu-j} \left(\frac{d^r}{d\omega^r} B_\alpha^\rho(t^j;\omega)\right)_{\omega=x} \\
&\quad + \left(\frac{d^r}{d\omega^r} B_\alpha^\rho(\psi(t,x))(t-x)^{r+2};\omega\right)_{\omega=x} \\
&:= J_1 + J_2, \ say.
\end{aligned}$$

Proceeding in a manner similar to the estimate of I_2 in Theorem 1, for each $x \in (0, \infty)$ we get $\alpha J_2 \to 0$ as $\alpha \to \infty$.

Next, we estimate J_1.

$$\begin{aligned}
J_1 &= \sum_{\nu=0}^{r-1} \frac{f^{(\nu)}(x)}{\nu!} \sum_{j=0}^{\nu} \binom{\nu}{j} (-x)^{\nu-j} \left(\frac{d^r}{d\omega^r} B_\alpha^\rho(t^j;\omega)\right)_{\omega=x} \\
&\quad + \frac{f^{(r)}(x)}{r!} \sum_{j=0}^{r} \binom{r}{j} (-x)^{r-j} \left(\frac{d^r}{d\omega^r} B_\alpha^\rho(t^j;\omega)\right)_{\omega=x} \\
&\quad + \frac{f^{(r+1)}(x)}{(r+1)!} \sum_{j=0}^{r+1} \binom{r+1}{j} (-x)^{r+1-j} \left(\frac{d^r}{d\omega^r} B_\alpha^\rho(t^j;\omega)\right)_{\omega=x} \\
&\quad + \frac{f^{(r+2)}(x)}{(r+2)!} \sum_{j=0}^{r+2} \binom{r+2}{j} (-x)^{r+2-j} \left(\frac{d^r}{d\omega^r} B_\alpha^\rho(t^j;\omega)\right)_{\omega=x}.
\end{aligned}$$

Making use of Lemma 2, we have

$$J_1 = f^{(r)}(x) + \alpha^{-1}\left(\sum_{\nu=1}^{r+2} Q(\nu, r, c, a, x) f^{(\nu)}(x) + o(1)\right).$$

Thus, from the estimates of J_1 and J_2, the required result follows.

The uniformity assertion follows as in the proof of Theorem 1. This completes the proof.

The next result provides an estimate of the degree of approximation in $B_\alpha^{\rho(r)}(f;x) \to f^{(r)}(x), r \in \mathbb{N}$.

Theorem 3 (Degree of approximation) *Let $r \le q \le r+2$, $f \in C_\gamma[0,\infty)$ and $f^{(q)}$ exist and be continuous on $(a-\eta, b+\eta)$ where $\eta > 0$ is sufficiently small. Then, for sufficiently large α*

$$\left\| \left(\frac{d^r}{d\omega^r} B_\alpha^\rho(f;\omega)\right)_{\omega=x} - f^{(r)}(x) \right\|_{C[a,b]} \le \max\{C_1 \alpha^{-(q-r)/2} \omega_{f^{(q)}}(\alpha^{-1/2}, (a-\eta, b+\eta)), C_2\, \alpha^{-1}\},$$

where $C_1 = C_1(r,c)$ and $C_2 = C_2(r,f,c)$.

Proof By our hypothesis we have,

$$f(t) = \sum_{i=0}^{q} \frac{f^{(i)}(x)}{i!}(t-x)^i + \frac{f^{(q)}(\xi) - f^{(q)}(x)}{q!}(t-x)^q \chi(t) + \phi(t,x)(1-\chi(t)), \tag{7}$$

where ξ lies between t and x and $\chi(t)$ is the characteristic function of $(a-\eta, b+\eta)$. The function $\phi(t,x)$ for $t \in [a,b]$ is bounded by $Me^{\gamma t}$ for some constant $M > 0$.

We operate $\frac{d^r}{d\omega^r} B_\alpha^\rho(.;\omega)$ on the equality (7) and break the right-hand side into three parts E_1, E_2 and E_3, say, corresponding to the three terms on the right-hand side of Eq. (7).

Now, treating E_1 in a manner similar to the treatment of J_1 of Theorem 2, we get $E_1 = f^{(r)}(x) + O(\alpha^{-1})$, uniformly in $x \in [a,b]$.
Making use of the inequality

$$|f^{(q)}(\xi) - f^{(q)}(x)| \le \left(1 + \frac{|t-x|}{\delta}\right)\omega_{f^{(q)}}(\delta),\ \delta > 0,$$

and Lemma 4, we get

$$|E_2| \le \frac{\omega_{f^{(q)}}(\delta)}{q!} \left\{ \sum_{k=1}^{\infty} \sum_{\substack{2i+j \le r \\ i,j \ge 0}} \frac{\alpha^i |k-\alpha x|^j |q_{i,j,r}(x,c)|}{(p(x,c))^r} p_{\alpha,k}(x,c) \right.$$
$$\times \int_0^\infty \theta_{\alpha,k}^\rho(t) \left(1 + \frac{|t-x|}{\delta}\right)|t-x|^q \chi(t)dt$$

$$+\left(x^q+\frac{x^{q+1}}{\delta}\right)\phi_{\alpha,c}^{(r)}(x)\Bigg\}$$
$$= E_4 + E_5.$$

Finally, let

$$S_1 = \sup_{x\in[a,b]} \sup_{\substack{2i+j\leq r\\ i,j\geq 0}} \frac{|q_{i,j,r}(x,c)|}{(p(x,c))^r},$$

then by applying Schwarz inequality, Lemmas 1 and 3, we obtain

$$\begin{aligned}
E_4 &\leq \frac{\omega_{f^{(q)}}(\delta)S_1}{q!}\sum_{\substack{2i+j\leq r\\ i,j\geq 0}}\alpha^{i+j}\left(\sum_{k=1}^{\infty}\left(\frac{k}{\alpha}-x\right)^{2j}p_{\alpha,k}(x,c)\right)^{1/2}\\
&\quad\times\left\{\left(\sum_{k=1}^{\infty}p_{\alpha,k}(x,c)\int_0^{\infty}\theta_{\alpha,k}^{\rho}(t)(t-x)^{2q}dt\right)^{1/2}\right.\\
&\quad\left.+\frac{1}{\delta}\left(\sum_{k=1}^{\infty}p_{\alpha,k}(x,c)\int_0^{\infty}\theta_{\alpha,k}^{\rho}(t)(t-x)^{2q+2}dt\right)^{1/2}\right\}\\
&\leq \omega_{f^{(q)}}(\delta)S_1\sum_{\substack{2i+j\leq r\\ i,j\geq 0}}\alpha^{i+j}\left(\upsilon_{\alpha,2j}(x)-x^{2j}\phi_{\alpha,c}(x)\right)^{1/2}\\
&\quad\times\left\{\left(B_{\alpha}^{\rho}((t-x)^{2q};x)-x^{2q}\phi_{\alpha,c}(x)\right)^{1/2}\right.\\
&\quad\left.+\frac{1}{\delta}\left(B_{\alpha}^{\rho}((t-x)^{2q+2};x)-x^{2q+2}\phi_{\alpha,c}(x)\right)^{1/2}\right\}\\
&= \omega_{f^{(q)}}(\delta)\sum_{\substack{2i+j\leq r\\ i,j\geq 0}}\alpha^{i+j}\{O(\alpha^{-j})+O(\alpha^{-s_1})\}^{1/2}\\
&\quad\times\{(O(\alpha^{-q})+O(\alpha^{-s_2})\}^{1/2}+\frac{1}{\delta}\{(O(\alpha^{-(q+1)})\\
&\quad+O(\alpha^{-s_3}))\}^{1/2},\quad \textit{for any } s_1,s_2,s_3>0.
\end{aligned}$$

Choosing s_1, s_2, s_3 such that $s_1 > j, s_2 > q, s_3 > q + 1$, we have

$$|E_4| = \omega_{f^{(q)}}(\delta)\sum_{\substack{2i+j\leq r\\ i,j\geq 0}}\alpha^{i+j}O\left(\frac{1}{\alpha^{j/2}}\right)\left\{O\left(\frac{1}{\alpha^{q/2}}\right)+\frac{1}{\delta}O\left(\frac{1}{\alpha^{(q+1)/2}}\right)\right\}.$$

Now, on choosing $\delta = \alpha^{-1/2}$, we get

$$|E_4| \leq C_1 \alpha^{-(q-r)/2} \omega_{f^{(q)}}(\alpha^{-1/2}, (a-\eta, b+\eta)).$$

Next, proceeding in a manner similar to the estimate of I_8 in Theorem 1, we have $E_5 = O(\alpha^{-p})$, for any $p > 0$. Choosing $p > 1$, we have $E_5 = O(\alpha^{-1})$, as $\alpha \to \infty$. Finally, proceeding along the lines of the estimate of I_{10} of Theorem 2, we obtain $E_3 = o(\alpha^{-1})$ as $\alpha \to \infty$.

On combining the estimates of $E_1 - E_5$, we get the required result.

3.2 Rate of Convergence

In this section, we shall estimate the rate of convergence for the generalized hybrid operators B_α^ρ for functions with derivatives of bounded variation. In recent years, several researchers have obtained results in this direction for different sequences of linear positive operators. We refer the reader to some of the related papers (cf. [1, 4, 6, 7, 9], etc.).

Let $f \in DBV_\gamma[0, \infty)$, $\gamma \geq 0$ be the class of all functions defined on $[0, \infty)$, having a derivative that coincides, a.e. with a function of bounded variation on every finite subinterval of $[0, \infty)$ and $|f(t)| \leq Mt^\gamma$, $\forall\ t > 0$.
It turns out that for $f \in DBV_\gamma[0, \infty)$, we may write

$$f(x) = \int_0^x g(t)dt + f(0),$$

where $g(t)$ is a function of bounded variation on each finite subinterval of $[0, \infty)$.

Lemma 5 *For all $x \in (0, \infty)$, $\lambda > 1$ and α sufficiently large, we have*

$$(i)\ \lambda_\alpha^\rho(x, t) = \int_0^t K_\alpha^\rho(x, u)du \leq \frac{1}{(x-t)^2}\frac{\lambda x(1+cx)}{\alpha},\ 0 \leq t < x;$$

$$(ii)\ 1 - \lambda_\alpha^\rho(x, z) = \int_z^\infty K_\alpha^\rho(x, u)du \leq \frac{1}{(z-x)^2}\frac{\lambda x(1+cx)}{\alpha},\ x < z < \infty.$$

Proof First we prove (i).

$$\lambda_\alpha^\rho(x,t) = \int_0^t K_\alpha^\rho(x,u)du \leq \int_0^t \left(\frac{x-u}{x-t}\right)^2 K_\alpha^\rho(x,u)du$$
$$\leq \frac{1}{(x-t)^2} B_\alpha^\rho((u-x)^2;x)$$
$$\leq \frac{1}{(x-t)^2}\frac{\lambda x(1+cx)}{\alpha}.$$

The proof of (ii) is similar.

Theorem 4 *Let $f \in DBV_\gamma[0,\infty)$, $\gamma \geq 0$. Then for every $x \in (0,\infty)$, $r(\in \mathbb{N}) > 2\gamma$ and sufficiently large α, we have*

$$|B_\alpha^\rho(f;x) - f(x)| \leq \left|\frac{f'(x+)-f'(x-)}{2}\right|\left\{\frac{\lambda x(1+cx)}{\alpha}\right\}^{1/2}$$
$$+\frac{x}{\sqrt{\alpha}}\bigvee_{x-\frac{x}{\sqrt{\alpha}}}^{x+\frac{x}{\sqrt{\alpha}}}(f_x') + \frac{\lambda(1+cx)}{\alpha}\sum_{m=1}^{\sqrt{[\alpha]}}\bigvee_{x-\frac{x}{m}}^{x+\frac{x}{m}}(f_x')$$
$$+|f'(x+)|\left\{\frac{\lambda x(1+cx)}{\alpha}\right\}^{1/2}$$
$$+|f(2x)-f(x)-xf'(x+)|\frac{\lambda(1+cx)}{\alpha x}$$
$$+M'\frac{A(r,x)}{\alpha^{\gamma/2}} + |f(x)|\frac{\lambda(1+cx)}{\alpha x},$$

where

$$f_x'(t) = \begin{cases} f'(t)-f'(x+), & x<t<\infty \\ 0 & t=x \\ f'(t)-f'(x-), & 0\leq t<x \end{cases},$$

$\bigvee_a^b(f'(x))$ is the total variation of f_x' on $[a,b]$, $A(r,x)$ is a constant depending on r and x and M' is a constant depending on f and γ.

Proof By the hypothesis, we may write

$$f'(t) = \frac{1}{2}\Big(f'(x+)+f'(x-)\Big) + f_x'(t)$$
$$+\frac{1}{2}\Big(f'(x+)-f'(x-)\Big)sgn(t-x)$$
$$+\delta_x(t)\left(f'(t)-\frac{1}{2}\Big(f'(x+)+f'(x-)\Big)\right), \tag{8}$$

where

$$\delta_x(t) = \begin{cases} 1 & t = x \\ 0 & t \neq x. \end{cases}$$

From Eqs. (2) and (8), we have

$$\begin{aligned} B_\alpha^\rho(f;x) - f(x) &= \int_0^\infty K_\alpha^\rho(x,t) f(t)dt - f(x) = \int_0^\infty (f(t) - f(x)) K_\alpha^\rho(x,t)dt \\ &= \int_0^x (f(t) - f(x)) K_\alpha^\rho(x,t)dt + \int_x^\infty (f(t) - f(x)) K_\alpha^\rho(x,t)dt \\ &= -\int_0^x \left(\int_t^x f'(u)du \right) K_\alpha^\rho(x,t)dt + \int_x^\infty \left(\int_x^t f'(u)du \right) K_\alpha^\rho(x,t)dt \\ &= I_1(x) + I_2(x), \ say. \end{aligned}$$

Using Eq. (8), we get

$$\begin{aligned} I_1(x) = \int_0^x \Bigg\{ \int_t^x \frac{1}{2}\Big(f'(x+) + f'(x-)\Big) + f'_x(u) + \frac{1}{2}\Big(f'(x+) - f'(x-)\Big) sgn(u - x) \\ + \delta_x(u)\Big(f'(u) - \frac{1}{2}\Big(f'(x+) + f'(x-)\Big)\Big)du \Bigg\} K_\alpha^\rho(x,t)dt. \end{aligned}$$

Since $\int_x^t \delta_x(u)du = 0$, we have

$$\begin{aligned} I_1(x) = \frac{1}{2}\Big(f'(x+) + f'(x-)\Big) \int_0^x (x - t) K_\alpha^\rho(x,t)dt + \int_0^x \left(\int_x^t f'_x(u)du \right) K_\alpha^\rho(x,t)dt \\ -\frac{1}{2}\Big(f'(x+) - f'(x-)\Big) \int_0^x |x - t| K_\alpha^\rho(x,t)dt. \end{aligned} \tag{9}$$

Proceeding similarly, we find that

$$\begin{aligned} I_2(x) = \frac{1}{2}\Big(f'(x+) + f'(x-)\Big) \int_x^\infty (t - x) K_\alpha^\rho(x,t)dt + \int_x^\infty \left(\int_x^t f'_x(u)du \right) K_\alpha^\rho(x,t)dt \\ +\frac{1}{2}\Big(f'(x+) - f'(x-)\Big) \int_x^\infty |t - x| K_\alpha^\rho(x,t)dt. \end{aligned} \tag{10}$$

By combining (9) and (10), we get

$$\begin{aligned} B_\alpha^\rho(f;x) - f(x) &= \frac{1}{2}\Big(f'(x+) + f'(x-)\Big)\int_0^\infty (t-x)K_\alpha^\rho(x,t)dt \\ &+ \frac{1}{2}\Big(f'(x+) - f'(x-)\Big)\int_0^\infty |t-x|K_\alpha^\rho(x,t)dt \\ &- \int_0^x \Big(\int_t^x f_x'(u)du\Big)K_\alpha^\rho(x,t)dt + \int_x^\infty \Big(\int_x^t f_x'(u)du\Big)K_\alpha^\rho(x,t)dt. \end{aligned}$$

Hence

$$\begin{aligned} |B_\alpha^\rho(f;x) - f(x)| &\le \left|\frac{f'(x+) + f'(x-)}{2}\right| |B_\alpha^\rho(t-x;x)| + \left|\frac{f'(x+) - f'(x-)}{2}\right| B_\alpha^\rho(|t-x|;x) \\ &+ \left|\int_0^x \Big(\int_t^x f_x'(u)du\Big)K_\alpha^\rho(x,t)dt\right| + \left|\int_x^\infty \Big(\int_x^t f_x'(u)du\Big)K_\alpha^\rho(x,t)dt\right|. \end{aligned} \tag{11}$$

On application of Lemma 5 and integration by parts, we obtain

$$\int_0^x \Big(\int_t^x f_x'(u)du\Big)K_\alpha^\rho(x,t)dt = \int_0^x \Big(\int_t^x f_x'(u)du\Big)\frac{\partial}{\partial t}\lambda_\alpha^\rho(x,t)dt = \int_0^x f_x'(t)\lambda_\alpha^\rho(x,t)dt.$$

Thus,

$$\begin{aligned} \left|\int_0^x \Big(\int_t^x f_x'(u)du\Big)K_\alpha^\rho(x,t)dt\right| &\le \int_0^x |f_x'(t)|\lambda_\alpha^\rho(x,t)dt \\ &\le \int_0^{x-\frac{x}{\sqrt{\alpha}}} |f_x'(t)|\lambda_\alpha^\rho(x,t)dt + \int_{x-\frac{x}{\sqrt{\alpha}}}^x |f_x'(t)|\lambda_\alpha^\rho(x,t)dt. \end{aligned}$$

Since $f_x'(x) = 0$ and $\lambda_\alpha^\rho(x,t) \le 1$, we get

$$\int_{x-\frac{x}{\sqrt{\alpha}}}^{x} |f_x'(t)|\lambda_\alpha^\rho(x,t)dt = \int_{x-\frac{x}{\sqrt{\alpha}}}^{x} |f_x'(t) - f_x'(x)|\lambda_\alpha^\rho(x,t)dt \le \int_{x-\frac{x}{\sqrt{\alpha}}}^{x} \bigvee_t^x (f_x')dt$$

$$\le \bigvee_{x-\frac{x}{\sqrt{\alpha}}}^{x} (f_x') \int_{x-\frac{x}{\sqrt{\alpha}}}^{x} dt = \frac{x}{\sqrt{\alpha}} \bigvee_{x-\frac{x}{\sqrt{\alpha}}}^{x} (f_x').$$

Similarly, using Lemma 5 and putting $t = x - \frac{x}{u}$, we get

$$\int_0^{x-\frac{x}{\sqrt{\alpha}}} |f_x'(t)|\lambda_\alpha^\rho(x,t)dt \le \frac{\lambda x(1+cx)}{\alpha} \int_0^{x-\frac{x}{\sqrt{\alpha}}} |f_x'(t)| \frac{dt}{(x-t)^2}$$

$$\le \frac{\lambda x(1+cx)}{\alpha} \int_0^{x-\frac{x}{\sqrt{\alpha}}} \bigvee_t^x (f_x') \frac{dt}{(x-t)^2}$$

$$= \frac{\lambda(1+cx)}{\alpha} \int_1^{\sqrt{\alpha}} \bigvee_{x-\frac{x}{u}}^{x} (f_x')du \le \frac{\lambda(1+cx)}{\alpha} \sum_{m=1}^{[\sqrt{\alpha}]} \bigvee_{x-\frac{x}{m}}^{x} (f_x').$$

Consequently,

$$\left|\int_0^x \left(\int_t^x f_x'(u)du\right) K_\alpha^\rho(x,t)dt\right| \le \frac{x}{\sqrt{\alpha}} \bigvee_{x-\frac{x}{\sqrt{\alpha}}}^{x} (f_x') + \frac{\lambda(1+cx)}{\alpha} \sum_{m=1}^{[\sqrt{\alpha}]} \bigvee_{x-\frac{x}{m}}^{x} (f_x'). \tag{12}$$

Also, we have

$$\left|\int_x^\infty \left(\int_x^t f_x'(u)du\right) K_\alpha^\rho(x,t)dt\right| \le \left|\int_x^{2x} \left(\int_x^t f_x'(u)du\right) \frac{\partial}{\partial t}(1 - \lambda_\alpha^\rho(x,t))dt\right|$$

$$+ \left|\int_{2x}^\infty \left(\int_x^t f_x'(u)du\right) K_\alpha^\rho(x,t)dt\right|$$

$$\le \left|\int_{2x}^\infty (f(t) - f(x)) K_\alpha^\rho(x,t)dt\right|$$

$$+ |f'(x+)| \left|\int_{2x}^\infty (t-x) K_\alpha^\rho(x,t)dt\right|$$

$$+\left|\int_x^{2x} f_x'(u)du\right||1-\lambda_\alpha^\rho(x,2x)|$$
$$+\int_x^{2x} |f_x'(t)|(1-\lambda_\alpha^\rho(x,t))dt.$$

Applying Lemma 5, we get

$$\begin{aligned}\left|\int_x^\infty\left(\int_x^t f_x'(u)du\right)K_\alpha^\rho(x,t)dt\right| &\le M\int_{2x}^\infty t^\gamma K_\alpha^\rho(x,t)dt + |f(x)|\int_{2x}^\infty K_\alpha^\rho(x,t)dt\\ &+|f'(x+)|\left\{\frac{\lambda x(1+cx)}{\alpha}\right\}^{1/2}\\ &+\frac{\lambda(1+cx)}{\alpha x}|f(2x)-f(x)-xf'(x+)|\\ &+\frac{x}{\sqrt{\alpha}}\bigvee_x^{x+\frac{x}{\sqrt{\alpha}}}(f_x')+\frac{\lambda(1+cx)}{\alpha}\sum_{m=1}^{[\sqrt{\alpha}]}\bigvee_x^{x+\frac{x}{m}}(f_x').\end{aligned} \tag{13}$$

We note that we can choose $r\in\mathbb{N}$ such that $2r>\gamma$.

Since $t\le 2(t-x)$ and $x\le t-x$ when $t\ge 2x$, using Hölder's inequality and Lemma 3, we obtain

$$M\int_{2x}^\infty t^\gamma K_\alpha^\rho(x,t)dt + |f(x)|\int_{2x}^\infty K_\alpha^\rho(x,t)dt$$

$$\begin{aligned}&\le 2^\gamma M\int_{2x}^\infty (t-x)^\gamma K_\alpha^\rho(x,t)dt + \frac{|f(x)|}{x^2}\int_{2x}^\infty (t-x)^2K_\alpha^\rho(x,t)dt\\ &\le 2^\gamma M\left(\int_0^\infty (t-x)^{2r}K_\alpha^\rho(x,t)dt\right)^{\gamma/2r} + |f(x)|\frac{\lambda(1+cx)}{\alpha x}\\ &\le M'\frac{A(r,x)}{\alpha^{\gamma/2}} + |f(x)|\frac{\lambda(1+cx)}{\alpha x},\ \text{where } M'=2^\gamma M.\end{aligned} \tag{14}$$

Using Lemma 3 and combining (11), (12), (13) and (14), we get the required result.

Acknowledgments The authors are extremely grateful to the reviewers for careful reading of the manuscript and for making valuable comments leading to better presentation of the paper. The first author is thankful to the "Council of Scientific and Industrial Research" India for financial support to carry out the above research work.

References

1. Acar, T., Gupta, V., Aral, A.: Rate of convergence for generalized Szasz operators. Bull. Math. Sci. **1**(1), 99–113 (2011)
2. Gupta, V.: A new class of Durrmeyer operators. Adv. Stud. Contemp. Math. (Kyungshang) **23**(2), 219–224 (2013)
3. Gupta, V., Agrawal, P.N., Goyal, M.: Approximation by certain genuine hybrid operators. Positivity (Under review)
4. Gupta, V., Agarwal, R.P.: Convergence Estimates in Approximation Theory. Springer, New York (2014)
5. Gupta, V., Rassias, ThM: Lupaş Durmeyer operators based on Polya distribution. Banach J. Math. Anal. **8**(2), 146–155 (2014)
6. Ispir, N.: Rate of convergence of generalized rational type Baskakov operators. Math. Comput. Model. **46**, 625–631 (2007)
7. Karsli, H.: Rate of convergence of new gamma type operators for functions with derivatives of bounded variation. Math. Comput. Model. **45**, 617–624 (2007)
8. Kasana, H.S., Prasad, G., Agrawal, P.N., Sahai, A.: Modified Szasz operators. In: Proceedings of the International Conference on Mathematical Analysis and its Applications, Kuwait (1985)
9. Özarslan, M.A., Duman O., Kaanoğlu, C.: Rates of convergence of certain King-type operators for functions with derivative of bounded variation. Math. Comput. Model. **52**, 334–345 (2010)
10. Păltănea, R.: Modified Szász-Mirakjan operators of integral form. Carpathian J. Math. **24**, 378–385 (2008)
11. Păltănea, R.: Simultaneous approximation by a class of Szász-Mirakjan operators. J. Appl. Funct. Anal. **9**(3–4), 356–368 (2014)
12. Sinha, R.P., Agrawal, P.N., Gupta, V.: On simultaneous approximation by modified Baskakov operators. Bull. Soc. Math. Belg. **43**, 217–231 (1991)

Frames in Semi-inner Product Spaces

N.K. Sahu and Ram N. Mohapatra

Abstract The objective of this paper is to study the theory of frames in semi-inner product spaces. Several researchers have studied frames in Banach spaces by using the bounded linear functionals. Application of semi-inner product is a new approach to investigate the theory of frames. The notion of semi-frame is introduced in this new aspect.

Keywords Frames · Semi-frames · Semi-inner product

1 Introduction

The theory of frames plays a fundamental role in signal processing, image processing, data compression, sampling theory and has found considerable applications in many more fields. Mathematically, the frame is equivalent to a spanning set in a vector space, but it may not be minimal. It may have more number of vectors than a basis. One of the main advantages in using frames in signal transmission over a basis is that if in the process of transmission, signal along a frame is lost, it is possible to reconstruct completely due to the built in redundancy which is not possible while using a basis. In applications one determines "optimal frames with erasers" (see Han and Sun [12], Pehlivan et al. [15] and the references there in).

The main objective of this paper is to describe Frames in Hilbert space, Banach space, Hilbert C^*-module and then define semi-inner product space, and Frames and semi-frames in that context. The advantage of using semi-inner product is to facilitate

N.K. Sahu (✉)
Dhirubhai Ambani Institute of Information and Communication Technology, Gandhinagar, India
e-mail: nabindaiict@gmail.com

R.N. Mohapatra
University of Central Florida, Orlando, FL 32816, USA
e-mail: ramm@mail.ucf.edu

P.N. Agrawal et al. (eds.), *Mathematical Analysis and its Applications*,
Springer Proceedings in Mathematics & Statistics 143,
DOI 10.1007/978-81-322-2485-3_11

calculations in uniformly convex smooth Banach spaces and obtain results that pose difficulties.

Frames in Hilbert Spaces

Frames for a Hilbert space were formally defined by Duffin and Schaeffer [5] in 1952. Frames in Hilbert space have been well investigated. For Hilbert space frames one can refer to Christensen [3] and the references there in.

Definition 1 Let H be a Hilbert space and I be an index set. A sequence $\{f_i\}_{i\in I}$ of elements in H is called a Bessel sequence for H if there exists a real constant $B > 0$ such that

$$\sum_{i\in I} |\langle f, f_i\rangle|^2 \leq B\|f\|^2 \text{ for all } f \in H. \tag{1}$$

B is called the Bessel bound for the Bessel sequence $\{f_i\}$.

Definition 2 A sequence $\{f_i\}_{i\in I}$ of elements in a Hilbert space H is called a frame for H, if there exist real constants A, B with $0 < A \leq B < \infty$ such that

$$A\|f\|^2 \leq \sum_{i\in I} |\langle f, f_i\rangle|^2 \leq B\|f\|^2 \text{ for all } f \in H. \tag{2}$$

Here A and B are called lower and upper frame bounds, respectively.

The largest number A and the smallest number B satisfying the frame inequality (2) for all $f \in H$ are called optimal frame bounds. If $A = B$, we call the frame $\{f_i\}$, a tight frame. When $A = B = 1$, the frame is called a Parseval frame. If all the frame elements have the same norm, then the frame is called equal norm frame, and if all the frame elements are of unit norm, then it is called unit norm frame. A frame is exact if it ceases to be a frame when any one of its element is removed. A frame is exact if and only if it is a Riesz basis. A non-exact frame is called over complete in the sense that if at least one vector is removed, the remaining ones still constitute a frame.

Frames in Banach Spaces

While constructing frames in Hilbert space we need the sequence space l^2. Similarly, while constructing frames in Banach space one needs a Banach space of scalar-valued sequences (BK-space).

A Banach space of scalar valued sequences (BK-space) is a linear space of sequences with a norm which makes it a Banach space and for which the coordinate functionals are continuous. In a BK-space, the unit vectors are defined by $e_i(j) = \delta_{ij}$ (Kronecker delta). Gröchenig [7] first generalized the concept of frames to Banach spaces and called them atomic decompositions.

Definition 3 Let X be a Banach space with norm $\|.\|_X$ and X_d be an associated BK-space with norm $\|.\|_{X_d}$. Let $\{f_i\}$ be a sequence of elements in X^*, the dual space of X, and $\{x_i\}$ be a sequence of elements in X. If

(i) $\{(x, f_i)\} \in X_d$, for all $x \in X$, where (x, f_i) denotes the value of the functional f_i at the point x,

(ii) the norms $\|x\|_X$ and $\|\{(x, f_i)\}\|_{X_d}$ are equivalent,

(iii) $x = \sum_i (x, f_i)x_i$, for all $x \in X$, then the pair $\big(\{f_i\}, \{x_i\}\big)$ is called an atomic decomposition of X with respect to X_d.

With a more general setting Gröchenig defined Banach frames as follows:

Definition 4 Let X be a Banach space with norm $\|.\|_X$ and X_d be an associated BK-space with norm $\|.\|_{X_d}$. Let $\{f_i\}$ be a sequence of elements in X^* and an operator $S : X_d \to X$ be given. If

(i) $\{(x, f_i)\} \in X_d$, for all $x \in X$,

(ii) the norms $\|x\|_X$ and $\|\{(x, f_i)\}\|_{X_d}$ are equivalent,

(iii) S is bounded and linear, and $S((x, f_i)) = x$ for each $x \in X$, then $\big(\{f_i\}, S\big)$ is a Banach frame for X with respect to X_d.

There is considerable research on frames in Banach spaces and for details on frames in Banach spaces one may refer to Christensen and Heil [4], Stoeva [17], Casazza and Christensen [2], Koushik [13].

Frames in Hilbert C^*-module

In recent years, many mathematicians generalized the frame theory in Hilbert spaces to frame theory in Hilbert C^*-modules and got significant results which enrich the theory of frames.

Definition 5 Let $\mathscr{A}$ be a unital C^*-algebra and $\mathbb{J}$ be a finite or countable index set. A sequence $\{x_j\}_{j\in\mathbb{J}}$ of elements in a Hilbert $\mathscr{A}$-module $\mathscr{H}$ is said to be a frame if there exists two real constants $A, B > 0$ such that

$$A\langle x, x\rangle \le \sum_{j\in\mathbb{J}} \langle x, x_j\rangle\langle x_j, x\rangle \le B\langle x, x\rangle \tag{3}$$

for every $x \in \mathscr{H}$. The optimal constants (maximal for A and minimal for B) are called frame bounds.

The frame $\{x_j\}_{j\in\mathbb{J}}$ is said to be a tight frame if $A = B$, and said to be a Parseval frame if $A = B = 1$.

Wu Jing in his doctoral dissertation to University of Central Florida gave an equivalent formulation of modular frames, and derived many interesting results. Details about these can be found in Han et al. [8–11].

Due to lack of inner product structure in general Banach spaces, people studied the theory of frames by taking the help of bounded linear functionals, that is by taking the help of the dual space. Many of the results on classical frame theory have been generalized to Banach spaces, in this way. The use of arbitrary bounded linear functionals is not always a convenient way to study these notions. It is also difficult to construct examples to verify the established theoretical results. So, in this paper we have put some effort to study the theory of frames in Banach spaces in a different way. We have seen frames being defined in semi-inner product spaces (see Zhang and Zhang [18]). We use the notion of semi-inner product to study more into the theory of frames. In the next section, we give a brief introduction of semi-inner product space. It is worth mentioning that this approach will result for frames in l^p and L^p spaces for $1 < p < \infty$.

2 Semi-inner Product Space

Definition 6 (Lumer [14]) Let X be a vector space over the field F of real or complex numbers. A functional $[.,.] : X \times X \to F$ is called a semi-inner product if it satisfies the following:

1. $[x + y, z] = [x, z] + [y, z], \quad \forall x, y, z \in X$;
2. $[\lambda x, y] = \lambda[x, y], \quad \forall \lambda \in F$ and $x, y \in X$;
3. $[x, x] > 0$, for $x \neq 0$;
4. $|[x, y]|^2 \leq [x, x][y, y]$.

   The pair $(X, [.,.])$ is called a semi-inner product space.

We observe that $\|x\| = [x, x]^{\frac{1}{2}}$ is a norm on X. Hence every semi-inner product space is a normed linear space. On the other hand, in a normed linear space, one can generate semi-inner product in infinitely many different ways. Giles [6] had proved that if the underlying space X is a uniformly convex smooth Banach space then it is possible to define a semi-inner product, uniquely. Also the unique semi-inner product has the following nice properties:

(i) $[x, y] = 0$ if and only if y is orthogonal to x, that is if and only if $\|y\| \leq \|y + \lambda x\|$, for all scalars λ.
(ii) Generalized Riesz representation theorem: If f is a continuous linear functional on X then there is a unique vector $y \in X$ such that $f(x) = [x, y]$, for all $x \in X$.
(iii) The semi-inner product is continuous, that is for each $x, y \in X$, we have $Re[y, x + \lambda y] \to Re[y, x]$ as $\lambda \to 0$.

The sequence space l^p, $p > 1$ and the function space L^p, $p > 1$ are uniformly convex smooth Banach spaces. So one can define semi-inner product on these spaces, uniquely.

Example 1 The real sequence space l^p for $1 < p < \infty$ is a semi-inner product space with the semi-inner product defined by

$$[x, y] = \frac{1}{\|y\|_p^{p-2}} \sum_i x_i y_i |y_i|^{p-2}, \quad x, y \in l^p.$$

Example 2 (Giles [6]) The real Banach space $L^p(X, \mu)$ for $1 < p < \infty$ is a semi-inner product space with the semi-inner product defined by

$$[f, g] = \frac{1}{\|g\|_p^{p-2}} \int_X f(x)|g(x)|^{p-1} \, sgn(g(x)) d\mu, \quad f, g \in L^p.$$

3 Frames in Semi-inner Product Spaces

Recently, Zhang and Zhang [18] investigated the theory of frames in Banach spaces by applying the notion of semi-inner product. They generalized the classical theory on frames and Riesz bases in this new perspective. They have defined frames in the following way:

Definition 7 Let X be a Banach space with a compatible semi-inner product $[., .]$ and norm $\|.\|_X$. Let X_d be an associated BK-space (sequence space with continuous coordinate linear functionals) with norm $\|.\|_{X_d}$. A sequence of elements $\{f_j\} \subseteq X$ is called an X_d-frame for X if $\{[f, f_j]\} \in X_d$, for all $f \in X$ and there exist two positive constants A, B such that

$$A\|f\|_X \leq \|\{[f, f_j]\}\|_{X_d} \leq B\|f\|_X \text{ for all } f \in X.$$

They have also defined frames for the dual space X^* of the Banach space X.

Definition 8 Let X be a Banach space with a compatible semi-inner product $[., .]$ and norm $\|.\|_X$. Let X^* be the dual space of X. Let X_d be an associated BK-space with norm $\|.\|_{X_d}$, and X_d^* be the dual space of X_d. A sequence of elements $\{f_j^*\} \subseteq X^*$ is an X_d^*-frame for X^* if $\{[f_j, f]\} \in X_d^*$, for all $f \in X$ and there exist two positive constants A, B such that

$$A\|f\|_X \leq \|\{[f_j, f]\}\|_{X_d^*} \leq B\|f\|_X \text{ for all } f \in X.$$

The notion of frame is too restrictive, in the sense that one cannot satisfy both upper and lower frame bounds simultaneously. Thus there is a scope for two natural generalizations, named as upper semi-frame and lower semi-frame. The notion of semi-frame in Hilbert space was studied by Antoine and Balazs [1]. In this paper we define the notion of semi-frame in Banach spaces by using the semi-inner product.

Definition 9 Let X be a Banach space with a compatible semi-inner product $[., .]$ and norm $\|.\|_X$. Let X_d be an associated BK-space with norm $\|.\|_{X_d}$. A sequence of elements $\{f_j\} \subseteq X$ is called upper semi-X_d-frame for X if

(i) $\{f_j\}$ is total in X;
(ii) $\{[f, f_j]\} \in X_d$, for all $f \in X$;
(iii) there exists a positive constant B such that

$$0 \leq \|\{[f, f_j]\}\|_{X_d} \leq B\|f\|_X \text{ for all } f \in X.$$

Definition 10 Let X be a Banach space with a compatible semi-inner product $[.,.]$ and norm $\|.\|_X$. Let X_d be an associated BK-space with norm $\|.\|_{X_d}$. A sequence of elements $\{f_j\} \subseteq X$ is called lower semi-X_d-frame for X if

(i) $\{f_j\}$ is total in X;
(ii) $\{[f, f_j]\} \in X_d$, for all $f \in X$;
(iii) there exists a positive constant A such that

$$A\|f\|_X \leq \|\{[f, f_j]\}\|_{X_d} \text{ for all } f \in X.$$

Similarly, we define upper semi-X_d^*-frame and lower semi-X_d^*-frame for the dual space X^*.

Definition 11 Let X be a Banach space with a compatible semi-inner product $[.,.]$ and norm $\|.\|_X$. Let X^* be the dual space of X. Let X_d be an associated BK-space with norm $\|.\|_{X_d}$, and X_d^* be the dual space of X_d. A sequence of elements $\{f_j^*\} \subseteq X^*$ is upper semi-X_d^*-frame for X^* if

(i) $\{f_j^*\}$ is total in X^*;
(ii) $\{[f_j, f]\} \in X_d^*$, for all $f \in X$;
(iii) there exists a positive constant B such that

$$0 \leq \|\{[f_j, f]\}\|_{X_d^*} \leq B\|f\|_X \text{ for all } f \in X.$$

Definition 12 Let X be a Banach space with a compatible semi-inner product $[.,.]$ and norm $\|.\|_X$. Let X^* be the dual space of X. Let X_d be an associated BK-space with norm $\|.\|_{X_d}$, and X_d^* be the dual space of X_d. A sequence of elements $\{f_j^*\} \subseteq X^*$ is lower semi-X_d^*-frame for X^* if

(i) $\{f_j^*\}$ is total in X^*;
(ii) $\{[f_j, f]\} \in X_d^*$, for all $f \in X$;
(iii) there exists a positive constant A such that

$$A\|f\|_X \leq \|\{[f_j, f]\}\|_{X_d^*} \text{ for all } f \in X.$$

Zhang and Zhang [18] established the reconstruction property for X_d-frame and X_d^*-frame in a semi-inner space X. They defined the operator (so-called analysis operator) $U : X \to X_d$ by $U(f) = \{[f, f_j]\}$. They proved that

Theorem 1 *([18]) If $\{f_j\}$ is an X_d-frame for X and $Rng(U)$ has an algebraic complement in X_d, then there exists an X_d^*-frame $\{g_j^*\}$ for X^* such that*

$$f = \sum_{j \in I} [f, f_j] g_j \text{ for all } f \in X$$

and

$$f^* = \sum_{j \in I} [g_j, f] f_j^* \quad \text{for all } f \in X.$$

Based on the above theorem, we formulate the following definition.

Definition 13 Let $\{f_j\}$ be an X_d-frame for X. If there exists an X_d^*-frame $\{g_j^*\}$ for X^* such that

$$f = \sum_{j \in I} [f, f_j] g_j \quad \text{for all } f \in X$$

and

$$f^* = \sum_{j \in I} [g_j, f] f_j^* \quad \text{for all } f \in X.$$

Then $\{f_j\}$ and $\{g_j^*\}$ are called dual frame pair.

Now we are in a position to propose the following theorem.

Theorem 2 *Let $\{f_j\}$ be an upper semi-X_d-frame for X with bound B. If $\{g_j^*\}$ is a sequence of element in X^* such that $\{f_j\}$ and $\{g_j^*\}$ are dual frame pair, then $\{g_j^*\}$ is a lower semi-X_d^*-frame for X^* with bound $\frac{1}{B}$.*

Proof Since $\{f_j\}$ is an upper semi-X_d-frame for X with bound B, we have

$$0 \leq \|\{[f, f_j]\}\|_{X_d} \leq B\|f\|_X \quad \text{for all } f \in X.$$

Now

$$\begin{aligned}
\|f\|_X^2 = [f, f] &= \left[\sum [f, f_j] g_j, f\right] \\
&= \sum [[f, f_j] g_j, f] \\
&= \sum [f, f_j][g_j, f] \\
&\leq \|\{[f, f_j]\}\|_{X_d} \|\{[g_j, f]\}\|_{X_d^*} \\
&\leq B\|f\|_X \|\{[g_j, f]\}\|_{X_d^*}
\end{aligned}$$

$$\Rightarrow \frac{1}{B}\|f\|_X \leq \|\{[g_j, f]\}\|_{X_d^*} \quad \text{for all } f \in X.$$

That is, $\{[g_j, f]\}$ is a lower semi-X_d^*-frame for X^*.

Similarly, we can easily prove the following theorem.

Theorem 3 *Let $\{g_j\}$ be an upper semi-X_d^*-frame for X^* with bound B. If $\{f_j\}$ is a sequence of element in X such that $\{f_j\}$ and $\{g_j^*\}$ are dual frame pair, then $\{f_j\}$ is a lower semi-X_d-frame for X with bound $\frac{1}{B}$.*

Proof Since $\{g_j\}$ is an upper semi-X_d^*-frame for X^* with bound B, we have

$$0 \leq \|\{[g_j, f]\}\|_{X_d^*} \leq B\|f\|_X \text{ for all } f \in X.$$

Now for any $f \in X$, we have

$$\begin{aligned} \|f\|_X^2 = [f, f] &= \left[\sum [f, f_j]g_j, f\right] \\ &= \sum [[f, f_j]g_j, f] \\ &= \sum [f, f_j][g_j, f] \\ &\leq \|\{[f, f_j]\}\|_{X_d} \|\{[g_j, f]\}\|_{X_d^*} \\ &\leq \|\{[f, f_j]\}\|_{X_d} B\|f\|_X \\ \Rightarrow \frac{1}{B}\|f\|_X &\leq \|\{[f, f_j]\}\|_{X_d} \text{ for all } f \in X. \end{aligned}$$

That is, $\{[f, f_j]\}$ is a lower semi-X_d-frame for X.

4 Frames for $l^p (1 < p < \infty)$ Spaces

In this section, we define frames in $l^p(1 < p < \infty)$ spaces.

We know that l^p $(1 < p < \infty)$ spaces are uniformly convex smooth Banach spaces. It is seen that those spaces are semi-inner product spaces with uniquely defined semi-inner product (see Giles [6]). For the rest of this section, we assume that X is the real sequence space $l^p(1 < p < \infty)$ with norm $\|.\|_p$ and semi-inner product $[.,.]$. The following definitions of Bessel sequence and frame can be found in Sahu and Nahak [16].

Definition 14 A set of elements $f = \{f_i\}_{i=1}^{\infty} \subseteq X$ is called a Bessel sequence if there exists a constant $B > 0$ such that

$$\sum_{i=1}^{\infty} |[f_i, x]|^q \leq B(\|x\|_p)^q, \ \forall x \in X,$$

where $1 < p, q < \infty$ and $\frac{1}{p} + \frac{1}{q} = 1$. The number B is called Bessel bound.

Definition 15 A sequence of elements $\{f_i\}_{i=1}^{\infty}$ in X is called a frame if there exist positive constants A and B such that

$$A(\|x\|_p)^q \leq \sum_{i=1}^{\infty} |[f_i, x]|^q \leq B(\|x\|_p)^q, \ \forall x \in X,$$

where $1 < p, q < \infty$ and $\frac{1}{p} + \frac{1}{q} = 1$. A and B are called lower and upper frame bounds, respectively.

If $A = B$ then the frame is called a tight frame, and if $A = B = 1$ then the frame is called a Parseval frame. A frame is called a normalized frame if each frame element has unit norm.

Example 3 Consider the set $\{e_i\}_{i=1}^{\infty} \in l^p$, where $e_i = (0, 0, ..., 1, 0, 0..)$, where 1 is at the ith coordinate and 0 at the other coordinates.

(i) $\{e_1, 0, e_2, 0, e_3, 0, ...\}$ is a Parseval frame.
(ii) $\{e_1, e_1, e_2, e_2, ...\}$ is a tight frame with bound 2.
(iii) $\left\{\frac{e_1}{\sqrt{2}}, \frac{e_1}{\sqrt{2}}, \frac{e_2}{\sqrt{2}}, \frac{e_2}{\sqrt{2}}, ...\right\}$ is a tight frame with bound $\frac{2}{(\sqrt{2})^{\frac{p}{p-1}}}$.
(iv) $\{ne_n\}_{n=1}^{\infty}$ is a lower semi-frame but not a frame.
(v) $\left\{\frac{1}{n}e_n\right\}_{n=1}^{\infty}$ is an upper semi-frame but not a frame.

Some of the classical frame theory results in Hilbert spaces can be generalized to l^p spaces in this new approach. The reconstruction formula naturally holds true for Parseval frames and tight frames. In this connection, we state the following two theorems.

Theorem 4 *A set of elements* $\{f_i\}_{i=1}^{\infty}$ *is a Parseval frame for* X *if and only if*

$$x = \sum_{i=1}^{\infty} \frac{|[f_i, x]|^{q-2}}{\|\{[f_i, x]\}\|^{q-2}} [f_i, x] f_i, \quad \forall x \in X. \tag{4}$$

Theorem 5 *A set of elements* $\{f_i\}_{i=1}^{\infty}$ *is a tight frame with bound* A *for* X *if and only if*

$$x = \sum_{i=1}^{\infty} \frac{1}{A^{\frac{2}{q}}} \frac{|[f_i, x]|^{q-2}}{\|\{[f_i, x]\}\|^{q-2}} [f_i, x] f_i \quad \forall x \in X. \tag{5}$$

Acknowledgments We thank the referees for their comments which have been used in the final presentation.

References

1. Antoine, J.P., Balazs, P.: Frames and semi-frames. J. Phys. A: Math. Theor. **44**(20), 205201 (2011)
2. Casazza, P.G., Christensen, O.: The reconstruction property in Banach spaces and perturbation theorem. Canad. Math. Bull. **51**, 348–358 (2008)
3. Christensen, O.: An Introduction to Frames and Riesz Bases. Birkhauser, Boston (2002)
4. Christensen, O., Heil, C.: Perturbations of Banach frames and atomic decompositions. Math. Nachr. **158**, 33–47 (1997)

5. Duffin, R.J., Schaeffer, A.C.: A class of nonharmonic Fourier series. Trans. Amer. Math. Soc. **72**, 341–366 (1952)
6. Giles, J.R.: Classes of semi-inner product spaces. Trans. Amer. Math. Soc. **129**, 436–446 (1967)
7. Gröchenig, K.: Localization of frames, Banach frames, and the invertibility of the frame operator. J. Four. Anal. Appl. **10**, 105–132 (2004)
8. Han, D., Jing, W., Mohapatra, R.N.: Structured Parseval frames in Hilbert C^* modules. Contemp. Math. (American Math. Soc.) **414**, 275–288 (2006)
9. Han, D., Jing, W., Larson, D., Mohapatra, R.N.: Riesz bases and their dual modular frames in Hilbert C^* modules. J. Math. Anal. Appl. **343**, 246–256 (2008)
10. Han, D., Jing, W., Mohapatra, R.N.: Perturbation of frames and Riesz bases in Hilbert C^* modules. Linear Algebra Appl. **431**, 746–759 (2009)
11. Han, D., Jing, W., Larson, D., Li, P., Mohapatra, R.N.: Dilation of dual frame pairs in Hilbert C^* modules. Resultate der Mathematik **63**, 241–250 (2013)
12. Han, D., Sun, W.: Reconstruction of signals from frame coefficients with erasures at unknown locations. IEEE Trans. Inform. Theor. **60**(7), 4013–4025 (2014)
13. Kaushik, S.K.: A generalization of frames in Banach spaces. J. Contemp. Math. Anal. **44**, 212–218 (2009)
14. Lumer, G.: Semi-inner product spaces. Trans. Amer. Math. Soc. **100**, 29–43 (1961)
15. Pehlivan, S., Han, D., Mohapatra, R.N.: Linearly connected sequences and spectrally optimal dual frames for erasures. J. Funct. Anal. **265**(11), 2855–2876 (2013)
16. Sahu, N.K., Nahak, C.: Semi-inner product: application to frame theory and numerical range of operators. Springer Proceedings in Mathematics & Statistics, pp. 77–90. Springer, New York (2014)
17. Stoeva, D.T.: On p-frames and reconstruction series in separable Banach spaces. Integ. Transforms Spec. Functions **17**, 127–133 (2006)
18. Zhang, H., Zhang, J.: Frames, Riesz bases, and sampling expansions in Banach spaces via semi-inner products. Appl. Comput. Harmon. Anal. **31**, 1–25 (2011)

Applications of Generalized Monotonicity to Variational-Like Inequalities and Equilibrium Problems

N.K. Mahato and R.N. Mohapatra

Abstract In this paper, we introduce the concept of relaxed $(\rho\text{-}\theta)$-η-invariant monotonicity to establish the existence of solutions for variational-like inequality problems in reflexive Banach spaces. Again we introduce the concept of $(\rho\text{-}\theta)$-monotonicity for bifunctions. The existence of solution for equilibrium problem with $(\rho\text{-}\theta)$-monotonicity is established by using the KKM technique.

Keywords Variational-like inequality problem · Relaxed $(\rho\text{-}\theta)$-η-invariant monotonicity · Equilibrium problem · $(\rho\text{-}\theta)$-monotonicity · KKM mappping

1 Introduction

Let K be a nonempty subset of a real reflexive Banach space X, and X^* be the dual space of X. Consider the operator $T : K \to X^*$ and the bifunction $\eta : K \times K \to X$. Then the variational-like inequality problem (in short, VLIP) is to find $x \in K$, such that

$$\langle Tx, \eta(y, x)\rangle \geq 0, \forall y \in K, \tag{1}$$

where $\langle ., .\rangle$ denote the pairing between X and X^*.

If we take $\eta(x, y) = x - y$, then (1) becomes to find $x \in K$, such that

$$\langle Tx, y - x)\rangle \geq 0, \forall y \in K, \tag{2}$$

N.K. Mahato
Indian Institute of Information Technology, Design and Manufacturing Jabalpur, 482 005 Jabalpur, India
e-mail: nihar@iiitdmj.ac.in

R.N. Mohapatra (✉)
University of Central Florida, Orlando 32816, FL, USA
e-mail: ramm@mail.ucf.edu

P.N. Agrawal et al. (eds.), *Mathematical Analysis and its Applications*, Springer Proceedings in Mathematics & Statistics 143,
DOI 10.1007/978-81-322-2485-3_12

which are variational inequality problems (VIP) [1, 2]. Variational inequalities have been studied by many authors [1–5] in both finite- and infinite-dimensional spaces. When we deal with variational inequalities, the most common assumption for the operator T is monotonicity. Recently, many authors have established the existence of solutions for variational inequalities with various types of generalized monotonicity assumptions (see [3, 5–8] and the references therein). Fang and Huang [5] defined the concept of relaxed η-α monotonicity and obtained the existence of solutions for variational-like inequalities. Bai et al. [3] extended the idea of relaxed η-α monotonicity to relaxed η-α pseudomonotonicity. Yang et al. [9] defined several kinds of invariant monotone maps and generalized invariant monotone maps. Behera et al. [10] defined various concepts of generalized $(\rho$-$\theta)$-η-invariant monotonicity to generalized concepts of Yang et al. [9]. Very recently, Mahato and Nahak [11] introduced relaxed $(\rho$-$\theta)$-η-invariant pseudomonotonicity to study variational-like inequalities and $(\rho$-$\theta)$-pseudomonotonicity to study equilibrium problems. But in [11], authors did not consider the concepts such as relaxed $(\rho$-$\theta)$-η-invariant monotone mappings, and $(\rho$-$\theta)$-monotone bifunctions. Therefore, we organized this article to consider these monotonicity concepts and study the variational-like inequality problems and equilibrium problem.

Inspired and motivated by [5, 9–11], in this paper, we introduce the concept of relaxed $(\rho$-$\theta)$-η-invariant monotone mappings to establish the existence of solutions for variational-like inequality problems. We also introduce the notion of $(\rho$-$\theta)$-monotonicity for bifunctions. By using the KKM technique we have studied the existence of solutions of equilibrium problem with $(\rho$-$\theta)$-monotone mappings in reflexive Banach spaces.

2 Preliminaries

We begin with the definition of relaxed $(\rho$-$\theta)$-η-invariant monotone mappings. For this consider the function $\theta : K \times K \to \mathbb{R}$ and $\rho \in \mathbb{R}$.

Definition 1 The operator $T : K \to X^*$ is said to be relaxed $(\rho$-$\theta)$-η-invariant monotone with respect to θ, if for any pair of distinct points $x, y \in K$, we have

$$\langle Tx, \eta(y, x)\rangle + \langle Ty, \eta(x, y)\rangle + \rho|\theta(x, y)|^2 \leq 0, \; where \; \theta(x, y) = \theta(y, x). \quad (3)$$

Remark 1 (i) If we take $\rho = 0$ then from (3) it follows that $\langle Tx, \eta(y, x)\rangle + \langle Ty, \eta(x, y)\rangle \leq 0, \forall x, y \in K$, and T is said to be invariant monotone, see [9].

(ii) If we take $\rho = 0$, and $\eta(x, y) = x - y$, then (3) reduces to $\langle Tx - Ty, x - y\rangle \geq 0, \forall x, y \in K$, and T is said to be monotone map.

From the above definitions, it is clear that **invariant monotonicity $\Rightarrow$ relaxed $(\rho$-$\theta)$-η-invariant monotonicity**. However, in general a relaxed $(\rho$-$\theta)$-η-invariant monotone map may not be an invariant monotone map.

Example 1 Let $K = [1, 5]$ and $T : [1, 5] \to \mathbb{R}$ be defined by $Tx = x^2 + 1$. Let the functions η and θ be defined by $\eta(x, y) = x^2+y^2, \theta(x, y) = (x^2+y^2)(x^2+y^2+5)$. Now, $\langle Tx, \eta(y, x)\rangle + \langle Ty, \eta(x, y)\rangle = (x^2 + y^2)(x^2 + y^2 + 2)$, which is not less than 0. Therefore, T is not invariant monotone. But, T is relaxed $(\rho\text{-}\theta)$-η-invariant monotone with respect to θ for any $\rho < 1$.

Definition 2 [5] The operator $T : K \to X^*$ is said to be η-hemicontinuous if for any fixed $x, y \in K$, the mapping $f : [0, 1] \to \mathbb{R}$ defined by $f(t) = \langle T(x + t(y - x)), \eta(y, x)\rangle$ is continuous at 0^+.

3 Relaxed $(\rho\text{-}\theta)$-η-Invariant Monotonicity and (VLIP)

In this section, we establish the existence of the solution for (VLIP), using relaxed $(\rho\text{-}\theta)$-η-invariant monotonicity. Consider the following problems:

$$\text{find } x \in K \text{ such that } \langle Ty, \eta(x, y)\rangle + \rho|\theta(x, y)|^2 \le 0, \forall y \in K. \tag{4}$$

Theorem 1 *Let K be a closed convex subset of a reflexive Banach space X. Assume that $T : K \to X^*$ is η-hemicontinuous and relaxed $(\rho\text{-}\theta)$-η-invariant monotone with the following conditions:*

(i) $\eta(x, y) + \eta(y, x) = 0, \forall x, y \in K$;

(ii) $\lim_{t\to 0} \frac{|\theta(x, x_t)|^2}{t} = 0$, where $x_t = ty + (1 - t)x, \forall x, y \in K$;

(iii) for a fixed $z, y \in K$, the mapping $x \mapsto \langle Tz, \eta(x, y)\rangle$ is convex.

Then the Problems (1) and (4) are equivalent.

Proof Let x be a solution of (1). From the definition of relaxed $(\rho\text{-}\theta)$-η-invariant monotonicity of T, we get $\langle Ty, \eta(x, y)\rangle + \rho|\theta(x, y)|^2 \le -\langle Tx, \eta(y, x)\rangle \le 0$.
Conversely, suppose that $x \in K$ is a solution of (4), i.e.,

$$\langle Ty, \eta(x, y)\rangle + \rho|\theta(x, y)|^2 \le 0, \forall y \in K. \tag{5}$$

Choose any point $y \in K$ and consider $x_t = ty + (1 - t)x, t \in (0, 1]$, then $x_t \in K$. Therefore, from (5) we have

$$\begin{aligned} &\langle Tx_t, \eta(x, x_t)\rangle + \rho|\theta(x, x_t)|^2 \le 0;\\ \Rightarrow &\langle Tx_t, \eta(x_t, x)\rangle - \rho|\theta(x, x_t)|^2 \ge 0;\\ \Rightarrow &\langle Tx_t, \eta(x_t, x)\rangle \ge \rho|\theta(x, x_t)|^2. \end{aligned} \tag{6}$$

Now, $\langle Tx_t, \eta(x_t, x)\rangle \le t\langle Tx_t, \eta(y, x)\rangle + (1-t)\langle Tx_t, \eta(x, x)\rangle = t\langle Tx_t, \eta(y, x)\rangle$. (7)

From (6) and (7) we have
$\langle Tx_t, \eta(y,x)\rangle \geq \rho\frac{|\theta(x,x_t)|^2}{t}$.
Since T is η-hemicontinuous and taking $t \to 0$ we get
$\langle Tx, \eta(y,x)\rangle \geq 0, \forall y \in K$.

Definition 3 Let $f : K \to 2^X$ be a set-valued mapping. Then f is said to be KKM mapping if for any $\{y_1, y_2, \ldots, y_n\}$ of K we have $co\{y_1, y_2, \ldots, y_n\} \subset \bigcup_{i=1}^{n} f(y_i)$, where $co\{y_1, y_2, \ldots, y_n\}$ denotes the convex hull of $y_1, y_2, \ldots, y_n$.

Lemma 1 ([12]) *Let M be a nonempty subset of a Hausdorff topological vector space X and let $f : M \to 2^X$ be a KKM mapping. If $f(y)$ is closed in X, for all $y \in M$ and compact for some $y \in M$, then $\bigcap_{y\in M} f(y) \neq \emptyset$.*

Theorem 2 Let K be a nonempty bounded closed convex subset of a real reflexive Banach space X. Assume that $T : K \to X^*$ is η-hemicontinuous and relaxed (ρ-θ)-η-invariant monotone. Let the following hold:

(i) $\eta(x, y) + \eta(y, x) = 0, \forall x, y \in K$;
(ii) $\lim_{t\to 0} \frac{|\theta(x, x_t)|^2}{t} = 0$, where $x_t = ty + (1-t)x$, $\forall x, y \in K$; and θ is lower semicontinuous in the first argument;
(iii) for a fixed $z, y \in K$, the mapping $x \mapsto \langle Tz, \eta(x, y)\rangle$ is convex and lower semicontinuous.

Then the Problem (1) has a solution.

Proof Consider the set-valued mapping $F : K \to 2^X$ such that
$F(y) = \{x \in K : \langle Tx, \eta(y,x)\rangle \geq 0\}, \forall y \in K$.
It is easy to see that $\overline{x} \in K$ solves the (VLIP) if and only if $\overline{x} \in \cap_{y\in K} F(y)$. We claim that F is a KKM mapping. If possible, let F not be a KKM mapping. Then there exists $\{x_1, x_2, \ldots, x_m\} \subset K$ such that $co\{x_1, x_2, \ldots, x_m\}$ not contained in $\cup_{i=1}^m F(x_i)$, that means there exists a $x_0 \in co\{x_1, x_2, \ldots, x_m\}$, $x_0 = \sum_{i=1}^m t_i x_i$ where $t_i \geq 0, i = 1, 2, \ldots, m, \sum_{i=1}^m t_i = 1$, but $x_0 \notin \cup_{i=1}^m F(x_i)$.

Hence, $\langle Tx_0, \eta(x_i, x_0)\rangle < 0$; for $i = 1, 2, \ldots, m$. From (i) and (iii) it follows that

$$0 = \langle Tx_0, \eta(x_0, x_0)\rangle \leq \sum_{i=1}^m t_i \langle Tx_0, \eta(x_i, x_0)\rangle < 0,$$

which is a contradiction. Hence F is a KKM mapping.

Assume $G : K \to 2^X$ such that $G(y) = \{x \in K : \langle Ty, \eta(x,y)\rangle + \rho|\theta(x,y)|^2 \leq 0\}, \forall y \in K$.

From the relaxed $(\rho\text{-}\theta)$-η-invariant monotonicity of T it follows that $F(y) \subset G(y)$, $\forall y \in K$. Therefore, G is also a KKM mapping.

Since K is closed bounded and convex, it is weakly compact. From the assumptions, we know that $G(y)$ is weakly closed for all $y \in K$. In fact, because $x \mapsto \langle Tz, \eta(x, y)\rangle$ and $x \mapsto \rho|\theta(x, y)|^2$ are lower semicontinuous. Therefore, $G(y)$ is weakly compact in K, for each y $\in$ K.

Therefore, from Lemma 1 and Theorem 1 it follows that $\bigcap_{y\in K} F(y) = \bigcap_{y\in K} G(y) \neq \emptyset$.

So there exists $\overline{x} \in K$ such that $\langle T\overline{x}, \eta(y, \overline{x})\rangle \geq 0, \forall y \in K$, i.e., the Problem (1) has a solution.

Theorem 3 Let K be a nonempty unbounded closed convex subset of a real reflexive Banach space X. Suppose that $T : K \to X^*$ is η-hemicontinuous and relaxed $(\rho\text{-}\theta)$-η-invariant monotone. Let the following hold:

(i) $\eta(x, y) + \eta(y, x) = 0, \forall x, y \in K$;

(ii) $\lim_{t\to 0} \dfrac{|\theta(x, x_t)|^2}{t} = 0$, where $x_t = ty + (1-t)x$, $\forall x, y \in K$; and θ is lower semicontinuous in the first argument;

(iii) for a fixed $z, y \in K$, the mapping $x \mapsto \langle Tz, \eta(x, y)\rangle$ is convex and lower semicontinuous;

(iv) T is weakly η-coercive, i.e., there exits $x_0 \in K$ such that $\langle Tx, \eta(x, x_0)\rangle > 0$, whenever $\|x\| \to \infty$ and $x \in K$.

Then the Problem (1) has solution.

Proof Since the proof of this theorem is very similar to Theorem 3 in [11], hence it is omitted.

4 $(\rho\text{-}\theta)$-Monotonicity and Equilibrium Problem

The equilibrium problem (in short, EP) for the bifunction $f : K \times K \to \mathbb{R}$ is to find $\overline{x} \in K$, such that

$$f(\overline{x}, y) \geq 0, \forall y \in K. \tag{8}$$

Problems like (8) were initially studied by Fan [13]. Later on Blum and Oettli [4] discussed that equilibrium problem contains many problems as particular cases for example, mathematical programming problems, complementary problems, variational inequality problems, fixed-point problems, and minimax inequality problems. Inspired and motivated by [11, 14], we introduced the concept of $(\rho\text{-}\theta)$-monotonicity to establish the existence of solution of equilibrium problem over bounded as well as unbounded domain.

Let K be a nonempty subset of a real reflexive Banach space X. Consider the function $f : K \times K \to \mathbb{R}$ and $\theta : K \times K \to \mathbb{R}$ and $\rho \in \mathbb{R}$.

Definition 4 The function $f : K \times K \to \mathbb{R}$ is said to be (ρ-θ)-monotone with respect to $\theta : K \times K \to \mathbb{R}$ if, for all $x, y \in K$, we have

$$f(x, y) + f(y, x) \leq \rho|\theta(x, y)|^2.$$

Remark 2 In the above definition,
(i) for $\rho > 0$ and $\theta(x, y) = \|x - y\|$, f is weakly monotone;
(ii) for $\rho = 0$, f is monotone;
(iii) for $\rho < 0$ and $\theta(x, y) = \|x - y\|$, f is strongly monotone.

We now give an example to show that (ρ-θ)-monotonicity is a generalization of monotonicity.

Example 2 Let $K = [1, 10]$. Let the functions f and θ be defined by

$$f(x, y) = x^2 + y^2 \textit{ and } \theta(x, y) = 2(x^2 + y^2) + 4.$$

$$\begin{aligned} f(x, y) + f(y, x) &= 2(x^2 + y^2) \\ &\leq \rho(2x^2 + 2y^2 + 4)^2, \text{ for any } \rho \geq 1. \end{aligned}$$

Therefore, f is (ρ-θ)-monotone with respect to θ. But f is not monotone.

Theorem 4 *Let K be a nonempty convex subset of a real reflexive Banach space X. Suppose $f : K \times K \to \mathbb{R}$ is (ρ-θ)-monotone with respect to θ and is hemicontinuous in the first argument with the following conditions:*
(i) $f(x, x) = 0$, $\forall x \in K$;
(ii) for fixed $z \in K$, the mapping $x \mapsto f(z, x)$ is convex;
(iii) $\lim_{t \to 0} \frac{|\theta(x, x_t)|^2}{t} = 0$, where $x_t = ty + (1 - t)x$, $\forall x, y \in K$.
Then $x \in K$ is a solution of (8) if and only if

$$f(y, x) \leq \rho|\theta(x, y)|^2, \forall y \in K. \tag{9}$$

Proof Let x is a solution of (8), i.e., $f(x, y) \geq 0$, $\forall y \in K$. Therefore, from the definition of (ρ-θ)-monotonicity of f it follows that

$$f(y, x) \leq \rho|\theta(x, y)|^2 - f(x, y) \leq \rho|\theta(x, y)|^2, \forall y \in K. \tag{10}$$

Conversely, suppose $x \in K$ satisfying (9), i.e.,

$$f(y, x) \leq \rho|\theta(x, y)|^2, \forall y \in K. \tag{11}$$

Choose any point $y \in K$ and $x_t = ty + (1-t)x$, $t \in (0, 1]$, then $x_t \in K$. Therefore, from (11) we have

$$f(x_t, x) \leq \rho|\theta(x, x_t)|^2, \forall y \in K. \tag{12}$$

Now conditions (i) and (ii) imply that,

$$0 = f(x_t, x_t) \leq tf(x_t, y) + (1-t)f(x_t, x)$$

$$\Rightarrow t[f(x_t, x) - f(x_t, y)] \leq f(x_t, x). \tag{13}$$

From (12) and (13) we have

$$f(x_t, x) - f(x_t, y) \leq \rho\frac{|\theta(x,x_t)|^2}{t}, \forall y \in K.$$

Since f is hemicontinuous in the first argument and taking $t \to 0$, it implies that $f(x, y) \geq 0, \forall y \in K$. Hence x is a solution of (8).

Theorem 5 Let K be a nonempty bounded convex subset of a real reflexive Banach space X. Suppose $f : K \times K \to \mathbb{R}$ is (ρ-θ)-monotone with respect to θ and is hemicontinuous in the first argument with the following conditions:
(i) $f(x, x) = 0,\ \forall x \in K$;
(ii) for fixed $z \in K$, the mapping $x \mapsto f(z, x)$ is convex and lower semicontunuous;
(iii) $\lim\limits_{t\to 0} \dfrac{|\theta(x, x_t)|^2}{t} = 0$, where $x_t = ty + (1-t)x$, $\forall x, y \in K$, and θ is upper semicontinuous in the first argument.
Then the Problem (8) has a solution.

Proof Consider the two set-valued mappings $F : K \to 2^X$ and $G : K \to 2^X$ such that

$$F(y) = \{x \in K : f(x, y) \geq 0\}, \forall y \in K,$$

and

$$G(y) = \{x \in K : f(y, x) \leq \rho|\theta(x, y)|^2\}, \forall y \in K.$$

It is easy to see that $\overline{x} \in K$ solves the equilibrium Problem (8) if and only if $\overline{x} \in \bigcap\limits_{y\in K} F(y)$. First to show that F is a KKM mapping. If possible, let F not be a KKM mapping. Then there exists $\{x_1, x_2, \ldots, x_m\} \subset K$ such that $co\{x_1, x_2, \ldots, x_m\}$ is not contained in $\bigcup\limits_{i=1}^{m} F(x_i)$, that means there exists a $x_0 \in co\{x_1, x_2, \ldots, x_m\}$, $x_0 = \sum\limits_{i=1}^{m} t_i x_i$ where $t_i \geq 0, i = 1, 2, \ldots, m, \sum\limits_{i=1}^{m} t_i = 1$, but $x_0 \notin \bigcup\limits_{i=1}^{m} F(x_i)$.
Hence, $f(x_0, x_i) < 0$; for $i = 1, 2, \ldots, m$. From (i) and (ii) it follows that

$$0 = f(x_0, x_0) \leq \sum_{i=1}^{m} t_i f(x_0, x_i) < 0,$$

which is a contradiction. Hence F is a KKM mapping.

From the (ρ-θ)-monotonicity of f we will show that $F(y) \subset G(y)$, $\forall y \in K$. For any given $y \in K$, let $x \in F(y)$, then

$$f(x, y) \geq 0.$$

From the (ρ-θ)-monotonicity of f, it follows that

$$f(y, x) \leq \rho|\theta(x, y)|^2 - f(x, y) \leq \rho|\theta(x, y)|^2.$$

Therefore $x \in G(y)$, i.e., $F(y) \subset G(y)$, $\forall y \in K$. This implies that G is also a KKM mapping.

Since K is closed bounded and convex, it is weakly compact. From the assumptions, we know that $G(y)$ is weakly closed for all $y \in K$. In fact, because $x \mapsto f(z, x)$ is lower semicontinuous and $x \mapsto \rho|(\theta(x, z)|^2$ is upper semicontinuous. Therefore, $G(y)$ is weakly compact in K, for each y $\in$ K.

Therefore from Lemma 1 and Theorem 4 it follows that $\bigcap_{y\in K} F(y) = \bigcap_{y\in K} G(y) \neq \emptyset$.

So there exists $\overline{x} \in K$ such that $f(\overline{x}, y) \geq 0, \forall y \in K$, i.e., (8) has a solution.

Theorem 6 Let K be a nonempty unbounded closed convex subset of a real reflexive Banach space X. Suppose $f : K \times K \to \mathbb{R}$ is (ρ-θ)-monotone with respect to θ and is hemicontinuous in the first argument and satisfy the following assumptions:

(i) $f(x, x) = 0, \ \forall x \in K$;
(ii) for fixed $z \in K$, the mapping $x \mapsto f(z, x)$ is convex and lower semicontinuous;
(iii) $\lim_{t\to 0} \frac{|\theta(x, x_t)|^2}{t} = 0$, where $x_t = ty + (1 - t)x, \ \forall x, y \in K$, and is upper semicontinuous in the first argument;
(iv) f is weakly coercive, that is there exists $x_0 \in K$ such that $f(x, x_0) < 0$, whenever $\|x\| \to +\infty$ and $x \in K$.

Then (8) has a solution.

Proof Since the proof of this theorem is very similar to Theorem 4.9. in [11], hence it is omitted.

5 Application to Fixed-Point Problems

Let $X = X^*$ be a Hilbert space. Let $T : K \to K$ be a given mapping. Then the fixed-point problem states that find $\overline{x} \in K$ such that

$$T\overline{x} = \overline{x}.$$

Now, by the setting $f(x, y) = \langle x - Tx, y - x\rangle$ we can show that if $\overline{x}$ solves the equilibrium problem (8) then $\overline{x}$ is also a solution of the above fixed-point problem.

Indeed, let $\overline{x}$ is a solution of the equilibrium problem, i.e., $f(\overline{x}, y) \geq 0,\ \forall y \in K$. Let us choose $y = T\overline{x}$, then

$$f(\overline{x}, y) = f(\overline{x}, T\overline{x}) = -\|T\overline{x} - \overline{x}\| \geq 0 \ \Rightarrow \ T\overline{x} = \overline{x},$$

which shows that $\overline{x}$ is a fixed point of T.

In this case, notice that $f(x, y)$ is $(\rho\text{-}\theta)$-monotone if and only if T is $(\rho\text{-}\theta)$-monotone. Since by Theorems 5 and 6, the equilibrium problem has solution, hence by the above result the fixed-point problem also has solution.

6 Conclusions

In this study the existence of solutions for variational-like inequality problems under a new concept relaxed $(\rho\text{-}\theta)$-η-invariant monotone maps in reflexive Banach spaces have been established. We have also obtained the existence of solutions of variational inequality and equilibrium problems with $(\rho\text{-}\theta)$-monotone mappings. This leads to the natural question of making sensitivity analysis and obtaining results using ε-efficiency conditions as in [15, 16]. We plan to pursue these as our subsequent research works.

References

1. Browder, F.E.: Nonlinear monotone operators and convex sets in banach spaces. Bull. Amer. Math. Soc **71**(5), 780–785 (1965)
2. Lions, J.L., Stampacchia, G.: Variational inequalities. Commun. Pure Appl. Math. **20**(3), 493–519 (1967)
3. Bai, M.R., Zhou, S.Z., Ni, G.Y.: Variational-like inequalities with relaxed η-α pseudomonotone mappings in Banach spaces. Appl. Math. Lett. **19**(6), 547–554 (2006)
4. Blum, E., Oettli, W.: From optimization and variational inequalities to equilibrium problems. Math. Student-India **63**(1), 123–145 (1994)
5. Fang, Y.P., Huang, N.J.: Variational-like inequalities with generalized monotone mappings in Banach spaces. J. Optim. Theory Appl. **118**(2), 327–338 (2003)
6. Lee, B.S., Lee, G.M.: Variational inequalities for (η, θ)-pseumonotone operators in nonreflexive banach spaces. Appl. Math. Lett. **12**(5), 13–17 (1999)
7. Luc, D.T.: Existence results for densely pseudomonotone variational inequalities. J. Math. Anal. Appl. **254**(1), 291–308 (2001)
8. Hadjisavvas, N., Schaible, S.: Quasimonotone variational inequalities in Banach spaces. J. Optim. Theory Appl. **90**(1), 95–111 (1996)
9. Yang, X.M., Yang, X.Q., Teo, K.L.: Generalized invexity and generalized invariant monotonicity. J. Optim. Theory Appl. **117**(3), 607–625 (2003)
10. Behera, N., Nahak, C., Nanda, S.: Generalized (ρ, θ)-η-invexity and generalized (ρ, θ)-η-invariant-monotonicity. Nonlinear Anal. Theory Meth. Appl. **68**(8), 2495–2506 (2008)
11. Mahato, N.K., Nahak, C.: Variational-like inequalities and equilibrium eroblems with generalized monotonicity in Banach spaces. Adv. Oper. Res. **2012**, 15pp. (2012)

12. Fan, K.: Some properties of convex sets related to fixed point theorems. Mathematische Annalen **266**(4), 519–537 (1984)
13. Fan, K.: A minimax inequality and applications. Inequalities **3**, 103–113 (1972)
14. Mahato, N.K., Nahak, C.: Mixed equilibrium problems with relaxed α-monotone mapping in banach spaces. Rendiconti del Circolo Matematico di Palermo (2013). doi: 10.1007/s12,215-013-0103-0
15. Mohapatra, R.N., Verma, R.U.: Sensitivity analysis for cocoercively monotone variational inclusions and (a, η)-maximal monotonicity. J. Appl. Math. Comput. **26**(1–2), 281–293 (2008)
16. Verma, R.U., Mohapatra, R.N.: The ε-efficiency conditions for multiobjective fractional programming problems. Dyn. Contin. Discrete Impuls. Syst. Ser. A-Math. Anal. **19**(1), 641–660 (2012)

Simultaneous Approximation Properties of q-Modified Beta Operators

Asha Ram Gairola, Girish Dobhal and Karunesh Kumar Singh

Abstract We establish some approximation properties in simultaneous approximation for a q-analogue of the q-modified Beta operators $B_n^q(f, x)$ introduced by Gupta and Kim. The convergence properties of the q-derivatives of these operators are discussed. Using the estimates for q-moments, the rate of approximation in simultaneous approximation is obtained in terms of modulus of continuity.

Keywords q-Modified beta operator · q-integers · Rate of approximation · Modulus of smoothness · Convergence

1 Introduction

In [21] Phillips introduced the q-Bernstein polynomials for a continuous function on $[0, 1]$. The approximation properties of these operators have been studied by several authors (see [13, 20–26]). In the last decade the q-analogue of the Meyer-König and Zeller operators, Baskakov operators, Szász operators, etc. have also been introduced and studied (see [1, 9–11, 17, 21]). In this paper we study the convergence properties of the q-derivatives of a q-analogue of modified Beta operators introduced by Gupta and Kim [12].

First, we provide the notations and definitions of the q-calculus used in this paper. Let $\mathbb{N}$ be the set of positive integers.
For $n \in \mathbb{N}$, and $q \in (0, 1)$ the q-analogue of a non-negative integer n is defined by

$$[n]_q = 1 + q + q^2 + \cdots + q^{n-1}.$$

A.R. Gairola (✉) · G. Dobhal
University of Petroleum and Energy Studies, Dehradun, India
e-mail: ashagairola@gmail.com

G. Dobhal
e-mail: girish.dobhal@gmail.com

K.K. Singh
Government Polytechnic, Rampur, India
e-mail: kks_iitr@gmail.com

P.N. Agrawal et al. (eds.), *Mathematical Analysis and its Applications*,
Springer Proceedings in Mathematics & Statistics 143,
DOI 10.1007/978-81-322-2485-3_13

Clearly, $[n]_q = n$ for $q = 1$. The q-factorial $[n]_q!$ is given by $[n]_q! = \prod_{j=0}^{n-1}[n-j]_q$ for $n \in \mathbb{N}$ and $[0]_q! = 1$. The q-binomial coefficients $\begin{bmatrix} n \\ k \end{bmatrix}_q$ are defined by

$$\begin{bmatrix} n \\ k \end{bmatrix}_q = \frac{[n]_q!}{[k]_q![n-k]_q!}, 0 \leqslant k \leqslant n$$

and the q-rising product $(a+b)_q^n$ is defined by $(a+b)_q^n = \prod_{j=0}^{n-1}(a+q^j b)$. The q-Jackson integrals and q-improper integrals are given by

$$\int_0^a f(x)\, d_q x = (1-q)\, a \sum_{n=0}^{\infty} f(aq^n)\, q^n$$

and

$$\int_0^{\infty/A} f(x)\, d_q x = (1-q) \sum_{n=-\infty}^{\infty} f\left(\frac{q^n}{A}\right) \frac{q^n}{A},\ A > 0,$$

respectively. Here the sums are assumed to be absolutely convergent (see [14, 18]). The q-derivative $D_q f(t)$ of a real function $f : \mathbb{R} \to \mathbb{R}$ is defined as

$$D_q f(t) = \frac{f(t) - f(qt)}{(1-q)t} \text{ if } t \neq 0,$$

$D_q f(0) = f'(0)$ provided $f'(0)$ exists. The product formula for q-differentiation is given by

$$D_q(f(x)g(x)) = f(qx) D_q(g(x)) + g(x) D_q(f(x)).$$

We remark that the q-analogue of the integration by part formula is given by

$$\int_a^b f(x) d_q g(x) = f(b)g(b) - f(a)g(a) - \int_a^b g(qx) d_q f(x).$$

For further details we refer to [2, 15, 22].

In [12] Gupta and Kim defined a q-analogue of modified Beta operators $B_n^q(f,x)$ as follows:

$$B_n^q(f,x) = \frac{[n-1]_q}{[n]_q} \sum_{k=0}^{\infty} b_{n,k}^q(x) \int_0^{\infty/A} q^k p_{n,k}^q(t) f(t)\, d_q t,$$

$$b_{n,k}^q(x) = \frac{q^{k(k-1)/2} x^k}{B_q(k+1,n)(1+x)_q^{n+k+1}},\quad p_{n,k}^q(x) = \begin{bmatrix} n+k-1 \\ k \end{bmatrix}_q \frac{q^{k(k-1)/2} x^k}{(1+x)_q^{n+k}}.$$

Here $f \in C_B[0,\infty)$, the class of the continuous and bounded functions on $[0,\infty)$. The space $C_B[0,\infty)$ is a Banach space with respect to the uniform norm $\|f\| = \sup_{0\leqslant x<\infty}|f(x)|$. The operators $B_n^q(f,x)$ are linear, positive and reproduce constant functions. The modulus of continuity $\omega(f,\delta)$ for $f\in C_B[0,\infty)$ is defined by

$$\omega(f,\delta)=\sup_{0<h\leqslant\delta}\ \sup_{x\in[0,\infty)}|f(x+h)-f(x)|.$$

We have denoted the set of non-negative integers by the symbol N_0. Let $\lambda\geqslant 0$. We define the class

$$C_\lambda^r[0,\infty)=\{f|D_q^1f, D_q^2f,\ldots D_q^rf\in C^1[0,\infty), D_q^rf(t)=O(t^\lambda) \text{ as } t\to\infty\},$$

where $C^1[0,\infty)$ is the class of continuously differentiable functions on $[0,\infty)$. It was shown in [12] that the operators $B_n^q(f,x)$ do not converge to any arbitrary real function $f(x)$ in case of a fixed q in $(0,1)$. In this paper we will show that this property is inherited to the q-derivatives of these operators, i.e. the q-derivatives of the operators $B_n^q(f,x)$ do not converge to the corresponding q-derivatives of the function f for a fixed q. In what follows, we shall use the notations $\varphi^2(x)=x(1+x)$ and throughout this paper C is a constant independent of f and n but may depend on q. Further, C is not necessarily same at each occurrence.
The paper is organized as follows:

Section 2 contains some estimates of the moments and expressions for the rth q-derivative of the operators B_n^q. Finally, Sect. 3 studies convergence properties of the derivatives $D_q^r\big(B_n^q(f,x)\big)$ and the error estimates for the functions of specified smoothness in terms of the modulus of continuity $\omega(f,\delta)$.

2 Moment Estimates

Remark 1 For q-differentiation, we have the formula (see [3])

$$D_q\frac{x^\alpha}{(1+x)_q^\beta}=[\alpha]\frac{x^{\alpha-1}}{(1+x)_q^{\beta+1}}-([\beta]-[\alpha])\frac{x^\alpha}{(1+x)_q^{\beta+1}}, \tag{1}$$

where D_q denotes the q-derivative operator and α,β are arbitrary real numbers. Using (1) for the weight $p_{n,k}^q(x)$ we obtain

$$\begin{aligned}D_q[p_{n,k}^q(x)]&=\begin{bmatrix}n+k-1\\k\end{bmatrix}_q q^{k(k-1)/2}\left(\frac{[k]x^{k-1}}{(1+x)_q^{n+k+1}}-\frac{([n+k]-[k])x^k}{(1+x)_q^{n+k+1}}\right)\\&=\begin{bmatrix}n+k-1\\k\end{bmatrix}_q q^{k(k-1)/2}\left(\frac{[k]}{x}-([n+k]-[k])\right)\frac{(qx)^k}{q^k(1+x)(1+qx)_q^{n+k}}.\end{aligned}$$

Since, we have $[n+k]-[k]=q^k[n]$ we get

$$D_q[p_{n,k}^q(x)] = \begin{bmatrix} n+k-1 \\ k \end{bmatrix}_q q^{k(k-1)/2}\left(\frac{[k]}{x}-q^k[n]\right)\frac{(qx)^k}{q^k(1+x)(1+qx)_q^{n+k}}$$
$$= \left(\frac{[k]}{q^k}-[n]\right)\frac{p_{n,k}^q(qx)}{\varphi^2(x)}.$$

Thus, we get the identity

$$q^k\varphi^2(x)D_q[p_{n,k}^q(x)] = \Big([k]_q - q^k[n]_q x\Big)p_{n,k}^q(qx). \tag{2}$$

Similarly, we have

$$q^k\varphi^2(x)D_q[b_{n,k}^q(x)] = \Big([k]_q - q^k[n+1]_q x\Big)b_{n,k}^q(qx). \tag{3}$$

Lemma 1 *For the functions $S_{n,m}(x) = B_n^q((t-x)_q^m, x)$ we have*

$$S_{n,0}(x) = 1,\ \ S_{n,1}(x) = \frac{[n+1]_q x}{q^2[n-2]_q} + \frac{1}{q[n-2]_q}$$

and there holds the recurrence relation

$$q^2\varphi^2(x)\left(D_q S_{n,m}(x) + [m]_q S_{n,m-1}(x)\right) + q^{m+2}[m]_q x(1+[2]_q x q^{m+2})S_{n,m-1}(qx)$$
$$= \left(\left([n]_q q^{m+2} - [n+1]_q q^2\right)x - [m+1]_q q(1+[2]_q x q^{m+2})\right) S_{n,m}(qx)$$
$$+([n]_q q - [m+2]_q)S_{n,m+1}(qx).$$

Proof $S_{n,0}(x)$ and $S_{n,1}(x)$ follow from Lemma 3, [12]. Now, using product formula for q-differentiation and Remark 1, we get

$$D_q(t-x)_q^m b_{n,k}^q(x)$$
$$= -[m]_q(t-x)_q^{m-1}b_{n,k}^q(x) + (t-qx)_q^m D_q b_{n,k}^q(x)$$
$$= -[m]_q(t-x)_q^{m-1}b_{n,k}^q(x) + \frac{(t-qx)_q^m([k]_q - q^k[n+1]_q x)}{q^k\varphi^2(x)}b_{n,k}^q(qx)$$

Hence,

$$\varphi^2(x)\left(D_q S_{n,m}(x) + [m]_q S_{n,m-1}(x)\right)$$
$$= \frac{[n-1]_q}{[n]_q}\sum_{k=0}^{\infty}\frac{([k]_q - q^k[n]_q t/q)}{q^k}b_{n,k}^q(qx)\int_0^{\infty/A} q^k p_{n,k}^q(t)(t-qx)_q^m\, d_q t$$

$$+\frac{[n-1]_q}{[n]_q}\sum_{k=0}^{\infty}\frac{(q^k[n]_q t/q - q^k[n+1]_q x)}{q^k} b_{n,k}^q(qx)\int_0^{\infty/A} q^k p_{n,k}^q(t)(t-qx)_q^m\, d_q t$$
$$= J_1 + J_2,\ \text{say.}$$

Using (2), with the transformation $t = qu$, we get

$$J_1 = \frac{[n-1]_q}{[n]_q}\sum_{k=0}^{\infty} b_{n,k}^q(qx)\int_0^{\infty/A} \varphi^2(u) D_q(p_{n,k}^q(u))(qu-qx)_q^m q\, du$$

We split $\varphi^2(u)$ as

$$\varphi^2(u) = q^m x\left(1+[2]_q x q^{m+2}\right) + \frac{1}{q}\left(1+[2]_q x q^{m+2}\right)\left(qu - q^{m+1}x\right) + \frac{1}{q^2}\left(qu - q^{m+1}x\right)\left(qu - q^{m+2}x\right)$$

and use in J_1. This yields

$$J_1 = \frac{[n-1]_q}{[n]_q}\sum_{k=0}^{\infty} b_{n,k}^q(qx)\Bigg(\int_0^{\infty/A} q^m x\left(1+[2]_q x q^{m+2}\right)(qu-qx)_q^m$$
$$+\int_0^{\infty/A}\frac{1}{q}\left(1+[2]_q x q^{m+2}\right)(qu-qx)_q^{m+1}$$
$$+\int_0^{\infty/A}\frac{1}{q^2}(qu-qx)_q^{m+2}\Bigg) D_q(p_{n,k}^q(u)) q\, du$$
$$= K_1 + K_2 + K_3,\ \text{say.}$$

Now, integration by parts gives

$$K_1 = -[m]q^{m+1}x\left(1+[2]_q x q^{m+2}\right)\frac{[n-1]_q}{[n]_q}\sum_{k=0}^{\infty} b_{n,k}^q(qx)$$
$$\times\int_0^{\infty/A} p_{n,k}^q(qu)(qu-qx)_q^{m-1}\, d_q u$$
$$= -[m]q^m x\left(1+[2]_q x q^{m+2}\right)\frac{[n-1]_q}{[n]_q}\sum_{k=0}^{\infty} b_{n,k}^q(qx)$$

$$\times \int_0^{\infty/A} p_{n,k}^q(t)(t-qx)_q^{m-1}\, d_q u$$

$$= -[m]q^m x\left(1+[2]_q x q^{m+2}\right) S_{n,m-1}(qx).$$

Similarly, we obtain

$$K_2 = -\frac{[m+1]_q}{q}\left(1+[2]_q x q^{m+2}\right) S_{n,m}(qx)$$

and

$$K_3 = -\frac{[m+2]_q}{q^2} S_{n,m+1}(qx).$$

By simple calculations we obtain

$$J_2 = ([n]_q q^m - [n+1])x) S_{n,m}(qx) + \frac{[n]_q}{q} S_{n,m+1}(qx).$$

Combining J_1 and J_2 the required relation follows.

Corollary 1 *For the functions $S_{n,m}(x)$ we have (i) $S_{n,m}(x)$ are polynomials in x of degree exactly m; (ii) there holds the order $S_{n,m}(x) = O\left([n]_q^{-m+1}\right)$, for all $x \in [0,\infty)$.*

Lemma 2 [8] *Let $m \in N_0$, $0 < q < 1$. There exists a constant $C = C(q,m) > 0$ independent of x and n such that*

$$B_n^q\left((t-x)_q^m, x\right) \leqslant C\left(\frac{1}{[n]_q^{\lfloor (m+1)/2 \rfloor}}\right).$$

Following is a Lorentz-type lemma (see [4, p. 112]) for the q-differentiation of $b_{n,k}^q(x)$.

Lemma 3 *For the functions $b_{n,k}^q(x)$ there holds*

$$q^{r-1}x^r\left(1+q^{n+k+1}x\right)_q^r D_q^r\left(b_{n,k}^q(x)\right) = \sum_{\substack{2i+j\leqslant r\\ i,j\geqslant 0}} \alpha_j(x)[n+1]_q^i\left([k]_q - q^k[n+1]_q x\right)^j b_{n,k}^q(x),$$

where $\alpha_i(x)$ are polynomials in x independent of $[n]_q$.

Proof The proof follows by an induction on r.

Lemma 4 *For the functions $Q_{r,l}(x)$ defined by*

$$Q_{r,l}(x) = \sum_{k=0}^{\infty} \left([k]_q - q^{k+l}[n+1]_q x\right)^r b_{n,k}^q(x),$$

there holds the order $Q_{2i,l}(x) = O\left([n]_q^{2i}\right)$.

Proof The identity $\sum_{k=0}^{\infty} \frac{q^{k(k-1)/2} t^k [n+k]_q!}{[k]_q![n-1]_q!(1+t)_q^{n+k+1}} = 1$ will be frequently used. We shall apply an induction on i. For $i = 1$ we write $\left([k]_q - q^{k+l}[n+1]_q x\right)^2 = [k]_q^2 - 2[k]_q[n+1]_q q^{k+l} x + q^{2k+2l}[n+1]_q^2 x^2$ and obtain three terms, namely D_1, D_2 and D_3. Now,

$$\begin{aligned}
D_1 &= \sum_{k=0}^{\infty} [k]_q^2 b_{n,k}^q(x) \\
&= \sum_{k=0}^{\infty} \frac{(1+q[k-1]_q) q^{k(k-1)/2} x^k [n+k]_q!}{[k-1]_q![n-1]_q!(1+x)_q^{n+k+1}} \\
&= \sum_{k=0}^{\infty} \frac{q^{k(k-1)/2} x^k [n+k]_q!}{[k-1]_q![n-1]_q!(1+x)_q^{n+k+1}} + q \sum_{k=0}^{\infty} \frac{q^{k(k-1)/2} x^k [n+k]_q!}{[k-2]_q![n-1]_q!(1+x)_q^{n+k+1}} \\
&= \frac{[n]_q x}{1+x} + x^2 q^2 \sum_{k=0}^{\infty} \frac{q^{k(k-1)/2} \left(q^2 x\right)^k [n+k+2]_q!}{[k]_q![n-1]_q!(1+x)(1+qx)\left(1+q^2 x\right)_q^{n+k+1}} \\
&= \frac{[n]_q x}{1+x} + \frac{[n+1]_q [n]_q x^2 q^2}{(1+x)(1+qx)} \sum_{k=0}^{\infty} \frac{q^{k(k-1)/2} \left(q^2 x\right)^k [n+k+2]_q!}{[k]_q![n-1]_q! \left(1+q^2 x\right)_q^{n+k+3}} (1+q^{n+k+2}x)_q^2.
\end{aligned}$$

This gives

$$\begin{aligned}
|D_1| &= \frac{[n]_q x}{1+x} + \frac{[n+1]_q [n]_q x^2 q^2 (1+x)^2}{(1+x)(1+qx)} \left| \sum_{k=0}^{\infty} \frac{q^{k(k-1)/2} (q^2 x)^k [n+k+2]_q!}{[k]_q![n-1]_q!(1+q^2 x)_q^{n+k+3}} \right| \\
&\leqslant \frac{[n]_q x}{1+x} + \frac{[n+1]_q [n]_q x^2 q^2 (1+x)}{(1+qx)}.
\end{aligned}$$

Next, we have

$$\begin{aligned}
D_2 &= -2xq^l [n+1]_q \sum_{k=0}^{\infty} \frac{q^k q^{k(k-1)/2} x^k [n+k]_q!}{[k-1]_q![n-1]_q!(1+x)_q^{n+k+1}} \\
&= \frac{-2xq^l [n+1]_q [n]_q}{(1+x)(1+qx)} \sum_{k=0}^{\infty} \frac{q^{2k} q^{k(k-1)/2} x^{k+1} [n+k+1]_q!}{[k]_q![n]_q!(1+q^2 x)_q^{n+k+2}} (1+q^{n+k+1}x)_q^3
\end{aligned}$$

Therefore, $|D_2| \leqslant \frac{2q^{l+1} x [n+1]_q [n]_q (1+x)^2}{(1+qx)}$. Similarly, we have

$$\begin{aligned}|D_3| &= \frac{(q^l[n+1]_q x)^2}{(1+x)(1+qx)}\left|\sum_{k=0}^{\infty}\frac{q^{2k}q^{k(k-1)/2}x^k[n+k]_q!}{[k]_q![n-1]_q!(1+q^2x)_q^{n+k+1}}(1+q^{n+k+1}x)_q^2\right|\\ &\leqslant \frac{(q^l[n+1]_q x)^2(1+x)}{(1+qx)}.\end{aligned}$$

Combining the estimates for D_1–D_3, we obtain

$$\begin{aligned}\left|Q_{2,l}(x)\right| &\leqslant \frac{[n]_q x}{1+x}+\frac{[n+1]_q[n]_q x^2 q^2(1+x)}{(1+qx)}+\frac{2q^{l+1}x[n+1]_q[n]_q(1+x)^2}{(1+qx)}\\ &\quad+\frac{(q^l[n+1]_q x)^2(1+x)}{(1+qx)}=O\left([n]_q^2\right),\ \text{for all}\, x\in(0,\infty).\end{aligned}$$

Let the lemma be true for a certain i. By q-differentiation we get

$$\begin{aligned}D_q Q_{2i,l}(x) &= -[2i]_q[n]_q q^l\sum_{k=0}^{\infty}q^k\left([k]_q-q^{k+l+1}[n+1]_q x\right)^{2i-1}b_{n,k}^q(qx)\\ &\quad+\sum_{k=0}^{\infty}\frac{\left([k]_q-q^{k+l+1}[n+1]_q x\right)^{2i}}{q^k\varphi(x)}b_{n,k}^q(qx)\left[\left([k]_q-q^k[n+1]_q x\right)\right]\end{aligned}$$

Rearrangement of the terms gives

$$\begin{aligned}&\sum_{k=0}^{\infty}\frac{\left([k]_q-q^{k+l+1}[n+1]_q x\right)^{2i+1}}{q^k}b_{n,k}^q(qx)\\ &\quad=\varphi^2(x)D_q Q_{2i,l}(x)\\ &\qquad+\varphi^2(x)[2i]_q[n]_q q^l\sum_{k=0}^{\infty}q^k\left([k]_q-q^{k+l+1}[n+1]_q x\right)^{2i-1}b_{n,k}^q(qx)\\ &\qquad-[n+1]_q x\left(q^l-1\right)\sum_{k=0}^{\infty}\left([k]_q-q^{k+l+1}[n+1]_q x\right)^{2i}b_{n,k}^q(qx).\end{aligned}\tag{4}$$

Therefore, from the definition of $Q_{2i+1,l}(x)$ and (4) we get

$$\begin{aligned}\left|Q_{2i+1,l}(x)\right| &\leqslant \left|\sum_{k=0}^{\infty}\frac{\left([k]_q-q^{k+l+1}[n+1]_q x\right)^{2i+1}}{q^k}b_{n,k}^q(qx)\right|\\ &\leqslant C[n]_q^{2i}+C'[n]_q^{2i+1}=C[n]_q^{2i+1}.\end{aligned}$$

This completes the proof.

Lemma 5 *Let $f\in C_\lambda^r[0,\infty)$. Then*

$$D_q^r\left(B_n^q(f,x)\right)=\prod_{m=0}^{r}\left(\frac{[n+m-1]_q}{[n-m]_q}\right)\sum_{k=0}^{\infty}b_{n+r,k}^q(x)\int_0^{\infty/A}q^k p_{n-r,k+r}^q(q^r t)D_q^r f(t)\,d_qt.$$

Proof First, we prove the lemma for $r=1$. By product rule for q-differentiation we obtain

$$D_q\left(\frac{x^l}{(1+x)_q^{l+m}}\right)=\frac{[l]_q x^{l-1}}{(1+x)_q^{l+m}}-\frac{[l+m]_q(qx)^l}{(1+x)_q^{l+m+1}}.$$

This gives

$$D_q b_{n,k}^q(x)=[n]_q q^{k-1}\left(b_{n+1,k-1}^q(x)-q\,b_{n+1,k}^q(x)\right)$$
$$D_q p_{n,k}^q(t)=[n]_q q^{k-1}\left(p_{n+1,k-1}^q(t)-q\,p_{n+1,k}^q(t)\right).$$

Next q-integration by parts gives

$$\begin{aligned}
&D_q\left(B_n^q(f,x)\right)\\
&=[n-1]_q\sum_{k=0}^{\infty}q^{k-1}\left(b_{n+1,k-1}^q(x)-b_{n+1,k}^q(x)\right)\\
&\quad\times\int_0^{\infty/A}q^k p_{n,k}^q(t)f(t)\,d_qt\\
&=[n-1]_q\sum_{k=0}^{\infty}q^k b_{n+1,k}^q(x)\int_0^{\infty/A}q^k\left(qp_{n,k+1}^q(t)-p_{n,k}^q(t)\right)f(t)\,d_qt\\
&=\sum_{k=0}^{\infty}q^k b_{n+1,k}^q(x)\int_0^{\infty/A}-D_q\left(p_{n-1,k+1}^q(t)\right)f(t)\,d_qt\\
&=\sum_{k=0}^{\infty}q^k b_{n+1,k}^q(x)\int_0^{\infty/A}p_{n-1,k+1}^q(t)D_q\left(f(t)\right)d_qt.
\end{aligned}$$

Hence, the result holds for $r=1$. Now the lemma follows by induction on r and straightforward calculations.

For $f\in C_\lambda^r[0,\infty)$ we define the operators $\widehat{B}_{n,q,r}(f,x)$ by

$$\widehat{B}_{n,q,r}(f,x)=\sum_{k=0}^{\infty}b_{n+r,k}^q(x)\int_0^{\infty/A}q^k p_{n-r,k+r}^q(q^r t)f(t)\,d_qt.$$

Lemma 6 *For the functions* $U_{n,m,r}(x) = \widehat{B}_{n,q,r}(t^m, x)$, *we have*

$$U_{n,0,r}(x) = \frac{1}{q^{2r}[n-r-1]_q},$$

$$U_{n,1,r}(x) = \frac{q^{-1}[n+r+1]_q x + [2]_q q^{r-2} + [r]_q}{[n-r+1]_q q^{2r}\left([n-r+1]_q - [2]_q q^{-2}\right)}$$

and there holds the relation

$$\begin{aligned} &q^{2r}\left([n-r+1]_q - [m+2]_q q^{-m-2}\right) U_{n,m+1,r}(qx) \\ &= \varphi^2(x) D_q U_{n,m,r}(x) + [n+r+1]_q x U_{n,m,r}(qx) \\ &+ \left([m+1]_q q^{r-m-1} + [r]_q\right) U_{n,m,r}(qx). \end{aligned} \tag{5}$$

Proof We have

$$U_{n,m,r}(x) = \sum_{k=0}^{\infty} b_{n+r,k}^q(x) \int_0^{\infty/A} q^k p_{n-r,k+r}^q(q^r t) t^m \, d_q t.$$

Using (2) we get

$$\begin{aligned} &\varphi^2(x) D_q U_{n,m,r}(x) + [n+r+1]_q x U_{n,m,r}(qx) \\ &= \sum_{k=0}^{\infty} q^{-k}[k]_q b_{n+r,k}^q(qx) \int_0^{\infty/A} q^k p_{n-r,k+r}^q(q^r t) t^m \, d_q t \end{aligned} \tag{6}$$

We write $q^{-k}[k]_q = q^r\left(q^{-k-r}[k+r]_q - [n-r+1]_q t q^r\right) + [n-r+1]_q t q^r - [r]_q$ and substitute in (6). This gives three terms I_1, I_2 and I_3. Using the transformation $t \to u/q^r$ we obtain

$$\begin{aligned} I_1 &= \frac{1}{q^{mr}} \sum_{k=0}^{\infty} q^k b_{n+r,k}^q(qx) \int_0^{\infty/A} \left(q^{-k-r}[k+r]_q - [n-r+1]_q u\right) p_{n-r,k+r}^q(u) \, d_q u \\ &= \frac{1}{q^{mr}} \sum_{k=0}^{\infty} q^k b_{n+r,k}^q(qx) \int_0^{\infty/A} \left(u^{m+1} + u^{m+2}\right) p_{n-r,k+r}^q(u) \, d_q u \\ &= J_1 + J_2, \text{ say.} \end{aligned}$$

Now, integration by parts and the inverse transformation $u \to tq^r$ give

$$\begin{aligned} J_1 &= \frac{-[m+1]_q}{q^{mr}} \sum_{k=0}^{\infty} q^k b^q_{n+r,k}(qx) \int_0^{\infty/A} p^q_{n-r,k+r}(u) \frac{u^m}{q^{m+1}}\, d_q u \\ &= -[m+1]_q q^{r-m-1} U_{n,m,r}(qx). \end{aligned}$$

Similarly, $J_2 = -[m+2]_q q^{2r-m-1} U_{n,m+1,r}(qx)$. From definition we get $I_2 = q^{2r}[n-r+1]_q U_{n,m+1,r}(qx)$ and $I_3 = -[r]_q U_{n,m,r}(qx)$. Combining these estimates, we obtain the required recurrence relation.

Corollary 2 *Since $\widehat{B}_{n,q,r}\left((t-x)^2, x\right)$ is a quadratic polynomial, we write $\widehat{B}_{n,q,r}\left((t-x)^2, x\right) = \alpha_0 + \alpha_1 x + \alpha_2 x^2$. Using the recurrence relation (5) the coefficients α_i are given as*

$$\alpha_0 = \frac{\left([2]_q q^{r-2} + [r]_q\right)^2}{q^{4r}\left([n-r+1]_q - [2]_q q^{-2}\right)\left([n-r+1]_q - [3]_q q^{-3}\right)},$$

$$\alpha_1 = \frac{\left(q^{-1} + 2\left([2]_q q^{r-2} + [r]_q\right)\right)[n+r+1]_q - 2q^{2r}\left([2]_q q^{r-2} + [r]_q\right)\left([n-r+1]_q - [3]_q q^{-3}\right)}{q^{4r}\left([n-r+1]_q - [2]_q q^{-2}\right)\left([n-r+1]_q - [3]_q q^{-3}\right)}$$

and

$$\begin{aligned} \alpha_2 &= \frac{[n+r+1]_q\left(q^{-2} + [n+r+1]_q - 2q^{2r-1}\left([n-r+1]_q - [3]_q q^{-3}\right)\right)}{q^{4r}\left([n-r+1]_q - [2]_q q^{-2}\right)\left([n-r+1]_q - [3]_q q^{-3}\right)} \\ &+ \frac{q^{2r}\left([n-r+1]_q - [2]_q q^{-2}\right)\left([n-r+1]_q - [3]_q q^{-3}\right)}{q^{4r}\left([n-r+1]_q - [2]_q q^{-2}\right)\left([n-r+1]_q - [3]_q q^{-3}\right)}. \end{aligned}$$

Now, it follows from straightforward calculations that $|\alpha_0| \leqslant \frac{C}{[n-r+1]_q^2}$, $|\alpha_1| \leqslant \frac{C'}{[n-r+1]_q}$ where $C, C' = C(r, a)$ and $|\alpha_2| \leqslant \frac{(1+q^{2r}-2q^{2r-1})}{q^{4r}} + O\left(\frac{1}{[n-r+1]_q}\right)$. Consequently, $\widehat{B}_{n,q,r}\left((t-x)^2, x\right) \leqslant C\left(\frac{(1+q^{2r}-2q^{2r-1})}{q^{4r}} x^2 + O\left(\frac{1}{[n-r+1]_q}\right)\right)$, for all $x \in [0, \infty)$.

3 Simultaneous Approximation Using q-Moments

Theorem 1 *If $f \in C^r_\lambda[0, \infty)$ and q_n be a sequence in $(0, 1)$ with $q_n \uparrow 1$, then there exists a number $0 < \hat{q}_n < 1$ such that the sequence $D^r_{q_n}\left(B^{q_n}_n(f, x)\right)$ converges to $D^r_{q_n}\left(f(x)\right)$ pointwise for $q_n \in (\hat{q}_n, 1)$.*

Proof There exists $\hat{q}_n \in (0, 1)$ (see [7]) such that for all $q_n \in (\hat{q}_n, 1)$ we have

$$f(t) = \sum_{l=0}^{r} \frac{(D^l_{q_n} f)(x)}{[k]_{q_n}!}(t-x)^l_q + \frac{(D^{r+1}_{q_n} f)(\xi)}{[r+1]_{q_n}!}(t-x)^{r+1}_q.$$

Since, $B_n^{q_n}\left((t-x)_{q_n}^m, x\right)$ are polynomials of degree exactly m, (see [8]), we obtain

$$\begin{aligned} & D_{q_n}^r\left(B_n^{q_n}(f,x)\right) - D_{q_n}^r\left(f(x)\right) \\ &= \frac{1}{[r+1]_{q_n}!}(B_n^{q_n})^{(r)}\left((D_{q_n}^{r+1}f)(\xi)(t-x)_{q_n}^{r+1}, x\right) \\ &= \frac{[n-1]_{q_n}}{[n]_{q_n}}\sum_{k=0}^{\infty}\sum_{\substack{2i+j\leqslant r\\ i,j\geqslant 0}} \alpha_j(x)[n+1]_{q_n}^i\left([k]_{q_n} - q_n^k[n+1]_q x\right)^j b_{n,k}^{q_n}(x) \\ &\times \int_0^{\infty/A} q_n^k p_{n,k}^{q_n}(t)\frac{(D_{q_n}^{r+1}f)(\xi)}{[r+1]_{q_n}!}(t-x)_{q_n}^{r+1}\,d_{q_n}t. \end{aligned}$$

Let $T_{i,j}$ be a typical term of the sum over i, j. Using Hölder's inequality first for integration and then for summation we obtain

$$\begin{aligned} |T_{i,j}| &\leqslant C\|D_{q_n}^{r+1}f\|\frac{[n-1]_{q_n}}{[n]_{q_n}}\sum_{\substack{2i+j\leqslant r\\ i,j\geqslant 0}}[n+1]_{q_n}^i\sum_{k=0}^{\infty}\left|[k]_{q_n} - q_n^k[n+1]_q x\right|^j b_{n,k}^{q_n}(x) \\ &\quad\times\left|\int_0^{\infty/A} q_n^k p_{n,k}^{q_n}(t)(t-x)_q^{r+1}\,d_{q_n}t\right| \\ &\leqslant C\|D_{q_n}^{r+1}f\|\sum_{\substack{2i+j\leqslant r\\ i,j\geqslant 0}}[n+1]_{q_n}^i\frac{[n-1]_{q_n}}{[n]_{q_n}}\sum_{k=0}^{\infty}\left|[k]_{q_n} - q_n^k[n+1]_q x\right|^j b_{n,k}^{q_n}(x) \\ &\quad\times\prod_{j=0}^{r}\left(\int_0^{\infty/A} q_n^k p_{n,k}^{q_n}(t)|t-q_n^j x|^{r+1}\,d_{q_n}t\right)^{\frac{1}{r+1}} \\ &\leqslant C\|D_{q_n}^{r+1}f\|\sum_{\substack{2i+j\leqslant r\\ i,j\geqslant 0}}[n+1]_{q_n}^i\left(\sum_{k=0}^{\infty}\left([k]_{q_n} - q_n^k[n+1]_q x\right)^{2j} b_{n,k}^{q_n}(x)\right)^{1/2} \\ &\quad\times\left(\frac{[n-1]_{q_n}}{[n]_{q_n}}\sum_{k=0}^{\infty}\prod_{j=0}^{r} b_{n,k}^{q_n}(x)\left(\int_0^{\infty/A} q_n^k p_{n,k}^{q_n}(t)|t-q_n^j x|^{r+1}\,d_{q_n}t\right)^{2/r+1}\right)^{1/2} \\ &\leqslant C\|D_{q_n}^{r+1}f\|\sum_{\substack{2i+j\leqslant r\\ i,j\geqslant 0}}[n+1]_{q_n}^i\left(\sum_{k=0}^{\infty}([k]_{q_n} - q_n^{k+s}[n+1]_q x)^{2j} b_{n,k}^{q_n}(x)\right)^{1/2} \end{aligned}$$

$$\times\left(\prod_{j=0}^{r}\left(\frac{[n-1]_{q_n}}{[n]_{q_n}}\sum_{k=0}^{\infty}b_{n,k}^{q_n}(x)\int_0^{\infty/A}q_n^k p_{n,k}^{q_n}(t)\left(t-q_n^j x\right)^{2r+2}d_{q_n}t\right)^{\frac{1}{r+1}}\right)^{1/2}.$$

Now, we use the relation (see [19])

$$(t-x)^{2r+2}=\sum_{s=1}^{2r+2}\alpha_{2r+2,s}\left(\frac{1-q_n^n}{[n]_{q_n}}\right)^{2r+2-s}x^{2r+2-s}(t-x)_q^s,$$

where the constants $\alpha_{2r+2,s}$ are independent of x, q_n and n. Using Corollary 1 in above relation we get

$$\begin{aligned}\left|B_n^{q_n}((t-x)^{2r+2},x)\right| &\leqslant \sum_{s=1}^{2r+2}\left|\alpha_{2r+2,s}\right|\left(\frac{1-q_n^n}{[n]_{q_n}}\right)^{2r+2-s}x^{2r+2-s}\left|B_n^{q_n}\left((t-x)_q^s,x\right)\right|\\ &\leqslant \sum_{s=1}^{2r+2}\left|\alpha_{2r+2,s}\right|\left(\frac{1-q_n^n}{[n]_{q_n}}\right)^{2r+2-s}x^{2r+2-s}\left(\frac{1}{[n]_{q_n}}\right)^{\lfloor\frac{(s+1)}{2}\rfloor}\\ &= O\left(\frac{1}{[n]_{q_n}}\right)^{r+1}.\end{aligned}$$

Therefore, using $(t-q_n^j x)^{2r+2}=\sum_{l=0}^{2r+2}\binom{2r+2}{l}(t-x)^l\left(x(1-q_n^j)\right)^{2r+2-l}$ and Hölder's inequality we obtain

$$\left|B_n^{q_n}((t-q_n^j x)^{2r+2},x)\right|\leqslant\sum_{l=0}^{2r+2}\binom{2r+2}{l}\left(x(1-q_n^j)\right)^{2r+2-l}\left(\frac{1}{[n]_{q_n}^{l/2}}\right).$$

Let $(1-q_n^j)=O\left(\frac{1}{[n]_{q_n}^{\rho}}\right)$, $\rho\geqslant 0$, for all j. This implies $(1-q_n^r)=O\left(\frac{1}{[n]_{q_n}^{\rho}}\right)$. Consequently, we get

$$\begin{aligned}|T_{i,j}| &\leqslant C\|D_{q_n}^{r+1}f\|\sum_{\substack{2i+j\leqslant r\\ i,j\geqslant 0}}[n+1]_{q_n}^{i+j}\left(\prod_{j=0}^{r}\left(\sum_{l=0}^{2r+2}\binom{2r+2}{l}\left(x(1-q_n^j)\right)^{2r+2-l}\left(\frac{1}{[n]_{q_n}^{l/2}}\right)\right)^{\frac{1}{r+1}}\right)^{1/2}\\ &\leqslant C\|D_{q_n}^{r+1}f\|\sum_{\substack{2i+j\leqslant r\\ i,j\geqslant 0}}[n+1]_{q_n}^{i+j}\left(\prod_{j=0}^{r}\left(\frac{[n]_{q_n}^{(2r+2)(\rho-1/2)}}{[n]_{q_n}^{(2r+2)\rho}}\right)^{\frac{1}{r+1}}\right)^{1/2}\\ &\leqslant C\|D_{q_n}^{r+1}f\|\frac{1}{[n]_{q_n}^{1/2}}.\end{aligned}$$

Since, this inequality is independent of ρ, it follows that $D^r_{q_n}\left(B^{q_n}_n(f,x)\right)$ converges pointwise to $D^r_{q_n}\left(f(x)\right)$ as $n\to\infty$ for $q_n\in(\hat{q_n},1)$.

Theorem 2 *Let $f\in C^r_\lambda[0,\infty)$ and $q\in(0,1)$. Then there exists $C>0$ independent of f and n such that*

$$\left|q^{2r}[n-r-1]_q\prod_{m=0}^{r}\left(\frac{[n-m]_q}{[n+m-1]_q}\right)D^r_q\left(B^q_n(f,x)\right)-D^r_q\left(f(x)\right)\right|$$
$$\leqslant C\omega\left(D^r_q f,[n-r-1]_q\sqrt{(1+q^{2r}-2q^{2r-1})x}\right),$$

for all $x\in[0,\infty)$.

Proof We have

$$\frac{[n-r-1]_q}{q^{-2r}}\sum_{k=0}^{\infty}b^q_{n+r,k}(x)\int_0^{\infty/A}q^k p^q_{n-r,k+r}(q^r t)\,d_qt=1.$$

This gives

$$\left|q^{2r}[n-r-1]_q\prod_{m=0}^{r}\left(\frac{[n-m]_q}{[n+m-1]_q}\right)D^r_q\left(B^q_n(f,x)\right)-D^r_q\left(f(x)\right)\right|$$
$$\leqslant q^{2r}[n-r-1]_q\sum_{k=0}^{\infty}b^q_{n+r,k}(x)\int_0^{\infty/A}q^k p^q_{n-r,k+r}(q^r t)\left|D^r_q f(t)-D^r_q f(x)\right|d_qt$$
$$\leqslant q^{2r}[n-r-1]_q\omega\left(D^r_q f,\delta\right)\sum_{k=0}^{\infty}b^q_{n+r,k}(x)\int_0^{\infty/A}q^k p^q_{n-r,k+r}(q^r t)\left(1+\frac{|t-x|}{\delta}\right)d_qt$$
$$=T_1+T_2,\ \text{say}.$$

We have $T_1=\omega\left(D^r_q f,\delta\right)$. Using Schwarz's inequality and Corollary 2, we obtain

$$T_2\leqslant\frac{q^{2r}[n-r-1]_q}{\delta}\left(\sum_{k=0}^{\infty}b^q_{n+r,k}(x)\int_0^{\infty/A}q^k p^q_{n-r,k+r}(q^r t)(t-x)^2\,d_qt\right)^{1/2}$$
$$\leqslant C\frac{q^{2r}[n-r-1]_q}{\delta}\sqrt{\left(\frac{(1+q^{2r}-2q^{2r-1})}{q^{4r}}x^2+O\left(\frac{1}{[n-r+1]_q}\right)\right)}.$$

Now, Choosing $\delta=[n-r-1]_q\sqrt{(1+q^{2r}-2q^{2r-1})x}$ we get

$$\left| q^{2r}[n-r-1]_q \prod_{m=0}^{r} \left(\frac{[n-m]_q}{[n+m-1]_q} \right) D_q^r \left(B_n^q(f,x) \right) - D_q^r \left(f(x) \right) \right|$$
$$\leqslant C\omega \left(D_q^r f, [n-r-1]_q \sqrt{(1+q^{2r}-2q^{2r-1})x} \right).$$

This completes the proof.

Remark 2 Let (q_n) be a sequence in (0, 1) such that $q_n \uparrow 1$ as $n \to \infty$ with the rate $1-q_n = O(1/[n]_q^\gamma)$, $\gamma > 2$. Then it follows that $(1+q^{2r}-2q^{2r-1}) = O(1/[n]_q^\gamma)$. Consequently, we obtain

$$\left| q^{2r}[n-r-1]_q \prod_{m=0}^{r} \left(\frac{[n-m]_q}{[n+m-1]_q} \right) D_q^r \left(B_n^q(f,x) \right) - D_q^r \left(f(x) \right) \right|$$
$$\leqslant C\omega \left(D_q^r f, \frac{x}{[n]_q^{(\gamma-2)/2}} \right).$$

The right-hand side tends to 0 on every finite compact subinterval of $[0, \infty)$.

Remark 3 From Theorem 1 it follows that $D_q^r \left(B_n^q(f) \right)$ do not converge to $D_q^r f$ unless we take a sequence (q_n) in (0, 1) such that $q_n \uparrow 1$. Similarly in Theorem 2, we observe that for a fixed q the right-hand side do not tend to 0 as $n \to \infty$. Therefore, the operators $B_n^q(f,x)$ fail to possess simultaneous approximation properties.

Acknowledgments The authors are indebted to anonymous referees for many valuable comments especially in connection with Lemma 1 which helped to improve the paper.

References

1. Aral, A., Gupta, V.: The q-derivative and applications to q-Szász Mirakyan operators. CALCOLO **43**, 151–170 (2006)
2. Aral, A., Gupta, V., Agarwal, R.P.: Applications of q-Calculus in Operator Theory. Springer, New York (2013)
3. De Sole, A., Kac, V.G.: On integral representations of q-gamma and q-beta functions. Rend. Mat. Acc. Lincei, s. 9. **16**, 11–29 (2005)
4. DeVore, R.A, Lorentz, G.G.: Constructive Approximation, Springer-Verlag, Berlin, New York (1993)
5. Doğru, O., Duman, O.: Statistical approximation of Meyer-König and Zeller operators based on the q-integers. Publ. Math. Debrecen **68**, 199–214 (2006)
6. Durrmeyer, J.L.: Une formula d'inversion, de la transformee de Laplace: Application a la theorie des Moments, These de 3e Cycle, Faculte des Sciences de l'universite de Paris (1967)
7. Ernst, T.: A new notation for q-calculus and a new q-Taylor formula, Uppsala University Report. Depart. Math. 1–28 (1999)
8. Gairola, A.R., Agrawal, P.N., Dobhal, G., Singh, K.K.: Moments of a q-Baskakov-beta operators in case $0 < q < 1$. J. Classical Anal. **2**(1), 9–22 (2013)

9. Gupta, V.: Some approximation properties of q-Durrmeyer operators. Appl. Math. Comput. **197**(1), 172–178 (2008)
10. Gupta, V.: On certain q-Durrmeyer type operators. Appl. Math. Comput. **209**, 415–420 (2009)
11. Gupta, V., Heping, W.: The rate of convergence of q-Durrmeyer operators for $0 < q < 1$. Math. Meth. Appl. Sci. **31**, 1946–1955 (2008)
12. Gupta, V., Kim, T.: On the rate of approximation by q-modified Beta operators. J. Math. Anal. Appl. **377**, 471–480 (2011)
13. Il'inskii, A., Ostrovska, S.: Convergence of generalized Bernstein polynomials. J. Approx. Theor. **116**(1), 100–112 (2002)
14. Jackson, F.H.: On a q-definite integrals. Quarterly J. Pure Appl. Math. **41**, 193–203 (1910)
15. Kac, V.G., Cheung, P.: Quantum Calculus. Universitext, Springer-Verlag, New York (2002)
16. Kantorovich, L.V.: Sur certain developments suivant les polynomes de la forms de S. Bernstein, Dokl. Akad. Nauk. SSSR. **1–2**, 563–568, 595–600 (1930)
17. Karsli, H., Gupta, V.: Some approximation properties of q-Chlodowsky operators. Appl. Math. Comput. **195**, 220–229 (2008)
18. Koornwinder, T.H.: q-Special Functions, a tutorial. In: Gerstenhaber, M., Stasheff, J. (eds.) Deformation and Qauntum Groups with Applications of Mathematical Physics, Contemp. Math., 134 (1992), Amer. Math. Soc. (1992)
19. Mahmudov, N.: The moments for q-Bernstein operators in the case $0 < q < 1$. Numer. Algor. 43–450 (2010)
20. Ostrovska, S.: q-Bernstein polynomials and their iterates. J. Approx. Theor. **123**(2), 232–255 (2003)
21. Phillips, G.M.: Bernstein polynomials based on the q-integers. Ann. Numer. Math. **4**, 511–518 (1997)
22. Phillips, G.M.: Interpolation and Approximation by Polynomials, CMS Books in Mathematics, vol. 14. Springer, Berlin (2003)
23. Wang, H., Meng, F.: The rate of convergence of q-Bernstein polynomials for $0 < q < 1$. J. Approx. Theor. **136**(2), 151–158 (2005)
24. Wang, H., Wu, X.: Saturation of convergence of q-Bernstein polynomials in the case $q > 1$. J. Math. Anal. Appl. **337**(1), 744–750 (2008)
25. Wang, H.: Voronovskaya-type formulas and saturation of convergence for q-Bernstein polynomials for $0 < q < 1$. J. Approx. Theor. **145**, 182–195 (2007)
26. Wang, H.: Properties of convergence for q-Bernstein polynomials. J. Math. Anal. Appl. **340**(2), 1096–1108 (2008)

Nonlinear Mixed Variational-Like Inequality with Respect to Weakly Relaxed $\eta - \alpha$ Monotone Mapping in Banach Spaces

Gayatri Pany and Sabyasachi Pani

Abstract In this paper, we have studied Nonlinear Mixed Variational-like inequality with respect to weakly relaxed $\eta - \alpha$ monotone mapping, involving a nonlinear bifunction, in Banach space. Significance of weakly relaxed $\eta - \alpha$ monotonicity is illustrated through an example. Existence of the solution to the problem is established using KKM (Knaster, Kuratowski and Mazurkiewicz) technique. Also we have proposed an iterative algorithm using auxiliary principle technique, which involves formulation of an auxiliary minimizing problem and then characterizing it by an auxiliary variational inequality problem. Solvability of the auxiliary variational inequality problem is established. Finally convergence of the iterates to the exact solution is proved.

Keywords Weakly relaxed $\eta - \alpha$ monotone mapping · KKM technique · Auxiliary principle technique · Iterative algorithm

1 Introduction

Study of variational inequality problem (in short VIP) mainly involves twofold aspect, namely qualitative and numerical. Qualitative aspect includes study of existence and uniqueness of the solution of the corresponding problem, while constructing iterative algorithm to find approximate solutions to the actual solution, study of convergence criteria, obtaining error bounds come under the numerical aspect. These two aspects of VIP provide an elegant framework of study in various fields like optimization, economics, transportation, oceanography, fluid flow through porous media, pure, applied and engineering sciences. These applications lead to generalizations of variational inequality theory in various directions. Variational-like inequalities problems (VLIP),

G. Pany (✉) · S. Pani
IIT Bhubaneswar, Odisha 751007, India
e-mail: gp11@iitbbs.ac.in

S. Pani
e-mail: spani@iitbbs.ac.in

P.N. Agrawal et al. (eds.), *Mathematical Analysis and its Applications*,
Springer Proceedings in Mathematics & Statistics 143,
DOI 10.1007/978-81-322-2485-3_14

generalized mixed variational-like inequalities are some of the importantgeneralization of variational inequality. Generalized mixed variational-like inequalities find applications in the areas like optimization theory (see [10, 12]), structural analysis [11], and economics [7].

The concept of monotonicity plays an important role for proving existence of solutions of VIP and VLIP. Chen [2] introduced semimonotonicity, using the combination of compactness and monotonicity and studied the corresponding variational inequality in Banach space. Huang and Deng [8] studied set-valued strongly nonlinear mixed variational-like inequality under strongly $\eta - \alpha$ monotonicity in Hilbert space. They have provided an iterative algorithm using auxiliary principle technique. There are substantial number of results on existence and uniqueness of variational inequalities under Hilbert space setting. These concepts were also generalized into Banach space setting. Fang and Huang [6] studied variational-like inequalities with respect to relaxed $\eta - \alpha$ monotone mapping in reflexive Banach space by introducing a new concept of relaxed $\eta - \alpha$ monotonicity. Later, this work was extended by Bai et al. [1], introducing the concept of relaxed $\eta - \alpha$ pseudomonotonicity. Recently, Kutbi and Sintunavarat [9] proved some existence results for variational-like inequalities involving a single operator using weakly relaxed $\eta - \alpha$ monotone mapping. But there was no discussion on the numerical aspect. Motivated by these works, we have extended these concepts to nonlinear mixed variational-like inequality with respect to weakly relaxed $\eta - \alpha$ monotone mapping, given by

$$\text{Find } w \in K, \langle N(w, y), \eta(v, w)\rangle + b(w, v) - b(w, w) \geq 0, \forall y, v \in K,$$

where $N : E \times E \to E^*$ is η-hemicontinuous and $b : E \times E \to \mathbb{R}$ is a convex lower semicontinuous function in second variable. Carrying out our study on the numerical aspect we have proposed an iterative algorithm to approximate the exact solution using auxiliary principle technique which is due to Glowinski et al. [11]. Following the ideas of Ding [4], we have discussed the covergence criteria for the proposed iterative algorithm.

2 Preliminaries

Definition 1 If $T : K \to E^*$ and $\eta : K \times K \to E^*$, then T is $\eta-$hemicontinuous if $f(t) = \langle T(x + t(y - x)), \eta(y, x)\rangle$ is continuous at 0, where $f : [0, 1] \to (-\infty, +\infty)$.

Definition 2 N and η are said to have $0-$diagonally concave relation, if the function $\phi : K \times K \to R$ defined by

$$\phi(w, v) = \langle N(w, v), \eta(w, v)\rangle$$

is 0−diagonally concave in v, i.e., for any finite set $\{v_1, \cdots, v_m\} \subset K$ and for any convex combination of v_i, $\sum_{i=1}^{m} \lambda_i \phi(w, v_i) \leq 0$. N and η are said to have 0−diagonally convex relation on K if $-N$ and η have 0−diagonally concave relation.

Definition 3 N is $\eta-$monotone with respect to first argument if

$$\langle N(w_1, v) - N(w_2, v), \eta(w, v)\rangle \geq 0, \ \forall w, v \in K.$$

Definition 4 N is $\eta-$antimonotone with respect to second argument if

$$\langle N(w, v_1) - N(w, v_2), \eta(w, v)\rangle \leq 0, \ \forall w, v \in K.$$

Definition 5 $F : K \to E^*$ is KKM mapping if, for any $\{x_1, \cdots, x_n\} \subset K$, $\text{co}\{x_1, \cdots, x_n\} \subset \bigcup_{i=1}^{n} F(x_i)$.

Definition 6 If $S = \{x_1, \cdots, x_n\}$, $\text{co}\{x_1, \cdots, x_n\} = \left\{ \sum_{i=1}^{n} \alpha_i x_i : \sum_{i=1}^{n} \alpha_i = 1, \forall \alpha_i \geq 0 \right\}$.

Definition 7 $f : K \to (-\infty, +\infty]$ is lower semicontinuous at x_0 if $f(x_o) \leq \liminf_{x \to x_0} f(x)$.

Definition 8 $T : K \to E^*$ is Lipschitz continuous if $\exists\, \alpha > 0$, such that $\|Tx - Ty\| \leq \alpha \|x - y\|, \forall\, x, y \in K$.

Definition 9 T is $\eta-$coercive with respect to a proper function $f : K \to (-\infty, +\infty]$ if there exists $x_0 \in K$ such that

$$\frac{\langle Tx - Tx_0, \eta(x, x_0)\rangle + f(x) - f(x_0)}{\|\eta(x, x_0)\|} \to \infty.$$

Definition 10 If $T : K \to E^*$, $\eta : K \times K \to E$, $\alpha : E \to \mathbb{R}$, $t > 0, z \in E$ and $p > 1$ is a constant, then T is

- weakly relaxed $\eta - \alpha$ monotone if $\langle Tx - Ty, \eta(y, x)\rangle \geq \alpha(x - y), \ \forall x, y \in K$, where $\lim_{t \to 0} \alpha(tz) = 0, \ \lim_{t \to 0} \frac{d}{dt}\alpha(tz) = 0$.
- relaxed $\eta - \alpha$ monotone if $\alpha(tz) = t^p \alpha(z)$ and $\langle Tx - Ty, \eta(y, x)\rangle \geq \alpha(x - y), \ \forall x, y \in K$.
- strongly $\eta - \alpha$ monotone if $\langle Tx - Ty, \eta(y, x)\rangle \geq c\|x - y\|^2, \ \forall x, y \in K$. Here $\alpha(tz) = c\|z\|^2$; $c > 0$ is a constant,

Remark 1 From the definitions it is obvious that

- Strongly $\eta - \alpha$ monotonicity is a special case of relaxed $\eta - \alpha$ monotonicity.
- Relaxed $\eta - \alpha$ monotonicity implies weakly relaxed $\eta - \alpha$ monotonicity.

But weakly relaxed $\eta - \alpha$ monotonicity does not always imply relaxed $\eta - \alpha$ monotonicity.

Example 1 If $T : \left[0, \frac{\pi}{2}\right] \to \mathbb{R}$, $Tx = \sin x$, $\eta(x, y) = x - y$, $\alpha(z) = -\sin^2 z$, then $\lim_{t \to 0} \alpha(tz) = 0$ and $\lim_{t \to 0} \frac{d}{dt}\alpha(tz) = 0$.

$$\langle Tx - Ty, \eta(y, x)\rangle = \langle \sin x - \sin y, x - y\rangle \geq -\sin^2(x - y) = \alpha(x - y).$$

So T is weakly relaxed $\eta - \alpha$ monotone, but not relaxed $\eta - \alpha$ monotone as $\alpha(tz) \neq t^p\alpha(z)$.

The following Lemma is used to prove the existence result for our problem.

Lemma 1 ([5]) *If M is a nonempty subset of a Hausdorff topological vector space X, $F : M \to 2^X$ is a KKM mapping, $F(x)$ is closed in X, $\forall x \in K$ and compact for some $x \in K$, then $\bigcap_{x \in M} F(x) \neq \phi$.*

Another useful result that we have used to prove the existence of solution for the auxiliary variational inequality is the following:

Lemma 2 ([3]) *Let K be a nonempty convex subset of a topological vector space and let $\phi : K \times K \to R$ be such that:*

1. *for each $x \in K$, $y \to \phi(x, y)$ is lower semicontinuous on each nonempty compact subset of K,*
2. *for each nonempty finite set $\{x_1, \cdots, x_m\} \subset K$ and for each $y = \sum_{i=1}^{m} \lambda_i x_i$, $(\lambda_i \geq 0, \sum_{i=1}^{m} = 1)$, $min_{1<i<m}\phi(x_i, y) \leq 0$,*
3. *there exists nonempty compact convex subset X_0 of K and a nonempty compact subset D of K such that for each $y \in K/D$, there is an $x \in co(X_0 \cup \{y\})$ with $\phi(x, y) > 0$.*

Then there exists an $\hat{y} \in D$ such that $\phi(x, \hat{y}) \leq 0$, for all $x \in K$.

3 Results

3.1 Existence Result

We have first established an equivalence between the problems (1) and (2) given below. Next we have shown that the set-valued mapping $F : K \to 2^E$ is a KKM mapping. Then applying Lemma 1, solvability is proved. In Theorem 3 adding an extra condition of $\eta-$coercivity to the mapping N, solvability is established in the case, where K is unbounded.

Theorem 1 *Let K be a nonempty compact convex subset of a real reflexive Banach space E and E^* be the dual space of E. Let $N : E \times E \to E^*$ be $\eta-$hemicontinuous and weakly relaxed $\eta - \alpha$ monotone and $b : E \times E \to \mathbb{R}$ be a convex lower semicontinuous function in second variable, such that,*

1. $\eta(w, w) = 0, \forall w \in K,$
2. $v \to \langle N(w, y), \eta(v, w)\rangle$ *is convex for any* $w, y \in K$.

Then the following problems are equivalent:

$$w \in K, \langle N(w, y), \eta(v, w)\rangle + b(w, v) - b(w, w) \geq 0, \forall v \in K, \tag{1}$$

$$w \in K, \langle N(v, y), \eta(v, w)\rangle + b(w, v) - b(w, w) \geq \alpha(w - v), \forall v \in K. \tag{2}$$

Proof Let $w \in K$ be a solution of (1). As N is weakly relaxed $\eta - \alpha$ monotone, we have,

$$\begin{aligned}\langle N(v, y), \eta(v, w)\rangle + b(w, v) - b(w, w) &\geq \langle N(w, y), \eta(v, w)\rangle + \alpha(w - v)\\ &+ b(w, v) - b(w, w) \geq \alpha(w - v), \forall v \in K.\end{aligned}$$

So w is a solution of (2). Conversely let $w \in K$ be a solution of (2). Let $v_t = (1 - t)w + tv, t \in (0, 1)$. So $v_t \in K$. As $w \in K$ is a solution of (2), we have,

$$\langle N(v_t, y), \eta(v_t, w)\rangle + b(w, v_t) - b(w, w) \geq \alpha\big(t(w - v)\big), \text{ and}$$

$$b(w, v_t) - b(w, w) \leq t(b(w, v) - b(w, w)).$$

Using these results, we get,

$$\langle N(w + t(v - w), y), \eta(v, w)\rangle + b(w, v) - b(w, w) \geq \frac{\alpha\big(t(w - v)\big)}{t}, \forall v \in K.$$

Since N is $\eta-$hemicontinuous and $\lim_{t\to 0} \frac{d}{dt}\alpha(tz) = 0$, letting $t \to 0$ and applying L'Hospital's rule, we get,

$$\langle N(w, y), \eta(v, w)\rangle + b(w, v) - b(w, w) \geq 0, \forall v \in K.$$

This completes the proof. □

Theorem 2 *If $N : E \times E \to E^*$ is $\eta-$ hemicontinuous and weakly relaxed $\eta - \alpha$ monotone and $b : E \times E \to \mathbb{R}$ is a convex lower semicontinuous function in second variable and the following assumptions*

1. $\eta(v, v) = 0, \ \forall y \in K,$
2. $v \to \langle N(w, y), \eta(v, w)\rangle$ *is convex and lower semicontinuous for any* $w, y \in K$,
3. *For any* v_β, v_β *converging to* v, $\alpha(v) \leq \liminf \alpha(v_\beta)$,

hold, then problem (1) is solvable.

Proof Let $F, G : K \to 2^E$ be defined by,

$$F(v) = \{w \in K, \langle N(w, y), \eta(v, w)\rangle + b(w, v) - b(w, w) \geq 0\}, \forall v \in K,$$

$$G(v) = \{w \in K, \langle N(v, y), \eta(v, w)\rangle + b(w, v) - b(w, w) \geq \alpha(w - v)\}, \forall v \in K.$$

We claim that F is a KKM mapping, if not, then there exists $\{v_1, \cdots, v_n\} \subset K$ and $t_i > 0, i = 1, 2, \cdots, n$, such that, $\sum_{i=1}^{n} t_i = 1$, $v = \sum_{i=1}^{n} t_i v_i \notin \bigcup_{i=1}^{n} F(v_i)$. Then by definition of F, $\langle N(v, y), \eta(v_i, v)\rangle + b(w, v_i) - b(w, v) < 0$, for $i = 1, 2 \cdots, n$. By our assumption

$$\begin{aligned} 0 &= \langle N(v, y), \eta(v, v)\rangle \\ \Longrightarrow \langle N(v, y), \eta\left(\sum_{i=1}^{n} t_i v_i, v\right)\rangle &\leq \sum_{i=1}^{n} t_i \langle N(v, y), \eta(v_i, v)\rangle \\ &< \sum_{i=1}^{n} t_i (b(w, v) - b(w, v_i)) \\ &= b(w, v) - \sum_{i=1}^{n} t_i b(w, v_i) \\ &\leq b(w, v) - b(w, v) = 0. \end{aligned}$$

This is a contradiction. So F is a KKM mapping. We now claim that $F(v) \subset G(v), \forall v \in K$. Let $w \in F(v)$, then

$$\langle N(w, y), \eta(v, w)\rangle + b(w, v) - b(w, w) \geq 0, \text{ for any } v \in K.$$

Using weakly relaxed $\eta - \alpha$ monotonicity of N, we have $w \in G(v)$. So $F(v) \subset G(v), \forall v \in K$. So G is a KKM mapping. As α is weakly lower semicontinuous, $G(v)$ is weakly closed. $v \to \langle N(w, y), \eta(v, w)\rangle$ and b are convex and lower semicontinuous and hence weakly lower semicontinuous. As K is bounded closed and convex in the reflexive Banach space E, it is weakly compact. Since $G(v)$ is weakly closed for all $v \in K$, it is weakly compact and hence the family $\{G(v)\}$ has finite intersection property, i.e., $\bigcap_{v \in K} G(v) \neq \Phi$. So conditions of Lemma 1 are satisfied, hence we have,

$$\bigcap_{v \in K} F(v) = \bigcap_{v \in K} G(v) \neq \Phi.$$

So there exists $w \in K$ such that

$$\langle N(w, y), \eta(v, w)\rangle + b(w, v) - b(w, w) \geq 0, \forall v \in K.$$

□

Next we prove the existence result for an unbounded set K.

Theorem 3 *If $N : E \times E \to E^*$ is $\eta-$hemicontinuous and weakly relaxed $\eta - \alpha$ monotone, $b : E \times E \to \mathbb{R}$ is a convex lower semicontinuous function in second variable, K is a nonempty closed convex unbounded subset of E and the following assumptions*

1. *for $w_0 \in K$, $[\langle N(w, y) - N(w_0, y), \eta(v, w_0)\rangle + b(w, v) - b(w_0, v)]/\|\eta(v, w_0)\| \to \infty$, as $\|v\| \to \infty$; i.e., N is $\eta-$coercive with respect to b in the second variable.*
2. *$\eta(w, v) + \eta(v, w) = 0, \ \forall w, v \in K$,*
3. *$v \to \langle N(w, y), \eta(v, w)\rangle$ is convex and lower semicontinuous for any $w, y \in K$,*
4. *for any v_β, v_β converging to v, $\alpha(v) \leq \liminf \alpha(v_\beta)$,*

hold, then problem (1) is solvable.

Proof Consider the problem, find $w_r \in K \cap B_r$ such that

$$\langle N(w_r, y), \eta(v, w_r)\rangle + b(w_r, v) - b(w_r, w_r) \geq 0, \ \forall v \in K \cap B_r \tag{3}$$

where $B_r = \{v \in E : \|v\| \leq r\}$.

By Theorem 2 (3) has a solution $w_r \in K \cap B_r$. Choosing $\|w_0\| < r$, we can put w_0 in place of v in (3), so we have,

$$\langle N(w_r, y), \eta(w_0, w_r)\rangle + b(w_r, w_0) - b(w_r, w_r) \geq 0.$$

Now,

$$\begin{aligned}
&\langle N(w_r, y), \eta(w_0, w_r)\rangle + b(w_r, w_0) - b(w_r, w_r)\\
&= -\langle N(w_r, y), \eta(w_r, w_0)\rangle + \langle N(w_0, y), \eta(w_0, w_r)\rangle + \langle N(w_0, y), \eta(w_r, w_0)\rangle\\
&+ b(w_r, w_0) - b(w_r, w_r)\\
&= -\langle N(w_r, y) - N(w_0, y), \eta(w_r, w_0)\rangle + b(w_r, w_0) - b(w_r, w_r) + \langle N(w_0, y), \eta(w_0, w_r)\rangle\\
&\leq \|\eta(w_r, w_0)\|\left(\frac{-\langle N(w_r, y) - N(w_0, y), \eta(w_r, w_0)\rangle + b(w_r, w_r) - b(w_r, w_0)}{\|\eta(w_r, w_0)\|} + \|N(w_0, y)\|\right).
\end{aligned}$$

If $\|w_r\| = r$ and $r \to \infty$, then by $\eta-$coercivity of N with respect to b in the second variable, the above inequality reduces to

$$\langle N(w_r, y), \eta(w_0, w_r)\rangle + b(w_r, w_0) - b(w_r, w_r) < 0.$$

This is a contradiction as $\langle N(w_r, y), \eta(w_0, w_r)\rangle + b(w_r, w_0) - b(w_r, w_r) \geq 0$. So $\|w_r\| < r$. Now for any $v \in K$, we choose $0 < \varepsilon < 1$, such that,

$$w_r + \varepsilon(v - w_r) \in K \cap B_r.$$

By assumption 3 and convexity of b, we have from Eq. (3),

$$\begin{aligned}0 &\le \langle N(w_r, y), \eta(w_r + \varepsilon(v - w_r), w_r)\rangle + b(w_r + \varepsilon(v - w_r)) - b(w_r, w_r)\\ &\le (1-\varepsilon)\langle N(w_r, y), \eta(w_r, w_r)\rangle + \varepsilon\langle N(w_r, y), \eta(v, w_r)\rangle + (1-\varepsilon)b(w_r, w_r) + \varepsilon b(v, w_r)\\ &\quad - b(w_r, w_r)\\ &= \varepsilon\langle N(w_r, y), \eta(v, w_r)\rangle + \varepsilon b(w_r, v) - \varepsilon b(w_r, w_r).\end{aligned}$$

So,

$$\langle N(w_r, y), \eta(v, w_r)\rangle + b(w_r, v) - b(w_r, w_r) \ge 0, \forall y \in K$$

and $w_r \in K$ is a solution of problem (1). Hence problem (1) is solvable. □

3.2 Iterative Algorithm and Convergence Analysis

In this section, we have obtained an iterative algorithm for finding approximate solutions to the nonlinear mixed variational-like inequality problem (1), using auxiliary principle technique. This technique involves first formulating an auxiliary minimizing problem and then characterizing it by an auxiliary variational inequality problem.

In the formulation of the auxiliary minimizing problem, the differentiable convex functional $\alpha : E \to \mathbb{R}$ is considered as an auxiliary differentiable convex functional. The auxiliary minimizing problem is defined as follows:

$$\min_{w \in K}\{\alpha(w) + \langle \rho N(v, y), \eta(w, v)\rangle - \langle \alpha'(v), w\rangle + \rho b(v, w)\}, \tag{4}$$

where $w \in E, v \in K$ and ρ is a positive constant. If $w \mapsto \langle N(v, y), \eta(w, v)\rangle$ is convex, then (4) is equivalent to following auxiliary variational inequality problem in the sense that solution to both the problems are same. The auxiliary variational inequality problem is given by,

$$\langle \alpha'(w) - \alpha'(v), u - w\rangle \ge -\rho\langle N(v, y), \eta(u, w)\rangle + \rho b(v, w) - \rho b(v, u), \text{ for all } u \in K. \tag{5}$$

Note 1 If $w = v$, then v is a solution of (1).

Keeping in view these results we propose an iterative algorithm as follows:

1. Let v_0 be the initial approximation for $n = 0$.
2. At the nth step solve the auxiliary minimizing or auxiliary variational inequality problem with $v^* = v_n$. Let v_{n+1} be the solution.
3. If $\|v_{n+1} - v_n\| \le \varepsilon;\ \varepsilon > 0$, stop, otherwise repeat 2.

Next, we prove the result that gurantees the existence of solution of (4) or (5).

Theorem 4 *Let E be a reflexive Banach space with dual space E^*, $N : E \times E \to E^*$, $b : E \times E \to \mathbb{R}$ be $\eta-$hemicontinuous, convex lower semicontinuous function in second variable, linear in the first argument, bounded and $b(u, v) - b(u, w) \le b(u, v - w)$ and $\alpha : E \to \mathbb{R}$ be a differentiable convex functional, such that*

1. $v \to \langle N(w, y), \eta(v, w)\rangle$ is convex and lower semicontinuous for any $w, y \in K$,
2. N is weakly lower semicontinuous, weakly relaxed $\eta - \alpha$ monotone and $\eta-$convex with respect to first argument for any $w, y \in K$,
3. N is strongly $\eta-$monotone in first argument, $\eta-$antimonotone and weakly relaxed monotone in second argument,
4. N is Lipschitz continuous in first and second argument with respect to constants σ_1 and σ_2 respectively,
5. $\eta(v, w) = \eta(v, u) + \eta(u, w)$, for any $u, v, w \in K$,
6. $\eta(v, u) + \eta(u, w) = 0$, for any $u, v, w \in K$,
7. η is Lipschitz continuous with respect to constant $\delta > 0$,
8. N and η have $0-$diagonally convex relation with respect to first argument,
9. $w \to \alpha'(w)$ is continuous from weak to strong topology and α' is strongly monotone,
10. $\alpha \neq 2\mu$, $(\sigma_1 + \sigma_2)\delta + \mu > 0$ and $0 < \rho < \frac{\rho(\sigma_1+\sigma_2)\delta}{\alpha-2\mu}$.

Then there exists a solution $w \in K$ of the problem (1) and for each $\rho > 0$, there exists a solution $w_{n+1} \in K$ of problems (4) or (5) and the approximate solutions converge strongly to the exact solution.

Proof As N is $\eta-$convex with respect to first argument for any $w, y \in K$, it is easy to check that condition 2 of Theorem 2 is satisfied. By condition 6, $\eta(v, v) = 0$. Hence condition 1 of Theorem 2 is satisfied. So all the conditions of Theorem 2 are satisfied. Hence solution to problem (1) exists. Now to prove the second part of the conclusion, we have to show that all the conditions of Lemma 2 are satisfied. For this purpose we define $\phi : K \times K \to R$, by,

$$\phi(u, w) = \langle \alpha'(v_n) - \alpha'(w), u - w\rangle - \rho\langle N(v_n, y_n), \eta(u, w)\rangle + \rho b(v_n, w) - \rho b(v_n, u).$$

As $w \to \alpha'(w)$ is continuous from weak to strong topology, the function $w \to \langle \alpha'(w), w\rangle$ is weak continuous on K. So $w \to \phi(u, w)$ is weakly lower semicontinuous. So condition 1 is satisfied. To prove the second condition we assume the contrary. So there exists $\{u_1, \cdots, u_n\} \subset K$ and w which is a convex combination of u_i, such that $\phi(u_i, w) > 0$. From this we get,

$$\sum_{i=1}^{n} \lambda_i \langle \alpha'(v_n) - \alpha'(w), u_i - w\rangle - \rho\langle N(v_n, y_n), \eta(u_i, w)\rangle + \rho b(v_n, w) - \rho \sum_{i=1}^{n} \lambda_i b(v_n, u_i) > 0.$$

As b is convex in the second argument we have

$$\sum_{i=1}^{n} \lambda_i \langle \alpha'(v_n) - \alpha'(w), u_i - w\rangle - \rho\langle N(v_n, y_n), \eta(u_i, w)\rangle > 0.$$

This contradicts condition 8. So condition 2 of Lemma 2 holds. Now considering a set $D = \{v \in K : \|v - u^*\| \leq \theta\}$, where $\theta = \frac{1}{\alpha}[\mu\|u^*\|] + \delta\|N(u^*, y)\|$ and using condition 2, 3, 6, 8 and conditions on b, condition 3 is proved. Hence all the conditions of Lemma reflem2 are satisfied. So there exists $w_0 \in K$ such that

$$\langle \alpha'(w_0) - \alpha'(v_n), u - w_0 \rangle \geq -\rho \langle N(v_n, y_n), \eta(u, w_0) \rangle + \rho b(v_n, w_0) - \rho b(v_n, u) \quad (6)$$

for all $u \in K$. Hence there exists a solution.

Now for convergence analysis we consider the following functional $\Gamma : K \to (-\infty, +\infty]$, defined by,

$$\Gamma(v) = \alpha(v_0) - \alpha(v) - \langle \alpha'(v), v_0 - v \rangle,$$

where v_0 is assumed to be the unique solution of problem (1). By strong monotonicity of α', we have,

$$\Gamma(v) = \alpha(v_0) - \alpha(v) - \langle \alpha'(v), v_0 - v \rangle \geq \frac{\sigma}{2} \|v - v_0\|^2.$$

Putting $w_0 = v_{n+1}$, $u = v_0$ in (6) and by antisymmetricity of η, strong monotonicity of α', we get,

$$\begin{aligned}
\Gamma(v_n) - \Gamma(v_{n+1}) &\geq \frac{\sigma}{2} \|v_n - v_{n+1}\|^2 + \rho \langle N(v_n, y_n), \eta(v_{n+1}, v_0) \rangle + \rho b(v_n, v_{n+1}) \\
&\quad - \rho b(v_n, v_0) = \frac{\sigma}{2} \|v_n - v_{n+1}\|^2 + \rho \langle N(v_n, y_n) - N(v_0, y_0), \eta(v_{n+1}, v_0) \rangle \\
&\quad + \rho \langle N(v_0, y_0), \eta(v_{n+1}, v_0) \rangle + \rho b(v_n, v_{n+1}) - \rho b(v_n, v_0).
\end{aligned}$$

v_0 being a solution of (1), it follows that,

$$\begin{aligned}
\Gamma(v_n) - \Gamma(v_{n+1}) \geq & \frac{\sigma}{2} \|v_n - v_{n+1}\|^2 + \rho \langle N(v_n, y_n) - N(v_0, y_0), \eta(v_{n+1}, v_0) \rangle \\
& + \rho [b(v_0, v_0) - b(v_0, v_{n+1}) + b(v_n, v_{n+1}) - b(v_n, v_0)] \\
& - \frac{\sigma}{2} \|v_n - v_{n+1}\|^2 + M.
\end{aligned}$$

Now using the conditions on b and the conditions 3–5 and 7, we get,

$$\begin{aligned}
M = & \rho \langle N(v_n, y_n) - N(v_0, y_0), \eta(v_{n+1}, v_0) \rangle \\
& - \rho [b(v_n - v_0, v_0) - b(v_n - v_0, v_{n+1}) + b(v_n - v_0, v_n) - b(v_n - v_0, v_n)] \\
& \geq \rho [\langle N(v_n, y_n) - N(v_0, y_0), \eta(v_{n+1}, v_n) \rangle + \langle N(v_n, y_n) - N(v_0, y_0), \eta(v_n, v_0) \rangle] \\
& - \rho [b(v_n - v_0, v_0 - v_n) + b(v_n - v_0, v_n - v_{n+1})] \\
& \geq \rho [\langle N(v_n, y_n) - N(v_0, y_n), \eta(v_n, v_0) \rangle \\
& + \langle N(v_0, y_n) - N(v_0, y_0), \eta(v_n, v_0) \rangle \\
& + \langle N(v_n, y_n) - N(v_0, y_n), \eta(v_{n+1}, v_n) \rangle \\
& + \langle N(v_0, y_n) - N(v_0, y_0), \eta(v_{n+1}, v_n) \rangle] \\
& - \rho \mu [\|v_n - v_0\|^2 + \|v_n - v_0\| \|v_n - v_{n+1}\|] \\
& \geq \rho \alpha \|v_n - v_0\|^2 - \rho \sigma_1 \delta \|v_n - v_0\| \|v_{n+1} - v_n\| \\
& - \rho \sigma_2 \delta \|v_n - v_0\| \|v_{n+1} - v_n\| - \rho \mu [\|v_n - v_0\|^2 + \|v_n - v_0\| \|v_n - v_{n+1}\|].
\end{aligned}$$

From this we have,

$$\Gamma(v_n) - \Gamma(v_{n+1}) \geq \rho[\alpha - (2\mu + (\sigma_1 + \sigma_2)\delta)]\|v_n - v_0\|^2.$$

Condition 10 implies that $\{\Gamma(v_n)\}$ is a strictly decreasing sequence and it is nonnegative by the strong monotonicity property and hence converges. So $\{v_n\}$ converges to v_0 strongly as $n \to \infty$. This completes the proof. □

4 Concluding Remarks

In this work, we have studied the existence of the solution of nonlinear mixed variational-like inequality with respect to weakly realaxed $\eta - \alpha$ monotone mapping in case of both bounded and unbounded sets. We have obtained an iterative algorithm using auxiliary principle technique and we have shown that the iterates approximate to the exact solution strongly. Further we are trying to frame this problem for nonconvex setting using hemivariational inequality concept.

Acknowledgments The first author wishes to thank *DST-INSPIRE* (Code No. IF110762) for the grant of research fellowship and to *IIT Bhubaneswar* for providing the research facilities.

References

1. Bai, M.R., Zhou, S.Z., Ni, G.Y.: Variational-like inequalities with relaxed pseudomonotone mappings in banach spaces. Appl. Math. Lett. **19**(6), 547–554 (2006)
2. Chen, Y.Q.: On the semi-monotone operator theory and applications. J. Math. Anal. Appl. **231**(1), 177–192 (1999)
3. Ding, X., Tan, K.: A minimax inequality with applications to existence of equilibrium point and fixed point theorems. Colloq. Math. **63**, 233–247 (1992)
4. Ding, X.P.: General algorithm for nonlinear variational-like inequalities in reflexive banach spaces. Indian J. Pure Appl. Math. **29**(2), (1998)
5. Fan, K.: Some properties of convex sets related to fixed point theorems. Math. Ann. **266**(4), 519–537 (1984)
6. Fang, Y., Huang, N.: Variational-like inequalities with generalized monotone mappings in banach spaces. J. Optim. Theory Appl. **118**(2), 327–338 (2003)
7. Huang, N.J.: On the generalized implicit quasivariational inequalities. J. Math. Anal. Appl. **216**(1), 197–210 (1997)
8. Huang, N.J., Deng, C.X.: Auxiliary principle and iterative algorithms for generalized set-valued strongly nonlinear mixed variational-like inequalities. J. Math. Anal. Appl. **256**(2), 345–359 (2001)

9. Kutbi, M.A., Sintunavarat, W.: On the solution existence of variational-like inequalities problems for weakly relaxed monotone mapping. In: Abstract and Applied Analysis, vol. 2013. Hindawi Publishing Corporation (2013)
10. Tian, G.: Generalized quasi-variational-like inequality problem. Math. Oper. Res. **18**(3), 752–764 (1993)
11. Tremolieres, R., Lions, J.L., Glowinski, R.: Numerical analysis of variational inequalities. Elsevier (2011)
12. Yao, J.C.: The generalized quasi-variational inequality problem with applications. J. Math. Anal. Appl. **158**(1), 139–160 (1991)

Pál Type (0; 1)-Interpolation on Mixed Tchebycheff Abscissas-II

Neha Mathur and Pankaj Mathur

Abstract In this paper, we have considered the interpolation problem when function values are prescribed on the zeros of (n-1)th Tchebycheff polynomial of second kind and weighted first derivatives are prescribed on the zeros of nth Tchebycheff polynomial of first kind. It has been shown that such an interpolation exists when n is even, the explicit representation of which has been obtained. The convergence theorem for the interpolatory polynomial has also been dealt with.

Keywords Interpolation · Tchebycheff polynomials · Inter-scaled zeros

1 Introduction

Let $\{x_{2i,2n+1}\}_{i=1}^{n}$ and $\{x_{2i+1,2n+1}\}_{i=1}^{n-1}$ be two distinct point systems in the interval $[-1, 1]$, which are interscaled such that

$$-1 = x_{2n+1,2n+1} < x_{2n,2n+1} < \cdots < x_{1,2n+1} = 1 \tag{1}$$

where the points $\{x_{2i,2n+1}\}_{i=1}^{n}$ are the distinct zeros of $T_n(x)$ the nth Tchebycheff polynomial of first kind and $\{x_{2i+1,2n+1}\}_{i=1}^{n-1}$ are the distinct zeros of $U_{n-1}(x)$ the nth Tchebycheff polynomial of second kind. Further, let $\{\alpha_{i,2n+1}\}_{i=1}^{2n+1}$ be arbitrary given numbers. We seek to find a polynomial $S_n(x)$ of degree $\leq 2n$ satisfying the conditions:

$$S_n(x_{2i+1,2n+1}) = \alpha_{2i+1,2n+1}; \quad i = 0, 1, 2, \ldots, n \tag{2}$$

$$\left(\sqrt{1-x^2}\,S_n\right)'(x_{2i,2n+1}) = \alpha_{2i,2n+1}; \quad i = 1, 2, \ldots, n \tag{3}$$

N. Mathur
Department of Mathematics, Career Convent College, Lucknow, India
e-mail: neha_mathur13@yahoo.com

P. Mathur (✉)
Department of Mathematics and Astronomy, Lucknow University, Lucknow, India
e-mail: pankaj_mathur14@yahoo.co.in

P.N. Agrawal et al. (eds.), *Mathematical Analysis and its Applications*,
Springer Proceedings in Mathematics & Statistics 143,
DOI 10.1007/978-81-322-2485-3_15

In 1975, Pál [11] introduced an interpolation process on an interscaled set of points

$$-\infty < x_n < y_{n-1} < \cdots < x_2 < y_1 < x_1 < \infty. \tag{4}$$

where $\{x_i\}_{i=1}^{n}$ and $\{y_i\}_{i=1}^{n-1}$ are the distinct real zeros of (say)

$$W_n(x) = \prod_{l=1}^{n}(x - x_i).$$

and of $W_n'(x)$, respectively. Pál proved that for given arbitrary numbers $\{\alpha_i^*\}_{i=1}^{n}$ and $\{\beta_i^*\}_{i=1}^{n-1}$ there exists a unique polynomial of degree $\leq 2n - 1$ satisfying the conditions:

$$R_n(x_i) = \alpha_i^* \ , \quad i = 1, 2, \ldots, n, \tag{5}$$

$$R_n'(y_i) = \beta_i^* \ , \quad i = 1, 2, \ldots, n - 1 \tag{6}$$

and an initial condition

$$R_n(a) = 0$$

where a is a given point, different from the nodal points (4). After which many mathematicians have taken up this problem on different sets of nodes. For more details, one is referred to [1–3, 5–10, 12–15, 17, 18], etc.

In this paper, we have considered the converse problem considered in [8]. We have shown that for n even, there exist a unique polynomial $S_n(x)$ of degree $\leq 2n$ satisfying the conditions (2)–(3). The explicit representation of $S_n(x)$ is obtained and the estimates of the fundamental polynomials leading to the convergence theorem has also been dealt with.

In Sect. 2, we give preliminaries. Existence, uniqueness, and the explicit representation of the interpolatory polynomials have been dealt with in Sect. 3. Section 4 is devoted to the estimation of the fundamental polynomials and the proof of the convergence theorem.

2 Preliminaries

We characterize the points

$$x_{2i} = \cos\left(i - \frac{1}{2}\right)\frac{\pi}{n}, \qquad i = 1(1)n \tag{7}$$

as the zeros of $T_n(x) = \cos n\theta$, $x = cos\theta(-1 < x < 1)$, nth Tchebycheff polynomial of first kind and

$$x_{2i+1} = \cos\left(\frac{i\pi}{n}\right), \qquad i = 1(1)n-1 \tag{8}$$

as the zeros of $U_{n-1}(x)$ the $(n-1)$th Tchebycheff polynomial of second kind. Obviously, $x = 0$ either in $\{x_{2i}\}_{i=1}^{n}$ or in $\{x_{2i+1}\}_{i=1}^{n-1}$ according as n is even or odd. Also $x_i = -x_{2n+2-i}$, $i = 1(1)n$. The differential equation satisfied by $T_n(x)$ [16] is

$$(1-x^2)T_n''(x) - xT_n'(x) + n^2T_n(x) = 0 \tag{9}$$

and that $U_{n-1}(x)$ is

$$(1-x^2)U_{n-1}''(x) - 3xU_{n-1}'(x) + (n^2-1)U_{n-1}(x) = 0. \tag{10}$$

We have that for $i = 1, 2, \ldots, n$

$$\ell_{2i}(x) = \frac{T_n(x)}{(x - x_{2i}T_n'(x_{2i}))} = \frac{1}{n} + \frac{2}{n}\sum_{r=1}^{n} T_r(x_{2i})T_r(x) \tag{11}$$

and for $i = 1, 2, \ldots, n-1$

$$\begin{aligned}\ell_{2i+1}(x) &= \frac{U_{n-1}(x)}{(x - x_{2i+1}U_{n-1}'(x_{2i+1}))}\\ &= \frac{2(1-x_{2i+1}^2)}{n}\sum_{r=1}^{n-1} U_r(x_{2i+1})U_r(x).\end{aligned} \tag{12}$$

Also,

$$\sum_{i=1}^{n} |\ell_{2i}(x)| = O(\log n), \tag{13}$$

$$\sum_{i=1}^{n-1} |\ell_{2i+1}(x)| = O(n), \tag{14}$$

$$\sum_{k=1}^{n}\left(1-x_k^2\right)^{-m} = \begin{cases} O(n\log n), & m = 1/2\\ O(n^2), & m = 1\end{cases} \tag{15}$$

and

$$c_1\frac{k}{n} \le \sqrt{1-x_k^2} \le c_2\frac{k}{n} \tag{16}$$

where $x_k's$ are the zeros of $T_n(x)$ or $U_{n-1}(x)$.

3 Existence, Uniqueness, and Explicit Representation of the Interpolatory Polynomials

We shall prove the following:

Theorem 1 *Let n be even and the (2n + 1) points in [−1, 1] be given by Eq.* (1) *together with* (7) *and* (8)*, then there exist a unique polynomial* $S_n(x)$ *of degree* $\leq 2n$ *satisfying the conditions* (2) *and* (3)*. But if n is odd then there exists, in general, no polynomial of degree* $\leq 2n$ *and if the polynomials exist, they are infinitely many.*

For n even, the interpolatory polynomial $S_n(x)$ satisfying the conditions (2) and (3) can be represented as:

$$S_n(x) = \sum_{i=0}^{n} \alpha_{2i+1} A_{2i+1}(x) + \sum_{i=1}^{n} \alpha_{2i} B_{2i}(x) \tag{17}$$

where $\{A_{2i+1}(x)\}_{i=0}^{n}$ and $\{B_{2i}(x)\}_{i=1}^{n}$ are called the fundamental polynomials of first and second kind, respectively, each of degree $\leq 2n$, which are uniquely determined by the following conditions: for $i = 0, 1, 2, \ldots, n$

$$\begin{cases} A_{2i+1}(x_{2j+1}) = \delta_{ij}, & j = 0, 1, 2, \ldots, n \\ \left(\sqrt{1-x^2} A_{2i+1}\right)'(x_{2j}) = 0, & j = 1, 2, \ldots, n \end{cases} \tag{18}$$

and for $i = 1, 2, \ldots, n$

$$\begin{cases} B_{2i}(x_{2j+1}) = 0, & j = 0, 1, 2, \ldots, n \\ \left(\sqrt{1-x^2} B_{2i}\right)'(x_{2j}) = \delta_{ij}, & j = 1, 2, \ldots, n \end{cases} \tag{19}$$

The explicit forms of fundamental polynomials are given in the following:

Theorem 2 *For n even, the fundamental polynomials* $\{B_{2i}(x)\}_{i=1}^{n}$ *satisfying the conditions* (19) *can be represented as*

$$B_{2i}(x) = \frac{U_{n-1}(x)}{\sqrt{1-x_{2i}^2}\, U_{n-1}(x_{2i})} \left[\int_{-1}^{x} \ell_{2i}(x)dx + c_{2i} \int_{-1}^{x} T_n(x)dx \right] \tag{20}$$

where $\ell_{2i}(x)$*are defined by* (11) *and*

$$c_{2i} = -\left(\int_{-1}^{1} T_n(x)dx\right)^{-1} \int_{-1}^{1} \ell_{2i}(x)dx.$$

Theorem 3 *For n even, the fundamental polynomials* $\{A_{2i+1}(x)\}_{i=0}^{n}$ *satisfying the conditions* (18) *can be represented as for* $i = 1, 2, \ldots, n-1$

$$A_{2i+1}(x) = \frac{(1-x^2)T_n(x)\ell_{2i+1}(x)}{(1-x_{2i+1}^2)T_n(x_{2i+1})} - \frac{U_{n-1}(x)}{(1-x_{2i+1}^2)T_n(x_{2i+1})U'_{n-1}(x_{2i+1})}\Big[\int_{-1}^{x}\frac{(1-x^2)T'_n(x)}{x-x_{2i+1}}dx + c_{3i}\int_{-1}^{x}T_n(x)dx\Big] \tag{21}$$

where $\ell_{2i+1}(x)$ *are given by* (12) *and*

$$c_{3i} = -\left(\int_{-1}^{1}T_n(x)dx\right)^{-1}\int_{-1}^{1}\frac{(1-x^2)T'_n(x)}{x-x_{2i+1}}dx.$$

For i = 0, n;

$$A_1(x) = c_3U_{n-1}(x)\int_{-1}^{x}T_n(x)dx$$

where

$$c_3 = \left(n\int_{-1}^{1}T_n(x)dx\right)^{-1}.$$

Similarly we have

$$A_{2n+1}(x) = c_4U_{n-1}(x)\int_{1}^{x}T_n(x)dx$$

where $c_4 = -c_3$.

The polynomial $S_n(x)$, for n even satisfies the following quantitative estimate:

Theorem 4 *Let* $f(x)$ *have a continuous derivative in* $[-1, 1]$*, then*

$$S_n(f, x) = \sum_{i=0}^{n} f(x_{2i+1})A_{2i+1}(x) + \sum_{i=1}^{n} f'(x_{2i})B_{2i}(x) \tag{22}$$

satisfies the relation:

$$|S_n(x) - f(x)| = O(1)\left[(1-x^2)|T_n(x)| + |U_{n-1}(x)|\log n + \frac{\sqrt{1-x^2}}{n}\right]\omega\left(f', \frac{\sqrt{1-x^2}}{n}\right)$$

where $\delta_n(y) = \frac{\sqrt{1-y^2}}{n}$ *and* $c_r(r = 1, 2, 3)$ *are constants, independent of* f, n *and* x.

We will prove only our main Theorem 4 in this chapter, as the proof of other Theorems is quite similar to that of theorems in [5]. In order to prove the theorem, we shall need the estimates of the fundamental polynomials.

4 Estimation of the Fundamental Polynomials

Lemma 1 *For n even and* $x \in [-1, 1]$, *we have*

$$\left|\int_{-1}^{x} T_n(x)dx\right| \le 2(n^2-1)^{-1} \tag{23}$$

and

$$\left|\int_{-1}^{x} \ell_{2i}(x)dx\right| \le O\left(\frac{\log n}{n}\right) \tag{24}$$

where $\ell_{2i}(x)$ *are given by* (11).

Proof On integrating the left-hand side of (23), we get

$$\int_{-1}^{x} T_n(x)dx = \frac{1}{2}\left[\frac{T_{n-1}(x)}{n-1} - \frac{T_{n+1}(x)}{n+1} + \frac{2}{n^2-1}\right]$$

from which (23) follows. For (24), we have by (11)

$$\begin{aligned}
\left|\int_{-1}^{x} \ell_{2i}(x)dx\right| &= \left|\int_{-1}^{x}\left\{\frac{1}{n} + \frac{2}{n}\sum_{r=1}^{n} T_r(x_{2i})T_r(x)\right\}dx\right| \\
&\le \frac{2}{n} + \frac{1}{n}\left|\sum_{r=1}^{n} T_r(x_{2i})\int_{-1}^{x} T_r(x)dx\right| \\
&\le \frac{2}{n} + \frac{1}{2n}\left[\frac{|x^2-1|}{2} + \sum_{r=2}^{n}\left|\frac{\cos(r-1)\theta}{r-1} - \frac{\cos(r+1)\theta}{r+1} + \frac{2}{r^2-1}\right|\right] \\
&= O\left(\frac{\log n}{n}\right).
\end{aligned}$$

Thus (24) follows. □

Lemma 2 *For n even and* $x \in [-1, 1]$, *we have*

$$\left|\int_{-1}^{x} \frac{(1-x^2)T_n'(x)}{x-x_{2i+1}}dx\right| \le \frac{4n\log n}{\sqrt{1-x_{2i+1}^2}}. \tag{25}$$

Proof Since $T_n'(x) = nU_{n-1}(x)$, we have

$$\left| \int_{-1}^{x} \frac{(1-x^2)T_n'(x)}{x - x_{2i+1}} dx \right| \leq n \left| U_{n-1}'(x_{2i+1}) \right| \left| \int_{-1}^{x} (1-x^2)\ell_{2i+1}(x)dx \right|. \tag{26}$$

By (12), we have

$$\left| \int_{-1}^{x} \left(1-x^2\right) \ell_{2i+1}(x)dx \right| \leq \frac{2\left(1-x_{2i+1}^2\right)}{n} \left| \sum_{r=0}^{n-1} U_r(x_{2i+1}) \int_{-1}^{x} \left(1-x^2\right) U_r(x)dx \right| \tag{27}$$

Since

$$\int_{-1}^{x} (1-x^2)U_r(x)dx = \left[\frac{T_{r-1}(x)}{4(r-1)} + \frac{T_{r+3}(x)}{4(r+3)} - \frac{T_{r+1}(x)}{2(r+1)} \right] + \frac{(-1)^r}{4} \left[\frac{2r^2+8r+1}{(r-1)(r+3)(r+1)} \right] \tag{28}$$

hence by using (27) and (28) in (26) the lemma follows. □

Lemma 3 *For n even,* $i = 1, 2, 3, \ldots, n$ *and* $x \in [-1, 1]$*, we have*

$$|B_{2i}(x)| \leq |U_{n-1}(x)| \ \frac{\log n}{n}$$

Proof The proof of this Lemma follows by (20) and Lemma 1. □

Lemma 4 *For n even,* $i = 1, 2, 3, \ldots, n-1$ *and* $x \in [-1, 1]$*, we have*

$$|A_{2i+1}(x)| \leq \frac{\left|\left(1-x^2\right) T_n(x)\ell_{2i+1}(x)\right|}{\left|\left(1-x_{2i+1}^2\right)\right|} + \frac{4\,|U_{n-1}(x)| \log n}{\sqrt{1-x_{2i+1}^2}}$$

Proof Using (23) and Lemma 2 in (21) the Lemma follows. □

5 Proof of the Main Theorem

In order to prove our main Theorem 4, we need the following important result of Gopengaus [4]: Let $f \in C^r[-1, 1]$, then for $n \geq 4r + 5$, there exists a polynomial $Q_n(x)$ of degree atmost n such that for all $x \in [-1, 1]$ and for $k = 0, 1, \ldots, r$

$$\left| f^{(k)}(x) - Q_n^{(k)}((x) \right| \leq c_k \left(\delta_n(x)\right)^{r-k} \omega\left(f^{(r)}, \delta_n(x)\right), \tag{29}$$

where $\delta_n(x) = \frac{\sqrt{1-x^2}}{n}$ and $c_k's$ are constants independent of f, n, and x.

Proof of Theorem 4 From the uniqueness of $S_n(x)$ in (22), it follows that every polynomial $Q_n(x)$ of degree $\leq 2n$ with the property (29) satisfies the relation

$$Q_n(x) = \sum_{i=0}^{n} Q_n(x_{2i+1})A_{2i+1}(x) + \sum_{i=1}^{n} Q_n'(x_{2i})B_{2i}(x).$$

Therefore,

$$\begin{aligned}
|S_n(x) - f(x)| &\leq |S_n(x) - f(x)| + |S_n(x) - f(x)| \\
&\leq \sum_{i=0}^{n} |f(x_{2i+1}) - Q_n(x_{2i+1})| \; |A_{2i+1}(x)| \qquad (30)\\
&+ \sum_{i=1}^{n} |f'(x_{2i+1}) - Q_n'(x_{2i+1})||B_{2i}(x)| + |Q_n(x) - f(x)| \\
&\equiv \Sigma_1 + \Sigma_2 + |Q_n(x) - f(x)| \qquad (31)
\end{aligned}$$

Now by (29) and Lemma 4, we have

$$\Sigma_1 \leq \sum_{i=1}^{n-1} \frac{c_8\sqrt{1 - x_{2i+1}^2}}{n} \omega\left(f', \frac{\sqrt{1 - x_{2i+1}^2}}{n}\right)\left[\frac{|(1 - x^2)T_n(x)\ell_{2i+1}(x)|}{|(1 - x_{2i+1}^2)|} + \frac{4\,|U_{n-1}(x)|\log n}{\sqrt{1 - x_{2i+1}^2}}\right]$$

Using the property of the modulus of continuity $\omega\left(f', \lambda\delta\right) \leq (1 + \lambda)\omega\left(f', \delta\right)$, we have

$$\begin{aligned}
\Sigma_1 &\leq \sum_{i=1}^{n-1} \frac{c_8}{n}\omega\left(f', \frac{\sqrt{1 - x^2}}{n}\right)\left(1 + \frac{\sqrt{1 - x_{2i+1}^2}}{\sqrt{1 - x^2}}\right)\left[\frac{|(1 - x^2)T_n(x)\ell_{2i+1}(x)|}{\sqrt{1 - x_{2i+1}^2}} + 4|(1 - x^2)U_{n-1}(x)|\log n\right] \\
&\leq \frac{c_8}{n}\omega\left(f', \frac{\sqrt{1 - x^2}}{n}\right)\left[\left|\left(1 - x^2\right) T_n(x)\right| \sum_{i=1}^{n-1} \frac{\ell_{2i+1}(x)}{\sqrt{1 - x_{2i+1}^2}} + 4n|U_{n-1}(x)|\log n \right.\\
&\left. + \left|\sqrt{(1 - x^2)}T_n(x)\right| \sum_{i=1}^{n-1} \ell_{2i+1}(x) + \frac{4|U_{n-1}(x)|\log n}{\sqrt{1 - x^2}} \sum_{i=1}^{n-1} \sqrt{1 - x_{2i+1}^2}\right]
\end{aligned}$$

Thus by Schwartz inequality, (14), (15) and (16), we have

$$\Sigma_1 \leq c_9 \omega\left(f', \frac{\sqrt{1-x^2}}{n}\right)\left[\left|(1-x^2)T_n(x)\right| + c_{10}\left|U_{n-1}(x)\right| \log n\right] \tag{32}$$

Also by (29) and Lemma 3, we have

$$\Sigma_2 \leq \sum_{i=1}^{n} c_{11} \omega\left(f', \frac{\sqrt{1-x_{2i}^2}}{n}\right)\left|U_{n-1}(x)\right| \frac{\log n}{n}.$$

Again by the property of the modulus of continuity, we have

$$\Sigma_2 \leq c_{11} \frac{\log n}{n} \omega\left(f', \frac{\sqrt{1-x^2}}{n}\right) \sum_{i=1}^{n-1}\left(1 + \frac{\sqrt{1-x_{2i}^2}}{\sqrt{1-x^2}}\right)$$

Thus by (15), we have

$$\Sigma_2 \leq c_{12}\left|U_{n-1}(x)\right| \log n \omega\left(f', \frac{\sqrt{1-x^2}}{n}\right). \tag{33}$$

Hence using (29), (32) and (33) in (30), Theorem 4 follows.

References

1. Akhlagi, M.R., Sharma, A.: Some Pál Type Interpolation Problems, Approx., Optimization and Computing Theory and Applications (A.G. Law and C.L. Wang Ceds.), pp. 37–40. Elsevier Science Publisher B.V. (North Holland), IMACS (1990)
2. Datta, S., Mathur, P.: On Pál type weighted (0,2;0)-interpolation on infinite interval $(-\infty, \infty)$. Approx. Theory Appl. China **17**(4), 1–10 (2001)
3. Datta, S., Mathur, P.: Modified weighted (0; 0, 2)-interpolation on infinite interval $(-\infty, \infty)$, Ann. Univ. Sci. Budapest **44**, 39–52 (2001)
4. Gopengauz, I.E.: On the Theorem of A.F. Timan on approximation of continuous functions on a line segment. Math. Zametski **2**, 163–172 (1967)
5. Joó, I.: On Pál interpolation. Ann. Univ. Sci. Budapest, Sect. Math. **37**, 247–262 (1994)
6. Joó, I., Pál, L.G.: Lacunary (0; 0, 1)-interpolation on the roots of Jacobi polynomials and their derivatives, respectively I (existence, explicit formulae, unicity). Anal. Math. **22**, 289–298 (1996)
7. Mathur, P.: (0, 1; 0)-interpolation on infinite interval $(-\infty, +\infty)$. Anal. Theory Appl. **22**(2), 105–113 (2006)
8. Mathur, N., Mathur, P.: Pál type (0; 1)-interpolation on mixed Tchebycheff abscissas-I. Int. J. Math. Arch. **4**(8), 225–235 (2013)
9. Mathur, K.K., Srivastava, R.: Pál type interpolation on infinite interval. J. Math. Anal. Appl. **192**, 346–359 (1995)
10. Muneer, Y.E.: On the convergence of (0, 1, 2) interpolation. Serdica Math. J. **31**, 87–94 (2005)
11. Pál, L.G.: A new modification of H-F interpolation. Anal. Math. **I**, 197–205 (1975)

12. Pál, L.G.: A general Lacunary (0; 0, 1)-interpolation process. Anal. Univ. Budapest, Sect. Comp. **16**, 291–301 (1996)
13. Sebestyen, Z.F.: Pál type interpolation on the roots of Hermite polynomials. Pure Math. Appl. **9**(3–4), 429–439 (1998)
14. Srivastava, R., Mathur, K.K.: Weighted (0; 0, 2)-interpolation on the roots of Hermite polynomials. Acta. Math. Hung. **70**(1–2), 57–73 (1996)
15. Szabados, J., Varma, A.K.: On a convergent Pál-type (0,2)-interpolation process. Acta. Math. Hungar. **66**(4), 301–326 (1995)
16. Szegö, G.: Orthogonal polynomials. Am. Math. Soc. Coll. Publ. (1959)
17. Szili, L.: A convergence theorem for the Pál, method of interpolation on the roots of Hermite polynomials. Anal. Math. **11**, 75–84 (1985)
18. Xie, T.F.: On Pás problem. Chinese Quart. J. Math. **7**, 48–52 (1992)

Degree of Approximation of $f \in L[0, \infty)$ by Means of Fourier–Laguerre Series

Soshal Saini and Uaday Singh

Abstract In this paper, we determine the degree of approximation of functions belonging to $L[0, \infty)$ by the Hausdorff means of its Fourier–Laguerre series at $x = 0$. Our theorem extends some of the recent results of Nigam and Sharma [A study on degree of approximation by $(E, 1)$ summability means of the Fourier–Laguerre expansion, Int. J. Math. Math. Sci. (2010), Art. ID 351016, 7], Krasniqi [On the degree of approximation of a function by $(C, 1)(E, q)$ means of its Fourier–Laguerre series, International Journal of Analysis and Applications 1 (2013), 33–39] and Sonker [Approximation of Functions by $(C, 2)(E, q)$ means of its Fourier–Laguerre series, Proceeding of ICMS-2014 ISBN 978-93-5107-261-4:125–128.] in the sense that the summability methods used by these authors have been replaced by the Hausdorff matrices.

Keywords Degree of approximation · Hausdorff means · Fourier–Laguerre series

1 Introduction

Let f be a function belonging to $L[0, \infty)$ in the sense that f is Labesgue integrable in the interval $[0, \infty)$. The Fourier–Laguerre expansion of f is given by

$$f(x) \sim \sum_{n=0}^{\infty} a_n L_n^{(\alpha)}(x), \tag{1}$$

S. Saini (✉) · U. Singh
Department of Mathematics, Indian Institute of Technology Roorkee, Roorkee 247667, India
e-mail: sainisoshal25@gmail.com

U. Singh
e-mail: usingh2280@yahoo.co.in

P.N. Agrawal et al. (eds.), *Mathematical Analysis and its Applications*, Springer Proceedings in Mathematics & Statistics 143, DOI 10.1007/978-81-322-2485-3_16

where

$$\Gamma(\alpha+1)\binom{n+\alpha}{n}a_n=\int_0^\infty e^{-x}x^\alpha f(x)L_n^{(\alpha)}(x)dx \tag{2}$$

and $L_n^{(\alpha)}(x)$ denotes the nth Laguerre polynomial of order $\alpha > -1$, defined by the generating function

$$\sum_{n=0}^{\infty} L_n^{(\alpha)}(x)\omega^n=(1-\omega)^{-\alpha-1}\exp\left(\frac{-x\omega}{1-\omega}\right). \tag{3}$$

When $x=0$,

$$L_n^{(\alpha)}(0)=\binom{n+\alpha}{n} \text{ [9]}.$$

The nth partial sums of (1) are defined by

$$s_n(f;x)=\sum_{k=0}^{n} a_k L_k^{(\alpha)}(x). \tag{4}$$

The Cesàro means of order λ of the Fourier–Laugerre series are defined by

$$C_n^\lambda(f;x)=\frac{1}{\binom{n+\lambda}{n}}\sum_{k=0}^{n}\binom{\lambda+n-k-1}{n-k}s_k(f;x).$$

The Euler means of the Fourier–Laugerre series are defined by

$$E_n^q(f;x)=\frac{1}{(1+q)^n}\sum_{k=0}^{n}\binom{n}{k}q^{n-k}s_k(f;x),\quad q>0.$$

The Hausdorff matrix $H\equiv(h_{n,k})$ is an infinite lower triangular matrix defined by

$$h_{n,k}=\begin{cases}\binom{n}{k}\Delta^{n-k}\mu_k\ , 0\le k\le n,\\ 0 \qquad\qquad\quad , \ k>n,\end{cases}$$

where Δ is the forward difference operator defined by $\Delta\mu_n=\mu_n-\mu_{n+1}$ and $\Delta^{k+1}\mu_n=\Delta^k(\Delta\mu_n)$. If H is regular, then $\{\mu_n\}$, known as moment sequence, has the representation

$$\mu_n=\int_0^1 u^n d\gamma(u),$$

where $\gamma(u)$, known as mass function, is continuous at $u = 0$ and belongs to $BV[0, 1]$ such that $\gamma(0) = 0, \gamma(1) = 1$; and for $0 < u < 1$, $\gamma(u) = [\gamma(u + 0) + \gamma(u - 0)]/2$ [11].

The Hausdorff means of the Fourier–Laugerre series are defined by

$$H_n(f; x) := \sum_{k=0}^{n} h_{n,k} s_k(f; x), \quad n = 0, 1, 2, ... \tag{5}$$

The Fourier–Laugerre series is said to be summable to s by the Hausdorff means, if $H_n(f; x) \to s \text{ as } n \to \infty$, [3].

For the examples of Hausdorff matrices, one can see [7, 8, 11] and references therein.

In this paper, the class of all regular Hausdorff matrices with moment sequence $\{\mu_n\}$ associated with mass function $\gamma(u)$ having constant derivative, is denoted by H_1.

We also write

$$\varphi(y) = \frac{e^{-y} y^{\alpha}(f(y) - f(0))}{\Gamma(\alpha + 1)},$$

and

$$g(u, y) = \sum_{k=0}^{n} \binom{n}{k} u^k (1 - u)^{n-k} L_k^{(\alpha+1)}(y).$$

2 Known Results

Gupta [2] obtained the degree of approximation of $f \in L[0, \infty)$ by Cesàro means of order k of the Fourier–Laguerre series at the point $x = 0$, where $k > \alpha + 1/2$. Nigam and Sharma [5] have used $(E, 1)$ means of the Fourier–Laguerre series for $-1 < \alpha < 1/2$ which is more appropriate range from the application point of view. The authors have proved the following result:

Theorem A *If*

$$E_n^1 = \frac{1}{2^n} \sum_{k=0}^{n} \binom{n}{k} s_k \quad \to \infty \text{ as } n \to \infty,$$

then the degree of approximation of Fourier–Laguerre expansion at the point $x = 0$ by $(E, 1)$ means E_n^1 is given by

$$E_n^1(0) - f(0) = o(\xi(n)), \tag{6}$$

provided that

$$\Phi(t) = \int_0^t |\varphi(y)|dy = o\left(t^{\alpha+1}\xi(1/t)\right), \quad t \to 0, \tag{7}$$

$$\int_\delta^n e^{y/2} y^{-((2\alpha+3)/4)} |\varphi(y)|dy = o\left(n^{-((2\alpha+1)/4)}\xi(n)\right), \tag{8}$$

$$\int_n^\infty e^{y/2} y^{-1/3} |\varphi(y)|dy = o(\xi(n)), \quad n \to \infty, \tag{9}$$

where δ is a fixed positive constant and $\alpha \in (-1, -1/2)$, and $\xi(t)$ is a positive monotonic increasing function of t such that $\xi(n) \to \infty$ as $n \to \infty$.

Following, Nigam and Sharma [5], Krasniqi [4] has used the $(C, 1)(E, q)$ means of the Fourier–Laguerre series to obtain the degree of approximation of $f \in L[0, \infty)$ at point $x = 0$ and has proved the following result:

Theorem B *The degree of approximation of the Fourier–Laguerre expansion at the point $x = 0$ by the $[(C, 1)(E, q)]_n$ means is given by*

$$[(C, 1)(E, q)]_n(0) - f(0) = o(\xi(n)), \tag{10}$$

provided that the conditions (7)–(9) given in Theorem A are satisfied.

Recently, Sonker [10] has also proved the same result using $[(C, 2)(E, q)]_n$ means of the Fourier–Laguerre series for the same range of α as follows:

Theorem C *The degree of approximation of the Fourier–Laguerre expansion at the point $x = 0$ by the $[(C, 2)(E, q)]_n$ means is given by*

$$[(C, 2)(E, q)]_n(0) - f(0) = o(\xi(n)), \tag{11}$$

provided that the conditions (7)–(9) given in Theorem A are satisfied.

Remark 1 We observe that Krasniqi [4, p. 37] has optimized $\sum_{k=0}^{\nu} \binom{\nu}{k} q^k k^{(2\alpha+1)/4}$ by its maximum value $(1+q)^\nu \nu^{(2\alpha+1)/4}$. This is possible only when $\alpha > -1/2$. But the author has used $-1 < \alpha < 1/2$ [4, p. 35, Theorem 2.1]. Similar error can also be seen in [5, 10].

3 Main Results

In this paper, we extend the above results using the Hausdorff means, which is a more general summability method, for an appropriate range of α. More precisely, we prove the following:

Theorem 1 *The degree of approximation of $f \in L[0, \infty)$ at the point $x = 0$ by the Hausdorff means of the Fourier–Laguerre series generated by $H \in H_1$ is given by*

$$H_n(f; 0) - f(0) = o(\xi(n)) \tag{12}$$

where $\xi(t)$ is a positive increasing function such that $\xi(t) \to \infty$ as $t \to \infty$ and satisfies the following conditions

$$\Phi(y) = \int_0^t |\varphi(y)|dy = o\left(t^{\alpha+1}\xi(1/t)\right), \quad t \to 0, \tag{13}$$

$$\int_\delta^n e^{y/2}\, y^{-((2\alpha+3)/4)}|\varphi(y)|dy = o\left(n^{-((2\alpha+1)/4)}\xi(n)\right), \tag{14}$$

and

$$\int_n^\infty e^{y/2}\, y^{-1/3}|\varphi(y)|dy = o(\xi(n)), \quad n \to \infty, \tag{15}$$

where δ is a fixed positive constant and $\alpha > -1/2$.

For the proof of our theorem, we need the following lemmas:

Lemma 1 [9, *p*. 177]. *Let α be an arbitrary real number, c and δ be fixed positive constants. Then*

$$L_n^{(\alpha)}(x) = \begin{cases} O\left(n^{(\alpha)}\right), & \text{if } 0 \le x \le \frac{c}{n}, \\ O\left(x^{-(2\alpha+1)/4} n^{(2\alpha-1)/4}\right), & \text{if } \frac{c}{n} \le x \le \delta, \end{cases} \tag{16}$$

as $n \to \infty$.

Lemma 2 [9, *p*. 240]. *Let α be an arbitrary real number, $\delta > 0$ and $0 < \eta < 4$. Then*

$$\max e^{-x/2} x^{(\alpha/2+1/4)}|L_n^{(\alpha)}(x)| = \begin{cases} O\left(n^{(\alpha/2-1/4)}\right), & \text{if } \delta \le x \le (4-\eta)n, \\ O\left(n^{(\alpha/2-1/12)}\right), & \text{if } x \ge \delta, \end{cases} \tag{17}$$

as $n \to \infty$.

Lemma 3 *For $0 < u < 1$ and $0 \le y \le \delta$,*

$$\left|\int_0^1 g(u, y)d\gamma(u)\right| = \begin{cases} O\left(n^{(\alpha+1)}\right), & \text{if } 0 \le y \le \frac{1}{n}, \\ O\left(y^{-(2\alpha+3)/4} n^{(2\alpha+1)/4}\right), & \text{if } \frac{1}{n} \le y \le \delta, \end{cases} \tag{18}$$

as $n \to \infty$.

Proof The $g(u, y)$ can be written as

$$g(u,y) = (1-u)^n \sum_{k=0}^{n} \binom{n}{k} \left(\frac{u}{1-u}\right)^k L_k^{(\alpha+1)}(y).$$

Then

$$\begin{aligned}\left|\int_0^1 g(u,y)d\gamma(u)\right| &= \left|\int_0^1 (1-u)^n \sum_{k=0}^{n} \binom{n}{k} \left(\frac{u}{1-u}\right)^k L_k^{(\alpha+1)}(y)d\gamma(u)\right| \\ &= \left|M\int_0^1 (1-u)^n \sum_{k=0}^{n} \binom{n}{k} \left(\frac{u}{1-u}\right)^k L_k^{(\alpha+1)}(y)du\right|\end{aligned}$$

Now, using Lemma 1 for $0 \le y \le \frac{1}{n}$ (taking $\alpha+1$ for α and $c = 1$), we have

$$\begin{aligned}\left|\int_0^1 g(u,y)d\gamma(u)\right| &= \int_0^1 (1-u)^n \sum_{k=0}^{n} \binom{n}{k} \left(\frac{u}{1-u}\right)^k O(k^{\alpha+1})du \\ &= O\left(n^{\alpha+1}\int_0^1 (1-u)^n \sum_{k=0}^{n} \binom{n}{k} \left(\frac{u}{1-u}\right)^k du\right) \\ &= O\left(n^{\alpha+1}\int_0^1 (1-u)^n du\right) \\ &= O\left(n^{\alpha+1}\right). \end{aligned} \tag{19}$$

Again, using Lemma 1 for $\frac{1}{n} \le y \le \delta$, we have

$$\begin{aligned}\left|\int_0^1 g(u,y)d\gamma(u)\right| &= \int_0^1 (1-u)^n \sum_{k=0}^{n} \binom{n}{k} \left(\frac{u}{1-u}\right)^k O\left(y^{-(2\alpha+3)/4}k^{(2\alpha+1)/4}\right)du \\ &= O\left(y^{-(2\alpha+3)/4}n^{(2\alpha+1)/4}\int_0^1 (1-u)^n \sum_{k=0}^{n} \binom{n}{k} \left(\frac{u}{1-u}\right)^k du\right) \\ &= O\left(y^{-(2\alpha+3)/4}n^{(2\alpha+1)/4}\right). \end{aligned} \tag{20}$$

Collecting (19) and (20), the proof of Lemma 3 is completed.

Lemma 4 *For* $0 < u < 1$,

$$\left|\int_0^1 g(u,y)d\gamma(u)\right| = \begin{cases} O\left(e^{y/2}y^{-(2\alpha+3)/4}n^{(2\alpha+1)/4}\right), & \text{if } \delta \le y \le n, \\ O\left(e^{y/2}y^{-(3\alpha+5)/6}n^{(\alpha+1)/2}\right), & \text{if } y \ge \delta, \end{cases} \tag{21}$$

as $n \to \infty$.

Proof Following the Lemma 3, we have

$$\left|\int_0^1 g(u, y)d\gamma(u)\right| = \left|\int_0^1 (1-u)^n \sum_{k=0}^{n} \binom{n}{k} \left(\frac{u}{1-u}\right)^k L_k^{(\alpha+1)}(y)du\right|$$

Now, using Lemma 2 for $\delta \leq y \leq n$ (taking $\alpha + 1$ for α and $\eta = 3$), we have

$$\begin{aligned}
\left|\int_0^1 g(u, y)d\gamma(u)\right| &= \left|\int_0^1 e^{(y/2)} y^{-(2\alpha+3)/4} (1-u)^n \sum_{k=0}^{n} \binom{n}{k} \left(\frac{u}{1-u}\right)^k e^{-(y/2)} y^{(2\alpha+3)/4} L_k^{(\alpha+1)}(y)du\right| \\
&= \int_0^1 e^{y/2} y^{-(2\alpha+3)/4} (1-u)^n \sum_{k=0}^{n} \binom{n}{k} \left(\frac{u}{1-u}\right)^k O\left(k^{(2\alpha+1)/4}\right) du \\
&= O\left(e^{y/2} y^{-(2\alpha+3)/4} n^{(2\alpha+1)/4}\right). \qquad (22)
\end{aligned}$$

Again, using Lemma 2 for $y \geq n$, we have

$$\begin{aligned}
\left|\int_0^1 g(u, y)d\gamma(u)\right| &= \left|\int_0^1 e^{(y/2)} y^{-(3\alpha+5)/6} (1-u)^n \sum_{k=0}^{n} \binom{n}{k} \left(\frac{u}{1-u}\right)^k e^{-(y/2)} y^{(3\alpha+5)/6} L_k^{(\alpha+1)}(y)du\right| \\
&= \int_0^1 e^{y/2} y^{-(3\alpha+5)/6} (1-u)^n \sum_{k=0}^{n} \binom{n}{k} \left(\frac{u}{1-u}\right)^k O\left(k^{(\alpha+1)/2}\right) du \\
&= O\left(e^{y/2} y^{-(3\alpha+5)/6} n^{(\alpha+1)/2}\right). \qquad (23)
\end{aligned}$$

Collecting (22) and (23), the proof of Lemma 4 is completed.

Proof of Theorem 1 We have

$$\begin{aligned}
s_n(0) &= \sum_{k=0}^{n} a_k L_k^{(\alpha)}(0) \\
&= \sum_{k=0}^{n} \frac{1}{\Gamma(\alpha+1)\binom{n+\alpha}{n}} \left(\int_0^\infty e^{-y} y^\alpha f(y) L_k^{(\alpha)}(y)dy\right) L_k^{(\alpha)}(0) \\
&= \frac{1}{\Gamma(\alpha+1)} \int_0^\infty e^{-y} y^\alpha f(y) \sum_{k=0}^{n} L_k^{(\alpha)}(y)dy \\
&= \frac{1}{\Gamma(\alpha+1)} \int_0^\infty e^{-y} y^\alpha f(y) L_n^{(\alpha+1)}(y)dy,
\end{aligned}$$

so that

$$\begin{aligned}
H_n(f; 0) &= \sum_{k=0}^{n} h_{n,k} s_k(0) \\
&= \sum_{k=0}^{n} \binom{n}{k} \Delta^{n-k} \mu_k \frac{1}{\Gamma(\alpha+1)} \int_0^\infty e^{-y} y^\alpha f(y) L_k^{(\alpha+1)}(y)dy.
\end{aligned}$$

Thus

$$\begin{aligned}
H_n(f;0)-f(0) &= \sum_{k=0}^{n}\binom{n}{k}\Delta^{n-k}\mu_k\left(\frac{1}{\Gamma(\alpha+1)}\int_0^\infty e^{-y}y^\alpha f(y)L_k^{(\alpha+1)}(y)dy-f(0)\right)\\
&= \sum_{k=0}^{n}\binom{n}{k}\Delta^{n-k}\mu_k\frac{1}{\Gamma(\alpha+1)}\int_0^\infty e^{-y}y^\alpha (f(y)-f(0))L_k^{(\alpha+1)}(y)dy\\
&= \sum_{k=0}^{n}\binom{n}{k}\Delta^{n-k}\mu_k\int_0^\infty \varphi(y)L_k^{(\alpha+1)}(y)dy\\
&= \int_0^\infty \varphi(y)\left(\sum_{k=0}^{n}\binom{n}{k}\Delta^{n-k}\mu_k L_k^{(\alpha+1)}(y)\right)dy\\
&= \int_0^\infty \varphi(y)\left(\sum_{k=0}^{n}\binom{n}{k}\int_0^1 u^k(1-u)^{n-k}d\gamma(u)L_k^{(\alpha+1)}(y)\right)dy\\
&= \int_0^\infty \varphi(y)\left(\int_0^1\sum_{k=0}^{n}\binom{n}{k}u^k(1-u)^{n-k}L_k^{(\alpha+1)}(y)d\gamma(u)\right)dy\\
&= \int_0^\infty \varphi(y)\left(\int_0^1 g(u,y)d\gamma(u)\right)dy
\end{aligned}$$

and

$$\begin{aligned}
|H_n(f;0)-f(0)| &= \left|\int_0^\infty \varphi(y)\left(\int_0^1 g(u,y)d\gamma(u)\right)dy\right|\\
&\le \int_0^\infty |\varphi(y)|\left|\int_0^1 g(u,y)d\gamma(u)\right|dy\\
&= \left(\int_0^{1/n}+\int_{1/n}^{\delta}+\int_\delta^n+\int_n^\infty\right)\left(|\varphi(y)|\left|\int_0^1 g(u,y)d\gamma(u)\right|dy\right)\\
&= I_1+I_2+I_3+I_4. \qquad (24)
\end{aligned}$$

Now, using Lemma 3 for $0\le y\le \frac{1}{n}$, we have

$$\begin{aligned}
I_1 &= \int_0^{1/n}|\varphi(y)|\left|\int_0^1 g(u,y)d\gamma(u)\right|dy\\
&= O\left(n^{\alpha+1}\right)\int_0^{1/n}|\varphi(y)|dy\\
&= O(n^{(\alpha+1)})o\left(\left(\frac{1}{n}\right)^{\alpha+1}\xi(n)\right)\\
&= o(\xi(n)), \qquad (25)
\end{aligned}$$

in view of condition (13).

Further, using Lemma 3 for $\frac{1}{n} \le y \le \delta$, we have,

$$I_2 = \int_{1/n}^{\delta} |\varphi(y)| O\left(y^{-(2\alpha+3)/4} n^{(2\alpha+1)/4}\right) dy$$
$$= O\left(n^{(2\alpha+1)/4}\right)\left(\int_{1/n}^{\delta} y^{-(2\alpha+3)/4} |\varphi(y)| dy\right).$$

Following [5, p. 6], we have

$$I_2 = o(\xi(n)), \tag{26}$$

in view of condition (13).

Now, using Lemma 4 for $\delta \le y \le n$, we have

$$I_3 = \int_{\delta}^{n} |\varphi(y)| \left| \int_0^1 g(u, y) d\gamma(u) \right| dy$$
$$= \int_{\delta}^{n} O\left(e^{y/2} y^{-((2\alpha+3)/4)} n^{(2\alpha+1)/4}\right) |\varphi(y)| dy$$
$$= O\left(n^{(2\alpha+1)/4}\right)\left(\int_{\delta}^{n} e^{y/2} y^{-((2\alpha+3)/4)} |\varphi(y)| dy\right)$$
$$= O\left(n^{(2\alpha+1)/4}\right) o\left((n^{-(2\alpha+1)/4})\xi(n)\right)$$
$$= o(\xi(n)), \tag{27}$$

in view of condition (14).

Further, using Lemma 4, we have

$$I_4 = \int_{n}^{\infty} |\varphi(y)| \left| \int_0^1 g(u, y) d\gamma(u) \right| dy$$
$$= \int_{n}^{\infty} |\varphi(y)| O\left(e^{y/2} y^{-(3\alpha+5)/6} n^{(\alpha+1)/2}\right) dy$$
$$= O\left(n^{(\alpha+1)/2}\right)\left(\int_{n}^{\infty} \frac{e^{y/2} y^{-1/3} |\varphi(y)|}{y^{(\alpha+1)/2}} dy\right)$$
$$= o\left((\xi(n)) n^{(\alpha+1)/2} \left(n^{-(\alpha+1)/2}\right)\right)$$
$$= o(\xi(n)), \tag{28}$$

in view of condition (15).

Collecting (24)–(28), we have

$$H_n(f; 0) - f(0) = o(\xi(n)).$$

Hence the proof of Theorem 1 is completed.

4 Corollaries

The following corollaries can be derived from our Theorem 1.

Corollary 1 *As discussed in* [7, *p.* 306, *Lemma* 1] *and* [11, *p.* 38], *if we take the mass function* $\gamma(u)$ *given by*

$$\gamma(u) = \begin{cases} 0, & 0 \leq u \leq a, \\ 1, & a \leq u \leq 1, \end{cases}$$

where $a = \frac{1}{(1+q)}$, $q > 0$, *the Hausdorff matrix* H *reduces to Euler matrix* (E, q), $q > 0$ *and defines the corresponding* (E, q) *means given by*

$$E_q^n(f; x) = \frac{1}{(1+q)^n} \sum_{k=0}^{n} \binom{n}{k} q^{n-k} s_k(f; x), \quad q > 0.$$

Hence the Theorem 1 *reduces to Theorem A (result proved by Nigam and Sharma* [5, *p.* 3, *Theorem* 2.1]*).*

Corollary 2 *As discussed in* [1, *p.* 400] *and* [6, *p.* 2747], *the Cesàro matrix of order* λ, *is also a Hausdorff matrix obtained by mass function* $\gamma(u) = 1 - (1-u)^{\lambda}$ *and the corresponding Cesàro means are given by*

$$C_n^{\lambda}(f; x) = \frac{1}{\binom{n+\lambda}{n}} \sum_{k=0}^{n} \binom{\lambda + n - k - 1}{n - k} s_k(f; x).$$

Further, Rhoades [7, *p.* 308] *and Rhoades et al.* [8, *p.* 6869] *has mentioned that the product of two Hausdorff matrices is again a Hausdorff matrix. Hence the Theorem B and Theorem C (results proved by Krasniqi* [4, *p.* 35, *Theorem* 2.1] *and Sonker* [10, *p.* 126, *Theorem* 1]*) are also particular cases of our Theorem* 1.

Remark 2 This is an open problem to associate the above discussed results with the L^p-spaces.

Acknowledgments The authors express their sincere gratitude to the reviewers for their valuable suggestions for improving the paper. This research is supported by the Council of Scientific and Industrial Research (CSIR), New Delhi, India (Award No.- 09/143(0821)/2012-EMR-I) in the form of fellowship to the first author.

References

1. Garabedian, H.L.: Hausdorff matrices. Am. Math. Monthly **46**(7), 390–410
2. Gupta, D.P.: Degree of approximation by Cesàro means of Fourier-Laguerre expansions. Acta Sci. Math. (Szeged) **32**, 255–259 (1971)
3. Hardy, G.H.: Divergent Series. Oxford at the Clarendon Press (1949)
4. Krasniqi, X.Z.: On the degree of approximation of a function by $(C, 1)(E, q)$ means of its Fourier-Laguerre series. Int. J. Anal. Appl. **1**, 33–39 (2013)
5. Nigam, H.K, Ajay, S.: A study on degree of approximation by $(E, 1)$ summability means of the Fourier-Laguerre expansion. Int. J. Math. Math. Sci. (Art. ID 351016), 7 (2010)
6. Rhoades, B.E.: Commutants for some classes of Hausdorff matrices. Proc. Am. Math. Soc. **123**(9), 2745–2755 (1995)
7. Rhoades, B.E.: On the degree of approximation of functions belonging to the weighted $(L^p, \xi(t))$ class by Hausdorff means. Tamkang J. Math. **32**(4), 305–314 (2001)
8. Rhoades, B.E., Kevser, O., Albayrak Inc.: On the degree of approximation of functions belonging to a Lipschitz class by Hausdorff means of its Fourier series. Appl. Math. Comput. **217**(16), 6868–6871 (2011)
9. Szegö, G.: Orthogonal polynomials. Am. Math. Soc. Colloquium Publ. **23**, (1939)
10. Sonker, S.: Approximation of Functions by $(C, 2)(E, q)$ means of its Fourier-Laguerre series. Proceeding of ICMS-2014, (2014). ISBN:978-93-5107-261-4:125–128
11. Singh, U., Sonker, S.: Trigonometric approximation of signals (functions) belonging to weighted $(L^p, \xi(t))$-class by Hausdorff means. J. Appl. Funct. Anal. **8**(1), 37–44 (2013)

Sparse Approximation of Overdetermined Systems for Image Retrieval Application

M. Srinivas and R. Ramu Naidu

Abstract The recent developments in the field of compressed sensing (CS) have been shown to have tremendous potential for applications such as content-based image retrieval. The underdetermined framework present in CS requires some implicit assumptions on the image database or needs the projection (or downsampling) of database members into lower dimensional space. The present work, however, poses the problem of image retrieval in overdetermined setting. The main feature of the proposed method is that it does not require any downsampling operation or implicit assumption on the databases. Our experimental results demonstrate that our method has potential for such applications as content-based image retrieval.

Keywords Overdetermined Systems · K-SVD · Image retrieval · LASSO · Underdetermined System

1 Introduction

Content-based image retrieval (CBIR) from large image databases has been an active area of research for long due to its applications in various fields like satellite imaging, medicine, etc. CBIR systems extract features from the raw images and calculate an associative measure (similarity or dissimilarity) between a query image and database images based on these features. Several CBIR systems based on wavelets, Gabor transform have been proposed in the literature ([1] and the references therein).

In recent years, sparse representations have received a lot of attention from the signal- and image-processing communities. Sparse coding involves representation of an image as a linear combination of a few atoms of a given dictionary [2]. It is a

M. Srinivas (✉)
Department of Computer Science and Engineering, IIT Hyderabad, Hyderabad, India
e-mail: cs10p002@iith.ac.in

R.R. Naidu
Department of Mathematics, IIT Hyderabad, Hyderabad, India
e-mail: ma11p003@iith.ac.in

P.N. Agrawal et al. (eds.), *Mathematical Analysis and its Applications*,
Springer Proceedings in Mathematics & Statistics 143,
DOI 10.1007/978-81-322-2485-3_17

powerful tool for efficiently processing data in unconventional ways. This is mainly due to the fact that signals and images of interest admit sparse representations in some dictionary. The dictionaries can be composed of wavelet or Fourier basis functions or can be learned from data. It has been observed that the dictionaries learned directly [2–4] from data provide better representation, and hence improve the performance on many practical applications such as classification. Several algorithms for learning dictionaries have been developed for example, the K-SVD [5] and the method of optimal directions (MOD) [6]. These techniques are used in many applications such as image restoration, denoising, and texture classification.

The image retrieval or classification based on the sparse approximations typically works under some assumptions implicitly on the databases. One often encounters the situations such as:

- the database may not be big enough so that the sparsity promoting underdetermined setting could be efficiently deployed
- when the classification of images is unsupervised, there is no guarantee that a cluster has enough members, and consequently the dictionary learning involving underdetermined setting may not be useful effectively.

One way to overcome the stated problems is to downsample the images (or project them to lower dimensional spaces) or to extract some relevant features so that the classification problem could be addressed in underdetermined setting.

The present work aims at proposing a novel CBIR algorithm by posing the problem in the form of an overdetermined framework. The salient features of our method are twofold: (1) even if the database is relatively smaller in size and images are bigger in size, the method could be useful and (2) the method per se does not require downsampling of images.

We realize our objective using the LASSO [7] at two stages. In the first step, we identify the most relevant clusters of the database by finding sparse approximation to the system $q \approx \Phi x$. Where q is query image and Φ is the matrix whose columns are the cluster centers of the data. While in the second step, we obtain the desired retrieval performance by obtaining sparse approximation to the overdetermined system $q \approx \Psi y$, where Ψ is the matrix whose columns are the images belonging to the relevant clusters.

The paper is organized into several sections. In Sects. 2 and 3, we present, respectively, sparse approximations to overdetermined systems and motivation for the present work in more detail. While in Sects. 4 and 5, we present image retrieval through overdetermined systems and simulation results, respectively. In the last section, we provide our concluding remarks.

2 Sparse Approximation to Overdetermined Linear System

The focus in this paper is on systems which can be represented in the form $y \approx Ax$, where the dimensions of the objects A, y, x are $m \times n$, $m \times 1$, $n \times 1$, respectively. It is assumed that $m > n$, which is typically the case in our image retrieval problem. When y comes from query image and A is generated using database members consisting of members relevant to y, one may suspect that the system y has sparse representation in A. We can obtain such a sparse approximation using the following optimization problem:

$$\|Au^k - y\|_2 = \inf\{\|Ax - y\|_2 \mid x \in \mathscr{R}^n, \|x\|_0 \leq k\},$$

where $\|x\|_0 = |\{i|x_i \neq 0\}|$, the number of nonzero components in x. The vector u^k is called as a best k-sparse approximation to $y \approx Ax$ and it contains atmost k nonzero terms.

A simple but computationally costly approach to solve this problem is by using brute force search, that is, randomly pick $(n - k)$ elements of x to be zero and find the remaining using the standard least squares method. It can be easily seen that this method becomes intractable for high values of n.

The main culprit in the problem is the l_0 norm which makes the problem computationally costly. Tibshirani [7] made use of the l_1 norm in the place of the l_0 norm and solved the following (modified) problem (for a fixed λ):-

$$\text{find } x \text{such that } \|Ax - y\|_2^2 + \lambda\|x\|_1 \text{is minimized.}$$

The main ideas behind using the l_1 norm are:

- convex optimization methods can be used to solve the above-modified problem
- minimizing l_1 norm typically provides sparse solutions

The minimizing algorithm called the *LASSO* is obtained by solving a series of quadratic programming problems [7]. Usually, the parameter λ is obtained by cross-validation.

3 Motivation for Present Work

The recent sparsity-based methods are found to be useful for applications in image processing [2, 3, 8]. In this section, we quickly review the relevant methods and present our motivation for the current work.

Given sufficient training samples of ith class, A_i (whose columns are samples) for $i = 1, 2, \ldots K$, it is shown in [8] that the class label of unknown object y may be obtained by solving

$$y = \underbrace{[A_1|A_2|\ldots|A_K]}_{A} x_0 \tag{1}$$

for sparse solution, which is obtained from

$$x_0 = \arg\min_{\alpha} \|\alpha\|_1 \quad \text{subject to} \quad A\alpha = y. \tag{2}$$

The index $\hat{i}$, defined by

$$\hat{i} = \arg \min_{i=1,2,\cdots,K} \|\mathbf{y} - \mathbf{A}\delta_i(x_0)\|_2^2, \tag{3}$$

is taken as the class label of y. Here, δ_i is a characteristic function that selects the coefficients associated with the ith labeled samples. This method being supervised provides excellent classification[1] provided the labeled data for each class are sufficient. The dictionary-based methods, however, train class-specific dictionaries using labeled data and then assign each testing image to the class for which the best reconstruction is obtained [2], which rely on the premise that two signals belonging to the same cluster have decomposition in terms of similar atoms. After identifying the class label, one may retrieve most relevant images from the ith class using some similarity metric.

These methods are effective when the database and labeled data for each class are sufficiently big enough to accommodate underdetermined framework. In the absence of the database or class-specific labeled data being sufficiently big enough; however, one may need to consider features of images or downsampled images or random projections of images into lower dimensional spaces meeting the theoretical restrictions posed by the Lemma [9]. The reduced dimension of data may have bearing on the retrieval performance. In addition, when the classification is unsupervised, sparse recovery based on underdetermined setting may not always be useful. This is because a cluster may not have enough members. Motivated by these considerations, we consider proposing an unsupervised method that involves the sparse approximations to overdetermined systems. We believe the overdetermined setting could involve very little restriction on the database being used.

4 Overdetermined Setting for Image Retrieval

Inspired by the ideas from [8], one may pose the retrieval problem in overdetermined setting as

$$q \approx \underbrace{[I_1 I_2 \ldots I_N]}_{\Phi} \alpha_0, \tag{4}$$

where, for $j = 1, 2, \ldots, N$, I_j is vectorized version of jth database image and q is query image. The problem of retrieving the L best matches of query image q may be obtained from

$$\alpha_0^L = \arg \min_{x \in \mathscr{R}^N} \{\|\Phi x - q\|_2 \mid \|x\|_0 \leq L\}, \tag{5}$$

[1] Viewing the problem of image retrieval as an extension of classification, in this paper, we use the words "classification" and "retrieval" synonymous. This is of course a slight abuse of convention.

which could be solved using LASSO. For large databases consisting of somewhat bigger images, Φ has bigger size and the computational cost in using Φ could be more, making thereby the method less useful. In view of this problem, instead of searching for the relevance in the entire database, we search for the relevance of q to few clusters, which is followed by further search within the relevant clusters through LASSO again.

After dividing the database into K clusters, namely $C_1, C_2, \ldots, C_K$ with cluster centers $e_1, e_2, \ldots, e_K$, we identify the K_1 most relevant clusters from

$$e^{K_1} = \arg \min_{x_K \in \mathscr{R}^K} \{\| \underbrace{[e_1 e_2 \ldots e_K]}_{\Psi} x_K - q \|_2 \mid \|x_K\|_0 \leq K_1\}, \tag{6}$$

The number K_1 may be heuristically determined as the significant number of cluster centers in whose spanning space q results in minimum misfit. Suppose $C_{i_1}, C_{i_2}, \ldots, C_{i_{K_1}} \in \{C_i\}_{i=1}^K$ are the clusters whose indices are those of the nonzero entries of e^{K_1}. We then search for the L most relevant images within the pruned clusters from

$$\theta^L = \arg \min_{\theta \in \mathscr{R}^{L_1}} \{\| \underbrace{[C_{i_1} C_{i_2} \ldots C_{i_{K_1}}]}_{\Gamma} \theta - q \|_2 \mid \|\theta\|_0 = L\}, \tag{7}$$

In the above equation, $L_1 := \sum_{l=1}^{K_1} |C_{i_l}|$, which is the column size of Γ. We determine the degree of relevance of retrieved images by computing the error:

$$\text{error}_{\text{rel}} = \min_{i \in \text{support}\{\theta^L\}} \|\mathbf{q} - \Gamma \delta_i(\theta^L)\|_2^2. \tag{8}$$

The images that result in small errors are considered to be more relevant. The distribution of data into several clusters is in general unknown. If the images within each cluster are closer to cluster center and the cluster centers are wide apart, the proposed method is expected to work very well. The block diagram of the method is shown in Fig. 1.

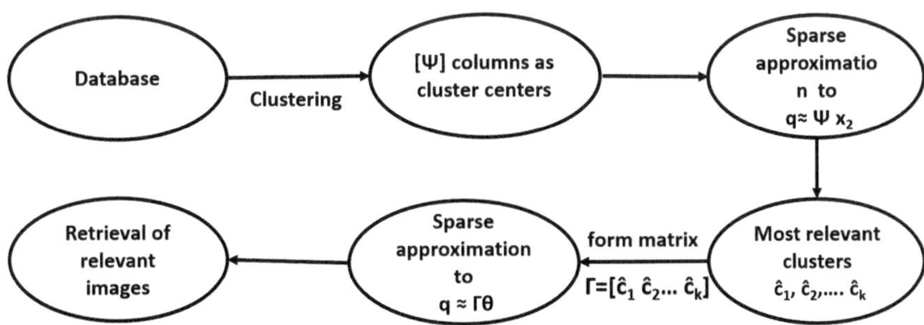

Fig. 1 Block diagram of proposed method

5 Simulation Results

In this section, we demonstrate the performance of the proposed overdetermined (OD) method on Image Retrieval in Medical Applications (IRMA) medical image database[2] and compare our results to the ones obtained by the underdetermined (UD) setting as stated in (4). In implementing the UD-based method, we use the orthogonal matching pursuit (OMP) [3] algorithm. We analyze the performances of both methods using the standard precision and recall, which are defined as

$$\text{Precision} = \frac{\text{Number of relevant images retrieved}}{\text{Total number of images retrieved}}$$

$$\text{Recall} = \frac{\text{Number of relevant images retrieved}}{\text{Total number of relevant images}}$$

To begin with, we execute our method on a small medical database wherein each image is of size 120×120. This database consists of 311 images of skull, breast, chest, hand, etc. Of them, we consider 11 as testing images. As in (4), if we form Φ with database members as columns, we get a matrix whose size is $14{,}400 \times 300$, a very tall and slim matrix that occupies huge memory space on computer and involves high computational cost. One may reduce the row dimension of this matrix by projecting the images to lower dimension spaces. But the dimension of reduced size in OD case is not dictated by the resulting size as it is in the UD case. To be able to apply the powerful theory of UD framework, one needs to downsample (or project) the database members to a space of dimension less than 300, which can impact retrieval performance. The dimension of projected space could be further small if one wishes to use dictionary-based approach [4].

Using K-means algorithm, we divided the database into 10 clusters (that is, $K = 10$). It is to be mentioned here that the bigger value for K might slightly increase computational complexity, but it can have no significant bearing on performance. This is because we consider more than one cluster when searching for relevant images. We form the matrix system (6) and obtain sparse approximation with $K_1 = 3$. This selection is based on the observation that the choice of $K_1 = 3$ is good enough to result relevant images in three clusters, which is supported by the plots in Figs. 2 and 3. In our simulation work, in order to deal with overdetermined matrix of reasonable size, we have downsampled the database members to 60×60. Therefore, the size of Φ becomes 3600×300. In UD setting, we have projected images to a space of lower dimension 256 (the associated matrix is of size 256×300). As stated already, this downsampling operation is not at all mandatory in OD case while it is necessary in UD case. The plots in Figs. 2 and 3 show that the average performance by OD method is better than that given by UD method.

[2] www.irma-project.org.

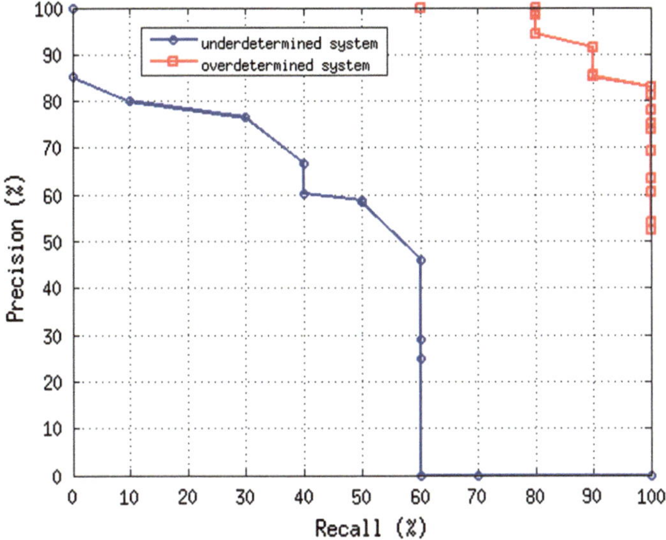

Fig. 2 Precision–Recall graph with $K_1 = 3$ on the medical database

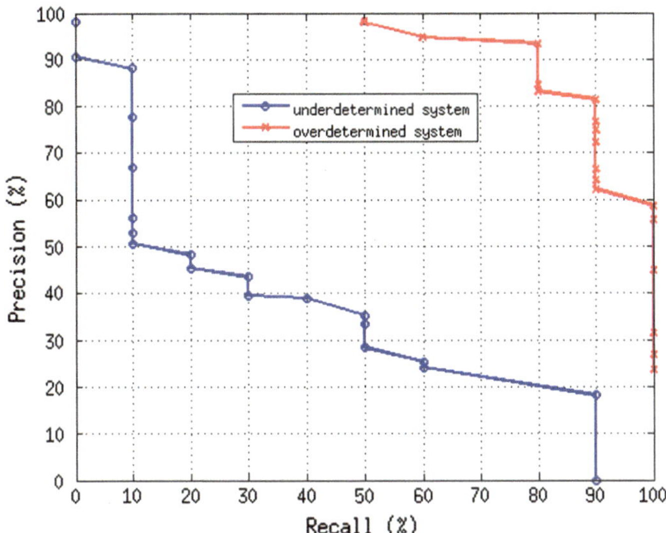

Fig. 3 Precision–Recall graph with $K_1 = 5$ on the medical database

Using the clusters whose indices correspond to the nonzero components of e^{K_1} (defined in (6)), we form the matrix Γ, and solve (7) for θ^L. From the nonzero locations of θ^L, we obtain the best matches from the database for a given query image. The performance of our method is shown in Fig. 4 for several query images. The precision–recall plots for the cases of $K_1 = 3$ and $K_1 = 5$ are shown in Figs. 2

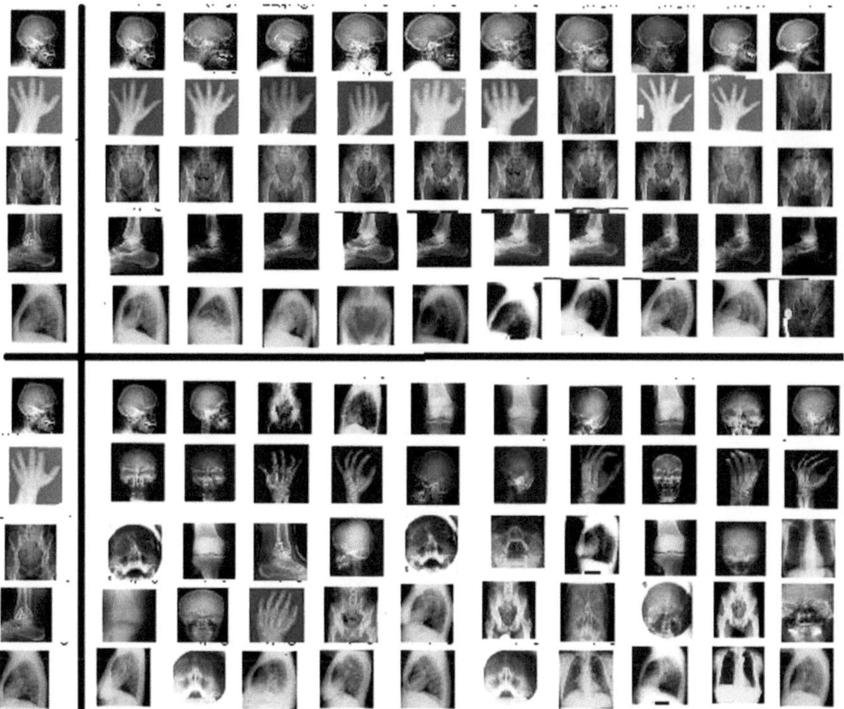

Fig. 4 Retrieval performance of the proposed method. The *first image on each row* is the query image and the remaining correspond to those retrieved by the method. The *upper* and *lower* parts of the figure, respectively, correspond to the performances of over- and underdetermined frameworks

and 3, respectively. In Fig. 4, on every row, the first image refers to the query image, the next five images correspond to the retrieval performance of OD setting, while the remaining five correspond to the retrieval by UD-based framework.

6 Conclusions

The present work has proposed an overdetermined framework for CBIR. The motivation for the present work comes from the fact that the sparsity promoting classification methods that involve the use of underdetermined matrix equations work on some implicit assumptions on the databases or project data into lower dimension spaces to accommodate the ideas from the theory of compressed sensing. The present work, however, does not need any such requirement. The preliminary simulation results reported in this paper demonstrate that the overdetermined framework has potential for image retrieval problems. The medical database members contain some scale variation. The retrieval performance on these databases can be improved by incorporating rotation and scale invariant features in the retrieval process. Our future efforts shall address this aspect.

Acknowledgments Authors are thankful to Dr. C. Krishna Mohan, Dr. Phanindra Jampana and Dr. C.S. Sastry for fruitful discussions. Authors would like to thank Dr. T.M. Deserno, Department of Medical Informatics, RWTH Aachen, Germany for making the original IRMA Database available for research purposes.

References

1. Chen, Y., Sastry, C.S., Patel, V., Philips, J., Chellappa, R.: In-Plane rotation and scale invariant simultaneous dictionary learning and clustering. IEEE Trans. Image Process. **22**(6), 2166–2180 (2013)
2. Patel, V.M., Wu, T., Biswas, S., Philips, P.J., Chellappa, R.: Dictionary-based face recognition under variable lighting and pose. IEEE Trans. Inform. Forensics Security **7**(3), 954–965 (2012)
3. Elad, M.: Sparse and Redundant Representations. Springer, New York (2010)
4. Sprechmann, P., Sapiro, G.: Dictionary learning and sparse coding for unsupervised clustering. In Proceedings of ICASSP (2010)
5. Aharon, M., Elad, M., Bruckstein, A.M.: The k-svd: an algorithm for designing of over-complete dictionaries for sparse representation. IEEE Trans. Signal Process. **54**(11), 4311–4322 (2006)
6. K-Delgado, K., Murray, J.F., Rao, B.D., Engan, K., Lee, T.W., Sejnowski, T.J.: Dictionary learning algorithms for sparse representation. Neural Comput. **15**, 349–396 (2003)
7. Tibshirani, R.: Regression shrinkage and selection via the LASSO. J. Statis. Soft. **33**(1), 1 (2010)
8. Wright, J., Yang, A.Y., Ganesh, A., Sastry, S.S., Ma, Y.: Robust face recognition via sparse representation. IEEE Trans. Pattern Anal. Mach. Intell. **31**(2), 210–227 (2009)
9. Baraniuk, R., Davenport, M., DeVore, R., Wakin, M.: A simple proof of the restricted isometry property for random matrices. Constr. Approximation **28**(3), 253–263 (2008)

kth Order Kantorovich Modification of Linking Baskakov-Type Operators

Margareta Heilmann and Ioan Raşa

Abstract In 1957 Baskakov introduced a general method for the construction of positive linear operators depending on a real parameter c. The so-called genuine Baskakov–Durrmeyer-type operators form a class of operators reproducing the linear functions, interpolating at (finite) endpoints of the interval, and having other nice properties. In this paper we consider a nontrivial link between Baskakov-type operators and genuine Baskakov–Durrmeyer-type operators. We establish explicit representations for the images of monomials and for the moments; they are useful, e.g., in studying asymptotic formulas.

Keywords Baskakov-and-Durrmeyer-type operators · Linking operators · Kantorovich-type modifications · Moments · Images of monomials

2010 AMS Subject Classification: 41A36 · 41A10 · 41A60

1 Introduction and Definition of the Operators

In 1957 Baskakov [1] introduced a general method for the construction of positive linear operators depending on a real parameter c including the classical Bernstein, Szász-Mirakjan, and Baskakov operators as special cases. All these Baskakov-type operators preserve linear functions and interpolate at (finite) endpoints of the corresponding interval. The so-called Bernstein–Durrmeyer operators were introduced by Durrmeyer in [2] and independently developed by Lupaş [9]. Afterwards, this construction was carried over to many other classical operators; for instance see

M. Heilmann (✉)
Faculty of Mathematics and Natural Sciences, University of Wuppertal,
Gaußstraße 20, Wuppertal 42119, Germany
e-mail: heilmann@math.uni-wuppertal.de

I. Raşa
Department of Mathematics, Technical University,
Str. Memorandumului 28, Cluj-napoca 400114, Romania
e-mail: Ioan.Rasa@math.utcluj.ro

P.N. Agrawal et al. (eds.), *Mathematical Analysis and its Applications*,
Springer Proceedings in Mathematics & Statistics 143,
DOI 10.1007/978-81-322-2485-3_18

[10, 16] and in the general setting for so-called Baskakov–Durrmeyer-type operators [6]. These operators have a lot of nice properties; they commute, they commute with certain differential operators, they are self-adjoint but they only reproduce constants.

The consideration of so-called genuine Baskakov–Durrmeyer-type operators leads to a class of operators again reproducing the linear functions and interpolating at (finite) endpoints of the corresponding interval. These operators are related to the Baskakov–Durrmeyer-type operators in the same way as the Baskakov-type operators to their corresponding Kantorovich variants.

In [11, 12] Păltănea introduces operators depending on a parameter $\rho \in \mathbb{R}^+$, which constitute a nontrivial link between the Bernstein and Szász-Mirakjan operators, respectively, and their genuine Durrmeyer modifications. Further results can also be found in [3, 4, 13].

In this paper we consider a nontrivial link between Baskakov-type operators and genuine Baskakov–Durrmeyer-type operators. Moreover, we investigate the kth order Kantorovich modification of them; for $k = 1$ this means a link between the Kantorovich modification of Baskakov-type and Baskakov–Durrmeyer-type operators.

In what follows for $c \in \mathbb{R}$ we use the notations

$$a^{c,\overline{j}} := \prod_{l=0}^{j-1}(a+cl), \ a^{c,\underline{j}} := \prod_{l=0}^{j-1}(a-cl), \ j \in \mathbb{N}; \quad a^{c,\overline{0}} = a^{c,\underline{0}} := 1$$

which can be considered as a generalization of rising and falling factorials. Note that $a^{-c,\overline{j}} = a^{c,\underline{j}}$ and $a^{c,\overline{j}} = a^{-c,\underline{j}}$. This notation enables us to state the results for the different operators in a unified form.

In a recent paper [8] we already considered the linking operators between the kth order Kantorovich modification of the Bernstein and the genuine Bernstein–Durrmeyer operators. Comparison of the results in [8] with the outcomes of the present paper shows that all the representations for the moments and the images of monomials are also valid for the Bernstein case by setting $c = -1$ in the subsequent theorems.

In the following definitions of the operators we omit the parameter c in the notations in order to reduce the necessary sub- and superscripts.

Let $c \in \mathbb{R}$, $c \geq 0$, $n \in \mathbb{R}$, $n > c$, $\rho \in \mathbb{R}^+$, $j \in \mathbb{N}_0$, $x \in [0, \infty)$. Then the basis functions are given by

$$p_{n,j}(x) = \begin{cases} \frac{n^j}{j!}x^j e^{-nx} & , \ c = 0, \\ \frac{n^{c,\overline{j}}}{j!}x^j(1+cx)^{-(\frac{n}{c}+j)} & , \ c > 0. \end{cases}$$

In the following definition we assume that $f : [0, \infty) \longrightarrow \mathbb{R}$ is given in such a way that the corresponding integrals and series are convergent.

Definition 1 The operators of Baskakov type are defined by

$$B_n(f,x) = \sum_{j=0}^{\infty} p_{n,j}(x) f\left(\frac{j}{n}\right), \tag{1}$$

the genuine Baskakov–Durrmeyer-type operators are denoted by

$$B_{n,1}(f,x) = f(0)p_{n,0}(x) + \sum_{j=1}^{\infty} p_{n,j}(x) \int_0^{\infty} p_{n+2c,j-1}(t) f(t)dt, \tag{2}$$

and for $\rho \in \mathbb{R}^+$ the linking operators are given by

$$B_{n,\rho}(f,x) = \sum_{j=0}^{\infty} F_{n,j}^{\rho}(f) p_{n,j}(x) \tag{3}$$

$$= f(0)p_{n,0}(x) + \sum_{j=1}^{\infty} p_{n,j}(x)(n+c) \int_0^{\infty} \mu_{n,j}^{\rho}(t) f(t)dt, \tag{4}$$

where

$$\mu_{n,j}^{\rho}(t) = \begin{cases} \frac{(n\rho)^{j\rho}}{\Gamma(j\rho)} t^{j\rho-1} e^{-n\rho t} & , c = 0, \\ \frac{c^{j\rho}}{B(j\rho, \frac{n}{c}\rho+1)} t^{j\rho-1} (1+ct)^{-(\frac{n}{c}+j)\rho-1} & , c > 0. \end{cases}$$

Setting $c = 0$ in (2) leads to the Phillips operators [14], $c > 0$ was investigated in [18]. To the best of our knowledge the case $c = 0$ in (3) was first considered in [12].

As in [8] for the Bernstein case we also consider the kth order Kantorovich modification of the operators $B_{n,\rho}$, i.e.,

$$B_{n,\rho}^{(k)} := D^k \circ B_{n,\rho} \circ I_k \tag{5}$$

where D^k denotes the kth order ordinary differential operator and

$$I_k f = f, \text{ if } k = 0, \text{ and } I_k(f,x) = \int_0^x \frac{(x-t)^{k-1}}{(k-1)!} f(t)dt, \text{ if } k \in \mathbb{N}.$$

For $k = 0$ we omit the superscript (k) as indicated by the definition above.

This general definition contains many known operators as special cases. For $c = 0$ we get the linking operators considered in [13]. For $\rho = 1$ we get the genuine Baskakov–Durrmeyer-type operators $B_{n,1}$, for $\rho = 1$, $k \in \mathbb{N}$ the Baskakov–Durrmeyer-type operators $B_{n,1}^{(1)}$ (see [6, (1.3)], named M_{n+c} there) and the auxiliary operators $B_{n,1}^{(k)}$ considered in [7, (3.5)], (named $M_{n+c,k-1}$ there) with the explicit representation

$$(B_{n,1}^{(k)}f)(x) = \frac{n^{c,\overline{k}}}{n^{c,\underline{k-1}}} \sum_{j=0}^{\infty} p_{n+ck,j}(x) \int_0^{\infty} p_{n-c(k-2),j+k-1}(t) f(t) dt.$$

For an arbitrary sequence of linear operators, the images of monomials and the moments are important, e.g., in studying the asymptotic behavior. In this paper we establish explicit representations for the images of the monomials and for the moments of the investigated operators. Corresponding recursion formulas and further results will be given in a forthcoming paper.

Below we will use the following basic formulas.

$$\int_0^{\infty} \mu_{n,j}^{\rho}(t) dt = B\left(j\rho, \frac{n}{c}\rho + 1\right), \tag{6}$$

$$\sum_{j=0}^{\infty} p_{n,j}(x) = 1, \tag{7}$$

$$\frac{j}{n} p_{n,j}(x) = x p_{n+c,j-1}(x), \tag{8}$$

$$x(1+cx) p_{n,j}'(x) = (j - nx) p_{n,j}(x), \tag{9}$$

with the convention $p_{n,l}(x) = 0$, if $l < 0$. As usual, empty products are defined to be one.

2 Explicit Formulas for the Images of Monomials

In this section we prove general explicit formulas for the images of the monomials of the operators $B_{n,\rho}^{(k)}$. In what follows we denote by $e_{\nu}(t) = t^{\nu}$, $\nu \in \mathbb{N}_0$, the monomials and by

$$\Delta_h^l f(x) = \sum_{\kappa=0}^{l} (-1)^{l-\kappa} \binom{l}{\kappa} f(x + \kappa h) \tag{10}$$

the lth order forward difference of a function f with step h and define

$$p_{\nu}^{\rho}(\xi) := \prod_{l=1}^{\nu-1} \left(\xi + \frac{l}{\rho}\right), \ \nu \in \mathbb{N}.$$

This can be rewritten as

$$p_\nu^\rho(\xi) = \sum_{i=0}^{\nu-1} \frac{\Delta_1^i p_\nu^\rho(1)}{i!} \prod_{l=1}^{i} (\xi - l), \tag{11}$$

which can be derived by using the Newton representation of the interpolation polynomial of p_ν^ρ for the equidistant knots $1, 2, \ldots, \nu$.

We first consider the images of the monomials for the case $k = 0$, i.e., for the operators $B_{n,\rho}$.

Theorem 1 *Let $n \in \mathbb{R}$, $n\rho > c(\nu - 1)$, $\rho \in \mathbb{R}_+$, $\nu \in \mathbb{N}_0$, $\nu \leq n$. Then*

$$(B_{n,\rho} e_0)(x) = 1, \tag{12}$$

$$(B_{n,\rho} e_\nu)(x) = \frac{\rho^\nu}{(n\rho)^{c,\underline{\nu}}} \sum_{i=1}^{\nu} \frac{n^{c,\overline{i}}}{(i-1)!} \left(\Delta_1^{i-1} p_\nu^\rho(1)\right) x^i, \; \nu \in \mathbb{N}. \tag{13}$$

Proof (12) follows immediately from (6) and (7).
In order to prove (13) we take into account that for $c = 0$

$$\frac{(n\rho)^{j\rho}}{\Gamma(j\rho)} \int_0^\infty t^\nu t^{j\rho-1} e^{-n\rho t} dt = \frac{1}{(n\rho)^\nu} \cdot \frac{\Gamma(j\rho + \nu)}{\Gamma(j\rho)} = \frac{1}{n^\nu} \prod_{l=0}^{\nu-1} \left(j + \frac{l}{\rho}\right)$$

and for $c > 0$

$$\frac{c^{j\rho}}{B\left(j\rho, \frac{n}{c}\rho + 1\right)} \int_0^\infty t^\nu t^{j\rho-1} (1 + ct)^{-(\frac{n}{c}+j)\rho-1} dt$$

$$= c^{-\nu} \frac{\Gamma(j\rho + \nu)\Gamma\left(\frac{n}{c}\rho + 1 - \nu\right)}{\Gamma(j\rho)\Gamma\left(\frac{n}{c}\rho + 1\right)} = \frac{\rho^\nu}{(n\rho)^{c,\underline{\nu}}} \prod_{l=0}^{\nu-1} \left(j + \frac{l}{\rho}\right).$$

Thus we get for $\nu \geq 1$ with (8) and (11)

$$\begin{aligned} (B_{n,\rho} e_\nu)(x) &= \frac{\rho^\nu}{(n\rho)^{c,\underline{\nu}}} \sum_{j=1}^{\infty} p_{n,j}(x) \prod_{l=0}^{\nu-1} \left(j + \frac{l}{\rho}\right) \\ &= \frac{\rho^\nu}{(n\rho)^{c,\underline{\nu}}} nx \sum_{j=1}^{\infty} p_{n+c,j-1}(x) p_\nu^\rho(j) \\ &= \frac{\rho^\nu}{(n\rho)^{c,\underline{\nu}}} nx \sum_{j=1}^{\infty} p_{n+c,j-1}(x) \sum_{i=0}^{\nu-1} \frac{\Delta_1^i p_\nu^\rho(1)}{i!} \prod_{l=1}^{i} (j - l) \\ &= \frac{\rho^\nu}{(n\rho)^{c,\underline{\nu}}} nx \sum_{i=0}^{\nu-1} \frac{\Delta_1^i p_\nu^\rho(1)}{i!} \sum_{j=i+1}^{\infty} p_{n+c,j-1}(x) \prod_{l=1}^{i} (j - l). \end{aligned} \tag{14}$$

Applying (8) for $j \geq i+1$ we have

$$p_{n+c,j-1}(x)\prod_{l=1}^{i}(j-l) = p_{n+c(i+1),j-i-1}(x)x^i\prod_{l=1}^{i}(n+cl).$$

Hence with (7)

$$(B_{n,\rho}e_\nu)(x) = \frac{\rho^\nu}{(n\rho)^{c,\underline{\nu}}}\sum_{i=1}^{\nu}\frac{n^{c,\bar{i}}}{(i-1)!}\left(\Delta_1^{i-1}p_\nu^\rho(1)\right)x^i. \qquad \square$$

Remark 1 Using (10), the representation (13) can be rewritten as

$$(B_{n,\rho}e_\nu)(x) = \frac{\rho^\nu}{(n\rho)^{c,\underline{\nu}}}\sum_{i=1}^{\nu}n^{c,\bar{i}}x^i\sum_{\kappa=0}^{i-1}(-1)^{i-1-\kappa}\frac{1}{\kappa!(i-1-\kappa)!}p_\nu^\rho(1+\kappa).$$

Now we consider the special cases $\rho = 1$ and $\rho \to \infty$.
$\underline{\rho = 1}$: Then with [5, (3.48)] (see [8, p. 323])

$$\Delta_1^{i-1}p_\nu^1(1) = (\nu-1)!\binom{\nu}{i}.$$

Thus

$$(B_{n,1}e_\nu)(x) = \frac{1}{n^{c,\underline{\nu}}}\sum_{i=1}^{\nu}n^{c,\bar{i}}\frac{(\nu-1)!}{(i-1)!}\binom{\nu}{i}x^i,$$

which coincides with the formula given in [18, Lemma 1.11] and [7, (4.3)] with $s = -1$ and taking $n+c$ instead of n there.
$\underline{\rho \to \infty}$: Then $\dfrac{\rho^\nu}{(n\rho)^{c,\underline{\nu}}} \to \dfrac{1}{n^\nu}$, and (see [8, p. 323])

$$\Delta_1^{i-1}p_\nu^\infty(1) = (i-1)!\sigma_\nu^i,$$

where σ_ν^j denote the Stirling numbers of second kind. Thus

$$(B_{n,\infty}e_\nu)(x) = \frac{1}{n^\nu}\sum_{i=1}^{\nu}n^{c,\bar{i}}\sigma_\nu^i x^i,$$

which coincides with the corresponding result for the classical Baskakov-type operators which can be calculated directly from the definition of the operators by using (8).

Next, we consider the images of the monomials for the case $k \in \mathbb{N}$.

Theorem 2 *Let $n \in \mathbb{R}$, $k \in \mathbb{N}$, $\rho \in \mathbb{R}_+$, $\nu \in \mathbb{N}_0$, $n\rho > c(\nu + k - 1)$. Then*

$$(B_{n,\rho}^{(k)} e_\nu)(x) = \frac{\nu!}{(\nu+k)!} \frac{\rho^{\nu+k}}{(n\rho)^{c,\underline{\nu+k}}} \sum_{i=0}^{\nu} \frac{n^{c,\overline{i+k}}}{i!} (i+k) \left(\mathbf{\Delta}_1^{i+k-1} p_{\nu+k}^{\rho}(1) \right) x^i. \quad (15)$$

Proof By using $B_{n,\rho}^{(k)} e_\nu = \frac{\nu!}{(\nu+k)!} D^k B_{n,\rho} e_{\nu+k}$ we get from (13) for $k \in \mathbb{N}$

$$\begin{aligned}(B_{n,\rho}^{(k)} e_\nu)(x) &= \frac{\nu!}{(\nu+k)!} \frac{\rho^{\nu+k}}{(n\rho)^{c,\underline{\nu+k}}} \sum_{i=k}^{\nu+k} \frac{n^{c,\overline{i}}}{(i-1)!} \left(\mathbf{\Delta}_1^{i-1} p_{\nu+k}^{\rho}(1) \right) \frac{i!}{(i-k)!} x^{i-k} \\ &= \frac{\nu!}{(\nu+k)!} \frac{\rho^{\nu+k}}{(n\rho)^{c,\underline{\nu+k}}} \sum_{i=0}^{\nu} \frac{n^{c,\overline{i+k}}}{i!} (i+k) \left(\mathbf{\Delta}_1^{i+k-1} p_{\nu+k}^{\rho}(1) \right) x^i. \quad \square\end{aligned}$$

Remark 2 Using again (10), the representation (15) can be rewritten as

$$\begin{aligned}(B_{n,\rho}^{(k)} e_\nu)(x) = &\frac{\nu!}{(\nu+k)!} \frac{\rho^{\nu+k}}{(n\rho)^{c,\underline{\nu+k}}} \sum_{i=0}^{\nu} n^{c,\overline{i+k}} \frac{(i+k)!}{i!} x^i \\ &\times \sum_{\kappa=0}^{i+k-1} (-1)^{i+k-1-\kappa} \frac{1}{\kappa!(i+k-1-\kappa)!} p_{\nu+k}^{\rho}(1+\kappa).\end{aligned}$$

Again we consider the special cases $\rho = 1$ and $\rho \to \infty$.
$\underline{\rho = 1}$: Then again with [5, (3.48)]

$$\mathbf{\Delta}_1^{i+k-1} p_{\nu+k}^{1}(1) = (\nu+k-1)! \binom{\nu+k}{i+k}.$$

Thus

$$(B_{n,1}^{(k)} e_\nu)(x) = \frac{1}{n^{c,\underline{\nu+k}}} \sum_{i=0}^{\nu} n^{c,\overline{i+k}} \frac{(\nu+k-1)!}{(i+k-1)!} \binom{\nu}{i} x^i.$$

This coincides with the corresponding result in [7, Satz 4.2] for the auxiliary operators with the notation $B_{n,\rho}^{(k)} = M_{n+c,k-1}$ there.
$\underline{\rho \to \infty}$: Then $\dfrac{\rho^{\nu+k}}{(n\rho)^{c,\underline{\nu+k}}} \to \dfrac{1}{n^{\nu+k}}$ and

$$\mathbf{\Delta}_1^{i+k-1} p_{\nu+k}^{\infty}(1) = (i+k-1)! \sigma_{\nu+k}^{i+k}.$$

Thus

$$(B_{n,\infty}^{(k)}e_\nu)(x) = \frac{\nu!}{(\nu+k)!}\frac{1}{n^{\nu+k}}\sum_{i=0}^{\nu}\frac{n^{c,\overline{i+k}}}{i!}(i+k)!\sigma_{\nu+k}^{i+k}x^i.$$

From the explicit representations of the images of the monomials we can deduce the following result concerning the limit of the operators $B_{n,\rho}^{(k)}$ when $\rho \to \infty$.

Corollary 1 *For each polynomial p we have*

$$\lim_{\rho\to\infty} B_{n,\rho}^{(k)}p(x) = B_n^{(k)}p(x)$$

uniformly on every compact subinterval of $[0,\infty)$.

For the evaluation of $B_{n,\rho}^{(k)}e_\nu$, $k \in \mathbb{N}$, for special values of ν, we use the representation

$$p_{\nu+k}^{\rho}(\xi) = \sum_{l=0}^{\nu+k-1}\rho^{-l}\sigma_l(1,2,\ldots,\nu+k-1)\xi^{\nu+k-1-l},$$

with the notation $\sigma_j(x_0,x_1,\ldots,x_n)$, $j \in \mathbb{N}$, for the symmetric function which is the sum of all products of j distinct values from the set $\{x_0,x_1,\ldots x_n\}$ and $\sigma_0(x_0,x_1,\ldots,x_n) := 1$.

For the monomial e_m, it is known (see, e.g., [15, Theorem 1.2.1]) that

$$\Delta_1^{j+k-1}e_m(1) = \begin{cases} 0, & m < j+k-1, \\ (j+k-1)!\tau_{m-(j+k-1)}(1,2,\ldots,j+k), & 0 \le j+k-1 \le m, \end{cases}$$

with the complete symmetric function $\tau_j(x_0,x_1,\ldots,x_n)$ which is the sum of all products of $x_0,x_1,\ldots,x_n$ of total degree j, $j \in \mathbb{N}$, and $\tau_0(x_0,x_1,\ldots,x_n) := 1$.

Thus we can rewrite $(B_{n,\rho}^{(k)}e_\nu)$ as

$$\begin{aligned}(B_{n,\rho}^{(k)}e_\nu)(x) &= \frac{\nu!}{(\nu+k)!}\frac{\rho^{\nu+k}}{(n\rho)^{c,\underline{\nu+k}}}\sum_{i=0}^{\nu}\frac{n^{c,\overline{i+k}}(i+k)!}{i!}x^i \\ &\quad\times\sum_{l=0}^{\nu-i}\rho^{-l}\sigma_l(1,2,\ldots,\nu+k-1)\tau_{\nu-l-j}(1,2,\ldots,i+k).\end{aligned} \tag{16}$$

As a corollary we present the results for $\nu = 0,\ 1,\ 2$.

Corollary 2 *For $k \in \mathbb{N}_0$ the images for the first monomials are given by*

$$(B_{n,\rho}^{(k)}e_0)(x) = \frac{\rho^k}{(n\rho)^{c,\underline{k}}} \cdot n^{c,\overline{k}},$$

$$(B_{n,\rho}^{(k)}e_1)(x) = \frac{\rho^{k+1}}{(n\rho)^{c,\underline{k+1}}} \cdot n^{c,\overline{k}} \left[\frac{1}{2}k\left(1+\frac{1}{\rho}\right) + (n+ck)x\right],$$

$$(B_{n,\rho}^{(k)}e_2)(x) = \frac{\rho^{k+2}}{(n\rho)^{c,\underline{k+2}}} \cdot n^{c,\overline{k}} \left[\frac{1}{2}k\left(\frac{3k+1}{6}+\frac{k+1}{\rho}+\frac{3k+5}{6\rho^2}\right) + (n+ck)\left((k+1)\left(1+\frac{1}{\rho}\right)x + (n+c(k+1))x^2\right)\right].$$

Proof For $k = 0$ the identities follow from Theorem 1. For $k \in \mathbb{N}$ we derive the proposition by using the representation (16) and the fact that for $m \in \mathbb{N}$

$$\sigma_0(1,\ldots,m) = \tau_0(1,\ldots,m) = 1,$$
$$\sigma_1(1,\ldots,m) = \tau_1(1,\ldots,m) = \frac{1}{2}m(m+1),$$
$$\sigma_2(1,\ldots,m) = \frac{1}{24}(m-1)m(m+1)(3m+2),$$
$$\tau_2(1,\ldots,m) = \frac{1}{24}m(m+1)(m+2)(3m+1). \qquad \square$$

In the following theorem we state a representation of $B_{n,\rho}^{(k)}e_\nu$ in terms of the images of monomials of the operators $B_n^{(k)}$. This underlines the close relationship beween the linking operators $B_{n,\rho}^{(k)}$ and the kth order Kantorovich modification of the classical operators B_n.

Theorem 3 *The images of the monomials under $B_{n,\rho}^{(k)}$ can be expressed as*

$$(B_{n,\rho}^{(k)}e_\nu)(x) = \frac{\nu!}{(\nu+k)!}\frac{1}{(n\rho)^{c,\underline{\nu+k}}}\sum_{i=0}^{\nu} s_{\nu+k}^{i+k}(\rho n)^{i+k}\frac{(i+k)!}{i!}(B_n^{(k)}e_i)(x),\ k \in \mathbb{N}_0,$$

where $s_{\nu+k}^{i+k}$ denote the Stirling numbers of first kind.

Proof For $\nu \in \mathbb{N}$ and $k = 0$ we derive from (14)

$$\begin{aligned}(B_{n,\rho}e_\nu)(x) &= \frac{1}{(n\rho)^{c,\underline{\nu}}}\sum_{j=1}^{\infty} p_{n,j}(x)\prod_{l=0}^{\nu-1}(j\rho+l) \\ &= \frac{1}{(n\rho)^{c,\underline{\nu}}}\sum_{i=0}^{\nu} s_\nu^i(\rho n)^i \sum_{j=1}^{\infty} p_{n,j}(x)\left(\frac{j}{n}\right)^i \\ &= \frac{1}{(n\rho)^{c,\underline{\nu}}}\sum_{i=0}^{\nu} s_\nu^i(\rho n)^i (B_n e_i)(x).\end{aligned}$$

For $k \in \mathbb{N}$ the conclusion follows by using $(B_{n,\rho}^{(k)} e_\nu) = \frac{\nu!}{(\nu+k)!} D^k(B_{n,\rho} e_{\nu+k})$ and $D^k(B_n e_i) = \frac{i!}{(i-k)!}(B_n^{(k)} e_{i-k})$, respectively. □

For the case $k = 0$ a corrresponding result for the Bernstein operators can be found in [17, Theorem 3.2.1].

3 Explicit Formulas for the Moments

Next, we consider the moments of $B_{n,\rho}$ and $B_{n,\rho}^{(k)}$. For abbreviation, we use the notation

$$M_{n,\rho,m}^{(k)}(x) = \left[B_{n,\rho}^{(k)}(e_1 - x e_0)^m \right](x), \ m \in \mathbb{N}_0, \ x \in [0, \infty) \tag{17}$$

where we again omit the superscript (k) in case $k = 0$. We use the fact that $M_{n,\rho,m}^{(k)}(x) = \sum_{\nu=0}^{m} \binom{m}{\nu} (-x)^{m-\nu} (B_{n,\rho}^{(k)} e_\nu)(x)$.
Again, we first treat the case $k = 0$.

Theorem 4 *Let $n \in \mathbb{R}$, $\rho \in \mathbb{R}_+$, $m \in \mathbb{N}_0$, $n\rho > c(m-1)$. Then*

$$M_{n,\rho,0}(x) = 1, \tag{18}$$

$$M_{n,\rho,1}(x) = 0, \tag{19}$$

$$\begin{aligned} M_{n,\rho,m}(x) = (-x)^m + \sum_{i=1}^{m} (-x)^i \sum_{\nu=1}^{i} \frac{\rho^{\nu+m-i}}{(n\rho)^{c,\underline{\nu+m-i}}} (-1)^\nu \binom{m}{i-\nu} \\ \times \frac{n^{c,\overline{\nu}}}{(\nu-1)!} \mathbf{\Delta}_1^{\nu-1} p_{\nu+m-i}^{\rho}(1), \ m \geq 2. \end{aligned} \tag{20}$$

Proof Equations (18) and (19) follow immediately from Corollary 2.

In order to prove (20) we apply Theorem 1. With the index transform $i \to i - m + \nu$, changing the order of summation and applying the index transform $\nu \to \nu + m - i$, we derive

$$\begin{aligned} &M_{n,\rho,m}(x) \\ &= (-x)^m + \sum_{\nu=1}^{m} \binom{m}{\nu} (-x)^{m-\nu} \frac{\rho^\nu}{(n\rho)^{c,\underline{\nu}}} \sum_{i=1}^{\nu} \frac{n^{c,\overline{i}}}{(i-1)!} \left(\mathbf{\Delta}_1^{i-1} p_\nu^\rho(1) \right) x^i \\ &= (-x)^m + \sum_{\nu=1}^{m} \binom{m}{\nu} (-1)^{m-\nu} \frac{\rho^\nu}{(n\rho)^{c,\underline{\nu}}} \\ &\quad \times \sum_{i=m-\nu+1}^{m} \frac{n^{c,\overline{i-m+\nu}}}{(i-m+\nu-1)!} \left(\mathbf{\Delta}_1^{i-m+\nu-1} p_\nu^\rho(1) \right) x^i \end{aligned}$$

$$= (-x)^m + \sum_{i=1}^{m} x^i \sum_{\nu=m+1-i}^{m} \binom{m}{\nu} (-1)^{m-\nu} \frac{\rho^\nu}{(n\rho)^{c,\underline{\nu}}}$$
$$\times \frac{n^{c,\overline{i-m+\nu}}}{(i-m+\nu-1)!} \left(\mathbf{\Delta}_1^{i-m+\nu-1} p_\nu^\rho(1) \right)$$
$$= (-x)^m + \sum_{i=1}^{m} (-x)^i \sum_{\nu=1}^{i} \binom{m}{i-\nu} (-1)^\nu \frac{\rho^{\nu+m-i}}{(n\rho)^{c,\underline{\nu+m-i}}}$$
$$\times \frac{n^{c,\overline{\nu}}}{(\nu-1)!} \left(\mathbf{\Delta}_1^{\nu-1} p_{\nu+m-i}^\rho(1) \right). \qquad \square$$

Remark 3 Analogously as for the images of monomials, (20) can be rewritten as

$$M_{n,\rho,m}(x) = (-x)^m + \sum_{i=1}^{m} (-x)^i \sum_{\nu=1}^{i} \frac{\rho^{\nu+m-i}}{(n\rho)^{\nu,\underline{\nu+m-i}}} n^{c,\overline{\nu}} \binom{m}{i-\nu}$$
$$\times \sum_{\kappa=0}^{\nu-1} (-1)^{\kappa+1} \frac{1}{\kappa!(\nu-1-\kappa)!} p_{\nu+m-i}^\rho(1+\kappa).$$

Next, we consider the special cases $\rho = 1$ and $\rho \to \infty$.
$\underline{\rho = 1}$: With [5, (3.48)]

$$\mathbf{\Delta}_1^{\nu-1} p_{\nu+m-i}^1(1) = (\nu+m-i-1)! \binom{\nu+m-i}{\nu}.$$

we get

$$M_{n,\rho,m}(x) = (-x)^m + \sum_{i=1}^{m} (-x)^i \frac{m!}{i!} \sum_{\nu=1}^{i} (-1)^\nu \frac{n^{c,\overline{\nu}}}{n^{c,\underline{\nu+m-i}}}$$
$$\times \binom{i}{\nu} \binom{\nu+m-i-1}{\nu-1},$$

which coincides with the result in [18, Korollar 1.12] and with [7, Korollar 4.4] with $s = -1$ and $n + c$ instead of n there.
$\underline{\rho \to \infty}$: Then $\dfrac{\rho^{\nu+m-i}}{(n\rho)^{c,\underline{\nu+m-i}}} \to \dfrac{1}{n^{\nu+m-i}}$ and

$$\mathbf{\Delta}_1^{\nu-1} p_{\nu+m-i}^\infty(1) = (\nu-1)! \sigma_{\nu+m-i}^\nu.$$

Thus

$$M_{n,\infty,m}(x) = (-x)^m + \sum_{i=1}^{m}(-x)^i \sum_{\nu=1}^{i} \frac{n^{c,\overline{\nu}}}{n^{\nu+m-i}} \binom{m}{i-\nu} (-1)^\nu \sigma_{\nu+m-i}^{\nu}.$$

In our next theorem, we evaluate the moments for the case $k \in \mathbb{N}$.

Theorem 5 *Let $n \in \mathbb{R}$, $\rho \in \mathbb{R}_+$, $k \in \mathbb{N}$, $m \in \mathbb{N}_0$, $n\rho > c(m+k-1)$. Then*

$$\begin{aligned} M_{n,\rho,m}^{(k)}(x) &= \sum_{i=0}^{m}(-x)^i \sum_{\nu=0}^{i} \frac{\rho^{\nu+m-i+k}}{(n\rho)^{c,\underline{\nu+m-i+k}}} (-1)^\nu \binom{m}{i-\nu} \\ &\quad \times \frac{(\nu+m-i)!}{(\nu+m-i+k)!} \frac{(\nu+k)}{\nu!} n^{c,\overline{k+\nu}} \boldsymbol{\Delta}_1^{\nu+k-1} p_{\nu+m-i+k}^{\rho}(1). \end{aligned} \tag{21}$$

Proof The result can be proved by using Theorem 2 and carrying out the same steps as in the proof of Theorem 4. □

Remark 4 With (10), we can rewrite the representation (21) as

$$\begin{aligned} &M_{n,\rho,m}^{(k)}(x) \\ &= \sum_{i=0}^{m}(-x)^i \sum_{\nu=0}^{i} \frac{\rho^{\nu+m-i+k}}{(n\rho)^{c,\underline{\nu+m-i+k}}} \binom{m}{i-\nu} \frac{(\nu+m-i)!}{(\nu+m-i+k)!} \\ &\quad \times n^{c,\overline{\nu+k}} \frac{(\nu+k)!}{\nu!} \sum_{\kappa=0}^{\nu+k-1} (-1)^{k+1+\kappa} \frac{1}{\kappa!(\nu+k-1-\kappa)!} p_{\nu+m-i+k}^{\rho}(1+\kappa). \end{aligned}$$

From Theorem 5 we derive the following identity for the special cases $\rho = 1$ and $\rho \to \infty$.

$\underline{\rho = 1}$: With [5, (3.48)] we have

$$\boldsymbol{\Delta}_1^{\nu+k-1} p_{\nu+m-i+k}^{1}(1) = (\nu+m-i+k-1)! \binom{\nu+m-i+k}{\nu+k}.$$

Thus

$$M_{n,1,m}^{(k)}(x) = \sum_{i=0}^{m}(-x)^i \frac{m!}{i!} \sum_{\nu=0}^{i} (-1)^\nu \frac{n^{c,\overline{\nu+k}}}{n^{c,\underline{\nu+m-i+k}}} \binom{i}{\nu} \binom{\nu+m-i+k-1}{\nu+k-1}.$$

This coincides with the result [7, Korollar 4.4] for the moments of the auxiliary operators named $M_{n+c,k-1}$ there.

$\underline{\rho \to \infty}$: Then $\dfrac{\rho^{\nu+m-i+k}}{(n\rho)^{c,\underline{\nu+m-i+k}}} \to \dfrac{1}{n^{\nu+m-i+k}}$ and

$$\mathbf{\Delta}_1^{\nu+k-1} p_{\nu+m-i+k}^{\infty}(1) = (\nu+k-1)!\sigma_{\nu+m-i+k}^{\nu+k}.$$

Thus

$$M_{n,\infty,m}^{(k)}(x) = \sum_{i=0}^{m}(-x)^i \sum_{\nu=0}^{i} \frac{n^{c,\overline{\nu+k}}}{n^{\nu+m-i+k}} \binom{m}{i-\nu} (-1)^{\nu} \frac{(\nu+m-i)!(\nu+k)!}{(\nu+m-i+k)!\nu!} \sigma_{\nu+m-i+k}^{\nu+k}.$$

With the same notations and arguments used for Corollary 2, the moments (20) and (21) can be computed by using

$$\mathbf{\Delta}_1^{\nu+k-1} p_{\nu+m-j+k}^{\rho}(1) = (\nu+k-1)! \sum_{l=0}^{m-j} \rho^{-l} \sigma_l(1,2,\ldots,\nu+m-j+k-1)\tau_{m-j-l}(1,2,\ldots,\nu+k).$$

Corollary 3 *For $k \in \mathbb{N}_0$ the first moments are given by*

$$M_{n,\rho,0}^{(k)}(x) = \frac{\rho^k}{(n\rho)^{c,\underline{k}}} n^{c,\overline{k}}, \quad M_{n,\rho,1}^{(k)}(x) = \frac{\rho^{k+1}}{(n\rho)^{c,\underline{k+1}}} n^{c,\overline{k}} \frac{1}{2} k \left(1+\frac{1}{\rho}\right)(1+2cx),$$

$$M_{n,\rho,2}^{(k)}(x) = \frac{\rho^{k+2}}{(n\rho)^{c,\underline{k+2}}} n^{c,\overline{k}} \left(1+\frac{1}{\rho}\right) \left\{ \left[n + c\left(1+\frac{1}{\rho}\right) k(k+1) \right] x(1+cx) + \frac{k}{12}\left[(3k+1)\left(1+\frac{1}{\rho}\right) + \frac{3k+5}{\rho} \right] \right\}.$$

References

1. Baskakov, V.A.: An instance of a sequence of positive linear operators in the space of continuous functions. Dokl. Akad. Nauk SSSR **113**(2), 249–251 (1957)
2. Durrmeyer, J.L.: Une formule d'inversion de la transformée de Laplace: applications à la théorie des moments, Thèse de 3e cycle, Faculté des Sciences de l'Université de Paris, 1967
3. Gonska, H., Păltănea, R.: Simultaneous approximation by a class of Bernstein-Durrmeyer operators preserving linear functions. Czechoslovak Math. J. **60**(135), 783–799 (2010)
4. Gonska, H., Păltănea, R.: Quantitative convergence theorems for a class of Bernstein-Durrmeyer operators preserving linear functions. Ukrainian Math. J. **62**(7), 1061–1072 (2010)
5. Gould, H.W.: Combinatorial Identities: A Standardized Set of Tables Listing 500 Binomial Coefficient Summations. Henry W. Gould, Morgantown (1972)
6. Heilmann, M.: Direct and converse results for operators of Baskakov-Durrmeyer operators. Approx. Theory Appl. **5**(1), 105–127 (1989)
7. Heilmann, M.: Erhöhung der Konvergenzgeschwindigkeit bei der Approximation von Funktionen mit Hilfe von Linearkombinationen spezieller positiver linearer Operatoren, Habilitationschrift Universität Dortmund (1992)
8. Heilmann, M., Raşa, I.: *k*-th order Kantorovich type modification of the operators U_n^{ρ}. J. Appl. Funct. Anal. **9**(3–4), 320–334 (2014)

9. Lupaş, A.: Die Folge der Betaoperatoren, Dissertation, Universität Stuttgart (1972)
10. Mazhar, S.M., Totik, V.: Approximation by modified Szász operators. Acta Sci. Math. **49**, 257–269 (1985)
11. Păltănea, R.: A class of Durrmeyer type operators preserving linear functions. Ann. Tiberiu Popoviciu Semin. Funct. Equ. Approx. Convexity (Cluj-Napoca) **5**, 109–117 (2007)
12. Păltănea, R.: Modified Szász-Mirakjan operators of integral form. Carpathian J. Math. **24**(3), 378–385 (2008)
13. Păltănea, R.: Simultaneous approximation by a class of Szász-Mirakjan operators. J. Appl. Funct. Anal. **9**(3–4), 356–368 (2014)
14. Phillips, R.S.: An inversion formula for Laplace transforms and semi-groups of linear operators. Ann. Math. **59**(2), 325–356 (1954)
15. Phillips, G.M.: Interpolation and Approximation by Polynomials. Springer (2003)
16. Sahai, A., Prasad, G.: On simultaneous approximation by modified Lupas operators. J. Approx. Theory **45**, 122–128 (1985)
17. Stanila, E.D.: On Bernstein-Euler-Jacobi operators, PhD Thesis University of Duisburg-Essen (2014)
18. Wagner, M.: Quasi-Interpolanten zu genuinen Baskakov-Durrmeyer-Typ Operatoren, Disssertation Universität Wuppertal (2013)

Rate of Convergence of Modified Schurer-Type q-Bernstein Kantorovich Operators

Manjari Sidharth and P.N. Agrawal

Abstract Lin (J. Inequal. Appl. **465**, 2014 [10]) introduced a new modified Schurer-type q-Bernstein Kantorovich operators and discussed a local approximation theorem and the statistical convergence of these operators. In this paper we study the rate of convergence by means of the first-order modulus of continuity, Lipschitz class function, the modulus of continuity of the first-order derivative and the Voronovskaja-type theorem.

Keywords Schurer type q-Bernstein Kantorovich operators · Rate of convergence · Modulus of continuity · Lipschitz class function

1 Introduction

In 1987, q-analogue of classical Bernstein polynomials was introduced by Lupas [11]. After a decade, Phillips [13] proposed another generalization of these polynomials based on q-integers and discussed the rate of convergence and Voronovskaja-type asymptotic formula. The q-analogue of Bernstein polynomials due to Phillips was studied by several researchers, e.g. Ostrovska [14, 15], Kim [9], Wang [17], etc. Subsequently, some other generalizations based on q-integers of the other well-known positive linear operators were proposed and studied.

In 2005, Derriennic [5] introduced the q-analogue of Bernstein Durmeyer polynomials with Jacobi weights and studied some approximation properties. Later, Gupta [7] introduced the q-analogue of the Bernstein Durmeyer operators which was inves-

M. Sidharth (✉) · P.N. Agrawal
Department of Mathematics, Indian Institute of Technology Roorkee,
Roorkee 247667, India
e-mail: manjarisidharth93@gmail.com

P.N. Agrawal
e-mail: pna_iitr@yahoo.co.in

P.N. Agrawal et al. (eds.), *Mathematical Analysis and its Applications*,
Springer Proceedings in Mathematics & Statistics 143,
DOI 10.1007/978-81-322-2485-3_19

tigated later by Finta and Gupta [6] and several other researchers. Dalmanoglu [4] introduced the Kantorovich-type modification of q-Bernstein polynomials and established some approximation results. Subsequently, Radu investigated the statistical convergence results of these operators.

Muraru [12] introduced Bernstein–Schurer polynomials based on q-integers and established the rate of convergence in terms of modulus of the continuity. Agrawal et al. [1] considered the Stancu varient of these operators and discussed some local and global direct results. Later, Agrawal et al. [2] proposed Durmeyer-type modification of these operators and discussed some local direct results and studied the rate of convergence of modified limit q-Bernstein–Schurer-type operators.

Very recently, Lin [10] introduced a new kind of modified Schurer-type q-Bernstein Kantorovich operators as follows:

Let $p \in \mathbb{N}^0$ (the set of non-negative integers) be arbitrary but fixed and α, β be integers satisfying $0 \leq \alpha \leq \beta$. For $f \in C[0, 1+p]$, he defined

$$K_{n,q}^{(\alpha,\beta)}(f;x) = \sum_{k=0}^{n+p} \bar{p}_{n,k}(q;x) \int_0^1 f\left(\frac{t}{[n+1+\beta]_q} + \frac{q[k+\alpha]_q}{[n+1+\beta]_q}\right) d_q t, x \in [0,1]$$

where $\bar{p}_{n,k}(q;x) = \binom{n}{k}_q x^k \prod_{s=0}^{n+p+k-1} (1-q^s x)$. It is clear that $K_{n,q}^{(\alpha,\beta)}(f;x)$ is a linear positive operator. It is remarked that when $\alpha = \beta = 0$, it reduces to the operator discussed in [16].

In the present paper, we continue the work done by Lin by discussing the rate of convergence in terms of the modulus of continuity, elements of Lipschitz-type space and Voronovskaja-type theorem. Throughout this paper, we consider $0 < q < 1$. For the properties of the q-calculus, we refer to [3, 8].

2 Preliminaries

In this section, we give some basic results which will be used in the sequel.

Lemma 1 ([10]) *For* $K_{n,q}^{(\alpha,\beta)}(t^m;x)$, $m = 0, 1, 2$, *we have*

(i) $K_{n,q}^{(\alpha,\beta)}(1;x) = 1$,

(ii) $K_{n,q}^{(\alpha,\beta)}(t;x) = \frac{[n+p]_q}{[n+1+\beta]_q} q^{\alpha+1} x + \frac{1}{[n+1+\beta]_q}\left(\frac{1}{[2]_q} + q[\alpha]_q\right)$,

(iii) $K_{n,q}^{(\alpha,\beta)}(t^2;x) = \frac{[n+p]_q[n+p-1]_q}{[n+1+\beta]_q^2} q^{2\alpha+3} x^2$

$$+\frac{[n+p]_q}{[n+1+\beta]_q^2}\left(\frac{2}{[2]_q}q^{\alpha+1}+q^{2+\alpha}(2[\alpha]_q+q^\alpha)\right)x$$
$$+\frac{1}{[n+1+\beta]_q^2}\left(\frac{1}{[3]_q}+\frac{2q[\alpha]_q}{[2]_q}+q^2[\alpha]_q^2\right).$$

Remark 1 For the modified Schurer-type q-Bernstein Kantorovich operators, we have

(i) $\lim_{n\to\infty}[n]_{q_n}(K_{n,q_n}^{(\alpha,\beta)}((t-x);x)=\left(\frac{1+2\alpha}{2}-(\alpha+1)x\right),$

(ii) $\lim_{n\to\infty}[n]_{q_n}(K_{n,q_n}^{(\alpha,\beta)}((t-x)^2;x)=x(1-x).$

3 Main Results

3.1 Rate of Convergence

The first modulus of continuity of $f\in C[0,1+p]$ for $\delta>0$, is given by

$$\omega(f,\delta)=\max_{0<|h|<\delta,x,x+h\in[0,1+p]}|f(x+h)-f(x)|.$$

We observe that for all $f\in C[0,1+p]$, we have

$$\lim_{\delta\to0^+}\omega(f,\delta)=0$$

and for any $\delta>0$,

$$|f(x)-f(y)|\le\omega(f,\delta)\left(\frac{|x-y|}{\delta}+1\right). \tag{1}$$

Theorem 1 *For $f\in C[0,1+p]$, we have*

$$|K_{n,q}^{(\alpha,\beta)}(f;x)-f(x)|\le 2\omega\left(f;\sqrt{\delta_{n,q}^{(\alpha,\beta)}}\right),$$

where $\omega(f,.)$ is the modulus of continuity of f and $\delta_{n,q}^{(\alpha,\beta)}:=K_{n,q}^{(\alpha,\beta)}((t-x)^2;x)$.

Proof Using the linearity and positivity of the operator $K_{n,q}^{(\alpha,\beta)}(f;x)$ in view of (1), we get

$$|K_{n,q}^{(\alpha,\beta)}(f;x)-f(x)|=\left|\sum_{k=0}^{n+p}\bar{p}_{n,k}(q;x)\int_0^1\left(f\left(\frac{t}{[n+1+\beta]_q}+\frac{q[k+\alpha]}{[n+1+\beta]_q}\right)-f(x)\right)d_qt\right|$$

$$\leq \sum_{k=0}^{n+p} \bar{p}_{n,k}(q;x) \int_0^1 \left| f\left(\frac{t}{[n+1+\beta]_q} + \frac{q[k+\alpha]_q}{[n+1+\beta]_q}\right) - f(x) \right| d_q t$$

$$\leq \sum_{k=0}^{n+p} \bar{p}_{n,k}(q;x) \int_0^1 \left(\frac{\left| \frac{t}{[n+1+\beta]_q} + \frac{q[k+\alpha]_q}{[n+1+\beta]_q} - x \right|}{\delta} + 1 \right) \omega(f,\delta) d_q t$$

$$\leq \omega(f,\delta) \sum_{k=0}^{n+p} \bar{p}_{n,k}(q;x) + \frac{\omega(f,\delta)}{\delta} \left(\sum_{k=0}^{n+p} \bar{p}_{n,k}(q;x) \int_0^1 \left| \frac{t}{[n+1+\beta]_q} + \frac{q[k+\alpha]_q}{[n+1+\beta]_q} - x \right| d_q t \right).$$

On applying the Cauchy–Schwarz inequality, we have

$$\int_0^1 \left| \frac{t}{[n+1+\beta]_q} + \frac{q[k+\alpha]_q}{[n+1+\beta]_q} - x \right| d_q t \leq \left\{ \int_0^1 \left(\frac{t}{[n+1+\beta]_q} + \frac{q[k+\alpha]_q}{[n+1+\beta]_q} - x \right)^2 d_q t \right\}^{1/2} = \sqrt{a_{n,k}^{(\alpha,\beta)}(x)}.$$

Hence,

$$|K_{n,q}^{(\alpha,\beta)}(f;x) - f(x)| \leq \omega(f,\delta) + \frac{\omega(f,\delta)}{\delta} \sum_{k=0}^{n+p} \{a_{n,k}^{(\alpha,\beta)}\}^{1/2} \bar{p}_{n,k}(q;x).$$

Again applying the Cauchy–Schwarz inequality, we get

$$\begin{aligned}
&|K_{n,q}^{(\alpha,\beta)}(f;x) - f(x)| \\
&\leq \omega(f,\delta) + \frac{\omega(f,\delta)}{\delta} \left\{ \sum_{k=0}^{n+p} a_{n,k}^{(\alpha,\beta)} \bar{p}_{n,k}(q;x) \right\}^{1/2} \left\{ \sum_{k=0}^{n+p} \bar{p}_{n,k}(q;x) \right\}^{1/2} \\
&= \omega(f,\delta) + \frac{\omega(f,\delta)}{\delta} \left\{ \sum_{k=0}^{n+p} \bar{p}_{n,k}(q;x) \int_0^1 \left(\frac{t}{[n+1+\beta]_q} + \frac{q[k+\alpha]_q}{[n+1+\beta]_q} - x \right)^2 d_q t \right\}^{1/2} \\
&= \omega(f,\delta) + \frac{\omega(f,\delta)}{\delta} \left\{ K_{n,q}^{(\alpha,\beta)}((t-x)^2;x) \right\}^{1/2}.
\end{aligned}$$

Choosing $\delta := \delta_{n,q}^{(\alpha,\beta)} = K_{n,q}^{(\alpha,\beta)}((t-x)^2;x)$, we have

$$|K_{n,q}^{(\alpha,\beta)}(f;x) - f(x)| \leq 2\omega\left(f, \sqrt{\delta_{n,q}^{(\alpha,\beta)}(x)}\right).$$

Hence, we get the desired result.

Corollary 1 *Let $f \in Lip_M(\xi)$ for $0 < \xi \leq 1$, then*

$$|K_{n,q}^{(\alpha,\beta)}(f;x) - f(x)| \leq 2M\left(\delta_{n,q}^{(\alpha,\beta)}(x)\right)^{\xi/2},$$

where $\delta_{n,q}^{(\alpha,\beta)}(x) = K_{n,q}^{(\alpha,\beta)}((t-x)^2;x)$.

Proof Since $f \in Lip_M(\xi)$, we have $\omega(f,\delta) \leq M\delta^{\xi}$ for any $\delta > 0$. Hence the result follows from Theorem 1.

Theorem 2 *If $f(x)$ has a continuous derivative $f'(x)$ and $\omega(f',\delta)$ is the modulus of continuity of $f'(x)$ on $[0, 1+p]$, then*

$$|K_{n,q}^{(\alpha,\beta)}(f;x) - f(x)| \leq M\,|\mu_{n,q,p}^{(\alpha,\beta)}| + \omega(f',\delta)\left(1 + \sqrt{\delta_{n,q}^{(\alpha,\beta)}(x)}\right),$$

where M is a positive constant such that $|f'(x)| \leq M$ and

$$\mu_{n,q,p}^{(\alpha,\beta)}(x) = \left(\frac{q^{\alpha+1}[n+p]_q}{[n+1+\beta]_q} - x\right)x + \frac{1}{[n+1+\beta]_q}\left(\frac{1}{[2]_q} + q[\alpha]_q\right). \quad (2)$$

Proof On applying the mean value theorem, we get

$$\begin{aligned} f\left(\frac{t}{[n+1+\beta]_q} + \frac{q[k+\alpha]_q}{[n+1+\beta]_q}\right) - f(x) &= \left(\frac{t}{[n+1+\beta]_q} + \frac{q[k+\alpha]_q}{[n+1+\beta]_q} - x\right)f'(\xi) \\ &= \left(\frac{t}{[n+1+\beta]_q} + \frac{q[k+\alpha]_q}{[n+1+\beta]_q} - x\right)f'(x) \\ &+ \left(\frac{t}{[n+1+\beta]_q} + \frac{q[k+\alpha]_q}{[n+1+\beta]_q} - x\right)(f'(\xi) - f'(x)), \end{aligned}$$

where ξ lies between $\left(\frac{t}{[n+1+\beta]_q} + \frac{q[k+\alpha]_q}{[n+1+\beta]_q}\right)$ and x.

Hence, we get

$$\begin{aligned} |K_{n,q}^{(\alpha,\beta)}(f;x) - f(x)| = \Bigg| & f'(x)\sum_{k=0}^{n+p}\int_0^1\left(\frac{t}{[n+1+\beta]_q} + \frac{q[k+\alpha]_q}{[n+1+\beta]_q} - x\right) \\ & \times \binom{n}{k}_q x^k \prod_{s=0}^{n+p+k-1}(1-q^s x)d_q t \\ & + \sum_{k=0}^{n+p}\int_0^1\left(\frac{t}{[n+1+\beta]_q} + \frac{q[k+\alpha]_q}{[n+1+\beta]_q} - x\right) \\ & \times (f'(\xi) - f'(x))\binom{n}{k}_q x^k \prod_{s=0}^{n+p+k-1}(1-q^s x)d_q t\Bigg| \end{aligned}$$

$$\leq |f'(x)||K_{n,q}^{(\alpha,\beta)}((t-x);x)| + \sum_{k=0}^{n+p}\int_0^1 \left|\frac{t}{[n+1+\beta]_q} + \frac{q[k+\alpha]_q}{[n+1+\beta]_q} - x\right|$$
$$\times |f'(\xi) - f'(x)| \binom{n}{k}_q x^k \prod_{s=0}^{n+p+k-1}(1-q^s x) d_q t$$
$$\leq M|\mu_{n,q,p}^{(\alpha,\beta)}| + \sum_{k=o}^{n+p}\int_0^1 \omega(f',\delta)\left(\frac{\left|\frac{t}{[n+1+\beta]_q} + \frac{q[k+\alpha]_q}{[n+1+\beta]_q} - x\right|}{\delta} + 1\right)$$
$$\times \left|\frac{t}{[n+1+\beta]_q} + \frac{q[k+\alpha]_q}{[n+1+\beta]_q} - x\right| \binom{n}{k}_q x^k \prod_{s=0}^{n+p+k-1}(1-q^s x) d_q t$$
$$\leq M\,|\mu_{n,q,p}^{(\alpha,\beta)}| + \omega(f',\delta)\sum_{k=o}^{n+p}\int_0^1 \left|\frac{t}{[n+1+\beta]_q} + \frac{q[k+\alpha]_q}{[n+1+\beta]_q} - x\right|$$
$$\times \binom{n}{k}_q x^k \prod_{s=0}^{n+p+k-1}(1-q^s x) d_q t$$
$$+ \frac{\omega(f',\delta)}{\delta}\sum_{k=0}^{n+p}\int_0^1 \left(\frac{t}{[n+1+\beta]_q} + \frac{t}{[n+1+\beta]_q} - x\right)^2$$
$$\times \binom{n}{k}_q x^k \prod_{s=0}^{n+p+k-1}(1-q^s x) d_q t,$$

where $\mu_{n,q,p}^{(\alpha,\beta)}$ is given by (2).

Now, applying Cauchy–Schwarz inequality in second term of the right side of the inequality, we have

$$|K_{n,q}^{(\alpha,\beta)}(f;x) - f(x)|$$
$$\leq M\,|\mu_{n,q,p}^{(\alpha,\beta)}| + \omega(f',\delta)\left(\sum_{k=0}^{n+p}\int_0^1 \left(\frac{t}{[n+1+\beta]_q} + \frac{q[k+\alpha]_q}{[n+1+\beta]_q} - x\right)^2\right.$$
$$\left.\times \binom{n}{k}_q x^k \prod_{s=0}^{n+p+k-1}(1-q^s x) d_q t\right)^{1/2}$$
$$+ \frac{\omega(f',\delta)}{\delta}\sum_{k=0}^{n+p}\int_0^1 \left(\frac{t}{[n+1+\beta]_q} + \frac{t}{[n+1+\beta]_q} - x\right)^2 \binom{n}{k}_q x^k \prod_{s=0}^{n+p+k-1}(1-q^s x) d_q t,$$
$$\leq M\,|\mu_{n,q,p}^{(\alpha,\beta)}| + \omega(f',\delta)\sqrt{K_{n,q}^{(\alpha,\beta)}((t-x)^2;x)} + \frac{\omega(f',\delta)}{\delta}K_{n,q}^{(\alpha,\beta)}((t-x)^2;x).$$

Choosing $\delta := \delta_{n,q}^{(\alpha,\beta)} = K_{n,q}^{(\alpha,\beta)}((t-x)^2;x)$, we have

$$|K_{n,q}^{(\alpha,\beta)}(f;x) - f(x)| \leq M|\mu_{n,q,p}^{(\alpha,\beta)}| + \omega(f',\delta_{n,q}^{(\alpha,\beta)}(x))(1+\sqrt{\delta_{n,q}^{(\alpha,\beta)}(x)}).$$

Hence, we get the desired result.

3.2 Asymptotic Result

Theorem 3 *Let $f \in C[0, 1+p]$, $0 < q_n < 1$ be a sequence such that $q_n \to 1$ and $\frac{1}{[n]_{q_n}} \to 0$, as $n \to \infty$. Suppose that $f''(x)$ exist at a point $x \in [0, 1]$, then we have*

$$\lim_{n\to\infty} [n]_{q_n} \left(K_{n,q_n}^{(\alpha,\beta)}(f; x) - f(x) \right) = \left(\frac{1+2\alpha}{2} - (\alpha+1)x \right) f'(x) + \frac{1}{2} x(1-x) f''(x).$$

Proof By Taylor's expansion we have

$$f(t) = f(x) + (t-x) f'(x) + \frac{1}{2} f''(x)(t-x)^2 + r(t,x)(t-x)^2, \tag{3}$$

where $r(t, x)$ is the Peano form of the remainder and $\lim_{t\to x} r(t, x) = 0$.
On applying $K_{n,q_n}^{(\alpha,\beta)}(.; x)$ on both sides of Eq. (3), we get

$$K_{n,q_n}^{(\alpha,\beta)}(f; x) - f(x) = f'(x) K_{n,q_n}^{(\alpha,\beta)}((t-x); x) + \frac{1}{2} f''(x) K_{n,q_n}^{(\alpha,\beta)}((t-x)^2; x) + K_{n,q_n}^{(\alpha,\beta)}((t-x)^2 r(t,x); x).$$

Taking the limit as $n \to \infty$ on both sides of the above equation, we get

$$\begin{aligned} \lim_{n\to\infty} [n]_{q_n} (K_{n,q_n}^{(\alpha,\beta)}(f; x) - f(x)) = \lim_{n\to\infty} [n]_{q_n} f'(x) \Bigg(K_{n,q_n}^{(\alpha,\beta)}((t-x); x) \\ + \lim_{n\to\infty} [n]_{q_n} \frac{f''(x)}{2} (K_{n,q_n}^{(\alpha,\beta)}((t-x)^2; x \Bigg) \\ + \lim_{n\to\infty} [n]_{q_n} K_{n,q_n}^{(\alpha,\beta)}((t-x)^2 r(t,x); x). \end{aligned} \tag{4}$$

From the Remark 1, we have

$$\lim_{n\to\infty} [n]_{q_n} (K_{n,q_n}^{(\alpha,\beta)}((t-x); x) = \left(\frac{1+2\alpha}{2} - (\alpha+1)x \right), \tag{5}$$

uniformly in [0,1], and

$$\lim_{n\to\infty} [n]_{q_n} (K_{n,q_n}^{(\alpha,\beta)}((t-x)^2; x) = x(1-x) \text{ , uniformly in } [0, 1]. \tag{6}$$

Hence in order to prove the result it is sufficient to show that

$$[n]_{q_n} K_{n,q_n}^{(\alpha,\beta)}((t-x)^2 r(t,x); x) \to 0 \text{ as } n \to \infty \text{ , uniformly in } [0, 1].$$

By using the Cauchy–Schwarz inequality, we have

$$K_{n,q_n}^{(\alpha,\beta)}((t-x)^2 r(t,x);x) \leq \sqrt{K_{n,q_n}^{(\alpha,\beta)}(r^2(t,x);x)}\sqrt{K_{n,q_n}^{(\alpha,\beta)}((t-x)^4;x)}. \quad (7)$$

We observe that $r^2(x,x) = 0$, and from the Basic convergence theorem [10], we have

$$\lim_{n\to\infty} K_{n,q_n}^{(\alpha,\beta)}(r^2(t,x);x) = r^2(x,x) = 0. \quad (8)$$

Hence, from (7) and (8), we get

$$\lim_{n\to\infty} [n]_{q_n} K_{n,q_n}^{(\alpha,\beta)}((t-x)^2 r(t,x);x) = 0, \text{ uniformly in } [0,1], \quad (9)$$

in view of the fact that

$$K_{n,q_n}^{(\alpha,\beta)}((t-x)^4;x) = O\left(\frac{1}{n^2}\right) \ as\ n \to \infty, \text{ uniformly in } [0,1].$$

Now, combining (4)–(6) and (9), we get the required result.
This completes the proof of the theorem.

Now, we consider the following two-parameter Lipschitz-type space:

$$Lip_M^{(a,b)}(r) := \Bigg\{ f \in C[0,1+p] : |f(t)-f(x)| \leq M\frac{|t-x|^r}{(t+ax+bx^2)^{r/2}}; x, \in (0,1], t \in [0,1+p] \Bigg\},$$

where M is a positive constant and $r \in (0,1]$.

Theorem 4 *Let $f \in Lip_M^{(a,b)}(r)$. Then $\forall x \in (0,1)$, we have*

$$|K_{n,q}^{(\alpha,\beta)}(f;x) - f(x)| \leq M\left(\frac{\delta_{n,q}^{(\alpha,\beta)}(x)}{ax+bx^2}\right)^{r/2},$$

where $\delta_{n,q}^{(\alpha,\beta)}(x) = K_{n,q}^{(\alpha,\beta)}((t-x)^2;x)$.

Proof First, we prove the result for $r = 1$.

$$|K_{n,q}^{(\alpha,\beta)}(f;x) - f(x)| \leq \sum_{k=0}^{n+p} \bar{p}_{n,k}(q;x) \int_0^1 \left| f\left(\frac{t}{[n+1+\beta]_q} + \frac{q[k+\alpha]_q}{[n+1+\beta]_q}\right) - f(x)\right| d_q t$$

$$\leq M \sum_{k=0}^{n+p} \bar{p}_{n,k}(q;x) \int_0^1 \frac{\left|\frac{t}{[n+1+\beta]_q} + \frac{q[k+\alpha]_q}{[n+1+\beta]_q} - x\right|}{\sqrt{\frac{t}{[n+1+\beta]_q} + \frac{q[k+\alpha]_q}{[n+1+\beta]_q} + ax + bx^2}} d_q t.$$

Since

$$\frac{1}{\sqrt{\frac{t}{[n+1+\beta]_q}+\frac{q[k+\alpha]_q}{[n+1+\beta]_q}+ax+bx^2}}<\frac{1}{\sqrt{ax+bx^2}},$$

the last inequality implies that

$$\begin{aligned}|K_{n,q}^{(\alpha,\beta)}(f;x)-f(x)| &\leq \frac{M}{\sqrt{ax+bx^2}}\sum_{k=0}^{n+p}\bar{p}_{n,k}(q;x)\int_0^1\left|\frac{t}{[n+1+\beta]_q}+\frac{q[k+\alpha]_q}{[n+1+\beta]_q}-x\right|d_qt\\ &= \frac{M}{\sqrt{ax+bx^2}}K_{n,q}^{(\alpha,\beta)}(|t-x|;x)\\ &= M\sqrt{\frac{\delta_{n,q}^{(\alpha,\beta)}(x)}{ax+bx^2}},\end{aligned}$$

on applying the Cauchy–Schwarz inequality. Hence, the result is true for $r=1$.

Now, we prove the result for $r\in(0,1)$. Applying the Hölder's inequality with $p=\frac{1}{r}$ and $q=\frac{1}{1-r}$, we get

$$\begin{aligned}|K_{n,q}^{(\alpha,\beta)}(f;x)-f(x)| &\leq \sum_{k=0}^{n+p}\bar{p}_{n,k}(q;x)\int_0^1\left|f\left(\frac{t}{[n+1+\beta]_q}+\frac{q[k+\alpha]_q}{[n+1+\beta]_q}\right)-f(x)\right|d_qt\\ &\leq \left\{\sum_{k=0}^{n+p}\bar{p}_{n,k}(q;x)\left(\int_0^1\left|f\left(\frac{t}{[n+1+\beta]_q}+\frac{q[k+\alpha]_q}{[n+1+\beta]_q}\right)-f(x)\right|d_qt\right)^{1/r}\right\}^r.\end{aligned}$$

Again applying Hölder's inequality with $p=\frac{1}{r}$ and $q=\frac{1}{1-r}$, we get

$$\begin{aligned}|K_{n,q}^{(\alpha,\beta)}(f;x)-f(x)| \leq \Bigg\{\sum_{k=0}^{n+p}\bar{p}_{n,k}(q;x)\int_0^1\Bigg|f\left(\frac{t}{[n+1+\beta]_q}+\frac{q[k+\alpha]_q}{[n+1+\beta]_q}\right)\\ -f(x)\Bigg|^{1/r}d_qt\Bigg\}^r.\end{aligned}$$

Since $f\in Lip_M^{(a,b)}(r)$, we have

$$|K_{n,q}^{(\alpha,\beta)}(f;x)-f(x)|\leq M\left\{\sum_{k=0}^{n+p}\bar{p}_{n,k}(q;x)\int_0^1\frac{\left|\frac{t}{[n+1+\beta]_q}+\frac{q[k+\alpha]_q}{[n+1+\beta]_q}-x\right|}{\sqrt{\frac{t}{[n+1+\beta]_q}+\frac{q[k+\alpha]_q}{[n+1+\beta]_q}+ax+bx^2}}d_qt\right\}^r$$

$$\leq \frac{M}{\sqrt[r]{ax+bx^2}}\left\{\sum_{k=0}^{n+p}\bar{p}_{n,k}(q;x)\int_0^1\left|\frac{t}{[n+1+\beta]_q}+\frac{q[k+\alpha]_q}{[n+1+\beta]_q}-x\right|d_qt\right\}^r$$
$$\leq \frac{M}{\sqrt[r]{ax+bx^2}}\left(K_{n,q}^{(\alpha,\beta)}(|t-x|;x)\right)^r.$$

Thus, on applying Cauchy–Schwarz inequality, we have

$$|K_{n,q}^{(\alpha,\beta)}(f;x)-f(x)|\leq M\left(\frac{\delta_{n,q}^{(\alpha,\beta)}(x)}{ax+bx^2}\right)^{r/2}.$$

Hence, we get the desired result.

Acknowledgments The authors are extremely grateful to the reviewers for their valuable comments leading to a better presentation of the paper.

References

1. Agrawal, P.N., Gupta, V., Sathish, A.: Kumar, On q-analogue of Bernstein-Schurer-Stancu operators. Appl. Math. Comput. **219**(14), 7754–7764 (2013)
2. Agrawal, P.N., Sathish, A., Kumar, Sinha, T.A.K.: Stancu type generalization of modified Schurer operators based on q-integers. Appl. Math. Comput. **226**, 765–776 (2014)
3. Aral, A., Gupta, V., Agarwal, R.P.: Applications of q-Calculus in Operator Theory. Springer (2013)
4. Dalmanoglu, Ö., Dogru, O.: On statistical approximation properties of Kantorovich type q-Bernstein operators. Math. Comput. Modell. **52**, 760–771 (2010)
5. Derriennic, M.M.: Modified bernstein polynomials and jacobi polynomials in q-calculus. Rend. Circ. Mat. Palermo Serie II (Suppl. 76), 269–290 (2005)
6. Finta, Z., Gupta, V.: Approximation properties of q-Baskakov operators. Cent. Eur. J. Math. **8**(1), 199–211 (2010)
7. Gupta, V.: Some approximation properties of q-Durmeyer operators. Appl. Math. Comput. **197**(1), 172–178 (2008)
8. Kac, V., Cheung, P.: Quantum Calculus. Universitext, Springer, New York (2002)
9. Kim, T.: A note on q-Bernstein polynomials. Russ. J. Math. Phys. **18**(1), 73–82 (2011)
10. Lin, Q.: Statistical approximation of modified Schurer type q-Bernstein Kantorovich operators. J. Inequal. Appl. **465**, (2014)
11. Lupas, A.: A q-analogue of the Bernstein operators. Semin. Numer. Stat. Calc. University of Cluj-Napoca **9**, 85–92 (1987)
12. Muraru, C.V.: Note on q-Bernstein-Schurer operators. Stud. Univ. Babes-Bolyai Math. **56**, 489–495 (2011)
13. Phillips, G.M.: Bernstein polynomials based on q-integers. Ann. Numer. Math. **4**(1–4), 511–518 (1997)
14. Ostrovska, S.: q-Bernstein polynomials and their iterates. J. Approx. Theory **123**(2), 232–255 (2003)

15. Ostrovska, S.: On Lupas q-analogue of the Bernstein operator. Rocky Mountain J. Math. **36**(5), 1615–1629 (2006)
16. Vedi, T.: Some Schurer Type q-Bernstein operators. Dissertation. Eastern Mediterranean University (2011)
17. Wang, H.: Properties of convergence for q Bernstein polynomials. J. Math. Anal. Appl. **340**(2), 1096–1108 (2008)

Operators of Durrmeyer Type with Respect to Arbitrary Measure

Elena E. Berdysheva

Abstract In this paper, we give an overview of operators of Durrmeyer type with respect to arbitrary measure. Our construction includes the Bernstein–Durrmeyer operator, the Szász–Mirakjan–Durrmeyer operator, and the Baskakov–Durrmeyer operator with respect to arbitrary measure. We are particularly interested in the convergence of the operators. We discuss the uniform and the pointwise convergence as well as convergence in the corresponding weighted L^p-spaces. A new result is the statement on the L^p-convergence of the Szász–Mirakjan–Durrmeyer operator and the Baskakov–Durrmeyer operator without additional restrictions on the measure.

Keywords Bernstein–Durrmeyer operator · Szász–Mirakjan–Durrmeyer operator · Baskakov–Durrmeyer operator · Uniform convergence · Pointwise convergence · L^p-convergence

2010 AMS Subject Classification: 41A36

1 Introduction

In this paper, we consider a class of positive linear operators of Durrmeyer type. Let $c \in \mathbb{R}$. Depending on the value of c, we consider the intervals $I_c = \left[0, -\frac{1}{c}\right]$ for $c < 0$ and $I_c = [0, \infty)$ for $c \geq 0$. For $n > 0$, $k \in \mathbb{N}_0 := \mathbb{N} \cup \{0\}$ and $x \in I_c$ we define the basis functions by the formulae

$$\begin{aligned} p_{n,k}^{[c]}(x) &:= (-1)^k \binom{-\frac{n}{c}}{k} (cx)^k \, (1 + cx)^{-\frac{n}{c}-k}, \quad && c \neq 0, \\ p_{n,k}^{[0]}(x) &:= \lim_{c \to 0} p_{n,k}^{[c]}(x) = \frac{(nx)^k}{k!} e^{-nx}, && c = 0. \end{aligned} \tag{1}$$

E.E. Berdysheva (✉)
Department of Mathematics and Science, German University of Technology in Oman, PC 130 Athaibah, P.O. Box 1816 Muscat, Sultanate of Oman
e-mail: elena.berdysheva@gutech.edu.om

P.N. Agrawal et al. (eds.), *Mathematical Analysis and its Applications*,
Springer Proceedings in Mathematics & Statistics 143,
DOI 10.1007/978-81-322-2485-3_20

The functions $p_{n,k}^{[c]}$ satisfy the property

$$\sum_{k=0}^{\infty} p_{n,k}^{[c]}(x) = 1.$$

The basis functions are nonnegative on I_c, i.e.,

$$p_{n,k}^{[c]}(x) \geq 0, \quad x \in I_c, \quad \text{for all } k \in \mathbb{N}_0,$$

if the following conditions are fulfilled: $n > 0$ if $c \geq 0$, and $n = -c\ell$, $\ell \in \mathbb{N}$, if $c < 0$. In the latter case $p_{n,k}^{[c]} \equiv 0$ for $k > -\frac{n}{c}$, $k \in \mathbb{N}_0$. It is not difficult to see that

$$p_{n,k}^{[c]}(x) = p_{\frac{n}{c},k}^{[1]}(cx), \quad c > 0,$$
$$p_{n,k}^{[c]}(x) = p_{-\frac{n}{c},k}^{[-1]}(-cx), \quad c < 0.$$

Thus, there are only three significantly different cases, namely, $c = -1$, $c = 0$, and $c = 1$. We will restrict our consideration to these three cases.

The basis functions (1) are traditionally used to define positive linear operators for functions on I_c. The first and the most well known of these operators is the operator of the form

$$\mathbf{B}_n^{[c]} f := \sum_{k=0}^{\infty} f\left(\frac{k}{n}\right) p_{n,k}^{[c]}, \tag{2}$$

where f is a continuous function on I_c. This is a positive linear operator that reproduces linear functions. In the case $c = -1$, the operator is the famous Bernstein operator. It was introduced by Bernstein [8] to give a constructive proof of the Weierstrass Approximation Theorem. The work of Bernstein initiated a study of numerous modifications and generalizations of this operator by many authors. We will call operator (2) when $c = 0$ the Szász–Mirakjan operator. It was considered independently by several authors, including Mirakjan [18], Favard [12], and Szász [22]. In the case $c = 1$, operator (2) was first defined by Baskakov [1], who also introduced a general frame which includes the operators $\mathbf{B}_n^{[c]}$ with $c \in \mathbb{R}$.

Aiming to have a similar construction for integrable functions, the so-called Durrmeyer variant of operator (2) was introduced as

$$\mathbf{M}_n^{[c]} f := \sum_{k=0}^{\infty} \frac{\int_{I_c} f(t)\, p_{n,k}^{[c]}(t)\, dt}{\int_{I_c} p_{n,k}^{[c]}(t)\, dt}\, p_{n,k}^{[c]}, \tag{3}$$

where $f \in L_p(I_c)$, $1 \leq p < \infty$, or $f \in C(I_c)$. This is a positive linear operator that reproduces constant functions. The Bernstein–Durrmeyer operator (that corresponds to the case $c = -1$) was defined independently by Durrmeyer [11] and Lupaş [16] and became known due to Derriennic (e.g., [9]). It can be naturally generalized for functions of several variables defined on the d-dimensional simplex. The Szász–Mirakjan–Durrmeyer operator (case $c = 0$) is due to Mazhar and Totik [17]. The Baskakov–Durrmeyer operator (case $c = 1$) was defined by Sahai and Prasad [21] and, independently, by Heilmann [13].

In what follows, we will use the following function spaces and norms. For a compact set A, we denote by $C(A)$ the space of continuous functions on A with the norm

$$\|f\|_{C(A)} := \max_{x\in A} |f(x)|.$$

The space $L^\infty(I_c, \rho)$ is the space of essentially bounded functions on I_c with respect to a measure ρ with the norm

$$\|f\|_{L^\infty(I_c,\rho)} := \text{ess sup}_{x\in I_c} |f(x)|.$$

The spaces $L^p(I_c, \rho)$, $1 \le p < \infty$, are the spaces of functions f for which $|f|^p$ is integrable on I_c with the norm

$$\|f\|_{L^p(I_c,\rho)} := \left(\int_{I_c} |f(x)|^p \, d\rho(x)\right)^{\frac{1}{p}}.$$

2 Durrmeyer-Type Operators with Respect to Arbitrary Measure

We generalize the operators of Durrmeyer-type (3) in the following way: the new operator has the same form, but the integration is taken not with respect to the Lebesgue measure dx but with respect to some measure $d\rho(x)$. The exact definition of the operator is as follows. Let ρ be a nonnegative locally bounded Borel measure on I_c. Then, in particular, ρ is regular (being a nonnegative bounded Borel measure on a metric space), and thus polynomials are dense in the spaces $L^p(I_c, \rho)$, $1 \le p < \infty$, and in $C(I_c)$. Furthermore, we suppose that

$$\operatorname{supp} \rho \ne \partial I_c \tag{4}$$

(where ∂I_c denotes the boundary of I_c). This condition guarantees that $\int_{I_c} p_{n,k}^{[c]}(t) \, d\rho(t) \ne 0$ for all n and k. In the cases $c = 0$ and $c = -1$, we need to additionally take care about the convergence. This requirement implies the following further conditions on the measure ρ: let

$$\int_0^\infty e^{-\gamma t} \, d\rho(t) < \infty \quad \text{for some} \quad \gamma > 0 \quad \text{if} \quad c = 0 \tag{5}$$

and

$$\int_0^\infty (1+t)^{-\gamma}\, d\rho(t) < \infty \quad \text{for some} \quad \gamma > 0 \quad \text{if} \quad c = 1. \tag{6}$$

Definition 1 Let ρ be a nonnegative locally bounded Borel measure on I_c that satisfies (4), (5), and (6). Let $f \in L^p(I_c, \rho)$, $1 \le p \le \infty$. The Durrmeyer-type operator with respect to the measure ρ is defined by the formula

$$\mathbf{M}_{n,\rho}^{[c]} f := \sum_{k=0}^{\infty} \frac{\int_{I_c} f(t)\, p_{n,k}^{[c]}(t)\, d\rho(t)}{\int_{I_c} p_{n,k}^{[c]}(t)\, d\rho(t)}\, p_{n,k}^{[c]} \tag{7}$$

for $n \in \mathbb{N}$ if $c = -1$, $n > \gamma \frac{p-1}{p}$ if $c = 0$, and $n \ge \gamma \frac{p-1}{p}$ if $c = 1$.

Please note that the Bernstein–Durremer operator ($c = -1$) can also be similarly designed for functions of several variables defined on the d-dimensional simplex. All results presented in this paper remain valid also in this case. However, we restrict our presentation to the one-dimensional case, for the sake of simplicity.

The Bernstein–Durrmeyer operator ($c = -1$) with respect to an arbitrary measure was for the first time studied in full generality in [5], to our knowledge. However, the Bernstein–Durrmeyer operator in a special case of a measure ρ different from the Lebesgue measure, namely, for the Jacobi measure $d\rho(x) = x^\alpha (1-x)^\beta dx$, $\alpha, \beta > -1$, is well known and very well studied. It was introduced by Păltănea [19], see also paper [7] by Berens and Xu. The multidimensional case was considered, for example, in [10]. A more general operator than the Bernstein–Durrmeyer operator with respect to Jacobi measure was considered by Păltănea in [20, Section 5.2]: he studied Bernstein–Durrmeyer operators of the form (7) with $d\rho(x) = x^\alpha (1-x)^\beta h(x)\, dx$, where $h \in C[0, 1]$, $h(t) > 0$ for all $t \in [0, 1]$, $\alpha, \beta > -1$. The Szász–Mirakjan–Durrmeyer operators (case $c = 0$) and the Baskakov–Durrmeyer operators (case $c = 1$) with respect to arbitrary measure were introduced and studied in [4].

We introduced an arbitrary measure in the construction having in mind applications in learning theory. Indeed, Jetter and Zhou [14] applied the Bernstein–Durrmeyer operator with respect to arbitrary measure, in the one-dimensional case, to bias-variance estimates for vector support machine classifiers. Li [15] used the Bernstein–Durrmeyer operators with respect to arbitrary measure in the multidimensional case in study of learning rates of least-squares regularized linear regression with polynomial kernels.

Just to give the reader a feeling how the Bernstein–Durrmeyer operators can be applied in the frames of learning theory, we give a short description of the problem considered by Li in [15]. Let X be a compact set in $\mathbb{R}^d$, $Y = \mathbb{R}$, and σ be a Borel probability measure on $Z := X \times Y$. We denote by σ_X the marginal distribution of σ on X. For a function $f : X \to Y$, the least-squares error is

$$\mathcal{E}(f) := \int_Z (f(x) - y)^2\, d\sigma.$$

The function

$$f_\sigma(x) := \int_Y y\, d\sigma(y|x),$$

where $\sigma(y|x)$ is the conditional probability induced by σ, minimizes the least-squares error. It is called the regression function. The aim is to find a good approximation of the regression function based on a random sample $z := \{(x_i, y_i)\}_{i=1}^m \subset Z$ of size m.

For $x, t \in X$, put $K_n(x,t) := (1 + x \cdot t)^n$, where $x \cdot t = x_1 t_1 + \cdots + x_d t_d$. This function is a Mercer kernel, and we denote by $\mathcal{H}_{K_n}$ the corresponding reproducing kernel Hilbert space, which is in this case the space of algebraic polynomials of degree at most n endowed with the corresponding norm. The least-squares regularized regression algorithm with the polynomial kernel K_n is the minimization problem

$$f_{z,n,\lambda} := \arg \min_{f \in \mathcal{H}_{K_n}} \left\{ \frac{1}{m} \sum_{i=1}^{m} (f(x_i) - y_i)^2 + \lambda \, \|f\|_{\mathcal{H}_{K_n}}^2 \right\},$$

where $\lambda > 0$ is the regularization parameter. One usually takes $\lambda = \lambda(m)$ with $\lim_{m\to\infty} \lambda(m) = 0$. We expect that $f_{z,n,\lambda}$ gives a good approximation of the regression function f_σ.

Li considered the case when X is the d-dimensional simplex, and, in particular, gave an estimate for the rate of convergence $\|f_{z,n,\lambda} - f_\sigma\|_{L^2(X,\sigma_X)}$ of type $O(m^{-\beta})$ with some $\beta > 0$. One of the important steps in deriving this estimate is approximating of f_σ by a function from $\mathcal{H}_{K_n}$. This approximation is realized by $\mathbf{M}_{n,\rho}^{[-1]} f_\sigma$ (more exactly, by its multivariate version) with $\rho = \sigma_X$. The method employs the estimates for the rate of convergence of the operator $\mathbf{M}_{n,\rho}^{[-1]}$ as given in Theorem 6 below.

Following this description of an application of operators (7), we return to discussing their properties. Obviously, operators (7) are linear positive operators that reproduce constant functions. Our first result is about boundedness of the operators.

Theorem 1 *Let $1 \le p \le \infty$. Let ρ and n be as in Definition 1. Then the operator*

$$\mathbf{M}_{n,\rho}^{[c]} : L^p(I_c, \rho) \to L^p(I_c, \rho)$$

is well-defined. Moreover,

$$\|\mathbf{M}_{n,\rho}^{[c]}\|_{L^p(I_c,\rho)\to L^p(I_c,\rho)} = 1.$$

In what follows, we concentrate on the question of convergence of operators (7). For other properties of the operators under consideration see, e.g., [5].

3 Uniform and Pointwise Convergence

Uniform and pointwise convergence of the Bernstein–Durrmeyer operators ($c = -1$) with respect to arbitrary measure, in the multivariate case, was studied in [2, 3]. The methods of these papers were adopted to the case of an unbounded interval in [4], where corresponding results for the Szász–Mirakjan–Durrmeyer operators ($c = 0$) and the Baskakov–Durrmeyer operators ($c = 1$) with respect to arbitrary measure were obtained. Below we summarize the results on the uniform and the pointwise convergence (in the one-dimensional case).

Recall that a measure ρ on I_c is called strictly positive if $\rho(A \cap I_c) > 0$ for every open set $A \subset \mathbb{R}$ such that $A \cap I_c \neq \emptyset$. This is equivalent to the fact that supp $(\rho) = I_c$.

Our first statement gives necessary and sufficient conditions for the uniform convergence of the Bernstein–Durrmeyer operator with respect to a measure ρ on the compact set $I_{-1} = [0, 1]$ for every function $f \in C([0, 1])$.

Theorem 2 *Let $c = -1$, and ρ and n be as in Definition 1. Then*

$$\lim_{n\to\infty} \|f - \mathbf{M}_{n,\rho}^{[-1]} f\|_{C([0,1])} = 0 \quad \textit{for every} \quad f \in C([0, 1])$$

if and only if ρ is strictly positive on $[0, 1]$.

A statement about pointwise convergence can be proved in all three cases $c = -1$, $c = 0$, or $c = 1$.

Theorem 3 *Let $c = -1$, $c = 0$, or $c = 1$, and ρ and n be as in Definition 1 with $p = \infty$. Let $x \in$ supp ρ. Let $f \in L^\infty(I_c, \rho)$ and continuous at x. Then*

$$\lim_{n\to\infty} \left| f(x) - \left(\mathbf{M}_{n,\rho}^{[c]} f\right)(x) \right| = 0.$$

Moreover, the convergence is uniform on every compact subset of the interior of supp ρ. For a set B, we denote by B° the interior of B. We understand the interior relatively to the set I_c, i.e., the boundary points of I_c may belong to it.

Theorem 4 *Let $c = -1$, $c = 0$, or $c = 1$, and ρ and n be as in Definition 1 with $p = \infty$. Let A be a compact set, $A \subset (\mathrm{supp}\,\rho)^\circ$. Let $f \in L^\infty(I_c, \rho)$ and continuous on A. Then*

$$\lim_{n\to\infty} \|f - \mathbf{M}_{n,\rho}^{[c]} f\|_{C(A)} = 0.$$

A very interesting open question is to obtain estimates for rates of uniform or pointwise convergence of operators of Durrmeyer type with respect to arbitrary measure.

4 Convergence in the Weighted L^p-spaces

The first result on the convergence of operators (7) in the spaces $L^p(I_c, \rho)$, $1 \le p < \infty$, was obtained for the Bernstein–Durrmeyer operator ($c = -1$) with respect to arbitrary measure, in the multidimensional case, by Li [15]. She also obtained estimates for the rate of convergence in terms of a K-functional that are good in the cases when $p = 1$ or $p = 2$. Using the same idea, the estimates for other values of p ($1 \le p < \infty$) were improved in [6].

The methods of Li's paper [15] were transferred to the cases of the Szász–Mirakjan–Durrmeyer operators ($c = 0$) and the Baskakov–Durrmeyer operators ($c = 1$) in [4]. Also estimates for the rate of convergence were proved. However, results in [4] were obtained under a very restrictive assumption that the measure ρ on $[0, \infty)$ has finite moments up to a certain order, and, in particular, is bounded on $[0, \infty)$ (see Theorem 6 below). Note that this condition is not satisfied in the classical case of the Lebesgue measure.

Here we present a new result. Namely, we prove that the Szász–Mirakjan–Durrmeyer operators ($c = 0$) and the Baskakov–Durrmeyer operators ($c = 1$) with respect to arbitrary measure converge in $L^p(I_c, \rho)$, $1 \le p < \infty$, for every $f \in L^p(I_c, \rho)$, $1 \le p < \infty$, without additional assumptions on the measure ρ. Our method is a further development of the method of Li from [15]. This development allows to overcome difficulties arising when we work on an infinite interval. The result formulated below includes Li's result [15] for $c = -1$ and is new in the cases $c = 0$ and $c = 1$.

Theorem 5 *Let $1 \le p < \infty$. Let $c = -1$, $c = 0$, or $c = 1$, and ρ and n be as in Definition 1. Then*

$$\lim_{n\to\infty} \|f - \mathbf{M}_{n,\rho}^{[c]} f\|_{L^p(I_c,\rho)} = 0$$

for each $f \in L^p(I_c, \rho)$.

We finish the paper by giving a statement about rates of convergence. This result was proved in [6] for $c = -1$ and in [4] for $c = 0$ and $c = 1$. Denote by $C^1(I_c)$ the set of continuously differentiable functions g on I_c such that $\|g'\|_{C(I_c)} := \sup_{x\in I_c} |g'(x)| < \infty$. Consider the K-functional

$$\mathcal{K}(f, t)_{p,\rho} := \inf_{g\in C^1(I_c)} \left\{\|f - g\|_{L^p(I_c,\rho)} + t\, \|g'\|_{C(I_c)}\right\}, \quad 1 \le p < \infty.$$

Theorem 6 *Let $1 \le p < \infty$. Let $c = -1$, $c = 0$, or $c = 1$, and ρ and n be as in Definition 1. Assume, in addition, that*

$$\int_{I_c} x^s \, d\rho(x) < \infty \qquad \textit{for some even } s. \tag{8}$$

Let $1 \le p \le s$, and $f \in L^p(I_c, \rho)$. Then there is a constant $C_{p,\rho}$ that depends only on p and the measure ρ such that

$$\|f - \mathbf{M}_{n,\rho}^{[c]} f\|_{L^p(I_c,\rho)} \leq 2\mathcal{K}\left(f, \frac{C_{p,\rho}}{\sqrt{n}}\right)_{p,\rho}.$$

Note that condition (8) is automatically satisfied for all s if $c = -1$ since in this case $I_{-1} = [0, 1]$ is a bounded interval. In the cases $c = 0$ and $c = -1$, when $I_c = [0, \infty)$, this condition is quite restrictive. Condition (8) guarantees, in particular, that the set of continuously differentiable functions with finite support is dense in the spaces $L^p(I_c, \rho)$, $1 \leq p < \infty$. Thus, $\mathcal{K}(f, t)_{p,\rho} \to 0$ as $t \to 0$ for all $f \in L^p(I_c, \rho)$, $1 \leq p < \infty$.

An open question is to obtain estimates for the rates of convergence of the Szász–Mirakjan–Durrmeyer operators ($c = 0$) and the Baskakov–Durrmeyer operators ($c = 1$) with respect to arbitrary measure in $L^p(I_c, \rho)$, $1 \leq p < \infty$, in the general case.

References

1. Baskakov, V.A.: An instance of a sequence of linear positive operators in the space of continuous functions. Dokl. Akad. Nauk SSSR **113**, 249–251 (1957) [in Russian]
2. Berdysheva, E.E.: Uniform convergence of Bernstein-Durrmeyer operators with respect to arbitrary measure. J. Math. Anal. Appl. **394**, 324–336 (2012)
3. Berdysheva, E.E.: Bernstein-Durrmeyer operators with respect to arbitrary measure, II: pointwise convergence. J. Math. Anal. Appl. **418**, 734–752 (2014)
4. Berdysheva, E.E., Al-Aidarous, E.: Szász-Mirakjan-Durrmeyer and Baskakov-Durrmeyer operators with respect to arbitrary measure (submitted)
5. Berdysheva, E.E., Jetter, K.: Multivariate Bernstein-Durrmeyer operators with arbitrary weight functions. J. Approx. Theory **162**, 576–598 (2010)
6. Berdysheva, E.E., Li, B.-Z.: On L^p-convergence of Bernstein-Durrmeyer operators with respect to arbitrary measure. Publ. Inst. Math. Nouv. Sér. **96**(110), 23–29 (2014)
7. Berens, H., Xu, Y.: On Bernstein-Durrmeyer polynomials with Jacobi-weights. In: Chui, C.K. (ed.) Approximation Theory and Functional Analysis, pp. 25–46. Academic Press, Boston (1991)
8. Bernstein, S.: Démonstration du théorème de Weierstrass, fondée sur le calcul des probabilités. Commun. Soc. Math. Kharkow **13**(2), 1–2 (1912–13)
9. Derriennic, M.-M.: Sur l'approximation de fonctions intégrables sur [0, 1] par des polynômes de Bernstein modifiés. J. Approx. Theory **31**, 325–343 (1981)
10. Ditzian, Z.: Multidimensional Jacobi-type Bernstein-Durrmeyer operators. Acta Sci. Math. (Szeged) **60**, 225–243 (1995)
11. Durrmeyer, J.-L.: Une formule d'inversion de la transformée de Laplace: Applications à la théorie des moments, Thèse de 3e cycle, Faculté des Sciences de l'Université de Paris (1967)
12. Favard, J.: Sur le multiplicateurs d'interpolation. J. Math. Pures Appl. IX **23**, 219–247 (1944)
13. Heilmann, M.: Direct and converse results for operators of Baskakov-Durrmeyer type. Approx. Theory Appl. **5**(1), 105–127 (1989)
14. Jetter, K., Zhou, D.-X.: Approximation with polynomial kernels and SVM classifiers. Adv. Comput. Math. **25**, 323–344 (2006)
15. Li, B.-Z.: Approximation by multivariate Bernstein-Durrmeyer operators and learning rates of least-square regularized regression with multivariate polynomial kernels. J. Approx. Theory **173**, 33–55 (2013)
16. Lupaş, A.: Die Folge der Betaoperatoren, Dissertation, Universität Stuttgart (1972)

17. Mazhar, S.M., Totik, V.: Approximation by modified Szász operators. Acta Sci. Math. **49**, 257–269 (1985)
18. Mirakjan, G.M.: Approximation des fonctions continues au moyen de polynômes de la forme $e^{-nx} \sum_{k=0}^{m_n} C_{k,n}x^k$, C. R. (Dokl.) Acad. Sc. URSS **31**, 201–205 (1941)
19. Păltănea, R.: Sur un opérateur polynomial defini sur l'ensemble des fonctions intégrables, "Babeş-Bolyai" Univ. Fac. Math. Res. Semin. **2**, 101–106 (1983)
20. Păltănea, R.: Approximation Theory Using Positive Linear Operators. Birkhäuser-Verlag, Boston (2004)
21. Sahai, A., Prasad, G.: On simultaneous approximation by modified Lupas operators. J. Approx. Theory **45**, 122–128 (1985)
22. Szász, O.: Generalization of S. Bernstein's polynomials to the infinite interval. J. Res. Nat. Bur. Stand. **45**, 239–245 (1950)

Part II
Applications

Construction of Sparse Binary Sensing Matrices Using Set Systems

R. Ramu Naidu

Abstract Recent developments at the intersection of algebra and optimization theory—by the name of compressed sensing (CS)—aim at providing linear systems with sparse descriptions. The deterministic construction of the sensing matrices is now an active directions in CS. The sparse sensing matrix contributes to fast processing with low computational complexity. The present work attempts to relate the notion of set systems to CS. In particular, we show that the set system theory may be adopted to designing a binary CS matrix of high sparsity from the existing binary CS matrices.

Keywords Compressed sensing · Restricted isometry property · Deterministic construction · Set systems

1 Introduction

In recent years, sparse representations have become a powerful tool for efficiently processing data in nontraditional ways. Compressed sensing (CS) is an emerging area potential for sparsity-based representations. Since the problem of sparse recovery through l_0 norm minimization is generally NP-hard, Donoho et al. [1], Candes [2] and Cohen et al. [3] have made several pioneering contributions and have reposed the problem as an l_1-minimization problem. It is known that restricted isometry property (RIP) is a sufficient condition to ensure the equivalence between l_0 and l_1 norm problems. As verifying RIP is computationally hard, there is much interest in construction of RIP matrices.

Of late, the deterministic construction of binary CS matrices has attracted significant attention. Devore [4], Li et al. [5], Amini et al. [6], Indyk [7] have constructed deterministic binary sensing matrices using ideas from algebra, graph theory, and

R.R. Naidu (✉)
Department of Mathematics, Indian Institute of Technology, Medak,
Hyderabad 502205, A.P., India
e-mail: ma11p003@iith.ac.in

P.N. Agrawal et al. (eds.), *Mathematical Analysis and its Applications*,
Springer Proceedings in Mathematics & Statistics 143,
DOI 10.1007/978-81-322-2485-3_21

coding theory. Devore [4] has been first to construct deterministic binary sensing matrix of size $p^2 \times p^{r+1}$ with p number of ones in each column and coherence being at most $\frac{r}{p}$, for every fixed r and prime power p such that $r < p$. In the present work, using the results from set systems we construct a binary sensing matrix from a given binary sensing matrix in such a way that the resulting matrix is more sparser than the input matrix. Consequently, the new matrix has potential for resulting in fast algorithms.

The paper is organized into several sections. In Sect. 2, we present basic CS theory and the conditions that ensure the equivalence between l_0-norm problem and l_1-norm problem. In Sect. 3, we use the ideas from the set system theory and construct binary sensing matrices of higher sparsity from the existing ones. We present our concluding remarks in the last section.

2 Sparse Recovery from Linear Measurements

CS refers to the problem of reconstruction of an unknown vector $u \in \mathscr{R}^M$ from the linear measurements $y = (\langle u, \phi_1\rangle, \ldots, \langle u, \phi_M\rangle) \in \mathscr{R}^m$ with $\langle u, \phi_j\rangle$ being the innerproduct between u and ϕ_j. The basic objective in CS is to design a recovery procedure based on the sparsity assumption on u when the number of measurements m is much small compared to M. Sparse representations seem to have merit for various applications in areas such as image/signal processing and numerical computation.

A vector $u \in \mathscr{R}^M$ is said to be k-sparse, if it has at most k nonzero coordinates. One can find the sparse vector from its linear measurements by solving the following l_0-norm optimization problem:

$$\min_v \|v\|_0 \text{ subject to } \phi v = y. \quad (1)$$

Here, $\|v\|_0 = |\{i \mid v_i \neq 0\}|$. The l_0-norm problem (1) is an NP-hard problem [2]. Candes et al. [2] have proposed the following l_1-norm minimization problem instead of l_0-norm problem, making it computationally tractable LPP problem:

$$\min_v \|v\|_1 \text{ subject to } \phi v = y. \quad (2)$$

Here, $\|v\|_1$ denotes the l_1-norm of the vector $v \in \mathscr{R}^M$.

Donoho et al. [1] and Kashin et al. [8], have provided the conditions under which the solution to l_0-norm problem (1) is the same as that of l_1-norm problem (2). For later use, we denote the solution to l_1-norm problem by $f_\phi(y)$ and solution to l_0-norm problem by $u^0_\phi(y) \in \mathscr{R}^M$.

2.1 Equivalence Between l_0-Norm and l_1-Norm Problems

Definition 1 The mutual coherence $\mu(\phi)$ of a given matrix ϕ is the largest absolute normalized innerproduct between the pairs of columns of ϕ, that is,

$$\mu(\phi) = \max_{1 \le i, j \le M,\, i \ne j} \frac{|\phi_i^T \phi_j|}{\|\phi_i\|_2 \|\phi_j\|_2}, \tag{3}$$

where ϕ_i is the ith column of ϕ. It is known [1] that for μ-coherent matrices ϕ, one has

$$u_\phi^0(y) = f_\phi(y) = u, \tag{4}$$

provided u is k-sparse with $k < \frac{1}{2}\,(1 + \frac{1}{\mu})$.

Candes et al. ([2] and the references therein) have introduced the following isometry condition on matrices ϕ and have established its important role in CS. An $m \times M$ matrix ϕ is said to satisfy the restricted isometry property (RIP) of order k with constant δ_k if for all k-sparse vectors $x \in \mathscr{R}^M$, we have

$$(1 - \delta_k)\, \|x\|_{l_2}^2 \le \|\phi x\|_{l_2}^2 \le (1 + \delta_k)\, \|x\|_{l_2}^2 \,. \tag{5}$$

The following proposition [9] relates the RIP constant δ_k and μ.

Proposition 1 *Suppose that $\phi_1, \ldots, \phi_M$ are the unit-normed columns of the matrix ϕ with coherence μ. Then ϕ satisfies the RIP of order k with constant $\delta_k = (k-1)\mu$.*

Candes [2] has shown that whenever ϕ satisfies RIP of order $3k$ with $\delta_{3k} < 1$, the CS reconstruction error satisfies the following estimate

$$\left\|u - f_\phi(\phi u)\right\|_{l_2^M} \le C k^{\frac{-1}{2}} \sigma_k(u)_{l_1^M}, \tag{6}$$

where $\sigma_k(u)_{l_1^M}$ denotes the l_1 error of the best k—term approximation, and the constant C depends only on δ_{3k}. This implies that the bigger the value of k for which we can verify the RIP then better the guarantee we have on the performance of ϕ.

One of the important problems in CS theory deals with constructing CS matrices that satisfy the RIP for the largest possible range of k. It is known that the widest range possible is $k \le C \frac{m}{\log(\frac{M}{m})}$ [4, 10–12]. However, the only known matrices that satisfy the RIP for this range are based on random constructions [10]. To the best of our knowledge, designing the good deterministic constructions of RIP matrices is still an open problem.

Since the sparsity of the sensing matrix is key to minimizing the computational complexity associated with the matrix vector multiplication, it is desirable that the CS matrix has smaller density. The sparse sensing matrix may contribute to fast processing with low computational complexity in compressed sensing [13].

Definition 2 [14] The density of a matrix is the ratio of the number of its nonzero entries to the total number of its entries.

It may be noted that the density of the sensing matrix constructed by Devore [4] is $\frac{1}{p}$. The sensing matrix constructed by Li et al. [5] have $\frac{1}{q}$ as density. This matrix, a generalization of [4], is of size $|\mathscr{P}|q \times q^{\mathscr{L}(G)}$, where q is any prime power and $\mathscr{P}$ is the set of all rational points on algebraic curve $\mathscr{X}$ over finite field $\mathscr{F}_q$. Amini et al. [6] have constructed binary sensing matrices using OOC codes. The density of this matrix is $\frac{\lambda}{m}$, where m is row size and λ is the number of ones in each column.

Many data mining tasks can be concerned with identifying a small number of interesting items from a tremendously large group without exceeding certain resource constraints. Specific examples [15] include the sketching and monitoring of heavy hitters in high-volume data streams, source localization in sensor networks, multiplier-less data compression and tomography. Note that, all these applications naturally correspond to binary matrices. Furthermore, binary matrices with small density are generally better. Thus, we focus on designing sparse binary matrices herein.

The present work attempts to address the deterministic construction of new binary sensing matrix of smaller density from a given binary sensing matrix. Suppose ϕ is a binary CS matrix of size $m \times M$ with $\frac{m(m+1)}{2} < M$. In the next section, using the results from set systems, we construct a binary sensing matrix ψ from ϕ in such a way that the resulting matrix ψ is more sparse compared to the given matrix ϕ.

3 Construction of Binary CS Matrix of Smaller Density from Existing Binary Matrix

Before presenting the main result, we discuss the definitions and results [16] relevant to our construction methodology. Let $V = \{v_1, v_2, \ldots, v_m\}$ be a set of m elements (treated as "universe"). A set system $\mathscr{S}$ on V is simply some subset chosen from all of the subsets of V, that is, $\mathscr{S} \subset 2^V$, the power set. A hypergraph is a collection of several subsets of V, where some subsets may be present with a multiplicity greater than 1. A set system may, however, contain each subset of V at most once.

Definition 3 Let $\mathscr{H} = \{H_1, H_2, \ldots, H_M\}$ be a hypergraph of M sets over the universe V, and let $\phi = \{\phi_{ij}\}$ be the $m \times M$ binary sensing incidence matrix of hypergraph $\mathscr{H}$, that is, the columns of ϕ correspond to the sets of H. The characteristic vector on each H_j gives the jth column in ϕ, that is, $\phi_{ij} = 1$ if $x_i \in H_j$ otherwise $\phi_{ij} = 0$.

Definition 4 Let $A = \{a_{ij}\}$ and $B = \{b_{ij}\}$ be the two $m \times M$ matrices over a ring R. Their dream product is an $m \times M$ matrix $C = \{c_{ij}\}$, denoted by $A \odot B$, and is defined as $c_{ij} = a_{ij}b_{ij}$ for $1 \leq i \leq m$, $1 \leq j \leq M$.

Definition 5 Let $f(x_1, x_2, \ldots, x_m) = \sum_{I \subseteq \{1,2,\ldots,m\}} a_I x_I$ be a multilinear polynomial, where $x_I = \prod_{i \in I} x_i$. Let $w(f) = |\{a_I : a_I \neq 0\}|$ and let $L_1(f) = \sum_{I \subseteq \{1,2,\ldots,m\}} |a_I|$.

Definition 6 Let $\mathcal{H}$ be a set system on the universe X with $m \times M$ incidence matrix ϕ. Let $f(x_1, x_2, \ldots, x_m) = \sum_{I \subseteq \{1,2,\ldots,m\}} a_I x_I$ be a multilinear polynomial with nonnegative integer coefficients or coefficients from $\mathcal{Z}_r$. Then $f(\mathcal{H}_\phi)$ is a hypergraph on the $L_1(f)$-element vertex set, and its incidence matrix is the $L_1(f) \times M$ matrix ψ. The rows of ψ correspond to x_I's of f; there are a_I identical rows of ψ corresponding to the same x_I. The row, corresponding to x_I is defined as the dream product of those rows of ϕ that correspond to $v_i, i \in I$.

Lemma 1 [16] *Suppose in the Definition 6, the coefficients of $x_1, x_2, \ldots, x_m$ are nonzeros in f. Then the resulting Hypergraph $f(\mathcal{H}_\phi)$ is a set system [16].*

The most remarkable property of $f(\mathcal{H}_\phi)$ is given by the following theorem:

Theorem 1 [16] *Let $\mathcal{H} = \{H_1, H_2, \ldots, H_M\}$ be a set system and ϕ its $m \times M$ incidence matrix. Let f be a multilinear polynomial with nonnegative integer coefficients or coefficients from $\mathcal{Z}_r$. Let $f(\mathcal{H}_\phi) = \{\widehat{H}_1, \widehat{H}_2, \ldots, \widehat{H}_M\}$. Then for any $1 \le k \le M$ and for any $1 \le i_1 < i_2 < \cdots < i_k \le M$: $f(H_{i_1} \cap H_{i_2} \cap \ldots \cap H_{i_k}) = |\widehat{H}_{i_1} \cap \widehat{H}_{i_2} \cap \ldots \widehat{H}_{i_k}|$.*

Following theorem discusses the construction of a new set system from a given set system using the Definition 6, Lemma 1 and Theorem 1.

Theorem 2 [16] *Let f be an m—variable symmetric polynomial with nonnegative integer coefficients, and $\mathcal{H}$ a set system of size M on the m element universe with $m \times M$ incidence matrix ϕ. Suppose that*

$$L(\mathcal{H}) = \{|H_i \cap H_j|, H_i \neq H_j, H_i, H_j \in \mathcal{H}\} = \{l_1, l_2, \ldots, l_s\}.$$

Then one may construct in $O(L_1(f)mM)$ time a hypergraph $f(\mathcal{H}_\phi)$ of size M on the $L_1(f)$—vertex universe such that the sizes of the pairwise intersections of the sets of $f(\mathcal{H}_\phi)$ are

$$f(l_1), f(l_2), \ldots, f(l_s).$$

3.1 Set Systems for Designing CS Matrices

Using the afore-stated results [16] from set system theory, we construct a new binary sensing matrix from a given binary sensing matrix. The new matrix has small density as compared to the given one.

Theorem 3 *Suppose $f(x_1, x_2, \ldots, x_m) = x_1 + x_2 + \cdots + x_m + \sum_{i<j} x_i x_j$ is a symmetric polynomial. Let ϕ be a binary sensing matrix of size $m \times M$ such that $\frac{m(m+1)}{2} < M$ with the coherence being at most $\frac{r}{k}$. Here k represents the number of nonzero elements that each column of ϕ has. Then there exists a binary sensing matrix ψ of size $\frac{m(m+1)}{2} \times M$ whose coherence is at most $\frac{r+\binom{r}{2}}{k+\binom{k}{2}}$.*

Proof Define $\mathscr{H} = \{H_i : 1 \leq i \leq M, H_i = \text{supp}(\phi_i)$, where ϕ_i is ith column of $\phi\}$. Since all columns of ϕ are distinct, $\mathscr{H}$ is a set system. Let $L(\mathscr{H}) = \{|H_i \cap H_j|, H_i \neq H_j, H_i, H_j \in \mathscr{H}\} = \{l_1, l_2, \ldots, l_s\}$. Since the coherence of the matrix ϕ is at most $\frac{r}{k}$, the cardinality of overlap between the supports of any two columns is at most r. Consequently, $l_i \leq r$ for all i. Let $X = \{1, 2, \ldots, m\}$.

Since $f(x_1, x_2, \ldots, x_m) = x_1 + x_2 + \cdots + x_m + \sum_{i<j} x_i x_j$ is a symmetric polynomial and $\mathscr{H}$ is a set system, we have $L_1(f) = m + \binom{m}{2}$ and $f(\mathscr{H}_\phi)$ is a set system of size M on $L_1(f)$-element universe, from Theorem 2 and Lemma 1.

Define $(v_{ij})_{m\times 1}$ to be the characteristic vector on $H_i \cap H_j$ in the universe X. Since each $l_s = |H_i \cap H_j|$ for some $i \neq j$, $f(l_s) = f((v_{ij})) \leq f(r) = r + \binom{r}{2}$. It follows that $f(l_i) \leq r + \binom{r}{2}$ for all i. Therefore, the sizes of the pairwise intersections of the sets of $f(\mathscr{H}_\phi)$ is at most $r + \binom{r}{2}$. The incidence matrix ψ of the set system $f(\mathscr{H}_\phi)$ is of size $L_1(f) \times M$, that is, $(m + \binom{m}{2}) \times M$. From the hypothesis of the theorem $(m + \binom{m}{2}) < M$, so it is an underdetermined system and its first m rows are same as ϕ and remaining $\binom{m}{2}$ rows are the dream products of first m rows. Each column in ψ contains $k + \binom{k}{2}$ number of ones. The cardinality of overlap between any two columns is at most $r + \binom{r}{2}$. It follows that coherence of the matrix ψ is $\mu(\psi)$, which is at most equal to $\frac{r+\binom{r}{2}}{k+\binom{k}{2}}$.

The following theorem concludes that the matrix $\psi_0 = \frac{1}{\sqrt{k+\binom{k}{2}}}\psi$ defined is RIP compliant.

Theorem 4 *The matrix* $\psi_0 = \frac{1}{\sqrt{k+\binom{k}{2}}}\psi$ *has the RIP with* $\delta = (k-1)\left(\frac{r+\binom{r}{2}}{k+\binom{k}{2}}\right)$ *whenever* $k - 1 < \frac{k+\binom{k}{2}}{r+\binom{r}{2}}$.

Proof Proof follows from the Proposition 1 and Theorem 3 □

Remark 1 The density of the new matrix ψ is $\frac{k+\binom{k}{2}}{m+\binom{m}{2}}$, which is smaller than $\frac{k}{m}$, the density of ϕ.

4 Concluding Remarks

As the sensing matrices of higher sparsity (or lower density) have potential for fast processing, the construction of such matrices is of relevance. In the present work, we have used the ideas from the set system theory and have showed that a CS matrix of higher sparsity can be generated from a given binary CS matrix.

Acknowledgments The author gratefully acknowledges the support (Ref No. 20-6/2009(i)EU-IV) that he receives from UGC, Govt of India. Author is thankful to Dr. C. S. Sastry and Dr. Phanindra Varma for fruitful discussions.

References

1. Donoho, D.L., Elad, M., Temlyakov, V.N.: Stable recovery of sparse overcomplete representations in the presence of noise. IEEE Trans. Inf. Theory **52**, 6–18 (2006)
2. Candes, E.: The restricted isometry property and its implications for compressed sensing. Comptes Rendus Mathematique **346**, 589–592 (2008)
3. Cohen, A., Dahmen, W., DeVore, R.: Compressed sensing and best k-term approximation. J. Amer. Math. Soc. **22**(1) 211–231 (2009)
4. DeVore, R.A.: Deterministic constructions of compressed sensing matrices. J. Complex. **23**, 918–925 (2007)
5. Li, S., Gao, F., Ge, G., Zhang, S.: Deterministic construction of compressed sensing matrices via algebraic curves. IEEE Trans. Inf. Theory **58**, 5035–5041 (2012)
6. Amini, A., Marvasti, F.: Deterministic construction of binary, bipolar and ternary compressed sensing matrices. IEEE Trans. Inf. Theory **57**, 2360–2370 (2011)
7. Indyk, P.: Explicit constructions for compressed sensing matrices. In: Proceedings 19th Annual ACM-SIAM Symposium Discrete Algorithms, pp. 30–33 (2008)
8. Kashin, B.S., Temlyakov, V.N.: A remark on compressed sensing. Math. Notes **82**(6), 748–755 (2007)
9. Bourgain, J., Dilworth, S., Ford, K., Konyagin, S., Kutzarova, D.: Explicit constructions of RIP matrices and related problems. Duke Math. J. **159**, 145–185 (2011)
10. Baraniuk, R., Davenpor, M., De Vore, R., Wakin, M.: A simple proof of the restricted isometry property for random matrices. Constr. Approx. **28**(3), 253–263 (2008)
11. Garnaev, A., Gluskin, E.: The widths of a Euclidean ball. Dokl. Akad. Nauk USSR **277**, 1048–1052 (1984); English transl. in Soviet Math. Dokl. **30**, 200–204 (1984)
12. Kashin, B.S.: Widths of certain finite-dimensional sets and classes of smooth functions. Izv. Akad. Nauk SSSR, Ser. Mat. **41**, 334–351 (1977). English transl. in Math. USSR IZV. **11**, 317–333 (1978)
13. Gilbert, A., Indyk, P.: Sparse recovery using sparsematrices. Proc. IEEE **98**(6), 937–947 (2010)
14. Fickus, M., Mixon, D.G., Tremain, J.C.: Steiner equiangular tight frames. Lin. Alg. Applicat. **436**(5), 1014–1027 (2012)
15. Iwen, M.: Compressed sensing with sparse binary matrices: instance optimal error guarantees in near-optimal time. J. Complex. **30**(1), 1–15 (2014)
16. Grolmusz, V.: Constructing set-systems with prescribed intersection sizes. J. Algorithms **44**, 321–337 (2001)

Topological and Nontopological 1-Soliton Solution of the Generalized KP-MEW Equation

Amiya Das and Asish Ganguly

Abstract In this paper, we obtain the topological and nontopological 1-soliton solution of the generalized Kadomtsev–Petviashvili modified equal width (KP-MEW) equation. The use of solitary wave ansatz method in context of doubly periodic Jacobi elliptic functions is done, which leads to the exact topological and nontopological soliton solutions. The Jacobi elliptic function solution degenerates into solitary wave solution in the limiting case of the modulus parameter. We derive the power law nonlinearity parameter domain for the existence of soliton solution, which is different for the topological and nontopological soliton. Also we identify the parametric restriction on the coefficients for the existence of solitary wave solutions. Finally, the remarkable features of such solitons are demonstrated in several interesting figures.

Keywords Topologican soliton · Nontopological soliton · KP-MEW equation · Jacobi elliptic functions

1 Introduction

Theory of nonlinear evolution equations (NLEE's) has a remarkable interest in the area of science and engineering, particularly in fluid dynamics, nonlinear optics, biochemistry, geophysics, etc [1–3]. The studies of these nonlinear evolution equations have attracted a great deal of attention as their closed form analytical solutions are necessary to carry out further investigation to examine the physical properties of these solutions. A bunch of powerful techniques have been introduced to carry out

A. Das (✉) · A. Ganguly
IIT Kharagpur, Kharagpur 721302, India
e-mail: amiya620@gmail.com

A. Ganguly
e-mail: aganguly@maths.iitkgp.ernet.in

P.N. Agrawal et al. (eds.), *Mathematical Analysis and its Applications*,
Springer Proceedings in Mathematics & Statistics 143,
DOI 10.1007/978-81-322-2485-3_22

the integration of various NLEE's. A few of these techniques are tanh–sech method, sine–cosine method, homogeneous balance method, Jacobi elliptic function method, F-expansion method, (G'/G)-expansion method [4–11], etc.

In 1965, Zabusky and Kruskal [12] discovered that the Korteweg de-Vries (KdV) equation has a pulse-like solitary wave solution which interacts "elastically" with another such solution. They termed such type of solution as soliton. In mathematics and physics, a soliton is a self-reinforcing solitary wave (a wave packet or pulse) that preserves its shapes while traveling at constant speed.

The delicate balance between nonlinearity and dispersion effects in the medium causes envelope soliton which are stable nonlinear wave packets that maintains their shapes during the propagation in a nonlinear dispersive medium [13]. Two different types of envelope solitons, cnoidal (nontopological) and snoidal (topological) can propagate in nonlinear dispersive media. The nontopological soliton is a pulse on a zero intensity background which has no phase change for large spatial distance, whereas the topological soliton appears as a intensity dip in an infinitely extended constant background [14].

The prototype example of NLEE in the area of theoretical physics and applied mathematics is the Korteweg de-Vries (KdV) equation

$$q_t + aqq_x + q_{xxx} = 0, \qquad (a \text{ is a constant parameter})\,. \tag{1}$$

A two-dimensional generalization of the KdV equation is Kadomtsev–Petviashvili (KP) equation

$$(q_t + aqq_x + q_{xxx})_x + q_{yy} = 0, \tag{2}$$

which depicts the evolution of quasi-one-dimensional shallow-water waves when effects of the surface tension and the viscosity are negligible [15]. The modified equal width (MEW) equation, which appears in many physical applications [16, 17] is of the form

$$q_t + a\left(q^3\right)_x - bq_{xxt} = 0\,. \tag{3}$$

In this work, our motivation is to seek the topological and nontopological soliton solutions of the gKP-MEW equation, which is a generalized form of the MEW equation in the KP sense along with the power law nonlinearity [5]

$$\left(q_t + a\left(q^n\right)_x + bq_{xxt}\right)_x + cq_{yy} = 0, \tag{4}$$

where a, b and c are real-valued constants. The first term in (4) depicts the evolution term, while the second term stands as the nonlinear term with the power law indicated by the exponent n, and the third term utters the dispersion in the x-direction. The fourth term, i.e., the y-dependence term is considered as a weak dependence on the y-coordinate. The index n indicating the power law nonlinearity is a positive real number. In [18], the single peak solitary wave solutions of the generalized KP-MEW (2, 2) equation is studied under an inhomogeneous boundary condition and different form of solutions like peakons, compactons, cuspons, loop solitons, and smooth solitons are analyzed by phase portrait analysis.

The rest of the paper is organized as follows. In Sect. 2, the nontopological soliton solution in context of the doubly periodic Jacobi elliptic functions are obtained for the gKP-MEW equation. In Sect. 3, the topological soliton solution in terms of the doubly periodic Jacobi elliptic functions is obtained. We conclude in Sect. 4.

2 Nontopological Soliton

Nontopological solitons are also known as bell-shaped solitons or cnoidal waves as there is no phase change for large spatial distance . The nontopological solitons are used to impart loads of information across trans-continental and trans-oceanic distances. In order to find the nontopological 1-soliton solution of (4), the solitary wave ansatze is assumed in the form of Jacobi elliptic functions as follows [19–23]:

$$q(x, y, t) = A\,\mathrm{cn}^p(B_1 x + B_2 y - vt) \tag{5}$$

where $\mathrm{sn}\tau \equiv \mathrm{sn}(\tau;\mathrm{k})$, $\mathrm{cn}\tau \equiv \mathrm{cn}(\tau;\mathrm{k})$, $\mathrm{dn}\tau \equiv \mathrm{dn}(\tau;\mathrm{k})$ are three Jacobian elliptic functions of real modulus $\mathrm{k}^2 (0 < \mathrm{k}^2 < 1)$ and $\mathrm{k}'^2 = 1 - \mathrm{k}^2$ is the complementary modulus. Here A indicates the soliton amplitude, with B_1 and B_2 as the inverse widths of the soliton along the x- and y- direction, respectively. Also v depicts the velocity of the soliton and the unknown exponent p will be calculated during tracking the solution of (4). To make the structure of the soliton more general, the inverse widths of the soliton in the x- and y-directions should be taken different, namely $B_1 \neq B_2$, in general. From (5), using the notation

$$\tau = B_1 x + B_2 y - vt, \tag{6}$$

it is possible to obtain q_{xt}, $(q^n)_{xx}$, q_{xxxt}, q_{yy} and substituting them in the KPMEW equation (4) yields

$$\begin{aligned}
&AB_1 v\left\{-p(p-1)\left(1-k^2\right)\mathrm{cn}^{p-2} - p^2\left(2k^2-1\right)\mathrm{cn}^p + p(p+1)k^2\mathrm{cn}^{p+2}\right\}\\
&+aA^n B_1^2\left\{np(np-1)\left(1-k^2\right)\mathrm{cn}^{np-2} + n^2p^2\left(2k^2-1\right)\mathrm{cn}^{np} - np(np+1)k^2\mathrm{cn}^{np+2}\right\}\\
&+bAB_1^3 v\left\{p(p-1)(p-2)(p-3)\left(1-k^2\right)^2\mathrm{cn}^{p-4} - 2p(p-1)\left(p^2-2p+2\right)\right.\\
&\times\left(1-k^2\right)\left(2k^2-1\right)\mathrm{cn}^{p-2} - \left[12pk^2\left(p^2-p+1\right)\left(k^2-1\right) + p^4\left(2k^2-1\right)^2\right]\mathrm{cn}^p\\
&\left.+2p(p+1)\left(p^2+2p+2\right)k^2\left(2k^2-1\right)\mathrm{cn}^{p+2} - p(p+1)(p+2)(p+3)k^4\mathrm{cn}^{p+4}\right\}\\
&+cAB_2^2\left\{p(p-1)\left(1-k^2\right)\mathrm{cn}^{p-2} + p^2\left(2k^2-1\right)\mathrm{cn}^p - p(p+1)k^2\mathrm{cn}^{p+2}\right\} = 0.
\end{aligned} \tag{7}$$

The value of p can be obtained by equating the exponents np and $p+2$ as

$$p = \frac{2}{n-1}, \tag{8}$$

which can also be achieved by equating the exponents $np+2$ and $p+4$. Finally, equating the coefficients of the linearly independent functions $\mathrm{cn}^{p+j}\tau$ for $j = 0, 2, 4$ value of v and A are obtained as follows:

$$v = \frac{c(n-1)^2\left(2k^2-1\right)B_2^2}{B_1\left\{(n-1)^2\left(2k^2-1\right)+4bB_1^2+bB_1^2k^2\left(k^2-1\right)\left(5n^3-32n^2+65n-26\right)\right\}}, \tag{9}$$

$$v = \frac{(n-1)^2\left\{ck^2B_2^2(n+1)-2\left(2k^2-1\right)an^2B_1^2A^{n-1}\right\}}{B_1(n+1)k^2\left\{(n-1)^2+4bB_1^2\left(n^2+1\right)\left(2k^2-1\right)\right\}}, \tag{10}$$

and

$$A = \left[\frac{bcB_2^2k^2(n+1)\left\{k^2\left(k^2-1\right)\left(5n^3-48n^2-42+65n\right)-4n^2\right\}}{2an^2\left\{(n-1)^2\left(2k^2-1\right)+4bB_1^2+bB_1^2k^2\left(k^2-1\right)\left(5n^3-32n^2+65n-26\right)\right\}}\right]^{\frac{1}{n-1}}. \tag{11}$$

Now from (9) to (11), it is clear that the restrictions that must be imposed on the parameters for the formation of the nontopological soliton solution are $c > 0$, and $ab < 0$. It is required to be noticed that equating the coefficient of $\mathrm{cn}^{p+4}\tau$ leads to the relation between the amplitude and the inverse widths of the soliton given in (11) and the two values of the velocity of the soliton are calculated equating the coefficients of $\mathrm{cn}^{p}\tau$ and $\mathrm{cn}^{p+2}\tau$ consequently. Now it is interesting to note that the relation between amplitude and the inverse widths of the soliton (11) can also be obtained by equating the two velocities of the soliton given by (9) and (10), which sustains the consistency of the method of solution. Thus the elliptic periodic solution of the gKP-MEW equation with power law nonlinearity is given by

$$q(x,y,t) = A\,\mathrm{cn}^{\frac{2}{n-1}}(B_1x+B_2y-vt), \tag{12}$$

where the amplitude A is given by (11) dependent on the inverse widths B_1 and B_2 and the velocity v of the soliton, given by (9) or (10).

Finally, consider the limits k $\to$ 1, $\mathrm{cn}\tau \to \mathrm{sech}\tau$, the Jacobi elliptic periodic solution (12) degenerate into the nontopological 1-soliton solution

$$q(x,y,t) = \frac{A}{\cosh^{\frac{2}{n-1}}(B_1x+B_2y-vt)}, \tag{13}$$

where the relation between the amplitude and inverse widths and the velocity of the soliton becomes

$$v = \frac{c(n-1)^2 B_2^2}{B_1 \left\{(n-1)^2 + 4bB_1^2\right\}}, \tag{14}$$

$$v = \frac{(n-1)^2 \left\{cB_2^2(n+1) - 2an^2 B_1^2 A^{n-1}\right\}}{B_1(n+1)\left\{(n-1)^2 + 4bB_1^2(n^2+1)\right\}}, \tag{15}$$

and

$$A = \left[-\frac{2bc(n+1)B_2^2}{a\left\{(n-1)^2 + 4bB_1^2\right\}}\right]^{\frac{1}{n-1}}. \tag{16}$$

The numerical simulation of the solution (13) with particular choice of parameters is shown in Fig. 1.

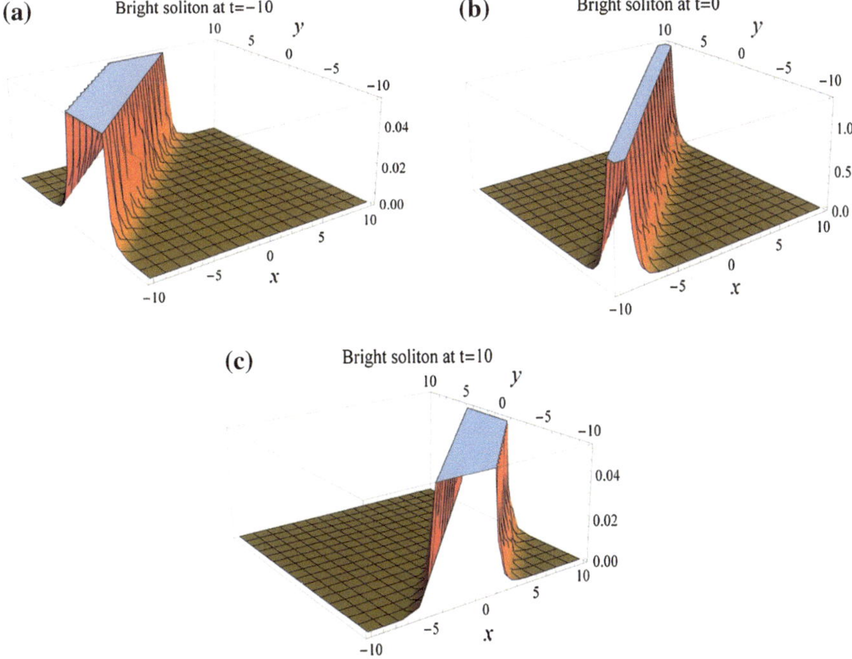

Fig. 1 Profile of the nontopological solitary wave at **a** $t = -10$, **b** $t = 0$, **c** $t = 10$ for $B_1 = 1$, $B_2 = -1$, $a = 0.2$, $b = -0.1$, $c = 0.5$, and $n = 2$

3 Topological Soliton

The topological solitons which are also known as snoidal waves are supported by the gKP-MEW equation under certain restictions. In case of topological solitons, the phase changes for large spatial distance. One of the essential difference between nontopological and topological solitons consists of the existence of multiple bound states that can form nontopological solitons in clear contrast with topological solitons [25]. To start off finding the topological soliton solution of the gKP-MEW equation, the starting hypothesis is taken in context of Jacobi elliptic function as follows [22–24]:

$$q(x, y, t) = A\ \mathrm{sn}^p(B_1x + B_2y - vt) \tag{17}$$

where A, B_1 and B_2 represent the soliton amplitude and the inverse widths of the soliton, respectively, and v depicts the velocity of the soliton. The value of the unknown exponent p will be determined during the course of derivation of the topological soliton solution of (4). From (17), by utilizing the notation

$$\tau = B_1x + B_2y - vt \tag{18}$$

one can easily obtain q_{xt}, $(q^n)_{xx}$, q_{xxxt}, q_{yy} and substituting them into (4) gives

$$\begin{aligned}
&AB_1v\left\{-p(p-1)\mathrm{sn}^{p-2} + p^2\left(1+k^2\right)\mathrm{sn}^p - p(p+1)k^2\mathrm{sn}^{p+2}\right\}\\
&+aA^nB_1^2\left\{np(np-1)\mathrm{sn}^{np-2} - n^2p^2\left(1+k^2\right)\mathrm{sn}^{np} + np(np+1)\mathrm{sn}^{np+2}\right\}\\
&+bAB_1^3v\left\{-p(p-1)(p-2)(p-3)\mathrm{sn}^{p-4} + 2\left(1+k^2\right)p(p-1)\left(p^2-2p+2\right)\mathrm{sn}^{p-2}\right.\\
&-p^2\left(p^2+4p^2k^2+p^2k^4+10k^2\right)\mathrm{sn}^p + 2k^2\left(1+k^2\right)p(p+1)\left(p^2+2p+2\right)\mathrm{sn}^{p+2}\\
&\left.-k^4p(p+1)(p+2)(p+3)\right\} + cAB_1^2\left\{-p^2\left(1+k^2\right)\mathrm{sn}^p + p(p-1)\mathrm{sn}^{p-2}\right.\\
&\qquad\left.+p(p+1)k^2\mathrm{sn}^{p+2}\right\} = 0
\end{aligned} \tag{19}$$

Equating the coefficients np and $p+2$ or $np+2$ and $np+4$, the value of p is obtained as

$$p = \frac{2}{n-1}. \tag{20}$$

The same value of p can also be obtained by equating the exponents np and $p+2$ and the exponent pairs $np-2$ and p. Note that, there are five linearly independent functions are there in (19), which are namely, $\mathrm{sn}^{p+j}\tau$ for $j = -4, -2, 0, 2, 4$. Hence, each of the coefficients of these linearly independent functions of (19) must be zero.

It is to be noticed that the function $\mathrm{sn}^{p-4}\tau$ stands alone linearly independent and its coefficient must vanish. Also, the common factor of all the coefficients of $\mathrm{sn}^{p-2}\tau$ is $(p-1)$. Hence, the coefficients of these two linearly independent functions produces

$$p = 1, \tag{21}$$

and

$$n = 3\,. \tag{22}$$

Thus from (22), it can be concluded that for the gKP-MEW equation, topological soliton exist only for $n = 3$, which is the KP-MEW equation. Hence, if $n \neq 3$, topological soliton do not exist for gKP-MEW equation, an important observation. Finally, setting the coefficients of the other linearly independent functions namely $\mathrm{sn}^{p+j}\tau$ for $j = 0, 2, 4$ to zero, we obtain

$$v = \frac{cB_2^2\left(1+k^2\right) - 6aA^2B_1^2}{B_1^2\left(1+k^2\right) - bB_1^3\left(1+14k^2+k^4\right)}, \tag{23}$$

$$v = \frac{9aA^2B_1^2\left(1+k^2\right) - 2ck^2B_2^2}{20bB_1^3k^2\left(1+k^2\right) - 2k^2B_1}, \tag{24}$$

$$v = \frac{aA^2}{2bB_1k^4}. \tag{25}$$

and the value of A is obtained as

$$A = \sqrt{\frac{2bck^2B_2^2}{a\left(1 - bB_1^2 - bB_1^2k^2\right)}}, \tag{26}$$

by equating the values of v in (23) and (24), which shows the relation between the amplitude and the inverse widths of the soliton. From (26), it is clear that the constraint relation

$$ab > 0 \quad \text{and} \quad c > 0 \tag{27}$$

has to be sustain for the existence of the Jacobi elliptic periodic solution. Hence the doubly periodic Jacobi elliptic solution for the KP-MEW equation

$$(q_t + 3aq^2q_x + bq_{xxt})_x + cq_{yy} = 0 \tag{28}$$

is given by

$$q(x, y, t) = A\,\mathrm{sn}(B_1x + B_2y - vt) \tag{29}$$

where the velocity of the soliton is given by (23) or (24) or (25). The relation between the free parameter and the inverse widths of the soliton is shown in (26) and the restriction on the coefficients of the dispersion term in x-direction and the weak dependence term along the y-coordinate for the existence of the doubly periodic Jacobi elliptic solution of the KP-MEW equation is conveyed in (27).

Finally, consider the limits $\mathrm{k} \to 1$, $\mathrm{sn}\tau \to \tanh\tau$, the elliptic periodic solution (29) degenerate into the topological 1-soliton solution

$$q(x, y, t) = A \tanh(B_1 x + B_2 y - vt), \tag{30}$$

with velocity as

$$v = \frac{cB_2^2 - 3aA^2B_1^2}{B_1 - 8bB_1^3}, \tag{31}$$

$$v = \frac{9aA^2B_1^2 - cB_2^2}{20bB_1^3 - B_1}, \tag{32}$$

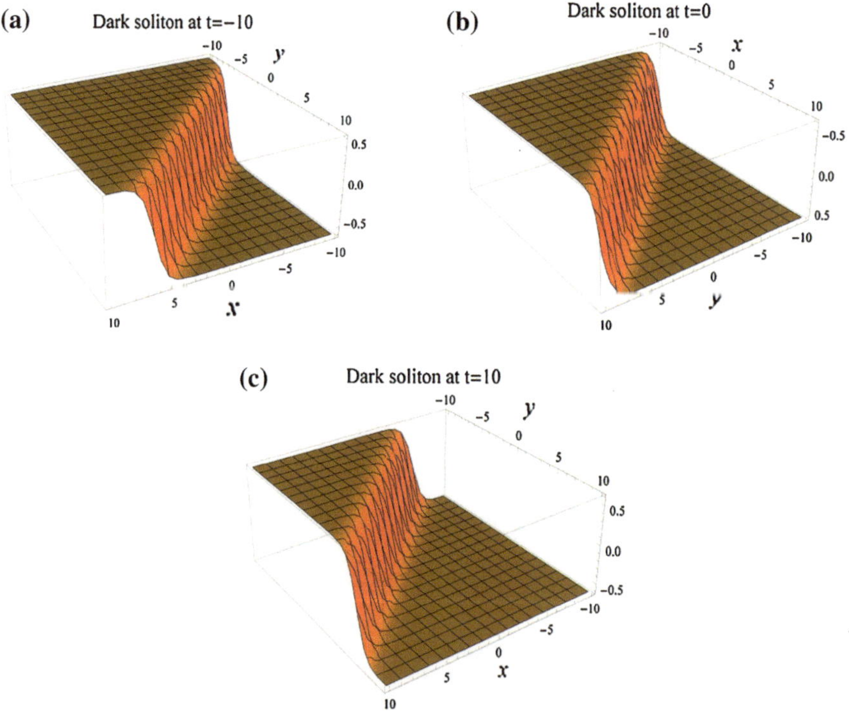

Fig. 2 Profile of the topological solitary wave at **a** $t = -10$, **b** $t = 0$, **c** $t = 10$ for $B_1 = 1$, $B_2 = -1$, $a = 0.2$, $b = 0.1$, $c = 0.3$

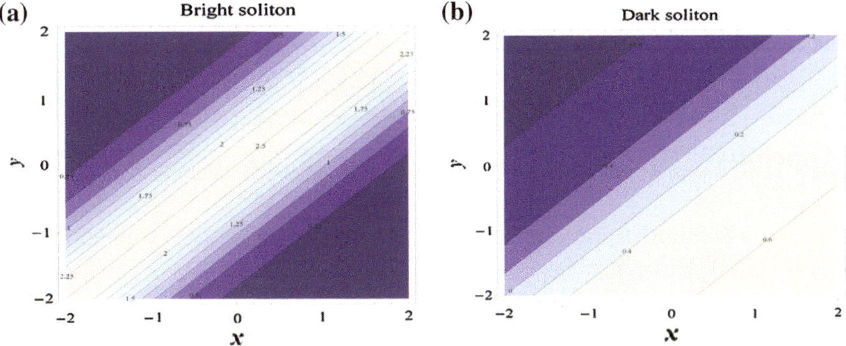

Fig. 3 Contour plot of the **a** nontopological and **b** topological soliton with same parameters at $t = 0$

and

$$v = \frac{aA^2}{2bB_1}. \tag{33}$$

Now, equating the three values of the velocity given by (31), (32) and (33), the relation between the free parameters A and B_1, B_2 is obtained as follows

$$A = \sqrt{\frac{2bcB_2^2}{a\left(1 - 2bB_1^2\right)}}. \tag{34}$$

The existence of the soliton solution requires the same constraint relation (27). The numerical simulation of the solution (30) with particular choice of parameters is shown in Fig. 2. The contour plot corresponding to the nontopological and topological soliton solutions are demonstrated in Fig. 3.

4 Conclusions

In this paper, the solitary wave ansatz method is exploited in the context of doubly periodic Jacobi elliptic functions to carry out the 1-soliton solution of the gKP-MEW equation. The elliptic function solution degenerates into the solitary wave solution in the limiting case of the elliptic modulus parameter. Both topological (snoidal) and nontopological (cnoidal) solitons are studied. An interesting fact is observed that the topological soliton for the gKP-MEW equation exist only for $n = 3$, or in other words for the KP-MEW equation only, which is reported in the literature for the first time as far as we know. In future, further aspects of this problem can be studied; such as addition of perturbation term, self-steeping term, soliton–soliton interaction and others.

Acknowledgments The authors acknowledge anonymous referees for critical comments and mentioning some useful references.

Appendix

Here we will furnish a primary introduction about elliptic functions (for more details, see [26, 27]). Three Jacobian elliptic functions are defined as

$$\mathrm{sn}(x,k) = \sin\varphi, \qquad \mathrm{cn}(x,k) = \cos\varphi, \qquad \mathrm{dn}(x,k) = \mathrm{d}\varphi/\mathrm{d}x \tag{35}$$

where the amplitude function $\varphi(\mathrm{z}, k)$ is defined by the integral

$$\mathrm{z}(\varphi,k) = \int_0^{\varphi} \frac{\mathrm{d}\tau}{\sqrt{1-k^2\sin^2\tau}} \tag{36}$$

The square of the real number k is called elliptic modulus parameter and $k^2 \in (0, 1)$. Also $k'^2 = 1-k^2$ is called complementary modulus parameter. To avoid the complexity, in the text we inhibit the explicit modular dependence and write $\mathrm{sn}x$, $\mathrm{cn}x$, $\mathrm{dn}x$ etc. These are doubly periodic functions of periods $4K$, $2iK'$; $4K$, $4iK'$, and $2K$, $4iK'$, respectively, where the quarter-periods K and K' are the real numbers given by

$$K(k) \equiv K = \mathrm{z}(\pi/2, k), \qquad K'(k) \equiv K' = K(k') \tag{37}$$

K is called complete elliptic integral of second kind. Some useful relations are

$$\mathrm{sn}^2x + \mathrm{cn}^2x = 1, \qquad \mathrm{dn}^2x + k^2\mathrm{sn}^2x = 1, \qquad k^2\left(\mathrm{cn}^2x - 1\right) = \mathrm{dn}^2x - 1 \tag{38}$$

the rules of differentiation are

$$\mathrm{sn}'x = \mathrm{cn}x\ \mathrm{dn}x, \qquad \mathrm{cn}'x = -\mathrm{sn}x\ \mathrm{dn}x, \qquad \mathrm{dn}'x = -k^2\mathrm{sn}x\ \mathrm{cn}x \tag{39}$$

and

$$\mathrm{sn}(x,k) \xrightarrow[k\to0]{k\to1} \begin{cases} \tanh x \\ \sin x \end{cases}, \quad \mathrm{cn}(x,k) \xrightarrow[k\to0]{k\to1} \begin{cases} \mathrm{sech}x \\ \cos x \end{cases}, \quad \mathrm{dn}(x,k) \xrightarrow[k\to0]{k\to1} \begin{cases} \mathrm{sech}x \\ 1 \end{cases} \tag{40}$$

References

1. Pelinovsky, D.: Intermediate nonlinear Schrödinger equation for internal waves in a fluid of finite depth. Phys. Lett. A **197**, 401–406 (1995)
2. Hasegawa, A.: Plasma Instabilities and Nonlinear Effects. Springer, Berlin (1975)
3. Gray, P., Scott, S.: Chemical Oscillations and Instabilities. Clarendon, Oxford (1990)

4. Malfliet, W., Hereman, W.: The tanh method. I: exact solutions of nonlinear evolution and wave equations. Phys. Scr. **54**, 563–568 (1996)
5. Wazwaz, A.M.: The tanh method for travelling wave solutions of nonlinear equations. Appl. Math. Comput. **154**, 713–723 (2004)
6. Fan, E., Zhang, H.: A note on the homogeneous balance method. Phys. Lett. A **246**, 403–406 (1998)
7. Fan, E., Zhang, J.: Application of the Jacobi elliptic function method to special-type nonlinear equations. Phys. Lett. A **305**, 383–392 (2002)
8. Bagchi, B., Das, S., Ganguly, A.: New exact solutions of a generalized shallow water wave equation. Phys. Scr. **82**, 025003 (2010)
9. Wang, M.L., Li, X.Z., Zheng, J.L.: The (G'/G)-expansion method and travelling wave solutions of nonlinear evolution equations in mathematical physics. Phys. Lett. A **372**, 417–423 (2008)
10. Das, A., Ganguly, A.: Some new exact solutions of the (2+1) dimensional potential Kadomstev-Petviashvili equation by the extended (G'/G)-expansion method. Int. J. Nonlinear Sci. **14**, 86–94 (2012)
11. Das, A., Ganguly, A.: A new variation of the (G'/G)-expansion method: traveling wave solutions to nonlinear equations. Int. J. Nonlinear Sci. **17**, 268–280 (2014)
12. Zabusky, N.J., Kruskal, M.D.: Interactions of solitons in a collisionless plasma and the recurrence of initial states, Phys. Rev. Lett. **15**, 240–243 (1965)
13. Bagada, A.V., Melkov, G.A., Serga, A.A., Slavin, A.N.: Parametric interaction of dipolar spin wave solitons with localized electromagnetic pumping. Phys. Rev. Lett. **79**, 21372140 (1997)
14. Scott, M.M., Kostylev, M.P., Kalinikos, B.A., Patton, C.E.: Excitation of bright and dark envelope solitons for magnetostatic waves with attractive nonlinearity. Phys. Rev. B **71**, 1–4 (2005)
15. Kadomtsev, B.B., Petviashvili, V.I.: On the stability of solitary waves in weakly dispersive media. Soviet. Phys. Dokl. **15**, 539–541 (1970)
16. Biswas, A.: Soliton perturbation theory for the generalized equal width equation. Pac. J. Appl. Math. **1**, 99–103 (2008)
17. Wazwaz, A.M.: The tanh and sine-cosine methods for a reliable treatment of the modified equal width equation and its variants. Commun. Nonlinear Sci. Num. Simul. **11**, 148–160 (2006)
18. Wei, M., Tang, S., Fu, H., Chen, G.: Single peak solitary wave solutions for the generalized KP-MEW (2, 2) equation under boundary condition. Appl. Math. Comput. **219**, 8979–8990 (2013)
19. Ismail, M.S., Petkovic, M.D., Biswas, A.: 1-soliton solution of the generalized KP equation with generalized evolution. Appl. Math. Comput. **216**, 2220–2225 (2010)
20. Biswas, A.: 1-soliton solution of the generalized camassaHolm Kadomtsev-Petviashvili equation. Commun. Nonlinear Sci. Num. Simul. **14**, 2524–2527 (2009)
21. Biswas, A., Ranasinghe, A.: 1-soliton solution of Kadomtsev-Petviashvili equation with power law nonlinearity. Appl. Math. Comput. **214**, 645–647 (2009)
22. Ebadi, G., Fard, N.Y., Bhrawy, A.H., Kumar, S., Triki, H., Yildirim, A., Biswas, A.: Solitons and other solutions to the $(3+1)$-dimensional extended Kadomtsev-Petviashvili equation with power law nonlinearity. Rom. Rep. Phys. **65**, 27–62 (2013)
23. Biswas, A., Triki, H., Labidi, M.: Bright and dark solitons of the Rosenau-Kawahara equation with power law nonlinearity. Phys. Wave Phenom. **19**, 24–29 (2011)
24. Biswas, A., Ranasinghe, A.: Topological 1-soliton solution of Kadomtsev-Petviashvili equation with power law nonlinearity. Appl. Math. Comput. **217**, 1771–1773 (2010)
25. Palacios, S.L., Fernandez-Diaz, J.M.: Black optical solitons for media with parabolic nonlinearity law in the presence of fourth order dispersion. Opt. Commun. **178**, 457–460 (2000)
26. Gradshteyn, I.S., Ryzhik, I.M.: Tables of Integrals, Series and Products. Academic Press, New York (1963)
27. Abramowitz, M., Stegun, I.A. (eds.): Handbook of Mathematical Functions. Dover Publications INC, New York (1965)

Mild Solutions for Impulsive Functional Differential Equations of Order $\alpha \in (1, 2)$

Ganga Ram Gautam and Jaydev Dabas

Abstract In this research paper, first we develop the definition of mild solutions for impulsive fractional differential equations of order $\alpha \in (1, 2)$. Second, we study the uniqueness result of mild solutions for impulsive fractional differential equation with state-dependent delay by applying fixed point theorem and solution operator. At last, we present an example to illustrate the uniqueness result using fractional partial derivatives.

Keywords Fractional order differential equation · Functional differential equations · Impulsive conditions · Fixed point theorem

1 Introduction

In this research paper, we consider the following impulsive fractional differential equation with state-dependent delay of the form

$$^C D_t^\alpha u(t) = Au(t) + f(t, u_{\rho(t,u_t)}),\ t \in J = [0, T],\ t \neq t_k, \tag{1}$$

$$u(t) = \phi(t),\ t \in (-\infty, 0],\ u'(0) = u_1 \in X, \tag{2}$$

$$\Delta u(t_k) = I_k(u(t_k^-)),\ \Delta u'(t_k) = Q_k(u(t_k^-)),\quad k = 1, 2, ...m, \tag{3}$$

where $^C D_t^\alpha$ is the Caputo's fractional derivative of order $\alpha \in (1, 2)$, u' is ordinary derivative with respect to t and J is operational interval. $A : D(A) \subset X \to X$ is the sectorial operator defined on a complex Banach space X. The functions

G.R. Gautam (✉) · J. Dabas
Department of Applied Science and Engineering, IIT Roorkee,
Saharanpur Campus, 247001 Saharanpur, India
e-mail: gangaiitr11@gmail.com

J. Dabas
e-mail: jay.dabas@gmail.com

P.N. Agrawal et al. (eds.), *Mathematical Analysis and its Applications*,
Springer Proceedings in Mathematics & Statistics 143,
DOI 10.1007/978-81-322-2485-3_23

$f : J \times \mathscr{B}_h \to X$, $\rho : J \times \mathscr{B}_h \to (-\infty, T]$ and $\phi \in \mathscr{B}_h$ are given and satisfies some assumptions, where $\mathscr{B}_h$ is introduced in Sect. 2. The history function $u_t : (-\infty, 0] \to X$ is defined by $u_t(\theta) = u(t+\theta)$, $\theta \in (-\infty, 0]$ belongs to $\mathscr{B}_h$. Here $0 = t_0 < t_1 < \cdots < t_m < t_{m+1} = T < \infty$ and the functions $I_k, Q_k \in C(X, X)$, $k = 1, 2, ...m$, are bounded. We have $\Delta u(t_k) = u(t_k^+) - u(t_k^-)$ where $u(t_k^+)$ and $u(t_k^-)$ represent the right- and left-hand limits of $u(t)$ at $t = t_k$, also we take $u(t_k^-) = u(t_k)$. Furthermore, $\Delta u'(t_k) = u'(t_k^+) - u'(t_k^-)$ where $u'(t_k^+)$ and $u'(t_k^-)$ represent the right- and left-hand limits of $u'(t)$ at $t = t_k$, also we take $u'(t_k^-) = u'(t_k)$, respectively.

Impulsive differential equations with fractional order (see for fractional calculus [15, 16, 18–20]) are paying attention by many researchers because the model processes which are subjected to abrupt changes cannot described by ordinary differential equations, so such type equations are modeled in term of impulse. The most important applications of these equations are in the ecology, mechanics, electrical, and medicine biology. On the other hand, functional differential equations originate in several branches of engineering, applied mathematics, and science. Recently, fractional functional differential equations with state-dependent delay seems frequently in many fields as modeling of equations, panorama of natural phenomena, and porous media. See for more details of the relevant development theory in the cited papers [1, 2, 4–9, 11, 13].

In our survey, we found that Feckan et al. [12] gave the new concept of solution for impulsive nonlinear fractional differential equation order $\alpha \in (0, 1)$. Wang et al. [22] defined the definition of mild solution using the probability density function for the impulsive fractional evolution equation of order $\alpha \in (0, 1)$. By Motivated work [22], Dabas and Chauhan [10] defined the mild solution for neutral impulsive fractional functional differential equation of order $\alpha \in (0, 1)$ using analytic operator theory. Wang et al. [23] extended the problem, consider in paper [12] for of order $\alpha \in (1, 2)$. Shu et al. [21] introduced the definition of mild solution for fractional differential equations with nonlocal conditions of order $\alpha \in (1, 2)$ without impulse. We found that there is no literature available on mild solution for impulsive fractional functional differential equation of order $\alpha \in (1, 2)$.

To fill this gap and inspired by the above-mentioned work [10, 12, 21–23], we develop the definition of mild solution for the problem (1)–(3) and show the existence result. For further details, this work has four sections, Sect. 2 provides some basic definitions, preliminaries, theorems, and lemmas. The Sect. 3 is equipped with main results for the considered problem (1)–(3) and in Sect. 4 an example is considered.

2 Preliminaries and Background Martials

Let $(X, \|\cdot\|_X)$ be a complex Banach space of functions with the norm $\|u\|_X = \sup_{t \in J}\{|u(t)| : u \in X\}$ and $L(X)$ denotes the Banach space of bounded linear operators from X into X equipped with norm is denoted by $\|\cdot\|_{L(X)}$.

For the analysis of the infinite delay, we shall use abstract phase space $\mathscr{B}_h$ as defined in [14] details are as follow:

Let $h : (-\infty, 0] \to (0, \infty)$ be a continuous function with $l = \int_{-\infty}^{0} h(s)ds < \infty$, $s \in (-\infty, 0]$. For any $a > 0$, we define space

$$\mathscr{B} = \{\psi : [-a, 0] \to X \text{ such that } \psi(t) \text{ is bounded and measurable}\},$$

equipped with the norm $\|\psi\|_{[-a,0]} = \sup_{s\in[-a,0]} \|\psi(s)\|_X$, $\forall\, \psi \in \mathscr{B}$. Let us define abstract space as

$$\mathscr{B}_h = \left\{\psi : (-\infty, 0] \to X, \text{ s.t. for any } a \geq c > 0, \psi\mid_{[-c,0]} \in \mathscr{B} \int_{-\infty}^{0} h(s)\|\psi\|_{[s,0]}ds < \infty\right\}.$$

If $\mathscr{B}_h$ is endowed with the norm $\|\psi\|_{\mathscr{B}_h} = \int_{-\infty}^{0} h(s)\|\psi\|_{[s,0]}ds$, $\forall\, \psi \in \mathscr{B}_h$, then it is clear that $(\mathscr{B}_h, \|\cdot\|_{\mathscr{B}_h})$ is a complete Banach space. Let

$$C_t^1([0, T], X) = C^1([0, t]; X), \ 0 < t \leq T < \infty,$$

be a Banach space of all functions $u : [0, T] \to X$ such that u is continuously differentiable on $[0, T]$ endowed with the norm

$$\|u\|_{C_t^1} = \sup_{t\in[0,T]} \left\{\sum_{j=0}^{1} \|u^j(t)\|_X, u \in C_t^1\right\}.$$

To use the impulsive condition with infinite delay, we consider a Banach space

$$\mathscr{B}_h' := PC^1((-\infty, T]; X), \ T < \infty,$$

formed by all functions $u : (-\infty, T] \to X$ such that u is continuously differentiable on $[0, T]$ except for a finite number of points $t_i \in (0, T)$, $i = 1, 2, \ldots,$ N, at which $u'(t_i^+)$ and $u'(t_i^-) = u'(t_i)$ exist and endowed with the seminorm $\|\cdot\|_{B_h'}$ in $\mathscr{B}_h'$

$$\|u\|_{\mathscr{B}_h'} = \sup\{\|u\|_{C_t^1} : 0 \leq t \leq T\} + \|\phi\|_{\mathscr{B}_h}, u \in \mathscr{B}_h'.$$

For a function $u \in \mathscr{B}_h'$ and $i \in \{0, 1, \ldots, N\}$, we introduce the function $\bar{u}_i \in C^1((t_i, t_{i+1}]; X)$ given by

$$\bar{u}_i(t) = \begin{cases} u'(t), & \text{for } t \in (t_i, t_{i+1}], \\ u'(t_i^+), & \text{for } t = t_i. \end{cases}$$

Let $u : (-\infty, T] \to X$ be the function such that $u_0 = \phi$, $u\mid_{J_k} \in C^1(J_k, X)$ then for all $t \in J_k$, the following conditions hold:

(C_1) $u_t \in \mathscr{B}_h$.
(C_2) $\|u(t)\|_X \leq H\|u_t\|_{\mathscr{B}_h}$.
(C_3) $\|u_t\|_{\mathscr{B}_h} \leq K(t)\sup\{\|u(s)\| : 0 \leq s \leq t\} + M(t)\|\phi\|_{\mathscr{B}_h}$, where $H > 0$ is constant; $K, M : [0,\infty) \to [0,\infty)$, $K(\cdot)$ is continuous, $M(\cdot)$ is locally bounded and K, M are independent of $u(t)$.
(C_{4_ϕ}) The function $t \to \phi_t$ is well-defined and continuous from the set

$$\Re(\rho^-) = \{\rho(s,\psi) : (s,\psi) \in [0,T] \times \mathscr{B}_h\}$$

into $\mathscr{B}_h$ and there exist a continuous and bounded function $J^\phi : \Re(\rho^-) \to (0,\infty)$ such that $\|\phi_t\|_{\mathscr{B}_h} \leq J^\phi(t)\|\phi\|_{\mathscr{B}_h}$ for every $t \in \Re(\rho^-)$.

Lemma 1 ([5]) *Let $u : (-\infty, T] \to X$ be function such that $u_0 = \phi, u\mid_{J_k} \in C^1(J_k, X)$ and if (C_{4_ϕ}) hold, then*

$$\|u_s\|_{\mathscr{B}_h} \leq (M_b + J^\phi)\|\phi\|_{\mathscr{B}_h} + K_b \sup\left\{\|u(\theta)\|;\ \theta \in [0, \max\{0,s\}]\right\},\ s \in \Re(\rho^-) \cup J_k,$$

where $J^\phi = \sup_{t\in\Re(\rho^-)} J^\phi(t),\ M_b = \sup_{s\in[0,T]} M(s)$ and $K_b = \sup_{s\in[0,T]} K(s)$.

Definition 1 Caputo's derivative of order $\alpha > 0$ with lower limit a, for a function $f : [a,\infty) \to \mathrm{R}$ such that $f \in C^n([a,\infty),\mathrm{R})$ is defined as

$${}_a^C D_t^\alpha f(t) = \frac{1}{\Gamma(n-\alpha)} \int_a^t (t-s)^{n-\alpha-1} f^{(n)}(s)ds =_a J_t^{n-\alpha} f^{(n)}(t),$$

where $a \geq 0,\ n-1 < \alpha < n,\ n \in \mathrm{N}$.

Definition 2 The Riemann Liouville fractional integral operator of order $\alpha > 0$ with lower limit a, for a continuous function $f : [a,\infty) \to \mathrm{R}$ such that $f \in L^1_{loc}([a,\infty),\mathrm{R})$ is defined by

$${}_aJ_t^0 f(t) = f(t),\ {}_aJ_t^\alpha f(t) = \frac{1}{\Gamma(\alpha)} \int_a^t (t-s)^{\alpha-1} f(s)ds, \quad t > 0,$$

where $a \geq 0$ and $\Gamma(\cdot)$ is the Euler gamma function.

Definition 3 ([21]) Let $A : D(A) \subseteq X \to X$ be a densely defined, closed, and linear operator in X. A is said to be sectorial of the type (M,θ,α,μ) if there exist $\mu \in \mathrm{R},\ \theta \in (\frac{\pi}{2},\pi),\ M > 0$, such that such that the α-resolvent of A exists outside the sector and following two conditions are satisfied

(1) $\mu + S_\theta = \{\mu + \lambda^\alpha : \lambda \in \mathscr{C},\ |Arg(-\lambda^\alpha)| < \theta\}$,
(2) $\|(\lambda^\alpha I - A)^{-1}\|_{L(X)} \leq \frac{M}{|\lambda^\alpha - \mu|},\ \lambda \notin \mu + S_\theta$,

where X is the complex Banach space with norm denoted $\|\cdot\|_X$.

Definition 4 ([19]) A two parameter function of the Mittag-Leffler type is defined by the series expansion and integral form

$$E_{\alpha,\beta}(y) = \sum_{k=0}^{\infty} \frac{y^k}{\Gamma(\alpha k + \beta)} = \frac{1}{2\pi\iota} \int_c \frac{\mu^{\alpha-\beta} e^{\mu}}{\mu^{\alpha} - y} d\mu, \ \alpha, \ \beta > 0, \ y \in C,$$

where c is a contour which starts and ends at $-\infty$ and encircles the disk $|\mu| \leq |y|^{\frac{1}{\alpha}}$ counter clockwise.

The Laplace integral of this function given by

$$\int_0^{\infty} e^{-\lambda t} t^{\beta-1} E_{\alpha,\beta}(\omega t^{\alpha}) dt = \frac{\lambda^{\alpha-\beta}}{\lambda^{\alpha} - \omega}, \ Re\lambda > \omega^{\frac{1}{\alpha}}, \ \omega > 0.$$

From paper [17], putting $\beta = 1$, $\omega = A$ and using the sign $\div$ for the juxtaposition of a function depending on t with its Laplace transform depending on λ, we get the following Laplace transform pairs

$$S_{\alpha}(t) = E_{\alpha}(At^{\alpha}) \div \frac{\lambda^{\alpha-1}}{\lambda^{\alpha} I - A}, \ Re\lambda > A^{\frac{1}{\alpha}}.$$

More general Laplace transform pairs with integral

$$_0J_t^j S_{\alpha}(t) \div \frac{\lambda^{\alpha-j-1}}{\lambda^{\alpha} I - A}, \ j = 0, 1.$$

Definition 5 ([2]) Let A be a closed and linear operator with the domain $D(A)$ defined in a Banach space X and $\alpha > 0$. We say that A is the generator of a solution operator if there exist $\omega \geq 0$ and a strongly continuous function $S_{\alpha} : \mathrm{R}^{+} \to L(X)$, such that $\{\lambda^{\alpha} : Re\lambda > \omega\} \subset \rho(A)$ and

$$\frac{\lambda^{\alpha-1}}{\lambda^{\alpha} I - A} x = \int_0^{\infty} e^{\lambda t} S_{\alpha}(t) x dt, \quad Re\lambda > \omega, x \in X.$$

In this case, $S_{\alpha}(t)$ is called the solution operator generated by A.

Definition 6 ([3]) Let A be a closed and linear operator with domain $D(A)$ defined on a Banach space X. Let $\rho(A)$ be the resolvent set of A, we call A is the generator of an α-resolvent family if there exists $\omega \geq 0$ and a strongly continuous function $T_{\alpha} : \mathrm{R}^{+} \to L(X)$ such that $\{\lambda^{\alpha} : Re\lambda > \omega\} \subset \rho(A)$ and

$$(\lambda^{\alpha} I - A)^{-1} x = \int_0^{\infty} e^{-\lambda t} T_{\alpha}(t) x dt, \ Re\lambda > \omega, \ x \in X.$$

In this case, $T_{\alpha}(t)$ is called α-resolvent family generated by A.

Lemma 2 *Let f be a continuous function and A be a sectorial operator of the type (M, θ, α, μ). Consider following differential equation of order $\alpha \in (1, 2)$*

$$ {}^C D_t^\alpha u(t) = Au(t) + f(t), \ t \in J = [0, T], t \neq t_k, \tag{4} $$
$$ u(0) = u_0 \in X, \ u'(0) = u_1 \in X, \tag{5} $$
$$ \Delta u(t_k) = I_k(u(t_k^-)), \Delta u'(t_k) = Q_k(u(t_k^-)), \ t \neq t_k, \quad k = 1, 2, ...m. \tag{6} $$

Then a function $u(t) \in PC^1([0, T], X)$ is a solution of the system (4)–(6) if it satisfies following integral equation

$$ u(t) = \begin{cases} S_\alpha(t)u_0 + u_1 \int_0^t S_\alpha(s)ds + \int_0^t T_\alpha(t-s)f(s)ds, t \in (0, t_1] \\ S_\alpha(t)u_0 + K_\alpha(t)u_1 + \sum_{i=1}^k S_\alpha(t-t_i)I_i(u(t_i^-)) \\ + \sum_{i=1}^k Q_i\left(u(t_i^-)\right) \int_{t_i}^t S_\alpha(s-t_i)ds + \int_0^t T_\alpha(t-s)f(s)ds, t \in (t_k, t_{k+1}], \end{cases} $$

where $S_\alpha(t)$ and $T_\alpha(t)$ are operators generated by A and defined as

$$ S_\alpha(t) = \frac{1}{2\pi i}\int_\Gamma e^{\lambda t}\lambda^{\alpha-1}(\lambda^\alpha I - A)^{-1}d\lambda; \ T_\alpha(t) = \frac{1}{2\pi i}\int_\Gamma e^{\lambda t}(\lambda^\alpha I - A)^{-1}d\lambda, $$

and Γ is a suitable path such that $\lambda^\alpha \notin \mu + S_\theta$ for $\lambda \in \Gamma$.

Proof If $t \in (0, t_1]$, we have following problem

$$ {}^C D_t^\alpha u(t) = Au(t) + f(t), \tag{7} $$
$$ u(0) = u_0, \ u'(0) = u_1. \tag{8} $$

By Lemma 3.1 in [23], the solution of Eqs. (7)–(8), we get

$$ u(t) = u_0 + u_1 t + \int_0^t \frac{(t-s)^{\alpha-1}}{\Gamma(\alpha)}Au(s)ds + \int_0^t \frac{(t-s)^{\alpha-1}}{\Gamma(\alpha)}f(s)ds. \tag{9} $$

If $t \in (t_k, t_{k+1}], k = 1, 2, ...m$, we have the following problem

$$ {}^C D_t^\alpha u(t) = Au(t) + f(t), \tag{10} $$
$$ u(t_k^+) = u(t_k^-) + I_k(u(t_k^-)), \tag{11} $$
$$ u'(t_k^+) = u'(t_k^-) + Q_k(u(t_k^-)). \tag{12} $$

By Lemma 3.1 in [23] the solution of Eqs. (10)–(12), we get

$$ u(t) = u_0 + u_1 t + \sum_{i=1}^k I_i(u(t_i^-)) + \sum_{i=1}^k Q_i(u(t_i^-))(t - t_i) $$
$$ + \int_0^t \frac{(t-s)^{\alpha-1}}{\Gamma\alpha}Au(s)ds + \int_0^t \frac{(t-s)^{\alpha-1}}{\Gamma(\alpha)}f(s)ds. \tag{13} $$

Summarizing Eqs. (9) and (13) to $t \in (0, T]$, we get

$$u(t) = u_0 + u_1 t + \sum_{i=1}^{m} \chi_{t_i}(t) I_i(u(t_i^-)) + \sum_{i=1}^{m} \chi_{t_i}(t) Q_i(u(t_i^-))(t - t_i) + \int_0^t \frac{(t-s)^{\alpha-1}}{\Gamma\alpha} Au(s)ds + \int_0^t \frac{(t-s)^{\alpha-1}}{\Gamma(\alpha)} f(s)ds, \tag{14}$$

where

$$\chi_{t_i}(t) = \begin{cases} 0 & t \le t_i \\ 1 & t > t_i. \end{cases}$$

By taking the Laplace transformation on Eq. (14), we have

$$L\{u(t)\} = \frac{u_0}{\lambda} + \frac{u_1}{\lambda^2} + \sum_{i=1}^{m} \frac{e^{-\lambda t_i}}{\lambda} I_i(u(t_i^-)) + \sum_{i=1}^{m} \frac{e^{-\lambda t_i}}{\lambda^2} Q_i(u(t_i^-)) + \frac{A}{\lambda^\alpha} L\{u(t)\} + \frac{1}{\lambda^\alpha} L\{f(t)\}. \tag{15}$$

On simplifying Eq. (15), we get

$$L\{u(t)\} = \frac{\lambda^{\alpha-1}(u_0)}{(\lambda^\alpha I - A)} + \frac{\lambda^{\alpha-2}(u_1)}{(\lambda^\alpha I - A)} + \sum_{i=1}^{m} \frac{\lambda^{\alpha-1}}{(\lambda^\alpha I - A)} e^{-\lambda t_i} I_i(u(t_i^-)) + \sum_{i=1}^{m} \frac{\lambda^{\alpha-2}}{(\lambda^\alpha I - A)} e^{-\lambda t_i} Q_i(u(t_i^-)) + \frac{1}{(\lambda^\alpha I - A)} L\{f(t)\}. \tag{16}$$

Now, taking the inverse Laplace transformation of Eq. (16), we have

$$u(t) = S_\alpha(t)u_0 + u_1 \int_0^t S_\alpha(s)ds + \sum_{i=1}^{m} \chi_{t_i}(t) I_i(u(t_i^-)) S_\alpha(t - t_i) + \sum_{i=1}^{m} \chi_{t_i}(t) Q_i(u(t_i^-)) \int_{t_i}^t S_\alpha(s - t_i)ds + \int_0^t T_\alpha(t-s) f(s)ds, \quad t \in J.$$

This complete the proof of the lemma.

Now, we state the definition of mild solutions of problem (1)–(3) by Lemma 2.

Definition 7 A function $u : (-\infty, T] \to X$ such that $u \in \mathscr{B}'_h$, $u(0) = \phi(0)$, $u'(0) = u_1$, is called a mild solution of problem (1)–(3) if it satisfies the following integral equation

$$u(t) = \begin{cases} S_\alpha(t)\phi(0) + u_1 \int_0^t S_\alpha(s)ds + \int_0^t T_\alpha(t-s) f(s, u_{\rho(s,u_s)})ds, & t \in (0, t_1] \\ S_\alpha(t)\phi(0) + u_1 \int_0^t S_\alpha(s)ds + \sum_{i=1}^k I_i(u(t_i^-))S_\alpha(t-t_i) \\ + \sum_{i=1}^k Q_i(u(t_i^-)) \int_{t_i}^t S_\alpha(s-t_i)ds + \int_0^t T_\alpha(t-s) f(s, u_{\rho(s,u_s)})ds, & t \in (t_k, t_{k+1}]. \end{cases}$$

3 Uniqueness Result of Mild Solution

In this section, we prove the existence of mild solutions for the problem (1)–(3) with a non-convex valued right-hand side . If A sectorial operator of the type (M, θ, α, μ) then the strongly continuous functions $\|S_\alpha(t)\| \leq M$, $\|T_\alpha(t)\| \leq M$. To prove our results, we shall assume the function ρ is continuous. Our result is based on contraction fixed point theorem, for this we have following assumptions

(H_1) The function f is continuous and there exists $l_f \in L^1(J, \mathrm{R}^+)$ such that

$$\|(f(t, \psi) - f(t, \xi))\|_X \leq l_f(t)\|\psi - \xi\|_{\mathscr{B}_h} \text{ for every } \psi, \xi \in \mathscr{B}_h.$$

(H_2) The functions I_k, Q_k are continuous and there exist $l_i, l_j \in L^1(J, \mathrm{R}^+)$ such that

$$\|I_k(x) - I_k(y)\|_X \leq l_i(t)\|x - y\|_X; \; \|Q_k(x) - Q_k(y)\|_X \leq l_j(t)\|x - y\|_X,$$

for all $x, y \in X$ and $k = 1, \ldots, m$.

Theorem 1 *Let the assumption (H_1) and (H_2) hold and the constant*

$$\Delta = M\left[m\|l_i\|_{L^1(J,\mathrm{R}^+)} + mT\|l_j\|_{L^1(J,\mathrm{R}^+)} + K_b \int_0^T l_f(s)ds\right] < 1.$$

Then problem (1)–(3) has a unique mild solutions u on J.

Proof We convert the problem (1)–(3) in to fixed point problem. Let $\bar{\phi} : (-\infty, T) \to X$ be the extension of ϕ to $(-\infty, T]$ such that $\phi(\bar{t}) = \phi(0)$ on J. Consider the space Banach $\mathscr{B}_h'' = \{u \in \mathscr{B}_h' : u(0) = \phi(0), u'(0) = u_1\}$ and define the operator $P : \mathscr{B}_h'' \to \mathscr{B}_h''$ as

$$Pu(t) = \begin{cases} S_\alpha(t)\phi(0) + u_1 \int_0^t S_\alpha(s)ds + \int_0^t T_\alpha(t-s) f(s, \bar{u}_{\rho(s,\bar{u}_s)})ds, & t \in (0, t_1] \\ S_\alpha(t)\phi(0) + u_1 \int_0^t S_\alpha(s)ds + \sum_{i=1}^k I_i(\bar{u}(t_i^-))S_\alpha(t-t_i) \\ + \sum_{i=1}^k Q_i(\bar{u}(t_i^-)) \int_{t_i}^t S_\alpha(s-t_i)ds + \int_0^t T_\alpha(t-s) f(s, \bar{u}_{\rho(s,\bar{u}_s)})ds, & t \in (t_k, t_{k+1}], \end{cases}$$

where $\bar{u} : (-\infty, T] \to X$ is such that $u(\bar{0}) = \phi$ and $\bar{u} = u$ on J. It is clear that u is unique mild solution of the problem (1)–(3) if and only if u is a solution of the operator equation $Pu = u$. Let $u, u^* \in \mathscr{B}_h''$, for $t \in (0, t_1]$ we have

$$\|Pu - Pu^*\|_X \le \int_0^t \|T_\alpha(t-s)\|_{L(X)} \|f\left(s, \bar{u}_{\rho(s,\bar{u}_s)}\right) - f\left(s, \bar{u}^*_{\rho(s,\bar{u}^*_s)}\right)\|_X ds$$
$$\|Pu - Pu^*\|_{\mathscr{B}''_h} \le MK_b\left[\int_0^T l_f(s)ds\right]\|u - u^*\|_{\mathscr{B}''_h}.$$

Now, without lose of generality we consider the subinterval $(t_k, t_{k+1}]$ to prove our result. Let $u, u^* \in \mathscr{B}''_h$ for $(t_k, t_{k+1}]$, we have

$$\begin{aligned}
\|Pu - Pu^*\|_X \le & \sum_{i=1}^{k} \|S_\alpha(t-t_i)\|_{L(X)} \|I_i\left(\bar{u}\left(t_i^-\right)\right) - I_i\left(\bar{u}^*\left(t_i^-\right)\right)\|_X \\
& + \sum_{i=1}^{k} \int_{t_i}^{t} \|S_\alpha(s-t_i)\|_{L(X)} ds \|Q_i\left(\bar{u}\left(t_i^-\right)\right) - Q_i\left(\bar{u}^*\left(t_i^-\right)\right)\|_X \\
& + \int_0^t \|T_\alpha(t-s)\|_{L(X)} \|f\left(s, \bar{u}_{\rho(s,\bar{u}_s)}\right) - f\left(s, \bar{u}^*_{\rho(s,\bar{u}^*_s)}\right)\|_X ds \\
\|Pu - Pu^*\|_{\mathscr{B}''_h} \le & M\left[m\|l_i\|_{L^1(J,\mathbb{R}^+)} + mT\|l_j\|_{L^1(J,\mathbb{R}^+)} + K_b\int_0^T l_f(s)ds\right]\|u - u^*\|_{\mathscr{B}''_h} \\
\le & \Delta\|u - u^*\|_{\mathscr{B}''_h}.
\end{aligned}$$

Since $\Delta < 1$, which implies that P is contraction map. Hence P has a unique fixed point, which is the mild solutions of problem (1)–(3) on J. This completes the proof of the theorem.

4 Application

Consider the following impulsive fractional partial differential equation of the form

$$\frac{\partial^\alpha}{\partial t^\alpha}u(t,x) = \frac{\partial^2}{\partial y^2}u(t,x) + \int_{-\infty}^{t} e^{2(s-t)}\frac{u(s - \rho_1(s)\rho_2(\|u\|), x)}{81}ds, t \ne \frac{1}{2}, \quad (17)$$
$$u(t,0) = u(t,\pi) = 0; u'(t,0) = u'(t,\pi) = 0 \quad t \ge 0, \quad (18)$$
$$u(t,x) = \phi(t,x), u'(t,x) = 0, t \in (-\infty, 0], x \in [0,\pi], \quad (19)$$
$$\Delta u|_{t=\frac{1}{2}} = \frac{\|u\left(\frac{1^-}{2}\right)\|}{36 + \|u\left(\frac{1^-}{2}\right)\|}, \quad \Delta u'|_{t=\frac{1}{2}} = \frac{\|u\left(\frac{1^-}{2}\right)\|}{49 + \|u\left(\frac{1^-}{2}\right)\|}, \quad (20)$$

where $\frac{\partial^\alpha}{\partial t^\alpha}$ is Caputo's fractional derivative of order $\alpha \in (1,2)$, $0 < t_1 < t_2 < \cdots < t_n < T$ are prefixed numbers and $\phi \in \mathscr{B}_h$. Let $X = L^2[0,\pi]$ and define the operator $A : D(A) \subset X \to X$ by $Aw = w''$ with the domain $D(A) := \{w \in X : w, w'$ are absolutely continuous, $w'' \in X, w(0) = 0 = w(\pi)\}$. Then

$$Aw = \sum_{n=1}^{\infty} n^2 (w, w_n) w_n, \quad w \in D(A),$$

where $w_n(x) = \sqrt{\frac{2}{\pi}} \sin(nx), \ n \in \mathrm{N}$ is the orthogonal set of eigenvectors of A. It is well known that A is the infinitesimal generator of an analytic semigroup $\{T(t)\}_{t \geq 0}$ in X given by

$$T(t)\omega = \sum_{n=1}^{\infty} e^{-n^2 t} (\omega, \omega_n) \omega_n, \ \text{ for all } \omega \in X, \ \text{ and every } t > 0.$$

By subordination principle of solution operator, we have $\|S_\alpha(t)\|_{L(X)} \leq M$ for $t \in J$. Let $h(s) = e^{2s}, \ s < 0$ then $l = \int_{-\infty}^{0} h(s) ds = \frac{1}{2} < \infty$, for $t \in (-\infty, 0]$ and define

$$\|\phi\|_{\mathscr{B}_h} = \int_{-\infty}^{0} h(s) \sup_{\theta \in [s,0]} \|\phi(\theta)\|_{L^2} ds.$$

Hence for $(t, \phi) \in [0, 1] \times \mathscr{B}_h$, where $\phi(\theta)(x) = \phi(\theta, x), \ (\theta, x) \in (-\infty, 0] \times [0, \pi]$. We assume that $\rho_i : [0, \infty) \to [0, \infty), \ i = 1, 2$, are continuous functions.

Set $u(t)(x) = u(t, x)$, and $\rho(t, \phi) = \rho_1(t)\rho_2(\|\phi(0)\|)$, we have

$$f(t, \phi)(x) = \frac{\phi}{81}, \ I_k(u) = \frac{\|u\|}{36 + \|u\|}, \ J_k(u) = \frac{\|u\|}{49 + \|u\|},$$

then with these settings the problem (17)–(20) can be written in the abstract form of Eqs. (1)–(3). It is obvious that the maps f, I_k, J_k following the assumption H_1, H_2. This implies that there exists a unique mild solutions of problem (17)–(20) on J.

References

1. Abbas, S., Benchohra, M.: Impulsive partial hyperbolic functional differential equations of fractional order with state-dependent delay. An Int. J. Theory Appl. (2010). ISSN 1311–0454
2. Agarwal, R.P., Andrade, B.D.: On fractional integro-differential equations with state-dependent delay. Comput. Math. Appl. **62**, 1143–1149 (2011)
3. Araya, D., Lizama, C.: Almost automorphic mild solutions to fractional differential equations. Nonlinear Anal. TMA **69**, 3692–3705 (2008)
4. Benchohra, M., Litimein, S., N'Guerekata, G.: On fractional integro-differential inclusions with state-dependent delay in Banach spaces. Appl. Anal. **2011**, 1–16 (2011)
5. Benchohra, M., Berhoun, F.: Impulsive fractional differential equations with state dependent delay. Commun. Appl. Anal. **14**(2), 213–224 (2010)
6. Carvalho dos Santos, J.P., Arjunan, M.M.: Existence results for fractional neutral integro-differential equations with state-dependent delay. Comp. and Math. with Appl. **62**, 1275–1283 (2011)

7. Chauhan, A., Dabas, J., Kumar, M.: Integral boundary-value problem for impulsive fractional functional integro-differential equation with infinite delay. Electr. J. Diff. Eqn. **2012**(229), 1–13 (2012)
8. Chauhan, A., Dabas, J.: Existence of mild solutions for impulsive fractional order semilinear evolution equations with nonlocal conditions. Electr. J. Diff. Eqn. **2011**(107), 1–10 (2011)
9. Chauhan, A., Dabas, J.: Local and global existence of mild solution to an impulsive fractional functional integro-differential equation with nonlocal condition. Commun. Nonlinear Sci. Num. Simul. **19**, 821–829 (2014)
10. Dabas, J., Chauhan, A.: Existence and uniqueness of mild solution for an impulsive neutral fractional integro-differential equations with infinity delay. Math. Comput. Mod.**57**, 754–763 (2013)
11. Dabas, J., Gautam, G.R.: Impulsive neutral fractional integro-differential equation with state dependent delay and integral boundary condition. Electr. J. Diff. Eqn. **2013**(273), 1–13 (2013)
12. Feckan, M., Zhou, Y., Wang, J.: On the concept and existence of solution for impulsive fractional differential equations. Commun. Nonlinear Sci. Numer. Simul. **17**, 3050–3060 (2012)
13. Fu, X., Huang, R.: Existence of solutions for neutral integro-differential equations with state-dependent delay. Appl. Math. Comput. **224**, 743–759 (2013)
14. Hale, J.K., Kato, J.: Phase space for retarded equations with infinite delay. Funkcialaj Ekvacioj **21**, 11–41 (1978)
15. Kilbas, A.A., Srivastava, H.M., Trujillo, J.J.: Theory and Applications of Fractional Differential Equations, vol. 204. North-Holland Mathematics Studies, Elsevier Science, Amsterdam (2006)
16. Lakshmikantham, V., Leela, V.S., Vasundhara Devi, J.: Theory of Fractional Dynamic Systems. Cambridge Scientific Publishers, Cambridge (2009)
17. Mainardi, F., Gorenflo, R.: On Mittag-Leffer-type functions in fractional evolution processes. J. Comput.Appl. Math. **118**, 283–299 (2000)
18. Miller, K.S., Ross, B.: An Introduction to the Fractional Calculus and Differential Equations. Wiley, New York (1993)
19. Podlubny, I.: Fractional Differential Equation. Academic Press, San Diego (1999)
20. Samko, S.G., Kilbas, A.A., Marichev, O.I.: Fractional Integrals and Derivatives Theory and Applications. Gordon and Breach, Yverdon (1993)
21. Shu, X., Wang, Q.: The existence and uniqueness of mild solutions for fractional differential equations with nonlocal conditions of order $1 < \alpha < 2$. Comput. Math. Appl. **64**, 2100–2110 (2012)
22. Wang, J., Fekckan, M., Zhou, Y.: On the new concept of solutions and existence results for mpulsive fractional evolution equations. Dyn. PDE **8**(4), 345–361 (2011)
23. Wang, J., Li, X., Wei, W.: On the natural solution of an impulsive fractional differential equation of order $q \in (1, 2)$. Commun. Nonlinear. Sci. Numer. Simul. **17**, 4384–4394 (2012)

An $O(N^{-1}lnN)^4$ Parameter Uniform Difference Method for Singularly Perturbed Differential-Difference Equations

Komal Bansal and Kapil K. Sharma

Abstract In this paper, we propose an $O(N^{-1}lnN)^4$ parameter uniform numerical scheme for singularly perturbed differential-difference equations (SPDDE) having both delay and advance arguments in reaction term. These types of problems are ubiquitous in many mathematical models of physical and biological phenomena. Piecewise uniform fitted mesh with improved fourth-order numerov method is used. A parameter uniform error estimate of order $O(N^{-1}lnN)^4$ is proved. We calculate numerical solution of some examples using the proposed method to establish the higher order parameter uniform estimates.

Keywords Singular perturbation · Parameter uniform error estimate · Differential-difference equations · Piecewise uniform fitted mesh · Numerov method

AMS subject classifications: 34K26 · 65L12 · 34K28

1 Introduction

Mathematical models for many real-life phenomena involve differential equations having nonsmooth solution with singularities related to boundary layer. In 1994, Lange and Miura [7] proposed a mathematical model for first-exit time problem in the modeling of the activation of neuronal variability. The problem for expected first-exit time, given initial membrane potential, can develop as a general boundary-value problem for linear second-order differential-difference equation [11, 12].

K. Bansal
Department of Mathematics, Panjab University, Chandigarh 160014, India
e-mail: bansalkomal87@gmail.com

K.K. Sharma (✉)
Department of Mathematics, South Asian University (SAU), Akbar Bhawan, Chanakyapuri, New Delhi 110021, India
e-mail: kapil.sharma@sau.ac.in

P.N. Agrawal et al. (eds.), *Mathematical Analysis and its Applications*, Springer Proceedings in Mathematics & Statistics 143,
DOI 10.1007/978-81-322-2485-3_24

Consider a linear singularly perturbed differential-difference equation of mixed type

$$\varepsilon y''(x)+\alpha(x)y(x-\delta)+\omega(x)y(x)+\beta(x)y(x+\eta) = f(x), \quad x \in \Omega = (0,1) \quad (1)$$

and $0 < \varepsilon \ll 1$ subject to the interval and boundary conditions

$$y(x) = \phi(x) \quad on \quad -\delta \leq x \leq 0, \quad y(x) = \gamma(x) \quad \text{on} \quad 1 \leq x \leq 1+\eta, \qquad (2)$$

where $\alpha(x)$, $\omega(x)$, $\beta(x)$, f(x), $\phi(x)$, and $\gamma(x)$ are smooth functions, δ and η are the small shifting parameters of $o(\varepsilon)$. It is assumed that $(\alpha(x) + \beta(x) + \omega(x))$ satisfies the condition

$$(\alpha(x) + \beta(x) + \omega(x)) \leq -b^* < 0, \qquad \forall x \in \overline{\Omega} = [0,1] \qquad (3)$$

where b^* is a positive constant. Because of the assumption (3) solution of the boundary-value problem considered here has no oscillation and will exhibit boundary layer behavior.

In 2002, Kadalbajoo and Sharma [4, 6] initiated the numerical study of such type of boundary-interval value problems, since than a lot of work had been carried out. Rao and Chakravarthy [1, 10] proposed some higher order methods but their methods are not parameter uniform. In [5] Kadalbajoo and Sharma established parameter uniform numerical methods but the order of convergence is less than one. In [9], Patidar and Sharma constructed nonstandard finite difference methods (NSFDMs) which are parameter uniform with second order of convergence. In this paper, we develop and analyze a parameter uniform difference method with $O(N^{-1}lnN)^4$ based on numerov finite difference scheme which utilizes a piecewise uniform mesh condensed in the boundary layer regions to capture the singular behavior of the solution in the layer regions.

Notations and Terminology: We used standard notations and symbols. Further, C will denote a positive constant which may take different values in different equations and inequalities and is always independent of ε and step size h. Here $||\,.||_\infty$ represent the standard supremum norm.

2 Some Deducible Estimates

Let us consider Taylor series expansion of the terms $y(x-\delta)$ and $y(x+\eta)$ in (1), we have

$$y(x-\delta) \approx y(x) - \delta y'(x), \qquad (4)$$

$$y(x+\eta) \approx y(x) + \eta y'(x), \qquad (5)$$

using (4)–(5) in (1)–(2), we obtain

$$-\varepsilon y''(x) + A_\varepsilon(x)y'(x) + B(x)y(x) \approx F(x), \tag{6}$$

$$y(0) \approx \phi(0),\ \ y(1) \approx \gamma(1); \tag{7}$$

where $\quad A_\varepsilon(x) \equiv A_{\varepsilon,\delta,\eta}(x) = \alpha(x)\delta - \beta(x)\eta,$
$B(x) = -(\alpha(x) + w(x) + \beta(x))\ \text{ and }\ F(x) = -f(x).$
Since (6)–(7) is an approximation of (1)–(2), it is important to have the knowledge of the error which may occur in neglecting the higher order terms in (4)–(5). As delay (δ) and advance (η) are sufficiently small, as a result the solution u of the new problem (resulting from (6)–(7))

$$-\varepsilon u''(x) + A_\varepsilon(x)u'(x) + B(x)u(x) = F(x), \tag{8}$$

$$u(0) = \phi(0) = \phi_0, \qquad u(1) = \gamma(1) = \gamma_1, \tag{9}$$

which differ from the original problem (1)–(2) by O$(\delta^2 u'', \eta^2 u'')$ terms will give a good approximation to the solution y of problem (1)–(2).

Remark 1 In the differential equations of the type considered in this paper when both the delay and the advance arguments are present, and the fact that δ and η are of the order of ε then to find the layer behavior of the solution, one has to observe how large is the coefficient of convection term as compared to ε, which solely depends on the values of ε. If ε is not very small, say, e.g., $\varepsilon \in [10^{-2}, 1]$, then $A_\varepsilon(x)$ contributes a lot and we have only one boundary layer in the solution. Boundary layer will be on the left- or right side of the domain depends upon the sign of $A_\varepsilon(x)$. When ε is very small then $A_\varepsilon(x)$ does not contribute significantly, and therefore contribution of the coefficient of the reaction term has much influence which results into two boundary layers one at each end.

As we are interested in developing the numerical scheme in the case when ε is very small, therefore, in view of the above remark, we dropout the term $A_\varepsilon(x)$ from (8) and consider the problem

$$\mathcal{L}_* u \equiv -\varepsilon(x)u''(x) + B(x)u(x) = F(x), \tag{10}$$

$$u(0) = \phi_0,\ u(1) = \gamma_1. \tag{11}$$

Operator of the continuous problem (10)–(11) will satisfy continuous maximum principle, this implies that the solution is unique and since the problem under consideration is linear, the existence of the solution is implied by its uniqueness.

Lemma 1 *The solution $u_\varepsilon(x)$ of the constant coefficient problem corresponding to (10)–(11) satisfies*

$$|u_\varepsilon| \leq C(1 + S(x, b^*))$$

where

$$S(x, b^*) = \exp\left(-x\sqrt{b^*/\varepsilon}\right) + \exp\left(-(1-x)\sqrt{b^*/\varepsilon}\right)$$

and

$$|u_\varepsilon^{(k)}| \leq C\varepsilon^{-(k/2)} S(x, b^*) \quad \forall\ k \geq 1$$

Lemma 2 *The solution $u(x)$ of (10)–(11) has the decomposition*

$$u(x) := v_\varepsilon(x) + w_L(x) + w_R(x)$$

where the regular(smooth) component $v_\varepsilon(x)$ satisfies

$$|v_\varepsilon(x)| \leq C(1 + S(x, b^*))$$

$$\left|v_\varepsilon^{(k)}(x)\right| \leq C\left[1 + \varepsilon^{-(k-2)/2} S(x, b^*)\right], \quad \forall k \geq 1.$$

The singular components w_L, w_R satisfy

$$\left|w_L^{(k)}(x)\right| \leq C\varepsilon^{-(k/2)} \exp\left(-x\sqrt{b^*/\varepsilon}\right), \quad \forall k \geq 0$$

$$\left|w_R^{(k)}(x)\right| \leq C\varepsilon^{-(k/2)} \exp\left(-(1-x)\sqrt{b^*/\varepsilon}\right), \quad \forall k \geq 0$$

Note: For the proofs of Lemma 1 and Lemma 2, the reader can refer to [5, 9].

Remark 2 Lemma 2 and the fact that the shift arguments are sufficiently small implies that the Eq. (8) is a good approximation to Eq. (1). This is because of the fact that

$$u^{''}(\xi) \leq C\left[1 + \frac{S(\xi, b^*)}{\varepsilon}\right]$$

with $\delta = o(\varepsilon)$, $\eta = o(\varepsilon)$ and therefore

$$\delta^2 u^{''} \approx C\left[\delta^2 + \delta^2 \frac{S(\xi, b^*)}{\varepsilon}\right].$$

Similarly, we can approximate $\eta^2 u^{''}(\xi)$ and $S(\xi, b^*) \to 0$ as $\varepsilon \to 0$ which leads to the fact that the terms $\delta^2 u^{''}$, $\eta^2 u^{''}$ are exponentially small. By similar arguents, one can show that higher order terms of Taylor series (4) and (5) are exponentially small.

3 The Discrete Problem

The fitted piecewise uniform mesh on the interval (0,1) is constructed by partitioning the interval in to three subintervals (0,ν), (ν, 1 − ν), and (1 − ν,1). Assuming that N= 2^l with $l \geq 3$ the intervals (0,ν) and (1 − ν,1) are each divided in to N/4 equal mesh elements while the interval (ν,1 − ν) is divided into N/2 equal mesh elements which guarantee that there is at least one point in the boundary layer region. The resulting piecewise uniform mesh Ω_ν^N depends on just one parameter ν where the transition parameter ν is defined as

$$\nu = \min\left(1/4, \frac{\sqrt{\varepsilon}}{b^*}\ln(N)\right). \tag{12}$$

We consider the following differential equation

$$\varepsilon U''(x) + (\beta(x)\eta - \alpha(x)\delta)U'(x) + (\alpha(x) + \omega(x) + \beta(x))U(x) = f(x), \quad x \in \Omega = (0, 1) \tag{13}$$

$$\varepsilon U''(x) = f(x) - (\beta(x)\eta - \alpha(x)\delta)U'(x) - (\alpha(x) + \omega(x) + \beta(x))U(x), \tag{14}$$

$$\varepsilon U''(x) = g(x, U, U'). \tag{15}$$

Now we consider fourth-order numerov method and fitted mesh [3, 8] to solve the Eq. (14) and this equation is approximated by the following scheme

$$\frac{2\varepsilon}{h_{i+1} + h_i}\left(\frac{U_{i+1} - U_i}{h_{i+1}} - \frac{U_i - U_{i-1}}{h_i}\right) = \frac{1}{12}[\overline{g_{i+1}} + 10\hat{g}_i + \overline{g_{i-1}}] \tag{16}$$

where $\hat{g}_i = g(x_i, U_i, \hat{U_i'})$

$$\overline{g_{i\pm1}} = g(x_{i\pm1}, U_{i\pm1}, \overline{U'_{i\pm1}}), \tag{17}$$

$$\overline{U'_i} = \frac{U_{i+1} - U_{i-1}}{h_{i+1} + hi}, \tag{18}$$

$$\overline{U'_{i+1}} = \frac{U_{i+1} - U_{i-1}}{h_{i+1} + hi} + \frac{2h_{i+1}}{h_{i+1} + h_i}\left(\frac{U_{i+1} - U_i}{h_{i+1}} - \frac{U_i - U_{i-1}}{h_i}\right), \tag{19}$$

$$\overline{U'_{i-1}} = \frac{U_{i+1} - U_{i-1}}{h_{i+1} + hi} - \frac{2h_i}{h_{i+1} + h_i}\left(\frac{U_{i+1} - U_i}{h_{i+1}} - \frac{U_i - U_{i-1}}{h_i}\right), \tag{20}$$

$$\hat{U_i'} = \overline{U'_i} - \frac{h_{i+1} + hi}{40}(\overline{g_{i+1}} - \overline{g_{i-1}}), \tag{21}$$

From the proposed scheme (16), we get the corresponding discrete operator L_ε^N.

Lemma 3 *Assume that the mesh function χ_i satisfies $\chi_0 \geq 0$, $\chi_N \geq 0$. Then for the discrete operator L_ε^N, if $L_\varepsilon^N \chi_i \geq 0$ for $1 \leq i \leq N-1$ implies that $\chi_i \geq 0$ for all $0 \leq i \leq N$.*

Lemma 4 *If Z_i is any mesh function such that $Z_0 = Z_n = 0$. Then*

$$|Z_i| \leq \frac{1}{b^*} \max_{1 \leq j \leq N-1} \left| \mathcal{L}_\varepsilon^N Z_j \right|, \quad \forall\ 0 \leq i \leq N.$$

Theorem 1 *The fitted mesh numerov method (16) with the fourth-order numerov method and the piecewise uniform fitted mesh Ω_ν^N, condensing at the boundary points x=0 and x=1, is ε-uniform for the problem (1)–(2) provided that ν is chosen to satisfy condition (12) above. Moreover, the solution u_ε of (1)–(2) and the solution U_ε of (15) satisfy the following ε-uniform error estimate*

$$\sup_{0<\varepsilon\leq 1} ||U_\varepsilon - u_\varepsilon||_{\overline{\Omega}_\nu^N} \leq CN^{-4}(lnN)^4$$

where C is constant independent of ε and mesh size h.
Solution: The solution U_ε of the discrete problem are decomposed in a similar manner to the decomposition of the solution u_ε of (1)–(2). Thus

$$U_\varepsilon = V_\varepsilon + W_\varepsilon, \tag{22}$$

where V_ε is the solution of the following nonhomogeneous problem

$$L_\varepsilon^N V_\varepsilon = f, \qquad V_\varepsilon(0) = v(0), \quad V_\varepsilon(1) = v(1) \tag{23}$$

and W_ε is the solution of the following homogeneous problem

$$L_\varepsilon^N W_\varepsilon = 0, \qquad W_\varepsilon(0) = w(0), \quad W_\varepsilon(1) = w(1). \tag{24}$$

We can write the error in the form

$$U_\varepsilon - u_\varepsilon = (V_\varepsilon - v_\varepsilon) + (W_\varepsilon - w_\varepsilon) \tag{25}$$

Now the singular and smooth components of the error can be calculated separately.

We use the following classical argument [3, 8] to estimate the smooth component. From the differential and difference equations

$$L_\varepsilon^N (V_\varepsilon - v_\varepsilon)(x_i) = f - L_\varepsilon^N v_\varepsilon \tag{26}$$

$$= (L_\varepsilon - L_\varepsilon^N) v_\varepsilon \tag{27}$$

$$= \frac{h^4}{240} v_\varepsilon^6(\xi)\varepsilon. \tag{28}$$

Now by using the estimate of $v_\varepsilon^{(6)}$ and the fact that $\frac{1}{4} \leq \frac{\sqrt{\varepsilon}}{\beta}\ln(N)$ we get

$$\left|L_\varepsilon^N(V_\varepsilon - v_\varepsilon)(x_i)\right| \leq C\,(\ln N)^2 N^{-4}. \tag{29}$$

By using Lemma 4 we get

$$|(V_\varepsilon - v_\varepsilon)(x_i)| \leq C\,(\ln N)^2 N^{-4}. \tag{30}$$

Estimates of the singular component of the error depends upon whether $\nu = \frac{1}{4}$ or $\nu = \frac{\sqrt{\varepsilon}}{\beta}\ln(N)$.

Case 1:
In the case when $\nu = \frac{1}{4}$ the mesh is uniform and $\frac{1}{4} \leq \frac{\sqrt{\varepsilon}}{\beta}\ln(N)$. By using the same classical argument as for $V_\varepsilon - v_\varepsilon$ estimates of $w_\varepsilon^{(6)}$ and the fact that $\frac{1}{4} \leq \frac{\sqrt{\varepsilon}}{\beta}\ln(N)$ leads to

$$\left|L_\varepsilon^N(W_\varepsilon - w_\varepsilon)(x_i)\right| \leq C\,(\ln N)^4 N^{-4}. \tag{31}$$

Case 2:
In the case when $\frac{1}{4} > \frac{\sqrt{\varepsilon}}{\beta}\ln(N)$. In this case, mesh will be piecewise uniform with the mesh spacing $\frac{2(1-2\nu)}{N}$ in the subinterval $[\nu, 1-\nu]$ and $\frac{4\nu}{N}$ in each of the subintervals $[0, \nu]$ and $[1-\nu, 1]$

Subcase 1:
Estimates of the error in $(0, \nu)$ and $(1-\nu, 1)$

$$L_\varepsilon^N(W_\varepsilon - w_\varepsilon)(x_i) = (L_\varepsilon - L_\varepsilon^N)w_\varepsilon \tag{32}$$

$$= \frac{h^4}{240} w_\varepsilon^6(\xi)\varepsilon. \tag{33}$$

Now by using the estimate of $w_\varepsilon^{(6)}$ and the fact that step size i.e. $h = \frac{4\nu}{N}$ and $\nu = \frac{\sqrt{\varepsilon}}{b^*}\ln(N)$ we get

$$\left|L_\varepsilon^N(W_\varepsilon - w_\varepsilon)(x_i)\right| \leq C\,(lnN)^4 N^{-4}. \tag{34}$$

Subcase 2:
Estimates of the error in $(\nu, 1-\nu)$

$$L_\varepsilon^N(W_\varepsilon - w_\varepsilon)(x_i) = (L_\varepsilon - L_\varepsilon^N)w_\varepsilon \tag{35}$$

$$= \frac{h^4}{240} w_\varepsilon^6(\xi)\varepsilon. \tag{36}$$

In the interval $(\nu, 1-\nu)$ solution is smooth, this implies

$$\left|L_{\varepsilon}^{N}(W_{\varepsilon}-w_{\varepsilon})(x_i)\right| \leq CN^{-4}. \tag{37}$$

Combining (31), (34), (37) gives

$$\left|L_{\varepsilon}^{N}(W_{\varepsilon}-w_{\varepsilon})(x_i)\right| \leq CN^{-4}(lnN)^4, \tag{38}$$

for all $x_i \in (0, 1)$. Applying Lemma 4 to the mesh function $W_{\varepsilon} - w_{\varepsilon}$ leads the required estimate of the error in the singular component of the solution

$$|(W_{\varepsilon}-w_{\varepsilon})(x_i)| \leq C\ (lnN)^4 N^{-4}. \tag{39}$$

This concludes the proof of the theorem.

4 Numerical Experiments

Example 1 Consider the problem

$$\varepsilon y''(x) + 0.25y(x-\delta) - y(x) + 0.25y(x+\eta) = 1, \quad x \in \Omega = (0, 1)$$

subject to the interval and boundary conditions

$$y(x) = 1, -\delta \leq x \leq 0; \qquad y(x) = 0, \ \ 1 \leq x \leq 1+\eta.$$

Example 2 Consider the problem

$$\varepsilon y''(x) - y(x-\delta) + y(x) - 0.5y(x+\eta) = 0, \quad x \in \Omega = (0, 1)$$

subject to the interval and boundary conditions

$$y(x) = 1, -\delta \leq x \leq 0; \qquad y(x) = 1, \ \ 1 \leq x \leq 1+\eta.$$

We use the double-mesh principle [2] to calculate the maximum absolute error and order of convergence of the numerical scheme. The following estimates for maximum point wise error E_{ε}^{N} and order of convergence P_{ε}^{n} are computed using the double-mesh principle

$$E_{\varepsilon}^{N} = \max_{0\leq i\leq n-1}\left|Y_i^N - Y_{2i}^{2N}\right|, \quad P_{\varepsilon}^{n} = \frac{(E_{\varepsilon}^{N}/E_{\varepsilon}^{2N})}{log2} \tag{40}$$

The calculated maximum absolute error and order of convergence for Examples 1 and 2 are arranged in the form of Tables 1 and 2 repectively.

Table 1 Maximum pointwise error E_ε^N and numerical order of convergence P_ε^N for each value of ε and N are given, respectively, for example 1 for $\delta = 0.0001$, $\eta = 0.0001$

ε	$N = 8$	$N = 16$	$N = 32$	$N = 64$
2^{-0}	9.764395E-008	6.105717E-009	3.816677E-010	2.390149E-011
	3.999	3.999	3.997	
2^{-2}	3.755846E-006	2.351924E-007	1.470671E-008	9.194143E-010
	3.997	3.999	3.999	
2^{-4}	6.193481E-005	3.899850E-006	2.442115E-007	1.527083E-008
	3.989	3.997	3.999	
2^{-6}	7.038173E-004	4.509816E-005	2.839569E-006	1.778154E-007
	3.964	3.989	3.997	
2^{-8}	7.773966E-003	6.939454E-004	4.446212E-005	2.799506E-006
	3.485	3.964	3.989	
2^{-10}	8.666641E-003	2.138141E-003	3.463243E-004	4.446165E-005
	2.019	2.626	2.961	
2^{-12}	9.412269E-003	2.150507E-00	3.467433E-004	4.624679E-005
	2.129	2.632	2.906	
2^{-14}	9.194118E-003	2.150605E-003	3.470122E-004	6.159023E-005
	2.095	2.631	2.494	
2^{-16}	8.772140E-003	2.146675E-003	3.469423E-004	8.252287E-005
	2.030	2.629	2.071	
2^{-18}	8.635118E-003	2.143429E-003	3.467774E-004	7.052199E-005
	2.010	2.627	2.297	
2^{-20}	8.608975E-003	2.141461E-003	3.466508E-004	5.029886E-005
	2.007	2.627	2.784	

Table 2 Maximum pointwise error E_ε^N and numerical order of convergence P_ε^N for each value of ε and N are given, respectively, for example 2 for $\delta = 0.00005$, $\eta = 0.0001$

ε	$N = 8$	$N = 16$	$N = 32$	$N = 64$
2^{-0}	1.345542e-008	8.412097e-010	5.263479e-011	3.377632e-012
	3.999	3.998	3.961	
2^{-4}	1.751296e-005	1.098727e-006	6.873708e-008	4.297450e-009
	3.994	3.998	3.999	
2^{-8}	2.535968e-003	1.671284e-004	1.115936e-005	7.001167e-007
	3.923	3.904	3.994	
2^{-12}	9.214721e-003	2.089910e-003	3.358351e-004	4.485627e-005
	2.140	2.637	2.904	
2^{-16}	9.134400e-003	2.092756e-003	3.619246e-004	8.178389e-005
	2.125	2.531	2.145	
2^{-20}	8.521945e-003	2.085891e-003	3.360410e-004	6.383956e-005
	2.030	2.633	2.396	

5 Conclusion

We construct a numerical scheme using shishkin mesh technique with the improved numerov method to solve singularly perturbed problem having both delay and advance arguments in reaction terms. We used priori estimates on the solution and its derivatives to prove a parameter uniform error estimate. We divide the domain in to two regions, namely interior and outer regions, which are dealt separately. We made a matlab program of the proposed numerical scheme to validate the computational efficiency, consistency, and stability of the scheme.

References

1. Chakravarthy, P.: Pramod, Rao, R. Nageshwar: A modified Numerov method for solving singularly perturbed differential-difference equations arising in science and engineering. Results Phys. **2**, 100–103 (2012)
2. Doolan, E.P., Miller, J.J.H., Schilders, W.H.A.: Uniform Numerical Methods for Problems with Initial and Boundary Layers. Boole Press, Dublin (1980)
3. Jain, M.K., Lynger, S.R.K., Jain, R.K.: Numerical Methods for Scientific and Engineering Computation. New Age International Publishers, New Delhi (1984)
4. Kadalbajoo, M.K., Sharma, K.K.: Numerical analysis of boundary-value problems for singularly perturbed differential-difference equations with small shifts of mixed type. J. Optim. Theory Appl. **115**, 145–163 (2002)
5. Kadalbajoo, M.K., Sharma, K.K.: ε-Uniform fitted mesh method for singularly perturbed differential-difference equations: Mixed type of shifts with layer behavior. Int. J. Comput. Math. **81**, 49–62 (2004)
6. Kadalbajoo, M.K., Sharma, K.K.: Numerical treatment of boundary value problems for second order singularly perturbed delay differential equations. Comput. Appl. Math. **24**, 151–172 (2005)
7. Lange, C.G., Miura, R.M.: Singular perturbation analysis of boundary-value problems for differential-difference equations. V. Small shifts with layer behavior. SIAM J. Appl. Math. **54**, 72–249 (1994)
8. Miller, J.J.H., Riordan, E.O'., Shishkin, G.I.: Fitted numerical methods for singular perturbation problems. In: Error estimates in the maximum norm for linear problems in one and two dimensions, World Scientific Publishing Co., Inc, River Edge, NJ (1996)
9. Patidar, Kailash C., Sharma, Kapil K.: ε-Uniformly convergent non-standard finite difference methods for singularly perturbed differential-difference equations with delay and advance. Int. J. Numer. Methods Eng. **66**, 272–296 (2006)
10. Rao, R.: Nageshwar, Chakravarthy, P. Pramod: A finite difference method for singularly perturbed differential-difference equations arising from a model of neuronal variability. J. Taibah Univ. Sci. **7**, 128–136 (2013)
11. Stein, R.B.: A theoretical analysis of neuronal variability. Biophys. J. **5**, 173–194 (1965)
12. Stein, R.B.: Some models of neuronal variability. Biophys. J. **7**, 37–68 (1967)

Almost Periodicity of a Modified Leslie–Gower Predator–Prey System with Crowley–Martin Functional Response

Jai Prakash Tripathi and Syed Abbas

Abstract In this paper, we discuss a modified Leslie–Gower Lotka–Volterra system with Crowley–Martin type functional response. Crowley–Martin functional response is similar to the Beddington–DeAngelis functional response but contains an extra term describing mutual interference by predators at high prey density. The rates are assumed to be almost periodic, which generalizes the concept of periodicity. We discuss the permanence, existence, uniqueness, and asymptotic stability of an almost periodic solution of the model under consideration by applying comparison theorem of differential equations and constructing a suitable Lyapunov functional. The analytical results obtained in this paper are illustrated with the help of a numerical example.

Keywords Almost-periodicity · Asymptotic stability · Crowley-Martin · Permanence · Predator-prey · Lyapunov functional

1 Introduction

The rate of prey consumption by an average predator per unit time, i.e., predator's functional response is one of important feature of predator–prey relationship [1, 2]. The Crowley–Martin [3, 4] type functional response is given by, $p(x, y) = \dfrac{aX}{1 + bX + cY + bcXY}$. The parameters a and b stand for the effects of capture rate and handling time, respectively. Parameter c describes the magnitude of interference among predators. Crowley–Martin functional response predicts that interference affects on feeding rate remain important at high prey abundance.

J.P. Tripathi (✉) · S. Abbas
School of Basic Sciences, Indian Institute of Technology Mandi,
Mandi 175001, H.P, India
e-mail: jtripathi85@gmail.com

S. Abbas
e-mail: sabbas.iitk@gmail.com

P.N. Agrawal et al. (eds.), *Mathematical Analysis and its Applications*,
Springer Proceedings in Mathematics & Statistics 143,
DOI 10.1007/978-81-322-2485-3_25

A model must becomes nonautonomous when the temporal inhomogeneity of the environment is taken into account in the model. The consideration of periodic and almost periodic coefficients can be found in case of nonautonomous systems. A function $g : R \to R$ is called almost periodic if $g(x+\tau) = g(x)$ is satisfied with an arbitrary degree of accuracy by infinitely many values of τ, those values being spread over the whole range from $-\infty$ to $+\infty$ in such a way as not to leave empty intervals of arbitrary great length.

Almost periodic solutions of ecological models have received significant attention from researchers during last few decades [5]. Recently, Abbas et al. [6] discussed the existence of a unique almost periodic solution of a delayed phytoplankton model. In theoretical ecology, there are several papers [5, 7, 8] on permanence and almost periodic solutions of Lotka–Volterra system. However, no work has been done for modified Leslie–Gower type predator–prey model with Crowley–Martin functional response.

Motivated by above, in this work, we consider the following nonautonomous modified Leslie–Gower [9] type predator–prey model with Crowley–Martin functional response

$$\frac{dx(t)}{dt} = x(t)\Big(a_1(t) - b_1(t)x(t) - \frac{b_2(t)y(t)}{a(t)+b(t)x(t)+c(t)y(t)+d(t)x(t)y(t)}\Big),$$
$$\frac{dy(t)}{dt} = y(t)\Big(a_2(t) - \frac{e(t)y(t)}{x(t)+k(t)}\Big), \tag{1}$$

where $x(t)$ and $y(t)$ denote the density of prey and predator at time t, respectively; $a_1, a_2, b_1, b_2, a, b, c, d, e, k \in C(R, R^+)$ are nonnegative almost periodic functions of t.

Functions $a_i(t), b_i(t), a(t), b(t), c(t), d(t), e(t), k(t), (i = 1, 2)$ are continuous, bounded by positive numbers. Our main objective is to obtain sufficient conditions for the existence of a unique globally attractive almost periodic solution of the system (1).

2 Boundedness and Permanence

Let $g(t)$ be a continuous and bounded function on R. Let g^l and g^u denote $\inf_{t\in R} g(t)$ and $\sup_{t\in R} g(t)$, respectively. Then we have

$$\min_{i=1,2}\{a_i^l, b_i^l, a^l, b^l, c^l, d^l, e^l, k^l,\}>0,\ \max_{i=1,2}\{a_i^u, b_i^u, a^u, b^u, c^u, d^u, e^u, k^u,\} < \infty.$$

Lemma 1 [10] *If* $p > 0$, $q > 0$ *and* $\frac{du}{dt} \le (\ge) \quad u(t)\Big(q - pu(t)\Big)$, $u(0) > 0$, *then we have*

$$\limsup_{t\to+\infty} u(t) \le \frac{q}{p}\ \Big(\liminf_{t\to+\infty} u(t) \ge \frac{q}{p}\Big).$$

Lemma 2 *The positive cone is positively invariant with respect to the model system (1).*

Proof The proof is similar to the proof given in [1].

Theorem 1 *If* $a_1^l a^l > b_2^u M_2^\varepsilon$, *then*
$\kappa_\varepsilon := \{(x, y) \in R^2 | m_1^\varepsilon \le x \le M_1^\varepsilon, m_2^\varepsilon \le x \le M_2^\varepsilon\}$,
is positively invariant with respect to the model system (1). Here

$$M_1^\varepsilon := \frac{a_1^u}{b_1^l} + \varepsilon, \qquad m_1^\varepsilon := \frac{a_1^l a^l - b_2^u M_2^\varepsilon}{b_1^u a^l} - \varepsilon,$$
$$M_2^\varepsilon := \frac{a_2^u (M_1^\varepsilon + k^u)}{e^l}, \qquad m_2^\varepsilon := \frac{a_2^l (m_1^\varepsilon + k^l)}{e^u}.$$

Proof First equation of the model system (1) gives

$$\frac{dx(t)}{dt} \le x(t)(a_1^u - b_1^l S(t)) \tag{2}$$

Using Lemma 1, Eq. (2) implies that
$\limsup_{t\to\infty} x(t) \le \frac{a_1^u}{b_1^l} \equiv M_1$.
Thus for sufficiently small $\varepsilon > 0$, $\exists$ a positive real number T_1 such that
$x(t) \le M_1 + \varepsilon \equiv M_1^\varepsilon, \qquad \forall \qquad t \ge T_1$.
From the second equation of the model system (1), we obtain

$$\frac{dy(t)}{dt} \le y(t)\Big(a_2^u - \frac{e^l y(t)}{M_1^\varepsilon + k^u}\Big). \tag{3}$$

Again according to Lemma 1 and Eq. (3), one has
$\limsup_{t\to\infty} y(t) \le \frac{a_2^u (M_1^\varepsilon + k^u)}{e^l} \equiv M_2$.
Hence, for the above ε, $\exists$ a positive real number $T_2 \ge T_1 \ge 0$ such that
$y(t) \le \frac{a_2^u (M_1^\varepsilon + k^u)}{e^l} + \varepsilon \equiv M_2^\varepsilon \qquad \forall \qquad t \ge T_2$.
From the first equation of the model system (1), we have

$$\frac{dx(t)}{dt} \le x(t)\Big(a_1^l - b_1^u x(t) - \frac{b_2^u M_2^\varepsilon}{a^l}\Big). \tag{4}$$

Using Lemma 1, for $\varepsilon > 0$ $\exists$ a positive real number $T_3 \ge T_2 \ge 0$ such that
$x(t) \ge \frac{a_1^l a^l - b_2^u M_2^\varepsilon}{b_1^u a^l} - \varepsilon \equiv m_1^\varepsilon$.

In the similar fashion, from the second equation of the model system (1), one can also find that

$$\frac{dy(t)}{dt} \le y(t)\Big(a_2^l - \frac{e^u y(t)}{m_1^\varepsilon + k^l}\Big). \tag{5}$$

Applying the Lemma 1, for $\varepsilon > 0$ $\exists$ a positive real number $T_4 \geq T_3 \geq 0$ such that

$$y(t) \geq \frac{a_2^l(m_1^\varepsilon + k^l)}{e^u} - \varepsilon \equiv m_2^\varepsilon.$$

Thus we arrive at the following result:

Theorem 2 *If $a_1^l a^l > b_2^u M_2^\varepsilon$, then the model system (1) is permanent and the set κ_ε with $\varepsilon > 0$ is an ultimately bounded region of the model system (1).*

Let (S) be the collection of all solutions $X(t) = (x(t), y(t))^T$ of (1) on R.

Theorem 3 *(S) is nonempty.*

Proof Proof is similar to the proof given in [8].

3 Existence of a Unique Almost Periodic Solution

Definition 1 [7] A function $g(t, x)$, where g is an m-vector, t is a real scalar and x is an $n-$ vector, is said to be almost periodic in t uniformly with respect to $x \in X \subset R^n$, if $g(t, x)$ is continuous in $t \in R$ and $x \in X$, and if for any $\varepsilon > 0$, it is possible to find a constant $l(\varepsilon) > 0$ such that in any interval of length $l(\varepsilon)$ there exist a τ such that the inequality

$$||g(t+\tau, x) - g(t, x)|| = \sum_{i=1}^{m} |g_i(t+\tau, x) - g_i(t, x)| < \varepsilon$$

is satisfied for all $t \in R$, $x \in X$. The number τ is called an ε-translation number of $g(t, x)$.

Definition 2 [11] A bounded positive solution $X(t) = (\hat{x}(t), \hat{y}(t))$ of the model system (1) with $X(0) > 0$ is said to be globally attractive (globally asymptotically stable), if any other solution $Y(t) = (x(t), y(t))$ of the system (1) with $Y(0) > 0$, satisfies $\lim_{t\to+\infty} |X(t) - Y(t)| = 0$

Definition 3 [8] The upper right Dini derivative for a function $G : R \to R$ is defined as

$$D^+G(t) = \limsup_{h\to 0^+} \frac{G(t+h) - G(t)}{h}$$

Theorem 4 *If the condition of Theorem 2 and if*

$$b_1^l > \frac{b_2^u M_2(b^u + d^u M_2)}{(a^l + b^l m_1 + c^l m_2 + d^l m_1 m_2)^2} + \frac{e^u M_2}{(m_1 + k^l)^2}$$

$$\frac{e^l}{M_1 + k^u} > \frac{b_2^u}{(a^l + b^l m_1 + c^l m_2 + d^l m_1 m_2)} + \frac{b_2^u M_2(c^u + d^u M_2)}{(a^l + b^l m_1 + c^l m_2 + d^l m_1 m_2)^2}$$

hold. Then any two positive solutions $X(t) = (x(t), y(t))$ *and* $Y^*(t) = (x^*(t), y*(t))$ *of the model system (1) satisfies*
$\lim_{t\to\infty} |X(t) - Y^*(t)| = 0$

Proof Consider any two positive solutions $X(t) = (x(t), y(t))^T$ and $Y^*(t) = (x^*(t), y^*(t))$ of the model system (1). Theorem 2 gives that for an enough small $\varepsilon > 0 \,\exists$ a $T > 0$ such that

$$\begin{aligned} s_1 - \varepsilon < x(t) < S_1 + \varepsilon, &\qquad s_2 - \varepsilon < y(t) < S_2 + \varepsilon,\\ s_1 - \varepsilon < x^*(t) < S_1 + \varepsilon, &\qquad s_2 - \varepsilon < y^*(t) < S_2 + \varepsilon, \end{aligned} \tag{6}$$

for all $t \geq T$. Define $\Delta(t, x(t), y(t)) = a(t) + b(t)x(t) + c(t)y(t) + d(t)x(t)y(t)$.
Let $S_1(t) = |\ln x(t) - \ln x^*(t)|$.
The Dini derivative of $S_1(t)$ along the solution of (1) gives

$$\begin{aligned} D^+G_1(t) &= sgn(x(t) - x^*(t))\Big(\frac{\dot{x}(t)}{x(t)} - \frac{\dot{x}^*(t)}{x^*(t)}\Big)\\ &= sgn(x(t) - x^*(t))\Big[- b_1(t)(x(t) - x^*(t) - b_2(t)\Big(\frac{y(t)}{\Delta(t, x(t), y(t))}\\ &\quad - \frac{y^*(t)}{\Delta(t, x^*(t), y^*(t))}\Big)\Big]\\ &= -b_1(t)|x(t) - x^*(t)| - sgn(x(t) - x^*(t))b_2(t)\Big(\frac{y(t)}{\Delta(t, x(t), y(t))} - \frac{y^*(t)}{\Delta(t, x(t), y(t))}\\ &\quad + \frac{y^*(t)}{\Delta(t, x(t), y(t))} - \frac{y^*(t)}{\Delta(t, x^*(t), y^*(t))}\Big)\\ &\leq -b_1(t)|x(t) - x^*(t) + b_2(t)\left(\frac{|y(t) - y^*(t)|}{\Delta(t, x(t), y(t))}\right)\\ &\quad + \frac{y^*(t)\big[(b(t) + d(t)y^*(t))|x(t) - x^*(t)| + (c(t) + d(t)x(t))|y(t) - y^*(t)|\big]}{\Delta(t, x^*(t), y^*(t)).\Delta(t, x(t), y(t))} \end{aligned}$$

Furthermore, consider $S_2(t) = |\ln y(t) - \ln y^*(t)|$.
The upper right derivative of $S_2(t)$ is given by

$$\begin{aligned} D^+G_2(t) &= sgn(y(t) - y^*(t))\Big(\frac{\dot{y}(t)}{y(t)} - \frac{\dot{y}^*(t)}{y^*(t)}\Big)\\ &= sgn(y(t) - y^*(t))e(t)\Big(\frac{y^*(t)}{x^*(t) + k(t)} - \frac{y(t)}{x(t) + k(t)}\Big)\\ &= sgn(y(t) - y^*(t))e(t)\Big(\frac{y^*(t)(x(t) - x^*(t))}{(x^*(t) + k(t))(x(t) + k(t))} + \frac{y^*(t) - y(t)}{x(t) + k(t)}\Big)\\ &\leq \frac{e(t)y^*(t)|x(t) - x^*(t)|}{(x^*(t) + k(t))(x(t) + k(t))} - e(t)\frac{|y(t) - y^*(t)|}{x(t) + k(t)} \end{aligned}$$

Combining the two functions $G_i(t)$, $i = 1, 2$, we obtain $G(t) = G_1(t) + G_2(t)$. For $t \geq T$, we have

$$D^+G(t) \le -\Big[b_1(t) - \frac{b_2(t)(S_2+\varepsilon)\big(b(t)+d(t)(S_2+\varepsilon)\big)}{\Delta^2(t,s_1-\varepsilon,m-\varepsilon)} - \frac{e(t)(S_2+\varepsilon)}{(s_1-\varepsilon+k(t))^2}\Big]|x(t)-x^*(t)|$$
$$-\Big[\frac{e(t)}{S_1+\varepsilon+k(t)} - \frac{b_2(t)}{\Delta(t,s_1-\varepsilon,m-\varepsilon)} - \frac{b_2(t)(S_2+\varepsilon)\big(e(t)+d(t)(S_2+\varepsilon)\big)}{\Delta^2(t,s_1-\varepsilon,m-\varepsilon)}\Big]$$
$$|y(t)-y^*(t)|.$$

Define $\rho = \min\Big\{b_1^l - \frac{b_2^u S_2(b^u+d^u S_2)}{(a^l+b^l s_1+c^l s_2+d^l s_1 s_2)^2} - \frac{e^u S_2}{(s_1+k^l)^2}, \frac{e^l}{S_1+k^u} - \frac{b_2^u}{(a^l+b^l s_1+c^l s_2+d^l s_1 s_2)} - \frac{b_2^u S_2(c^u+d^u S_2)}{(a^l+b^l s_1+c^l s_2+d^l s_1 s_2)^2}\Big\}$. Choosing $\varepsilon \to 0$, the above inequality takes the following form:

$$D^+G(t) \le -\rho\big[|x(t)-x^*(t)|+|y(t)-y^*(t)|\big]. \tag{7}$$

Integrating the above relation (7) from T to t, we obtain

$$G(t)+\rho\int_T^t \Big[|x(s)-x^*(s)|+|y(s)-y^*(s)|\Big]ds < G(t) < +\infty,$$

which gives

$$\limsup_{t\to\infty}\int_T^t \Big[|x(s)-x^*(s)|+|y(s)-y^*(s)|\Big]ds < \frac{G(t)}{\rho} < +\infty.$$

One can easily observe that $|x(t)-x^*(t)|$ and $|y(t)-y^*(t)|$ are uniformly continuous on $[T,+\infty)$. Thus we have

$$\lim_{t\to\infty}|x(t)-x^*(t)| = 0, \ \lim_{t\to\infty}|y(t)-y^*(t)| = 0.$$

Hence the solution of the model system (1) is globally attractive.

Theorem 5 *Under the conditions of the Theorem 4, the model system (1) has a unique almost periodic solution.*

Proof Theorem 3, implies that ∃ a bounded positive solution $u(t) = (u_1(t), u_2(t))$ of the model system (1). Thus ∃ a sequence $\{t_n\}$, $t_n \to \infty$ as $n \to \infty$, such that $(u_1(t+t_n), u_2(t+t_n))^T$ satisfies

$$\frac{dx(t)}{dt} = x(t)\Big(a_1(t+t_n) - b_1(t+t_n)x(t) - \frac{b_2(t+t_n)y(t)}{a(t+t_n)+b(t+t_n)x(t)+c(t+t_n)y(t)+d(t+t_n)x(t)y(t)}\Big),$$
$$\frac{dy(t)}{dt} = y(t)\Big(a_2(t+t_n) - \frac{e(t+t_n)y(t)}{x(t)} + k(t+t_n)\Big). \tag{8}$$

Thus $\{u_i(t+t_n)\}$ for $i = 1, 2$ and $\{\dot{u_i}(t+t_n)\}$ for $i = 1, 2$ are uniformly bounded and equicontinuous. Hence the Ascoli's theorem implies that $\exists$ a uniformly convergent subsequence $\{u_i(t+t_m)\} \subset \{u_i(t+t_n)\}$ and for any $\varepsilon > 0$ there exist $k(\varepsilon) > 0$ such that

$$|u_i(t+t_k) - u_i(t+t_m)| < \varepsilon, \qquad i = 1, 2. \tag{9}$$

Thus one can deduce that $u_i(t)$ for $i = 1, 2$ are asymptotically almost periodic. Hence $\{u_i(t+t_m)\}$ can be written as sum of an almost periodic function $u_{i_1}(t+t_m)$ and a continuous function $u_{i_2}(t+t_m)$ $(i = 1, 2)$ defined on R, and we have

$$u_i(t+t_m) = u_{i_2}(t+t_m) + u_{i_1}(t+t_m) \qquad \forall \qquad t \in R.$$

We also have that $\lim_{m\to\infty} u_{i_2}(t+t_m) = 0$, $\quad \lim_{m\to\infty} u_{i_1}(t+t_m) = u_{i_1}(t)$, u_{i_1} is an almost periodic function. Thus we obtain that $\lim_{m\to\infty} u_i(t+t_m) = u_{i_1}(t)$ $(i = 1, 2)$. Furthermore,

$$\lim_{m\to\infty} \dot{u}_i(t+t_m) = \lim_{m\to\infty}\lim_{h\to 0} \frac{u_i(t+t_m+h) - u_i(t+t_m)}{h} = \lim_{h\to 0}\lim_{m\to\infty} \frac{u_i(t+t_m+h) - u_i(t+t_m)}{h}$$
$$\lim_{h\to 0} \frac{u_{i_1}(t+h) - u_{i_1}(t)}{h}. \tag{10}$$

Thus one can deduce that $\dot{u}_{i_1}$ exist for $i = 1, 2$. We have a sequence $\{t_n\}$ such that $t_n \to \infty$ as $n \to \infty$ for which

$$a_1(t+t_n) \to a_i(t), \quad b_i(t+t_n) \to b_i(t), \quad a(t+t_n) \to a(t), \quad b(t+t_n) \to b(t),$$
$$c(t+t_n) \to c(t), \quad d(t+t_n) \to d(t), \quad e(t+t_n) \to e(t), \quad k(t+t_n) \to k(t) \quad (i = 1, 2).$$

$$\begin{aligned}
\dot{u}_{1_1} &= \lim_{n\to\infty} \dot{u}_1(t+t_n)\\
&= \lim_{n\to\infty} u_1(t+t_n)\Big[a_1(t+t_n) - b_1(t+t_n)u_1(t+t_n)\\
&\quad - \frac{b_2(t+t_n)u_2(t+t_n)}{a(t+t_n) + b(t+t_n)u_1(t+t_n) + c(t+t_n)u_2(t+t_n) + d(t+t_n)u_1(t+t_n)u_2(t+t_n)}\Big]\\
&= u_{1_1}(t)\Big[a_1(t) - b_1(t)u_{1_1}(t) - \frac{b_2(t)u_{2_1}(t)}{a(t) + b(t)u_{1_1}(t) + c(t)u_{2_1}(t) + d(t)u_{1_1}(t)u_{2_1}(t)}\Big],\\
\dot{u}_{2_1} &= \lim_{n\to\infty} \dot{u}_2(t+t_n)\\
&= \lim_{n\to\infty} u_2(t+t_n)\Big[a_2(t+t_n) - \frac{e(t+t_n)u_2(t+t_n)}{u_1(t+t_n) + k(t+t_n)}\Big]\\
&= u_{2_1}(t)\Big[a_2(t) - \frac{e(t)u_{2_1}(t)}{u_{1_1} + k(t)}\Big].
\end{aligned}$$

Thus Theorem 4 implies that system (1) possess a unique positive almost periodic solution.

4 Numerical Example

Consider the following predator–prey model system:

$$\frac{dx(t)}{dt} = x(t)\Big(9.9 + \sin\sqrt{5}t - 10.9x(t) - \frac{(0.3 + 0.19\sin\sqrt{5}t)y(t)}{8 + \cos\sqrt{11}t + (10 + \sin\sqrt{3}t)x(t) + 5y(t) + 0.1x(t)y(t)}\Big),$$

$$\frac{dy(t)}{dt} = y(t)\Big(0.5 + 0.29\sin\sqrt{3}t - \frac{(12 + 0.2\sin\sqrt{13}t)y(t)}{x(t) + 2}\Big). \quad (11)$$

Here $a_1^l = 8.9, a_1^u = 10.9, b_1^l = b_1^u = 10.9, b_2^l = 0.11, b_2^u = 0.49, a^l = 7,$ $a^u = 9, b^l = 9, b^u = 11, c^l = c^u = 5, d^l = d^u = 0.1, e^l = 11.79,$ $e^u = 12.19, k^l = k^u = 2$. And so $S_1 = 1$, $M_2 = 0.2034$, $s_1 = 0.8167$, $s_2 = 0.0462$. Hence we have $b_1^l = 10.9 > \frac{b_2^u M_2(b^u + d^u M_2)}{(a^l + b^l s_1 + c^l s_2 + d^l s_1 s_2)^2} + \frac{e^u M_2}{(s_1 + k^l)^2} = 0.0057 + 0.9300 = 0.9357$ and $\frac{e^l}{S_1 + k^u} = 3.9333 > \frac{b_2^u}{(a^l + b^l s_1 + c^l s_2 + d^l s_1 s_2)} + \frac{b_2^u M_2(c^u + d^u M_2)}{(a^l + b^l s_1 + c^l s_2 + d^l s_1 s_2)^2} = 0.3428 + 0.0027 = 0.3455$. This confirms that the parametric values involved with the model system (11) satisfy all the sufficient conditions for the global attractivity of solutions obtained in Theorem 4. Hence the model system (11) admits a unique, globally attractive, positive, and almost periodic solution. The almost periodic coexistence have been depicted in Fig. 1.

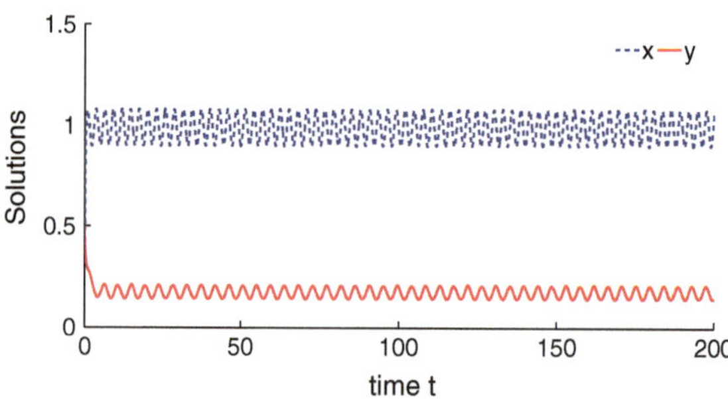

Fig. 1 Almost periodic solution of the nonautonomous system (11)

5 Concluding Remarks

In this present work, a Lotka–Volterra model with modified Leslie–Gower and Crowley–Martin functional response is considered. By constructing a suitable Lyapunov function, a set of sufficient conditions for the existence of a unique, globally attractive, almost periodic solution of the model system (1) is obtained. Furthermore, an example is given to substantiate our analytical findings. Our results and example generalize the results obtained in [8] and indicate that the dynamic behavior of the considered model (global attractivity of almost periodic solution) depend upon the mutual interference parameter d, i.e., the mutual interference among predators also affects the global attractivity of solution at high prey density. Another important task will be to consider similar type of predator–prey system with mutual interference among prey species.

Acknowledgments The research work of first author (J.P. Tripathi) is supported by the Council of Scientific and Industrial Research (CSIR) (No. 09/1058(0001)/2011-EMR-1), India.

References

1. Tripathi, J.P., Abbas, S., Thakur, M.: Local and global stability analysis of two prey one predator model with help. Commun. Nonlinear Sci. Numer. Simulat. **19**, 3284–3297 (2014)
2. Tripathi, J.P., Abbas, S., Thakur, M.: Stability analysis of two prey one predator model. AIP Conf. Proc. **1479**, 905–909 (2012)
3. Crowley, P.H., Martin, E.K.: Functional responses and interference within and between year classes of a dragonfly population. J. North Amer. Benth. Soc. **8**, 21–211 (1989)
4. Tripathi, J.P., Abbas, S.: Dynamical analysis of a prey predator model with Beddington DeAngelis type function response incorporating a prey refuge. Nonlinear Dyn. **21**, 99–111 (2014). doi:10.1007/s11071-014-1859-2
5. Chen, F., Shi, C.: Global attractivity in an almost periodic multi-species nonlinear ecological model. Appl. Math. Comput. **180**, 376–392 (2006)
6. Abbas, S., Sen, M., Banerjee, M.: Almost periodic solution of a non-autonomous model of phytoplankton allelopathy. Nonlinear Dynamics **67**, 203–214 (2012)
7. Chen, F., Cao, X.: Existence of almost periodic solution in a ratio-dependent Leslie system with feedback controls. J. Math. Anal. Appl. **341**, 1399–1412 (2008)
8. Lin, X., Chen, F.: Almost periodic solution for a Volterra model with mutual interference and Beddington-DeAngelis functional response. Appl. Math. Comput. **214**, 548–556 (2009)
9. Leslie, P.H.: Some further notes on the use of matrices in population mathematics. Biometrica **35**, 213–245 (1948)
10. Tripathi, J.P., Abbas, S., Thakur, M.: A density dependent delayed predator-prey model with Beddington-DeAngelis type Function Response incorporating a prey refuge. Commun. Nonlinear Sci. Numer. Simulat. (2014). doi:10.1016/j.cnsns.2014.08.018
11. Fan, M., Kuang, Y.: Dynamics of a nonautonomous predator-prey system with Beddington-DeAngelis functional response. J. Math. Anal. Appl. **295**, 15–39 (2004)

Existence of a Mild Solution for Impulsive Neutral Fractional Differential Equations with Nonlocal Conditions

Alka Chadha and Dwijendra N. Pandey

Abstract In the present work, we study the existence of a mild solution of a fractional-order differential equation with impulsive conditions in a Banach space X. We establish the existence and uniqueness of the mild solution by using some fixed-point theorems and resolvent semigroup theory.

Keywords Fractional calculus · Caputo derivative · Impulsive conditions · Resolvent operator · Neutral fractional differential equation

2010 Mathematics Subject Classification: 26A33 · 34K37 · 34K40 · 34K45 · 35R11 · 45J05 · 45K05

1 Introduction

In recent few decades, fractional calculus has received much attention of researchers mainly due to its demonstrated applications in widespread fields of science and engineering, e.g., fluid flow, rheology dynamical, mechanics, electrical engineering, modeling of many physical phenomena, and so on. Fractional calculus has been available and applicable to deal with real system characterized by power laws, anomalous diffusion process, etc. The nonlinear oscillations of earthquake are one of such important models. The deficiency of continuum traffic flow can be characterized by the fractional derivative. For more details on fractional calculus, we refer to [15–17, 20].

On the other hand, impulsive differential equations have played an important role in real-world problems for describing a process which at certain moments changes

A. Chadha (✉) · D.N. Pandey
Department of Mathematics, Indian Institute of Technology Roorkee, Roorkee 247667, India
e-mail: alkachadda23@gmail.com

D.N. Pandey
e-mail: dwij.iitk@gmail.com

P.N. Agrawal et al. (eds.), *Mathematical Analysis and its Applications*, Springer Proceedings in Mathematics & Statistics 143, DOI 10.1007/978-81-322-2485-3_26

their state rapidly and which cannot be described by using the classical differential problem. Such process is investigated in various fields such as biology, physics, control theory, population dynamics, medicine, and so on. Impulsive differential equations are an appropriate model to hereditary phenomena for which a delay argument arises in the modeling equations. For the general theory of such equations, we refer to monographs [1, 2] and papers [3–5, 8, 11].

In this paper, our main objective is to establish the existence and uniqueness of a solution for the fractional order neutral differential equation in a Banach space X with norm $\|\cdot\|_X$,

$$ {}^cD_t^\eta[u(t) - F(t, u(h_1(t)))] = A[u(t) - F(t, u(h_1(t)))] + G(t, u(h_2(t))), $$
$$ t \in [0, T], \quad t \neq t_k \quad 0 < T < \infty, \tag{1.1} $$
$$ \Delta u(t_k) = I_k(u(t_k^-)), \quad k = 1, 2, \dots m, \tag{1.2} $$
$$ u(0) = u_0 + g(u) \in X, \tag{1.3} $$

where ${}^cD_t^\eta$ is the Caputo fractional derivative of order $0 < \eta < 1$ and $A : D(A) \subset X \to X$ is a closed linear operator with dense domain $D(A)$ in a Banach space X and $I_k : X \to X$, $0 = t_0 < t_1 < \cdots < t_m < t_{m+1} = T$, $\Delta u|_{t=t_k} = u(t_k^+) - u(t_k^-)$, and $u(t_k^+) = \lim_{h\to 0+} u(t_k + h)$ and $u(t_k^-) = \lim_{h\to 0-} u(t_k + h)$ denote the right and left limits of $u(t)$ at $t = t_k$, respectively. The functions F, G, h_1, h_2, and g are appropriate continuous functions to be specified later.

The organization of the paper is as follows: Sect. 2 gives some basic definitions, Lemmas, and Theorems as preliminaries as these are useful for proving our results. Section 3 focuses on proving the existence result of mild solution to problem (1.1)–(1.3). Section 4 provides an example to illustrate the theory.

2 Preliminaries and Assumptions

In this section, we discuss some definitions and notations about sectorial operators, solution operator, and analytic solution operators required for establishing our results. Throughout this paper, X is a complex Banach space equipped with the norm $\|\cdot\|_X$. The symbol $C([0, T]; X)$ stands for the Banach space of all continuous functions from $[0, T]$ into X with supremum norm, i.e., $\|y\|_{[0,T]} = \sup_{t\in[0,T]} \|y(t)\|$. The notation $L(X, Y)$ denotes the Banach spaces of all bounded linear operators from X into Y with the operator norm denoted by $\|\cdot\|_{L(X,Y)}$ and when $X = Y$ then we write simply $L(X)$ and $\|\cdot\|_{L(X)}$. In addition, $PC([0, T], X)$ represents the Banach space of all the piecewise continuous functions from $[0, T]$ into X with the norm

$$ \|u\|_{PC} = \max\{ \sup_{t\in[0,T]} \|u(t + 0)\|_X, \ \sup_{t\in[0,T]} \| u(t - 0)\|_X\}, $$

and $B_r(x, X)$ denotes a closed ball with center at x and radius r in X.

To set the structure for our primary existence results, we recall the following definitions.

Definition 1 The definition of one parameter **Mittag-Leffler** function is given by

$$E_\alpha(z) = \sum_{k=0}^{\infty} \frac{z^k}{\Gamma(\alpha k + 1)},$$

it is the standard definition of Mittag-Leffler function of one parameter and two parameter function of Mittag-Leffler type is defined by

$$E_{\alpha,\beta}(z) = \sum_{k=0}^{\infty} \frac{z^k}{\Gamma(\alpha z + \beta)} = \frac{1}{2\pi i}\int_C \frac{\mu^{\alpha-\beta} e^{\mu}}{\mu^a - z} d\mu, \quad \alpha, \beta > 0, z \in \mathbb{C},$$

where C is a contour, which starts and ends at $-\infty$ and encircles the disk $|\mu| \leq |z|^{1/\alpha}$ counterclockwise. The Laplace transform of the Mittag-Leffler is defined as

$$L(t^{\beta-1}E_{\alpha,\,\beta}(-\rho^\alpha t^\alpha)) = \frac{\lambda^{\alpha-\beta}}{\lambda^\alpha + \rho^\alpha}, \quad \text{Re}\,\lambda > \rho^{1/\alpha}, \;\; \rho > 0.$$

For more details we refer to [15].

Definition 2 The Riemann–Liouville fractional integral operator $\mathscr{J}$ of order $\eta > 0$ is defined by

$$\mathscr{J}_t^\eta F(t) = \frac{1}{\Gamma(\eta)} \int_0^t (t-s)^{\eta-1} F(s) ds, \tag{2.1}$$

where $F \in L^1((0, T); X)$.

Definition 3 The Riemann–Liouville fractional derivative is given by

$$D_t^\eta F(t) = D_t^m J_t^{m-\eta} F(t), \;\; m-1 < \eta < m, \;\; m \in \mathbb{N}, \tag{2.2}$$

where $D_t^m = \frac{d^m}{dt^m}$, $F \in L^1((0, T); X)$, $J_t^{m-\eta} F \in W^{m,1}((0, T); X)$. Here the notation $W^{m,1}((0, T); X)$ stands for the Sobolev space defined by

$$W^{m,1}((0, T); X) = \{y \in X : \exists z \in L^1((0, T); X) : y(t) = \sum_{k=0}^{m-1} d_k \frac{t^k}{k!} \tag{2.3}$$

$$+ \frac{t^{m-1}}{(m-1)!} * z(t), \;\; t \in (0, T)\}.$$

Note that $z(t) = y^m(t)$, $d_k = y^k(0)$.

Definition 4 The Caputo fractional derivative is given by

$$ {}^{c}D_t^{\eta}F(t)=\frac{1}{\Gamma(m-\eta)}\int_0^t (t-s)^{m-\eta-1}F^{m}(s)ds,\quad m-1<\eta<m, \tag{2.4}$$

where $F\in C^{m-1}((0,T);X)\cap L^1((0,T);X)$ and the following holds

$$\mathscr{J}_t^{\eta}({}^{c}D_t^{\eta}F(t))=F(t)-\sum_{k=0}^{m-1}\frac{t^k}{k!}F^k(0). \tag{2.5}$$

The Laplace transform of the Caputo derivative of order $\eta > 0$ is given by

$$L[{}^{c}D_t^{\eta}u(t);\lambda]=\lambda^{\eta}L[u(t)]-\sum_{k=0}^{m-1}\lambda^{\eta-k-1}u^k(0),\quad m-1<\eta<m. \tag{2.6}$$

Definition 5 [3] An operator A, which is closed and linear, is called sectorial operator if there are constants $\omega\in\mathbb{R}$, $\theta\in[\pi/2,\pi]$, $M>0$ such that the following two conditions are satisfied:

(1) $\rho(A)\supset\sum_{(\theta,\,\omega)}=\{\lambda\in C:\lambda\neq\omega,\ |\arg(\lambda-\omega)|<\theta\}$,
(2) $\|R(\lambda,A)\|_{L(X)}\le\frac{M}{|\lambda-\omega|},\quad \omega\in\sum_{(\theta,\,\omega)}$,

where $\rho(A)$ be the resolvent set of A.

For more details we refer to [13]. Consider the following Cauchy problem for the fractional evolution equation

$$ {}^{c}D_t^{\eta}u(t)=Au(t),\quad t>0,\quad u(0)=x,\quad u^k(0)=0,\quad k=1,\ldots,m-1, \tag{2.7}$$

where $\eta>0$ and $m=\lceil\eta\rceil$.

Definition 6 [13] A family $\{S_\eta(t)\}_{t\ge 0}\subset L(X)$ is called a solution operator for (2.7) if the following conditions are satisfied:

(a) $S_\eta(t)$ is strongly continuous for $t\ge 0$ and $S_\eta(0)=I$;
(b) $S_\eta(t)D(A)\subset D(A)$ and $AS_\eta(t)x=S_\eta(t)Ax$, for all $x\in D(A)$, $t\ge 0$;
(c) $S_\eta(t)x$ is a solution of (2.7), for all $x\in D(A)$, $t\ge 0$.

The solution operator $S_\eta(t)$ of (2.7) is also defined by (see [13])

$$\lambda^{\eta-1}(\lambda^{\eta}I-A)^{-1}x=\int_0^{\infty}e^{-\lambda t}S_\eta(t)xdt,\quad \mathrm{Re}\,\lambda>\omega,\ x\in X, \tag{2.8}$$

where $\omega\ge 0$ and $\{\lambda^{\eta}:\ \mathrm{Re}\lambda>\omega\}\subset\rho(A)$.

An operator A is said to belong to $\mathscr{C}^{\eta}(X; M, \omega)$, or $\mathscr{C}^{\eta}(M, \omega)$ if the problem (2.7) has a solution operator $S_{\eta}(t)$ satisfying $\|S_{\eta}(t)\| \leq Me^{\omega t}$, $t \geq 0$. Denote $\mathscr{C}^{\eta}(\omega) = \bigcup\{\mathscr{C}^{\eta}(M, \omega);\ M \geq 1\}$, or $\mathscr{C}^{\eta} = \bigcup\{\mathscr{C}^{\eta}(\omega;\ \omega \geq 0)\}$ Bazhlekova [13].

Definition 7 [13] A solution operator $S_{\eta}(t)$ of (2.7) is said to be analytic if $S_{\eta}(t)$ admits an analytic extension to a sector $\sum_{\theta_0}$ for some $\theta_0 \in (0, \pi/2]$.

An analytic solution operator $S_{\eta}(t)$ is said to be of analyticity type (θ_0, ω_0) if for each $\theta < \theta_0$ and $\omega > \omega_0$ there exists a positive constant $M = M(\theta, \omega)$ such that $\|S_{\eta}(t)\| \leq Me^{\omega \operatorname{Re} t}$, for $t \in \sum_{\theta} = \{t \in C/\{0\} : |\arg t| < \theta\}$. Denote $\mathscr{A}^{\eta}(\theta_0, \omega_0) = \{\mathscr{A} \in \mathscr{C}^{\eta};\ \mathscr{A}$ generates analytic solution operator $S_{\eta}(t)$ of type $(\theta_0, \omega_0)\}$.

Lemma 1 [13, 14] *Let $\eta \in (0, 2)$. A linear closed densely defined operator A belongs to $\mathscr{A}^{\eta}(\theta_0, \omega_0)$ if and only if $\lambda^{\eta} \in \rho(A)$ for each $\lambda \in \sum_{\theta_0+\pi/2}(\omega_0)$, and for any $\omega > \omega_0$, $\theta < \theta_0$, there exists a constant $C = C(\theta, \omega)$ such that*

$$\|\lambda^{\eta-1}R(\lambda^{\eta}, A)\| \leq \frac{C}{|\lambda - \omega|}, \ \lambda \in \sum_{\theta+\pi/2}(\omega). \tag{2.9}$$

Now, we have following result for mild solution of Eqs. (2.12) and (2.13).

Theorem 1 *Suppose A is a sectorial operator and f satisfies the uniform Hölder condition with exponent $\beta \in (0, 1]$, then*

$$u(t) = S_{\eta}(t)x_0 + \int_0^t T_{\eta}(t-s)f(s)ds, \quad t \in [0, T], \tag{2.10}$$

where

$$\begin{aligned} S_{\eta}(t) &= \frac{1}{2\pi i}\int_{\Gamma} e^{\lambda t}\lambda^{\eta-1}R(\lambda^{\eta}, A)d\lambda, \\ T_{\eta}(t) &= \frac{1}{2\pi i}\int_{\Gamma} e^{\lambda t}R(\lambda^{\eta}, A)d\lambda, \end{aligned} \tag{2.11}$$

is the mild solution for fractional Cauchy problem

$$^{c}D_t^{\eta}u(t) = Au(t) + f(t), \quad 0 < \eta < 1, \ t \in [0, T], \tag{2.12}$$

$$u(0) = x_0 \in X, \tag{2.13}$$

where Γ is a suitable path lying on $\sum_{\theta,\omega}$. For $0 < \eta < 1$, $T_{\eta}(t)$ is the η-resolvent family and $S_{\eta}(t)$ is the solution operator, generated by A.

If $\eta \in (0, 1)$ and $A \in A^{\eta}(\theta_0, \omega_0)$, then for any $x \in X$ and $t > 0$, we have $S_{\eta}(t)x \in D(A)$ and

$$\|S_{\eta}(t)\|_{L(X)} \leq Me^{\omega t}, \quad \|T_{\eta}(t)\|_{L(X)} \leq Ce^{\omega t}(1 + t^{\eta-1}), \quad t > 0, \quad \omega > \omega_0.$$

Let

$$\widetilde{M}_S = \sup_{0\le t\le T} \|S_\eta(t)\|_{L(X)}, \quad \widetilde{M}_T = \sup_{0\le t\le T} Ce^{\omega t}(1+t^{1-\eta}).$$

Thus we have

$$\|S_\eta(t)\|_{L(X)} \le \widetilde{M}_S, \quad \|T_\eta(t)\|_{L(X)} \le t^{\eta-1}\widetilde{M}_T.$$

For more details about solution operators, we refer to [6, 14], and references cited in these papers.

Consider the set of functions

$$PC(I,\ X) = \{u : I \to X : u \in C((t_k,\ t_{k+1}], X), k = 0, 1, \ldots, m \text{ and } \exists \\ u(t_k^+) \text{ and } u(t_k^-), k = 1, \ldots, m \text{ with } u(t_k^-) = u(t_k)\}, \tag{2.14}$$

equipped with the norm

$$\|u\|_{PC} = \sup_{t\in I} \|u(t)\|_X,$$

which is a Banach space $(PC(I,\ X), \|\cdot\|_{PC})$.

Now, we assume following assumptions on F, G, h_1, h_2, and I_k which will be used later to establish main result.

(A1) For $0 < \beta < 1$, the function $A^\beta F : [0, T] \times X \to X$ is continuous and there exists a constant $L_F > 0$ such that

$$\|A^\beta F(t, x) - A^\beta F(s, y)\| \le L_F[|t - s| + \|x - y\|_X], \tag{2.15}$$

and

$$\| A^\beta F(t, x)\| \le L_1\| x\| + L_2, \tag{2.16}$$

for every $x,\ y \in X$ and $t,\ s \in [0, T]$ and L_1 and L_2.

(A2) The function $G : [0, T] \times X \to X$ is continuous and there exists a constant $L_G > 0$ such that

$$\|G(t, x) - G(t, y)\| \le L_G\|x - y\|_X, \tag{2.17}$$

for every $x,\ y \in X$ and $t, \in [0, T]$.

(A3) $I_k : X \to X$, where $k = 1, \ldots, m$ are continuous functions and there exists a constant $L > 0$ such that

$$\|I_k(x) - I_k(y)\| \le L\|x - y\|_X \tag{2.18}$$

for each $x,\ y \in X$.

(A4) There exists a constant $L_g > 0$ such that

$$\|g(x) - g(y)\| \le L_g\|x - y\|_X, \quad x, y \in X, \tag{2.19}$$

and

$$\|g(x)\| \leq \mathscr{C}_1\|x\| + \mathscr{C}_2, \ x \in X. \tag{2.20}$$

(A5) $h_i \in C([0,T],[0,T]), \ i = 1, 2.$

Next, we are giving the definition of the mild solution for the problem (1.1)–(1.3).

Definition 8 The function $u : [0, T] \to X$ is said to be a mild solution of Eqs. (1.1)–(1.3) if $u(\cdot) \in PC([0, T], X)$ satisfying the following integral equation

$$\begin{aligned} u(t) &= S_\eta(t)[u_0 + g(u) - F(0, u(h_1(0)))] + F(t, u(h_1(t))) \\ &+ \int_0^t T_\eta(t-s)G(s, u(h_2(s)))ds \\ &+ \sum_{i=1}^{m} S_\eta(t-t_i)I_i(u(t_i)), \end{aligned} \tag{2.21}$$

for each $t \in [0, T]$. and also satisfies the following impulsive conditions $\Delta u|_{t=t_i} = I_i(u(t_i^-)), \quad i = 1, \ldots, m.$

3 Existence Results

In this section, we establish the existence and uniqueness of a mild solution of Eqs. (1.1)–(1.3) using Banach fixed-point theorem.

Theorem 2 *Let (A1)–(A5) holds and*

$$\mathscr{R} = \widetilde{M}_S(L_g + L_F) + L_F\|A^{-\beta}\| + \widetilde{M}_T L_G \frac{T^\eta}{\eta} + m\widetilde{M}_S L < 1. \tag{3.1}$$

Then impulsive problem (1.1)–(1.3) has a unique mild solution $u \in X$.

Proof Let $u_0 \in X$ be fixed. Define a mapping $Q : PC([0, T]; X) \to PC([0, T]; X)$ such that

$$\begin{aligned} (Qu)(t) &= S_\eta(t)[u_0 + g(u) - F(0, u(h_1(0)))] + F(t, u(h_1(t))) \\ &+ \int_0^t T_\eta(t-s)G(s, u(h_2(s)))ds \\ &+ \sum_{i=1}^{m} S_\eta(t-t_i)I_i(u(t_i)), \end{aligned} \tag{3.2}$$

for each $t \in [0, T]$. Since F and G are continuous functions and $S_\eta(t), \ t \geq 0$ and $R_\eta(t), \ t \geq 0$ are compact, thus it is easy to show that the map Q is well defined on $PC([0,T];X)$. To establish the result, it is sufficient to show that the mapping Q is a contraction mapping on $PC([0,T];X)$. To this end, let $t\in[0,T]$ and $u^*,\ u^{**}\in PC([0,T];X)$. Thus, we obtain

$$
\begin{aligned}
&\|(Qu^*)(t)-(Qu^{**})(t)\|\\
&\qquad\le \|S_\eta(t)[g(u^*)-g(u^{**})-(F(0,u^*(h_1(0)))-F(0,u^{**}(h_1(0))))]\|\\
&\qquad\quad+\|F(t,u^*(h_1(t)))-F(t,u^{**}(h_1(t)))\|\\
&\qquad\quad+\int_0^t\|T_\eta(t-s)\|\times\|G(s,u^*(h_2(s)))-G(s,u^{**}(h_2(s)))\|ds\\
&\qquad\quad+\sum_{i=1}^m\|S_\eta(t-t_i)[I_i(u^*(t_i))-I_i(u^{**}(t_i))]\|,\\
&\quad\le(\widetilde{M}_S[L_g+L_F]+L_F\|A^{-\beta}\|)\|u^*-u^{**}\|_X+\widetilde{M}_T\int_0^t(t-s)^{\eta-1}L_G\|u^*-u^{**}\|_X ds\\
&\quad\ +m\widetilde{M}_S L\|u^*-u^{**}\|_X,\\
&\quad\le[\widetilde{M}_S(L_g+L_F)+L_F\|A^{-\beta}\|+\widetilde{M}_T L_G\frac{T^\eta}{\eta}+m\widetilde{M}_S L]\|u^*-u^{**}\|_X.
\end{aligned}
\tag{3.3}
$$

Since $\mathscr{R}<1$ by the inequality (3.1), it indicates that the map Q is a strict contraction on $PC([0,T];X)$. Hence, by Banach fixed-point theorem, there exists a unique fixed $u\in PC([0,T];X)$ such that $Qu(t)=u(t)$ which is a mild solution of the problem (1.1)–(1.3). Therefore, the proof of the theorem is completed.

Our second result is based on the Schaefer's fixed-point theorem. Schaefer's theorem is a special case of the far-reaching Leray–Schauder theorem which was discovered earlier by Juliusz Schauder and Jean Leray. The statement is as follows:

Theorem 3 *Let $Q: X\to X$ be a continuous and a compact map such that the set $\{x\in X: x=\lambda Px$ for some $0\le\lambda\le1\}$ is bounded, then Q has a fixed point.*

For this second result, we need to assume a new set of assumptions on G, F & I_k's, where $k=1,2,\ldots,m$.

Assumptions

(A6) $G:[0,T]\times X\to X$ is a continuous function and there exists a continuous function $m_G:[0,T]\to(0,\infty)$ and continuous nondecreasing function $W:[0,\infty)\to(0,\infty)$ such that

$$\|G(t,x)\|_X\le m_G(t)W(\|x\|),\quad (t,x)\in[0,T]\times X, \tag{3.4}$$

and $\int_0^\infty\frac{ds}{W(s)}<\infty$.

(A7) The functions $I_k: X\to X$ are completely continuous and there exist $\Omega>0$ such that

$$\Omega=\max_{1\le k\le m,\ x\in X}\{\|I_k(x)\|_X\}. \tag{3.5}$$

(A8) The function F is compact satisfying hypotheses $(A1)$.
(A9) The operator families $S_\eta(t), t \geq 0$ and $T_\eta(t), t \geq 0$ are compacts.

Theorem 4 *Suppose (A4)–(A9) holds. If* $A \in \mathscr{A}^\eta(\theta_0, \omega_0)$ *and* $L_F(1+\widetilde{M_S}) < 1$ *and* $\frac{\widetilde{M}_T T^\eta}{\eta(1-L_1(1+\widetilde{M}_S))} \int_0^T m_G(s)ds < \int_{\omega_1}^\infty \frac{ds}{W(s)}$, *where*

$$\omega_1 = \frac{\widetilde{M_S}}{1-L_1(1+\widetilde{M_S})} \| u_0 \| + \frac{\widetilde{M_S} m}{1-L_1(1+\widetilde{M_S})} \Omega(1+L_F) + \frac{1}{1-L_1(1+\widetilde{M_S})} L_2(1+\widetilde{M_S})$$

holds, then there exist at least one mild solution of the impulsive problem (1.1)–(1.3) on [0,T].

Proof Consider an operator $Q : PC([0,T]; X) \to PC([0,T]; X)$ as in Theorem 2. It can be easily proved that map Q is well defined on $PC([0,T]; X)$.

Step 1: We show the continuity of the map Q.
To prove the continuity, let u_n be sequence in $PC([0,T]; X)$ such that $\lim_{n\to\infty} u_n(t) = u(t)$, i.e., $u_n \to u$ as $n \to \infty$ in $PC([0,T]; X)$. Since G and F are continuous. Therefore, by the continuity of G, F, g, and I_i, $i = 1, \ldots, m$, we deduce that

$$G(t, u_n(h_2(t))) \to G(t, u(h_2(t))), \quad \text{as } n \to \infty, \tag{3.6}$$

$$F(t, u_n(h_1(t))) \to F(t, u(h_1(t))), \quad \text{as } n \to \infty, \tag{3.7}$$

$$g(u_n) \to g(u), \quad \text{as } n \to \infty, \tag{3.8}$$

$$I_i(u_n(t_i)) \to I_i(u(t_i)), \quad \text{as } n \to \infty. \tag{3.9}$$

Now for every $t \in [0,T]$, we have

$$\begin{aligned}
\| (Qu_n)(t) - (Qu)(t)\| &\leq \|S_\eta(t)[g(u_n) - g(u) - (F(0, u_n(h_1(0))) - F(0, u(h_1(0))))]\| \\
&+ \| F(t, u_n(h_1(t))) - F(t, u(h_1(t)))\| \\
&+ \int_0^t \| T_\eta(t-s)\| \cdot \| G(s, u_n(h_2(s))) - G(s, u(h_2(s)))\| ds \\
&+ \sum_{i=1}^m \|S_\eta(t-t_i)[I_i(u(t_i)) - I_i(u(t_i))]\|, \\
&\leq \widetilde{M}_S[\|g(u_n) - g(u)\| + \|F(0, u_n(h_1(0))) - F(0, u(h_1(0)))\|] \\
&+ \| F(t, u_n(h_1(t))) - F(t, u(h_1(t)))\| \\
&+ \widetilde{M}_T \int_0^t (t-s)^{\eta-1} \| G(s, u_n(h_2(s))) - G(s, u(h_2(s)))\| ds \\
&+ \widetilde{M}_S \sum_{i=1}^m [I_i(u(t_i)) - I_i(u(t_i))].
\end{aligned} \tag{3.10}$$

Then by the continuity of G, F, g, and I_i and dominated convergence theorem, we get that $Qu_n(t)$ converges to $Qu(t)$ in X, i.e., $\lim_{n\to\infty} Q_n u(t) = Qu(t)$ in X for each $t \in [0, T]$. Hence this proves the continuity of the map Q.

Step 2: Second, we show that Q maps bounded sets into bounded sets in $PC([0, T]; X)$. To prove the result, it is enough to show that for any $r > 0$ there exists $\gamma > 0$ such that $\| Qu\|_{PC} \leq \gamma$ for each $u \in B_r(PC) = \{u \in PC([0, T]; X) : \| u\|_{PC} \leq r\}$. Let $G_1 = \sup_{t\in I,\, u\in B_r} \| G(t, u(h_2(t)))\|$, then for any $u \in B_r(PC)$, $t \in [0, T]$, we have

$$\begin{aligned}
\| Qu(t)\|_X \leq\ & \widetilde{M_S}[\| u_0\| + \|g(u)\| + L_1 r + L_2] + \sup_{0\leq t\leq T,\, u\in B_r} \| F(t, u(h_1(t)))\| \\
& + \frac{\widetilde{M_T} T^\eta G_1}{\eta} + \sum_{i=1}^{m} \|S_\eta(t - t_i) I_i(u(t_i))\|, \\
& \leq \widetilde{M_S}[\| u_0\| + \mathscr{C}_1 r + \mathscr{C}_2 + L_1 r + L_2] + L_1 r + L_2 + \frac{\widetilde{M_T} T^\eta G_1}{\eta} + m\widetilde{M}_S \Omega, \\
& = \gamma.
\end{aligned} \tag{3.11}$$

Thus, we conclude that $\| Qu(t)\| \leq \gamma$.

Step 3: Q maps bounded sets into equicontinuous sets of $PC([0, T]; X)$. To this end, we show that $Q(B_r)$ is equicontinuous. Take $0 \leq \tau < t \leq T$ and $u \in C([0, T]; X)$, we have

$$\begin{aligned}
\| Qu(t) - Qu(\tau)\| \leq\ & \| [S_\eta(t) - S_\eta(\tau)](u_0 + g(u) - F(0, u(h_1(0))))\| \\
& + \| F(t, u(h_1(t))) - F(\tau, u(h_1(\tau)))\| \\
& + \| \int_0^t T_\eta(t - s) G(s, u(h_2(s)))ds - \int_0^\tau T_\eta(\tau - s) G(s, u(h_2(s)))ds\|, \\
& + \sum_{i=1}^{m} \|[S_\eta(t - t_i) - S_\eta(\tau - t_i)] I_i(u(t_i))\|,
\end{aligned}$$

$$\begin{aligned}
& \leq \| [S_\eta(t) - S_\eta(\tau)]\| [\| u_0\| + \| F(0, u(h_1(0)))\|] + L_F(|t - \tau|) \\
& + \widetilde{M_T} \int_\tau^t (t - s)^{\eta-1} \| G(s, u(h_2(s)))\| ds, \\
& + \widetilde{M_T} \int_0^t [(t - s)^{\eta-1} - (\tau - s)^{\eta-1}] \| G(s, u(h_2(s)))\| ds \\
& + \sum_{i=1}^{m} \|S_\eta(t - t_i) - S_\eta(\tau - t_i)\| \|I_i(u(t_i))\|.
\end{aligned} \tag{3.12}$$

Since $S_\eta(t), t > 0$ and $T_\eta(t), t > 0$ are compact, therefore $\| Qu(t) - Qu(\tau)\| \to 0$ as $t \to \tau$. Therefore, Qu is equicontinuous on $[0, T]$. Hence we conclude that $Qu(t)$ is equicontinuous on $[0, T]$

Step 4: Q maps B_r into a compact set in X.
For this, we decompose Q into Q_1 and Q_2, where

$$(Q_1u)(t) = S_\eta(t)[g(u) - F(0, u(h_1(0)))] + F(t, u(h_1(t))) + \int_0^t T_\eta(t-s)G(s, u(h_2(u(s))))ds,\ t \in [0, T], \tag{3.13}$$

$$(Q_2u)(t) = \sum_{i=1}^{m} S_\eta(t-t_i)I_i(u(t_i)),\ t \in [0, T]. \tag{3.14}$$

We now prove that $\{Q_1u(t),\ u \in B_r\}$ is relatively compact on X, for all $t \in [0, T]$. It is obvious that the set $\{Q_1u(t),\ u \in B_r\}$ is relatively compact in X for $t = 0$. Let $0 < t \leq T$ be fixed and $0 < \varepsilon < t$. For $u \in B_r$ define an operator $Q_{1,\varepsilon}$ by

$$Q_{1,\varepsilon}u(t) = S_\eta(t)[g(u) - F(0, u(h_1(0)))] + F(t, u(h_1(t))) + \int_0^{t-\varepsilon} T_\eta(t-s)G(s, u(h_2(u(s))))ds,\ t \in [0, T]. \tag{3.15}$$

Since $S_\eta(t)$ and $T_\eta(t)$ are compact, the set $\{Q_{1,\varepsilon}u(t),\ u \in B_r\}$ is relatively compact in X for every $\varepsilon,\ 0 < \varepsilon < t$. Moreover, for every $u \in B_r$, we have

$$\| Q_1u(t) - Q_{1,\varepsilon}u(t)\| \leq \int_{t-\varepsilon}^{t} \| T_\eta(t-s)\| \| G(s, u(h_2(s)))\|ds, \tag{3.16}$$

therefore, taking $\varepsilon \to 0$ we can easily see that there are relatively compact sets arbitrarily close to the set $\{Q_1u(t),\ u \in B_r\}$. Hence the set $\{Q_1u(t),\ u \in B_r\}$ is relatively compact in X and by $(A8)$ we conclude that Q_1 is compact for all $t \in [0, T]$. Next, we show that $\{Q_2u(t),\ u \in B_r\}$ is relatively compact in X, for all $t \in [0, T]$. For $t \in [0, T]$ we have $Q_2u(t) = \sum_{i=1}^{m} S_\eta(t-t_i)I_i(u(t_i))$ which is equicontinuous and bounded, by $(A9)$ it follows that $\{Q_2u(t),\ u \in B_r\}$ is relatively compact subset of X, for all $t \in [0, T]$. Hence, by $(A6)$–$(A8)$ and Arzela–Ascoli theorem, we conclude that Q_2 is compact for all $t \in [0, T]$. Therefore, $Q = Q_1 + Q_2$ is compact.

Step 5: (A priori bounds) We prove that the set $E = \{u \in PC([0, T]; X)$ such that $u = \lambda Qu$ for some $0 < \lambda < 1\}$ is bounded.
Let $u \in E$ with $u(t) = \lambda Qu(t)$ for some $0 < \lambda < 1$. Then for each $t \in [0, T]$,

$$\begin{aligned}\| u(t)\|_X &\leq \lambda[\widetilde{M_S}[\| u_0\| + \mathscr{C}_1\|u(t)\| + \mathscr{C}_2] + \widetilde{M_S}(L_1\| u(t)\| + L_2) + L_1\| u(t)\| + L_2 \\ &\quad + \widetilde{M_T}\int_0^t (t-s)^{\eta-1}\| G(s, u(h_2(s)))\|ds + m\widetilde{M}_S\Omega], \\ &\leq \lambda[\widetilde{M_S}\| u_0\| + [L_1(\widetilde{M_S}+1) + \widetilde{M}_S\mathscr{C}_1]\| u(t)\| + L_2(1+\widetilde{M_S}) + \widetilde{M}_S\mathscr{C}_2 \\ &\quad + \frac{\widetilde{M_T}T^\eta}{\eta}\int_0^t m_G(s)W(\| u(s)\|)ds + m\widetilde{M}_S\Omega].\end{aligned} \tag{3.17}$$

Therefore, for all $t \in [0, T]$, by the Young inequality ([13], p. 6), we get

$$\begin{aligned}\| u(t)\|_X &\leq \frac{\widetilde{M_S}}{1-L_1(1+\widetilde{M_S})-\widetilde{M}_S\mathscr{C}_1}\| u_0\| + \frac{\widetilde{M_S\mathscr{C}_2}}{1-L_1(1+\widetilde{M_S})-\widetilde{M}_S\mathscr{C}_1} \\ &+ \frac{L_2(1+\widetilde{M_S})+m\widetilde{M}_S\Omega}{1-L_1(1+\widetilde{M_S})-\widetilde{M}_S\mathscr{C}_1} \\ &+ \frac{\widetilde{M_T}T^\eta}{\eta(1-L_1(1+\widetilde{M_S}))-\widetilde{M}_S\mathscr{C}_1}\int_0^t m_G(s)W(\| u(s)\|)ds, \\ &\leq \omega_1 + \frac{\widetilde{M_T}T^\eta}{\eta(1-L_1(1+\widetilde{M_S})-\widetilde{M}_S\mathscr{C}_1)}\int_0^t m_G(s)W(\| u(s)\|)ds\end{aligned}$$

where $\omega_1 = \frac{\widetilde{M_S}}{1-L_1(1+\widetilde{M_S})-\widetilde{M}_S\mathscr{C}_1}\| u_0\| + \frac{\widetilde{M_S\mathscr{C}_2}}{1-L_1(1+\widetilde{M_S})-\widetilde{M}_S\mathscr{C}_1} + \frac{L_2(1+\widetilde{M_S})+m\widetilde{M}_S\Omega}{1-L_1(1+\widetilde{M_S})-\widetilde{M}_S\mathscr{C}_1}$. Then for all $t \in [0, T]$,

$$\| u(t)\| \leq \beta_\lambda(t) \triangleq \omega_1 + \frac{\widetilde{M_T}T^\eta}{\eta(1-L_1(1+\widetilde{M_S})-\widetilde{M}_S\mathscr{C}_1)}\int_0^t m_G(s)W(\| u(s)\|)ds.$$

Calculating $\beta'_\lambda(t)$ for $t \in [0, T]$, we obtain

$$\beta'_\lambda(t) \leq \frac{\widetilde{M_T}T^\eta}{\eta(1-L_1(1+\widetilde{M_S})-\widetilde{M}_S\mathscr{C}_1)} m_G(t)W(\| u(t)\|).$$

Thus we have

$$\frac{d\beta_\lambda(t)}{W(\| \beta_\lambda(t)\|)} \leq \frac{d\beta_\lambda(t)}{W(\| u(t)\|)} \leq \frac{\widetilde{M_T}T^\eta}{\eta(1-L_1(1+\widetilde{M_S})-\widetilde{M}_S\mathscr{C}_1)} m_G(t)dt. \tag{3.18}$$

Since $W(s)$ is positive and nondecreasing. Integrating both sides, we get

$$\int_{\omega_1}^{\beta_\lambda(t)} \frac{ds}{W(s)} \leq \frac{\widetilde{M_T}T^\eta}{\eta(1-L_1(1+\widetilde{M_S})-\widetilde{M}_S\mathscr{C}_1)}\int_0^T m_G(s)ds < \int_{\omega_1}^{\infty}\frac{ds}{W(s)}, \tag{3.19}$$

where, we have $\beta_\lambda(0) = \omega_1$, $\beta_\lambda(t)$ is positive and nondecreasing. Hence, from the above inequality, we obtain that the set of functions $\{\beta_\lambda : \lambda \in (0, 1)\}$ is bounded. This implies that set $\{u \in PC([0, T]; X) : u = \lambda Qu, \ 0 < \lambda < 1\}$ is bounded in X. Hence by Schaefer's fixed-point theorem, we get that Q has a fixed point on $[0, T]$. This completes the proof of the theorem.

4 Application

We consider the following fractional-order impulsive partial functional differential system of the form

$$\frac{\partial^\eta}{\partial t^\eta}\left[z(t,x)+\int_0^t b(t,\xi,x)\left[z(\sin t,\xi)+\frac{\partial}{\partial \xi}z(\sin t,\xi)\right]d\xi\right]$$
$$=\frac{\partial^2}{\partial x^2}\left[z(t,x)+\int_0^t b(t,\xi,x)\left[z(\sin t,\xi)+\frac{\partial}{\partial \xi}z(\sin t,\xi)\right]d\xi\right]+\chi\left(t,\frac{\partial}{\partial x}z(\sin t,x)\right),$$
$$0\le t\le 1,\quad 0\le x\le \pi, \tag{4.1}$$

$$z(t,0)=z(t,\pi)=0, \tag{4.2}$$
$$z(0,x)=z_0(x),\quad 0\le x\le \pi, \tag{4.3}$$
$$\Delta z|_{t_k}=z(t_{t_k}^{+})-z(t_{t_k}^{-})=I_k(z(t_{t_k}^{-})),\ k=1,\ldots,m, \tag{4.4}$$

where $0<\eta<1$ and $0<t_1<t_2<\cdots<t_m<1$ and $b:[0,1]\times[0,\pi]\times[0,\pi]\to\mathbb{R}$ and $\chi:[0,1]\times\mathbb{R}\to\mathbb{R}$ are continuous functions. Take $X=L^2[0,\pi]$ and let an operator A such that

$$Af=f'' \tag{4.5}$$

with the domain

$$D(A)=H^2([0,\pi])=\{f(\cdot)\in X: f',f''\in X \text{ and } f(0)=f(\pi)=0\}. \tag{4.6}$$

It implies that A generates a strongly continuous semigroup $T(\cdot)$ which is analytic and semi-adjoint and is given by $T(t)f=\sum_{n=1}^{\infty}e^{-n^2t}(f,z_n)z_n$. On the other hand, A has a discrete spectrum, the eigenvalues are $-n^2$, $n\in\mathbb{N}$ with the corresponding normalized eigenvectors $z_n(x)=\sqrt{\frac{2}{\pi}}\sin(nx)$. We have that if $f\in D(A)$ then $Af=\sum_{n=1}^{\infty}n^2(f,z_n)z_n$, for all $f\in X$ and $t>0$.

Now we assume following assumptions:

(i) $b:[0,1]\times[0,\pi]\times[0,\pi]\to\mathbb{R}$ is continuously differentiable with $b(t,\xi,0)=b(t,\xi,\pi)=0$.

(ii) The function b is measurable and

$$\sup_{0\le t\le 1}\int_0^\pi\int_0^\pi b^2(t,\xi,x)d\xi dx<\infty, \tag{4.7}$$

and function $\frac{\partial^2}{\partial x^2}$ is measurable and

$$K_1 = \sup_{0\le t\le 1}\left[\int_0^{\pi}\int_0^{\pi}\left(\frac{\partial^2}{\partial x^2}b(t,\xi,x)\right)^2 d\xi dx\right]^{1/2} < \infty. \tag{4.8}$$

(iii) $\chi : [0,1]\times\mathbb{R} \to \mathbb{R}$ is Lipschitz continuous with respect to the second argument and there exists positive constant a_0 such that

$$\| \chi(t,x_1) - \chi(t,x_2)\| \le a_0\| x_1 - x_2\|, \tag{4.9}$$

for $t \in [0, 1],\ x_1,\ x_2 \in \mathbb{R}$.

(iv) $I_k \in C(X,X),\ k = 1, 2, \ldots, m$ such that

$$\| I_k(x)\| \le \psi_k(\| x\|), \tag{4.10}$$

for $x \in X$, where $\psi_k \in (J, \mathbb{R}_+)$ is nondecreasing function.

Let $h_1(t) = h_2(t) = \sin t$, $F(t,z)(x) = \int_0^{\pi} b(t,\xi,x)[z(\xi), z'(\xi)]d\xi$, and $G(t,z)(x) = \chi(t, z'(\xi))$.
Therefore, the Eqs. (4.1) and (4.2) can be reformulated as

$$\begin{aligned}\frac{d^{\eta}}{dt^{\eta}}[u(t) + F(t, u(h_1(t)))] &= A[u(t) + F(t, u(h_1(t)))] + G(t, u(h_2(t))),\ 0 \le t \le 1,\\ u(0) &= u_0,\\ \Delta u|_{t_k} &= I_k(u(t_{t_k}^-)),\ k = 1, 2, \ldots, m.\end{aligned} \tag{4.11}$$

It is easy to verify that F and G satisfy the condition $A1$ and $A2$, respectively, and from (ii) it is clear that $F(t,z)$ is bounded linear operator on $\mathbb{R}$. Thus from Theorem 2, the system (4.11)–(4.11) admits a mild solution $[0,\ T]$ as well as (4.1)–(4.2).

Acknowledgments The authors would like to thank the referee for valuable comments and suggestions. The work of the first author is supported by the University Grants Commission (UGC), Government of India, New Delhi.

References

1. Benchohra, M., Henderson, J., Ntouyas, S.K.: Impulsive differential equations and inclusions. Contemporary Mathematics and Its Applications, vol. 2. Hindawi Publishing Corporation, New York (2006)
2. Lakshmikantham, V., Baǐnov, D., Simeonov, P.S.: Theory of Impulsive Differential Equations. World Scientific, Singapore-London (1989)
3. Shu, X.-B., Lai, Y., Chen, Y.: The existence of mild solutions for impulsive fractional partial differential equations. Nonlinear Anal.: TMA **74**, 2003–2011 (2011)

4. Zhang, X., Huang, X., Liu, Z.: The existence and uniqueness of mild solutions for impulsive fractional equations with nonlocal conditions and infinite delay. Noninear Anal. Hybrid Syst. **4**, 775–781 (2010)
5. Balachandran, K., Samuel, F.P.: Existence of mild solutions for quasilinear integrodifferential equations with impulsive conditions. Elect. J. Diff. Equ. **84**, 1–9 (2009)
6. Lizama, C.: Regularized solutions for abstract Volterra equations. J. Math. Anal. Appl. **243**, 278–292 (2000)
7. Araya, D., Lizama, C.: Almost automorphic mild solutions to fractional differential equations. Nonlinear Anal.: TMA. **69**, 3692–3705 (2008)
8. Abbas, M.I.: Existence for fractional order impulsive integrodifferential inclusions with nonlocal initial conditions. Int. J. Math. Anal. **6**, 1813–1828 (2012)
9. Ezzinbia, K., Fu, X., Hilal, K.: Existence and regularity in the α-norm for some neutral partial differential equations with nonlocal conditions. Nonlinear Anal.: TMA. **67**, 1613–1622 (2007)
10. Ezzinbia, K., Fu, X.: Existence and regularity of solutions for some neutral partial differential equations with nonlocal conditions. Nonlinear Anal.: TMA. **57**, 1029–1041 (2004)
11. Mophou, M.G.: Existence and uniqueness of mild solutions to impulsive fractional differential equations. Nonlinear Analy.: TMA. **72**, 1604–1615 (2010)
12. Pazy, A.: Semi-groups of Linear Operator and Applications of Partial Differential Equations. Springer verlag, New York (1983)
13. Bazhlekova, E.: Fractional evolution equations in Banach spaces, Ph.D. Thesis. Eindhoven University of Technology (2001)
14. Prüss, J.: Evolutionary Integral Equations and Applications (Monographs Math), vol. 87. Birkhauser-Verlag, New York (1993)
15. Podlubny, I.: Fractional Differential Equations. Academic press, New York (1993)
16. Miller, K.S., Ross, B.: An Introduction to the Fractional Calculus and Fractional Differential Equations. John Wiley and Sons Inc, New York (1993)
17. Samko, S.G., Kilbas, A.A., Marichev, O.I.: Fractional Integrals and Derivatives: Theory and Applications. Gordon and Breach Science Publisher, Yverdon (1993)
18. Dabas, J., Chauhan, A.: Existence and uniqueness of mild solution for an impulsive neutral fractional integro-differential equation with infinite delay. Math. Comp. Model. **57**, 754–763 (2013)
19. Li, K., Peng, J., Jia, J.: Cauchy problems for fractional differential equations with Riemann-Liouville fractional derivatives. J. Fract. Anal. **263**, 476–510 (2012)
20. Kilbas, A.A., Srivastava, H.M., Trujillo, J.J.: Theory and Applications of Fractional Differential Equations. Elsevier, Amsterdam (2006)
21. Wang, J., Fečkan, M., Zhou, Y.: On the new concept of solution and existence results for impulsive fractional evolution equations. Dyn. PDE **8**, 345–361 (2011)
22. Fečkan, M., Zhou, Y., Wang, J.: On the concept and existence of solution for impulsive fractional differential equations. Commun. Nonlinear Sci. Numer. Simulat. **17**, 3050–3060 (2012)
23. Wang, J., Li, X., Wei, W.: On the natural solution of an impulsive fractional differential equation of order q$\in$(1,2). Commun. Nonlinear Sci. Numer. Simulat. **17**, 4384–4394 (2012)
24. Granas, A., Dugundji, J.: Fixed Point Theory. Springer-Verlag, New York (2003)

Numerical Solution of Highly Oscillatory Nonlinear Integrals Using Quasi-Monte Carlo Methods

Nageswara Rao Narni

Abstract Highly oscillating integrals occur in many engineering applications. This paper discusses the quasi-Monte Carlo methods for calculation of the highly oscillating integrals using a low discrepancy sequence. We evaluated the highly oscillating integrals using a low discrepancy sequence known as Vander Corput sequence. The theoretical error bounds are calculated and are compared with analytical results. The reliability of the quasi-Monte Carlo methods is compared with He's homotopy perturbation method.

Keywords Highly oscillatory integrals · Quasi Monte-Carlo methods · Low discrepancy sequence

1 Introduction

Highly oscillatory functions arise in wide range of applications in science and engineering. The integration of high oscillatory functions is a challenging task from several years. Most of the techniques or analysis for integration of highly oscillatory functions are problem-oriented or technique-oriented. For example, integration of these functions occurs in solving the problems modeling of wave phenomena like diffraction of light, scattering of acoustic waves [8], scattering of electromagnetic waves [11], etc. The boundary element method also requires the evaluation of highly oscillatory integrals [3]. Explicit solution exists only for a few cases. So one needs to go for numerical methods.

The main goal of the present paper is on the analysis and computation of the integrals of the form

N.R. Narni (✉)
Rajiv Gandhi University of Knowledge Technologies,
RK Valley, YSR Dist., AP 516329, India
e-mail: narninrao@gmail.com

P.N. Agrawal et al. (eds.), *Mathematical Analysis and its Applications*,
Springer Proceedings in Mathematics & Statistics 143,
DOI 10.1007/978-81-322-2485-3_27

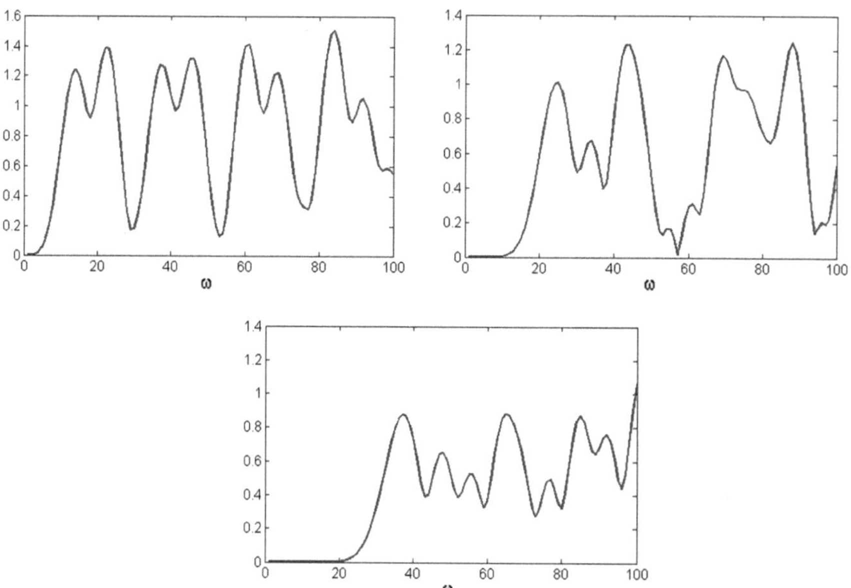

Fig. 1 Error in Gaussian quadrature with $f(x) = \cos(x)$, $g(x) = x^2$ for the quadrature points 5, 10, and 16

$$I[f, \Omega] = \int_{\Omega} f(x) e^{i\omega g(x)} dV, \tag{1}$$

where $\Omega \subset \mathbb{R}^n$ is bounded and open domain with piecewise smooth boundary. The functions $f, g \in C^\infty$ are smooth. For large values of $|\omega|$, the integral (1) oscillates rapidly as a function of ω.

A classical technique to compute (1) is *Gaussian quadrature* [6] method. If the integrand oscillates rapidly (for large values of ω), the Gaussian quadrature methods are not appropriate. For example, let us consider the following integral

$$\int_0^1 \cos(x) e^{i\omega x^2} dx. \tag{2}$$

The integral is evaluated using Gauss–Legendre quadrature rule at different quadrature points. One may observe that the Gauss–Legendre quadrature rule gives good results for small values of ω. As ω becomes large in comparison with quadrature points, high oscillations set in and the error becomes $\mathcal{O}(1)$. The absolute error with respect to ω is plotted in the Fig. 1.

There exists few techniques to calculate the highly oscillating integrals. Among all *Asymptotic expansion methods*, *Filon-type methods*, and *Levin-type methods* are most popular. The Asymptotic method in a straightforward manner is nothing but repeatedly applying the integration by parts. But the accuracy of the asymptotic expansion is limited due to the divergence of the series.

An even better approach is Filon [2]-type method. In this method instead of approximating whole integral, we approximate $f(x)$ of Eq. (1) at a set of quadrature nodes $c_1, \ldots, c_\nu$, by a polynomial $\tilde{f}$. Evaluation of the moments makes Filon-type methods difficult to certain type of applications.

In Levin [9]-type method we collocate the integrand at specific points. The Levin-type method is advantageous over Filon-type method and it is due to the fact that Levin-type method works easily on all types of domains and nonlinear oscillators. Most of the methods used in the current research are either Filon- or Levin-type methods or a modified form of these methods [1, 14]. He's homotopy Perturbation Method (HPM) is used in [10] for the numerical solution of the highly oscillating integrals.

In spite of all these methods, new applications continuously give rise to situations where straightforward application of these formulas are either inefficient or simply not possible. For example, if the function to be integrated contains critical point, then both Filon-type method and Levin collocation method are not accurate. So there is a need to device new methods for the integration of highly oscillatory functions.

The Monte Carlo method can be used to approximate the definite integral. This method gives the accuracy $\mathcal{O}\left(\frac{1}{\sqrt{n}}\right)$, which is not at all competitive with good algorithms, such as the Romberg method [6]. The present paper proposes the application of quasi-Monte Carlo methods for the numerical integration of highly oscillatory functions. In these methods selection of abscissas are based on Vander Corput sequence [7], which is a low discrepancy sequence.

In Sect. 2, an introduction to quasi-Monte Carlo methods, low discrepancy sequences, and Vander Corput sequence are presented. Section 3 gives the error bounds for quasi-Monte Carlo integration of the highly oscillating integrals. In Sect. 4, the quasi-Monte Carlo method with a Vander Corput sequence is applied to various problems. The efficiency of the QMC method is compared with other methods. Conclusions are drawn and are discussed in Sect. 5.

2 Quasi-Monte Carlo Methods

The only difference between the Monte Carlo and quasi-Monte Carlo methods is the selection of abscissa set $\{x_i\}$ (grid points). In Monte Carlo methods the abscissa are generated as a set of random number, whereas in quasi-Monte Carlo methods the quadrature nodes are calculated from deterministic algorithms.

2.1 Low Discrepancy Sequences

Now we introduce a quantity (the so-called discrepancy of the sequence) that measures the deviation of the sequence from an ideal distribution. This measure enables us to distinguish between *good* and *bad* sequences.

Definition 1 [7] For a nonempty set $\mathscr{M}$ of measurable subsets of C_N^+, the discrepancy $D_n^{\mathscr{M}}$ of the finite sequence $x_1, x_2, \ldots, x_n \in C_N^+$ with respect to $\mathscr{M}$ is defined by

$$D_n^{\mathscr{M}}(x_1, x_2, \ldots, x_n) := \sup_{E \in \mathscr{M}} \left| \frac{A(E; n)}{n} - \int_{C_N^+} C_E(x)dx \right|.$$

where $A(E; n) := \sum_{i=1}^{n} C_E(x_i)$ counts the number of points $x_i \in E$ and $E \subseteq C_N^+$.

2.2 Vander Corput Sequence

In this subsection, we are going to describe a low discrepancy sequence known as *Vander Corput sequence* [7]. In fact, this is the only infinite sequence having a uniformly smaller discrepancy than any other sequence exists upto now. Therefore, we have chosen this sequence for our numerical integration.

We define the so-called Vander Corput sequence $\{x_n\}$ as follows: For $n \geq 1$, let $n - 1 = \sum_{j=0}^{s} a_j 2^j$ be the dyadic expansion of $n - 1$. Then we set $x_n = \sum_{j=0}^{s} a_j 2^{-j-1}$. The sequence $\{x_n\}$ is then clearly contained in the unit interval. The following theorem gives the bounds for discrepancy of the Vander Corput sequence.

Theorem 1 *The discrepancy $D_N(\zeta)$ of the Vander Corput sequence $\zeta = \{x_N\}$ satisfies*

$$D_N(\zeta) \leq \frac{\ln(N+1)}{N \ln 2}$$

for N grid points.

Proof See [7].

3 Error Bounds for Quasi-Monte Carlo Methods

The selection of a numerical scheme is generally based on its accuracy (error bound), convergence, and computational cost. In this section, we are going to analyze these characters for quasi-Monte Carlo methods.

3.1 Variation of a Function

The variation of a univariate real function $f : [a, b] \to \mathbb{R}$ characterizes the regularity of f on the interval $[a, b]$. For a partition $\mathscr{P}$ of the interval $[a, b]$ into n subintervals,

$$\mathscr{P} : \{x_i : a = x_0 < x_1 < \cdots < x_{N-1} < x_N = b\},$$

the sum

$$V(f; \mathscr{P}) := \sum_{i=1}^{n} | f(x_i) - f(x_{i-1}) |$$

measures the discrete variation of f with respect to the specific partition $\mathscr{P}$. The continuous variation of f can be characterized by the supremum of all such discrete variations $V(f; \mathscr{P})$.

Definition 2 [12] **Variation of a univariate function**. The variation of a univariate function $f : [a, b] \to \mathbb{R}$ is defined as

$$V(f) := \sup\{V(f; \mathscr{P})\} = \sup_{\mathscr{P}} \left\{ \sum_{i=1}^{n} | f(x_i) - f(x_{i-1}) | \right\}.$$

If $V(f)$ is finite, f is said to be of bounded variation on $[a, b]$. If f is continuously differentiable then the relationship holds.

$$V(f) = \int_a^b | f'(x) | \, dx$$

The variation of nonlinear oscillator is obtained as follows:

Proposition 1 *Suppose $f(x)$, $g(x)$ are two continuously differentiable real valued functions with a finite bounded variation on $[a, b]$ and let*

$$I = \int_a^b f(x) e^{i\omega g(x)} dx,$$

then the bounded variation of the integrand is calculated as

$$BV\left[f(x)e^{i\omega g(x)}\right] \leq \int_a^b \left|i\omega g'(x) f(x) + f'(x)\right| dx$$

Proof Let the integrand be denoted as

$$\phi(x) = f(x) e^{i\omega g(x)}$$

Since $\phi(x)$ is piecewise smooth then the bounded variation is calculated as

$$
\begin{aligned}
BV(\phi) &= \int_a^b |\phi'(x)|dx \\
&= \int_a^b \left| i\omega g'(x) f(x) e^{i\omega g(x)} + f'(x) e^{i\omega g(x)} \right| dx
\end{aligned}
$$

$$
\begin{aligned}
\therefore \quad BV(\phi) &\leq \int_a^b \left| i\omega g'(x) f(x) + f'(x) \right| \left| e^{i\omega g(x)} \right| dx \\
&\leq \int_a^b \left| i\omega g'(x) f(x) + f'(x) \right| dx
\end{aligned}
$$

Corollary 1 *Suppose* $g(x) = $ *Constant, then*

$$
BV\left[f(x) e^{i\omega g(x)} \right] \leq \int_a^b \left| f'(x) \right| dx
$$

which corresponds to normal integration.

Corollary 2 *Suppose* $g(x) = x$, *i.e., linear oscillator, then*

$$
BV\left[f(x) e^{i\omega g(x)} \right] \leq \int_a^b \left| i\omega f(x) + f'(x) \right| dx
$$

But it is not always possible to calculate the variation of the functions.

3.2 Error Bounds

Now we discuss the error bounds for quasi-Monte Carlo approximation for more general integration domains. All these bounds depend on the variation of the integrand which involves the oscillatory parameter ω. A classical result is the following inequality of Koksma [7].

Theorem 2 *If* f *has bounded variation* $V(f)$ *on* $[0, 1]$, *then, for any sequence* $x_1, x_2, \ldots, x_N \in [0, 1]$, *we have*

$$
\left| \frac{1}{N} \sum_{n=1}^{N} f(x_n) - \int_0^1 f(u) du \right| \leq V(f) D_N(x_1, x_2, \ldots, x_N). \tag{3}
$$

Proof We can assume that $x_1 \leq x_2 \leq \cdots \leq x_N$. Put $x_0 = 0$ and $x_{N+1} = 1$. Using summation by parts and integration by parts, we obtain

$$\frac{1}{N}\sum_{n=1}^{N} f(x_n) - \int_0^1 f(u)du = -\sum_{n=0}^{N}\frac{n}{N}(f(x_{n+1}) - f(x_n)) + \int_0^1 udf(u)$$
$$= \sum_{n=0}^{N}\int_{x_n}^{x_{n+1}}\left(u - \frac{n}{N}\right)df(u).$$

For fixed n with $0 \leq n \leq N$, we have

$$\left|u - \frac{n}{N}\right| \leq D_N(x_1, x_2, \ldots, x_N) \quad \text{for } x_n \leq u \leq x_{n+1}$$

$$\therefore \quad \left|\frac{1}{N}\sum_{n=1}^{N} f(x_n) - \int_0^1 f(u)du\right| \leq D_N(x_1, x_2, \ldots, x_N)\sum_{n=0}^{N}\int_{x_n}^{x_{n+1}} |df(u)|$$
$$\leq D_N(x_1, x_2, \ldots, x_N)\sum_{n=1}^{N}|f(x_{n+1}) - f(x_n)|$$
$$\leq D_N(x_1, x_2, \ldots, x_N)V(f).$$

Hence we get the desired inequality.

The Koksma's inequality is applicable for C^{∞} functions also.

4 Numerical Experiments

In this section, we consider two different example problems. We evaluated the integrals using quasi-Monte Carlo methods with Vander Corput sequence and error bounds are calculated using Koksma's inequality.

Example 1 In this example, we consider the highly oscillating integrals of the form

$$I = \int_a^b e^{i\omega g(x)}dx$$

where $g'(0) = g''(0) = \cdots = g^{(r-1)}(0) = 0$ and $g^{(r)}(x) \neq 0$, for all $x \in [0, 1]$. This oscillator is known as irregular oscillator, where g is real.

In particular let us consider the integral

$$\int_0^1 e^{i\omega x^2}dx = \frac{\text{erf}(\sqrt{-i\omega})\sqrt{\pi}}{2\sqrt{-i\omega}}$$

It can be observed that the above integral has a unique critical point 0 of $g(x)$ in [0, 1]. Therefore, quasi-Monte Carlo method can be applied to evaluate the integral by using Vander Corput sequence.

Now we calculate the error bound for the given integral as follows: As we know $f(x) = 1$, and $g(x) = x^2$ and is continuous and differentiable. Therefore, the variation of $\phi(x)$ can be calculated as

$$V(\phi) = \int_0^1 |\phi'(x)|dx = \int_0^1 \left|2i\omega x e^{i\omega x^2}\right| dx$$

$$\therefore V(\phi) = 2\omega \int_0^1 \left|x e^{i\omega x^2}\right| dx$$

$$\leq 2\omega$$

By Koksma inequality (3), we get

$$\left|\frac{1}{N}\sum_{n=1}^{N} f(x_n) - \int_0^1 f(u)du\right| \leq \omega D_N(x_1, x_2, \ldots, x_N).$$

From Theorem 1, we know that

$$D_N(x_1, x_2, \ldots, x_N) \leq \frac{\ln(N+1)}{N \ln 2}$$

Therefore, the error bound for this integral is obtained as

$$\text{Error Bound} \leq \omega \frac{\ln(N+1)}{N \ln 2} \tag{4}$$

From this Eq. (4), we can observe that the error bound is dependent on both oscillating parameter ω and number of quadrature points N. The absolute error of the numerical scheme is plotted in Fig. 2 for $N = \omega$. We can observe that the error is of order $\ln(N)$.

Example 2 Consider the highly oscillatory integral [13]:

$$I(f) = \int_0^1 f(x) e^{i\omega \sin x} dx \tag{5}$$

where $f(x) = \cos(\sin x)\cos(x)$ and $g(x) = \sin x$. The bounded variation of the above integral is obtained as follows:

$$\begin{aligned}BV(f) &\leq \int_0^1 \left|i\omega g'(x) f(x) + f'(x)\right| dx \\ &\leq \int_0^1 |i\omega \cos(x)\cos(\sin x)\cos(x) - \cos(x)\sin(\sin(x))\cos(x) \\ &\quad - \cos(\sin(x))\sin(x)|\, dx \leq \omega\end{aligned}$$

The exact solution of the above integral (5) is obtained as

$$I(Q) = \frac{e^{i\omega \sin 1}}{1+\omega^2}\left[i\omega\cos(\sin 1) + \sin(\sin 1)\right] - \frac{1}{1+\omega^2} i\omega.$$

The approximate solution of the above integral (5) is calculated using homotopy perturbation method (HPM) in [10]. It is given as

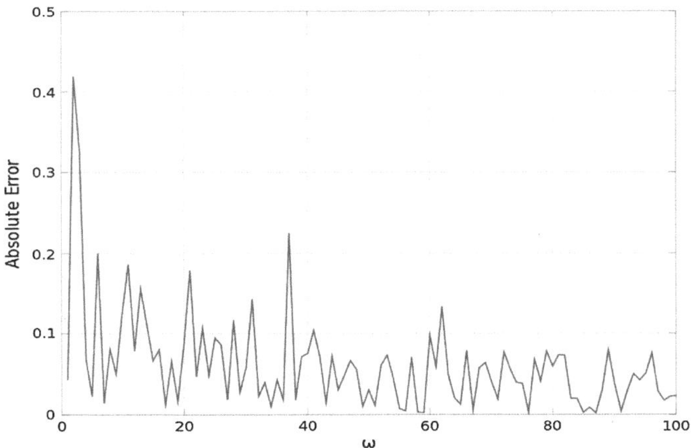

Fig. 2 Absolute error for $N = \omega$ and $0 \leq \omega \leq 100$

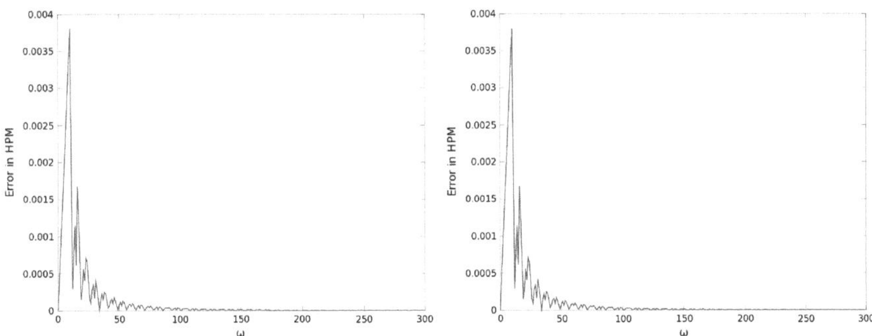

Fig. 3 The Absolute error of the integral (5) using HPM and QMC methods

$$I_{10}(Q) = \frac{1}{\omega^{10}}\left[i\omega(457+37\omega^2+5\omega^4+\omega^6+\omega^8)\right]+e^{0.8414709848 i\omega}(968.8821733 +169.63521533 i\omega+37.55809703\omega^2+17.513719298 i\omega^3 -2.7632852366\omega^4+2.3069869759 i\omega^5-1.4293570493\omega^6 +0.4328909146 i\omega^7-0.4028624431\omega^8-0.6663667454 i\omega^9)$$

and the absolute error between exact and HPM is plotted in Fig. 3.The numerical solution of the above integral (5) is calculated using quasi-Monte Carlo method and the corresponding absolute error with respect to the exact integral is plotted in Fig. 3.

We can observe that the numerical solution is same as the exact solution with very small difference as ω increases. The relative absolute error for the two methods are calculated. Corresponding to HPM method we have relative error 0.002522606 and corresponding quasi-Monte Carlo method gives the relative error 0.068065001. The relative error corresponding to HPM is less compared to quasi-Monte Carlo method due to the fact that HPM is a semiquantitive method and is applicable to few specific problems. This shows that the quasi-Monte Carlo methods are reliable and can be applied to a wide range of problems.

5 Conclusions

In this paper we are able to find the numerical solutions of highly nonlinear oscillatory integrals using quasi-Monte Carlo Methods. It is observed that the numerical solution satisfies the analytical error bounds. This shows the good agreement of results between analytical and numerical calculations. The work is under progress for the application of quasi-Monte Carlo methods to higher dimensional problems. This is due to the fact that as dimension of the problem increases the computational cost of traditional Gaussian-type methods increases. Therefore, Monte Carlo and quasi-Monte Carlo methods are the best choices.

References

1. Asheim, S.: A combined Filon/Asymptotic quadrature method for highly oscillatory problems. BIT Numer. Math. **48**, 425–448 (2008)
2. Filon, L.N.G.: On a quadrature formula for trigonometric integrals. Proc. Roy. Soc. Edinburgh **49**, 38–47 (1928)
3. Hsiao, G.C., Wendland, W.L.: Boundary Integral Equations. Springer-verlag, Berlin (2008)
4. Iserles, A., Norsett, S.P., Efficient quadrature of highly oscillatory integrals using derivatives, Proc. Roy. Soc. Edinburgh, A. 461,1383–1399 (2005)
5. Iserles, A., Norsett, S.P., Olver, S., Highly oscillating quadrature: The story so far
6. Jain, M.K., Iyengar, S.R.K., Jain, R.K.: Introduction to Numerical Analysis. Narosa Publishers, New Delhi (1993)
7. Kuipers, L., Niederreiter, H.: Uniform Distribution of Sequences. John-Wiley & Sons, Newyork (1974)

8. Kunik, M., Skrzpacz, P.: Diffraction of light revisited. Math. Methods. Appl. Sci. **31**(7), 793–820 (2007)
9. Levin, D.: Procedures for computing one-and-two dimensional integrals of functions with rapid irregular oscillations. Math. Comp. **38**, 531–538 (1982)
10. Molabahrami, A., Khani. F.: Numerical solutions of highly oscillatory integrals. Appl. Math. Comput. **198**, 657–664 (2008)
11. Oseledets, I.V., Stavtsev, S.L., Tyrtyshnikov, E.E.: Integration of oscillating functions in a quasi-three-dimensional electrodynamic problem. Comput. Math. Math. Phys. **49**(2), 292–303 (2009)
12. Walter R.: Real and complex analysis, 3rd edition, McGraw-Hill Company, New Delhi, 1987
13. Wu, X.Y., Xia, J.L.: Two low accuracy methods for stiff systems. Appl. Math. Comput. **123**, 141–153 (2001)
14. Xiang, S.: Efficient quadrature for highly oscillatory integrals involving critical points. J. Comput. Appl. Math. **206**, 688–698 (2006). doi:10.1016/j.cam.2006.08.018

Approximate Controllability of Semilinear Stochastic System with State Delay

Anurag Shukla, N. Sukavanam and D.N. Pandey

Abstract The objective of this paper is to present some sufficient conditions for approximate controllability of semilinear stochastic system with state delay. Sufficient conditions are obtained by separating the given semilinear system into two systems namely a semilinear deterministic system and a linear stochastic system. To prove our results, the Schauder fixed-point theorem is applied. At the end, an example is given to show the effectiveness of the result.

Keywords Approximate controllability · State delay · Stochastic system · Schauder fixed point theorem

1 Introduction

Controllability concepts play a vital role in deterministic control theory. It is well known that controllability of deterministic equation is widely used in many fields of science and technology. But in many practical problems such as fluctuating stock prices or physical system subject to thermal fluctuations, population dynamics, etc., some randomness appear, so the system should be modelled stochastic form.

In setting of deterministic systems: Kalman [1] introduced the concept of controllability for finite-dimensional deterministic linear control systems. Then Barnett [2] and Curtain [3] introduced the concepts of deterministic control theory in finite and infinite-dimensional spaces. Naito [4] established sufficient conditions for approximate controllability of deterministic semilinear control system dominated by the

A. Shukla (✉) · N. Sukavanam · D.N. Pandey
Department of Mathematics, Indian Institute of Technology Roorkee, Roorkee 247667, India
e-mail: anuragshukla259@gmail.com

N. Sukavanam
e-mail: nsukavanam@gmail.com

D.N. Pandey
e-mail: dwij.iitk@gmail.com

P.N. Agrawal et al. (eds.), *Mathematical Analysis and its Applications*, Springer Proceedings in Mathematics & Statistics 143,
DOI 10.1007/978-81-322-2485-3_28

linear part using Schauder's fixed-point theorem. In [5, 6], Wang extended the results of [4] and established sufficient conditions for delayed deterministic semilinear systems using same Schauder's fixed-point theorem. In [7] author provided more applications of Schauder's fixed-point theorem in nonlinear controllability problems.

In setting of stochastic systems: In [8, 9] Mahmudov established some results for controllability of linear stochastic systems in finite-dimensional and infinite-dimensional spaces, respectively. Sukavanam et al. in [10] obtained some sufficient conditions for s-controllability of an abstract first-order semilinear control system using Schauder's fixed-point theorem. Recently, Anurag et al. [11] obtained some sufficient conditions for approximate controllability of retarded semilinear stochastic system with nonlocal conditions using Banach fixed-point theorem.

The present paper is generalized form of the system taken in [10]. In this paper system is taken with finite delay in state which is not discussed up to now in the literature in best of my knowledge. The technique is adopted similar to discussed in [10, 12] with suitable modifications.

Let X and U be the Hilbert spaces and $Z = L_2[0, b; X]$, $Z_h = L_2[-h, b; X]$, $0 < h < b$, and $Y = L_2[0, b; U]$ be function spaces. $\mathbf{R}^k$ denotes k-dimensional real Euclidean space. Let (Ω, ζ, P) be the probability space with a probability measure P on Ω and a filtration $\{\zeta_t | t \in [0, b]\}$ generated by a Wiener Process $\{\omega(s) : 0 \leq s \leq t\}$.

We consider the semilinear stochastic control system of the form:

$$\begin{aligned} dx(t) &= [Ax(t) + Bu(t) + f(t, x_t)]dt + d\omega(t), \ t > 0 \\ x(t) &= \xi(t), \quad t \in [-h, 0] \end{aligned} \tag{1}$$

where the state function $x \in Z$; $A : D(A) \subseteq X \to X$ is a closed linear operator which generates a strongly continuous semigroup $S(t)$; $B : Y \to Z$ is a bounded linear operator; function $f : [0, b] \times X \to X$ is a nonlinear operator such that, f is measurable with respect to t, for all $x \in Z$ and continuous with respect to x for almost all $t \in [0, b]$; $x_t \in L_2([-h, 0], X) = \mathbf{C}$ (let)-valued stochastic processes and defined as $x_t(s) = \{x(t + s)| - h \leq s \leq 0|\}$; Control $u(t)$ takes values in U for each $t \in [0, b]$.

By splitting the system (1), we get the following pair of coupled systems

$$\begin{aligned} \frac{dy(t)}{dt} &= [Ay(t) + Bv(t) + f(t, (y + z)_t]; \quad 0 \leq t \leq b \\ y(t) &= \psi(t), \quad t \in [-h, 0] \end{aligned} \tag{2}$$

and

$$\begin{aligned} dz(t) &= [Az(t) + Bw(t)]dt + d\omega(t); \quad 0 \leq t \leq b \\ z(t) &= \xi(t) - \psi(t), \quad t \in [-h, 0] \end{aligned} \tag{3}$$

The system represented by (3) is linear stochastic system and for each realization $z(t)$ of system (3), the system given by (2) is a deterministic system. Thus the solution $y(t)$

of the semilinear system (2) depends on the solution $z(t)$ of linear stochastic system (3). The functions v and w are Y-valued control function, such that $u = v + w$.

It can be easily seen that, the solution $x(t)$ of the semilinear stochastic system (1) is given by $y(t) + z(t)$ where $y(t)$ and $z(t)$ are the solutions of the systems (2) and (3), respectively.

2 Preliminaries

In this section, some definitions are discussed which will be used in proof of main results.

The mild solution of the systems (1) can be written as

$$x(t) = \begin{cases} S(t)\xi(0) + \int_0^t S(t-s)\{Bu(s) + f(s, x_s)\}ds + \int_0^t S(t-s)d\omega(s), & t > 0 \\ \xi(t) & -h \leq t \leq 0 \end{cases} \tag{4}$$

the mild solution of the semilinear system (2) can be written as

$$y(t) = \begin{cases} S(t)\psi(0) + \int_0^t S(t-s)\{Bv(s) + f(s, (y+z)_s\}ds, & t > 0 \\ \psi(t) & -h \leq t \leq 0 \end{cases} \tag{5}$$

and the mild solution of the linear stochastic system (3) can be written as

$$z(t) = \begin{cases} S(t)(\xi(0) - \psi(0)) + \int_0^t S(t-s)Bw(s)ds + \int_0^t S(t-s)d\omega(s), & t > 0 \\ \xi(t) - \psi(t) & -h \leq t \leq 0 \end{cases} \tag{6}$$

Consider the linear system corresponding to the system (2) given by

$$\begin{aligned} \frac{dp(t)}{dt} &= Ap(t) + Br(t), \quad t > 0 \\ p(t) &= \psi(t) \quad t \in [-h, 0] \end{aligned} \tag{7}$$

The mild solution of the above linear system is expressed as

$$p(t) = \begin{cases} S(t)\psi(0) + \int_0^t S(t-s)Br(s)ds \; t > 0 \\ \psi(t) & -h \leq t \leq 0 \end{cases} \tag{8}$$

Definition 1 The set given by $K_T(f) = \{x(T) \in X : x \in Z_h\}$ where x is a mild solution of (1) corresponding to control $u \in Y$ is called Reachable set of the system (1).

Definition 2 The system (1) is said to be approximately controllable if $K_T(f)$ is dense in X, means $\overline{K_T(f)} = X$.

3 Basic Assumptions

In this section, some basic conditions and lemmas are assumed and discussed for obtaining the main results. Throughout this paper $D(A)$, $R(A)$, and $N_0(A)$ denote the domain, range, and null space of operator A, respectively.
The following conditions are assumed:
(H_1) For every $p \in Z$ there exists a $q \in \overline{R(B)}$ such that $Lp = Lq$ where the operator $L : Z \to X$ is defined as

$$Lx = \int_0^b S(b-s)x(s)ds$$

(H_2) The semigroup $\{S(t), t \geq 0\}$ generated by A is compact on X and there is a constant $M \geq 0$ such that $||S(t)|| \leq M$.
(H_3) $f(t, x)$ satisfies Lipschitz continuity on Z. i.e

$$||f(t, x_1) - f(t, x_2)|| \leq l_p||x_1 - x_2||, \quad l_p > 0$$

(H_4) $f(t, x)$ satisfies linear growth condition, that is,

$$||f(t, x)|| \leq a_1 + b_1||x||,$$

where a_1 and b_1 are constants.
(H_5) $Mbb_1(1 + c) < 1$
where the constants b and b_1 appear in the above conditions. The constant c is defined in Lemma 1.
Let $G : N_0^{\perp}(L) \to \overline{R(B)}$ be an operator defined as follows:

$$Ga = a_0$$

where $a \in N_0^{\perp}(L)$ and a_0 is the unique minimum norm element in the set $\{a + N_0(L)\} \bigcap \overline{R(B)}\}$ satisfying the following condition

$$||Ga|| = ||a_0|| = \min\left[||e|| : e \in \{a + N_0(L)\} \bigcap \overline{R(B)}\}\right] \tag{9}$$

The operator G is well defined, linear, and continuous (see [4], Lemma 1). From continuity of G, it follows that $||Ga|| \leq c||a||_Z$, for some constant $c \geq 0$.

Since $Z = N_0(L) + \overline{R(B)}$ as is evident from condition $(H1)$, any element $z \in Z$ can be expressed as

$$z = n + q : n \in N_0(L), q \in \overline{R(B)}$$

Lemma 1 *In* [12], *for $z \in Z$ and $n \in N_0(L)$, the following inequality holds*

$$||n||_Z \leq (1 + c)||z||_Z \tag{10}$$

where c is such that $||G|| \leq c$.

Let us introduce some operators in the following way:
$K : Z \to Z$ defined by

$$(Kz)(t) = \int_0^t S(t-s)z(s)ds$$

Now, let M_0 be the subspace of Z_h (see [11]) such that

$$M_0 = \begin{cases} m \in Z_h : m(t) = (Kn)(t), \;\; n \in N_0(L) & 0 \leq t \leq b \\ m(t) = 0, & -h \leq t \leq 0 \end{cases}$$

It can be noted that $m(b) = 0$, for all $m \in M_0$.

For each solution $p(t)$ of the system (7) with control r and for each realization $z(t)$ of the system (3), define the random operator $f_p : \overline{M_0} \to M_0$ as

$$f_p = \begin{cases} Kn, & 0 < t < b \\ 0, & -h \leq t \leq 0 \end{cases} \tag{11}$$

where n is given by the unique decomposition

$$F(p + z + m) = n + q : n \in N_0(L), \;\; q \in \overline{R(B)}, \tag{12}$$

where $F : L_2([0, b], \mathbf{C}) \to X$ given by

$$(Fx)(t) = f(t, x_t(.)); \qquad 0 \leq t \leq b$$

It is easy to see that F satisfies Lipschitz continuity (H_3) and linear growth conditions (H_4).

4 Main Results

In this section, approximate controllability of systems (2), (3) is proved. System (1) is splitted in systems (2), (3), so if systems (2), (3) are approximately controllable then system (1) is also approximately controllable.

The linear system (7) corresponding to system (2) is approximately controllable under the condition (H_1) (see [5]).
For approximate controllability of (3)

$$\begin{aligned} dz(t) &= [Az(t) + Bw(t)]dt + d\omega(t); \quad 0 \leq t \leq b \\ z(t) &= \xi(t) - \psi(t), \quad t \in [-h, 0] \end{aligned} \tag{13}$$

The mild solution of above system is

$$z(t) = \begin{cases} S(t)(\xi(0) - \psi(0)) + \int_0^t S(t-s)Bw(s)ds + \int_0^t S(t-s)d\omega(s), \ t > 0 \\ \xi(t) - \psi(t) \qquad\qquad -h \leq t \leq 0 \end{cases} \tag{14}$$

Define the operator $L_0^b : L_2[0, b; U] \to L_2[\Omega, \zeta_t, X]$, the controllability operator $\Pi_s^b : L_2[\Omega, \zeta_t, X] \to L_2[\Omega, \zeta_t, X]$ associated with (14), and the controllability operator $\Gamma_s^b : X \to X$ associated with the corresponding deterministic system of (14) as

$$L_0^b = \int_0^b S(b-s)Bw(s)ds \tag{15}$$

$$\Pi_s^b\{.\} = \int_s^b S(b-t)BB^*S^*(b-t)\mathbf{E}\{.|\zeta_t\}dt \tag{16}$$

$$\Gamma_s^b = \int_s^b S(b-t)BB^*S^*(b-t)dt \tag{17}$$

It is easy to see that the operators L_0^b, Π_s^b, Γ_s^b are linear-bounded operators, and the adjoint $(L_0^b)^* : L_2[\Omega, \zeta_t, X] \to L_2[0, b; U]$ of L_0^b is defined by

$$(L_0^b)^* = B^*S^*(b-t)\mathbf{E}\{z|\zeta_t\}\Pi_0^b = L_0^b(L_0^b)^*.$$

Before studying the approximate controllability of system (3), let us first investigate the relation between Π_s^b and Γ_s^b; $s \leq r < b$ and resolvent operator $R(\lambda, \Pi_s^b) = (\lambda I + \Pi_s^b)^{-1}$ and $R(\lambda, \Gamma_r^b) = (\lambda I + \Gamma_r^b)^{-1}$, $s \leq r < b$ for $\lambda > 0$, respectively.

Lemma 2 *For every $z \in L_2[\Omega, \zeta_t, X]$ there exists $\varphi(.) \in L_2^{\zeta}(0, b; \mathbf{L}(\mathbf{R}^k, X))$ such that*

1. $\mathbf{E}\{z|\zeta_t\} = \mathbf{E}\{z\} + \int_0^t \varphi(s)d\omega(s)$,
2. $\Pi_s^b z = \Gamma_s^b \mathbf{E}z + \int_s^b \Gamma_r^b \varphi(r)d\omega(r)$,
3. $R(\lambda, \Pi_s^b)z = R(\lambda, \Gamma_s^b)\mathbf{E}\{z|\zeta_t\} + \int_s^b \Gamma_r^b \varphi(r)d\omega(r)$.

Proof The proof is straightforward adaption of the proof of [10, Lemma 2.3]. □

Theorem 1 *The control system (3) is approximately controllable on* $[0, b]$ *if and only if one of the following conditions holds.*

1. $\Pi_0^b > 0$.
2. $\lambda R(\lambda, \Pi_0^b)$ *converges to the zero operator as* $\lambda \to 0^+$ *in the strong operator topology.*
3. $\lambda R(\lambda, \Pi_0^b)$ *converges to the zero operator as* $\lambda \to 0^+$ *in the weak operator topology.*

Proof The proof is straightforward adaption of the proof of [9, Theorem 2]. □

Lemma 3 *Under the conditions* (H_2), (H_4), *and* (H_5), *the operator* f_p *has a fixed point* $m_0 \in M_0$ *for each realization* $z(t)$ *of the system (3).*

Proof From the compactness of $S(t)$ the integral operator K is compact and hence f_p is compact for each p, (see [1]). Now let $||m|| \leq \tilde{r}$. Then from the condition (H_4) and from the inequality (10) and (12), we have

$$\begin{aligned}||f_p(m)||^2 &\leq \left|\left|\int_0^t S(t-s)n(s)ds\right|\right|^2 \\ &\leq \int_0^b \left|\left|\int_0^t S(t-s)n(s)ds\right|\right|^2 dt \\ &\leq M^2b^2(1+c)^2||F(p+z+m||_Z^2 \\ &\leq M^2b^2(1+c)^2\{a_1+b_1||p+z+m||_Z\}^2 \\ &\leq M^2b^2(1+c)^2\{a_1+b_1||p+z||+b_1\tilde{r}\}^2 \end{aligned} \tag{18}$$

Using Schauder's fixed-point theorem, it is clear from the compactness of f_p and (18) that f_p has a fixed point in M_0 in a ball of radius $\tilde{r} > 0$, if

$$\tilde{r} > \frac{Mb(1+c)(a_1+b_1||p+z||)}{1-Mb(1+c)b_1}$$

Thus $f_p(m_0) = m_0$

The approximate controllability of the semilinear system (2) is proved in following manner using the above lemma.

Lemma 4 *For each realization* $z(t)$ *of the system (3), the semilinear control system (2) is approximately controllable under the conditions* (H_1)–(H_4).

Proof From the Eq. (12), we have

$$F(p+z+m) = n+q$$

Operating K on both the sides at $m = m_0$ (fixed point of f_p) and using (11), we get

$$\begin{aligned}KF(p+z+m_0) &= Kn+Kq \\ &= m_0+Kq\end{aligned}$$

Adding p on both sides, we get

$$p + KF(p + z + m_0) = p + m_0 + Kq$$

Let $p + m_0 = y^*$, then the above equation is equivalent to

$$p + KF(y^* + z) = y^* + Kq$$

Since, from the Eq. (8)

$$p = S(t)\psi(0) + KBr$$

we have

$$S(t)\psi(0) + KBr + KF(y^* + z) = y^* + Kq$$
$$S(t)\psi(0) + K(Br - q) + KF(y^* + z) = y^*$$

Thus, it follows that $y^*(t)$ is a solution of the semilinear system

$$\frac{dy^*(t)}{dt} = Ay^*(t) + f(t, (y^* + z)_t) + Br(t) - q(t),$$
$$y^*(0) = \psi(0) \tag{19}$$

with control $(Br - q)$.
Moreover, since $y^*(t) = p(t) + m_0(t)$, it follows that

$$y^*(b) = p(b) + m_0(b),$$

as $m_0(b) = 0$ it follows that

$$y^*(b) = p(b) \tag{20}$$

From the Eqs. (19) and (20), it is clear that the reachable set of (19) is a superset of the reachable set of the system (7), which is dense in X.

Further $q \in \overline{R(B)}$ implies that for any given $\varepsilon_1 > 0$, there exists $v_1 \in Y$ such that $||q - Bv_1|| \leq \varepsilon_1$.

Now consider the equation

$$\frac{dy(t)}{dt} = Ay(t) + f(t, (y + z)_t) + B(r(t) - v_1(t)),$$
$$y(0) = \psi(0) \tag{21}$$

Let $y(t)$ be the solution of the system (21), corresponding to control $v = r - v_1$. Then $||y^*(b) - y(b)||$ can be made arbitrary small by choosing a suitable v_1, which implies that the reachable set of the system (21) is dense in the reachable set of the system (19), which in turn is dense in X. This proves that the system (2) is approximately controllable. □

5 Example

Consider the stochastic control system with delay governed by the semilinear heat equation

$$\partial y(t,x) = \left[\frac{\partial^2 y(t,x)}{\partial x^2} + Bu(t,x) + f(t, y(t+v,x))\right]\partial t + \partial\omega(t)$$
$$\text{for } \ 0 < t < \tau; \ v \in [-h,0]; \quad 0 < x < \pi$$
$$\text{with conditions} \quad y(t,0) = y(t,\pi) = 0, \quad 0 \le t \le \tau$$
$$y(t,x) = \xi(t,x), \qquad -h \le t \le 0,\ 0 \le x \le \pi \tag{22}$$

The system (22) can be written in the abstract form (1), by setting $X = L_2(0,\pi)$ and $A = \frac{d^2}{dx^2}$, with domain consisting of all $y \in X$ with $\left(\frac{d^2y}{dx^2}\right) \in X$ and $y(0) = 0 = y(\pi)$. Take $\phi_(x) = (2/\pi)^{1/2} sin(nx),\ 0 \le x \le \pi,\ n = 1,2,3,...$, then $\{\phi_n(x)\}$ is an orthonormal basis for X and ϕ_n ia an eigenfunction corresponding to the eigenvalue $\lambda_n = -n^2$ of the operator $A, n = 1,2,3, ...$ Then the C_0-semigroup $T(t)$ generated by A has $e^{\lambda_n t}$ as the eigenvalues and ϕ_n as their corresponding eigenfunctions.

Define an infinite-dimensional space U by

$$U = \left\{u : u = \sum_{n=2}^{\infty} u_n\phi_n \text{ with } \sum_{n=2}^{\infty} u_n^2 < \infty\right\}$$

The norm defined by

$$||u||_U = \left(\sum_{n=2}^{\infty} u_n^2\right)^{1/2}$$

$\xi(t,x)$ is known function.

Let B be a continuous linear operator from U to X defined as

$$Bu = 2u_2\phi_1 + \sum_{n=2}^{\infty} u_n\phi_n, \quad u = \sum_{n=2}^{\infty} u_n\phi_n \in U$$

The nonlinear operator f is assumed to satisfy conditions (H_3) and (H_4).

The approximate controllability of the corresponding semilinear deterministic heat equation of (22) was considered by Naito [4] and proved under the conditions

(H_1)–(H_4). Here approximate controllability of the stochastic semilinear heat control system (22) is considered.

The system (22) can be associated with two control systems under the initial and boundary conditions, as given below

$$\frac{\partial y(t,x)}{\partial t} = \frac{\partial^2 y(t,x)}{\partial x^2} + y(t-h,x) \\ +Bv(t,x) + f(t, y(t-h,x) + z(t-h,x)) \; t \in [0,b] \; x \in [0,\pi] \quad (23)$$

$$y(t,x) = \xi(t,x), \qquad -h \le t \le 0, \; 0 \le x \le \pi \\ \partial z(t,x) = \left[\frac{\partial^2 z(t,x)}{\partial x^2} + z(t-h,x) + Bw(t)\right]\partial t + \partial\omega(t) \quad (24)$$

The system (24) is a linear stochastic system and for each realization $z(t)$ of the system (24), the system (23) is a deterministic system.

From Lemma 4 and using the conditions (H_1)–(H_4), it is clear that for each realization $z(t)$ of the system (24), the system (23) is approximately controllable. The linear stochastic system (24) is approximately controllable from Lemma 3 corresponding to (23) and linear system corresponding to system (23) is approximately controllable from [4].

Acknowledgments The authors express their sincere gratitude to the reviewers for their valuable suggestions for improving the paper. This research is supported by the Council of Scientific and Industrial Research (CSIR) in the form of fellowship to the first author.

References

1. Kalman, R.E.: Controllability of linear systems. Contribut. Differ. Equ. **1**, 190–213 (1963)
2. Barnett, S.: Introduction to Mathematical Control Theory. Clarendon Press, Oxford (1975)
3. Curtain, R.F., Zwart, H.J.: An Introduction to Infinite Dimensional Linear Systems Theory. Springer, New York (1995)
4. Naito, K.: Controllability of semilinear control systems dominated by the linear part. SIAM J. Control Optim. **25**, 715–722 (1987)
5. Wang, L.W.: Approximate controllability of integrodifferential equations with multiple delays. J. Optim. Theory Appl. **143**, 185–206 (2009)
6. Wang, L.W.: Approximate controllability of delayed semilinear control of control system. J. Appl. Math. Stoch. Anal. **1**, 67–76 (2005)
7. Klamka, J.: Schuder fixed point theorem in nonlinear controllability problems. Control Cybernet. **29**(1), 153–165 (2000)
8. Mahmudov, N.I., Denker, A.: On controllability of linear stochastic systems. Int. J. Control **73**(2), 144–151 (2000)
9. Mahmudov, N.I.: Controllability of linear stochastic systems in hilbert spaces. J. Math. Anal. Appl. **259**(1), 64–82 (2001)
10. Sukavanam, N., Kumar, M.: S-controllability of an abstract first order semilinear control system. Numer. Funct. Anal. Optim. **31**(7–9), 1023–1034 (2010)

11. Shukla, A., Arora, U., Sukavanam, N.: Approximate controllability of retarded semilinear stochastic system with non local conditions. J. Appl. Math. Comput. doi:10.1007/s12190-014-0851-9
12. Sukavanam, N.: Approximate controllability of semilinear control systems with growing nonlinearit. In: Mathematical Theory of Control Proceedings of International Conference, pp. 353–357. Marcel Dekker, New York (1993)

Fourth-Order Derivative-Free Optimal Families of King's and Ostrowski's Methods

Ramandeep Behl, S.S. Motsa, Munish Kansal and V. Kanwar

Abstract In this paper, we present several new fourth-order optimal schemes that do not require any derivative evaluation for solving nonlinear equations, numerically. The presented approach of deriving these families is based on approximating derivatives by finite difference and weight function approach. The fourth-order derivative-free optimal families of King's and Ostrowski's methods are the main findings of the present work. Further, we have also shown that the families of fourth-order methods proposed by Petković et al., Appl Math Comput 217:1887–1895 (2010) [12] and Kung-Traub, J ACM 21:643–651 (1974) [8] are special cases of our proposed schemes. The proposed methods are compared with their closest competitors in a series of numerical experiments. All the methods considered here are found to be more effective to similar robust methods available in the literature.

Keywords Nonlinear equation · Newton's method · Order of convergence · Ostrowski's method · Simple root

1 Introduction

Multipoint iterative methods for solving nonlinear equations play a significant role in the field of iterative processes since they circumvent the drawbacks of one-point

R. Behl (✉) · S.S. Motsa
School of Mathematics, Statistics and Computer Science,
University of KwaZulu-Natal, Private Bag X01, Pietermaritzburg 3209, South Africa
e-mail: ramanbehl87@yahoo.in

S.S. Motsa
e-mail: sandilemotsa@gmail.com

M. Kansal · V. Kanwar
University Institute of Engineering and Technology, Panjab University,
Chandigarh 160 014, India
e-mail: mkmaths@gmail.com

V. Kanwar
e-mail: vmithil@yahoo.co.in

P.N. Agrawal et al. (eds.), *Mathematical Analysis and its Applications*,
Springer Proceedings in Mathematics & Statistics 143,
DOI 10.1007/978-81-322-2485-3_29

iterations, such as Newton's method. Such constructions occasionally possess a better order of convergence and efficiency index for finding the simple or multiple roots. In past and recent years, many fourth-order optimal multipoint iterative methods have been proposed and analyzed by King [7], Kanwar et al. [6], Behl et al. [1] and for detailed explanation of such methods, one can refer the excellent text books written by Traub [14], Ostrowski [9], and Petković et al. [11].

King's family [7], Jarratt's method [1], and Ostrowski's method [1, 6, 9] are one of the most efficient fourth-order multipoint methods known to date. However, all these multipoint methods require the evaluation of first-order derivative of function at each step. But, there are many practical situations in which the calculations of derivatives are expensive and/or it requires a great deal of time for them to be given or calculated. Therefore, derivative-free family of King's methods and Ostrowski's method are still needed.

In the last few years, many derivative-free fourth-order multipoint methods were proposed and analyzed by Petković et al. [12], Cordero et al. [4], Soleymani et al. [13], Cordero et al. [2, 3, 5], Zheng et al. [15], Peng et al. [10], and references cited therein. Some attempts have been made by Cordero et al. [4] to develop an optimal family of Ostrowski's method free from derivative. They obtained a fourth- and sixth-order multipoint family of Ostrowski's method in which the derivative is not required. But these proposed methods are not optimal in the sense of Kung-Traub conjecture.

Therefore, the construction of an optimal derivative-free family of King's method or Ostrowski's method having biquadratic convergence is an open and challenging problem in computational mathematics. With this aim, we proposed an optimal scheme of King's family in which there is no need to find the derivatives of functions. All the proposed methods considered here are found to be effective and comparable to the classical Ostrowski's method, King's method, and recently developed robust methods, respectively.

2 Construction of Novel Techniques Without Memory

In this section, we intend to develop derivative-free optimal families of King's method and Ostrowski's method. For this purpose, we consider the following scheme:

$$\begin{aligned} y_n &= x_n - \frac{f(x_n)}{f'(x_n)}, \\ x_{n+1} &= y_n - \frac{f(y_n)}{f'(y_n)}. \end{aligned} \tag{1}$$

This method has a fourth-order of convergence. But, the scheme (1) has two major drawbacks as it requires four functional evaluations per iteration and first-order derivative is computed at every iteration. Therefore, we consider here some suitable approximations of $f'(x_n)$ by forward approximation and $f'(y_n)$ by similar to

King's approximation [7], which are given by

$$f'(x_n) \approx f[x_n, u_n], \text{ where } u_n = x_n + \alpha f(x_n), \text{ and } \alpha \in \mathbb{R}\backslash\{0\},$$
and
$$f'(y_n) = f[x_n, u_n]\frac{f(x_n) + af(y_n)}{f(x_n) + \beta f(y_n)}, \text{ where } a,\ \beta \in \mathbb{R}, \tag{2}$$

respectively.
Using the above approximations of derivatives $f'(x_n)$ and $f'(y_n)$ in (1), we get

$$\begin{cases} y_n = x_n - \dfrac{f(x_n)}{f[x_n, u_n]}, \ u_n = x_n + \alpha f(x_n), \text{ where } \alpha \in \mathbb{R}\backslash\{0\}, \\ x_{n+1} = y_n - \dfrac{f(y_n)}{f[x_n, u_n]}\left[\dfrac{f(x_n) + \beta f(y_n)}{f(x_n) + af(y_n)}\right], \text{ where } a,\ \beta \in \mathbb{R}. \end{cases} \tag{3}$$

This is a modified family of King's method and satisfies the following error equation:

$$e_{n+1} = \left(\frac{a - \beta + 2 + \alpha(a - \beta + 1)c_1}{c_1^3}\right)(1 + \alpha c_1)c_2^3 e_n^3 + O(e_n^4), \tag{4}$$

where $e_n = x_n - r,\ c_n = \frac{f^{(n)}(r)}{n!},\ n = 1, 2, 3, \ \ldots$
But, according to the Kung-Traub conjecture [8], the above scheme is not optimal. Therefore, we shall now make use of weight function approach to build our optimal scheme based on (3) by a simple change in its second step. Therefore, we consider

$$\begin{cases} y_n = x_n - \dfrac{f(x_n)}{f[x_n, u_n]}, \ u_n = x_n + \alpha f(x_n), \text{ where } \alpha \in \mathbb{R}\backslash\{0\}, \\ x_{n+1} = y_n - \dfrac{f(y_n)}{f[x_n, u_n]}\dfrac{f(x_n) + \beta f(y_n)}{f(x_n) + \gamma f(y_n)}Q\left(\dfrac{f(y_n)}{f(u_n)}\right), \end{cases} \tag{5}$$

where γ and β are two free disposable parameters and $Q\left(\frac{f(y_n)}{f(u_n)}\right) \in \mathbb{R}$ is a real-valued weight function. Theorem 1 indicates that under what conditions on the weight function and disposable parameters in (5), the order of convergence will reach at the optimal level four.

3 Convergence Analysis

Theorem 1 *Let $f : I \subseteq \mathbb{R} \rightarrow \mathbb{R}$ be sufficiently differentiable function defined on an open interval I, enclosing a simple zero of $f(x)$ (say $x = r \in I$). Assume that initial guess $x = x_0$ is sufficiently close to $r \in I$. Then, our proposed scheme (5) free from* *derivatives, has an optimal fourth-order convergence when*

$$\gamma = \beta - 1, \quad Q(0) = 1, \quad Q'(0) = 1. \tag{6}$$

It satisfies the following error equation

$$e_{n+1} = \left(\frac{\beta + 3 + 2\alpha(\beta - 1)c_1 + \alpha^2 \beta c_1^2}{c_1^3} \right)(1 + \alpha c_1)c_2^3 e_n^4 + O(e_n^5), \tag{7}$$

where e_n and c_n are already defined in Eq. (4).

Proof Let $x = r$ be a simple zero of $f(x)$. Expanding $f(x_n)$ and $f'(x_n)$ about $x = r$ by the Taylor's series expansion, we have

$$f(x_n) = (c_1 e_n + c_2 e_n^2) + O(e_n^3), \tag{8}$$

and

$$f(x_n + \alpha f(x_n)) = (1 + \alpha c_1)c_1 e_n + (1 + 3\alpha c_1 + \alpha^2 c_1^2)c_2 e_n^2 + 2\alpha(\alpha c_1 + 1)c_2^2 e_n^3 + O(e_n^4), \tag{9}$$

respectively.
From Eqs. (8) and (9), we have

$$\frac{\alpha f^2(x_n)}{f(x_n + \alpha f(x_n)) - f(x_n)} = e_n - \left(\alpha + \frac{1}{c_1}\right) c_2 e_n^2 + \left(\frac{2 + 2\alpha c_1 + \alpha^2 c_1^2}{c_1^2} \right) c_2^2 e_n^3 - \left(\frac{4 + 5\alpha c_1 + 3\alpha^2 c_1^2 + \alpha^3 c_1^3}{c_1^3} \right) c_2^3 e_n^4 + O(e_n^5), \tag{10}$$

and

$$\begin{aligned} f(y_n) &= f\left(x_n - \frac{\alpha f^2(x_n)}{f(x_n + \alpha f(x_n)) - f(x_n)}\right), \\ &= \left(\frac{1 + \alpha c_1}{c_1}\right) c_2 e_n^2 - \left(\frac{2 + 2\alpha c_1 + 2\alpha^2 c_1^2}{c_1}\right) c_2^2 e_n^3 + \left(\frac{5 + 7\alpha c_1 + 4\alpha^2 c_1^2 + \alpha^3 c_1^3}{c_1^2}\right) c_2^3 e_n^4 + O(e_n^5). \end{aligned} \tag{11}$$

Furthermore, we have

$$\frac{\alpha f(x_n) f(y_n)(f(x_n) + \beta f(y_n))}{(f(u_n) - f(x))(f(x_n) + \gamma f(y_n))} = \left(\frac{1 + \alpha c_1}{c_1}\right) c_2 e_n^2 - \left(\frac{4 + \gamma - \beta + \alpha(5 + 2\gamma - 2\beta)c_1 + \alpha^2(\gamma + 2 - \beta)c_1^2}{c_1^2} \right) c_2^2 e_n^3 + O(e_n^4), \tag{12}$$

and

$$\frac{f(y_n)}{f(u_n)} = \left(\frac{c_2}{c_1}\right) e_n - \left(\frac{2\alpha c_1 + 3}{c_1^2}\right) c_2^2 e_n^2 + \left(\frac{8 + 8\alpha c_1 + 3\alpha^2 c_1^2}{c_1^3}\right) c_2^3 e_n^3 + O(e_n^4). \tag{13}$$

Since, it is clear from (13) that $\left(\frac{f(y_n)}{f(u_n)}\right)$ is of order e_n. Hence, we can consider the Taylor's expansion of the weight function Q in the neighborhood of zero. Therefore, we have

$$Q\left(\frac{f(y_n)}{f(u_n)}\right) = Q(0) + \left(\frac{f(y_n)}{f(u_n)}\right) Q'(0) + O(e_n^3). \tag{14}$$

Using (11), (12), (13) and (14) in scheme (5), we have the following error equation:

$$\begin{aligned}
e_{n+1} &= e_n - \frac{f(x_n)}{f[x_n, u_n]} - \frac{f(y_n)}{f[x_n, u_n]} \left(\frac{f(x_n) + \beta f(y_n)}{f(x_n) + \gamma f(y_n)}\right) Q\left(\frac{f(y_n)}{f(u_n)}\right), \\
&= -(Q(0) - 1)\left(\frac{\alpha c_1 + 1}{c_1}\right) c_2 e_n^2 + \frac{1}{c_1^2}\Big[-2 - Q'(0) + Q(0)(\gamma - \beta + 4) - \alpha(2 + Q'(0) \\
&\quad + Q(0)(2\beta - 2\gamma - 5))c_1 + \alpha^2(-1 + Q'(0)(\gamma - \beta + 2))c_1^2\Big] c_2^2 e_n^3 - \frac{1}{c_1^3}\Big[-4 - Q'(0)(\gamma + 7) \\
&\quad + Q(0)(13 + \gamma^2 - \gamma(\beta - 7) - 7\beta) + \beta Q'(0) + \alpha(-5 + Q(0)(20 + 3\gamma^2 - 3\gamma(\beta - 5) - 15\beta) \\
&\quad - 2Q'(0)(5 + \gamma - \beta))c_1 + \alpha^2(-3 + Q(0)(12 + 3\gamma^2 + \gamma(11 - 3\beta) - 11\beta) + Q'(0)(-4 - \gamma + \beta))c_1^2 \\
&\quad + \alpha^3(-1 + Q(0)(3 + \gamma^2 - \gamma(\beta - 3) - 3\beta))c_1^3\Big] c_2^3 e_n^4 + O(e_n^5).
\end{aligned} \tag{15}$$

For obtaining an optimal general class of fourth-order methods, the coefficients of e_n^2 and e_n^3 in the error Eq. (15) must be zero simultaneously. After simplifying the Eq. (15), we have the following equations involving γ, $Q(0)$ and $Q'(0)$

$$\begin{cases} Q(0) = 1, \\ -Q'(0) + Q(0)(\gamma - \beta + 4) = 2, \\ (2 + Q'(0) + Q(0)(2\beta - 2\gamma - 5)) = 0, \\ (-1 + Q'(0)(\gamma - \beta + 2)) = 0, \end{cases} \tag{16}$$

respectively.

After simplifying the Eq. (16), we get the following values of γ, $Q(0)$ and $Q'(0)$

$$\gamma = \beta - 1, \quad Q(0) = 1, \quad Q'(0) = 1. \tag{17}$$

Using the above conditions in scheme (5), we shall get the following error equation:

$$e_{n+1} = \left(\frac{\beta + 3 + 2\alpha(\beta - 1)c_1 + \alpha^2 \beta c_1^2}{c_1^3} \right) (1 + \alpha c_1) c_2^3 e_n^4 + O(e_n^5), \tag{18}$$

where $\alpha, \beta \in \mathbb{R}$ are two free disposable parameters.

This completes the proof of the Theorem 1. □

Remark 1 From computational point of view, the family of methods (5) which is totally derivative-free reaches the highest possible convergence and efficiency index using only three functional evaluations viz., $f(x_n),\ f(u_n),\ f(y_n)$ per full iteration. From the application point of view, methods in which there are derivative evaluations per full cycle are restricted, when the problem considered a bear massive time or load for computing the derivatives. For example, the nonlinear function $h(x) = \tan(\ln x) + \cos(x^4) \times \sqrt{(1/(2x)}$, has a first derivative that is hard to write. In fact in such cases, the derivative evaluation is expensive and/or occasionally takes a great deal of time. Such shortcomings lead us to study optimal iterative methods which are totally derivative-free per full iteration.

4 Special Cases

In this section, we shall consider some particular cases of our proposed scheme (5) and mention some weight functions $Q(x)$ that satisfy all the conditions of Theorem 1, as follow:

Case 1 Let us consider the following weight function:

$$Q(x) - x + 1. \tag{19}$$

Using the above weight function in scheme (5), we obtain

$$\begin{cases} y_n = x_n - \dfrac{f(x_n)}{f[x_n, u_n]}, \quad u_n = x_n + \alpha f(x_n), \ \text{where } \alpha \in \mathbb{R}\backslash\{0\}, \\ x_{n+1} = y_n - \dfrac{\alpha f(x_n) f(y_n)(f(u_n) + f(y_n))(f(x_n) + \beta f(y_n))}{f(u_n)(f(u_n) - f(x_n))(f(x_n) + (\beta - 1) f(y_n))}, \end{cases} \tag{20}$$

where $\beta \in \mathbb{R}$.

This is a new modified derivative-free optimal general class of fourth-order King's method and one can easily get many new families of methods by choosing different values of the disposable parameters α and β.

Subspecial Cases of Optimal Family (20)

(i) For $\beta = \frac{1}{2}$, family (20) reads as

$$\begin{cases} y_n = x_n - \dfrac{f(x_n)}{f[x_n, u_n]}, \; u_n = x_n + \alpha f(x_n), \\ x_{n+1} = y_n - \dfrac{\alpha f(x_n) f(y_n)(f(u_n + f(y_n))(2f(x_n) + f(y_n))}{f(u_n)(f(u_n) - f(x_n))(2f(x_n) - f(y_n))}. \end{cases} \tag{21}$$

This is a new modified derivative-free optimal family of King's method.
(ii) For $\beta = 0$, family (20) reads as

$$\begin{cases} y_n = x_n - \dfrac{f(x_n)}{f[x_n, u_n]}, \; u_n = x_n + \alpha f(x_n), \\ x_{n+1} = y_n - \dfrac{\alpha f^2(x_n) f(y_n)(f(u_n + f(y_n))}{f(u_n)(f(u_n) - f(x_n))(f(x_n) - f(y_n))}. \end{cases} \tag{22}$$

This is a new modified derivative-free optimal family of Ostrowski's method.

Case 2 Let us consider the following weight function:

$$Q(x) = \frac{1}{1-x}. \tag{23}$$

Using the above weight function in scheme (5), we obtain

$$\begin{cases} y_n = x_n - \dfrac{f(x_n)}{f[x_n, u_n]}, \; u_n = x_n + \alpha f(x_n), \; \text{where } \alpha \in \mathbb{R}\backslash\{0\}, \\ x_{n+1} = y_n - \dfrac{\alpha f(x_n) f(u_n) f(y_n)(f(x_n) + \beta f(y_n))}{(f(x_n) - f(u_n))(f(u_n) - f(y_n))(f(x_n) + (\beta - 1) f(y_n))}, \end{cases} \tag{24}$$

where $\beta \in \mathbb{R}$.
This is another new modified general class of fourth-order optimal King's method.

Subspecial Cases of Optimal Family (24)
(i) For $\beta = 1$, family (24) reads as

$$\begin{cases} y_n = x_n - \dfrac{f(x_n)}{f[x_n, u_n]}, \; u_n = x_n + \alpha f(x_n), \\ x_{n+1} = y_n - \dfrac{\alpha f(u_n) f(y_n)(f(x_n) + f(y_n))}{(f(x_n) - f(u_n))(f(u_n) - f(y_n))}. \end{cases} \tag{25}$$

This is a fourth-order optimal family of derivative-free methods independently derived by Petković et al. [12].
(ii) For $\beta = 0$, family (24) reads as

$$\begin{cases} y_n = x_n - \dfrac{f(x_n)}{f[x_n, u_n]}, \; u_n = x_n + \alpha f(x_n), \\ x_{n+1} = y_n - \dfrac{\alpha f^2(x_n) f(u_n) f(y_n)}{(f(x_n) - f(u_n))(f(u_n) - f(y_n))(f(x_n) - f(y_n))}. \end{cases} \tag{26}$$

This is another fourth-order optimal family of multipoint methods independently derived by Kung and Traub [8].

Remark 2 The first most striking feature of this contribution is that we have developed optimal derivative-free families of King's method and Ostrowski's method for the first time which will converge even though the guess is far from zero or the derivative is small in the vicinity of the required root. Therefore, these techniques can be used as an alternative to King's and Ostrowski's techniques or in the cases where King's and Ostrowski's techniques are not successful.

5 Numerical Experiments

In this section, we shall check the effectiveness of new optimal methods. We employ the present methods, namely (21), method (22), and family (20) for $\left(\beta = \frac{1}{100} \text{ and } \beta = \frac{101}{100}\right)$ denoted by MKM_4^1, MOM_4, MKM_4^2, and MKM_4^3, respectively, with $|\beta = 1|$ to solve nonlinear equations given in Table 1. We compare them with existing King's method for $\beta = \frac{1}{2}$ (KM_4), Ostrowski's method (OM_4), Cordero et al. method (4) (CM_4) [4], Petković et al. method (12) (PM_4) [12], and Kung-Traub method (KT_4) [8], respectively. For better comparison of our proposed methods, we have given three comparison tables in each example: one is corresponding to absolute error value of given nonlinear functions (with the same total number of functional evaluations = 12), the second is with respect to the number of iterations taken by each method to obtain the root correct up to 35 significant digits, and the last one is corresponding to computational order of convergence in Table 2, 3 and 4, respectively. All computations have been performed using the programming package *Mathematica* 9 with multiple precision arithmetic. We use $\varepsilon = 10^{-34}$ as a tolerance error. The following stopping criteria are used for computer programs:
(i) $|x_{n+1} - x_n| < \varepsilon$ and (ii) $|f(x_{n+1})| < \varepsilon$.

Table 1 Test functions

$f(x)$	r	[a, b]
$f_1(x) = \sin x$	0.000000000000000000000000000000000	$[-1.51, 1.51]$
$f_2(x) = e^{-x} + \sin x$	3.1830630119333635919391869956363946	$[1.3, 4]$
$f_3(x) = \sin^2 x - x^2 + 1$	1.4044916482153412260350868177868681	$[0.5, 1.8]$
$f_4(x) = \tan^{-1} x$	0.000000000000000000000000000000000	$[-1, 1]$
$f_5(x) = (x-1)^3 - 1$	2.000000000000000000000000000000000	$[1.5, 3]$
$f_6(x) = x^3 - \cos x + 2$	$-1.1725779647539700126733327148688486$	$[-3, 0]$
$f_7(x) = \tan(\log x) + \cos(x^3) \times \sqrt{1/2x}$	0.443260783556767073513472596321	$[0.38, 0.50]$

Table 2 Comparison of different fourth-order methods with the same total number of functional evaluation (TNFE = 12)

$f(x)$	x_0	KM_4	MKM_4^1	OM_4	MOM_4	CM_4	PM_4	MKM_4^2	KT_4	MKM_4^3
1.	−1.51	C	1.1.e−552	C	4.9e−557	C	1.4e−801	3.6e−800	4.9e−557	4.0e−917
	−1.31	C	5.4e−1118	C	7.4e−1126	C	5.9e−1397	2.3e−1396	7.4e−1126	7.3e−1452
	1.31	C	5.4e−1118	C	7.4e−1126	C	5.9e−1397	2.3e−1396	7.4e−1126	7.3e−1452
	1.51	C	1.1.e−552	C	4.9e−557	C	1.4e−801	3.6e−800	4.9e−557	4.0e−917
2.	1.5	C	2.1e−184	C	5.9e−183	C	4.0e−174	4.0e−174	5.9e−183	3.2e−180
	2.5	5.3e−182	3.6e−530	6.8e−187	3.7e−530	8.8e−153	1.1e−586	1.1e−586	3.7e−530	1.1e−586
	3.5	1.4e−274	3.5e−568	6.8e−278	3.5e−568	4.0e−245	9.2e−598	9.2e−598	3.5e−568	9.6e−598
	4	1.2e−128	1.1e−317	6.4e−140	9.0e−318	4.1e−127	2.5e−344	2.5e−344	9.0e−318	5.8e−345
3.	0.5	2.6e+0	1.8e−73	1.6e−16	1.9e−64	9.4e−130	2.1e−130	7.5e−131	1.9e−64	6.9e−107
	1.2	1.7e−160	1.4e−220	3.4e−197	4.5e−184	3.9e−256	1.5e−224	1.1e−225	4.5e−184	3.7e−169
	1.8	3.6e−141	9.5e−117	1.1e−159	6.2e−77	1.0e−195	1.0e−56	4.3e−51	6.2e−77	5.3e−78
4.	−1	1.4e−102	5.7e−1485	1.5e−134	1.2e−1490	7.0e−147	8.1e−1773	1.8e−1772	1.2e−1490	1.2e−1804
	1	1.4e−102	5.7e−1485	1.5e−134	1.2e−1490	7.0e−147	88.1e−1773	1.8e−1772	1.2e−1490	1.2e−1802
5.	1.5	3.7e−10	9.5e−125	9.7e−60	1.2e−61	8.7e−93	1.6e−136	1.5e−137	1.2e−61	1.6e−48
	2.5	1.4e−100	2.1e−43	1.2e−122	3.6e−35	D	5.6e−83	3.9e−81	1.4e−31	2.7e−34
6.	−1.3	5.9e−219	6.0e−231	1.2e−248	5.1e−146	1.4e−174	9.4e−184	4.4e−181	5.1e−146	2.5e−126
	−1	6.7e−153	1.3e−225	1.4e−194	1.8e−142	2.6e−173	8.6e−183	7.7e−181	1.8e−142	1.4e−134
	0	F	8.1e+0	F	6.2e+1	D	7.2e+0	7.1e+0	6.2e+1	6.4e−6
7.	0.44	3.8e−191	1.5e−130	0.1e−343	1.6e−345	*C*	2.3e−101	6.4e−105	1.6e−142	2.1e−100
	0.45	9.0e−396	1.0e−342	1.6e−560	1.0e−346	9.2e−317	1.92e−312	1.0e−330	1.0e−346	6.8e−312

C Stands for converge to undesired root, *D* Stands for divergent, *F* Stands for failure

Table 3 Comparison of different fourth-order methods with respect to number of iterations

$f(x)$	x_0	KM_4	MKM_4^1	OM_4	MOM_4	CM_4	PM_4	MKM_4^2	KT_4	MKM_4^3
1.	−1.51	C	4	C	4	C	4	4	4	4
	−1.31	C	4	C	4	C	4	4	4	4
	1.31	C	4	C	4	C	4	4	4	4
	1.51	C	4	C	4	C	4	4	4	4
2.	1.5	C	4	C	4	C	4	4	4	4
	2.5	4	4	4	4	4	3	3	4	3
	3.5	4	3	4	3	4	3	3	3	3
	4	5	4	4	4	5	4	4	4	4
3.	0.5	9	5	6	5	5	5	5	5	5
	1.2	4	4	4	4	4	4	4	4	4
	1.8	4	5	4	5	4	5	5	5	5
4.	−1	5	4	5	4	5	4	4	4	4
	1	5	4	5	4	5	4	4	4	4
5.	1.5	6	5	5	5	5	4	4	5	5
	2.5	5	6	5	5	D	5	5	6	5
6.	−1.3	4	4	4	4	4	4	4	4	5
	−1	4	4	4	4	4	4	4	4	4
	0	F	15	F	10	D	11	13	10	7
7.	0.40	4	4	4	4	C	5	4	4	4
	0.45	4	4	4	4	4	4	4	4	4

C Stands for converge to undesired root, *D* Stands for divergent, *F* Stands for failure

Table 4 Computational order of convergence of different fourth-order methods

$f(x)$	x_0	KM_4	MKM_4^1	OM_4	MOM_4	CM_4	PM_4	MKM_4^2	KT_4	MKM_4^3
1.	−1.51	na	5.0000	na	5.0000	na	4.0009	4.0009	4.0067	4.0004
	−1.31	na	4.0001	na	4.0001	na	4.0000	4.0000	4.0001	4.0000
	1.31	na	4.0001	na	4.0001	na	4.0000	4.0000	4.0001	4.0000
	1.5	na	5.0000	na	5.0000	na	4.0009	4.0009	9.0067	4.0004
2.	1.5	na	3.8024	na	3.8147	na	4.0026	4.0026	3.8147	4.0029
	2.5	4.0031	4.0000	4.0025	4.0000	4.0000	4.0000	4.0000	4.0000	4.0000
	3.5	3.9999	4.0000	3.9999	4.0000	3.9997	4.0000	4.0000	4.0000	4.0000
	4	4.0000	4.0000	4.0000	4.0000	4.0000	4.0000	4.0000	4.0000	4.0000
3.	0.5	4.0000	4.0000	4.0000	4.0000	4.0000	4.0000	4.0000	4.0000	4.0000
	1.2	3.9994	4.0001	3.9999	4.0002	4.0025	4.0001	4.0001	4.0002	4.0003
	1.8	3.9985	4.0000	3.9995	4.0000	3.8975	4.0000	4.0001	4.0000	4.0000
4.	−1	5.0000	4.0000	5.0000	4.0000	5.0000	4.0000	4.0000	4.0000	4.0000
	1	5.0000	4.0000	5.0000	4.0000	5.0000	4.0000	4.0000	4.0000	4.0000
5.	1.5	4.0000	4.0000	4.0000	4.0000	4.0000	4.0016	4.0016	4.0000	4.0001
	2.5	4.0000	4.0000	4.0000	3.9998	nd	4.0000	4.0000	4.0000	4.0013
6.	−1.3	4.0000	4.0000	4.0000	4.0008	3.9997	3.9998	3.9998	4.0007	4.0000
	−1	3.9992	4.0001	3.9999	4.0007	3.9997	3.9998	3.9998	4.0008	4.0014
	0	*	4.0000	*	3.9997	nd	4.0000	4.0000	3.9997	4.0000
7.	0.40	4.000	3.9999	4.0000	4.0001	na	3.9999	4.0333	4.0001	4.0000
	0.45	4.000	4.0000	4.0000	4.0001	4.000	4.0000	4.0000	4.0000	4.0000

na Not applicable in the case of converge to undesired root, *nd* Not defined in the case of divergent, * In the case of failure

Example 1 $\sin x = 0$.
This equation has an infinite number of roots. It can be seen that King's method (for $\beta = \frac{1}{2}$), Ostrowski's method, and Cordero et al. method do not necessarily converge to the root that is nearest to the starting value. For example, King's method (for $\beta = \frac{1}{2}$), Ostrowski's method, and Cordero et al. method with initial guess $x_0 = -1.51$ converge to $6.2831\ldots$, $9.4247\ldots$, and $-25.1327\ldots$, respectively, far away from the required root zero. Similarly, King's method (for $\beta = \frac{1}{2}$), Ostrowski's method, and Cordero et al. method with initial guess $x_0 = -1.31$ converge to $28.2743\ldots$, $-6.2831\ldots$, and $3.1415\ldots$, respectively, and so on. Our methods do not exhibit this type of behavior.

Example 2 $e^{-x} + \sin x = 0$.
Again, this equation has an infinite number of roots. It can be seen that King's method (for $\beta = \frac{1}{2}$), Ostrowski's method, and Cordero et al. method do not necessarily converge to the root that is nearest to the starting value. For example, King's method (for $\beta = \frac{1}{2}$) and Ostrowski's method with initial guess $x_0 = 1.5$ converge to $9.4248\ldots$, while Cordero et al. method converges to $25.1327\ldots$ far away from the required root $3.1830\ldots$. Similarly, King's method (for $\beta = \frac{1}{2}$), Ostrowski's method, and Cordero et al. method with initial guess $x_0 = 1.6$ converge to $12.566\ldots$, $18.8495\ldots$, and $6.2813\ldots$, respectively, and so on. Our methods do not exhibit this type of behavior. Similarly, we can check the behavior of all the methods on the text problems mentioned in Table 1.

6 Conclusions

In this paper, we have proposed several new modified derivative-free families of King's method, Ostrowski's method for solving nonlinear scalar equations. The main advantage of these methods is that they do not use the evaluation of any derivatives but an optimal order of convergence is nonetheless maintained. All the proposed families of methods require three functional evaluations, viz. $f(x_n)$, $f(u_n)$, and $f(y_n)$. In order to obtain an assessment of the efficiency index of our proposed families of methods, we shall make use of the efficiency index [14]. For our proposed iteration schemes, we find $p = 4$ and $d = 3$, yielding $E = \sqrt[3]{4} \cong 1.587$, which is better than those of most third-order methods $E \approx 1.442$, Newton's method and Steffensen's method $E \approx 1.414$. Furthermore, the numerical examples considered here show that in many cases all our proposed methods are efficient alternative to the existing methods.

Acknowledgments The authors wish to thank the anonymous referees for their valuable suggestions and comments.

References

1. Behl, R., Kanwar, V., Sharma, K.K.: Optimal equi-scaled families of Jarratt's method. Int. J. Comput. Math. **90**, 408–422 (2013)
2. Cordero, A., Hueso, J.L., Martínez, E., Torregrosa, J.R.: Generating optimal derivative free iterative methods for nonlinear equations by using polynomial interpolation. Math. Comput. Model. **57**, 1950–1956 (2013)
3. Cordero, A., Hueso, J.L., Martínez, E., Torregrosa, J.R.: A new technique to obtain derivative-free optimal iterative methods for solving nonlinear equations. J. Comput. Appl. Math. **252**, 95–102 (2013)
4. Cordero, A., Hueso, J.L., Martínez, E., Torregrosa, J.R.: Steffensen type methods for solving nonlinear equations. J. Comput. Appl. Math. **236**, 3058–3064 (2012)
5. Cordero, A., Torregrosa, J.R.: A class of Steffensen type methods with optimal order of convergence. Appl. Math. Comput. **217**, 7653–7659 (2011)
6. Kanwar, V., Behl, R., Sharma, K.K.: Simply constructed family of a Ostrowski's method with optimal order of convergence. Comput. Math. Appl. **62**, 4021–4027 (2011)
7. King, R.F.: A family of fourth order methods for nonlinear equations. SIAM J. Numer. Anal. **10**, 876–879 (1973)
8. Kung, H.T., Traub, J.F.: Optimal order of one-point and multi-point iteration. J. ACM **21**, 643–651 (1974)
9. Ostrowski, A.M.: Solutions of Equations and System of Equations. Academic Press, New York (1960)
10. Peng, Y., Feng, H., Li, Q., Zhang, X.: A fourth-order derivative-free algorithm for nonlinear equations. J. Comput. Appl. Math. **235**, 2551–2559 (2011)
11. Petković, M.S., Neta, B., Petković, L.D., Dzunić, J.: Multipoint Methods for Solving Nonlinear Equations. Academic Press, New York (2012)
12. Petković, M.S., Ilić, S., Džunić, J.: Derivative free two-point methods with and without memory for solving nonlinear equations. Appl. Math. Comput. **217**, 1887–1895 (2010)
13. Soleymani, F., Sharma, R., Li, X., Tohidi, E.: An optimized derivative-free form of the Potra-Pták method. Math. Comp. Model. **56**, 97–104 (2012)
14. Traub, J.F.: Iterative Methods for the Solution of Equations. Prentice-Hall, Englewood Cliffs, NJ (1964)
15. Zheng, Q., Zhao, P., Huang, F.: A family of fourth-order Steffensen-type methods with the applications on solving nonlinear ODEs. Appl. Math. Comput. **217**, 8196–8203 (2011)

Existence of Solution for Fractional Stochastic Integro-Differential Equation with Impulsive Effect

Mohd Nadeem and Jaydev Dabas

Abstract This paper is concerned with the existence and uniqueness of the solution for an impulsive fractional stochastic integro-differential equation. The existence and uniqueness results are shown using the fixed point technique on a Hilbert space.

Keywords Fractional order differential equation · Stochastic functional differential equations · Existence results · Impulsive conditions

1 Introduction

It is well known that the fractional calculus is a classical mathematical notion and is a generalization of ordinary differentiation and integration to arbitrary order. Nowadays, studying fractional calculus has become an active area of research field as it has gained considerable importance due to its numerous applications in various fields, such as physics, chemistry, viscoelasticity, engineering sciences, etc. For more details, one can see the cited papers [1–8, 14] and reference therein.

The deterministic models often fluctuate due to environmental noise. A natural extension of a deterministic model is stochastic model, where relevant parameters are modeled as suitable stochastic processes. Due to this fact that, most of the problems in a practical life situation are modeled by stochastic equations rather than deterministic. Therefore, it is of great significance to introduce stochastic effects in the investigation of differential equations [13]. For more details on stochastic differential equations see [10–12] and references therein.

However, it is known that the impulsive effects exist widely in different areas of real world such as mechanics, electronics, telecommunications, finance, economics,

M. Nadeem (✉) · J. Dabas
Department of Applied Science and Engineering, IIT Roorkee,
Saharanpur Campus, Saharanpur 247001, India
e-mail: mohdnadeem.jmi@gmail.com

J. Dabas
e-mail: jay.dabas@gmail.com

P.N. Agrawal et al. (eds.), *Mathematical Analysis and its Applications*,
Springer Proceedings in Mathematics & Statistics 143,
DOI 10.1007/978-81-322-2485-3_30

etc., for more detail see [9]. Due to this fact, the states of many evolutionary processes are often subject to instantaneous perturbations and experience abrupt changes at certain moments of time. The duration of these changes is very short and negligible in comparison with the duration of the process considered, and can be thought of as impulses. Therefore, it is important to consider the effect of impulses in the investigation of stochastic differential equations.

Wang et al. [16] considered the following impulsive fractional differential equation for order $q \in (1, 2)$

$$\begin{aligned} &{}^cD_t^q u(t) = f(t, u(t)), \quad t \in J' = [0, T], \ q \in (1, 2), \\ &\Delta u(t_k) = I_k(u(t_k^-)), \Delta u'(t_k) = J_k(u(t_k^-)), k = 1, 2, \dots, m, \\ &u(0) = u_0, \ u'(0) = \overline{u}_0, \end{aligned}$$

and discussed the existence and uniqueness of solutions with the help of Banach fixed point theorem and Krasnoselskii fixed point theorem.

Sakthivel et al. [15] considered the following impulsive fractional stochastic differential equations with infinite delay in the form

$$\begin{cases} D_t^\alpha x(t) = Ax(t) + f(t, x_t, B_1 x(t)) + \sigma(t, x_t, B_2 x(t)) \frac{dw(t)}{dt}, t \in [0, T], t \neq t_k, \\ \Delta x(t_k) = I_k(x(t_k)), \ k = 1, 2, \dots, m, \\ x(t) = \phi(t), \ \ \phi(t) \in \mathscr{B}_h, \end{cases}$$

and discussed the existence of mild solutions using Banach contraction principle, Krasnoselskii's fixed point theorem.

Motivated by the mentioned work [15, 16], in this article, we are concerned with the existence and uniqueness of solution for impulsive fractional functional integro-differential equation of the form:

$$\begin{aligned} {}^cD_t^\alpha x(t) = &f\left(t, x(t), x_t, \int_0^t K(t, s)x(s)ds\right) \\ &+ g\left(t, x(t), x_t, \int_0^t K(t, s)x(s)ds\right) \frac{dw(t)}{dt}, t \in J = [0, T], t \neq t_k, \end{aligned} \tag{1}$$

$$\Delta x(t_k) = I_k(x(t_k^-)), \Delta x'(t_k) = Q_k(x(t_k^-)), k = 1, 2, \dots, m, \tag{2}$$

$$x(t) = \phi(t), x'(0) = x_1, \ t \in [-d, 0], \tag{3}$$

where J is an operational interval and ${}^cD_t^\alpha$ denotes the Caputo's fractional derivative of order $\alpha \in (1, 2)$ and $x(\cdot)$ takes the value in the real separable Hilbert space $\mathscr{H}$; $f : J \times \mathscr{H} \times PC^0_{\mathscr{L}} \times \mathscr{H} \to \mathscr{H}$ and $g : J \times \mathscr{H} \times PC^0_{\mathscr{L}} \times \mathscr{H} \to \mathscr{L}(\mathscr{K}, \mathscr{H})$ and $I_k, Q_k : \mathscr{H} \to \mathscr{H}$ are appropriate functions; $\phi(t)$ is $\mathscr{F}_0$-measurable $\mathscr{H}$-valued random variables independent of w. Here let $0 = t_0 < t_1 < \cdots < t_m < t_{m+1} = T$, $\Delta x(t_k) = x(t_k^+) - x(t_k^-)$, $\Delta x'(t_k) = x'(t_k^+) - x'(t_k^-)$, $x(t_k^+)$ and $x(t_k^-)$ denote the right and left limits of x at t_k. Similarly, $x'(t_k^+)$ and $x'(t_k^-)$ denote the right and left limits of x' at t_k, respectively.

For further details, this work has three sections. Second section provides some basic definitions, preliminaries, theorems, and lemmas. Third section is equipped with main results for the considered problem (1)–(3).

2 Preliminaries

Let $\mathscr{H}, \mathscr{K}$ be two real separable Hilbert spaces and $\mathscr{L}(\mathscr{K}, \mathscr{H})$ be the space of bounded linear operators from $\mathscr{K}$ into $\mathscr{H}$. For convenience, we will use the same notation $\|\cdot\|$ to denote the norms in $\mathscr{H}$, $\mathscr{K}$ and $\mathscr{L}(\mathscr{K}, \mathscr{H})$, and use $(\cdot, \cdot)$ to denote the inner product of $\mathscr{H}$ and $\mathscr{K}$ without any confusion. Let $(\Omega, \mathscr{F}, \{\mathscr{F}_t\}_{t\geq 0}, \mathscr{P})$ be a complete filtered probability space satisfying that $\mathscr{F}_0$ contains all $\mathscr{P}$-null sets of $\mathscr{F}$. An $\mathscr{H}$-valued random variable is an $\mathscr{F}$-measurable function $x(t) : \Omega \to \mathscr{H}$ and a collection of random variables $S = \{x(t, \omega) : \Omega \to \mathscr{H} \setminus t \in J\}$ is called stochastic process. Usually we write $x(t)$ instead of $x(t, \omega)$ and $x(t) : J \to \mathscr{H}$ in the space of S. $\mathscr{W} = (\mathscr{W}_t)_{t\geq 0}$ be a $\mathscr{Q}$-Wiener process defined on $(\Omega, \mathscr{F}, \{\mathscr{F}_t\}_{t\geq 0}, \mathscr{P})$ with the covariance operator $\mathscr{Q}$ such that $Tr\mathscr{Q} < \infty$. We assume that there exists a complete orthonormal system $\{e_k\}_{k\geq 1}$ in $\mathscr{K}$, a bounded sequence of nonnegative real numbers λ_k such that $\mathscr{Q}e_k = \lambda_k e_k,\ k = 1, 2, \ldots,$ and a sequence of independent Brownian motions $\{\beta_k\}_{k\geq 1}$ such that

$$(w(t), e)_{\mathscr{K}} = \sum_{k=1}^{\infty} \sqrt{\lambda_k}(e_k, e)_{\mathscr{K}} \beta_k(t), e \in \mathscr{K}, t \geq 0.$$

Let $\mathscr{L}_0^2 = \mathscr{L}^2(\mathscr{Q}^{\frac{1}{2}}\mathscr{K}, \mathscr{H})$ be the space of all Hilbert Schmidt operators from $\mathscr{Q}^{\frac{1}{2}}\mathscr{K}$ to $\mathscr{H}$ with the inner product $< \varphi, \psi >_{\mathscr{L}_0^2} = Tr[\varphi\mathscr{Q}\psi*]$.

The collection of all strongly measurable, square integrable, $\mathscr{H}$-valued random variables, denoted by $\mathscr{L}^2(\Omega, \mathscr{F}, \{\mathscr{F}_t\}_{t\geq 0}, \mathscr{P}; \mathscr{H}) = \mathscr{L}^2(\Omega; \mathscr{H})$, is a Banach space equipped with norm $\|x(\cdot)\|^2_{\mathscr{L}^2} = E\|x(\cdot, w)\|^2_{\mathscr{H}}$, where E denotes expectation defined by $E(h) = \int_\Omega h(w)d\mathscr{P}$. An important subspace is given by $\mathscr{L}_0^2(\Omega; \mathscr{H}) = \{f \in \mathscr{L}^2(\Omega, \mathscr{H}) : f$ is $\mathscr{F}_0$- is measurable$\}$.

Let $PC^0_{\mathscr{L}} = C([-d, 0], \mathscr{L}^2(\Omega; \mathscr{H}))$ be a Banach space of all continuous map from $[-d, 0]$ into $\mathscr{L}^2(\Omega; \mathscr{H})$ satisfying the condition $\sup E\|\phi(t)\|^2 < \infty$ with norm

$$\|\phi\|_{PC^0_{\mathscr{L}}} = \sup_{t\in[-d,0]} \left\{E\|\phi(t)\|_{\mathscr{H}}, \phi \in PC^0_{\mathscr{L}}\right\}.$$

Consider $C^2(J, \mathscr{L}^2(\Omega; \mathscr{H}))$ be a Banach space of all continuously differentiable map from J into $\mathscr{L}^2(\Omega; \mathscr{H})$ satisfying the condition $\sup E\|x(t)\|^2 < \infty$ with norm defined

$$\|x\|_{C^2}^2 = \sup_{t\in J}\sum_{j=0}^{1}\left\{E\|x^j(t)\|_{\mathscr{H}}^2, x \in C^2(J, \mathscr{L}^2(\Omega; \mathscr{H}))\right\}.$$

To study the impulsive conditions, we consider

$$PC_{\mathscr{L}}^2 = PC^2([-d, T], \mathscr{L}^2(\Omega; \mathscr{H}))$$

a Banach space of all such continuous functions $x : [-d, T] \to \mathscr{L}^2(\Omega; \mathscr{H})$, which are continuously differentiable on $[0, T]$ except for a finite number of points $t_i \in (0, T),\ i = 1, 2, \ldots, \mathscr{N}$, at which $x'(t_i^+)$ and $x'(t_i^-) = x'(t_i)$ exist and are endowed with the norm

$$\|x\|_{PC_{\mathscr{L}}^2}^2 = \sup_{t\in J}\sum_{j=0}^{1}\left\{E\|x^j(t)\|_{\mathscr{H}}^2, x \in PC_{\mathscr{L}}^2\right\}.$$

Definition 1 The Reimann–Liouville fractional integral operator for order $\alpha > 0$, of a function $f : \mathscr{R}^+ \to \mathscr{R}$ and $f \in L^1(\mathscr{R}^+, X)$ is defined by

$$J_t^0 f(t) = f(t),\ J_t^\alpha f(t) = \frac{1}{\Gamma(\alpha)}\int_0^t (t-s)^{\alpha-1} f(s)ds, \quad \alpha > 0,\ t > 0,$$

where $\Gamma(\cdot)$ is the Gamma function.

Definition 2 Caputo's derivative of order $\alpha > 0$ for a function $f : [0, \infty) \to \mathscr{R}$ is defined as

$$D_t^\alpha f(t) = \frac{1}{\Gamma(n-\alpha)}\int_0^t (t-s)^{n-\alpha-1} f^{(n)}(s)ds = J^{n-\alpha} f^{(n)}(t),$$

for $n - 1 < \alpha < n,\ n \in N$. If $0 < \alpha < 1$, then

$$D_t^\alpha f(t) = \frac{1}{\Gamma(1-\alpha)}\int_0^t (t-s)^{-\alpha} f^{(1)}(s)ds.$$

Obviously, Caputo's derivative of a constant is equal to zero.

Lemma 1 *A measurable $\mathscr{F}_t$-adapted stochastic process $x : [-d, T] \to \mathscr{H}$ such that $x \in PC_{\mathscr{L}}^2$ is called a mild solution of the system (1)–(3) if $x(0) = \phi(0)$ and $x'(0) = x_1$ on $[-d, 0]$, $\Delta x|_{t=t_k} = I_k(x(t_k^-))$ and $\Delta x'|_{t=t_k} = Q_k(x(t_k^-))$, $k = 1, 2, \cdots, m$ the restriction of $x(\cdot)$ to the interval $[0, T)\backslash t_1, \cdots, t_m$ is continuous and $x(t)$ satisfies the following fractional integral equation*

$$x(t)=\begin{cases}\phi(0)+x_1t+\frac{1}{\Gamma(\alpha)}\int_0^t(t-s)^{\alpha-1}f\left(s,x(s),x_s,\int_0^t K(s,t)x(s)ds\right)ds\\ +\frac{1}{\Gamma(\alpha)}\int_0^t(t-s)^{\alpha-1}g\left(s,x(s),x_s,\int_0^t K(s,t)x(s)ds\right)dw(s), & t\in(0,t_1],\\ \phi(0)+x_1t+I_1(x(t_1^-))+Q_1(x(t_1^-))(t-t_1)\\ +\frac{1}{\Gamma(\alpha)}\int_0^t(t-s)^{\alpha-1}f\left(s,x(s),x_s,\int_0^t K(s,t)x(s)ds\right)ds\\ +\frac{1}{\Gamma(\alpha)}\int_0^t(t-s)^{\alpha-1}g\left(s,x(s),x_s,\int_0^t K(s,t)x(s)ds\right)dw(s), & t\in(t_1,t_2],\\ \cdots\\ \phi(0)+x_1t+\sum_{i=1}^k\left[I_i(x(t_i^-))+Q_i(x(t_i^-))(t-t_i)\right]\\ +\frac{1}{\Gamma(\alpha)}\int_0^t(t-s)^{\alpha-1}f\left(s,x(s),x_s,\int_0^t K(s,t)x(s)ds\right)ds\\ +\frac{1}{\Gamma(\alpha)}\int_0^t(t-s)^{\alpha-1}g\left(s,x(s),x_s,\int_0^t K(s,t)x(s)ds\right)dw(s), & t\in(t_k,t_{k+1}].\end{cases}$$

Further, we introduce the following assumptions to establish our results:

(H1) The nonlinear maps f and g are continuous and there exit constants $\mu_1,\mu_2,\mu_3,\nu_1,\nu_2,\nu_3>0$ such that

$$E\|f(t,x,\varphi,u)-f(t,y,\psi,v)\|^2_{\mathscr{H}}\le\mu_1\|x-y\|^2_{\mathscr{H}}+\mu_2\|\varphi-\psi\|_{PC^0_{\mathscr{L}}}+\mu_3\|u-v\|^2_{\mathscr{H}},$$
$$E\|g(t,x,\varphi,u)-g(t,y,\psi,v)\|^2_{\mathscr{H}}\le\nu_1\|x-y\|^2_{\mathscr{H}}+\nu_2\|\varphi-\psi\|_{PC^0_{\mathscr{L}}}+\nu_3\|u-v\|^2_{\mathscr{H}}$$

for all $x,y,u,v\in\mathscr{H}$, $t\in J$ and $\varphi,\psi\in PC^0_{\mathscr{L}}$.

(H2) The functions I_k,Q_k are continuous and there exists $L_I,L_Q>0$, such that

$$E\|I_k(x)-I_k(y)\|^2_{\mathscr{H}}\le L_IE\|x-y\|^2_{\mathscr{H}},$$
$$E\|Q_k(x)-Q_k(y)\|^2_{\mathscr{H}}\le L_QE\|x-y\|^2_{\mathscr{H}}$$

for all $x,y\in\mathscr{H}$ and $k=1,2,\cdots,m$.

3 Existence and Uniqueness Results

This result is based on Banach contraction fixed point theory.

Theorem 1 *Suppose that the assumptions (H1) and (H2) hold and*

$$\Theta=\left\{4(mL_I+mT^2L_Q)+\frac{4T^{2\alpha}}{\Gamma(\alpha)}\left[\frac{1}{\alpha^2}(\mu_1+\mu_2+\mu_3K^*)+\frac{1}{T(2\alpha-1)}(\nu_1+\nu_2+\nu_3K^*)\right]\right\}<1,$$

where $K^*=\sup_{t\in[0,t]}\int_0^t K(t,s)ds<\infty$. Then the system (1)–(3) has a unique solution.

Proof We convert the problem (1)–(3) into fixed point problem. We consider an operator $N : PC^2_{\mathscr{L}} \to PC^2_{\mathscr{L}}$ defined by

$$(Nx)(t) = \begin{cases} \phi(0) + x_1 t + \frac{1}{\Gamma(\alpha)} \int_0^t (t-s)^{\alpha-1} f\left(s, x(s), x_s, \int_0^t K(s,t)x(s)ds\right) ds \\ + \frac{1}{\Gamma(\alpha)} \int_0^t (t-s)^{\alpha-1} g\left(s, x(s), x_s, \int_0^t K(s,t)x(s)ds\right) dw(s), & t \in (0, t_1], \\ \phi(0) + x_1 t + I_1(x(t_1^-)) + Q_1(x(t_1^-))(t - t_1) \\ + \frac{1}{\Gamma(\alpha)} \int_0^t (t-s)^{\alpha-1} f\left(s, x(s), x_s, \int_0^t K(s,t)x(s)ds\right) ds \\ + \frac{1}{\Gamma(\alpha)} \int_0^t (t-s)^{\alpha-1} g\left(s, x(s), x_s, \int_0^t K(s,t)x(s)ds\right) dw(s), & t \in (t_1, t_2], \\ \cdots \\ \phi(0) + x_1 t + \sum_{i=1}^k \left[I_i(x(t_i^-)) + Q_i(x(t_i^-))(t - t_i)\right] \\ + \frac{1}{\Gamma(\alpha)} \int_0^t (t-s)^{\alpha-1} f\left(s, x(s), x_s, \int_0^t K(s,t)x(s)ds\right) ds \\ + \frac{1}{\Gamma(\alpha)} \int_0^t (t-s)^{\alpha-1} g\left(s, x(s), x_s, \int_0^t K(s,t)x(s)ds\right) dw(s), & t \in (t_k, t_{k+1}]. \end{cases}$$

Now we show that N is a contraction map. For this we take two points x, x^* such that for $t \in (0, t_1]$

$$\begin{aligned} E\|(Nx)(t) - (Nx^*)(t)\|^2_{\mathscr{H}} &\leq 2E\|\frac{1}{\Gamma(\alpha)} \int_0^t (t-s)^{\alpha-1} [f\left(s, x(s), x_s, \int_0^t K(s,t)x(s)ds\right) \\ &\quad - f\left(s, x^*(s), x_s^*, \int_0^t K(s,t)x^*(s)ds\right) ds\|^2_{\mathscr{H}} \\ &\quad + 2E\|\frac{1}{\Gamma(\alpha)} \int_0^t (t-s)^{\alpha-1} [g(s, x(s), x_s, \int_0^t K(s,t)x(s)ds) \\ &\quad - g\left(s, x^*(s), x_s^*, \int_0^t K(s,t)x^*(s)ds\right) dw(s)\|^2_{\mathscr{H}} \\ &\leq \frac{2T^{2\alpha}}{\Gamma(\alpha)} \left[\frac{1}{\alpha^2}(\mu_1 + \mu_2 + \mu_3 K^*)\right) \\ &\quad + \frac{1}{T(2\alpha - 1)}(\nu_1 + \nu_2 + \nu_3 K^*)\|x - x^*\|^2_{PC^2_{\mathscr{L}}}. \end{aligned}$$

When $t \in (t_1, t_2]$,

$$\begin{aligned} E\|(Nx)(t) - (Nx^*)(t)\|^2_{\mathscr{H}} &\leq 4E\|I_1(x(t_1^-)) - I_1(x^*(t_1^-))\|^2_{\mathscr{H}} \\ &\quad + 4E\| Q_1(x(t_1^-))(t - t_1) - Q_1(x^*(t_1^-))(t - t_1)\|^2_{\mathscr{H}} \\ &\quad + 4E\|\frac{1}{\Gamma(\alpha)} \int_0^t (t-s)^{\alpha-1} [f\left(s, x(s), x_s, \int_0^t K(s,t)x(s)ds\right) \\ &\quad - f\left(s, x^*(s), x_s^*, \int_0^t K(s,t)x^*(s)ds\right)]ds\|^2_{\mathscr{H}} \\ &\quad + 4E\|\frac{1}{\Gamma(\alpha)} \int_0^t (t-s)^{\alpha-1} [g(s, x(s), x_s, \int_0^t K(s,t)x(s)ds) \\ &\quad - g\left(s, x^*(s), x_s^*, \int_0^t K(s,t)x^*(s)ds\right)]dw(s)\|^2_{\mathscr{H}} \end{aligned}$$

$$\leq \left\{4(L_I + T^2 L_Q) + \frac{4T^{2\alpha}}{\Gamma(\alpha)}\left[\frac{1}{\alpha^2}(\mu_1 + \mu_2 + \mu_3 K^*) + \frac{1}{T(2\alpha - 1)}(\nu_1 + \nu_2 + \nu_3 K^*)\right]\right\} \|x - x^*\|^2_{PC^2_{\mathscr{L}}}.$$

Similarly for $t \in (t_k, t_{k+1}]$, $k = 2, 3, \ldots, m$,

$$E\|(Nx)(t) - (Nx^*)(t)\|^2_{\mathscr{H}} \leq \left\{4(mL_I + mT^2 L_Q) + \frac{4T^{2\alpha}}{\Gamma(\alpha)}\left[\frac{1}{\alpha^2}(\mu_1 + \mu_2 + \mu_3 K^*) \frac{1}{T(2\alpha - 1)}(\nu_1 + \nu_2 + \nu_3 K^*)\right]\right\} \|x - x^*\|^2_{PC^2_{\mathscr{L}}}$$
$$= \Theta\|x - x^*\|^2_{PC^2_{\mathscr{L}}}.$$

Since $\Theta < 1$, by the condition given in Theorem 1, N is a contraction map and therefore it has a unique fixed point $x \in PC^2_{\mathscr{L}}$ which is a solution of our equation (1)–(3) on J. This completes the proof of the theorem.

References

1. Chauhan, A., Dabas, J.: Existence of mild solutions for impulsive fractional-order semilinear evolution equations with nonlocal conditions. Electron. J. Differ. Equ. **2011**(107), 1–10 (2011)
2. Chauhan, A., Dabas, J.: Local and global existence of mild solution to an impulsive fractional functional integro-differential equation with nonlocal condition. Commun. Nonlinear. Sci. Numer. Simulat. **19**, 821–829 (2014)
3. Chauhan, A., Dabas, J., Kumar, M.: Integral boundary-value problem for impulsive fractional functional differential equations with infinite delay. Electron. J. Differ. Equ. **2012**(229), 1–13 (2012)
4. Dabas, J., Chauhan, A.: Existence and uniqueness of mild solution for an impulsive neutral fractional integro-differential equation with infinite delay. Math. Comput. Model. **57**, 754–763 (2013)
5. Dabas, J., Chauhan, A., Kumar, M.: Existence of the mild solutions for impulsive fractional equations with infinite delay. Intern. J. Differ. Equ. **793023**, 20 (2011)
6. Dabas, J., Gautam, G.R.: Impulsive neutral fractional integro-differential equations with state dependent delays and integral conditions. Electron. J. Differ. Equ. **2013**(273), 1–13 (2013)
7. Kilbas, A.A., Srivastava, H.M., Trujillo, J.J.: Theory and Applications of Fractional Differential Equations. Elsevier Science B.V, Amsterdam (2006)
8. Lakshmikantham, V.: Theory of fractional differential equations. Nonlinear Anal. **69**, 3337–3343 (2008)
9. Lakshmikantham, V., Bainov, D., Simeonov, P.: Theory of Impulsive Differential Equations. World Scientific Press, Singapore (1989)
10. Li, C., Sun, J., Sun, R.: Stability analysis of a class of stochastic differential delay equations with nonlinear impulsive effects. J. Franklin Inst. **347**, 1186–1198 (2010)
11. Lin, A., Ren, Y., Xia, N.: On neutral impulsive stochastic integro-differential equations with infinite delays via fractional operators. Math. Comput. Model. **51**, 413–424 (2010)
12. Longa, S., Teng, L., Xu, D.: Global attracting set and stability of stochastic neutral partial functional differential equations with impulses. Stat. Probab. Lett. **82**, 1699–1709 (2012)
13. Oksendal, B.: Stochastic Differential Equations. Springer, Berlin, Heidelberg (2005)

14. Podlubny, I.: Fractional Differential Equations. Acadmic Press, New York, USA (1993)
15. Sakthivel, R., Revathi, P., Ren, Y.: Existence of solutions for nonlinear fractional stochastic differential equations. Nonlilear Anal. **81**, 70–86 (2013)
16. Wang, J.R., Li, X., Wei, W.: On the natural solution of an impulsive fractional differential equation of order $\alpha \in (1, 2)$. Commun. Nonlinear Sci. Numer. Simulat. **17**, 4384–4394 (2012)

Singularly Perturbed Convection-Diffusion Turning Point Problem with Shifts

Pratima Rai and Kapil K. Sharma

Abstract In this paper, a class of singularly perturbed turning point problem with shifts (i.e., delay as well as advance) is considered. Presence of turning point results into twin boundary layers in the solution of the problem under consideration. For the numerical approximation of the problem, a finite difference scheme is proposed on a uniform mesh. Interpolation is used to tackle the terms containing shifts and to deal with the difficulty arising due to presence of the turning point a combination of backward and forward difference is used in the first derivative term. Convergence analysis is given for the proposed numerical scheme. Numerical results are presented which illustrate the theoretical results and depict the effect of shifts on the layer behavior of the solution.

Keywords Singular perturbation · Turning point · Positive shifts · Negative shifts · Finite difference scheme · Boundary layer

AMS Subject Classifications: 34K26 · 65L12 · 34K28

1 Introduction

In the recent past, numerical treatment of singularly perturbed differential equations with positive/negative shifts have attracted the attention of many researchers. These type of differential equations model a wide range of real-life phenomena including variety of models of physiological processes or diseases [5], first exit problem in neu-

P. Rai
Department of Mathematics, University of Delhi, Delhi 110007, India
e-mail: prai@maths.du.ac.in

K.K. Sharma (✉)
Department of Mathematics, South Asian University (SAU), Akbar Bhawan, Chanakyapuri 110021, New Delhi, India
e-mail: kapil.sharma@sau.ac.in

P.N. Agrawal et al. (eds.), *Mathematical Analysis and its Applications*,
Springer Proceedings in Mathematics & Statistics 143,
DOI 10.1007/978-81-322-2485-3_31

robiology [18, 19], population ecology, materials with thermal memory, variational problem in control theory [2, 4], etc.

In this paper, we consider a boundary value problem for singularly perturbed differential-difference equation of the form

$$\varepsilon u''(x) + a(x)u'(x) - b(x)u(x) + c(x)u(x-\delta) + d(x)u(x+\eta) = f(x) \quad x \in \Omega = (-1, 1) \tag{1}$$

subject to the interval and boundary conditions

$$u(x) = \varphi(x), \quad x \in \Omega_0; \quad u(x) = \gamma(x), \quad x \in \Omega_4 \tag{2}$$

where $0 < \varepsilon \ll 1$, $\bar{\Omega} = [-1, 1]$, $\Omega = \Omega_1 \cup \Omega_2 \cup \Omega_3$, $\Omega_1 = (-1, -1+\delta]$, $\Omega_2 = (-1+\delta, 1-\eta)$, $\Omega_3 = [1-\eta, 1)$, $\Omega_0 = [-1-\delta, -1]$, $\Omega_4 = [1, 1+\eta]$, δ, η are the shift arguments independent of ε; $a(x)$, $b(x)$, $c(x)$, $d(x)$, $\varphi(x)$, $\gamma(x)$, $f(x)$ are sufficiently smooth functions satisfying

$$a(0) = 0, \quad a'(0) \le 0, \tag{3}$$

$$|a'(x)| \ge |a'(0)/2| \quad \forall\, x \in \bar{\Omega}, \tag{4}$$

$$b(x) > 0, \quad b(x) - c(x) - d(x) \ge k > 0, \quad c(x) > 0, \quad d(x) > 0 \,\forall\, x \in \Omega. \tag{5}$$

For $u(x)$ to be smooth solution of the problem (1)–(2), it should satisfy boundary conditions, be continuous on $\bar{\Omega}$ and differentiable on Ω. When shift terms are zero (i.e., δ, $\eta = 0$), the above problem reduces to boundary value problem for singularly perturbed differential equation with turning point. In this case, as the perturbation parameter tends to zero there may be boundary or interior layer depending upon the argument $\beta = b(0)/a'(0)$. For $\beta > 0$, the solution of the problem exhibit interior layer around the turning point whereas for $\beta < 0$ there are twin boundary layers of exponential type in the solution of the problem.

For questions on existence and uniqueness of differential equations with shifts, one can see [6, 10–12] whereas numerical analysis is given in [3, 13]. Investigation of boundary value problems for the second-order differential equations with shifts was initiated by Lange and Miura [7–9] who extended the method of matched asymptotic expansions developed for ODEs. Numerical study of singularly perturbed differential equations with small shifts was considered in [14–16] where authors approximated the terms containing shifts by Taylor series expansion. Taylor's series holds good when shifts are small but give bad approximation when shifts are large. In [1, 17], authors constructed numerical schemes to deal with such type of problems when shifts are large but, their study is limited to the case when the convection coefficient has same sign throughout the domain, i.e., non-turning point case. In this paper, we study the turning point case with both positive (advance) as well as negative (delay) shifts.

An outline of the paper is as follows. In the next section, we discuss some properties of the exact solution. In Sect. 3, numerical scheme is described and convergence properties of the proposed scheme are analyzed. Numerical experiments are presented in Sect. 4 which validate the theoretical results computationally. The paper ends with a summary of the main conclusions.

We adopt certain conventions that throughout the paper C denotes a generic positive constant independent of N and ε, $||.||$ is maximum norm.

2 Some Properties of the Exact Solution

In this section, we analyze some properties of the solution and its derivatives for the problem (1)–(2) which are needed later on for the convergence analysis of the proposed numerical method.

Let L_ε denotes the differential operator occurring in problem (1)–(2) which is defined as

$$L_\varepsilon u(x) = \begin{cases} \varepsilon u''(x) + a(x)u'(x) - b(x)u(x) + d(x)u(x+\eta) = f(x) - c(x)\varphi(x-\delta) & x \in \Omega_1 \\ \varepsilon u''(x) + a(x)u'(x) - b(x)u(x) + c(x)u(x-\delta) + d(x)u(x+\eta) = f(x) & x \in \Omega_2 \\ \varepsilon u''(x) + a(x)u'(x) - b(x)u(x) + c(x)u(x-\delta) = f(x) - d(x)\gamma(x+\eta) & x \in \Omega_3. \end{cases} \tag{6}$$

L_ε satisfy following minimum principle on $\bar{\Omega}$

Lemma 1 *Let Ψ be a smooth function satisfying $\Psi(-1) \geq 0$, $\Psi(1) \geq 0$ and $L_\varepsilon \Psi(x) \leq 0$, $\forall\, x \in \Omega$. Then, $\Psi(x) \geq 0\, \forall\, x \in \bar{\Omega}$.*

□

Immediate consequence of the above minimum principle is the following stability estimate.

Lemma 2 *If $u(x)$ is solution of the problem (1)–(2) then for some positive constant C we have*

$$||u(x)|| \leq C\left[||f||/k + max\{||\varphi||,\ ||\gamma||\}\right]. \tag{7}$$

□

Next, we divide the domain $\bar{\Omega}$ into three regions, $D_1 = [-1, -\mu]$, $D_2 = (-\mu, \mu)$, $D_3 = [\mu, 1]$, $0 < \mu \leq 1/2$. Following theorem gives us bound on the derivatives of the solution $u(x)$ of the problem (1)–(2) in the intervals D_1, D_3.

Theorem 3 *Let $a(x)$, $b(x)$, $c(x)$, $d(x)$, $f(x) \in C^j(\bar{\Omega})$, $\varphi(x) \in C^j(\Omega_0)$, $\gamma(x) \in C^j(\Omega_4)$, $||a|| = M$, $|a(x)| \geq \alpha > 0$, $x \in D_1 \cup D_3$. Then, there exist a positive constant C such that the solution $u(x)$ of the problem (1)–(2) satisfies*

$$|u^{(i)}(x)| \leq \begin{cases} C\left(1 + \varepsilon^{-i} e^{-\alpha(1-x)/\varepsilon}\right),\ i = 1, \ldots j+1,\ x \in D_3 \\ C\left(1 + \varepsilon^{-i} e^{-\alpha(x+1)/\varepsilon}\right)\ i = 1, \ldots j+1,\ x \in D_1. \end{cases}$$

□

Theorem 3 gives us bound on the derivatives of the solution outside the turning point region. Therefore, we are left with obtaining the bound on the derivatives of the solution in the region D_2 which is given by the following theorem.

Theorem 4 *Let* $a(x),\ b(x),\ c(x),\ d(x),\ f(x) \in C^j(\bar{\Omega}), \varphi(x) \in C^j(\Omega_0),\ \gamma(x) \in C^j(\Omega_4)$ *and conditions* (3)–(5) *holds. Then there exist a positive constant* C *such that the solution* $u(x)$ *of the problem* (1)–(2) *satisfies*

$$|u^{(i)}(x)| \leq C, \quad i = 1, \ldots j, \quad x \in D_2. \tag{8}$$

□

3 Discretization and Convergence

In this section, we describe a upwind finite difference scheme on uniform mesh. Let $w = x_i = -1 + ih$, where $i = 1,\ 2, \ldots,\ N-1;\ h = 2/N$ be uniform mesh on $\Omega, \bar{w} = w \cup \{-1,\ 1\}$ and $x = 0$ be the turning point. To tackle the terms containing positive/negative shifts, we use interpolation. Earlier, authors [1, 17] constructed numerical schemes in which they considered a special type of mesh in which the term containing shifts lie at the nodal point. But this mesh selection has a drawback that it put restriction on the number of mesh generations. To overcome this drawback, we propose a scheme which works equally well in both the cases, i.e., whether the terms containing shifts lie at the node or not. If $x_i-\delta,\ x_i+\eta,\ i = 0, \ldots, N$ are not the nodal points then there exist $0 < m_0,\ m_1 < N$ such that $m_0 h < \delta < (m_0 + 1)h$ and $m_1 h < \eta < (m_1 + 1)h$. In our algorithm, interpolation is used to approximate the value of $x_i - \delta$ and $x_i + \eta$ in terms of neighboring nodal points.

We introduce certain notations for the mesh functions. For any mesh function $g(x)$, we have

$$g_i = g(x_i),\ D_+g_i = (g_{i+1} - g_i)/h,\ D_-g_i = (g_i - g_{i-1})/h$$
$$||g|| \equiv ||g||_\infty = \max_{0 \leq i \leq N} |g_i|,\ D_+D_-g_i = (g_{i+1} - 2g_i + g_{i-1})/h^2$$
$$s_i = x_i - \delta,\ r_i = x_i + \eta$$

$$D_* g_i = \begin{cases} D_+g_i, & a_i > 0 \\ D_-g_i, & a_i < 0. \end{cases}$$

The difference scheme for the boundary value problem (1)–(2) is given by

$$L^N U_i = \varepsilon D_+ D_- U_i + a_i D_* U_i - b_i U_i + \frac{c_i}{h}\left[(s_i - x_{i-m_0-1})U_{i-m_0} + (x_{i-m_0} - s_i)U_{i-m_0-1}\right] + \frac{d_i}{h}\left[(r_i - x_{i+m_1})U_{i+m_1+1} + (x_{i+1+m_1} - r_i)U_{i+m_1}\right] = f_i \quad (9)$$
$$U_i = \varphi_i, \quad i = -m_0, \ldots, 0$$
$$U_i = \gamma_i, \quad i = N, \ldots, N + m_1.$$

Lemma 5 *Suppose $\Psi_0 \geq 0$ and $\Psi_N \geq 0$. Then, $L_\varepsilon \Psi_i \leq 0$ for all $i = 1, 2, \ldots, N-1$ implies $\Psi_i \geq 0$ for all $i = 0, 1, \ldots, N$.*

□

Lemma 6 *Let U_i be the solution of the problem* (9). *Then*

$$||U|| \leq C\left[||f||/K + max\{||\varphi||, ||\gamma||\}\right]. \quad (10)$$

where $K = min_{1 \leq i \leq N-1}(b_i - c_i - d_i)$.

□

Theorem 7 *Assume $h < \varepsilon$. Then the error $e_i = L^N(u_i - U_i)$ between the solution $u(x_i)$ of the continuous problem* (1)–(2) *and the solution U_i of the discrete problem* (9) *satisfies the estimate*

$$|e_i| \leq C\frac{h}{h+\varepsilon}, \quad i = 0, \ldots N. \quad (11)$$

□

Lemma 6 and Theorem 7 gives us following theorem

Theorem 8 *Let u be solution of* (1) *and* (2) *and U be solution of* (9). *Then*

$$||U - u|| \leq C\frac{h}{h+\varepsilon}.$$

□

4 Numerical Results

In this section, we computationally verify the theoretical results obtained in the previous section. For this we study the performance of the proposed scheme (9) applied to the following test examples. For $\delta,\ \eta = 0$ the solution of the boundary value problem exhibit twin boundary layers. We also illustrate the effect of shift on the solution behavior. Since exact solutions are not known for the considered examples we use double mesh principle to estimate the accuracy in the maximum norm

$$E^N = \max_{0\le i\le N}|U_i^N - U_{2i}^{2N}|,$$

and the convergence rate

$$R^N = \log_2\left(\frac{E^N}{E^{2N}}\right).$$

Example 1

$$\varepsilon u''(x) + 4(1-2x)u'(x) - 4u(x) + u(x-\delta) + 2u(x+\eta) = 0, \quad x \in (0,1) \tag{12}$$

$$u(x) = 1, \ -\delta \le x \le 0, \quad u(x) = 1, 1 \le x \le 1+\eta. \tag{13}$$

5 Conclusion and Discussion

In this paper, a class of linear second-order singularly perturbed convection-diffusion turning point problem with shifts is considered. The solution of such type of differential equations have boundary or interior layers depending upon the sign of the convection and the reaction coefficients. Here, we considered the case where the presence of turning point results into twin boundary layers. There are also terms containing positive and negative shifts.

Difficulty arising due to the presence of the turning point is tackled using combination of forward and backward difference in the numerical approximation of the first derivative term. Interpolation is used to deal with the terms containing shifts. The Tables 1, 2 and 3 gives maximum pointwise error E^N and the rate of convergence R^N for the considered example for different nonzero values of δ, η. Table 4 gives maximum point wise error for the solution when $\delta = 0$, i.e., when we have only positive shifts, whereas Table 5 gives maximum point wise error for the solution when $\eta = 0$, i.e., when only negative shift is there. Graphs are plotted (Figs. 1, 2 and 3) for the considered example to illustrate the effect of shifts on the layer behavior of the solution. It is observed that as the value of the shift argument increases the thickness of the boundary layers increases and steepness decreases. Moreover, the magnitude of the shift depends upon the value of the positive/negative shifts and the coefficient of the terms containing shifts.

Table 1 Maximum pointwise error E^N and rate for convergence R^N for $\delta = 0.15,\ \eta = 0.15$

ϵ	$N = 128$	$N = 256$	$N = 512$	$N = 1024$	$N = 2048$
1	$7.56580E-4$	$3.78626E-4$	$1.12511E-4$	$5.61834E-5$	$4.73650E-5$
	0.99	1.7	1.0	0.25	1.0
2^{-1}	$3.89010E-3$	$1.97371E-3$	$6.22136E-4$	$3.11787E-4$	$2.49418E-4$
	0.98	1.7	1.0	0.32	1.0
2^{-2}	$1.40237E-2$	$7.20224E-3$	$2.49091E-3$	$1.25406E-3$	$9.17788E-4$
	0.96	1.5	0.99	0.45	1.0
2^{-3}	$2.29938E-2$	$1.20099E-2$	$5.11759E-3$	$2.59958E-3$	$1.56537E-3$
	0.94	1.2	0.98	0.73	0.99
2^{-4}	$3.46149E-2$	$1.93660E-2$	$9.90240E-3$	$5.13310E-3$	$2.69769E-3$
	0.84	0.97	0.95	0.93	0.99
2^{-5}	$5.30166E-2$	$3.31337E-2$	$1.87340E-2$	$1.00508E-2$	$5.24686E-3$
	0.68	0.82	0.9	0.94	0.97
2^{-6}	$8.01397E-2$	$5.30070E-2$	$3.31428E-2$	$1.88765E-2$	$1.01488E-2$
	0.6	0.68	0.81	0.9	0.95
2^{-7}	$8.60812E-2$	$7.98999E-2$	$5.30411E-2$	$3.32542E-2$	$1.89585E-2$
	1.1	0.59	0.67	0.81	0.9

Table 2 Maximum pointwise error E^N and rate for convergence R^N for $\delta = 0.1,\ \eta = 0.3$

$\varepsilon \downarrow$	$N = 128$	$N = 256$	$N = 512$	$N = 1024$	$N = 2048$
1	$7.55930E-4$	$1.71292E-4$	$8.54595E-5$	$6.45943E-5$	$4.23081E-5$
	2.1	1.0	0.4	0.6	2.1
2^{-1}	$3.44352E-3$	$8.93709E-4$	$4.49362E-4$	$4.38537E-4$	$2.19673E-4$
	1.9	0.99	0.35	1.0	1.9
2^{-2}	$1.12767E-2$	$3.20727E-3$	$1.62803E-3$	$1.46869E-3$	$7.37259E-4$
	1.8	0.97	0.15	0.99	1.8
2^{-3}	$2.09117E-2$	$7.08495E-3$	$3.66956E-3$	$2.83638E-3$	$1.42862E-3$
	1.6	0.95	0.37	1.0	1.6
2^{-4}	$3.39450E-2$	$1.48829E-2$	$7.98753E-3$	$5.12619E-3$	$2.60610E-3$
	1.2	0.9	0.64	0.98	1.3
2^{-5}	$5.03637E-2$	$2.78583E-2$	$1.58758E-2$	$9.46715E-3$	$4.89823E-3$
	0.85	0.81	0.75	0.95	1.1
2^{-6}	$7.59403E-2$	$4.57304E-2$	$2.86893E-2$	$1.72498E-2$	$9.26065E-3$
	0.73	0.67	0.73	0.9	1.0
2^{-7}	$8.26705E-2$	$6.93906E-2$	$4.64092E-2$	$2.99831E-2$	$1.70699E-2$
	0.25	0.58	0.63	0.81	0.93

Table 3 Maximum pointwise error E^N and rate for convergence R^N for $\delta = 0.3,\ \eta = 0.1$

$\varepsilon \downarrow$	$N = 128$	$N = 256$	$N = 512$	$N = 1024$	$N = 2048$
1	$4.79304E-4$	$2.38424E-4$	$1.74441E-4$	$8.71994E-5$	$2.98139E-5$
	1.0	0.45	1.0	1.5	1.0
2^{-1}	$2.45619E-3$	$1.23898E-3$	$8.91549E-4$	$4.47194E-4$	$1.56942E-4$
	0.99	0.47	0.99	1.5	1.0
2^{-2}	$9.00776E-3$	$4.62390E-3$	$3.20011E-3$	$1.61050E-3$	$5.94638E-4$
	0.96	0.53	0.99	1.4	1.0
2^{-3}	$1.75028E-2$	$9.23944E-3$	$5.64799E-3$	$2.86161E-3$	$1.21974E-3$
	0.92	0.71	0.98	1.2	0.99
2^{-4}	$3.06080E-2$	$1.74018E-2$	$1.00675E-2$	$5.20123E-3$	$2.46531E-3$
	0.81	0.79	0.95	1.1	0.99
2^{-5}	$4.97544E-2$	$3.11371E-2$	$1.84108E-2$	$9.87295E-3$	$4.94305E-3$
	0.68	0.76	0.9	1.0	0.97
2^{-6}	$7.57999E-2$	$5.02539E-2$	$3.21188E-2$	$1.82812E-2$	$9.64407E-3$
	0.59	0.65	0.81	0.92	0.95
2^{-7}	$8.16723E-2$	$7.60033E-2$	$5.11203E-2$	$3.20309E-2$	$1.80823E-2$
	1.0	0.57	0.67	0.82	0.9

Table 4 Maximum pointwise error E^N for $\delta = 0,\ \varepsilon = .01$

$\eta \downarrow$	$N = 128$	$N = 256$	$N = 512$	$N = 1024$	$N = 2048$
0.15	$9.06105E-2$	$7.49621E-2$	$4.76055E-2$	$2.88584E-2$	$1.59083E-2$
0.25	$9.09313E-2$	$7.27323E-2$	$4.61588E-2$	$2.77859E-2$	$1.52839E-2$
0.35	$2.48565E-1$	$5.79099E-1$	$7.31938E-1$	$9.10935E-1$	$9.15949E-1$
0.45	$5.47839E-2$	$5.15808E-2$	$3.26882E-2$	$1.83858E-2$	$1.01365E-2$

Table 5 Maximum pointwise error E^N for $\eta = 0,\ \varepsilon = .01$

$\delta \downarrow$	$N = 128$	$N = 256$	$N = 512$	$N = 1024$	$N = 2048$
0.15	$9.04784E-2$	$7.49711E-2$	$4.77148E-2$	$2.89146E-2$	$1.59173E-2$
0.25	$9.00035E-2$	$7.43240E-2$	$4.72919E-2$	$2.86399E-2$	$1.57649E-2$
0.35	$8.47844E-2$	$6.99147E-2$	$4.44824E-2$	$2.69308E-2$	$1.48238E-2$
0.45	$7.83714E-2$	$6.36476E-2$	$4.04636E-2$	$2.44278E-2$	$1.34427E-2$

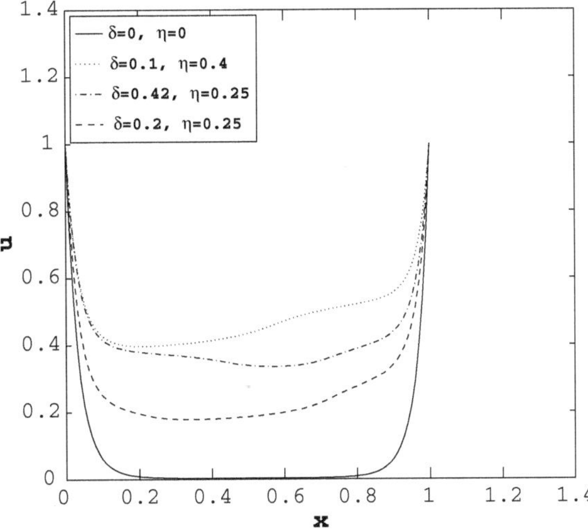

Fig. 1 The numerical solution for Example 1 when δ and η are nonzero ($\varepsilon = 0.1$)

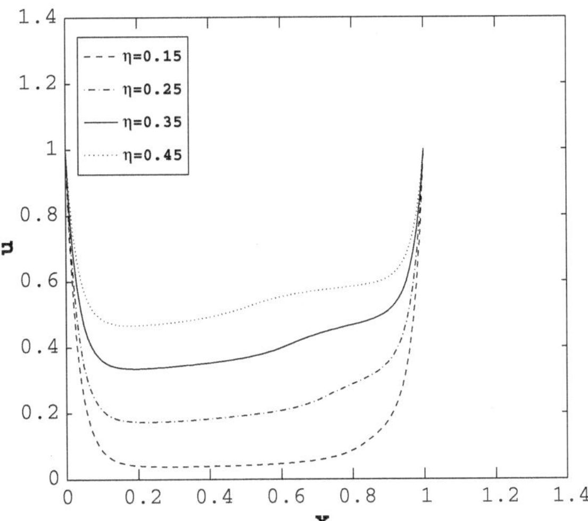

Fig. 2 The numerical solution for Example 1 when $\delta = 0$ ($\varepsilon = 0.1$)

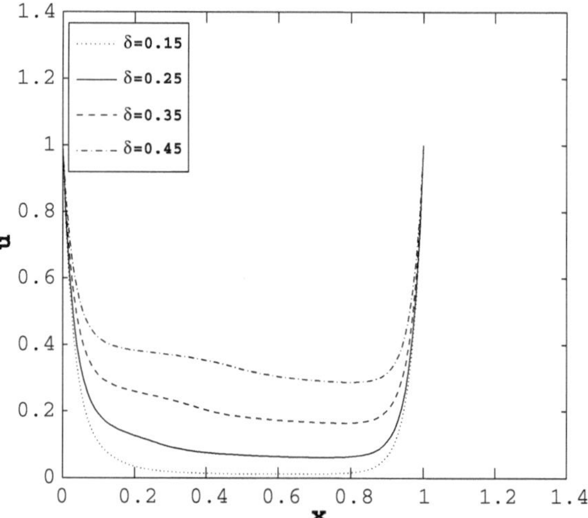

Fig. 3 The numerical solution for Example 1 when $\eta = 0$ ($\varepsilon = 0.1$)

Acknowledgments The first author wish to acknowledge Delhi University, India, 110007 for providing financial support for the above work.

References

1. Amiraliyev, G.M., Cimen, E.: Numerical method for a singularly perturbed convection-diffusion problem with delay. Appl. Math. Comput. **216**, 2351–2359 (2010)
2. Bellman, R., Cooke, K.L.: Differential Difference Equations. Academic Press, New York (1991)
3. Cryer, C.W.: Numerical methods for functional differential equations, in delay and functional differential equations. In: Schmitt, K. (ed.) Proceeding of Conference held at Park city, Utah, 1972
4. E'lsgolts, L.E.: Qualitative Methods in Mathematical Analysis, Translations of Mathematical Monographs, vol. 12. American Mathematical Society, Providence (1964)
5. Glass, L., Mackey, M.C.: Oscillation and chaos in physiological control system. Science **197**, 287–289 (1977)
6. Henderson, J.: Boundary Value Problems for Functional Differential Equations. World Scientific (1995)
7. Lange, C.G., Miura, R.M.: Singular perturbation analysis of boundary value problems for differential-difference equations. SIAM J. Appl. Math. **42**, 502–531 (1982)
8. Lange, C.G., Miura, R.M.: Singular perturbation analysis of boundary value problems for differential-difference equations II. Rapid oscillations and resonance. SIAM J. Appl. Math. **45**, 687–707 (1985)
9. Lange, C.G., Miura, R.M.: Singular perturbation analysis of boundary value problems for differential-difference equations V. Small shifts with layer behavior. SIAM J. Appl. Math. **54**, 249–272 (1994)
10. Lee, J.W., O'Regan, D.: Existence results for differential delay equations-II. Non-linear Anal. **17**, 683–702 (1991)
11. Lee, J.W., O'Regan, D.: Existence results for differential delay equations-I. J. Diff. Eqn. **102**, 342–359 (1993)

12. Ntouyas, S.K., Sficas, Y.G., Tsamatos, P.: An existence principle for boundary value problems for second order functional differential equations. Nonlinear Anal. **20**, 215–222 (1993)
13. Norkin, S.B.: Differential Equations of the Second Order with Retarded Arguments, Translations of Mathematical Monographs, vol. 31. American Mathematical Society, Providence (1972)
14. Ramos, J.I.: Exponential methods for singularly perturbed ordinary differential-difference equations. Appl. Math. Comput. **182**, 1528–1541 (2006)
15. Kadalbajoo, M.K., Kumar, D.: Fitted mesh B-spline collocation method for singularly perturbed differential-difference equations with small delay. Appl. Math. Comput. **204**, 90–98 (2008)
16. Kadalbajoo, M.K., Sharma, K.K.: Parameter- uniform fitted mesh method for singularly perturbed delay differential equations with layer behaviour. Elect. Trans. Numer. Anal. **23**, 180–201 (2006)
17. Kadalbajoo, M.K., Sharma, K.K.: A numerical method based on finite difference for boundary value problems for singularly perturbed delay differential equations. Appl. Math. Comput. **197**, 692–707 (2008)
18. Stein, R.B.: A theoretical analysis of neuronal variability. Biophys. J. **5**, 173–194 (1965)
19. Tuckwell, H.C.: On the first exit time problem for temporally homogeneous Markov process. J. Appl. Probab. **13**, 39–46 (1976)

High-Order Compact Finite Difference Method for Black–Scholes PDE

Kuldip Singh Patel and Mani Mehra

Abstract In this paper, Black–Scholes PDE is solved for European option pricing by high-order compact finite difference method using polynomial interpolation. Numerical results obtained are compared with standard finite difference method and error with the analytic solution is discussed.

Keywords Option pricing · European options · Black-Scholes PDE · Compact finite difference methods

1 Introduction

The seminal work in the area of option pricing was done by Black and Scholes in 1973 [1] for pricing European option by solving a parabolic partial differential equation (PDE), (commonly known as Black–Scholes PDE). Many computational techniques [2] such as finite difference methods, spectral methods, Fast Fourier transform techniques [3] have been extensively used for solving Black–Scholes PDE. In the era of advanced computational techniques, compact finite difference method [4] is highly recommended. Various compact finite difference techniques have already been studied for Black–Scholes PDE for pricing the European and American option [5, 6] and their convergence has also been studied [7].

An option, in finance, is a contract that gives its owner the right (but not the obligation) to buy or sell a prescribed amount of particular asset from the writer of the option for a prescribed fixed price (called the strike price) on or before the certain date (called maturity date). For various purpose, there are many kinds of options, such as, vanilla options (European call or put option, American call or put option), Asian option, Bermudan option, exotic option, look-back option, barrier option, etc.

K.S. Patel (✉) · M. Mehra
Indian Institute of Technology, Delhi, India
e-mail: kuldip@maths.iitd.ac.in

M. Mehra
e-mail: mmehra@maths.iitd.ac.in

P.N. Agrawal et al. (eds.), *Mathematical Analysis and its Applications*,
Springer Proceedings in Mathematics & Statistics 143,
DOI 10.1007/978-81-322-2485-3_32

[8, 9]. Options that can be exercised only on the maturity date are called European option, while options that can be exercised at any time up to the maturity date are called American option. If the option is to buy the asset it is a call option, if to sell the asset it is a put option.

Various finite difference methods have been studied for Black–Scholes PDE. A major disadvantage of the finite difference approach is the widening of the computational stencil as the order of the approximation is increased. High-order compact finite difference schemes [4] consider the value of the function and its first/higher derivatives as unknowns at each discretization point. Compared to standard explicit finite difference schemes, these schemes are implicit and give a higher order of accuracy for the same number of grid points and also provide high resolution characteristics [10]. This feature brings them closer to the spectral methods, while the freedom in choosing the mesh geometry and the boundary conditions is maintained.

The rest of the paper is organized as follows. In Sect. 1.1, some introduction is given about Black–Scholes PDE. In Sect. 1.2, a very short review of finite difference method is given. In Sect. 1.3, compact finite difference scheme for first and second derivative is discussed. In Sect. 2.1, analytical solution of Black–Scholes PDE is given. In Sect. 2.2, solution of Black–Scholes PDE by compact finite difference method is given and error with the analytic solution is discussed.

1.1 Black–Scholes PDE

It is not always easy to determine the value of option because of the stochastic nature of financial markets. Option pricing theory has made a great leap forward since the development of the Black–Scholes option pricing model by Black and Scholes in 1973 [1] and previously by Merton in 1973 [11]. The famous Black–Scholes PDE can be written as:

$$\frac{\partial V}{\partial t} + \frac{1}{2}\sigma^2 S^2 \frac{\partial^2 V}{\partial S^2} + rS\frac{\partial V}{\partial S} - rV = 0. \tag{1}$$

The solution of above famous Black–Scholes PDE provides both an option pricing for European call and put option and a hedging portfolio that replicates the contingent claim under the following assumptions [12]:

- The asset price S follows geometric Brownian motion, i.e. S satisfies the following stochastic differential equation:

$$dS = \mu S\,dt + \sigma S\,dW,$$

  where μ is the drift rate, σ is the volatility and dW is the increment of a standard Brownian motion.
- The drift μ (which measures the average rate of growth of the asset price), the volatility σ (which measures the standard deviation of the returns) and the

risk-free interest rate r are constant for $0 \leq t \leq T$ and no dividends are paid in that time period.
- The market is frictionless.
- There are no arbitrage opportunities.

Under the assumption discussed above the market is complete. The completeness of market implies that any derivative and any asset can be replicated or hedged with a portfolio of other assets in the market [13]. The parabolic PDE given in Eq. (1) can be transformed into the heat equation and solved analytically to price the option [14].

1.2 Finite Difference Method

In general, finite difference (FD) methods are used to numerically approximate the solutions of certain ordinary and partial differential equations. In the case of a bivariate, parabolic PDE, such as Eq. (1), we start by establishing a rectangular solution domain in the two variables, S and t. We then form finite difference approximations to each of the derivative terms in the PDE. Perhaps the most popular FD methods used in computational finance are:

- Explicit Euler,
- Implicit Euler and
- Crank–Nicolson method.

Using each of these three methods has its advantages and disadvantages. The easiest scheme among the above three methods to implement is the explicit Euler method. The main disadvantage in using explicit Euler is that it is unstable for certain choices of domain discretization. Though Implicit Euler and Crank–Nicolson methods involve solving linear systems of equations at each time step, they are each unconditionally stable with respect to the domain discretization. Crank–Nicolson exhibits the greatest accuracy among the above three methods for a given domain discretization.

1.3 Compact Finite Difference Method

High-order compact finite difference schemes [15] are developed conventionally using method of undetermined coefficient [4, 16]. Polynomial interpolation has also been used to derive arbitrarily high-order compact schemes for the first and second derivatives on non-uniform grids [17]. Boundary and near boundary schemes of the same order as the interior have also been developed using polynomial interpolation

Table 1 Compact finite difference scheme for first derivative

Index i	I_n, I_m	Uniform grid, $x_i = x_1 + h(i-1)$
1	$(3,4),(1,2)$	$f_1' + 3f_2' = \frac{-17}{6h}f_1 + \frac{3}{2h}f_2 + \frac{3}{2h}f_3 - \frac{1}{6h}f_4$
$2, 3, .., N-1$	$(i-1, i+1), (i)$	$\frac{1}{4}f_{i-1}' + f_i' + \frac{1}{4}f_{i+1}' = \frac{3}{4h}(f_{i+1} - f_{i-1})$
N	$(N-2, N-3), (N, N-1)$	$f_N' + 3f_{N-1}' = \frac{17}{6h}f_N - \frac{3}{2h}f_{N-1} - \frac{3}{2h}f_{N-2} + \frac{1}{6h}f_{N-3}$

Table 2 Compact finite difference scheme for second derivative

Index i	Uniform grid, $x_i = x_1 + h(i-1)$
1	$f_1'' + 44f_2'' = \frac{13}{h^2}f_1 - \frac{27}{h^2}f_2 + \frac{15}{h^2}f_3 - \frac{1}{h^2}f_4$
$2, 3, .., N-1$	$\frac{1}{10}f_{i-1}'' + f_i'' + \frac{1}{10}f_{i+1}'' = \frac{6}{5h^2}(f_{i+1} - f_{i-1}) - \frac{12}{5h^2}f_i$
N	$f_N'' + 44f_{N-1}'' = \frac{13}{h^2}f_N - \frac{27}{h^2}f_{N-1} + \frac{15}{h^2}f_{N-2} - \frac{1}{h^2}f_{N-3}$

[17]. It has been proved that polynomial interpolation is more efficient than the conventional method of undetermined coefficients [16] for finding coefficients of the scheme.

Consider a set of n points I_n on which values of the function and its first derivative have been specified and another set of m points I_m on which only function values have been specified. The independent variable representing the points is x_i, i being the index of the node and the function values are given by $f_i = f(x_i)$. First derivative is given by $f_i' = f'(x_i)$ and second derivative is given by $f_i'' = f''(x_i)$. Compact finite difference scheme for first and second derivative on uniform grid with step size h are given in Tables 1 and 2 respectively.

2 Numerical Results

2.1 Closed Form Solution of Black–Scholes PDE

Black–Scholes PDE can be written as

$$\frac{\partial V}{\partial t} + \frac{1}{2}\sigma^2 S^2 \frac{\partial^2 V}{\partial S^2} + rS\frac{\partial V}{\partial S} - rV = 0. \tag{2}$$

By the definition of the European option, it is clear that at expiry date T, the value of European option $V(S,t)$ (also called as pay-off function) is given by

$$V(S,T) = \begin{cases} \text{max(S-X,0)} & \text{for a European call,} \\ \text{max(X-S,0)} & \text{for a European put,} \end{cases} \tag{3}$$

and the solution $V(S,t)$ of the Black–Scholes PDE (Eq. (2)) with the above final condition (Eq. (3)) is given by

$$V(S,t) = \begin{cases} SN(d1) - XN(d2)e^{-r(T-t)} & \text{for a European call,} \\ XN(-d2)e^{-r(T-t)} - SN(-d1) & \text{for a European put,} \end{cases} \tag{4}$$

where X is the strike price, r is the interest rate,

$$N(x) = \frac{1}{\sqrt{2\pi}} \int_{-\infty}^{x} e^{\left(\frac{-y^2}{2}\right)} dy,$$

$$d1 = \frac{\log(S/X) + \left(r + \frac{1}{2}\sigma^2\right)(T-t)}{\sigma\sqrt{T-t}},$$

$$d2 = \frac{\log(S/X) + \left(r - \frac{1}{2}\sigma^2\right)(T-t)}{\sigma\sqrt{T-t}},$$

this implies

$$d2 = d1 - \sigma\sqrt{T-t}.$$

For example price of European put option (Fig. 1) for a non-dividend paying stock by the above formula is

$$V = 1.6306, \tag{5}$$

for the given stock price $S = 50$, strike price $X = 50$, interest rate $r = 0.1$, volatility $\sigma = 0.2$ and time to maturity $T = 5/12$.

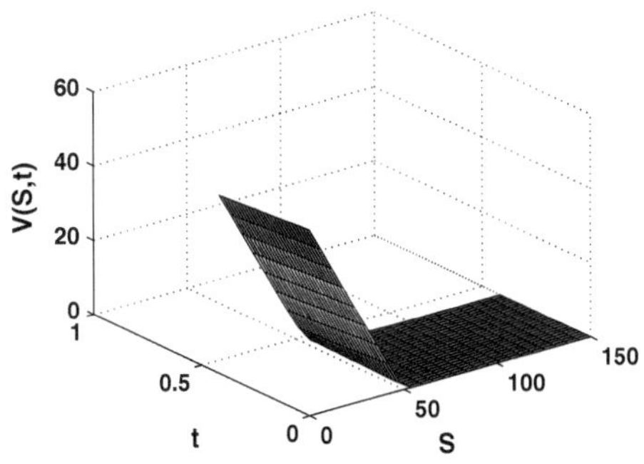

Fig. 1 Price of European put options calculated by closed form solution for $X = 50$, $S = 50$, $r = 0.10$, $\sigma = 0.20$ and $T = 5/12$

2.2 Compact Finite Difference Method for Black–Scholes PDE

In the following, Black–Scholes PDE is solved by compact finite difference method using polynomial interpolation to get the option price and the price of option obtained by closed form solution of Black–Scholes PDE will serve as benchmark for comparison. The initial and boundary conditions for a European call option are

$$V(S,T) = \max(S - X, 0),$$

$$V(0,t) = 0,$$

$$\lim_{S \to \infty} V(S,t) = S,$$

while the initial and boundary conditions for a European put option are

$$V(S,T) = \max(X - S, 0),$$

$$V(0,t) = Xe^{-r(T-t)},$$

$$\lim_{S \to \infty} V(S,t) = 0,$$

with X as the exercise price.

Black–Scholes PDE (Eq. (1)) can be written as:

$$V_t = LV \text{ where } L \equiv -\frac{1}{2}\sigma^2 S^2 \frac{\partial^2}{\partial S^2} - rS\frac{\partial}{\partial S} + rI. \tag{6}$$

If we write $\frac{\partial}{\partial S} \equiv D$ and $\frac{\partial^2}{\partial S^2} \equiv D^2$ (where D and D^2 are compact finite difference differentiation matrix for first and second derivative respectively), then

$$L \equiv -\frac{1}{2}\sigma^2 S^2 D^2 - rSD + rI,$$

where $D = A^{-1}B$ and $D^2 = P^{-1}Q$. Now time discretization can be done for Eq. (6) according to implicit Euler, explicit Euler and Crank–Nikolson method. Matrices A, B and P, Q (constructed from Tables 1 and 2 respectively) are given as:

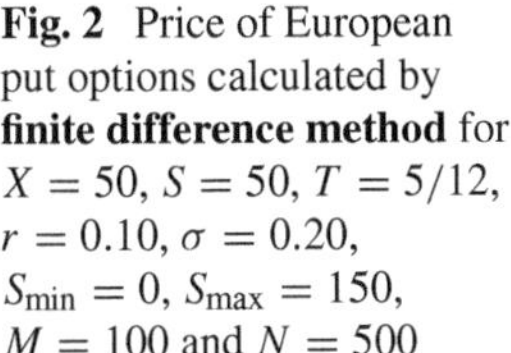

Fig. 2 Price of European put options calculated by **finite difference method** for $X = 50$, $S = 50$, $T = 5/12$, $r = 0.10$, $\sigma = 0.20$, $S_{\min} = 0$, $S_{\max} = 150$, $M = 100$ and $N = 500$

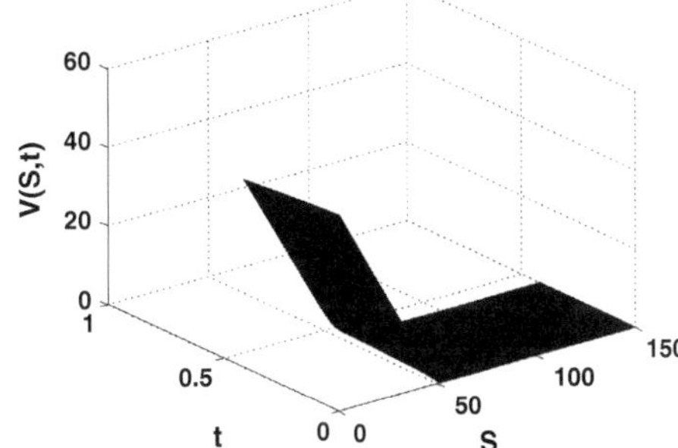

Fig. 3 Price of European put options calculated by **compact finite difference method** for $X = 50$, $S = 50$, $T = 5/12$, $r = 0.10$, $\sigma = 0.20$, $S_{\min} = 0$, $S_{\max} = 150$, $M = 100$ and $N = 500$

$$A = \begin{pmatrix} 1 & 3 & 0 & 0 & \cdots\cdots & 0 \\ \frac{1}{4} & 1 & \frac{1}{4} & 0 & & \vdots \\ \vdots & & & \ddots & & \vdots \\ \vdots & & & \ddots & \frac{1}{4} \quad 1 & \frac{1}{4} \\ 0 & \cdots\cdots & & & 3 & 1 \end{pmatrix} \qquad B = \begin{pmatrix} \frac{-17}{6h} & \frac{3}{2h} & \frac{3}{2h} & \frac{-1}{6h} & \cdots\cdots & 0 \\ \frac{-3}{4h} & 0 & \frac{3}{4h} & 0 & & \vdots \\ \vdots & & & & \ddots & \vdots \\ \vdots & & & \ddots & \frac{-3}{4h} \quad 0 & \frac{3}{4h} \\ 0 & \cdots\cdots & \frac{1}{6h} & \frac{-3}{2h} & \frac{-3}{2h} & \frac{17}{6h} \end{pmatrix} \tag{7}$$

$$P = \begin{pmatrix} 1 & 44 & 0 & 0 & \cdots\cdots & 0 \\ \frac{1}{10} & 1 & \frac{1}{10} & 0 & & \vdots \\ \vdots & & & \ddots & & \vdots \\ \vdots & & & \ddots & \frac{1}{10} \quad 1 & \frac{1}{10} \\ 0 & \cdots\cdots & & & 44 & 1 \end{pmatrix} \qquad Q = \begin{pmatrix} \frac{13}{h^2} & \frac{-27}{h^2} & \frac{15}{h^2} & \frac{-1}{h^2} & \cdots & \cdots & 0 \\ \frac{6}{5h^2} & \frac{-12}{5h^2} & \frac{6}{5h^2} & 0 & & & \vdots \\ \vdots & & & & \ddots & & \vdots \\ \vdots & & & & \ddots & \frac{6}{5h^2} \quad \frac{-12}{5h^2} & \frac{6}{5h^2} \\ 0 & \cdots & \cdots & \frac{-1}{h^2} & \frac{15}{h^2} & \frac{-27}{h^2} & \frac{13}{h^2} \end{pmatrix} \tag{8}$$

Price of European put option by closed form solution (Fig. 1), by standard finite difference method (Fig. 2) and by compact finite difference method (Fig. 3) for a non-dividend paying stock are given in Table 3. It can be noticed that explicit (Euler)

Table 3 Price of European put option for parameters $S = 50$, strike price $X = 50$, interest rate $r = 0.1$, volatility $\sigma = 0.2$, time to maturity $T = 5/12$, number of time steps $M = 100$ number of space steps $N = 500$, $S_{\min} = 0$ and $S_{\max} = 150$

Method	Standard finite difference method	Compact finite difference method	Closed form solution
Explicit Euler	$-7.2447e^{90}$	$6.1154e^{106}$	
Implicit Euler	1.6277	1.6279	1.6306
Crank–Nicolson method	1.6304	1.6307	

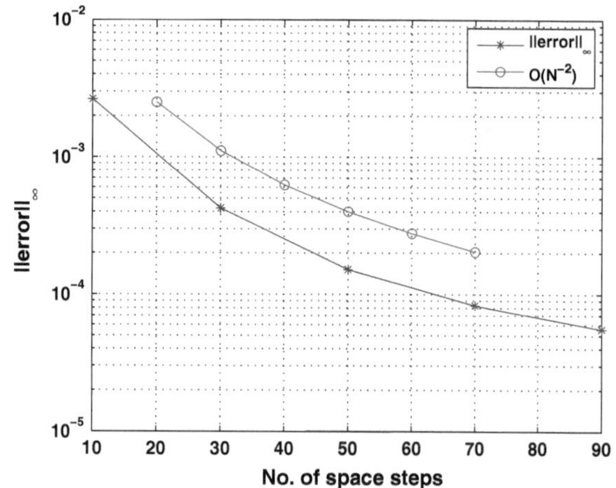

Fig. 4 Error versus number of space steps in **finite difference method** by Crank–Nicolson method for $X = 0.2$, $T = 5/12$, $r = 0.1$, $\sigma = 0.4$, $S_{\min} = 0$, $S_{\max} = 1$, and $M = 200$

method gives unstable results and Crank–Nicolson method gives the better results for the same parameter.

Error with the analytic solution is plotted for finite difference method (Fig. 4) and for compact finite difference method (Fig. 5) versus number of grid points. It can be seen from Figs. 4 and 5 that method presented in this paper has more order of convergence ($O(h^3)$) as compared to standard finite difference method ($O(h^2)$). From Table 4, it can be concluded that high order of accuracy is obtained by using compact finite difference method (up to 10^{-6} for $N = 200$) as compared to standard finite difference method (up to 10^{-4} for $N = 200$) for the same parameters.

In Figs. 6 and 7, the results for various time steps are compared and it is observed that three standard finite difference methods converge to identical precision for lager number of time steps. Errors are plotted as the difference between analytic Black–Scholes price and the finite difference price. In Figs. 8 and 9, the results for various time steps are compared and it is observed that three compact finite difference methods converge to identical precision for larger number of time steps. Errors are plotted as the difference between analytic Black–Scholes price and the compact finite difference price. It can be noticed that error is largest at the money.

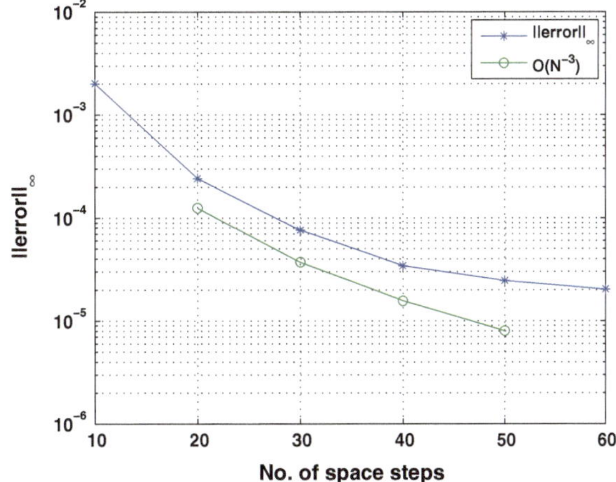

Fig. 5 Error versus number of space steps in **compact finite difference method** by Crank–Nicolson method for $X = 0.2, T = 5/12$, $r = 0.1, \sigma = 0.4, S_{\min} = 0$, $S_{\max} = 1$, and $M = 200$

Table 4 Comparison of errors for parameters $S = 50$, strike price $X = 50$, interest rate $r = 0.3$, volatility $\sigma = 0.05$, time to maturity $T = 5/12$, number of time steps $M = 100$, $S_{\min} = 0$ and $S_{\max} = 150$

Number of grid points difference method	Error in standard finite difference method	Error in compact finite difference method
N = 50	0.2712	0.0444
N = 100	0.0461	0.0045
N = 150	0.0033	8.8943e-06
N = 200	3.1888e-04	1.0864e-06

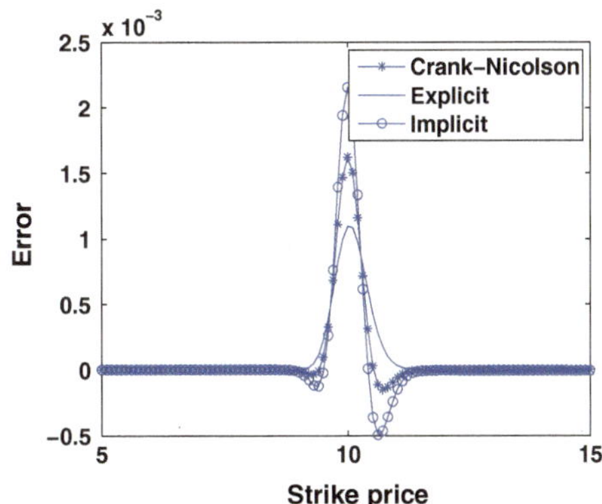

Fig. 6 Error versus strike price with 30 time steps in **finite difference method** for $X = 10,\ S = 10,\ T = 0.1,\ r = 0.05,\ \sigma = 0.10,\ S_{\min} = 5,\ S_{\max} = 15,\ N = 100,\ M = 30$

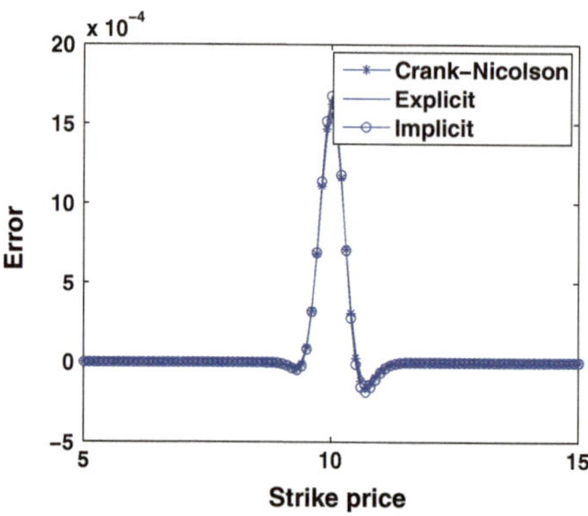

Fig. 7 Error versus strike price with 300 time steps in **finite difference method** for $X = 10,\ S = 10,\ T = 0.1,\ r = 0.05,\ \sigma = 0.10,\ S_{\min} = 5,\ S_{\max} = 15,\ N = 100,\ M = 300$

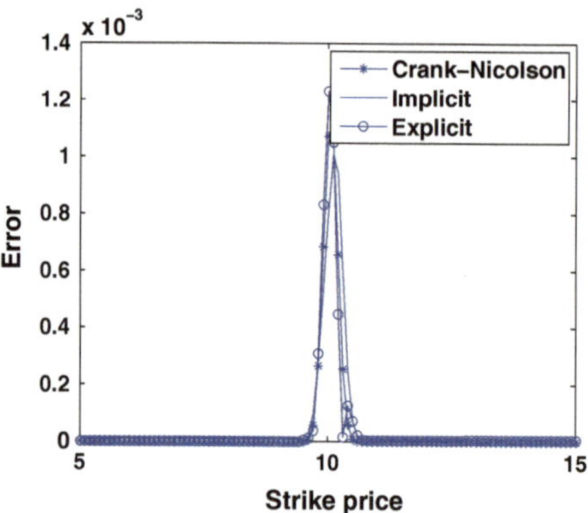

Fig. 8 Error versus strike price with 30 time steps in **compact finite difference method** for $X = 10,\ S = 10,\ T = 0.1,\ r = 0.05,\ \sigma = 0.05,\ S_{\min} = 5,\ S_{\max} = 15$

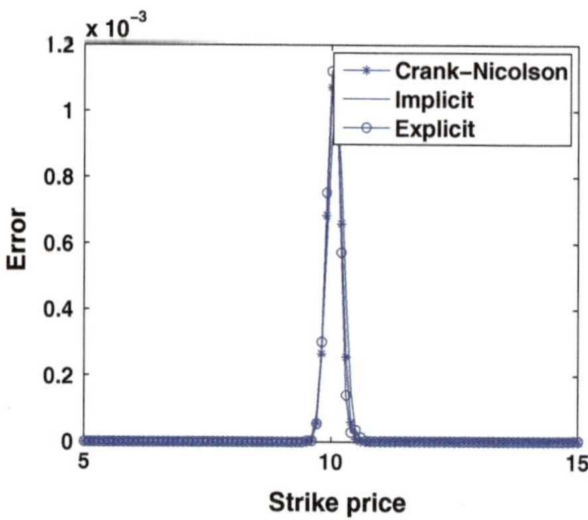

Fig. 9 Error versus strike price with 60 time steps in **compact finite difference method** for $X = 10,\ S = 10,\ T = 0.1,\ r = 0.05,\ \sigma = 0.05,\ S_{\min} = 5,\ S_{\max} = 15$

3 Discussion and Conclusion

In this paper, compact finite difference method using polynomial interpolation is used to solve Black–Scholes PDE for pricing European put option. It can be noticed from Table 4 that high-order accuracy is obtained by using compact finite difference method (up to 10^{-6} for $N = 200$) as compared to standard finite difference method (up to 10^{-4} for $N = 200$) for the same parameters. It is shown in Figs. 4 and 5 that more order of convergence is obtained by using compact finite difference method $(O(h^3))$ as compared to standard finite difference method $(O(h^2))$ for the same parameters. Also the results for various time steps are compared (Figs. 8 and 9 for compact finite difference method and Figs. 6 and 7 for standard finite difference method) and it is observed that the three method converge to identical precision for larger number of time steps.

References

1. Black, F., Scholes, M.: Pricing of options and corporate liabilities. J. Polit. Econ. **81**, 637–654 (1973)
2. Achdou, Y., Pironneau, O.: Computational Methods for Options Pricing. SIAM, Philadelphia (2005)
3. Carr, P., Madan, D.B., Smith, R.H.: Option valuation using the fast fourier transform. J. Comput. Finan. **2**, 61–73 (1999)
4. Lele, S.K.: Compact finite difference schemes with spectral-like resolution. J. Comput. Phys. **103**, 16–42 (1992)
5. Tangman, D.Y., Gopaul, A., Bhuruth, M.: A fast high-order finite difference algorithm for pricing american options. J. Comput. Appl. Math. **222**, 17–29 (2008)
6. Cen, Zhongdi, Le, Anbo: A robust and accurate finite difference method for a generalized black-scholes equation. J. Comput. Appl. Math. **235**, 3728–3733 (2011)
7. During, Bertram, Fournie, Michel, Jungel, Ansgar: Convergence of a high-order compact finite difference scheme for a non-linear black-scholes equation. Math. Model. Numer. Anal. **38**, 359–369 (2004)
8. Hull, J.C.: Options, Futures and Other Derivatives. Prentice Hall, Upper Saddle River (2006)
9. Ugur, O.: An introduction to Mathematical Finance. Imperial College Press, London (2008)
10. Adam, Y.: Highly accurate compact implicit methods and boundary conditions. J. Comput. Phys. **24**, 10–22 (1977)
11. Merton, R.C.: Theory of rational option pricing. Bell J. Econ. **4**, 141–183 (1973)
12. Seydel, R.: Tools for Computational Finance, 2nd edn. Springer, Berlin (2004)
13. Tavella, D., Randall, C.: Pricing Financial Instruments: The Finite Difference Method. Wiley, New York (2000)
14. Willmott, P., Howison, S., Dewynne, J.: Option Pricing: Mathematical Models and Computation. Oxford Financial Press, Oxford (1993)
15. Goedheer, W.J., Potters, J.H.H.M.: A compact finite difference scheme on a non-equidistant mesh. J. Comput. Phys. **61**, 269–279 (1985)
16. Gamet, L., Ducros, F., Nicoud, F., Poinsot, T.: Compact finite difference schemes on non-uniform meshes, application to direct numerical simulation of compressible flows. Int. J. Numer. Meth. Fluids **29**, 159–191 (1999)
17. Shukla, R.K., Zhong, X.: Derivation of high-order compact finite difference schemes for non-uniform grid using polynomial interpolation. J. Comput. Phys. **204**, 404–429 (2005)

On Stability of Steady-States for a Two-Dimensional Network Model of Ferromagnetic Nanowires

Sharad Dwivedi and Shruti Dubey

Abstract This article concerns with the mathematical study of stability properties of steady-states for a two-dimensional network model of ferromagnetic nanowires. We consider the finite network model of ferromagnetic nanowires of semi-infinite length. We derive a sufficient condition independent of the size of the network under which the relevant configurations (steady-states) of magnetization are shown to be asymptotically stable. To be precise, we establish the result under certain condition on the length between the two consecutive nanowires. We use perturbation technique and energy method to derive the result.

Keywords Ferromagnetic material · Landau–Lifschitz equation · Stability · Domain walls · Micromagnetics

AMS Subject Classification: Primary: 35B35 · 34D05 · 35Q60 · Secondary: 35K55

1 Introduction

Experimentally, it has been observed that below a critical temperature, ferromagnetic materials have a tendency to split up into a small uniformly magnetized regions called domains separated by a thin transition layer known as domain walls. Over the period of time, study of formation and motions of domain walls gained a lot of attention and became one of the most fascinating topic among researchers. It is due to the fact that the ferromagnetic materials are used on a wide scale in magnetic storage industry. In particular, ferromagnetic nanowires play a very dominant role

---

S. Dwivedi (✉) · S. Dubey
Department of Mathematics, Indian Institute of Technology Madras, Chennai 600036, India
e-mail: sharadiitm@gmail.com

S. Dubey
e-mail: sdubey@iitm.ac.in

P.N. Agrawal et al. (eds.), *Mathematical Analysis and its Applications*, Springer Proceedings in Mathematics & Statistics 143,
DOI 10.1007/978-81-322-2485-3_33

in nanoelectronic devices in which the information is encoded as magnetic domains separated by domain walls along the wire. For example, in case of racetrack memory, we obtain a three-dimensional storage device by using U-shaped nanowires normal to the plane of silicon wafer (see [1]). For the rigorous treatment of domains and its characteristics, we refer the reader to [2] and the references therein.

The model used to describe the magnetic behavior of ferromagnetic material is called micromagnetism and was introduced by Brown [3]. The evolution of magnetization inside the ferromagnetic medium is triggered by the Landau–Lifschitz equation which is parabolic and nonlinear. The relevant configurations of magnetization are minimizers of an energy functional, consisting of several components. We shall see that these relevant configurations of magnetization coincide with the steady-states of Landau–Lifschitz equation.

The general framework of the ferromagnetism is as follows. We consider a finite homogeneous ferromagnetic material which occupies a domain $\Omega \subset \mathbb{R}^3$. The time-varying magnetic moment u of a ferromagnetic material is a solution of the Landau–Lifschitz equation (LLE)

$$\frac{\partial u}{\partial t} = -u \times \mathcal{H}_{eff} - u \times \left(u \times \mathcal{H}_{eff}\right), \tag{1}$$

with the physical saturation constraint

$$|u(t,.)| = 1 \text{ for } (t,.) \in \mathbb{R}^+ \times \mathbb{R}^3 \; a.e., \tag{2}$$

where the abbreviation $a.e.$ stands for almost everywhere. The total effective field $\mathcal{H}_{eff} = -\nabla\mathcal{E}$ is derived from the micromagnetism energy $\mathcal{E}$ given by

$$\mathcal{E}(u) = \frac{A}{2}\int_{\Omega} |\nabla u|^2 + \frac{1}{2}\int_{\mathbb{R}^3} |\mathcal{H}_d(u)|^2 - \int_{\Omega} \mathcal{H}_a \cdot u, \tag{3}$$

where the term represents the exchange, stray field and external energy contribution respectively. The constant $A > 0$ is called the exchange constant. Also, $\mathcal{H}_a$ denotes an applied magnetic field and $\mathcal{H}_d(u)$ is the stray field which is characterized by the Maxwell equations:

$$\begin{cases} \text{curl } \mathcal{H}_d(u) = 0 \text{ in } \mathbb{R}^3, \\ \text{div } (\mathcal{H}_d(u) + \bar{u}) = 0 \text{ in } \mathbb{R}^3, \\ H_d(u) \text{ vanishes at infinity.} \end{cases}$$

where $\bar{u}$ is the extension of u in $\mathbb{R}^3$ by 0 outside of Ω. We obtain that,

$$\mathcal{H}_{eff} = \Delta u + \mathcal{H}_d(u) + \mathcal{H}_a.$$

We take the scalar product of (1) with $\mathscr{H}_{eff}$ and integrate in time (assuming time invariant applied field). Using $\mathscr{H}_{eff} = -\nabla\mathscr{E}$, we obtain (see [4, 5])

$$\frac{d}{dt}\mathscr{E}(u(t)) = -\int\limits_{\Omega} |\mathscr{H}_{eff}(u) - \left(\mathscr{H}_{eff}(u)) \cdot u\right) u|^2, \tag{4}$$

this denotes the dissipation of energy which is mainly due to the second term appears on the right hand side of (1). Furthermore, steady-states of (1) satisfy $u \times \mathscr{H}_{eff} = 0$ in domain Ω, which is exactly the Euler–Lagrange equations of the minimization problem for (3). Therefore, minimizers of (3), i.e., relevant physical configurations of the magnetization are nothing but the steady-state solutions of (1) under the constraint (2).

Existence results of weak solutions for the Landau–Lifschitz equation have been discussed in [6–8], whereas the strong solutions are considered in [9, 10] and known to exist locally in time. Numerical aspects of ferromagnetic materials have been investigated in [11, 12] and the references therein. Stability and controllability results related with ferromagnetic nanowires are studied in [5, 13, 14]. Higher dimensional models and network models of such materials can be found in [15, 16].

In the present article, we consider a two-dimensional finite network model of ferromagnetic nanowires of semi-infinite length. We assume the relevant configuration of the magnetization of the network is of the form $u^* = \mu\mathbf{e}_1$ where $\mu = (\mu_i)_{i\in I}$ with $\mu_i = \{-1, +1\}$. We prove that these relevant configurations are asymptotically stable in a long time behavior under certain condition on the distance between the consecutive nanowires. The organization of this article is as follows:

In Sect. 2, we present the schematics of the considered model and introduce the problem related to the stability of the steady-states in the absence of external magnetic field. In Sect. 3, we give the statement of the main result and establish some preliminary estimates to derive the Theorem.

2 Modeling of a Network Model

In this section, we present a schematics and modeling of a network model under consideration. We consider a two-dimensional finite network model of ferromagnetic nanowires of semi-infinite length. In which nanowires are supposed to have homogeneous geometry and to be placed on the plane $(\mathbf{e}_1, \mathbf{e}_2)$, where $(\mathbf{e}_1, \mathbf{e}_2, \mathbf{e}_3)$ is the canonical basis of $\mathbb{R}^3$. We represent the distance between the two consecutive nanowires by $\ell > 0$. Since, we consider the finite framework of a network model therefore the index i takes it values in the finite set $I = \{0, 1, 2, \ldots, N\}$. We denote the coordinates of a point on the ith nanowire by $(x_i, i\ell)$, where $0 \leq x_i < \infty$ with $i \in I$ (see Fig. 1). We use the following notations:

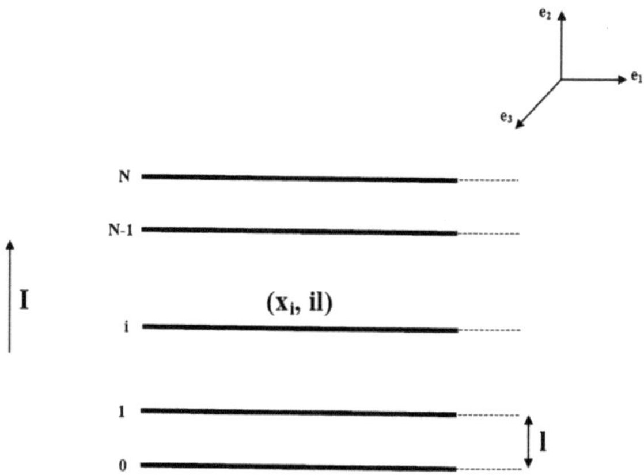

Fig. 1 Schema of network model

$$\begin{cases} \left(\mathbb{R}^3\right)^I = \left\{u = (u_i)_{i \in I}, \text{ such that } \forall\, i \in I, u_i \in \mathbb{R}^3\right\}, \\ \left(\mathbb{S}^2\right)^I = \left\{u = (u_i)_{i \in I} \in (\mathbb{R}^3)^I, \text{ such that } \forall\, i \in I, |u_i| = 1\right\}, \\ \|u\| = \sup\limits_i |u_i|, \text{ where } i \in \text{I and } |\cdot| \text{ is the euclidean norm in } \mathbb{R}^3. \end{cases}$$

where $\mathbb{S}^2$ represents the unit sphere in $\mathbb{R}^3$.

We assume that the magnetization on each nanowire is constant in the space variable, i.e., we deal with the ordinary differential model of micromagnetism. This can be justified by the assumption that the radius of the nanowires are very small as compared to ℓ. We denote $u_i = u_i(t)$ the magnetization at any point on the ith nanowire. Therefore, the unknown $u = (u_0, \ldots, u_N)$ is defined as $u : \mathbb{R}^+ \to (\mathbb{S}^2)^I$, i.e., $u = u(t) = (u_i(t))_{i \in I}$. Exchange field vanishes due to the aforementioned assumption renders the only contribution of demagnetizing (stray) field in total effective field.

We recall that the stray energy is connected with the magnetic field generated by the medium itself. We calculate the stray field for the entire network in the following fashion. On a fixed nanowire say j_0, we represent its stray field as $\mathscr{H}_d(u)(j_0)$ which consist of two parts: the stray field generated on j_0th nanowire by its own magnetization, i.e., by u_{j_0}, denoted by $\mathscr{H}_d^{int}(u)(j_0)$, and the field generated by the magnetization of other nanowires, denoted by $\mathscr{H}_d^{ext}(u)(j_0)$. We write the stray field as:

$$\mathscr{H}_d(u)(j_0) = \mathscr{H}_d^{int}(u)(j_0) + \mathscr{H}_d^{ext}(u)(j_0).$$

The demagnetizing field on j_0th nanowire due to its own magnetization is given by (see [13, 17])

$$\mathscr{H}_d^{int}(u)(j_0) = -u_{j_0}^2 \mathbf{e}_2 - u_{j_0}^3 \mathbf{e}_3. \tag{5}$$

The stray field generated by the i_0th nanowire on the j_0th nanowire is given by (see [18])

$$\mathscr{H}_{i_0,j_0}(u_{i_0})(\zeta) = -\frac{1}{4\pi}\int_0^\infty \frac{u_{i_0}}{|\zeta-\eta|^3}dy + \frac{3}{4\pi}\int_0^\infty \frac{(\zeta-\eta)}{|\zeta-\eta|^5} u_{i_0}\cdot(\zeta-\eta)dy.$$

with $\zeta = (x, j_0\ell)$ and $\eta = (y, i_0\ell)$, where x and y belongs to $[0, \infty)$.
On calculating the values of these integrals, we obtain:

$$\mathscr{H}_{i_0,j_0}(u_{i_0})(\zeta) = (\mathscr{H}_{i_0,j_0}^1, \mathscr{H}_{i_0,j_0}^2, \mathscr{H}_{i_0,j_0}^3),$$

where,

$$\mathscr{H}_{i_0,j_0}^1 = \frac{1}{4\pi\ell^2|i_0-j_0|^2}\left[-\frac{x\ell^2|i_0-j_0|^2}{(x^2+\ell^2|i_0-j_0|^2)^{3/2}}u_{i_0}^1 + \frac{\ell^3|i_0-j_0|^3}{(x^2+\ell^2|i_0-j_0|^2)^{3/2}}u_{i_0}^2\right].$$

$$\begin{aligned}\mathscr{H}_{i_0,j_0}^2 = \frac{1}{4\pi\ell^2|i_0-j_0|^2}&\left[\frac{\ell^3|i_0-j_0|^3}{(x^2+\ell^2|i_0-j_0|^2)^{3/2}}u_{i_0}^1\right.\\ &\left.+\left\{1+\frac{x}{(x^2+\ell^2|i_0-j_0|^2)^{3/2}}(x^2+2\ell^2|i_0-j_0|^2)\right\}u_{i_0}^2\right].\end{aligned}$$

$$\mathscr{H}_{i_0,j_0}^3 = -\frac{1}{4\pi\ell^2|i_0-j_0|^2}\left[1+\frac{x}{(x^2+\ell^2|i_0-j_0|^2)^{1/2}}\right]u_{i_0}^3.$$

Therefore, the total network exterior field at the j_0th nanowire is given by:

$$\mathscr{H}_d^{ext}(u)(j_0) = \sum_{i_0\neq j_0}\mathscr{H}_{i_0,j_0}(u(i_0)) = \begin{pmatrix}\Psi^1(u^1)(j_0)+\Psi^2(u^2)(j_0)\\ \Psi^2(u^1)(j_0)+\Psi^3(u^2)(j_0)\\ \Psi^4(u^3)(j_0)\end{pmatrix}, \tag{6}$$

where (u^1, u^2, u^3) are the coordinates of u and for $k = \{1, \dots, 4\}$, the linear operators $\Psi^k\colon (\mathbb{R}^3)^I \to (\mathbb{R}^3)^I$ is defined as, for all $u = (u_i)_{i\in I}$ in $(\mathbb{R}^3)^I$,

$$\Psi^1(u)(j_0)(\zeta) = -\frac{1}{4\pi\ell^2}\sum_{j\neq j_0}\frac{1}{|j-j_0|^2}\left[\frac{x\ell^2|j-j_0|^2}{(x^2+\ell^2|j-j_0|^2)^{3/2}}\right]u(j),$$

$$\Psi^2(u)(j_0)(\zeta) = \frac{1}{4\pi\ell^2}\sum_{j\neq j_0}\frac{1}{|j-j_0|^2}\left[\frac{\ell^3|j-j_0|^3}{(x^2+\ell^2|j-j_0|^2)^{3/2}}\right]u(j),$$

$$\Psi^3(u)(j_0)(\zeta) = \frac{1}{4\pi\ell^2}\sum_{j\neq j_0}\frac{1}{|j-j_0|^2}\left[1+\frac{x}{(x^2+\ell^2|j-j_0|^2)^{3/2}}(x^2+2\ell^2|j-j_0|^2)\right]u(j),$$

$$\Psi^4(u)(j_0)(\zeta) = -\frac{1}{4\pi\ell^2}\sum_{j\neq j_0}\frac{1}{|j-j_0|^2}\left[1+\frac{x}{(x^2+\ell^2|j-j_0|^2)^{1/2}}\right]u(j).$$

Hence, we study the following system:

$$\begin{aligned}\frac{du_i}{dt} &= -u_i \times (\mathscr{H}_{eff}(u))(i) - u_i \times (u_i \times (\mathscr{H}_{eff}(u))(i)) \\ \mathscr{H}_{eff}(u))(i) &= \mathscr{H}_d^{int}(u)(i) + \mathscr{H}_d^{ext}(u)(i)\end{aligned} \tag{7}$$

for $i \in I$ and $t \in \mathbb{R}^+$ with $u_i : \mathbb{R}^+ \to \mathbb{S}^2$.

We assume the relevant steady-states configurations of the magnetization distribution as:

$$u_i^* = \mu_i \mathbf{e}_1 \ \text{ with } \ \mu_i = \{-1, +1\}, \ \ \forall\, i \in I. \tag{8}$$

Experimentally, we relate these relevant configurations to the memory state in a magnetic storage device, where $\mu_i = 1$ corresponds to a bit 1 and $\mu_i = -1$ corresponds to a bit 0. Next, we introduce the problem related to stability of the relevant configurations under consideration.

Asymptotic stability of any relevant configuration.

In the absence of an external applied field, for any initial conditions in a vicinity of a given relevant configuration, the solution of the Landau–Lifschitz equation (7) *converges to the relevant configuration.*

We give the mathematical statement of the result in the following section:

3 Main Result

In order to state the result, we need to introduce the following notations. We observe that for $\rho \in \mathbb{S}^2$, if $0 < \rho_1 < 1$ (resp. $-1 < \rho_1 < 0$), the quantity $\rho_2^2 + \rho_3^2$ exhibits the distance between ρ and $+\mathbf{e}_1$ (resp. $-\mathbf{e}_1$). We have:

$$\frac{1}{2}|\rho - \mathbf{e}_1|^2 \le \rho_2^2 + \rho_3^2 \le |\rho - \mathbf{e}_1|^2.$$

For $\alpha > 0$ sufficiently small, we define $\mathscr{D}_{+1}(\alpha)$ and $\mathscr{D}_{-1}(\alpha)$ by

$$\mathscr{D}_{+1}(\alpha) = \left\{\rho \in \mathbb{S}^2,\ \rho_1 > 0 \text{ and } \rho_2^2 + \rho_3^2 < \alpha^2\right\},$$
$$\mathscr{D}_{-1}(\alpha) = \left\{\rho \in \mathbb{S}^2,\ \rho_1 < 0 \text{ and } \rho_2^2 + \rho_3^2 < \alpha^2\right\}.$$

For $\mu = (\mu_i)_{i \in I}$ with $\mu_i \in \{-1, +1\}$, we denote, for $\alpha > 0$,

$$\mathscr{D}_\mu(\alpha) = \left\{u \in (\mathbb{S}^2)^I,\ \ \forall\, i \in I,\ \ u_i \in \mathscr{D}_{\mu_i}(\alpha)\right\}. \tag{9}$$

Our main result about the asymptotic stability of any relevant position is the following:

Theorem 1 *Suppose u is the solution of the Landau–Lifschitz equation (7) with initial condition $u(0) = u^{init}$, where u^{init} satisfies the saturation condition (2). There exists β, a positive constant independent of the size of the network such that if*

$$\frac{1}{\ell^2} \leq \beta, \tag{10}$$

then there exist $\alpha_0 > 0$ and $\kappa > 0$, such that for all relevant configurations u^* (i.e., $u_i^* = \mu_i \mathbf{e}_1$ for all $i \in I$), for all $u^{init} \in \mathscr{D}_\mu(\alpha_0)$, u satisfies:

$$\|u(t) - u^*\| \to 0 \ \text{as}\ t \to \infty.$$

Proof We derive the stability result of a relevant configuration for the Landau–Lifschitz equation without an external magnetic source. For this we analyze the following system with unknown u defined as $u\colon \mathbb{R}^+ \to (\mathbb{S}^2)^I$,

$$\begin{aligned} \frac{du}{dt} &= -u \times \mathscr{H}_{eff}(u) - u \times (u \times \mathscr{H}_{eff}(u)) \\ \mathscr{H}_{eff}(u) &= \mathscr{H}_d^{int}(u) + \mathscr{H}_d^{ext}(u) \end{aligned} \tag{11}$$

The existence and uniqueness of a solution of (11) for any initial condition follows from the Cauchy–Lipschitz theorem. We assume that u^* be a fixed relevant configuration satisfies the saturation constraint (2), i.e, $u^* \in (\mathbb{S}^2)^I$ such that

$$u_i^* = \mu_i \mathbf{e}_1, \ \text{ with } \mu_i \in \{-1, +1\},\ \ \forall\, i \in I.$$

Because of the physical saturation constraint (2), we only deal with perturbations u of u^* satisfying:

$$|u_i(t)| = 1,\ \ \forall\, i \in I \ \text{ and } \forall\, t \geq 0.$$

We consider u as a small perturbation of u^* and describe it as:

$$u_i = \mu_i \mathbf{e}_1 + \gamma(\omega_i)\mu_i \mathbf{e}_1 + \omega_i^2 \mathbf{e}_2 + \omega_i^3 \mathbf{e}_3, \quad \forall\, i \in I \tag{12}$$

with $\omega_i = \left(\omega_i^2, \omega_i^3\right)$ and $\gamma : (\mathbb{R}^2)^I \to (\mathbb{R})^I$ is a smooth map defined as $\gamma(\omega_i) = \sqrt{1 - |\omega_i|^2} - 1$.

To obtain the transformed system of (11) in new variable $\omega \in C^1\left(\mathbb{R}^+; (\mathbb{R}^2)^I\right)$, we use the perturbation (12) of u^*. We substitute (12) in (11) and take the projection of the obtained expression along the direction of $\mathbf{e}_2$ and $\mathbf{e}_3$.

After a lengthy algebraic computations, it yields that u given by (12) satisfies (11) if and only if $\omega = (\omega^2, \omega^3)$ verifies the following system:

$$\frac{d\omega}{dt} = \begin{pmatrix} -1 & -\mu \\ \mu & -1 \end{pmatrix} \omega + \mathscr{A}(\mu) + \mathscr{B}(\omega) + \mathscr{C}(\omega), \tag{13}$$

where

$$\mathscr{A}(\mu) = \begin{pmatrix} \Psi^2(\mu) \\ -\mu\Psi^2(\mu) \end{pmatrix},$$

The linear term $\mathscr{B}(\omega)$ is given by:

$$\mathscr{B}(\omega) = \begin{pmatrix} -(\omega^3 + \mu\omega^2)\Psi^1(\mu) + \Psi^2(\mu\gamma) \\ +\Psi^3(\omega^2) + \mu\Psi^4(\omega^3) \\ \\ (\omega^2 - \mu\omega^3)\Psi^1(\mu) - \mu\Psi^2(\mu\gamma) \\ -\mu\gamma\Psi^2(\mu) - \mu\Psi^3(\omega^2) + \Psi^4(\omega^3) \end{pmatrix},$$

The nonlinear term $\mathscr{C}(\omega)$ is given by:

$$\mathscr{C}(\omega) = \begin{pmatrix} -\mu\gamma\omega^3 - (\omega^3 + \mu\omega^2 + \mu\gamma\omega^2)\Psi^1(\mu\gamma) \\ -\mu\gamma\omega^2\Psi^1(\mu) - (\omega^3 + \mu\omega^2 + \mu\gamma\omega^2)\Psi^2(\omega^2) \\ -(\Psi^2(\mu) + \Psi^2(\mu\gamma))(\omega^2)^2 - (\omega^2)^2\Psi^3(\omega^2) \\ +(\mu\gamma - \omega^2\omega^3)\Psi^4(\omega^3) + ((\omega^2)^2 + (\omega^3)^2)\omega^2 \\ \\ \mu\gamma\omega^2 + (\omega^2 - \mu\omega^3 - \mu\gamma\omega^3)\Psi^1(\mu\gamma) \\ -\mu\gamma\omega^3\Psi^1(\mu) + (\omega^2 - \mu\omega^3 - \mu\gamma\omega^3)\Psi^2(\omega^2) \\ -(\Psi^2(\mu) + \Psi^2(\mu\gamma))\omega^2\omega^3 - \mu\gamma\Psi^2(\mu\gamma) \\ -(\mu\gamma + \omega^2\omega^3)\Psi^3(\omega^2) - (\omega^3)^2\Psi^4(\omega^3) \\ +((\omega^2)^2 + (\omega^3)^2)\omega^3 \end{pmatrix}.$$

Our objective is to analyze the stability behavior of a relevant configuration u^* for LLE (11). Evidently, both the forms of Landau–Lifschitz equation, (11) and (13) are equivalent and the stability of zero solution for (13) renders the stability of u^* for (11). We state this in the following Proposition.

Proposition 1 *Let* $u \in C^1\left(\mathbb{R}^+; (\mathbb{S}^2)^I\right)$ *with* $|u| = 1$ *and verifies* (11). *Let* $\omega \in C^1\left(\mathbb{R}^+; (\mathbb{R}^2)^I\right)$ *defined by:*

$$u = \mu \mathbf{e}_1 + \gamma(\omega)\mu \mathbf{e}_1 + \omega^2 \mathbf{e}_2 + \omega^3 \mathbf{e}_3$$

Then u *is a solution to Landau–Lifschitz equation* (11) *if and only if* ω *is a solution to* (13). *Moreover,* u^* *is asymptotically stable for* (11) *if and only if* 0 *is asymptotically stable for* (13).

Proof We follow the similar technique used in partial differential equation framework in [13–15]. It is apparent that by taking the projection on both $\mathbf{e}_2$ and $\mathbf{e}_3$ axis, if u satisfies (11) then ω verifies (13). For the converse part, we write (11) on the form

$$\frac{du}{dt} = \mathscr{F}(u).$$

Furthermore $u \cdot \mathscr{F}(u) = 0$. Since ω satisfies (13), we have

$$\left(\frac{du}{dt} - \mathscr{F}(u)\right) \cdot \mathbf{e}_k = 0, \quad \forall\, k \in \{2, 3\}.$$

Using the constraint $|u| = 1$, which renders $u \cdot \dfrac{du}{dt} = 0$. we obtain

$$\mu\,(1 + \gamma)\left(\frac{du}{dt} - \mathscr{F}(u)\right) \cdot \mathbf{e}_1 = 0,$$

with $\mu \neq 0$ and $\gamma \neq -1$, implies that u satisfies (11). This completes the proof of Proposition 1.

Now we study the stability of zero solution for the transformed Landau–Lifschitz equation (13). First, we establish some preliminary estimates. We estimate the linear operators $\Psi^k\colon (\mathbb{R}^3)^I \to (\mathbb{R}^3)^I$ in the following fashion, we obtain for all $k = \{1, \ldots, 4\}$

$$\|\Psi^k(u)\| \le \frac{K_1}{\pi \ell^2}\left(\sum_{j \neq 0} \frac{1}{|j|^2}\right)\|u\|, \quad \forall\, u = (u_i)_{i \in I} \in (\mathbb{R}^3)^I. \tag{14}$$

Using (14), the operators $\mathscr{A}$, $\mathscr{B}$ and $\mathscr{C}$ appear on the right hand side of (13) are estimated with straightforward arguments in the following lemmas.

Lemma 1 *There exists a constant* K_2 *such that, for all* $\omega \in (\mathbb{R}^3)^I$ *with* $\|\omega\| < 1$, *we have*

$$\|\mathscr{B}(\omega)\| \le \frac{K_2}{\ell^2}\|\omega\|.$$

Lemma 2 *We assume that* $\frac{1}{\ell^2} \leq 1$. *There exist constants* K_3 *and* K_4 *such that, for all* $\omega \in (\mathbb{R}^3)^I$ *with* $\|\omega\| < 1$, *we have*

$$\|\mathscr{A}(\mu)\| \ \leq \ K_3 \ \textit{and} \ \|\mathscr{C}(\omega)\| \leq K_4\|\omega\|^2.$$

It is worth to mention that the constants K_2, K_3 and K_4 neither depend on ℓ nor on the size of the network. We notice that $\|\gamma(\omega)\| \leq \|\omega\|$ whenever $\|\omega\| < 1$. We have $\omega \in C^1\left(\mathbb{R}^+; (\mathbb{R}^2)^I\right)$ with,

$$|\omega_i(t)| = \left((\omega_i^2(t))^2 + (\omega_i^3(t))^2\right)^{\frac{1}{2}}$$

We notice that $u \in \mathscr{D}_\mu(\alpha)$ if and only if $|\omega| < \alpha$ (see (9)).
Taking the inner product of (13) with $\left(\omega_i^2, \omega_i^3\right)$, we obtain, for all $i \in I$,

$$\left(\omega_i^2 \frac{d}{dt}\omega_i^2 + \omega_i^3 \frac{d}{dt}\omega_i^3\right) + \left((\omega_i^2)^2 + (\omega_i^3)^2\right) = ((\mathscr{A}(\mu))_i + (\mathscr{B}(\omega))_i + (\mathscr{C}(\omega))_i) \cdot \left(\omega_i^2, \omega_i^3\right),$$

Using Lemmas 1 and 2, we have,

$$\frac{1}{2}\frac{d}{dt}\left(|\omega_i|^2\right) + |\omega_i|^2 \ \leq \ K_3\|\omega\| + \frac{K_2}{\ell^2}\|\omega\|^2 + K_4\|\omega\|^3. \tag{15}$$

We define β by,

$$\beta = \frac{1}{K_2} \tag{16}$$

Our goal is to show that zero solution is asymptotically stable for (13). We set the distance between the nanowires in such a way so that $\frac{1}{\ell^2}$ remains less than β.
Multiplying (15) by e^{2t} and integrate from 0 to t. We get, for all $i \in I$,

$$\left(|\omega_i(t)|^2\right) e^{2t} \ \leq \ \|\omega(0)\|^2 + 2K_3 \int_0^t \|\omega(v)\| e^{2v} dv + 2\frac{K_2}{\ell^2} \int_0^t \|\omega(v)\|^2 e^{2v} dv + 2K_4 \int_0^t \|\omega(v)\|^3 e^{2v} dv$$

We take the supremum on $i \in I$ and obtain that

$$\|\omega(t)\|^2 e^{2t} \leq \|\omega(0)\|^2 + 2K_3 \int_0^t \|\omega(v)\| e^{2v} dv + 2\frac{K_2}{\ell^2} \int_0^t \|\omega(v)\|^2 e^{2v} dv + 2K_4 \int_0^t \|\omega(v)\|^3 e^{2v} dv$$

We denote $\kappa = 1 - \frac{K_2}{\ell^2}$. Equation (16) together with condition $\frac{1}{\ell^2} < \beta$ implies $\kappa > 0$. Now while $\|\omega(v)\| \leq \frac{\kappa}{2K_4} e^{-2v} \leq \frac{\kappa}{2K_4}$, we have

$$\|\omega(t)\|^2 e^{2t} \leq \|\omega(0)\|^2 + \frac{K_3}{K_4}\kappa t + (2-\kappa) \int_0^t \|\omega(v)\|^2 e^{2v} dv, \quad \forall\, t \geq 0.$$

Using Gronwall lemma, while $\|\omega(v)\| \leq \frac{\kappa}{2K_4}$, we obtain

$$\|\omega(t)\|^2 \leq \left[\|\omega(0)\|^2 + \frac{K_3}{K_4}\kappa t \right] e^{-\kappa t}$$

It is evident that the term $te^{-\kappa t} \to 0$ as $t \to \infty$. Therefore, whenever $\|\omega(0)\| \leq \frac{\kappa}{2K_4}$, we obtain

$$\|\omega(t) - 0\| \to 0 \text{ as } t \to \infty.$$

We set $\alpha_0 = \frac{\kappa}{2K_4}$, and it shows that zero solution is asymptotically stable for the perturbed LLE (13) which in turn reflect the asymptotic behavior of relevant configurations for LLE (11). This completes the proof of Theorem 1.

References

1. Parkin, S.S.P., Hayashi, M., Thomos, L.: Magnetic domain-wall racetrack memory. Science **320**, 190–194 (2008)
2. Hubert, A., Schäfer, R.: Magnetic Domains: The Analysis of Magnetic Microstructures. Springer, Berlin (1998)
3. Brown, W.F.: Micromagnetics. Wiley, New York (1963)
4. Alouges, F.: Mathematical models in micromagnetism: An introduction. ESAIM Proc. **22**, 114–117 (2007)
5. Labbé, S., Privat, Y., Trélat, E.: Stability properties of steady-states for a network of ferromagnetic nanowires. J. Diff. Eqn. **253**, 1709–1728 (2012)
6. Alouges, F., Soyeur, A.: On global weak solutions for Landau-Lifshitz equations: Existence and nonuniqueness. Nonlinear Anal. Theory Meth. Appl. **18**, 1071–1084 (1992)
7. Carbou, G., Fabrie, P.: Time average in micromagnetism. J. Differ. Eqn. **147**, 383–409 (1998)

8. Visintin, A.: On Landau Lifschitz equation for ferromagnetism. Japan J. Appl. Math. **1**, 69–84 (1985)
9. Carbou, G., Fabrie, P.: Regular solutions for Landau-Lifschitz equation in a bounded domain. Differ. Integral Eqn. **14**, 213–229 (2001)
10. Carbou, G., Fabrie, P.: Regular solutions for Landau-Lifschitz equation in $\mathbb{R}^3$. Commun. Appl. Anal. **5**, 17–30 (2001)
11. Bañas, L., Bartels, S., Prohl, A.: A convergent implicit finite element discretization of the Maxwell-Landau-Lifshitz-Gilbert equation, SIAM J. Numer. Anal. **46**, 1399–1422 (2008)
12. Labbé, S.: Fast computation for large magnetostatic systems adapted for micromagnetism. SIAM J. Sci. Comput. **26**, 2160–2175 (2005)
13. Carbou, G., Labbé, S.: Stability for static walls in ferromagnetic nanowires. Discrete Contin. Dynam. Syst. Ser. B **6**, 273–290 (2006)
14. Carbou, G., Labbé, S., Trélat, E.: Control of travelling walls in a ferromagnetic nanowire. Discrete Contin. Dynam. Systems Ser. S **1**, 51–59 (2008)
15. Carbou, G.: Stability of static walls for a three-dimensional model of ferromagnetic material. J. Math. Pures Appl. **93**, 183–203 (2010)
16. Agarwal, S., Carbou, G., Labbé, S., Prieur, C.: Control of a network of magnetic ellipsoidal samples. Math. Control Relat. Fields **1**(2), 129–147 (2011)
17. Sanchez, D.: Behaviour of the LandauLifschitz equation in a ferromagnetic wire. Math. Meth. Appl. Sci. **32**(2), 167–205 (2009)
18. Griffiths, D.J.: Introduction to Electrodynamics, 3rd edn. PearsonBenjamin Cummings, San Francisco (2008)

Fractional Functional Impulsive Differential Equation with Integral Boundary Condition

Vidushi Gupta and Jaydev Dabas

Abstract In this article, we discuss the existence and uniqueness of solution for fractional order differential equation with integral boundary condition and fractional impulsive conditions. In our problem delay also include with finite domain. Some important fixed point theorems are the main tools to establish the existence and uniqueness results for the solution of the problem.

Keywords Fractional order differential equation · Impulsive conditions · Boundary value problem · Fixed point theorems

1 Introduction

The present work is related to study the existence and uniqueness of solution for impulsive fractional differential equations with some special boundary conditions given as follows:

$$^{c}D_t^{\alpha}x(t) = f(t, x(t), x(t-\tau)),\ t \in [0, T],\ t \neq t_k \tag{1}$$

$$\Delta x(t_k) = I_k(x(t_k^-)),\ k = 1, 2, \ldots, m, \tag{2}$$

$$\Delta(^{c}D^{q}x(t_k^-)) = J_k(x(t_k^-)),\ q \in (0, 1),\ k = 1, 2, \ldots, m, \tag{3}$$

$$x(t) = \phi(t),\ t \in [-\tau, 0] \tag{4}$$

$$x(0) - x(T) = \int_0^T p(x(s))ds, \tag{5}$$

V. Gupta (✉) · J. Dabas
Department of Applied Science and Engineering, IIT Roorkee,
Saharanpur Campus, 247001 Saharanpur, India
e-mail: rinkygupta00@gmail.com

J. Dabas
e-mail: jay.dabas@gmail.com

P.N. Agrawal et al. (eds.), *Mathematical Analysis and its Applications*,
Springer Proceedings in Mathematics & Statistics 143,
DOI 10.1007/978-81-322-2485-3_34

where ${}^cD_t^\alpha$ is Caputo's derivative and $\alpha \in (1,2)$, $T < \infty$. The function $f\colon [0,T] \times X \times PC_0 \to X$ and $p\colon X \to X$ are given continuous functions and satisfied some assumptions. The Eq. (2)–(3) are impulsive conditions having some properties, with $0 = t_0 < t_1 < \cdots < t_m < t_{m+1} = T$, $I_k, J_k \in C(X, X)$, $(k = 1, 2, \ldots, m)$, are bounded functions. We have $\Delta x(t_k) = x(t_k^+) - x(t_k^-)$ and $\Delta({}^cD^q x(t_k)) = ({}^cD^q x(t_k^+)) - ({}^cD^q x(t_k^-))$, $x(t_k^+) = \lim_{h\to 0} x(t_k + h)$ and $x(t_k^-) = \lim_{h\to 0} x(t_k - h)$ represents the right and left-hand limits of $x(t)$ at $t = t_k$ respectively with $x(t_i^-) = x(t_i)$.

Recently, the study of differential equations of the type of non-integer order has been an important tool in the area of research in mathematics. Its useful applications included mathematical modeling in many engineering and science discipline like physics, chemistry, biophysics, biology, etc. Its nonlocal behavior is the vital characteristic that makes it vary from its rival in classical calculus. For more details one can refer the books [1–3] and papers [4, 5] and the references therein.

In recent years, the theory of impulsive differential equations for integer order comes in various applications of mathematical modeling of phenomena and practical situations. For instance, the impulsive differential equations captured from real-world problems describe the dynamics of processes in which sudden, discontinuous jumps occurs. For more details one can see the papers [4, 7–9] and references therein.

Problems with integral boundary conditions arise naturally in thermal conduction problems, semiconductor problems, hydrodynamic problems, etc. However, periodic boundary value problems with impulsive fractional evolution equations have not been studied extensively. For more details one can see the papers [6, 7, 10–14, 20] and the references therein.

Differential equations with time delay are often used to model phenomena in economics, biology, medicine, ecology, and other fields of sciences. They take into account that in many applications, some time elapses between causes and their effects. These concerns, for instance, investments in economics and finance, typically yielding returns only after some time lag. Delay reaction also transpire in population dynamics, where individuals always need some time to mature, or in medicine, where contagious diseases have cross-infection. Presently, many authors [12, 15–20] are currently working on field of fractional delay differential equation with finite domain.

In [21] Ravichadran et al. proved the existence and uniqueness of mild solutions for the following impulsive fractional functional differential equations of the form:

$$D^\alpha x(t) = Ax(t) + f(t, x_t, \int_0^t h(t, s, x_s)ds),\ t \in [0, T],\ t \neq t_k,$$

$$x(t) = \phi(t),\ t \in [-d, 0],\ \Delta x(t_k) = Q_k(x(t_k^-)),\ k = 1, 2, \ldots, m,$$

Authors established their results by using some fixed point theorems.

In [20] Dabas et al. considered the following impulsive neutral fractional integro-differential equation with state-dependent delay subject to integral boundary condition

$$D_t^{\alpha}\left[x(t)+\int_0^t \frac{(t-s)^{\alpha-1}}{\Gamma(\alpha)} g(s, x_{\rho(s,x_s)})ds\right] = f(t, x_{\rho(t,x_t)}, B(x)(t)),\ t\in[0,T],\ t\neq t_k$$

$$\Delta x(t_k) = I_k(x(t_k^-)),\ \Delta x'(t_k) = Q_k(x(t_k^-)),\ k=1,2,\ldots,m,$$

$$x(t)=\phi(t),\ t\in[-d,0],\ ax'(0)+bx'(T)=\int_0^T q(x(s))ds.$$

The existence results are proved by applying the classical fixed point theorems.

In [8] Xi Fu et al. concerned with the fractional separated boundary value problem with fractional impulsive conditions. By using the Schaefer fixed point theorem, Banach fixed point theorem, and nonlinear alternative of Leray–Schauder type authors obtained the existence results. Chouhan et al. [12] studied the solution for the system with infinite delay and by using Banach contraction and Krasnoselkii fixed point theorems authors established the existence and uniqueness results. In [14] Yu et al. concerned periodic fractional impulsive boundary value problem. The existence and boundedness of piecewise continuous mild solutions and design parameter drift for periodic motion of linear problems are presented. Furthermore, the authors' established existence results of piecewise continuous mild solutions for semilinear impulsive periodic problems are shown by using the Schauder's fixed point theorem.

To the best of the authors' knowledge, no one has studied the existence and uniqueness of solutions for fractional boundary value problem (1)–(5) by using the Caputo derivative. The purpose of this paper is to fill in this gap. The organization of the paper is as follows. In Sect. 2, we give some basic preliminaries concerning the fractional integral, fractional derivative, and fixed point theorems. In Sect. 3, we present our main results.

2 Preliminary

Let $(X, \|\cdot\|_X)$ be a complex Banach space. In order to defined the solution of the problem (1), we consider the following space: $PC_T = PC([-\tau, T]; X) = \{\phi: [-\tau, T] \to X: \phi(t)$ is continuous every where except for a finite number of points $t_i, i = 1, 2, \ldots, m$, at which $\phi(t_i^+)$ and $\phi(t_i^-)$ exist and $\phi(t_i) = \phi(t_i^-)\}$. The space $PC_t = PC([-\tau, t]: X)$ for $t \in [-\tau, T]$ is the Banach space of all continuous functions except for a finite number of points $t_i, i = 1, 2, \ldots, m$, at which $\phi(t_i^+)$ and $\phi(t_i^-)$ exist and $\phi(t_i) = \phi(t_i^-)$ endowed with the norm

$$\|\phi\|_t = \sup_{-\tau\leq s\leq t}\{\|\phi(s)\|_X,\ \phi\in PC_t\},$$

where $\|\cdot\|_X$ is the norm in X. For delay we consider the Banach space $PC_0 = C([-\tau, 0]: X)$ endowed with the above sup-norm.

Definition 1 ([4]) The fractional integral of order α with lower limit zero for a function $f: [0, \infty) \to R$ of order is defined as

$$I_t^{\alpha} f(t) = \int_0^t \frac{(t-s)^{\alpha-1}}{\Gamma(\alpha)} f(s)ds, \ t > 0, \ \alpha > 0, \tag{6}$$

provided the right side is pointwise defined on $[0, \infty)$, where Γ is the gamma function.

Definition 2 ([4]) The Riemann–Liouville derivative of order α with the lower limit zero for a function $f: [0, \infty) \to R$ can be written as

$${}^L D_t^{\alpha} f(t) = \frac{1}{\Gamma(n-\alpha)} \Big(\frac{d}{dt}\Big)^n \int_0^t (t-s)^{n-\alpha-1} f(s)ds, \ t > 0, \ n-1 < \alpha < n. \tag{7}$$

Definition 3 ([4]) The Caputo's derivative of order α for a function $f: [0, \infty) \to R$ can be written as

$${}^c D_t^{\alpha} f(t) = {}^L D_t^{\alpha} \Big[f(t) - \sum_{k=0}^{n-1} \frac{t^k}{k!} f^{(k)}(0) \Big], \ t > 0, \ n-1 < \alpha < n. \tag{8}$$

Remark 1 ([4]) If $f(t) \in C^n[0, \infty)$, for order $n-1 < \alpha < n$ then

$${}^c D_t^{\alpha} f(t) = \frac{1}{\Gamma(n-\alpha)} \int_0^t \frac{f^{(n)}(s)}{(t-s)^{\alpha+1-n}} ds = I_t^{n-\alpha} f^{(n)}(t), \ t > 0. \tag{9}$$

The Caputo derivative of constant is equal to zero.

Lemma 1 ([5]) *Let $\alpha > 0$, then the differential equation*

$${}^c D^{\alpha} h(t) = 0 \tag{10}$$

has solutions $h(t) = c_0 + c_1 t + c_2 t^2 + \cdots + c_{n-1} t^{n-1}$ and $I^{\alpha} D^{\alpha} h(t) = h(t) + c_0 + c_1 t + c_2 t^2 + \cdots + c_{n-1} t^{n-1}$, where $c_i \in R$, $i = 0, 1, \ldots, n-1$, $n = [\alpha] + 1$.

Lemma 2 *Let $\alpha \in (1, 2)$ and $f: [0, T] \times X \times PC_0 \to X$ be continuously differentiable function. A piecewise continuous differential function $x(t): (-\tau, T] \to X$ is a solution of system (1)–(5) on $[0, T]$ iff $x(t)$ satisfied following integral equation,*

$$x(t) = \begin{cases} \phi(t), & t \in [-\tau, 0], \\ \int_0^t \frac{(t-s)^{\alpha-1}}{\Gamma(\alpha)} f(s, x(s), x(s-\tau))ds + \phi(0) \\ -\frac{t}{T}\Big[\int_0^T p(x(s))ds + T \sum_{i=1}^{m} \Big(\frac{\Gamma(2-q)}{t_i{}^{1-q}} J_i(x(t_i^-))\Big) \\ + \int_0^T \frac{(T-s)^{\alpha-1}}{\Gamma(\alpha)} f(s, x(s), x(s-\tau))ds\Big], & t \in [0, t_1), \\ \cdots \\ \int_0^t \frac{(t-s)^{\alpha-1}}{\Gamma(\alpha)} f(s, x(s), x(s-\tau))ds + \sum_{i=1}^{k} I_i(x(t_i^-)) + \phi(0) \\ -\frac{t}{T}\Big[\int_0^T p(x(s))ds + T \sum_{i=1}^{m} \Big(\frac{\Gamma(2-q)}{t_i{}^{1-q}} J_i(x(t_i^-))\Big) \\ + \int_0^T \frac{(T-s)^{\alpha-1}}{\Gamma(\alpha)} f(s, x(s), x(s-\tau))ds\Big] \\ + \sum_{i=1}^{k}(t - t_i)\Big(\frac{\Gamma(2-q)}{t_i{}^{1-q}} J_i(x(t_i^-))\Big), & t \in (t_k, t_{k+1}]. \end{cases}$$

Proof For $t \in [0, t_1]$, then

$$\begin{aligned} D_t^{\alpha} x(t) &= f(t, x(t), x(t-\tau)), \\ x(t) &= \phi(t), \ t \in [-\tau, 0]. \end{aligned} \tag{11}$$

Taking the Riemann–Liouville fractional integral on (11) and using the Lemma (1), then

$$x(t) = \int_0^t \frac{(t-s)^{\alpha-1}}{\Gamma(\alpha)} f(s, x(s), x(s-\tau))ds - c_0 - c_1 t, \tag{12}$$

using the condition $x(0) = \phi(0)$ we compute $c_0 = -\phi(0)$, then we have

$$x(t) = \int_0^t \frac{(t-s)^{\alpha-1}}{\Gamma(\alpha)} f(s, x(s), x(s-\tau))ds + \phi(0) - c_1 t. \tag{13}$$

Same way for $t \in (t_1, t_2]$, then

$$\begin{cases} D_t^{\alpha} x(t) = f(t, x(t), x(t-\tau)), \\ \Delta x(t_1) = I_1(x(t_1^-)), \ \Delta({}^c D^q x(t_1^-)) = J_1(x(t_1^-)), \end{cases} \tag{14}$$

Again applying the Riemann–Liouville fractional integral operator and using the Lemma (1), then

$$x(t) = \int_0^t \frac{(t-s)^{\alpha-1}}{\Gamma(\alpha)} f(s, x(s), x(s-\tau))ds - c_2 - c_3 t, \tag{15}$$

on applying first impulsive condition $\Delta x(t_1) = I_1(x(t_1^-))$, we get

$$-c_2 = I_1(x(t_1^-)) + c_3 t_1 + \phi(0) - c_1 t_1. \tag{16}$$

Substituting the value of c_2 in (15), we obtain

$$x(t) = \int_0^t \frac{(t-s)^{\alpha-1}}{\Gamma(\alpha)} f(s, x(s), x(s-\tau))ds + I_1(x(t_1^-)) \\ +\phi(0) - c_1 t_1 + c_3(t_1 - t). \tag{17}$$

From (17) and (13), we get

$$D^q x(t) = \frac{1}{\Gamma(\alpha-q)} \int_0^t (t-s)^{\alpha-q-1} f(s, x(s), x(s-\tau))ds - c_3 \frac{t^{1-q}}{\Gamma(2-q)}, \tag{18}$$

$$D^q x(t) = \frac{1}{\Gamma(\alpha-q)} \int_0^t (t-s)^{\alpha-q-1} f(s, x(s), x(s-\tau))ds - c_1 \frac{t^{1-q}}{\Gamma(2-q)}. \tag{19}$$

Using the second impulsive condition $\Delta(D^q x(t_1)) = J_1(x(t_1^-))$, then we have

$$c_3 = -\frac{\Gamma(2-q)}{{t_1}^{1-q}} J_1(x(t_1^-)) + c_1. \tag{20}$$

On putting the value of c_3 in (17), we prevail

$$x(t) = \int_0^t \frac{(t-s)^{\alpha-1}}{\Gamma(\alpha)} f(s, x(s), x(s-\tau))ds + I_1(x(t_1^-)) \\ +\phi(0) + (t - t_1)\frac{\Gamma(2-q)}{{t_1}^{1-q}} J_1(x(t_1^-)) - c_1 t. \tag{21}$$

Similarly for $t \in (t_k, t_{k+1}]$, then

$$x(t) = \int_0^t \frac{(t-s)^{\alpha-1}}{\Gamma(\alpha)} f(s, x(s), x(s-\tau))ds + \sum_{i=1}^{k} I_i(x(t_i^-)) + \phi(0) - c_1 t \\ + \sum_{i=1}^{k} (t - t_i)\left(\frac{\Gamma(2-q)}{{t_i}^{1-q}} J_i(x(t_i^-))\right). \tag{22}$$

Now using the boundary condition $x(0) - x(T) = \int_0^T p(x(s))ds$, we compute the following value of the constant c_1 given as:

$$c_1 = \frac{1}{T}\Big[\int_0^T p(x(s))ds + T\sum_{i=1}^{m}\left(\frac{\Gamma(2-q)}{{t_i}^{1-q}} J_i(x(t_i^-))\right) \\ + \int_0^T \frac{(T-s)^{\alpha-1}}{\Gamma(\alpha)} f(s, x(s), x(s-\tau))ds\Big]. \tag{23}$$

So that for $t \in (t_k, t_{k+1}]$, then we get

$$
\begin{aligned}
x(t) &= \int_0^t \frac{(t-s)^{\alpha-1}}{\Gamma(\alpha)} f(s, x(s), x(s-\tau))ds + \sum_{i=1}^{k} I_i(x(t_i^-)) + \phi(0) \\
&-\frac{t}{T}\Big[\int_0^T p(x(s))ds + T\sum_{i=1}^{m}\left(\frac{\Gamma(2-q)}{t_i{}^{1-q}} J_i(x(t_i^-))\right) \\
&+\int_0^T \frac{(T-s)^{\alpha-1}}{\Gamma(\alpha)} f(s, x(s), x(s-\tau))ds\Big] \\
&+\sum_{i=1}^{k}(t-t_i)\left(\frac{\Gamma(2-q)}{t_i{}^{1-q}} J_i(x(t_i^-))\right). \qquad (24)
\end{aligned}
$$

This is complete proof.

3 Existence and Uniqueness Results

Our first result is based on Banach contraction principle.

Theorem 1 *Suppose the following conditions holds:*

1. *There exist the positive constants* L_{f1}, L_{f2}, L_p *such that*

$$
\begin{aligned}
&\|f(t, x, \psi) - f(t, y, \chi)\|_X \le L_{f2}\|x - y\|_X + L_{f1}\|\psi - \chi\|_X, \\
&\|p(x) - p(y)\|_X \le L_p\|x - y\|_X, \ t \in [0, T], \ \forall \psi, \chi, x, y \in X.
\end{aligned}
$$

2. *The bounded continuous function* I_k, J_k *satisfied the condition for constant* L_I, L_J *such that*

$$
\|I_k(x) - I_k(y)\|_X \le L_I\|x - y\|_X, \ \|J_k(x) - J_k(y)\|_X \le L_J\|x - y\|_X, \ \forall x, y \in X.
$$

3. *And assume that* $\left(\frac{2T^\alpha}{\Gamma(\alpha+1)}(L_{f_1} + L_{f_2}) + mL_I + TL_p + 2T^q m\Gamma(2-q)L_J\right) < 1.$

Then system (1)–(5) has a unique solution.

Proof Let the space PC_T is closed convex set, invested with the uniform topology and the operator $P : PC_T \to PC_T$ is defined by

$$Px(t) = \begin{cases} \phi(t) & t \in [-\tau, 0], \\ \int_0^t \frac{(t-s)^{\alpha-1}}{\Gamma(\alpha)} f(s, x(s), x(s-\tau))ds + \phi(0) \\ -\frac{t}{T}\Big[\int_0^T p(x(s))ds + T\sum_{i=1}^m \left(\frac{\Gamma(2-q)}{t_i^{1-q}} J_i(x(t_i^-))\right) \\ + \int_0^T \frac{(T-s)^{\alpha-1}}{\Gamma(\alpha)} f(s, x(s), x(s-\tau))ds\Big], & t \in [0, t_1), \\ \ldots, \\ \int_0^t \frac{(t-s)^{\alpha-1}}{\Gamma(\alpha)} f(s, x(s), x(s-\tau))ds + \sum_{i=1}^k I_i(x(t_i^-)) + \phi(0) \\ -\frac{t}{T}\Big[\int_0^T p(x(s))ds + T\sum_{i=1}^m \left(\frac{\Gamma(2-q)}{t_i^{1-q}} J_i(x(t_i^-))\right) \\ + \int_0^T \frac{(T-s)^{\alpha-1}}{\Gamma(\alpha)} f(s, x(s), x(s-\tau))ds\Big] \\ + \sum_{i=1}^k (t-t_i)\left(\frac{\Gamma(2-q)}{t_i^{1-q}} J_i(x(t_i^-))\right), & t \in (t_k, t_{k+1}]. \end{cases} \tag{25}$$

Let $x, x^* \in PC_T$ and $t \in [0, t_1)$.

$$\|P(x) - P(x^*)\|_X \leq \int_0^t \frac{(t-s)^{\alpha-1}}{\Gamma(\alpha)} \|f(s, x(s), x(s-\tau)) - f(s, x^*(s), x^*(s-\tau))\|_X ds$$
$$+ \frac{|t|}{T}\Big[\int_0^T \|p(x(s)) - p(x^*(s))\|_X ds + T\sum_{i=1}^m \left(\frac{\Gamma(2-q)}{|t_i|^{1-q}} \|J_i(x(t_i^-)) - J_i(x^*(t_i^-))\|_X\right)$$
$$+ \int_0^T \frac{(T-s)^{\alpha-1}}{\Gamma(\alpha)} \|f(s, x(s), x(s-\tau)) - f(s, x^*(s), x^*(s-\tau))\|_X ds\Big]$$

Using the given conditions, we get

$$\|P(x) - P(x^*)\|_{PC_T} \leq \Big(\frac{2T^\alpha}{\Gamma(\alpha+1)}(L_{f_1} + L_{f_2}) + L_p T + T^q m \Gamma(2-q) L_J\Big)\|x - x^*\|_{PC_T}.$$

For $t \in (t_k, t_{k+1}]$, we have

$$\|P(x) - P(x^*)\|_X \leq$$
$$\int_0^t \frac{(t-s)^{\alpha-1}}{\Gamma(\alpha)} \|f(s, x(s), x(s-\tau)) - f(s, x^*(s), x^*(s-\tau))\|_X ds$$
$$+ \sum_{i=1}^k \|I_i(x(t_i^-)) - I_i(x^*(t_i^-))\|_X + \frac{|t|}{T}\Big[\int_0^T \|p(x(s)) - p(x^*(s))\|_X ds$$
$$+ T\sum_{i=1}^m \left(\frac{\Gamma(2-q)}{|t_i|^{1-q}} \|J_i(x(t_i^-)) - J_i(x^*(t_i^-))\|_X\right)$$
$$+ \int_0^T \frac{(T-s)^{\alpha-1}}{\Gamma(\alpha)} \|f(s, x(s), x(s-\tau)) - f(s, x^*(s), x^*(s-\tau))\|_X ds\Big]$$
$$+ \sum_{i=1}^k (|t - t_i|)\left(\frac{\Gamma(2-q)}{|t_i|^{1-q}} \|J_i(x(t_i^-)) - J_i(x^*(t_i^-))\|_X\right)$$

Again in the same way using the given assumptions of the Theorem (1), we have

$$\begin{aligned}&\|P(x)-P(x^*)\|_{PC_T}\\&\leq\Big(\frac{2T^\alpha}{\Gamma(\alpha+1)}(L_{f_1}+L_{f_2})+mL_I+TL_p+2T^q m\Gamma(2-q)L_J\Big)\|x-x^*\|_{PC_T}\\&\leq\Delta\|x-x^*\|_{PC_T}.\end{aligned}$$

Since $\Delta<1$, implies that the map P is a contraction map and has a unique fixed point $x\in PC_T$, which is a solution of the system (1)–(5) on $[-\tau,T]$. This is complete proof of theorem.

Our second result is based on Krasnoselkii's fixed point theorem [20].

Theorem 2 *Let I_k, J_k and p be the continuous functions and satisfy*

$$\|I_k(x)\|_X\leq C_1,\ \|J_k(x)\|_X\leq C_2,\ \|p(x)\|_X\leq C_3,\ C_1>0,\ C_2>0,\ C_3>0.$$

Further, $f:[0,T]\times X\times PC_0\to X$ is continuous function for every $t\in[0,T]$, and satisfy the condition

$$\|f(t,x,\psi)-f(t,y,\chi)\|_X\leq L_{f2}\|x-y\|_X+L_{f1}\|\psi-\chi\|_X,\ L_{f1},\ L_{f2}>0.$$

Then system (1)–(5) has at least one solution.

Proof Let $r\geq[mC_1+\|\phi(0)\|+2T^q mC_2+\frac{2T^\alpha}{\Gamma(\alpha+1)}(L_{f_1}+L_{f_2})r+C_3T]$.

Consider the space $PC_T^r=\{x\in PC_T:\|x\|_{PC_T}\leq r\}$, then PC_T^r is a bounded, closed convex subset in PC_T. For $t\in(t_k,t_{k+1}]$, the operators $N:PC_T^r\to PC_T^r$ and $P:PC_T^r\to PC_T^r$ are defined as

$$\begin{aligned}N(x)&=\sum_{i=1}^{k}I_i(x(t_i^-))+\phi(0)-t\sum_{i=1}^{m}\left(\frac{\Gamma(2-q)}{t_i{}^{1-q}}J_i(x(t_i^-))\right)\\&\quad+\sum_{i=1}^{k}(t-t_i)\left(\frac{\Gamma(2-q)}{t_i{}^{1-q}}J_i(x(t_i^-))\right)\end{aligned}\tag{26}$$

$$\begin{aligned}P(x)&=\int_0^t\frac{(t-s)^{\alpha-1}}{\Gamma(\alpha)}f(s,x(s),x(s-\tau))ds-\frac{t}{T}\Big[\int_0^T p(x(s))ds\\&\quad+\int_0^T\frac{(T-s)^{\alpha-1}}{\Gamma(\alpha)}f(s,x(s),x(s-\tau))ds\Big].\end{aligned}\tag{27}$$

Now the proof of the Theorem (2) is given in form of following steps:

Step 1. Let $x, x^* \in PC_T^r$ then

$$\begin{aligned}
&\|N(x)+P(x^*)\|_X\\
&\le \sum_{i=1}^{k}\|I_i(x(t_i^-))\|_X+\|\phi(0)\|+|t|\sum_{i=1}^{m}\left(\frac{\Gamma(2-q)}{|t_i|^{1-q}}\|J_i(x(t_i^-))\|_X\right)\\
&+\sum_{i=1}^{k}(|t-t_i|)\left(\frac{\Gamma(2-q)}{|t_i|^{1-q}}\|J_i(x(t_i^-))\|_X\right)\\
&+\int_0^t\frac{(t-s)^{\alpha-1}}{\Gamma(\alpha)}\|f(s,x^*(s),x^*(s-\tau))\|_X ds\\
&+\frac{|t|}{T}\Big[\int_0^T\|p(x^*(s))\|_X ds+\int_0^T\frac{(T-s)^{\alpha-1}}{\Gamma(\alpha)}\|f(s,x^*(s),x^*(s-\tau))\|_X ds\Big]\\
&\le mC_1+\|\phi(0)\|+2T^q mC_2+\frac{2T^\alpha}{\Gamma(\alpha+1)}(L_{f_1}+L_{f_2})r+C_3T\\
&\le r.
\end{aligned}$$

Which implies that $\|N(x)+P(x^*)\|_X\le r$. which means that $N(x)+P(x^*)\in PC_T^r$.

Step 2. Let $x_n\to x$ be sequence in PC_T^r then,

$$\begin{aligned}
&\|N(x_n)-N(x)\|_X\\
&\le\sum_{i=1}^{k}\|I_i(x_n(t_i^-))-I_i(x(t_i^-))\|_X+\|\phi(0)\|\\
&+|t|\sum_{i=1}^{m}\left(\frac{\Gamma(2-q)}{|t_i|^{1-q}}\|J_i(x_n(t_i^-))-J_i(x(t_i^-))\|_X\right)\\
&+\sum_{i=1}^{k}(|t-t_i|)\left(\frac{\Gamma(2-q)}{|t_i|^{1-q}}\|J_i(x_n(t_i^-))-J_i(x(t_i^-))\|_X\right).
\end{aligned}$$

Since the functions I_k and J_k, $k=1,2,\ldots,m$, are continuous, hence

$$\|N(x_n)-N(x)\|\to 0.$$

Which implies that the mapping N is continuous on PC_T^r.

Step 3. For each $t\in(t_k,t_{k+1}]$, $k=0,1,\ldots,m$ for each $x\in PC_T^r$, we have

$$\begin{aligned}
\|N(x)\|_X&\le\sum_{i=1}^{k}\|I_i(x(t_i^-))\|_X+\|\phi(0)\|+|t|\sum_{i=1}^{m}\left(\frac{\Gamma(2-q)}{|t_i|^{1-q}}\|J_i(x(t_i^-))\|_X\right)\\
&+\sum_{i=1}^{k}|(t-t_i)|\left(\frac{\Gamma(2-q)}{|t_i|^{1-q}}\|J_i(x(t_i^-))\|\right)\\
&\le mC_1+\|\phi(0)\|+2T^q mC_2.
\end{aligned}$$

Which implies that the mapping N is uniformly bounded.

Step 4. Now to show that N is equicontinuous. Let $l_1, l_2 \in (t_k, t_{k+1}],\ t_k \leq l_1 < l_2 \leq t_{k+1},\ k = 1, 2, \ldots, m,\ x \in PC_T^r$, we have

$$\begin{aligned}
&\|N(x)(l_2) - N(x)(l_1)\|_X \\
&\leq (l_2 - l_1)\sum_{i=1}^{m}\Big(\frac{\Gamma(2-q)}{|t_i|^{1-q}}\|J_i(x(t_i^-))\|_X\Big) + (l_2 - l_1)\sum_{i=1}^{k}\Big(\frac{\Gamma(2-q)}{|t_i|^{1-q}}\|J_i(x(t_i^-))\|_X\Big)
\end{aligned}$$

As $l_2 \to l_1$, then $\|N(x)(l_2) - N(x)(l_1)\| \to 0$. Which show that N is equicontinuous map. Combing Step 2 to Step 4 together with the Ascol's theorem, we conclude that the operator N is a compact.

Step 5. Now we show that P is a contraction mapping. Let $x, x^* \in PC_T^r$ and $t \in (t_k, t_{k+1}],\ k = 1, 2, \ldots, m,$ we have

$$\begin{aligned}
&\|P(x) - P(x^*)\|_X \\
&\leq \int_0^t \frac{(t-s)^{\alpha-1}}{\Gamma(\alpha)}\|f(s, x(s), x(s-\tau)) - f(s, x^*(s), x^*(s-\tau))\|_X ds \\
&+ \frac{|t|}{T}\Big[\int_0^T \|p(x(s)) - p(x^*(s))\|_X ds \\
&+ \int_0^T \frac{(T-s)^{\alpha-1}}{\Gamma(\alpha)}\|f(s, x(s), x(s-\tau)) - f(s, x^*(s), x^*(s-\tau))\|_X ds\Big]
\end{aligned}$$

$$\|P(x) - P(x^*)\|_{PC_T^r} \leq \Big(\frac{2T^\alpha}{\Gamma(\alpha+1)}(L_{f_1} - L_{f_2}) + TL_p\Big)\|x - x^*\|_{PC_T^r}$$

As $\Delta < 1$, therefore P is a contraction map. Thus all the assumptions of the Krasnoselkii's theorem are satisfied, which implies that the system (1)–(5) has at least one solution on $(\tau, T]$. This completes the proof of the theorem.

References

1. Kilbas, A.A., Srivastava, H.M., Trujillo, J.J.: Theory and Applications of Fractional Differential Equations. In: North-Holland Mathematical Studies, vol. 204. Elsevier, Amsterdam (2006)
2. Miller, K.S., Ross, B.: An Introduction to the Fractional Calculus and Fractional Differential Equations. Wiley, New York (1993)
3. Podlubny, I.: Fractional Differential Equations. Academic Press, New York (1999)
4. Feckan, M., Zhou, Y., Wang, J.R.: On the concept and existence of solution for impulsive solution for impulsive fractional differential equations. Commun. Nonlinear Sci. Numer. Simulat. **17**, 3050–3060 (2012)
5. Agarwal, R.P., Benchohra, M., Hamani, S.: A survey on existence results for boundary value problems of nonlinear fractional differential equations and inclusions. Acta Applicandae Mathematicae **109**, 973–1033 (2010)

6. Fu, X., Liu, X.: Existence results for fractional differential equations with separated boundary conditions and fractional impulsive conditions. Abstr. Appl. Anal. Article ID 785078, 9 (2013)
7. Ahmad, B., Sivasundaram, S.: Existence of solutions for impulsive integral boundary value problems of fractional order. Nonlinear Anal. Hybrid Syst. **4**, 134–141 (2010)
8. Guo, T.L., Jiang, W.: Impulsive fractional functional differential equations. Comput. Math. Appl. **64**(10), 3414–3424 (2012)
9. Feckan, M., Zhou, Y., Wang, J.: Response to "Comments on the concept of existence of solution of existence of solution for impulsive fractional differential equations [Commun Nonlinear Sci Numer Simul 2014;19:401–3.]", Commun. Nonlinear Sci. Numer. Simulat. **19**, 4213–4215 (2014)
10. Ahmad, B., Nieto, J.J.: Existence results for nonlinear boundary value problems of fractional integrodifferential equations with integral boundary conditions. Bound. Value Probl. **10**(Article ID 708576), 11 (2009)
11. Boucherif, A.: Second-order boundary value problems with integral boundary conditions. Nonlinear Anal. **70**, 364–371 (2009)
12. Chauhan, A., Dabas, J.: Integral boundary-value problem for impulsive fractional functional integrodifferential equation with infinite delay. Electron. J. Differ. Equ. **2012**, 1–13 (2012)
13. Wang, J., Zhou, Y., Feckan, M.: On recent developments in the theory of boundary value problems for impulsive fractional differential equations. Comp. Math. Appl. **64**, 3008–3020 (2012)
14. Yu, X., Wang, J.: Periodic BVPs for fractional order impulsive evolution equations. Bound. Value Probl. (2014)
15. Agarwal, R.P., Andrade, B.D.: On fractional integro-differential equations with state-dependent delay. Comput. Math. Appl. **62**, 1143–1149 (2011)
16. Belmekki, M., Mekhalfi, K., Ntouyas, S.K.: Semilinear functional differential equations with fractional order and finite delay. Malaya J. Matematika **1**(1), 73–81 (2012)
17. Benchohra, M., Berhoun, F.: Impulsive fractional differential equations with state dependent delay. Commun. Appl. Anal. **14**(2), 213–224 (2010)
18. Benchohra, M., Litimein, S., NGuerekata, G.: On fractional integro-differential inclusions with state-dependent delay in Banach spaces. Appl. Anal. 1–16 (2011)
19. Carvalho, J.P., Santos, D., Cuevas, C., Andrade, B.D.: Existence Results for a fractional equation with state-dependent delay. Adv. Differ. Equ. Article ID 642013, (2011)
20. Dabas, J., Gautam, G.R.: Impulsive neutral fractional integro-differential equations with state dependent delays and integral condition. Electron. J. Differ. Equ. **273**, 1–13 (2013)
21. Ravicharndran, C., Arjunan, M.M.: Existence result for impulsive fractional semilinear functional integrodifferential equations in Banach spaces. J. Fractional Calc. Appl. **1–11**, 3 (2012)

Controllability of Nonlinear Fractional Neutral Stochastic Dynamical Systems with Poisson Jumps

T. Sathiyaraj and P. Balasubramaniam

Abstract This paper is concerned with the controllability of fractional neutral stochastic dynamical systems with Poisson jumps in the finite dimensional space. Sufficient conditions for controllability results are obtained by using Krasnoselskii's fixed point theorem. The controllability Grammian matrix is defined by Mittag-Leffler matrix function.

Keywords Controllability · Fractional differential equation · Mittag-Leffler function · Neutral stochastic system · Poisson jumps

MSC: 93B05 · 26A33 · 34A08 · 34K50 · 60J65

1 Introduction

Fractional differential equations have recently been proved to be valuable tools in the modeling of many phenomena in various fields of science and engineering. It draws a great application in nonlinear oscillations of earthquakes, many physical phenomena such that seepage flow in porous media and in fluid dynamic traffic model. There has been a significant development in fractional differential equations in recent years (see [6, 9, 11, 12]).

It is well known that the concept of controllability plays an important role in engineering and control theory. The controllability results for linear and nonlinear integral order dynamical systems in finite-dimensional space have discussed extensively (see [4]). Local null controllability of nonlinear functional differential systems

T. Sathiyaraj · P. Balasubramaniam (✉)
Gandhigram Rural Institute-Deemed University,
Gandhigram 624 302, Tamil Nadu, India
e-mail: balugru@gmail.com

T. Sathiyaraj
e-mail: sathiyaraj133@gmail.com

P.N. Agrawal et al. (eds.), *Mathematical Analysis and its Applications*,
Springer Proceedings in Mathematics & Statistics 143,
DOI 10.1007/978-81-322-2485-3_35

in Banach space has been studied in [1]. Approximate controllability of fractional order semilinear systems with bounded delay has been studied (see [8]).

In recent years, the controllability problems for stochastic differential equations have become a field of increasing interest (see [2, 7, 10] and references therein). Stochastic differential equations have many applications in ecology, finance, and economics. The extensions of deterministic controllability concepts to stochastic system have been discussed only in a limited number of publications.

The Poisson jumps have become very popular in recent years, because it is extensively used to model many of the phenomena arising in areas such as economics, finance, physics, biology, medicine, and other science. For example, if a system jumps from a "normal state" to a "bad state," the strength of systems is random. It is natural and necessary to include a jump term in any dynamical system to make more realistic sysrems. Complete controllability of stochastic evolution equations with jumps has been studied in [13].

However, to the best of authors' knowledge, there are no relevant reports on the controllability of fractional neutral stochastic dynamical systems with Poisson jumps in the finite-dimensional space. Motivated by the above, in this article the controllability of fractional neutral stochastic dynamical systems is studied with Poisson jumps in finite-dimensional spaces. Sufficient conditions for controllability results are obtained by using Krasnoselskii's fixed point theorem with a Grammian matrix defined by Mittag-Leffler matrix function.

The paper is organized as follows: In Sect. 2, some well-known fractional operators and the solution representation of linear fractional stochastic differential equation with Poisson jumps are discussed. In Sect. 3, the linear and nonlinear fractional neutral stochastic differential equation with Poisson jumps are considered and the controllability conditions are established by using the controllability Grammian matrix which is defined by means of the Mittag-Leffler matrix function. Finally, concluding remarks are given in Sect. 4.

2 Preliminaries

Let p and q are some positive constants satisfying $n-1<q<n, n-1<p<n$ and $n \in \mathbb{N}$. Let $\mathbb{R}^m$ be the m-dimensional Euclidean space. The following notations and definitions are well known, for a suitable function $f \in L_1(\mathbb{R}_+), \mathbb{R}_+ = [0, \infty)$ for more details, (see [6]).

(a) Riemann–Liouville fractional operator:

$$(I_{0+}^q f)(x) = \frac{1}{\Gamma(q)} \int_0^x (x-t)^{q-1} f(t)dt$$

(b) Mittag-Leffler Function:

The most interesting properties of the Mittag-Leffler function are associated with their Laplace integral

$$\int_0^\infty e^{-st} t^{p-1} E_{q,p}(\pm a t^q) dt = \frac{s^{q-p}}{(s^q \mp a)},$$

That is,

$$\mathscr{L}\{t^{p-1} E_{q,p}(\pm a t^q)\}(s) = \frac{s^{q-p}}{(s^q \mp a)},$$

(see [12]) for more details.

(c) Solution representation:

Consider the linear fractional stochastic differential equation with Poisson jumps represented in the following form:

$$\begin{aligned} d\Big[J_t^{1-q}(x(t) - x_0)\Big] &= \Big[Ax(t) + Bu(t) + \int_0^t \sigma(s)dw(s)\Big]dt + \int_{-\infty}^{+\infty} h(t,\eta)\lambda(dt,d\eta), \\ & \quad s, t \in J := [0, T], \\ x(0) &= x_0, \end{aligned} \tag{1}$$

where $0 < q < 1$, J_t^{1-q} is the $(1-q)-$ order Riemann–Liouville fractional integral operator $x \in \mathbb{R}^n$, $u \in \mathbb{R}^m$, A, B are matrices of dimensions $n \times n$, $n \times m$ respectively and $\sigma : J \longrightarrow \mathbb{R}^{n \times n}$, $h : J \times J \longrightarrow \mathbb{R}^n$ are given functions.

Let $\{\overline{\lambda}(dt, d\eta),\ t, \eta \in J\}$ is a centered Poisson random measure with parameter $\pi(d\eta)dt$. Let $\int_{-\infty}^{+\infty} \pi(d\eta) < \infty$ and $\lambda(dt, d\eta) = \overline{\lambda}(dt, d\eta) - \pi(d\eta)dt$ is compensated Poisson random measure which is independent of $w(s)$.

Now applying the Riemann–Liouville fractional integral operator on both sides, we get

$$\begin{aligned} x(t) = x_0 &+ \frac{1}{\Gamma(q)} \int_0^t (t-s)^{q-1} Ax(s)ds + \frac{1}{\Gamma(q)} \int_0^t (t-s)^{q-1} Bu(s)ds \\ &+ \frac{1}{\Gamma(q)} \int_0^t (t-s)^{q-1} \int_0^s \sigma(\theta)dw(\theta)ds \\ &+ \frac{1}{\Gamma(q)} \int_0^t (t-s)^{q-1} \int_{-\infty}^{+\infty} h(s,\eta)\lambda(ds,d\eta). \end{aligned}$$

Taking the Laplace Transformation on both sides, we obtain

$$\hat{x}(s) = \frac{1}{s} x_0 + \frac{1}{s^q} A\hat{x}(s) + \frac{1}{s^q} B\hat{u}(s) + \frac{1}{s^q}\hat{\sigma}(s) + \frac{1}{s^q}\hat{h}(s).$$

Taking inverse Laplace Transformation on both sides, we get

$$x(t) = E_{q,1}(At^q)x_0 + \int_0^t (t-s)^{q-1} E_{q,q}(A(t-s)^q)\Big(Bu(s) + \int_0^s \sigma(\theta)dw(\theta)\Big)ds$$
$$+ \int_0^t (t-s)^{q-1} E_{q,q}(A(t-s)^q) \int_{-\infty}^{+\infty} h(s,\eta)\lambda(ds,d\eta). \tag{2}$$

Let $(\Omega, \mathscr{F}, P)$ be the complete probability space with a probability measure P on Ω and $w(t) = (w_1(t), w_2(t), \ldots, w_n(t))^T$ be an $n-$dimensional Wiener process defined on the probability space. Let $\{\mathscr{F}_t | t \in J\}$ be the filtration generated by $\{w(s) : 0 \leq s \leq t\}$ defined on the probability space $(\Omega, \mathscr{F}, P)$. Let $L_2(\Omega, \mathscr{F}_T, \mathbb{R}^n)$ denotes the Hilbert space of all $\mathscr{F}_T$ measurable square integrable random variables with values in $\mathbb{R}^n$. Let $L_2^{\mathscr{F}}(J, \mathbb{R}^n)$ be the Hilbert space of all square integrable and $\mathscr{F}_t$-measurable processes with values in $\mathbb{R}^n$. Let $\mathscr{B}$ is the Banach space of all square integrable and $\mathscr{F}_t$-adapted process $x(t)$ with norm

$$\|x\|^2 = \sup_{t \in J}\{\mathbb{E}\|x(t)\|^2\},$$

where $\mathbb{E}(\cdot)$ denotes the mathematical expectation operator of stochastic process with respect to the given probability measure P. Let $\mathscr{L}(\mathbb{R}^n, \mathbb{R}^m)$ be the space of all linear transformation from $\mathbb{R}^n$ to $\mathbb{R}^m$. Further, we assume that the set of admissible controls $\mathscr{U}_{ad} := L_2^{\mathscr{F}}(J, \mathbb{R}^m)$. Now let us introduce the following operators and sets. The linear bounded operator

$$\mathbb{L} \in \mathscr{L}(L_2^{\mathscr{F}}(J, \mathbb{R}^m), L_2(\Omega, \mathscr{F}_t, \mathbb{R}^n))$$

is defined by

$$\mathbb{L}u = \int_0^T (T-s)^{q-1} E_{q,q}(A(T-s)^q)Bu(s)ds$$

and its adjoint linear bounded operator

$$\mathbb{L}^* : L_2(\Omega, \mathscr{F}_T, \mathbb{R}^n) \longrightarrow L_2^{\mathscr{F}}(J, \mathbb{R}^m)$$

is defined by

$$(\mathbb{L}^*z)(t) = B^* E_{q,q}(A^*(T-t)^q)\mathbb{E}\{z|\mathscr{F}_t\},$$

and the set of all states attainable from x_0 in time $t > 0$ using admissible controls is defined by

$$\mathscr{R}_t(\mathscr{U}_{ad}) = \{x(t; x_0, u) \in L_2(\Omega, \mathscr{F}_T, \mathbb{R}^n) : u(\cdot) \in \mathscr{U}_{ad}\}.$$

The linear controllability operator $W_0^T \in \mathscr{L}(L_2(\Omega, \mathscr{F}_T, \mathbb{R}^n), L_2(\Omega, \mathscr{F}_T, \mathbb{R}^n))$ which is associated with the operator $\mathbb{L}$ is defined by

$$W_0^T = \mathbb{L}\mathbb{L}^*\{\cdot\} = \int_0^T (T-\tau)^{q-1}[E_{q,q}(A(T-\tau)^q)B][E_{q,q}(A(T-\tau)^q)B]^*\mathbb{E}\{(\cdot)|\mathscr{F}_t\}d\tau,$$

and the deterministic matrix $\Gamma_s^T \in \mathscr{L}(\mathbb{R}^n, \mathbb{R}^n)$ is

$$\Gamma_s^T = \int_s^T (T-\tau)^{q-1}[E_{q,q}(A(T-\tau)^q)B][E_{q,q}(A(T-\tau)^q)B]^*d\tau, \ s \in J.$$

Definition 1 The system (1) is said to be controllable on J if for every $x_0, x_1 \in \mathbb{R}^n$ there exists a stochastic control $u(t) \in \mathscr{U}_{ad}$ such that the solution of $x(t)$ of system (1) satisfies the conditions $x(0) = x_0$ and $x(T) = x_1$.

Definition 2 The system (1) is completely controllable on J if

$$\mathscr{R}_T(x_0) = L_2(\Omega, \mathscr{F}_T, \mathbb{R}^n),$$

that is, all points in $L_2(\Omega, \mathscr{F}_T, \mathbb{R}^n)$ can be exactly reached from an arbitrary initial condition $x_0 \in L_2(\Omega, \mathscr{F}_T, \mathbb{R}^n)$ at time T.

3 Controllability Results

In this section, we discuss the controllability criteria of linear and nonlinear stochastic system with Poisson jumps.

Lemma 1 ([10]) *If the linear system (1) is completely controllable, then for some* $\gamma > 0$,

$$\mathbb{E}\langle W_0^T z, z\rangle \geq \gamma \mathbb{E}\|z\|^2,$$

for all $z \in L_2(\Omega, \mathscr{F}_t, \mathbb{R}^n)$
and, consequently,

$$\mathbb{E}\|(W_0^T)^{-1}\|^2 \leq \frac{1}{\gamma} = l_2.$$

Lemma 2 ([10]) *Assume that the operator* W_0^T *is invertible. Then, for arbitrary* $x_1 \in L_2(\Omega, \mathscr{F}_T, \mathbb{R}^n)$, *the control*

$$u(t) = B^*E_{q,q}(A^*(T-t)^q)\mathbb{E}\Big\{(W_0^T)^{-1}\Big(x_1 - E_{q,1}(AT^q)x_0 - \int_0^T (T-s)^{q-1}E_{q,q}(A(T-s)^q)$$
$$\times \Big[\int_0^s \sigma(\theta)dw(\theta)\Big]ds - \int_0^T (T-s)^{q-1}E_{q,q}(A(T-s)^q)\int_{-\infty}^{+\infty} h(s,\eta)\lambda(ds,d\eta)\Big)\Big|\mathscr{F}_t\Big\}$$

transfers the system (1) form $x_0 \in \mathbb{R}^n$ *to* $x_1 \in \mathbb{R}^n$ *at time* T.

Proof Substituting the control $u(t)$ into the solution $x(t)$ in (2) and substituting $t = T$, one can easily verify that the control $u(t)$ steers the linear system $x(t)$ from x_0 to x_1.

Let us consider the nonlinear fractional neutral stochastic dynamical systems with Poisson jumps represented in the following form

$$\begin{aligned} d\Big[J_t^{1-q}(x(t) - g(t, x(t)) - x_0 - g(0, x_0))\Big] &= \Big[A\Big(x(t) - g(t, x(t))\Big) + Bu(t) \\ &\quad + J_t^{1-q} f(t, x(t)) + \int_0^t \sigma(s, x(s))dw(s)\Big]dt \\ &\quad + \int_{-\infty}^{+\infty} h(t, x(t), \eta)\lambda(dt, d\eta), \ s, t \in J, \\ x(0) &= x_0, \end{aligned} \tag{3}$$

where $0 < q < 1$, J_t^{1-q} is the $(1-q)$–order Riemann–Liouville fractional integral operator, A, B are the matrices of dimensions $n \times n, n \times m$ respectively and $f : J \times \mathbb{R}^n \longrightarrow \mathbb{R}^n$, $\sigma : J \times \mathbb{R}^n \longrightarrow \mathbb{R}^{n \times n}$ and $h : J \times \mathbb{R}^n \times \mathbb{R} \longrightarrow \mathbb{R}^n$, are given functions. Then the solution (3) is given by (see [3, 5])

$$\begin{aligned} x(t) = E_{q,1}(At^q)[x_0 + g(0, x_0)] + g(t, x(t)) + \int_0^t E_{q,1}(A(t-s)^q) f(s, x(s))ds \\ + \int_0^t (t-s)^{q-1} E_{q,q}(A(t-s)^q)\Big(Bu(s) + \int_0^s \sigma(\theta, x(\theta))dw(\theta)\Big)ds \\ + \int_0^t (t-s)^{q-1} E_{q,q}(A(t-s)^q) \int_{-\infty}^{+\infty} h(s, x(s), \eta)\lambda(ds, d\eta). \end{aligned}$$

Lemma 3 (Krasnoselskii's fixed point theorem) *Let E be a Banach space, let B be a bounded closed and convex subset of E and let Φ_1, Φ_2 be maps of B into E such that $\Phi_1 x, \Phi_2 y \in B$ for every pair $x, y \in B$. If Φ_1 is a contraction and Φ_2 is completely continuous, then the equation $\Phi_1 x + \Phi_2 x = x$ has a solution of B.*

In order to prove the main results we assume the following conditions hold:

(H1) The functions g, f, σ and h satisfy the following Lipschitz conditions and there exist some positive constants K, L, M and N such that

(i) $\|g(t, x) - g(t, y)\|^2 \le K\|x - y\|^2$
(ii) $\|f(t, x) - f(t, y)\|^2 \le L\|x - y\|^2$
(iii) $\|\sigma(t, x) - \sigma(t, y)\|^2 \le M\|x - y\|^2$
(iv) $\int_{-\infty}^{+\infty} \|h(t, x, \eta) - h(t, y, \eta)\|^2 \lambda(d\eta) \le N\|x - y\|^2$

(H2) The functions g, f, σ and h are continuous and satisfy the following linear growth conditions. That is, there exist some positive constants $\overline{K}, \overline{L}, \overline{M}$ and $\overline{N}$ such that

(i) $\|g(t,x)\|^2 \le \overline{K}(1+\|x\|^2)$
(ii) $\|f(t,x)\|^2 \le \overline{L}(1+\|x\|^2)$
(iii) $\|\sigma(t,x)\|^2 \le \overline{M}(1+\|x\|^2)$
(iv) $\int_{-\infty}^{+\infty} \|h(t,x,\eta)\|^2\lambda(d\eta) \le \overline{N}(1+\|x\|^2)$

(H3) The linear system (1) is completely controllable on J.

Now, define the nonlinear operator Φ from $\mathscr{B}$ to $\mathscr{B}$ as follows

$$\begin{aligned}(\Phi x)(t) &= E_{q,1}(At^q)[x_0+g(0,x_0)]+g(t,x(t))+\int_0^t E_{q,1}(A(t-s)^q)f(s,x(s))ds\\ &+\int_0^t (t-s)^{q-1}E_{q,q}(A(t-s)^q)\Big(Bu(s)+\int_0^s \sigma(\theta,x(\theta))dw(\theta)\Big)ds\\ &+\int_0^t (t-s)^{q-1}E_{q,q}(A(t-s)^q)\int_{-\infty}^{+\infty} h(s,x(s),\eta)\lambda(ds,d\eta)\end{aligned}$$

$$\begin{aligned}\mathbb{E}\|(\Phi x)(t)\|^2 \le \Delta := 36l_1l_2\|x_1\|^2+6(1+6l_1l_2)\Big(2S_1(\|x_0\|^2+\|g(0,x_0)\|^2)\\ +\Big[\overline{K}+T^2S_2\overline{L}+\frac{T^{2q+1}}{q^2}S_3M_\sigma\overline{M}+\frac{T^{2q}}{q^2}S_3\overline{N}\Big](1+\mathbb{E}\|x\|^2)\Big)\end{aligned}$$

where

$$\begin{aligned}u_x(t) = B^*E_{q,q}(A^*(T-t)^q)\mathbb{E}\Big\{(W_0^T)^{-1}\big[x_1-E_{q,1}(AT^q)[x_0+g(0,x_0)]-g(T,x(T))\\ -\int_0^T E_{q,1}(A(T-s)^q)f(s,x(s))ds-\int_0^T (T-s)^{q-1}E_{q,q}(A(T-s)^q)\\ \times\Big(\int_0^s \sigma(\theta,x(\theta))dw(\theta)\Big)ds-\int_0^T (T-s)^{q-1}E_{q,q}(A(T-s)^q)\\ \times\int_{-\infty}^{+\infty} h(s,x(s),\eta)\lambda(ds,d\eta)\big]\Big|\mathscr{F}_t\Big\}.\end{aligned}$$

Applying Lemma 3 we need to construct two mapping Φ_1 and Φ_2 such that

$$(\Phi x)(t) = (\Phi_1 x)(t)+(\Phi_2 x)(t)$$

where

$$\begin{aligned}(\Phi_1 x)(t) = \int_0^t E_{q,1}(A(t-s)^q)f(s,x(s))ds+\int_0^t (t-s)^{q-1}E_{q,q}(A(t-s)^q)\Big(Bu(s)\\ +\int_0^s \sigma(\theta,x(\theta))dw(\theta)\Big)ds+\int_0^t (t-s)^{q-1}E_{q,q}(A(t-s)^q)\\ \times\int_{-\infty}^{+\infty} h(s,x(s),\eta)\lambda(ds,d\eta),\end{aligned}$$

and

$$(\Phi_2 x)(t) = E_{q,1}(At^q)[x_0+g(0,x_0)]+g(t,x(t)).$$

For convenience, let us introduce the following notations:

$$l_1 = \max\{\|\Gamma_s^T\|^2\},\ S_1 = \|E_{q,1}(At^q)\|^2,\ S_2 = \|E_{q,1}(A(T-s)^q)\|^2,\ S_3 = \|E_{q,q}(A(T-s)^q)\|^2.$$

Theorem 1 *Assume that the conditions (H1)–(H3) are hold and if $\Delta < 1$ are satisfied, then the nonlinear system (3) is completely controllable on J.*

Proof In order to make more clear presentations, we divide the proof into the following three several steps.

Step I: For $t \in J$ and any $x, y \in \mathscr{B}$, we have

$$\begin{aligned}\mathbb{E}\|(\Phi_1 x)(t)\|^2 &\le 4\mathbb{E}\left\|\int_0^t E_{q,1}(A(t-s)^q)f(s,x(s))ds\right\|^2 + 4\mathbb{E}\left\|\int_0^t (t-s)^{q-1}E_{q,q}(A(t-s)^q)\right.\\ &\quad\left.\times Bu_x(s)ds\right\|^2 + 4\mathbb{E}\left\|\int_0^t (t-s)^{q-1}E_{q,q}(A(t-s)^q)\int_0^s \sigma(\theta,x(\theta))dw(\theta)ds\right\|^2\\ &\quad + 4\mathbb{E}\left\|\int_0^t (t-s)^{q-1}E_{q,q}(A(t-s)^q)\int_{-\infty}^{+\infty} h(s,x(s),\eta)\lambda(ds,d\eta)\right\|^2.\end{aligned}$$

Now, we have the following estimate

$$\begin{aligned}\mathbb{E}\left\|\int_0^t (t-s)^{q-1}E_{q,q}(A(t-s)^q)Bu_x(s)ds\right\|^2 &\le 6l_1l_2\Big[\|x_1\|^2 + 2S_1(\|x_0\|^2 + \|g(0,x_0)\|^2)\\ &\quad + \Big(\overline{K} + T^2S_2\overline{L} + \frac{T^{2q+1}}{q^2}S_3M_\sigma\overline{M} + \frac{T^{2q}}{q^2}S_3\overline{N}\Big)\\ &\quad \times (1+\mathbb{E}\|x\|^2)\Big].\end{aligned}$$

Thus

$$\begin{aligned}\mathbb{E}\|(\Phi_1 x)(t)\|^2 &\le 4\Big[T^2S_2\mathbb{E}\|f(t,x(t))\|^2 + 6l_1l_2\Big(\|x_1\|^2 + 2S_1(\|x_0\|^2 + \|g(0,x_0)\|^2)\\ &\quad + \Big(\overline{K} + T^2S_2\overline{L} + \frac{T^{2q+1}}{q^2}S_3M_\sigma\overline{M} + \frac{T^{2q}}{q^2}S_3\overline{N}\Big)(1+\mathbb{E}\|x\|^2)\Big)\\ &\quad + \frac{T^{2q+1}}{q^2}S_3M_\sigma\mathbb{E}\|\sigma(t,x(t))\|^2 + \frac{T^{2q}}{q^2}S_3\int_{-\infty}^{+\infty}\mathbb{E}\|h(t,x(t),\eta)\|^2\lambda(d\eta)\Big]\\ &\le 4\Big[6l_1l_2[\|x_1\|^2 + 2S_1(\|x_0\|^2 + \|g(0,x_0)\|^2)] + \Big(6l_1l_2\overline{K} + (1+6l_1l_2)\\ &\quad \times \Big(T^2S_2\overline{L} + \frac{T^{2q+1}}{q^2}S_3M_\sigma\overline{M} + \frac{T^{2q}}{q^2}S_3\overline{N}\Big)\Big)(1+\mathbb{E}\|x\|^2)\Big]\end{aligned}$$

and

$$\begin{aligned}\mathbb{E}\|(\Phi_2 y)(t)\|^2 &\le 2\|E_{q,1}(At^q)[x_0 + g(0,x_0)]\|^2 + 2\mathbb{E}\|g(t,y(t))\|^2\\ &\le 4S_1[\|x_0\|^2 + \|g(0,x_0)\|^2] + 2\overline{K}(1+\mathbb{E}\|y\|^2).\end{aligned}$$

By the condition $\varDelta < 1$, we can find a $r > 0$ such that

$$x, y \in \mathscr{B}_r = \{x \in \mathscr{B} : \mathbb{E}\|x\|^2 \le r\},\ \mathbb{E}\|\varPhi_1 x + \varPhi_2 y\|^2 \le r$$

that is $\varPhi_1 x + \varPhi_2 y \in \mathscr{B}_r$.

Step II: $\varPhi_1$ is a contraction mapping on $\mathscr{B}_r$. For any $x, y \in \mathscr{B}_r$ and $t \in J$, we have

$$\begin{aligned}
\mathbb{E}\|(\varPhi_1 x)(t) - (\varPhi_1 y)(t)\|^2 \le\ & 4\mathbb{E}\left\|\int_0^t E_{q,1}(A(t-s)^q)[f(s,x(s)) - f(s,y(s))]ds\right\|^2 \\
& + 4\mathbb{E}\left\|\int_0^t (t-s)^{q-1}E_{q,q}(A(t-s)^q)B[u_x(s) - u_y(s)]ds\right\|^2 \\
& + 4\mathbb{E}\Big\|\int_0^t (t-s)^{q-1}E_{q,q}(A(t-s)^q) \\
& \times \left(\int_0^s [\sigma(\theta,x(\theta)) - \sigma(\theta,y(\theta))]dw(\theta)\right) ds\Big\|^2 \\
& +4\mathbb{E}\Big\|\int_0^t (t-s)^{q-1}E_{q,q}(A(t-s)^q) \\
& \times \left(\int_{-\infty}^{+\infty} [h(s,x(s),\eta) - h(s,y(s),\eta)]\lambda(ds,d\eta)\right)\Big\|^2 \\
\le\ & 4\left[4l_1l_2K + (1+4l_1l_2)\left(T^2S_2L + \frac{T^{2q+1}}{q^2}S_3M_\sigma M + \frac{T^{2q}}{q^2}S_3N\right)\right] \\
& \times \mathbb{E}\|x(t) - y(t)\|^2 =: \varUpsilon\mathbb{E}\|x(t) - y(t)\|^2.
\end{aligned}$$

From the condition $\varDelta < 1$, we obtain $\varUpsilon < 1$, which implies that $\varPhi_1$ is a contraction mapping.

Step III: $\varPhi_2$ is a completely continuous operator.

Due to continuity of A and continuity of g, the operator is $\varPhi_2$ is continuous. Next, we will show that $\{\varPhi_2 x, x \in \mathscr{B}_r\}$ is relatively compact. It suffices to show that the family of function $\{\varPhi_2 x, x \in \mathscr{B}_r\}$ is uniformly bounded and equicontinuous for any $t \in J$ and $\{(\varPhi_2 x)(t), x \in \mathscr{B}_r\}$ is relatively compact. For any $x \in \mathscr{B}_r$, we have $\mathbb{E}\|\varPhi_2 x\|^2 \le r$ which implies that $\{\varPhi_2 x, x \in \mathscr{B}_r\}$ is uniformly bounded. In the following, we will show that $\{\varPhi_2 x, x \in \mathscr{B}_r\}$ is a family of equicontinuous functions. For any $x \in \mathscr{B}_r$ and $0 \le t_1 < t_2 \le T$, we have

$$\begin{aligned}
\mathbb{E}\|(\varPhi_2 x)(t_2) - (\varPhi_2 x)(t_1)\|^2 \le\ & 4\|E_{q,1}(At_2^q) - E_{q,1}(At_1^q)\|^2(\|x_0\|^2 + \|g(0,x_0)\|^2) \\
& + 2\mathbb{E}\|g(t_2,x(t_2)) - g(t_1,x(t_1))\|^2.
\end{aligned}$$

The right side of the above equation is independently of $x \in \mathscr{B}_r$ as $(t_2 - t_1) \longrightarrow 0$ which means that $\{\varPhi_2 x, x \in \mathscr{B}_r\}$ is equicontinuous. Therefore $\{\varPhi_2 x, x \in \mathscr{B}_r\}$ is relatively compact by Arzela–Ascoli theorem. The continuity of $\varPhi_2$ and relative compactness of $\{\Phi_2 x, x \in \mathscr{B}_r\}$ imply that Φ_2 is a completely continuous operator. By using Krasnoseskii's fixed point theorem we obtain that $\Phi_1 + \Phi_2$ has a fixed point on $\mathscr{B}_r$. Therefore the system (3) has atleast one fixed point on J.

4 Conclusion

This paper deal with the controllability of fractional neutral stochastic dynamical systems with Poisson jumps in the finite-dimensional space. Sufficient conditions for controllability results have been obtained by using Krasnoseskii's fixed point theorem . The controllability Grammian matrix is defined by Mittag-Leffler matrix function.

Acknowledgments The work of authors are supported by Council of Scientific and Industrial Research, Extramural Research Division, Pusa, New Delhi, India under the grant No. 25/(0217)/13/EMR-II.

References

1. Balachandran, K., Balasubramaniam, P., Dauer, J.: Local null controllability of nonlinear functional differential systems in Banach space. J. Optim. Theory Appl. **88**(1), 61–75 (1996)
2. Balasubramaniam, P., Vembarasan, V., Senthilkumar, T.: Approximate controllability of impulsive fractional integro-differential systems with nonlocal conditions in Hilbert space. Numer. Func. Anal. Opt. **35**(2), 177–197 (2014)
3. Chikriy, A.A., Matichin, I.I.: Presentation of solutions of linear systems with fractional derivatives in the sense of Riemann-Liouville, Caputo and Miller-Ross. J. Automat. Informat. Sc. **40**(6), 1–11 (2008)
4. Karthikeyan, S., Balachandran, K.: Constrained controllability of nonlinear stochastic impulsive systems. Int. J. Appl. Math. Comput. Sci. **21**(2), 307–316 (2011)
5. Kexue, L., Jigen, P.: Controllability of fractional neutral stochastic functional differential system. Z. Angew. Math. Phys. 1–19 (2013). doi:10.1007/s00033-013-0369-2
6. Kilbas, A.A.A., Srivastava, H.M., Trujillo, J.J.: Theory and Applications of Fractional Differential Equations. Elsevier Science Limited (2006)
7. Klamka, J.: Stochastic controllability of linear systems with delay in control. Tech. Sci. **55**(1), 23–29 (2007)
8. Kumar, S., Sukavanam, N.: Approximate controllability of fractional order semilinear systems with bounded delay. J. Differ. Equ. **252**(11), 6163–6174 (2012)
9. Lakshmikantham, V., Leela, S., Devi, J.V.: Theory of Fractional Dynamic Systems. Scientific Publishers, Cambridge (2009)
10. Mahmudov, N., Zorlu, S.: Controllability of nonlinear stochastic systems. Int. J. Control. **76**(2), 95–104 (2003)
11. Miller, K.S., Ross, B.: An Introduction to the Fractional Calculus and Fractional Differential Equations. Wiley, New York (1993)
12. Podlubny, I.: Fractional Differential Equations: An Introduction to Fractional Derivatives, Fractional Differential Equations, to Methods of Their Solution and Some of Their Applications. Academic Press (1998)
13. Sakthivel, R., Ren, Y.: Complete controllability of stochastic evolution equations with jumps. Rep. Math. Phys. **68**(2), 163–174 (2011)

Efficient Meshfree Method for Pricing European and American Put Options on a Non-dividend Paying Asset

Kailash C. Patidar and Abdelmgid O.M. Sidahmed

Abstract We develop efficient meshfree method based on radial basis functions (RBFs) to solve European and American option pricing problems arising in computational finance. The application of RBFs leads to system of differential equations which are then solved by a time integration θ-method. The main difficulty in pricing the American options lies in the fact that these options are allowed to be exercised at any time before their expiry. Such an early exercise right purchased by the holder of the option results into a free boundary problem. Following the approach of Nielsen et al. [B.F. Nielsen, O. Skavhaug and A. Tveito, Penalty methods for the numerical solution of American multi-asset option problems. J. Comput. Appl. Math. **222**, 3–16 (2008)], we use a small penalty term to remove the free boundary. The method is analyzed for stability. Numerical results describing the payoff functions and option values are also present. We also compute the two important Greeks, delta and gamma, of these options.

Keywords American and European Options · Meshfree Methods · Free Boundary Value Problems · Stability Analysis

1 Introduction

Options are frequently priced by means of partial differential equations (PDEs). These options can be categorized into standard and nonstandard options. The present work deals with the standard options (like European and American options). A large amount of work has already been done to solve the PDEs representing European options. However, the same for American options is not fully explored. For some

K.C. Patidar (✉) · A.O.M. Sidahmed
Department of Mathematics and Applied Mathematics, University of the Western Cape, Bellville 7535, South Africa
e-mail: kpatidar@uwc.ac.za

A.O.M. Sidahmed
e-mail: abdelmgid@gmail.com

P.N. Agrawal et al. (eds.), *Mathematical Analysis and its Applications*, Springer Proceedings in Mathematics & Statistics 143,
DOI 10.1007/978-81-322-2485-3_36

historical developments, readers may refer to classical works of Black and Scholes [1] and Merton [12, 13].

Researchers have attempted to solve these problems using a variety of techniques, see e.g., adaptive θ-methods for solving American options (Khaliq et al. [11]), compact finite difference methods (Zhao et al. [17]), generalized trapezoidal schemes (Chawla et al. [3]), etc. On the other hand, methods based on meshfree approximations have been used a lot for problems in other domains of science and engineering, see e.g., [8, 9, 14]. One of these popular meshfree methods are those based on the radial basis functions (RBFs). Wua and Hon [16] used such an approximation for solving diffusion-type problems under free boundary condition. In their work, the numerical solution of the Black–Scholes equation for pricing American options, which is a classical heat diffusion equation under free boundary value condition, is obtained and compared with the traditional binomial method for numerical verification.

In this work, we construct a meshfree method based on RBFs to solve European and American option pricing problems. For American put option we use the penalty method to remove the free boundary by adding a small penalty term. The basic idea behind the use of RBFs is to use interpolation with a linear combination of basis functions of the same type. A variety of RBFs are found in the literature. The two RBFs that we will use in this paper are Gaussian and Multiquadratic.

The rest of the paper is organized as follows. Two option pricing problems are described in Sect. 2. In Sect. 3 we discuss the application of radial basis functions to solve these problems. The stability analysis of the numerical methods is presented in Sect. 4. Finally, some numerical results along with a discussion on them are given in Sect. 5.

2 Problem Description

In this paper, we consider the mathematical models for pricing European and American options. While a European option can only be exercised on the expiration date, the American option can be exercised at any time before the expiration date.

The European option satisfies the following Black–Scholes equation

$$\frac{\partial V}{\partial t} + \frac{1}{2}\sigma^2 S^2 \frac{\partial^2 V}{\partial S^2} + rS\frac{\partial V}{\partial S} - rV = 0, \tag{1}$$

where r is the risk-free interest rate, σ is the volatility of the stock price, and $V(S,t)$ is the option value at time t for the stock's price S.

The initial condition is given by the terminal payoff

$$V(S,T) = \begin{cases} \max(X - S, 0) & \text{for put} \\ \max(S - X, 0) & \text{for call} \end{cases} \tag{2}$$

whereas the boundary conditions are given by

$$V(S,T)=\begin{cases} V(0,t)=Xe^{-r(T-t)}, & V(S,t)\to 0 \text{ as } S\to\infty \quad \text{for put} \\ V(0,t)=0, & V(S,t)\to S \text{ as } S\to\infty \quad \text{for call,} \end{cases} \tag{3}$$

where T is the maturity time and X is the strike price of the option.

The exact solution of Eq. (1) with the initial condition (2) and the boundary conditions (3) is given by [15]

$$V(S,T)=\begin{cases} Xe^{-r(T-t)}N(-d_2)-SN(-d_1) & \text{for put} \\ SN(d_1)-Xe^{-r(T-t)}N(d_2) & \text{for call} \end{cases} \tag{4}$$

where $N(\cdot)$ is the cumulative distribution function of the standard normal distribution with

$$d_1=\frac{\log(S/X)+\left(r+\frac{1}{2}\sigma^2\right)(T-t)}{\sigma\sqrt{T-t}} \tag{5}$$

and

$$d_2=\frac{\log(S/X)+\left(r-\frac{1}{2}\sigma^2\right)(T-t)}{\sigma\sqrt{T-t}}. \tag{6}$$

On the other hand, the American option problem takes the form of a free boundary problem. The early exercise constraint leads to the following model for the value $P(S,t)$ of an American put to sell the underlying asset [10]:

$$\left.\begin{array}{l} \frac{\partial P}{\partial t}+\frac{1}{2}\sigma^2S^2\frac{\partial^2 P}{\partial S^2}+rS\frac{\partial P}{\partial S}-rP=0, \quad S>S_f(t),\ 0\le t<T \\ P(S,T)=\max(E-S,0), \quad S\ge 0, \\ \frac{\partial P}{\partial S}(S_f,t)=-1;\ P(S_f(t),t)=E-S_f(t), \\ \lim_{S\to\infty}P(S,t)=0, \\ S_f(T)=E;\ P(S,t)=E-S, \quad 0\le S<S_f(t), \end{array}\right\} \tag{7}$$

where $S_f(t)$ represents the free boundary, σ is the volatility of the underlying asset, r is the risk-free interest rate, and E is the exercise price of the option. Since early exercise is permitted, the value P of the option must satisfy

$$P(S,t)\ge\max(E-S,0), \quad S\ge 0,\ 0\le t\le T. \tag{8}$$

3 Use of Radial Basis Functions in Pricing Options

3.1 European Options

We approximate the unknown function V (the value of the European option) using the radial basis functions as

$$V(S,t) \approx \sum_{j=1}^{N} a_j(t)\phi(|S-x_j|), \tag{9}$$

where a_j are unknown coefficients and $\phi(|S-x_j|)$ are the RBFs. We will use the following Gaussian radial basis functions for this problem

$$\phi(S) = e^{-|S-x_j|^2/c^2}, \tag{10}$$

where c is a positive parameter.

Collocating at the N points x_j $(j = 1, 2, \ldots, N)$, Eq. (1) becomes

$$\frac{\partial V(x_i,t)}{\partial t} + \frac{1}{2}\sigma^2 S_i^2 \frac{\partial^2 V(x_i,t)}{\partial S^2} + rS_i \frac{\partial V(x_i,t)}{\partial S} - rV(x_i,t) = 0. \tag{11}$$

Differentiating (9), we obtain

$$\frac{\partial V(x_i,t)}{\partial t} = \sum_{j=1}^{N} \frac{da_j(t)}{dt}\phi(|S-x_j|), \tag{12}$$

$$\frac{\partial V(x_i,t)}{\partial S} = \sum_{j=1}^{N} a_j(t)\frac{\partial \phi(|S-x_j|)}{\partial S}, \tag{13}$$

$$\frac{\partial^2 V(x_i,t)}{\partial S^2} = \sum_{j=1}^{N} a_j(t)\frac{\partial^2 \phi(|S-x_j|)}{\partial S^2}. \tag{14}$$

Using (10)–(11), and simplifying, we obtain

$$\Phi \frac{d\mathbf{a}}{dt} + \mathbf{Ra} = 0, \tag{15}$$

where

$$\Phi_{ij} = e^{-\|x_i-x_j\|^2/c^2} \tag{16}$$

and

$$\mathbf{R}_{ij} = \frac{1}{2}\sigma^2 x_i^2 \left(\frac{4(x_i - x_j)^2 - 2c^2}{c^4}\right) \Phi_{ij} + r x_i \left(\frac{-2(x_i - x_j)}{c^2}\right) \Phi_{ij} - r\Phi_{ij}. \quad (17)$$

To solve the system described by Eq. (15), we use a θ-method

$$\Phi \frac{a^{n+1} - a^n}{\Delta t} + \theta \mathbf{R} a^{n+1} + (1 - \theta)\mathbf{R} a^n = 0, \quad (18)$$

with the initial condition given by the first part of Eq. (2) and boundary conditions given by the first part of Eq. (3).

We can rewrite Eq. (18) as

$$[\Phi - (1 - \theta)\Delta t \mathbf{R}]a^n = [\Phi + \theta \Delta t \mathbf{R}]a^{n+1}. \quad (19)$$

$$\Rightarrow a^n = [\Phi - (1 - \theta)\Delta t \mathbf{R}]^{-1}[\Phi + \theta \Delta t \mathbf{R}]a^{n+1}. \quad (20)$$

Furthermore, Eq. (9) applied at all collocation points can be written in the matrix form as

$$V = \Phi \mathbf{a}. \quad (21)$$

Using Eq. (21), Eq. (20) can be written as

$$V^n = \Phi^{-1}[\Phi - (1 - \theta)\Delta t \mathbf{R}]^{-1}[\Phi + \theta \Delta t \mathbf{R}]\Phi V^{n+1}. \quad (22)$$

The above equation is solved along with (2) and the first part of Eq. (9) to obtain the numerical solution. Also the form of this equation should be read in context to the computing process because in the problems like those considered in this paper, we usually have a final boundary value problem rather than an initial-boundary value problem. To this end, note that the scheme given by (19) corresponding to $\theta = 0,\ 0.5,$ and 1 are implicit Euler, Crank–Nicolson and explicit Euler methods, respectively.

3.2 American Options

To solve the American option problem (7), which is a free boundary problem, we approximate the model by adding a penalty term. This leads to a nonlinear partial differential equation on a fixed domain. More precisely, we consider the initial-boundary value problem

$$\frac{\partial P_\varepsilon}{\partial t} + \frac{1}{2}\sigma^2 S^2 \frac{\partial^2 P_\varepsilon}{\partial S^2} + rS\frac{\partial P_\varepsilon}{\partial S} - rP_\varepsilon + \frac{\varepsilon C}{P_\varepsilon + \varepsilon - q(S)} = 0, \quad (23)$$

with the initial condition as the first part of Eq. (2) and the boundary conditions as

$$P_\varepsilon(0,t) = E, \qquad \lim_{S\to\infty} P_\varepsilon(S,t) = 0, \tag{24}$$

where $C \geq rE$, $q(S) = E - S$, and $0 < \varepsilon \ll 1$.

Using multiquadric radial basis functions, we find

$$\frac{\partial\phi(x_i - x_j)}{\partial S} = \frac{(x_i - x_j)}{\sqrt{(x_i - x_j)^2 + c^2}} \tag{25}$$

and

$$\frac{\partial^2\phi(x_i - x_j)}{\partial S^2} = \frac{c^2}{\sqrt{((x_i - x_j)^2 + c^2)^3}}. \tag{26}$$

Proceeding in the similar manner as in the previous case, we obtain

$$\Phi\frac{d\mathbf{a}}{dt} + \widetilde{\mathbf{R}}\mathbf{a} + Q(\mathbf{a}) = 0, \tag{27}$$

where

$$\Phi_{ij} = \sqrt{(x_i - x_j)^2 + c^2}, \quad i, j = 1, \ldots, N, \tag{28}$$

$$Q(\mathbf{a}) = \frac{\varepsilon C}{\Phi_i\mathbf{a} + \varepsilon - q(x_i)}, \quad i = 1, \ldots, N$$

and

$$\widetilde{\mathbf{R}}_{ij} = \frac{1}{2}\sigma^2 x_i^2\left(\frac{c^2}{\sqrt{((x_i - x_j)^2 + c^2)^3}}\right) + r x_i\left(\frac{(x_i - x_j)}{\sqrt{(x_i - x_j)^2 + c^2}}\right) - r\Phi_{ij}. \tag{29}$$

Using θ-method, Eq. (27) becomes

$$\Phi\frac{a^{n+1} - a^n}{\Delta t} + \theta\widetilde{\mathbf{R}}a^{n+1} + (1-\theta)\widetilde{\mathbf{R}}a^n + \theta Q(a^{n+1}) + (1-\theta)Q(a^n) = 0. \tag{30}$$

Consequently, the nonlinear penalty term gives rise to a nonlinear system of equations whose solution is typically found by a modified Newton method. By replacing a^n in the penalty term by a^{n+1}(as in [10]), the linearly implicit scheme corresponding to Eq. (30) is given by

$$\Phi\frac{a^{n+1}-a^n}{\Delta t}+\theta\widetilde{\mathbf{R}}a^{n+1}+(1-\theta)\widetilde{\mathbf{R}}a^n+Q(a^{n+1})=0, \tag{31}$$

with the initial condition given by the first part of Eq. (2) and boundary conditions given by Eq. (3).

4 Stability Analysis

To proceed with the stability analysis, let us define the error at the nth time level by

$$e^n = V^n_{\text{exact}} - V^n_{\text{app}}, \tag{32}$$

where V^n_{exact} and V^n_{app} are the exact and numerical solutions obtained by either (18) or (31), respectively.

For the scheme given by (21) the error equation at $(n+1)$th level can be written as

$$e^n = \mathbf{B}e^{n+1}, \tag{33}$$

where $\mathbf{B}$ is the amplification matrix is given by

$$\mathbf{B}=\Phi^{-1}[\Phi+\theta\Delta t\mathbf{R}][\Phi-(1-\theta)\Delta t\mathbf{R}]^{-1}\Phi.$$

The numerical method will be stable if $\rho(\mathbf{B})\leq 1$, where $\rho(\mathbf{B})$ is the spectral radius of $\mathbf{B}$.

Substituting the value of $\mathbf{B}$ in Eq. (33) and simplifying, we obtain

$$[\Phi-(1-\theta)\Delta t\mathbf{R}]\Phi^{-1}e^n=[\Phi+\theta\Delta t\mathbf{R}]\Phi^{-1}e^{n+1}. \tag{34}$$

This implies

$$[I-(1-\theta)\Delta tM]e^n=[I+\theta\Delta tM]e^{n+1} \tag{35}$$

where $M=\mathbf{R}\Phi^{-1}$ and I is an $N\times N$ identity matrix.

It is clear from Eq. (35) that the numerical scheme is stable if all the eigenvalues of the matrix $[I-(1-\theta)\Delta tM]^{-1}[I+\theta\Delta tM]$ are less than unity, which means that

$$\left|\frac{1+\theta\Delta t\lambda_M}{1-(1-\theta)\Delta t\lambda_M}\right|\leq 1, \tag{36}$$

where λ_M represent the eigenvalues of the matrix M.

To check the above, we consider different cases. Firstly, when $\theta = 1$, we have explicit Euler method. The above condition for stability in this case becomes

$$|1 + \Delta t \lambda_M| \leq 1. \tag{37}$$

Hence the explicit Euler method will be stable if $\Delta t \geq -2/\lambda_M$ and $\lambda_M \leq 0$. Secondly, when $\theta = 0$, we have implicit Euler method which is unconditionally stable as can be seen from (36) because $\lambda_M \leq 0$. Finally, when $\theta = 0.5$, we have the Crank–Nicolson's method. Even in this case, the inequality (36) will hold as long as $\lambda_M \leq 0$ and this does happen. Therefore, the Crank–Nicholson's method is unconditionally stable. Note that the stability analysis for (31) can be done along the similar lines.

5 Numerical Results and Discussion

Using the RBF approach, the resulting problems for European and American put options are solved via Crank–Nicolson's method (i.e., $\theta = 0.5$) with $\Delta t = 0.01$. Results are presented in Table 1.

The parameters used for the simulations for European put option problem are: $r = 0.05$, $\sigma = 0.2$, $D = 0$, $E = 10$, $t_0 = 0$, $T = 0.5$, $S_0 = 0$ and $S_{\max} = 30$. The first column in this table represents values of the asset price S, the second column represents the exact solution and the other three columns indicate the numerical values of the European put option that we obtain using the radial basis function approach with 21, 41 and 101 nodes, respectively.

For the American put options, we choose $r = 0.1$, $\sigma = 0.2$, $D = 0$, $E = 1$, $t_0 = 0$, $T = 1$, $\varepsilon = 0.01$, $S_0 = 0$, and $S_{\max} = 2$. We again use the Crank–Nicolson method with $\Delta t = 0.01$. Using the multiquadratic radial basis function

Table 1 Values of European put option using radial basis functions

S	Exact	RBF21	RBF41	RBF101
6	3.7532	3.7528	3.7529	3.7532
7	2.7568	2.7659	2.7594	2.7572
8	1.7987	1.8510	1.8080	1.8003
9	0.9880	1.0079	0.9908	0.9886
10	0.4420	0.5280	0.4628	0.4454
11	0.1606	0.2087	0.1754	0.1629
12	0.0483	0.0499	0.0504	0.0486
13	0.0124	0.0206	0.0147	0.0127
14	0.0028	0.0040	0.0035	0.0029

Table 2 Values of American put option using radial basis functions

S	RBF21	RBF41	RBF101
0.6	4.0000e-01	4.0000e-01	4.0000e-01
0.7	3.0011e-01	3.0011e-01	3.0012e-01
0.8	2.0198e-01	2.0202e-01	2.0204e-01
0.9	1.1657e-01	1.1687e-01	1.1695e-01
1.0	5.9688e-02	6.0169e-02	6.0295e-02
1.1	2.8756e-02	2.9196e-02	2.9330e-02
1.2	1.3656e-02	1.4005e-02	1.4147e-02
1.3	6.7494e-03	7.0219e-03	7.2231e-03
1.4	3.6229e-03	3.9109e-03	4.2547e-03

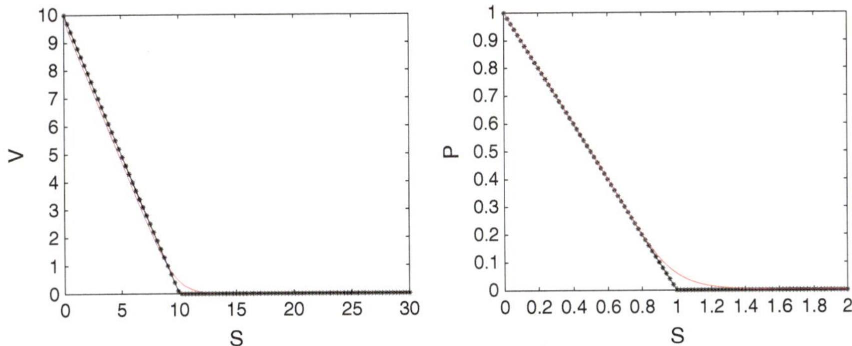

Fig. 1 *Left plot* Value of the European put at t_0 using 101 points and $r = 0.05$, $\sigma = 0.2$, $E = 10$, $t_0 = 0$, $T = 0.5$, $S_0 = 0$ and $S_{max} = 30$. The curve with '*' shows payoff whereas the *solid curve* represents the value of the option. *Right plot* Value of an American put at t_0 using 101 points and $r = 0.1, \sigma = 0.2, E = 1, T = 1, \varepsilon = 0.01$. The curve with '*' shows payoff whereas the *solid curve* represents the value of the option

$\sqrt{r^2 + c^2}$, we obtain reasonably accurate results in the sense that they are very close to those obtained by Fasshauer in [4]. This can be seen from Table 2.

Finally, in Fig. 1, we depict some special cases for European and American options as indicated in the figure caption.

The accuracy of the solution obtained by using meshfree methods depends on the choice of the shape parameter c. The choice of the optimal value of this parameter is still an open problem. Many researchers have chosen it as $c = 2h$, where $h = (S_{max} - S_0)/(N - 1)$. After some numerical experiments, we found the optimal value of this shape parameter using Gaussian RBFs as approximately 0.79.

Since the radial basis functions are infinitely differentiable, the computations of the derivatives of options are readily available from the derivatives of the basis functions. In Table 3, we present values of delta for European put options and compare them with their exact values. It is clear from the results presented in these tables

Table 3 Values of option's delta for European put

S	Analytic values of option's Δ	Numerical values of option's Δ
8.0000	−0.9083	−0.90665
9.0000	−0.6906	−0.6902
10.0000	−0.4023	−0.4031
11.0000	−0.1784	−0.1798
12.0000	−0.0622	−0.0625

Table 4 Comparison of delta for American put option

S	LUBA	EXP	QFK	RBFs
80	−1.0000	−1.0000	−1.0000	−0.9997
90	−0.6173	−0.6207	−0.6212	−0.6220
100	−0.3588	−0.3582	−0.3581	−0.3602
110	−0.2108	−0.2109	−0.2108	−0.2129
120	−0.1256	−0.1257	−0.1256	−0.1280

Table 5 Values of option's gamma for European put

S	Analytic values of option's Γ	Numerical values of option's Γ
6.0000	0.0016	0.0014
7.0000	0.0303	0.0315
8.0000	0.1455	0.1461
9.0000	0.2770	0.2767
10.0000	0.2736	0.2722
11.0000	0.1677	0.1678
12.0000	0.0722	0.0723

that the numerical values of the option's delta lie between -1 and 0 which is in agreement with what is mentioned in Hull [5]. Note that analytical solution for the Δ for American option is not available and therefore in Table 4, we compare them with some of those seen in the literature. We also calculate the Γ of a portfolio of options on an underlying asset which is the rate of change of the portfolio's delta with respect to the price of the underlying asset. It is the second partial derivative of the portfolio with respect to the asset price. If the absolute value of Γ is large, Δ is highly sensitive to the price of the underlying asset. Table 5 gives the values of Γ for European put options. The first column in this table represents the values of the asset price S, the second column represents the analytical values of option's Γ and the third column represents the numerical values of it using the proposed approach.

In Table 4, the acronyms LUBA, EXP, QFK and RBF, respectively, stand for Lower and Upper Bound Approximations [2], multipiece Exponention [6], Quadrature Formula of Kim equations [7], and Radial Basis Function approach proposed in this paper.

6 Conclusions

In this paper, we presented a meshfree method based on radial basis function approximations to solve European and American style option pricing problems. While the approach for European option problems was straightforward, we have to use a penalty approach to solve the problems for pricing American options. Proposed method is analyzed for stability. Numerous comparative results are present. It may be noted that the calculation of Greeks from our method was free of any spurious oscillations. Furthermore, the methods can be used to solve the problems where asset pays a dividend because the only difference in that case would be the fact that the asset can be less than the payoff which will not affect the performance of the method. Another important feature of the proposed meshfree method is the local adaptivity of the radial basis functions which allows for its possible extensions to multi-asset problems.

Acknowledgments We thank the anonyms referees for their valuable comments and suggestions. The research of KCP was supported by the South African National Research Foundation. AOMS acknowledges the financial support of AL-Neelain University, Sudan.

References

1. Black, F., Scholes, M.: Pricing of options and corporate liabilities. J. Pol. Eco. **81**(3), 637–654 (1973)
2. Broadie, M., Detemple, J.: American option valuation: new bounds, approximations, and a comparison of existing methods. Rev. Fin. Stud. **9**, 1211–1250 (1996)
3. Chawla, M.M., AL-Zanaidi, M.A., Evans, D.J.: Generralized trapezoidal formulas for valuing American options. Int. J. Comp. Math. **81**(3), 375–381 (2004)
4. Fasshauer, G.E., Khaliq, A.Q.M., Voss, D.A.: Using meshfree approximation for multi-asset American option. J. Chin. Inst. Eng. **27**(4), 563–571 (2004)
5. Hull, J.C.: Options, Futures, and Other Derivatives. Prentice Hall, Upper Saddle River (2009)
6. Ju, N.: Pricing an American option by approximating its early exercise boundary as a multipiece exponential function. Rev. Fin. Stud. **11**, 627–646 (1998)
7. Kallast, S., Kivinukk, A.: Pricing and hedging American options using approximations by Kim integral equations. Eur. Fin. Rev. **7**(361–383), 361–383 (2003)
8. Kansa, E.J.: Multiquadrics—a scatered data approximation scheme with applications to computional Fluid-Dynamics-I. Comput. Math. Appl. **19**, 127–145 (1990)
9. Kansa, E.J.: Multiquadrics—a scatered data approximation scheme with applications to computional Fluid-Dynamics-II. Comput. Math. Appl. **19**, 147–161 (1990)

10. Khaliq, A.Q.M., Voss, D.A., Kazmi, S.H.K.: A linearly implicit predictor-corrector scheme for pricing American options using a penalty method approach. J. Bank. Fin. **30**, 489–502 (2006)
11. Khaliq, A.Q.M., Voss, D.A., Kazmi, S.H.K.: Adaptive θ-methods for pricing American options. J. Comput. Appl. Math. **222**, 210–227 (2008)
12. Merton, R.C.: Theory of rational option pricing. Bell J. Eco. Man. Sci. **4**, 141–183 (1973)
13. Merton, R.C.: Option pricing when the underlying stocks are discontinuous. J. Fin. Eco. **5**, 125–144 (1976)
14. Tatari, M., Dehghan, M.: A method for solving partial differential equations via radial basis functions: application to the heat equation. Eng. Anal. Bound. Elem. **34**, 206–212 (2010)
15. Wilmott, P., Howison, S., Dewynne, J.: The Mathematical Financial Derivatives: A Student Introduction. Cambridge University Press, Oxford (1995)
16. Wua, Z., Hon, Y.C.: Convergence error estimate in solving free boundary diffusion problem by radial basis functions method. Eng. Anal. Bound. Elem. **27**, 73–79 (2003)
17. Zhao, J., Davison, M., Corless, R.M.: Compact finite difference method for American option pricing. J. Comput. Appl. Math. **206**, 306–321 (2007)

A Laplace Transform Approach for Pricing European Options

Edgard Ngounda and Kailash C. Patidar

Abstract In this paper we investigate two efficient numerical methods for solving the Black–Scholes equation for pricing European options. We use spectral methods to discretize the associated partial differential equation with respect to space (asset direction) and generate a system of ordinary differential equations in time. This system is then solved by applying the numerical inversion of the Laplace transform which is based on the Talbot's method [A. Talbot, The accurate numerical inversion of Laplace transforms, IMA J. Appl. Math. **23(1)**, 97–120 (1979)]. This involves an application of trapezoidal rule to approximate a Bromwich integral. Using Cauchy's integral theorem, we deform the Bromwich line into a contour which starts and ends in the left half plane. Comparative numerical results obtained by this and other three methods (Exponential Time Differencing Runge–Kutta Methods of order 4, MATLAB solver ode15s and Crank-Nicholson's method) are presented.

Keywords Option Pricing · Contour Integrals · Spectral Methods · Exponential Time Differencing Runge-Kutta Methods

1 Introduction

Since its development in the 1970s by F. Black and M. Scholes, the Black–Scholes equation has become a fundamental model for pricing financial derivatives [1]. A derivative security is a financial instrument whose value depends on the values of some other underlying variables, e.g. stocks, foreign currency. Among the most popular derivatives, options are actively traded on different financial markets over the world. An option gives its holder the right without any obligation to buy (call

E. Ngounda · K.C. Patidar (✉)
Department of Mathematics and Applied Mathematics, University of the Western Cape, Bellville 7535, South Africa
e-mail: kpatidar@uwc.ac.za

E. Ngounda
e-mail: edgardngounda@gmail.com

P.N. Agrawal et al. (eds.), *Mathematical Analysis and its Applications*, Springer Proceedings in Mathematics & Statistics 143,
DOI 10.1007/978-81-322-2485-3_37

option) or to sell (put option) the underlying asset by a certain date (maturity date) for a certain price (strike price). The European options can only be exercised at maturity.

The Black–Scholes partial differential equation can be used to model different types of options. However, a closed form solution cannot always be found and we must therefore resort to numerical methods to solve such a PDE. Some of the most popular methods used in the past to tackle these type of problems are those based on Monte Carlo simulations [2], binomial trees [3] and finite difference methods [7].

Finite difference methods are classical methods for solving PDEs and have been used extensively to price options since the advent of the financial mathematics. The authors in [9] used a grid stretching in combination with backward difference method of fourth order in time to solve the European options. In [11], Tangman et al. used a method based on the grid stretching to generate a high-order compact scheme to improve on the well-known second-order Crank–Nicolson method for solving these problems. In spite of the popularity of these time marching methods, one of their critical drawback is that they usually require as many time steps as spatial meshes to maintain their stability.

In this paper, we consider the application of Laplace transform which has recently been investigated by some researchers and is considered to be a valuable alternative method to finite differences methods for solving parabolic PDEs [4, 10, 15]. This has led to great applications in the financial world.

The rest of the paper is organized as follows. In Sect. 2 we give a full description of the Black–Scholes equation which is used to model the European put and call options. In Sect. 3, we introduce the spectral discretization method. Application of the Laplace integration method is discussed in Sect. 4. Finally, in Sect. 2, we present comparative numerical results.

2 Problem Description

We consider the following Black–Scholes (BS) equation to price European options

$$\frac{\partial V}{\partial t} + \frac{1}{2}\sigma^2 S^2 \frac{\partial^2 V}{\partial S^2} + rS\frac{\partial V}{\partial S} - rV = 0, \qquad S \in (0, \infty), \quad t \in (0, T). \tag{1}$$

Final and boundary conditions are given by

$$V(S, T) = \begin{cases} \max(S - K, 0) \text{ for call} \\ \\ \max(K - S, 0) \text{ for put} \end{cases} \tag{2}$$

and

$$\left.\begin{array}{l} V(0,t)=0, \quad V(S,t)\to S-Ke^{-r(T-t)} \quad \text{as} \quad S\to\infty \quad \text{for call,} \\ \\ V(0,t)=Ke^{-rt}, \quad V(S,t)\to 0 \quad \text{as} \quad S\to\infty \quad \text{for put.} \end{array}\right\} \tag{3}$$

In the above, $V(S,t)$ is the price of a call/put option for the underlying asset whose price is S at time t up to the expiry date T, r is the interest rate, σ is the volatility of the underlying asset, and K is the strike price.

We set $\tau = T - t$ to transform the backward formulation (1)–(3) to the following forward equation:

$$\frac{\partial V}{\partial \tau} - \frac{1}{2}\sigma^2 S^2 \frac{\partial^2 V}{\partial S^2} - rS\frac{\partial V}{\partial S} + rV = 0, \tag{4}$$

The initial condition is given by the terminal payoff

$$V(S,0) = \begin{cases} \max(S-K,0) \text{ for call} \\ \\ \max(K-S,0) \text{ for put} \end{cases} \tag{5}$$

and the boundary conditions are given by

$$\left.\begin{array}{l} V(0,\tau)=0, \quad V(S,\tau)\to S-Ke^{-r\tau} \quad \text{as} \quad S\to\infty \quad \text{for call,} \\ \\ V(0,\tau)=Ke^{-r\tau}, \quad V(S,\tau)\to 0 \quad \text{as} \quad S\to\infty \qquad \text{for put.} \end{array}\right\} \tag{6}$$

3 Spectral Discretization

To semi-discretize the PDE (1), we consider a spectral method. The basic idea behind the spectral methods is as follows. For a given set of points, we interpolate the unknown solution and differentiate the interpolating polynomial at these grid points. This discretization process leads to a system of equations which can then be solved using any state-of-the-art solvers.

The discretization using spectral method (in this paper) is based on the Chebyshev polynomial interpolation [13]. Methods such as finite elements or finite differences divide the domain into subdomains and use local polynomials of low degree. By contrast, spectral methods use global representations of high degree over the entire domain.

The implementation of spectral methods can be divided into three categories, namely, the Galerkin, tau, and the collocation (or pseudospectral) methods. The first two of these methods use the expansion coefficients of the global approximation and the latter can be viewed as a method of finding numerical approximations to derivatives at collocation points. In a manner similar to finite difference or finite element methods, the equation to be solved is satisfied in space at the collocation points. In this paper, we use the third one, i.e., the spectral collocation method.

The spectral process involves seeking the solution to a differential equation by polynomial interpolation. In order to review the concept of polynomial interpolation, we consider interpolating an arbitrary function $f(x)$ at $N+1$ distinct nodes $\{x_k\}_{k=0}^{N}$ in $[-1, 1]$.

Given a set of grid points $\{x_j\}_{j=0}^{N}$, an interpolating approximation to a function $f(x)$ is a polynomial $f_N(x)$ of degree N, determined by the requirement that the interpolant agrees with $f(x)$ at the set of interpolation points $\{x_j\}_{j=0}^{N}$, i.e.,

$$f_N(x_i) = f(x_i), \qquad i = 0, 1, ..., N. \tag{7}$$

We define by $L_k(x)$, the Lagrange polynomial of degree N,

$$L_k(x) = \prod_{\substack{j=0 \\ j \neq k}}^{N} \frac{x - x_j}{x_k - x_j}, \qquad k = 0, 1, ..., N.$$

Note that $L_k(x)$ satisfies $L_j(x_k) = \delta_{jk}$, where δ_{jk} is the Kronecker delta function. The interpolation polynomial $f_N(x)$ is then given by

$$f_N(x) = \sum_{k=0}^{N} f(x_k) L_k(x). \tag{8}$$

In this paper, we use the Chebyshev points as the grid points. These are given by

Chebyshev zeros: $x_j = \cos\left(\frac{2j+1}{2(N+1)}\pi\right), \qquad j = 0, ..., N,$

and

Chebyshev extrema: $x_j = \cos\left(\frac{j\pi}{N}\right), \qquad j = 0, ..., N.$

The Chebyshev points are often defined as the projection onto the interval $[-1, 1]$ of the roots of unity along the unit circle $|z| = 1$ in the complex plane [13]. For European options, since the payoff is nonsmooth, a direct application of the Chebyshev points for discretization leads to low-order approximation. To regain a high-order accuracy an alternative approach was proposed by Tangman [12]. The basic idea is to modify the Chebyshev points as

$$x = [x_k, x_\ell]^T, \tag{9}$$

where

$$x_k = S_{\min} + \left(\frac{K - S_{\min}}{2}\right)\left(1 - \cos\left(\frac{2\pi k}{N}\right)\right), \quad k = 0, 1, ..., \frac{N}{2}, \tag{10}$$

$$x_\ell = K + \left(\frac{S_{\max} - K}{2}\right)\left(1 - \cos\left(\frac{2\pi l}{N}\right)\right), \quad \ell = 1, 2, ..., \frac{N}{2}. \tag{11}$$

for N even. This discretization clusters grid nodes at the boundaries located at $S_{\min}$ and $S_{\max}$ as well as at the strike price K where the discontinuity of the payoff occurs. As we show in Sect. 5, it follows that local grid refinement improve accuracy of the spectral method at the payoff. Another advantage of this strategy is that it applies directly to the Eq. (4) without the need for transforming into the interval $[-1, 1]$.

Differentiation Matrices

The concept of collocation derivatives is associated with the interpolation polynomial $f_N(x)$ as described above. These are the derivatives of $f_N(x)$ at the collocation points $\{x_k\}_{k=0}^N$. Using (8), we can see that the mth order collocation derivative of $f_N(x)$ is given by

$$\frac{d^m f_N(x)}{dx^m} = \sum_{k=0}^{N} f(x_k) \frac{d^m L_k(x)}{dx^m}. \tag{12}$$

Nodal representation yields

$$\frac{d^m f_N(x_j)}{dx^m} = \sum_{k=0}^{N} f(x_k) \frac{d^m L_k(x_j)}{dx^m}, \qquad j = 0, ..., N, \tag{13}$$

which can be expressed by the matrix formula

$$\mathbf{f^{(m)}}_N = D_N^{(m)} \mathbf{f}_N, \tag{14}$$

where

$$\mathbf{f}_N = \begin{bmatrix} f_N(x_0) \\ \vdots \\ f_N(x_N) \end{bmatrix}, \quad \mathbf{f^{(m)}}_N = \begin{bmatrix} f_N^{(m)}(x_0) \\ \vdots \\ f_N^{(m)}(x_N) \end{bmatrix},$$

and $D_N^{(m)}$ is the $(N+1) \times (N+1)$ differentiation matrix of order m with entries

$$\left(D_N^{(m)}\right)_{j,k} = L_k^{(m)}(x_j), \qquad j, k = 0, ..., N. \tag{15}$$

The computation of these differentiation matrices for an arbitrary order m has been considered in [6, 13]. Following the approach in [16], Weideman and Reddy [14] developed a MATLAB algorithm (DMSUITE package) that computes the Chebyshev grid points as well as the differentiation matrices of an arbitrary order. The suite contains a function `chebdif` that computes the extreme points of the Chebyshev polynomial $T_N(x)$ and the differentiation matrix $D_N^{(m)}$. The code takes as input the size of the differentiation matrix N and the highest derivative order m and produces matrices $D_N^{(\ell)}$ of order $\ell = 1, 2, ..., m$.

Formulas for the computation of the entries of $D_N^{(1)}$, $N \geq 1$, let $i, j = 0, 1, ...N$, are (as given in [13]):

$$\left(D_N^{(1)}\right)_{00} = \frac{2N^2+1}{6}, \qquad \left(D_N^{(1)}\right)_{NN} = \frac{2N^2+1}{6}, \tag{16}$$

$$\left(D_N^{(1)}\right)_{jj} = \frac{-x_j}{2(1-x_j^2)}, \qquad j = 1, ..., N-1, \tag{17}$$

$$\left(D_N^{(1)}\right)_{ij} = \frac{c_i}{c_j}\frac{(-1)^{i+j}}{(x_i - x_j)}, \qquad i \neq j, \qquad i, j = 0, ..., N, \tag{18}$$

where

$$c_i = \begin{cases} 2, \; i = 0 \text{ or } N \\ 1, \quad \text{otherwise.} \end{cases}$$

Higher order derivatives are evaluated by recursions at a cost of $\mathcal{O}(N^2)$ operations [14, 16]. This turns out to be cost-effective as compared to $\mathcal{O}(N^3)$ if higher derivatives are obtained by taking powers of the first derivative [14].

Using the differentiation matrices as described above, we can rewrite (4) in matrix form as

$$\frac{\partial \mathbf{V}}{\partial \tau} - \frac{1}{2}\sigma^2 P D^{(2)}\mathbf{V} - rQD^{(1)}\mathbf{V} + r\mathbf{V} = 0, \tag{19}$$

where P and Q are the diagonal matrices with entries on the main diagonals as $(x_k+1)^2$ and (x_k+1), respectively, for $k = 0, ..., N$.

We will solve Eq. (19) using several time integration methods as indicated in the next two sections.

4 Application of the Laplace Transform to Price the European Call and Put Options

Applying the Laplace transform to Eq. (4), we obtain

$$z\bar{V} - \frac{1}{2}\sigma^2 S^2 \frac{\partial^2 \bar{V}}{\partial S^2} - rS\frac{\partial \bar{V}}{\partial S} + r\bar{V} = V_0. \tag{20}$$

The boundary conditions are given by

$$\left.\begin{array}{l}\bar{V}(0,z)=0,\quad \bar{V}(S,z)=\frac{S_{\max}}{z}-\frac{K}{(z+r)}\quad \text{for call,}\\[2ex] \bar{V}(0,z)=\frac{K}{(z+r)},\quad \bar{V}(S,z)=0\quad \text{for put.}\end{array}\right\}\tag{21}$$

The Eq. (19), therefore, becomes

$$z\bar{\mathbf{V}}-\frac{1}{2}\sigma^2 P D_N^{(2)}\bar{\mathbf{V}}+rQD_N^{(1)}\bar{\mathbf{V}}-r\bar{\mathbf{V}}=\mathbf{V}_0,$$

$$\left(z_k\mathbf{I}-\frac{1}{2}\sigma^2 P D_N^{(2)}+rQD_N^{(1)}-r\mathbf{I}\right)\bar{\mathbf{V}}_k=\mathbf{V}_0\qquad k=0,\ldots,N-1.\tag{22}$$

A straight forward application of the Laplace inversion formula [8] yields

$$\mathbf{V}(t)=\frac{h}{2\pi i}\int_{-\infty}^{\infty}e^{z(\ell)t}\bar{\mathbf{V}}z'(\ell)d\ell.\tag{23}$$

Using the symmetry, the trapezoidal approximation yields

$$\mathbf{V}_M(t)=\frac{h}{\pi}\sum_{k=0}^{M-1}e^{z_k t}\bar{\mathbf{V}}_{\mathbf{k}}z_k',\tag{24}$$

where

$$\bar{\mathbf{V}}_k=(z_k\mathbf{I}-A)^{-1}\mathbf{V}_0,\quad k=0,1,\ldots,N-1,\tag{25}$$

and

$$A=\frac{1}{2}\sigma^2 P D_N^{(2)}-rQD_N^{(1)}+r\mathbf{I}.\tag{26}$$

Now since the differentiation matrices $D_N^{(1)}$ and $D_N^{(2)}$ are not sparse, the Eq. (25) indicates the bulk of the computation in the trapezoidal rules (24). To speed up this computation, an Hessenberg decomposition can be computed once at the beginning as follows:

$$A=MHM^T,\tag{27}$$

where $H=(h_{ij})$ is an upper Hessenberg matrix, i.e., $h_{ij}=0,\ i>j+1$, and M an orthogonal matrix. Then for each z_k, $k=0,1,\ldots,M-1$, the Eq. (25) becomes

$$(z_k I-MHM^T)\mathbf{V}_k=\mathbf{V}_0.\quad k=0,1,\ldots,N-1.\tag{28}$$

From this we have

$$(z_k I-H)\mathbf{U}_k=M^T\mathbf{V}_0\quad k=0,1,\ldots,N-1,\tag{29}$$

Table 1 Parameters used in the contour over an interval $[t_0, \Lambda t_0]$

Λ	α	$A(\alpha)$	$\widetilde{\mu}\Lambda t_0/M$	$B(\alpha)$
1	1.1721	1.0818	4.4921	2.3157
5	1.0791	2.4578	1.5013	1.2570
10	1.0236	3.3744	0.8871	1.0888
50	0.9381	5.5582	0.3452	0.7152

The right column $B(\alpha)$ shows the convergence rate over the contour for each set parameters

where $\mathbf{U}_k = M^T \mathbf{V}_k$, so that

$$\mathbf{V}_k = M\mathbf{U}_k, \quad k = 0, 1..., N-1. \tag{30}$$

The solution $\mathbf{V}_k$ for each z_k, is obtained by the computation of an almost triangular system (29) and combining the result in (30) at only $O(N^2)$ operations [5]. During this process, the Hessenberg reduction (27) is only computed once beforehand.

For numerical implementation, we considered the following contour parameters defined over an interval $[t_0, \Lambda t_0]$ (as defined in [15])

$$z = \widetilde{\mu}(1 + \sin(iw - \alpha)), \tag{31}$$

where

$$A(\alpha) = \cosh^{-1}\left(\frac{(\pi - 2\alpha)\,\Lambda - \pi + 4\alpha}{(4\alpha - \pi)\sin\alpha}\right)$$

and

$$h = \frac{A(\alpha)}{M}, \; \widetilde{\mu} = \frac{4\alpha\pi - \pi^2}{A(\alpha)}\left(\frac{M}{\Lambda t_0}\right),$$

with $\Lambda \in N$ and $\widetilde{M}$ is the number of points in the trapezoidal rule. The convergence rate of the Laplace method on these contour is given by $\mathcal{O}\left(e^{-B(\alpha)M}\right)$ where

$$B(\alpha) = \frac{\pi^2 - 2\pi\alpha}{\cosh^{-1}\left(\frac{(\pi - 2\alpha)\Lambda + 4\alpha - \pi}{(4\alpha - \pi)\sin(\alpha)}\right)}.$$

Values of the above parameters are given in Table 1.

5 Numerical Results and Discussion

We compare the results obtained by using our Laplace transform method with those obtained by simulations that we perform using ETDRK-4 (Exponential Time Differencing Runge–Kutta Method of order 4) as well as the more conventional

Table 2 Comparison of the errors defined by (32), for the Crank–Nicolson's method, ETDRK4 and the Laplace inversion approach applied for a European call option

	ode15s		Crank–Nicolson		ETDRK4		Laplace inversion method	
N	Time (s)	Error	Time (s)	Error	Time (s)	Error	Time (s)	Error
20	12.5E-2	8.2E-2	6.0E-2	7.1E-3	5.2E-2	7.4E-3	1.1E-2	7.4E-3
30	14.8E-2	6.97E-4	4.47E-2	1.3E-3	5.4E-2	1.00E-3	5.1E-3	1.0E-3
40	19.3E-2	6.59E-5	10.9E-2	1.93E-4	7.5E-2	1.22E-4	7.0E-3	1.18E-4
50	22.3E-2	7.86E-6	13.3E-2	9.67E-5	9.5E-2	4.85E-5	9.1E-3	1.07E-5
60	24.2E-2	4.63E-6	1.63E-2	9.73E-5	12.6E-2	4.86E-5	1.2E-4	3.52E-6
80	31.0E-2	1.86E-5	26.1E-2	9.80E-5	21.8E-2	4.89E-5	2.0E-4	5.80E-7

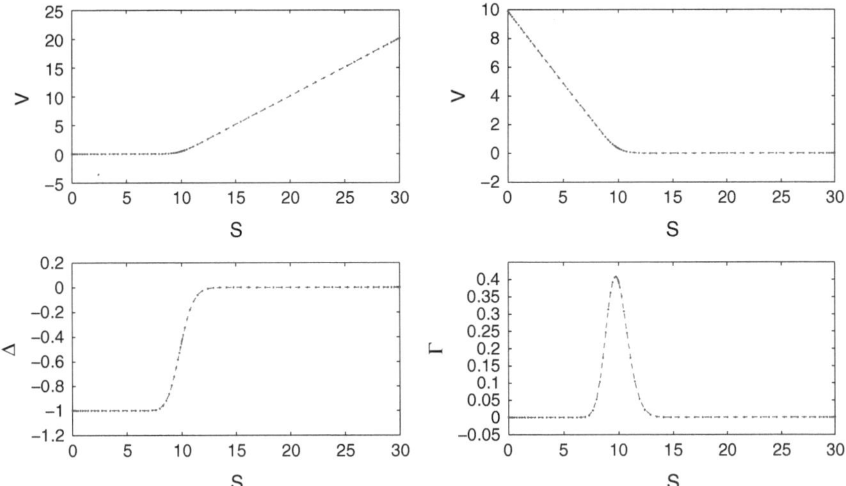

Fig. 1 *Top figures* Europeans call option (*left*), put (*right*). *Bottom figures* Δ (*left*) and Γ (*right*) for European put option. $K = 10, r = 0.05, \sigma = 0.2, S_{\max} = 3K, T = 0.25$ $N = 80$

time-marching methods such as Crank-Nicholson's method (with stepsize $2.5e-3$) and the well-known MATLAB solver *ode15s*. These results are presented in Table 2. For the numerical simulations, we fix spatial variable S at $S_{\max} = 3K$ to reduce the domain truncation error. Other parameters are chosen as follows $K = 15$, $\sigma = 0.2$, $r = 0.05$, $T = 0.25$. Maximum absolute errors are calculated using the formula

$$\text{error} = \max_{t\in[0,T]} |\mathbf{V}(t) - \mathbf{V}_M(t)|, \tag{32}$$

where $\mathbf{V}(t)$ is the analytical solution obtained by using the Black–Scholes formula and $\mathbf{V}_M(t)$ is the numerical solution obtained by any of the three methods as indicated in Table 2.

In Fig. 1, we plot values for Europeans call (and put) options as well as the Greeks Δ and Γ. We notice that both Greeks are free of oscillations.

It is worth mentioning here that even though in practice, the use of spectral methods for boundary value problems may be troublesome because the presence of boundaries often introduces stability conditions that are both highly restrictive and often difficult to analyze, one should note that for smooth solutions the results using spectral methods are of a degree of accuracy that local approximation methods cannot produce. For such solutions spectral methods can often achieve an exponential convergence rate as compared to the algebraic convergence rate of finite difference or finite element methods.

One may also think that the matrices in spectral methods are neither sparse nor symmetric, in contrast to the situation in finite differences or finite elements where the sparsity structure of the matrices simplifies the computation. However, the number of discretization points required to achieve the expected accuracy using the spectral method is much less than those required in finite difference or finite element methods, and therefore the spectral method is still very efficient as compared to these other two methods.

Acknowledgments E. Ngounda acknowledges the Agence National des Bourses du Gabon for the financial support. Patidar's research was supported by the South African National Research Foundation.

References

1. Black, F., Scholes, M.: The pricing of options and corporate liabilities. J. Pol. Eco. **81**, 637–659 (1973)
2. Boyle, P.: Options: a Monte-Carlo approach. J. Fin. Eco. **4**, 323–338 (1973)
3. Cox, J.C., Rubinstein, M.: Options Markets. Prentice-Hall, Englewood Cliffs (1985)
4. Duffy, D.G.: Transform Methods for Solving Partial Differential Equations, 2nd edn. Chapman & Hall/CRC, Boca Raton (2004)
5. Golub, G.H., van Loan, C.F.: Matrix Computations, 3rd edn. Johns Hopkins University Press, Baltimore (1996)
6. Huang, W.Z., Sloan, M.D.: The pseudospectral method for solving differential eigenvalue problems. J. Comput. Phys. **111**, 399–409 (1994)
7. Hull, J.C.: Options Futures and Others Derivatives, 4th edn. Prentice-Hall International, New Jersey (2000)
8. Ngounda, E., Patidar, K.C., Pindza, E.: Contour integral method for European options with jumps. Commun. Nonlinear Sci. Numer. Sim. **18**, 478–492 (2013)
9. Oosterlee, C.W., Leentvaar, C.C.W., Huang, X.: Accurate american option pricing by grid stretching and high-order finite differences. Delft University of Technology, The Netherlands, Technical Report (2005)
10. Talbot, A.: The accurate numerical inversion of Laplace transforms. IMA J. Appl. Math. **23**(1), 97–120 (1979)
11. Tangman, D.Y., Gopaul, A., Bhuruth, M.: Numerical pricing of option using high-order compact finite difference schemes. J. Comput. Appl. Math. **218**(2), 270–280 (2008)
12. Tangman, D.Y., Gopaul, A., Bhuruth, M.: Exponential time integration and Chebyshev discretization schemes for fast pricing of options. Appl. Numer. Math. **58**(9), 1309–1319 (2008)

13. Trefeten, L.N.: Spectral Method in MATLAB. SIAM, Philadelphia (2000)
14. Weideman, J.A.C., Reddy, S.C.: A MATLAB differentiation matrix suit. ACM Trans. Math. Soft. **26**, 465–519 (2000)
15. Weideman, J.A.C., Trefethen, L.N.: Parabolic and hyperbolic contours for computing the Bromwich integral. Math. Comput. **76**, 1341–1356 (2007)
16. Welfert, D.B.: Generation of pseudospectral differentiation matrices. I. SIAM J. Numer. Anal. **34**, 1640–1657 (1997)

Quintic Hermite Fractal Interpolation in a Strip: Preserving Copositivity

A.K.B. Chand and S.K. Katiyar

Abstract The notion of fractal interpolation provides a general framework which includes traditional nonrecursive splines as special cases. In this paper, we describe a procedure for the construction of quintic Hermite FIFs as α-fractal function corresponding to the classical quintic Hermite interpolant. In contrast to traditional piecewise nonrecursive quintic Hermite interpolant, its fractal version has a second derivative which is differentiable in a finite or dense subset of the interpolation interval. This scheme offers an additional freedom over the classical quintic Hermite interpolants due to the presence of scaling factors. The elements of the iterated function system are identified so that the class of α-fractal function f^α reflects the fundamental shape properties such as positivity, monotonicity, and convexity in addition to the regularity of f in the given interval. Using this general theory, an algorithm for positivity of quintic Hermite FIF is presented. Finally, the algorithm for a quintic Hermite fractal interpolants copositive with a given data set is prescribed.

Keywords Fractals · Iterated function system · Quintic Hermite fractal function · Positivity

MSC 28A80 · 41A20 · 65D10 · 26A48

1 Preamble

It is not ideal to use (piecewise) smooth interpolant with a desired precision when the data has very irregular structure, for instance, real-world signal presented by climate data, time series, financial series, and bioelectric recordings. Barnsley [1] introduced

---

A.K.B. Chand · S.K. Katiyar (✉)
Department of Mathematics, Indian Institute of Technology Madras,
Chennai 600036, India
e-mail: sbhkatiyar@gmail.com

A.K.B. Chand
e-mail: chand@iitm.ac.in

P.N. Agrawal et al. (eds.), *Mathematical Analysis and its Applications*,
Springer Proceedings in Mathematics & Statistics 143,
DOI 10.1007/978-81-322-2485-3_38

fractal interpolation function (FIF) based on the theory of iterated function system (IFS). A FIF is obtained as a fixed point of a suitable map defined on a space of continuous functions. Fractal interpolation function (FIF) captures the irregularity of data very effectively in comparison with the classical interpolants and provides an effective tool for modeling data sampled from real-world signals which are usually difficult using classical approach. FIFs are used to approximate naturally occurring functions which show some sort of self-similarity under magnification. The fractal continuation of an analytic function and fractal tiling from the attractor of one IFS to the attractor of another are studied recently in [3, 4], respectively.

By imposing appropriate conditions on the scaling factors, Barnsley et al. [2] observed that if the problem is of differentiable type, then the elements of the IFS may be suitably chosen so that the corresponding FIF is smooth. Smooth FIF constitutes an advance in the technique of approximation, since the classical methods of real data interpolation can be generalized by means of smooth fractal techniques. However, it is difficult to get all types of boundary conditions for fractal splines in this iterative construction. Fractal splines with general boundary conditions have been studied in [5–7] in simpler ways. By using suitable iterated function system (IFS), Barnsley and Navascués have provided a method to perturb a continuous function so as to yield a class of continuous functions f^α, where α is a free parameter, called scaling vector. For suitable values of scale vector α, the fractal functions f^α simultaneously interpolate and approximate f (see, for instance, [12–15, 19]). Further, the parameter α can be used to modify or preserve properties of f.

Apart from suitable degree of smoothness, it may be desirable or even necessary that the interpolant possesses some properties inherent in the data, depending on practical background of the problem. The problem of searching a sufficiently smooth function that preserves the qualitative shape property inherent in the data is generally referred to as shape preserving interpolation/approximation. The shape properties are mathematically expressed in terms of conditions such as positivity, monotony, and convexity. Recently, Chand and collaborators have developed the shape preserving aspects of the cubic Hermite fractal interpolation function and several rational cubic FIFs with shape parameters (see, for instance, [7–9, 18]). Though these FIFs can render shape preserving interpolants, the order of continuity is only $\mathscr{C}^1$. For enhancing the order of continuity to $\mathscr{C}^2$, the IFS parameters are to be selected so as to satisfy a linear system governing the global $\mathscr{C}^2$-continuity. It is not known whether the solution of this system is compatible with the shape preserving conditions. In the current article we construct quintic Hermite FIF which is $\mathscr{C}^2$, and then constrain the parameters (treating slopes and moments as parameters) so as to obtain positivity. By a slight modification of the proposed method we shall also obtain $\mathscr{C}^2$-continuous fractal spline copositive with given data.

2 Basic Facts

In this section we briefly recall requisite general material for our study. For a detailed exposition reader may refer to [1, 2, 10, 16].

2.1 *Quintic Hermite Interpolant*

For $r \in \mathbb{N}$, let $\mathbb{N}_r$ denote the subset $\{1, 2, \ldots, r\}$ of $\mathbb{N}$. Let a set of data points $\mathscr{D} = \{(x_i, y_i) \in I \times \mathbb{R} : i \in \mathbb{N}_N\}$ satisfying $x_1 < x_2 < \cdots < x_N$, where $I = [x_1, x_N]$ be given. The local mesh spacing is $h_i = x_{i+1} - x_i$, and the slope of the linear interpolant between the data points is $\Delta_i = \frac{y_{i+1} - y_i}{h_i}$. A quintic Hermite function $Q \in \mathscr{C}^2(I)$ is uniquely determined by y_i, d_i, and D_i, where $Q(x_i) = y_i$, $Q^{'}(x_i) = d_i$, and $Q^{''}(x_i) = D_i$, $i \in \mathbb{N}_N = \{1, 2, \ldots, N\}$. The quintic Hermite interpolation defined over the subinterval $I_i = [x_i, x_{i+1}]$ has the form:

$$
\begin{aligned}
Q_i(x) = & \left[\frac{D_{i+1} - D_i}{2h_i^3} + \frac{-3d_i - 3d_{i+1} + 6\Delta_i}{h_i^4}\right](x - x_i)^5 \\
& + \left[\frac{-2D_{i+1} + 3D_i}{2h_i^2} + \frac{8d_i + 7d_{i+1} - 15\Delta_i}{h_i^3}\right](x - x_i)^4 \\
& + \left[\frac{D_{i+1} - 3D_i}{2h_i} + \frac{-6d_i - 4d_{i+1} + 10\Delta_i}{h_i^2}\right](x - x_i)^3 \\
& + \frac{D_i}{2}(x - x_i)^2 + d_i(x - x_i) + y_i.
\end{aligned} \tag{1}
$$

2.2 *Nonnegativity/Nonpositivity of Quintic Hermite Interpolant*

The quintic Hermite interpolant preserves nonnegativity or positivity if

$$
\frac{-5\tau_i y_i}{h_i} \le \tau_i d_i \le \frac{5\tau_i y_i}{h_{i-1}} \; and \; \tau_i D_i \ge \tau_i max\left\{\frac{8d_i}{h_{i-1}} - \frac{20y_i}{h_{i-1}^2}, \frac{-8d_i}{h_i} - \frac{20y_i}{h_i^2}\right\}, \tag{2}
$$

where $\tau_i = sgn(y_i)$, $i = 2, 3, \ldots, N-1$, and endpoint derivatives are calculated by arithmetic mean method and it is addressed in [10].

2.3 IFS for Fractal Functions

Let a set of data points $\mathscr{D} = \{(x_i, y_i) \in \mathbb{R}^2 : i \in \mathbb{N}_N\}$ satisfying $x_1 < x_2 < \cdots < x_N$, $N > 2$, be given. Set $I = [x_1, x_N]$, $I_i = [x_i, x_{i+1}]$ for $i \in \mathbb{N}_{N-1}$. Suppose $L_i : I \to I_i, i \in \mathbb{N}_{N-1}$ be contraction homeomorphisms such that

$$L_i(x_1) = x_i,\ \ L_i(x_N) = x_{i+1}. \tag{3}$$

Let $0 < r_i < 1, i \in \mathbb{N}_{N-1}$, and $X := I \times \mathbb{R}$. Let $N-1$ continuous mappings $F_i : X \to \mathbb{R}$ be given satisfying:

$$F_i(x_1, y_1) = y_i,\ \ F_i(x_N, y_N) = y_{i+1},\ \ |F_i(x, y) - F_i(x, y^*)| \le r_i|y - y^*|, \tag{4}$$

where $(x, y), (x, y^*) \in X$. Define functions $w_i : X \to I_i \times \mathbb{R}$, $w_i(x, y) = (L_i(x), F_i(x, y))\ \forall\ i \in \mathbb{N}_{N-1}$. It is known [1] that there exists a metric on $\mathbb{R}^2$, equivalent to the Euclidean metric, with respect to which $w_i, i \in \mathbb{N}_{N-1}$, are contractions. The collection $\mathscr{I} = \{X; w_i : i \in \mathbb{N}_{N-1}\}$ is called an Iterated Function System (IFS). Associated with the IFS $\mathscr{I}$, there is a set valued Hutchinson map $W : H(X) \to H(X)$ defined by $W(B) = \bigcup_{i=1}^{N-1} w_i(B)$ for $B \in H(X)$, where $H(X)$ is the set of all nonempty compact subsets of X endowed with the Hausdorff metric h. The Hausdorff metric h completes $H(X)$. Further, W is a contraction map on the complete metric space $(H(X), h)$. By the Banach Fixed Point Theorem, there exists a unique set $G \in H(X)$ such that $W(G) = G$. This set G is called the attractor or deterministic fractal corresponding to the IFS $\mathscr{I}$. For any choices of L_i and F_i satisfying the conditions prescribed in (3)–(4) , the following theorem holds.

Proposition 1 (Barnsley [1]) *The IFS $\{X; w_i : i \in \mathbb{N}_{N-1}\}$ defined above admits a unique attractor G, and G is the graph of a continuous function $g : I \to \mathbb{R}$ which obeys $g(x_i) = y_i$ for $i \in \mathbb{N}_N$.*

Definition 1 The aforementioned function g whose graph is the attractor of an IFS is called a **Fractal Interpolation Function** (FIF) or a **Self-referential function** corresponding to the IFS $\{X;\ w_i : i \in \mathbb{N}_{N-1}\}$.

The above fractal interpolation function g is obtained as the fixed point of the Read-Bajraktarević (RB) operator T on a complete metric space $(\mathscr{G}, \rho)$ defined as

$$(Th)(x) = F_i\left(L_i^{-1}(x), h \circ L_i^{-1}(x)\right)\ \forall\, x \in I_i,\ i \in \mathbb{N}_{N-1},$$

where $\mathscr{G} := \{h : I \to \mathbb{R} \mid h \text{ is continuous on } I,\ h(x_1) = y_1, h(x_N) = y_N\}$ equipped with the metric $\rho(h, h^*) = \max\{|h(x) - h^*(x)| : x \in I\}$ for $h, h^* \in \mathscr{G}$. It can be seen that T is a contraction mapping on $(\mathscr{G}, \rho)$ with a contraction factor $r^* := \max\{r_i : i \in \mathbb{N}_{N-1}\} < 1$. The fixed point of T is the FIF g corresponding to the IFS $\mathscr{I}$. Therefore, g satisfies the functional equation:

$$g(x) = F_i\left(L_i^{-1}(x), g \circ L_i^{-1}(x)\right), \quad x \in I_i, \ i \in \mathbb{N}_{N-1}, \tag{5}$$

The most extensively studied FIFs in theory and applications so far are defined by the mappings:

$$L_i(x) = a_i x + b_i, \ F_i(x, y) = \alpha_i y + q_i(x), \ i \in \mathbb{N}_{N-1}. \tag{6}$$

Here $-1 < \alpha_i < 1$ and $q_i : I \to \mathbb{R}$ are suitable continuous functions satisfying (4). The parameter α_i is called a scaling factor of the transformation w_i, and $\alpha = (\alpha_1, \alpha_2, \ldots, \alpha_{N-1})$ is the scale vector corresponding to the IFS. Let $f \in \mathscr{C}(I)$ be a continuous function and consider the case:

$$q_i(x) = f \circ L_i(x) - \alpha_i b(x). \tag{7}$$

Here $b : I \to \mathbb{R}$ is a continuous map that fulfills the conditions $b(x_1) = f(x_1)$, $b(x_N) = f(x_N)$, and $b \neq f$. This case is proposed by Barnsley [1] and Navascués [11] as generalization of any continuous function. Here the interpolation data set is $\{(x_i, f(x_i)) : i \in \mathbb{N}_N\}$. We define the α-fractal function corresponding to f in the following:

Definition 2 The continuous function $f^\alpha : I \to \mathbb{R}$ whose graph is the attractor of the IFS defined by (6)–(7) is referred to as α-**fractal function** associated with f, with respect to b and the partition $\mathscr{D}$.

According to (5), f^α satisfies the functional equation:

$$f^\alpha(x) = f(x) + \alpha_i[(f^\alpha - b) \circ L_i^{-1}(x)] \ \forall x \in I_i, \ i \in \mathbb{N}_{N-1}. \tag{8}$$

Note that for $\alpha = 0$, $f^\alpha = f$. Thus the aforementioned equation may be treated as an entire family of functions f^α with f as its germ. By this method one can define fractal analogues of any continuous function.

2.4 Differentiable FIFs (Fractal Splines)

For a prescribed data set, a FIF with $\mathscr{C}^r$-continuity is obtained as the fixed point of IFS (6), where the scaling factors α_i and the functions q_i are chosen according to the following proposition.

Proposition 2 (Barnsley and Harrington [2]) *Let $\{(x_i, y_i) : i \in \mathbb{N}_N\}$ be given interpolation data with strictly increasing abscissae. Let $L_i(x) = a_i x + b_i, i \in \mathbb{N}_{N-1}$, satisfy (3) and $F_i(x, y) = \alpha_i y + q_i(x), i \in \mathbb{N}_{N-1}$, satisfy (4). Suppose that for some integer $r \geq 0$, $|\alpha_i| \leq \kappa a_i^r$, $0 < \kappa < 1$, and $q_i \in \mathscr{C}^r(I)$, $i \in \mathbb{N}_{N-1}$. Let*

$$F_{i,k}(x, y) = \frac{\alpha_i y + q_i^{(k)}(x)}{a_i^k}, \quad y_{1,k} = \frac{q_1^{(k)}(x_1)}{a_1^k - \alpha_1}, \quad y_{N,k} = \frac{q_{N-1}^{(k)}(x_N)}{a_{N-1}^k - \alpha_{N-1}}, \quad k = 1, 2, \ldots, r.$$

If $F_{i-1,k}(x_N, y_{N,k}) = F_{i,k}(x_1, y_{1,k})$ *for* $i = 2, 3, \ldots, N-1$ *and* $k = 1, 2, \ldots, r$, *then the IFS* $\{X; (L_i(x), F_i(x, y)) : i \in \mathbb{N}_{N-1}\}$ *determines a FIF* $g \in \mathscr{C}^r[x_1, x_N]$, *and* $g^{(k)}$ *is the FIF determined by the IFS* $\{X; (L_i(x), F_{i,k}(x, y)) : i \in \mathbb{N}_{N-1}\}$ *for* $k = 1, 2, \ldots, r$.

The equality proposed in the above proposition demands the resolution of systems of equations. Sometimes the system has no solution, mainly whenever some boundary conditions are imposed on the function (see [2]). However, for the special class of IFS used to construct α-fractal function f^{α}, the procedure can be easily carried out.

Assume $|\alpha_i| < a_i^r$, for all $i \in \mathbb{N}_{N-1}$. Then to obtain a fractal perturbation $f^{\alpha} \in \mathscr{C}^r(I)$ corresponding to a given $f \in \mathscr{C}^r(I)$, it is enough to find the conditions on function b so that the IFS defined by (6) fulfills the conditions of the above theorem. Assuming a uniform partition, Navascués and Sebastián have undertaken this in Ref. [17] which is improved by Viswanathan et al. in Ref. [20] by allowing nonuniform partition and unequal scale factors in different subintervals.

Proposition 3 *Let* $f \in \mathscr{C}^r(I)$. *Suppose* $\mathscr{D} = \{x_1, x_2, \ldots, x_N\}$ *be an arbitrary partition on* I *satisfying* $x_1 < x_2 < \cdots < x_N$. *Let* $|\alpha_i| < a_i^r$, *for all* $i \in \mathbb{N}_{N-1}$. *Further suppose that* $b \in \mathscr{C}^r(I)$ *fulfills* $b^{(k)}(x_1) = f^{(k)}(x_1)$, $b^{(k)}(x_N) = f^{(k)}(x_N)$ *for* $k = 0, 1, \ldots, r$. *Then the corresponding fractal function* f^{α} *is* r*-smooth, and* $(f^{\alpha})^{(k)}(x_i) = f^{(k)}(x_i)$ *for* $i \in \mathbb{N}_N$ *and* $k = 0, 1, \ldots, r$.

3 $\mathscr{C}^2$ Quintic Hermite FIF Preserving Copositivity

In this section, first we shall find the strip condition for rth derivative of α-fractal function, then we develop an algorithm for copositivity of $\mathscr{C}^2$-quintic Hermite FIF.

3.1 Strip Condition for rth Derivative of α-Factal Function

In this subsection we shall provide conditions on the parameters so as to ensure that rth derivative of a $\mathscr{C}^r$-continuous fractal function lies in a rectangle whenever its classical counterpart f does so. Our proof is a modification of that given in [20] where we take into account new requirements. However, we work with a slightly more general obstacle for the rth derivative where $r \in \mathbb{N} \cup \{0\}$ is arbitrary, whereas [20] deals with constraining graph of f^{α} or its first two derivatives within a rectangle $I \times [0, M]$. For a succinct presentation of the theorem, let us introduce the following notation for a continuous function g defined on a compact interval J:

$$m(g; J) = min\ \{g(x) : x \in J\}, \qquad M(g; J) = max\ \{g(x) : x \in J\}.$$

Theorem 1 *Let $f \in \mathscr{C}^r(I)$ be such that $M_1 \le f^{(r)}(x) \le M_2$, for all $x \in I$, some suitable constants M_1 and M_2. The α-fractal function f^α (cf. (5)) corresponding to f satisfies $M_1 \le (f^\alpha)^{(r)}(x) \le M_2$, for all $x \in I$ provided the base function $b \in \mathscr{C}^r(I)$ satisfies $b^{(k)}(x_1) = f^{(k)}(x_1)$, $b^{(k)}(x_N) = f^{(k)}(x_N)$ for $k = 0, 1, \ldots, r$, and the scaling factors $|\alpha_i| < a_i^r$, for all $i \in \mathbb{N}_{N-1}$ obey the following additional conditions:*

$$max\left\{\frac{a_i^r[M_1 - m(f^{(r)}; I_i)]}{M_2 - m(b^{(r)}; I)}, \frac{-a_i^r[M_2 - M(f^{(r)}; I_i)]}{M(b^{(r)}; I) - M_1}\right\} \le \alpha_i$$

$$\le min\left\{\frac{a_i^r[m(f^{(r)}; I_i) - M_1]}{M(b^{(r)}; I) - M_1}, \frac{a_i^r[M_2 - M(f^{(r)}; I_i)]}{M_2 - m(b^{(r)}; I)}\right\}.$$

Proof With the stated conditions on the scale factors and the function b we can ensure from Proposition 3 that corresponding fractal function f^α is r-smooth. Note that $(f^\alpha)^{(r)}$ is a fractal function corresponding to the IFS $\{X; (L_i(x), F_{i,r}(x, y)) : i \in \mathbb{N}_{N-1}\}$ (see Proposition (3)), $(f^\alpha)^{(r)}(x_i) = f^{(r)}(x_i)$ and $(f^\alpha)^{(r)}$ is constructed iteratively using the following functional equation:

$$(f^\alpha)^{(r)}(L_i(x)) = F_{i,r}(x, (f^\alpha)^{(r)}(x)) = f^{(r)}(L_i(x)) + \frac{\alpha_i}{a_i^r}\{(f^\alpha)^{(r)} - b^{(r)}\}(x).$$

Therefore, to prove $M_1 \le (f^\alpha)^{(r)}(x) \le M_2$, for all $x \in I$, by the property of the attractor of the IFS, it is enough to prove that $M_1 \le y \le M_2$ implies $M_1 \le F_{i,r}(x, y) \le M_2 \; \forall i \in \mathbb{N}_{N-1}$.

First, let $0 \le \alpha_i < a_i^r$. We note that $M_1 \le y \le M_2$ implies $\frac{\alpha_i}{a_i^r} M_1 + f^{(r)}(L_i(x)) - \frac{\alpha_i}{a_i^r} b^{(r)}(x) \le \frac{\alpha_i}{a_i^r} y + f^{(r)}(L_i(x)) - \frac{\alpha_i}{a_i^r} b^{(r)}(x) \le \frac{\alpha_i}{a_i^r} M_2 + f^{(r)}(L_i(x)) - \frac{\alpha_i}{a_i^r} b^{(r)}(x)$. Therefore, our target $M_1 \le \frac{\alpha_i}{a_i^r} y + f^{(r)}(L_i(x)) - \frac{\alpha_i}{a_i^r} b^{(r)}(x) \le M_2$ is achieved if $M_1(1 - \frac{\alpha_i}{a_i^r}) \le f^{(r)}(L_i(x)) - \frac{\alpha_i}{a_i^r} b^{(r)}(x) \le M_2(1 - \frac{\alpha_i}{a_i^r})$.

Note that $f^{(r)}(L_i(x)) \ge m(f^{(r)}; I_i)$ and $b^{(r)}(x) \le M(b^{(r)}; I)$ is true for all $x \in I$. Therefore, $M_1(1 - \frac{\alpha_i}{a_i^r}) \le f^{(r)}(L_i(x)) - \frac{\alpha_i}{a_i^r} b^{(r)}(x)$ holds if $M_1(1 - \frac{\alpha_i}{a_i^r}) \le m(f^{(r)}; I_i) - \frac{\alpha_i}{a_i^r} M(b^{(r)}; I)$. This inequality holds if $\alpha_i \le \frac{a_i^r[m(f^{(r)}; I_i) - M_1]}{M(b^{(r)}; I) - M_1}$. Similarly, $f^{(r)}(L_i(x)) \le M(f^{(r)}; I_i)$ and $b^{(r)}(x) \ge m(b^{(r)}; I)$ is true for all $x \in I$. Therefore, $f^{(r)}(L_i(x)) - \frac{\alpha_i}{a_i^r} b^{(r)}(x) \le M_2(1 - \frac{\alpha_i}{a_i^r})$ holds if $M(f^{(r)}; I_i) - \frac{\alpha_i}{a_i^r} m(b^{(r)}; I) \le M_2(1 - \frac{\alpha_i}{a_i^r})$, which in turn holds if $\alpha_i \le \frac{a_i^r[M_2 - M(f^{(r)}; I_i)]}{M_2 - m(b^{(r)}; I)}$.

Now assume $-a_i^r < \alpha_i \le 0$. In this case, $M_1 \le y \le M_2$ implies that $\frac{\alpha_i}{a_i^r} M_2 + f^{(r)}(L_i(x)) - \frac{\alpha_i}{a_i^r} b^{(r)}(x) \le \frac{\alpha_i}{a_i^r} y + f^{(r)}(L_i(x)) - \frac{\alpha_i}{a_i^r} b^{(r)}(x) \le \frac{\alpha_i}{a_i^r} M_1 + f^{(r)}(L_i(x)) - \frac{\alpha_i}{a_i^r} b^{(r)}(x)$. Consequently, for $M_1 \le F_{i,r}(x, y) \le M_2$, it is sufficient to

verify $M_1 - \frac{\alpha_i}{a_i^r} M_2 \le f^{(r)}(L_i(x)) - \frac{\alpha_i}{a_i^r} b^{(r)}(x) \le M_2 - \frac{\alpha_i}{a_i^r} M_1$. By appropriately using the definition of $m(b^{(r)}; I)$, $M(b^{(r)}; I)$, $m(f^{(r)}; I_i)$, $M(f^{(r)}; I_i)$ on lines similar to the first part, we get
$\alpha_i \ge \dfrac{a_i^r[M_1 - m(f^{(r)}; I_i)]}{M_2 - m(b^{(r)}; I)}$ and $\alpha_i \ge \dfrac{-a_i^r[M_2 - M(f^{(r)}; I_i)]}{M(b^{(r)}; I) - M_1}$. Combination of the obtained conditions on the scale factors completes the proof.

Consequence. If $f : I = [a, b] \to \mathbb{R}$ is a $\mathscr{C}^r$-continuous r-convex function (i.e., $f^{(r)} \ge 0$), then using the foregoing theorem we can select a scale vector α such that $f^\alpha \in \mathscr{C}^r(I)$ and f^α preserves r-convexity of f. Note that for $r = 0, 1, 2$, r-convexity reduces to positivity, monotonicity, and convexity, respectively.

Remark 1 Let us confine to the positivity case (i.e., $r = 0$ and $M_1 = 0$). If it is enough to consider the nonnegative scale factors, then the following condition on the scale factors ensures the nonnegativity of f^α: $0 \le \alpha_i \le \frac{m(f; I_i)}{M(b; I)}$, where $|\alpha_i| < 1$ is assumed.

Remark 2 If $f \in \mathscr{C}(I)$ is nonpositive (i.e., $f(x) \le 0$, for all $x \in I$), then we may construct f^α satisfying $f^\alpha(x) \le 0$, for all $x \in I$ by employing Theorem 1. Taking $r = 0$ and $M_2 = 0$ then the condition for $M_1 \le f^\alpha \le 0$ can be obtained as: $|\alpha_i| < 1$ and $\max\left\{-\dfrac{M_1 - m(f; I_i)}{m(b; I)}, \dfrac{-M(f; I_i)}{M_1 - M(b; I)}\right\} \le \alpha_i \le \min\left\{\dfrac{M_1 - m(f; I_i)}{M_1 - M(b; I)}, \dfrac{M(f; I_i)}{m(b; I)}\right\}$.

Remark 3 The aforementioned fractal scheme can be modified and extended to produce piecewise-defined α-fractal function which is copositive with the given $f \in \mathscr{C}(I)$. For this, the interval I has to be subdivided into subintervals, say I_j, $j = 1, 2, \ldots, r$ in such a way that the function f is positive or negative throughout the subinterval I_j. In each of these subintervals I_j, we take base function b_j, and a scaling factor α^j so as to meet the specification in Theorem 1. Consequently, we can get the fractal functions $f_j^{\alpha^j}$ that retain the nonpositivity/nonnegativity of the functions $f_j = f|I_j$, $j = 1, 2, \ldots, r$. Denoting α to be the r-rowed matrix whose rows are the scaling factors α^j, we define f^α in a piecewise manner as follows: $f^\alpha|I_j = f_j^{\alpha^j}$.

3.2 *Quintic Hermite FIF as α-Fractal Function*

Consider a set of data points $\mathscr{D} = \{(x_i, y_i, d_i, D_i) : i \in \mathbb{N}_N\}$, where y_i denotes the function value, d_i, D_i denote the first derivative and second derivative value of an unknown function Ψ at the knot point x_i, respectively. To construct the $\mathscr{C}^2$-quintic Hermite FIF corresponding to $\mathscr{D}$, one may employ the general theory given in Sect. 2. The traditional nonrecursive $\mathscr{C}^2$-quintic Hermite interpolant corresponding to $\mathscr{D}$ can be represented as:

$$
\begin{aligned}
Q(L_i(x)) = {} & \left[\frac{h_i^2(D_{i+1}-D_i)-6h_i(d_i+d_{i+1})+12(y_{i+1}-y_i)}{2}\right]\left(\frac{x-x_1}{x_N-x_1}\right)^5\\
& + \left[\frac{h_i^2(-2D_{i+1}+3D_i)+h_i(16d_i+14d_{i+1})-30(y_{i+1}-y_i)}{2}\right]\left(\frac{x-x_1}{x_N-x_1}\right)^4\\
& + \left[\frac{h_i^2(D_{i+1}-3D_i)-h_i(12d_i+8d_{i+1})+20(y_{i+1}-y_i)}{2}\right]\left(\frac{x-x_1}{x_N-x_1}\right)^3\\
& + \frac{h_i^2 D_i}{2}\left(\frac{x-x_1}{x_N-x_1}\right)^2 + h_i d_i\left(\frac{x-x_1}{x_N-x_1}\right)+y_i. \qquad (9)
\end{aligned}
$$

For $x \in [x_i, x_{i+1}]$, using $\frac{L_i^{-1}(x)-x_1}{x_N-x_1} = \frac{x-x_i}{x_{i+1}-x_i}$, one can see that the above expression coincides with the classical quintic Hermite interpolant (cf. (1)). According to prescription in Theorem 1, we have to select the function b so as to obtain a $\mathscr{C}^2$-quintic Hermite fractal perturbation for $Q \in \mathscr{C}^2(I)$. A natural choice of b is the two-point quintic Hermite interpolants (with knots at x_1 and x_N) corresponding to Q. That is,

$$
\begin{aligned}
b(x) = {} & \left[\frac{(x_N-x_1)^2(D_N-D_1)-6(x_N-x_1)(d_1+d_N)+12(y_N-y_1)}{2}\right]\left(\frac{x-x_1}{x_N-x_1}\right)^5\\
& + \left[\frac{(x_N-x_1)^2(-2D_N+3D_1)+(x_N-x_1)(16d_1+14d_N)-30(y_N-y_1)}{2}\right]\left(\frac{x-x_1}{x_N-x_1}\right)^4\\
& + \left[\frac{(x_N-x_1)^2(D_N-3D_1)-(x_N-x_1)(12d_1+8d_N)+20(y_N-y_1)}{2}\right]\left(\frac{x-x_1}{x_N-x_1}\right)^3\\
& + \frac{D_1}{2}(x-x_1)^2+d_1(x-x_1)+y_1. \qquad (10)
\end{aligned}
$$

Using (9)–(10), we obtain the fractal function $Q^\alpha \in \mathscr{C}^2(I)$ corresponding to $Q \in \mathscr{C}^2(I)$ as:

$$
\begin{aligned}
Q^\alpha(L_i(x)) &= \alpha_i Q^\alpha(x) + Q(L_i(x)) - \alpha_i b(x),\\
&= \alpha_i Q^\alpha(x) + \tfrac{A_i}{2}\left(\tfrac{x-x_1}{x_N-x_1}\right)^5 + \tfrac{B_i}{2}\left(\tfrac{x-x_1}{x_N-x_1}\right)^4 + \tfrac{C_i}{2}\left(\tfrac{x-x_1}{x_N-x_1}\right)^3 \qquad (11)\\
&\quad + \tfrac{D_i}{2}\left(\tfrac{x-x_1}{x_N-x_1}\right)^2 + E_i\left(\tfrac{x-x_1}{x_N-x_1}\right) + F_i, \quad \forall x \in I, i \in \mathbb{N}_{N-1},
\end{aligned}
$$

where

$$
\begin{aligned}
A_i &= h_i^2(D_{i+1}-D_i)-6h_i(d_i+d_{i+1})+12(y_{i+1}-y_i)\\
&\qquad - \alpha_i[(x_N-x_1)^2(D_N-D_1)-6(x_N-x_1)(d_1+d_N)+12(y_N-y_1)],\\
B_i &= h_i^2(-2D_{i+1}+3D_i)+h_i(16d_i+14d_{i+1})-30(y_{i+1}-y_i)\\
&\qquad - \alpha_i[(x_N-x_1)^2(-2D_N+3D_1)+(x_N-x_1)(16d_1+14d_N)-30(y_N-y_1)],\\
C_i &= h_i^2(D_{i+1}-3D_i)-h_i(12d_i+8d_{i+1})+20(y_{i+1}-y_i)
\end{aligned}
$$

$$- \alpha_i[(x_N - x_1)^2(D_N - 3D_1) - (x_N - x_1)(12d_1 + 8d_N) + 20(y_N - y_1)],$$
$$D_i = h_i^2 D_i - \alpha_i D_1(x_N - x_1)^2, \quad E_i = h_i d_i - \alpha_i d_1(x_N - x_1), \quad F_i = y_i - \alpha_i y_1.$$

Note that the function $Q^\alpha : I \to \mathbb{R}$ enjoys the interpolation conditions $Q^\alpha(x_i) = y_i$, $(Q^\alpha)'(x_i) = d_i$ and $(Q^\alpha)''(x_i) = D_i$.

Remark 4 When $\alpha_i = 0$, for all $i \in J$, Q^α reduces to the classical $\mathscr{C}^2$-quintic Hermite interpolant Q given in (9). The diversity of options α in Q^α allows us to choose the best if a problem combined with approximation and optimization is to be approached. The function $(Q^\alpha)'' : I \to \mathbb{R}$ may be nondifferentiable in a dense subset of I or a finite subset of I depending on the values of α. Thus the perturbation allows new geometric possibility: The graph of $(Q^\alpha)''$ owns a fractal dimension which may be used as an index for experimental signals.

3.3 Algorithm for Positivity of $\mathscr{C}^2$-Quintic Hermite FIF

Here we shall find conditions for the positivity of $\mathscr{C}^2$-quintic Hermite FIF. Recall that we have viewed $\mathscr{C}^2$-quintic Hermite FIF as α-fractal function Q^α corresponding to quintic Hermite interpolant Q. Therefore, it is not hard to see that Theorem 1 in conjunction with positivity condition for Q gives the following algorithm.

An Algorithm for positive $\mathscr{C}^2$-quintic Hermite FIF

Step 1: Compute the approximate derivative values $d_i, D_i, i \in \mathbb{N}_N$ and check if they satisfy conditions given in (2).

Step 2: To get positive quintic Hermite interpolant by the modified derivative values obtained in Step 1. If not, modify according to $\frac{-5y_i}{h_i} \le d_i \le \frac{5y_i}{h_{i-1}}$ *and* $D_i \ge \max\left\{\frac{8d_i}{h_{i-1}} - \frac{20y_i}{h_{i-1}^2}, \frac{8d_i}{h_i} - \frac{20y_i}{h_i^2}\right\}$ for $i = 2, 3, \ldots, N-1$, and endpoint derivatives are calculated by arithmetic mean method.

Step 3: Denote the derivative values obtained at the end of Step 2 by d_i, D_i for $i \in \mathbb{N}_N$. For Q and b, compute the constants $m(b; I) = \min_{x \in I} b(x)$, $M(b; I) = \max_{x \in I} b(x)$, $m(f; I_i) = \min_{x \in I_i} f(x)$, $M(f; I_i) = \max_{x \in I_i} f(x)$. Choose $|\alpha_i| < a_i$ and $\max\left\{\frac{M_1 - m(f; I_i)}{M_2 - m(b; I)}, -\frac{M_2 - M(f; I_i)}{M(b; I) - M_1}\right\} \le \alpha_i \le \min\left\{\frac{m(f; I_i) - M_1}{M(b; I) - M_1}, \frac{M_2 - M(f; I_i)}{M_2 - m(b; I)}\right\}$ according to the prescription in Theorem 1.

Step 4: Input these derivative values chosen in Step 2 and the scaling parameters as prescribed by Step 3 in the functional equation represented by (11) whereupon the points of the graph of Q^α are computed.

Remark 5 On similar lines we can get an algorithm for a nonpositive $\mathscr{C}^2$-quintic Hermite FIF.

Next, we shed some light on to get interpolation of positive/negative data in adjacent intervals by considering Remark 3. Let us discuss this with an example. Consider a continuous function Φ defined on data set $\{(x_i, y_i) : i = 1, 2, \ldots, 7\}$

having zeros at the points x_3, x_6. It is nonnegative on $[x_1, x_3]$, nonpositive on $[x_3, x_6]$, and nonnegative on $[x_6, x_7]$. To get the desired interpolant, subdivide the interpolation domain into three subintervals $I_1 = [x_1, x_3]$, $I_2 = [x_3, x_6]$, and $I_3 = [x_6, x_7]$ for our convenience. We can apply the developed algorithm to obtain positivity preserving quintic Hermite FIF $Q_1^{\alpha^1}$ on I_1. With proper renaming of the data points if necessary, the negative quintic Hermite FIF algorithm can be applied to obtain a negative quintic Hermite FIF $Q_2^{\alpha^2}$ on I_2. Since I_3 does not contain sufficient number of knot points, iterations of the IFS code cannot produce any new points. To overcome this problem, we introduce a new node say, (x_6^*, y_6^*) in such a way $x_6 < x_6^* < x_7$. We apply the developed positivity preserving quintic Hermite FIF algorithm with an arbitrary but shape consistent extra node to obtain a positive quintic Hermite FIF $Q_3^{\alpha^3}$. Define a quintic Hermite FIF Q^α in a piecewise manner such that $Q^\alpha|I_j = Q_j^{\alpha^j}$ for $j = 1, 2, 3$. Then Q^α is copositive with the original data.

4 Numerical Illustration

In this section, we will illustrate positivity preserving $\mathscr{C}^2$-quintic Hermite FIF with some simple examples. Let us take a set of positive data $\mathscr{D} = \{(0, 0.1), (0.4, 1), (0.75, 2), (1, 5)\}$. Note that for the implementation of the $\mathscr{C}^2$-quintic Hermite FIF one requires in input the values of the derivatives at the knot points. Therefore, in the absence of other conditions/information, estimates of derivatives are necessary. Values (rounded off to two decimal places) of d_i, D_i, $i = 1, 2, 3, 4$ estimated using the arithmetic mean method are $d_1 = 1.92$, $d_2 = 2.57$, $d_3 = 8.19$, $d_4 = 15.80$, $D_1 = -6.07$, $D_2 = 9.31$, $D_3 = 8.19$, and $D_4 = 36.48$. The nonpositive $\mathscr{C}^2$-quintic Hermite FIF $Q^\alpha(L_i(x))$ is displayed in Fig. 1a. This illustrates the importance of the positivity preserving $\mathscr{C}^2$-quintic Hermite FIF algorithm developed in the previous section. Now improve the derivative values as prescribed in (2), i.e., $d_2 = 10.50$, $d_3 = 20.57$, $D_2 = -73.60$, $D_3 = -139.31$, and endpoint derivatives as in Fig. 1a.

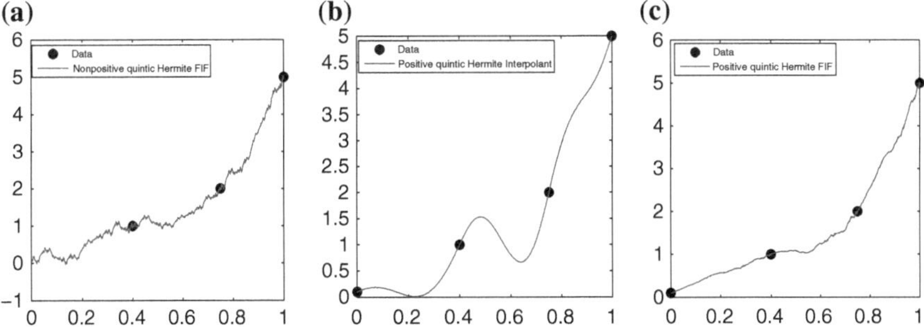

Fig. 1 $\mathscr{C}^2$-Quintic Hermite FIF (the interpolating data points are given by the *circles* and the relevant $\mathscr{C}^2$-quintic Hermite FIF by the *solid lines*)

We generate positive $\mathscr{C}^2$-quintic Hermite interpolant displayed in Fig. 1b by taking these modified derivative values and assuming the values of scaling factors $\alpha_i = 0$, for all $i \in \{1, 2, 3\}$. To illustrate our algorithm, we first construct the positive $\mathscr{C}^2$-quintic Hermite interpolant Q as in Fig. 1b. After computing the maximum and minimum value of Q and b, we select scaling factors according to the prescription in Theorem 1. Input the derivative values and parameters values in the functional equation represented by (11) whereupon the points of the graph of Q^α are computed in Fig. 1c.

5 Concluding Remarks

In this paper, we have developed methods to identify the elements of the IFS so that the α-fractal function f^α retains fundamental shape property, namely, positivity, and order of continuity inherent in the function f. For a data with prescribed or estimated slopes and moments at knot points, the quintic Hermite FIF is constructed, which generalizes the classical quintic Hermite interpolant. The considerable flexibility and diversity offered by quintic Hermite FIF (see Remark 2) can be exploited in fields of applications such as CAE, computer graphics, animation, visual simulation, and image processing.

Acknowledgments The first author is thankful to the Department of Science & Technology, India for the SERC DST Project No. SR/S4/MS: 694/10

References

1. Barnsley, M.F.: Fractal functions and interpolation. Constr. Approx. **2**(4), 303–329 (1986)
2. Barnsley, M.F., Harrington, A.N.: The calculus of fractal functions. J. Approx. Theory **57**(1), 14–34 (1989)
3. Barnsley, M., Vince, A.: Fractal continuation. Constr. Approx. **38**(1), 311–337 (2013)
4. Barnsley, M., Vince, A.: Fractal tilings from iterated function system. Discrete Comput. Geom. **51**(1), 729–752 (2014)
5. Chand, A.K.B., Kapoor, G.P.: Generalized cubic spline fractal interpolation functions. SIAM J. Numer. Anal. **44**(2), 655–676 (2006)
6. Chand, A.K.B., Navascués, M.A.: Generalized Hermite fractal interpolation. Rev. R. Acad. de ciencias. Zaragoza **64**(2), 107–120 (2009)
7. Chand, A.K.B., Viswanathan, P.: A constructive approach to cubic Hermite fractal interpolation function and its constrained aspects. BIT Numer. Math. **53**(4), 841–865 (2013)
8. Chand, A.K.B., Katiyar, S.K., Saravana Kumar, G.: A new class of rational fractal function for curve fitting. Proceeding of Computer Aided Engineering CAE 2013, pp. 78–93, ISBN:No-80689-17-3 (2013)
9. Chand, A.K.B., Vijender, N., Navascués, M.A.: Shape preservation of scientific data through rational fractal splines. Calcolo **51**, 329–362 (2013)
10. Dougherty, R.L., Edelman A.S., Hyman, J.M.: Nonnegativity-, monotonicity-, or convexity-preserving cubic and quintic Hermite interpolation. Math. Comput. **52**, 471–494 (1989)
11. Navascués, M.A.: Fractal polynomial interpolation. Z. Anal. Anwend. **25**(2), 401–418 (2005)

12. Navascués, M.A.: Fractal approximation. Complex Anal. Oper. Theory **4**, 953–974 (2010)
13. Navascués, M.A.: Reconstruction of sampled signals with fractal functions. Acta. Appl. Math. **110**, 1199–1210 (2010)
14. Navascués, M.A.: Fractal Haar system. Nonlinear Anal. **74**, 4152–4165 (2011)
15. Navascués, M.A., Chand, A.K.B.: Fundamental sets of fractal functions. Acta Appl. Math. **100**, 247–261 (2008)
16. Navascués, M.A., Chand, A.K.B., Viswanathan, P., Sebastián, M.V.: Fractal interpolation functions: a short survey. Appl. Math. **5**, 1834–1841 (2014)
17. Navascués, M.A., Sebastián, M.V.: Smooth fractal interpolation. J. Inequal. Appl. (Article ID 78734), 20 (2004)
18. Viswanathan, P., Chand, A.K.B.: A fractal procedure for monotonicity preserving interpolation. Appl. Math. Comput. **247**, 190–204 (2014)
19. Viswanathan, P., Chand, A.K.B.: Fractal rational functions and their approximation properties. J. Approx. Theory **185**, 31–50 (2014)
20. Viswanathan, P., Chand, A.K.B., Navascués, M.A.: Fractal perturbation preserving fundamental shapes: bounds on the scale factors. J. Math. Anal. Appl. **419**, 804–817 (2014)

Existence Result for Semilinear Fractional Stochastic Evolution Inclusions Driven by Poisson Jumps

P. Tamilalagan and P. Balasubramaniam

Abstract In this manuscript, the sufficient conditions are established for the existence of mild solutions of semilinear fractional stochastic evolution inclusions driven by Poisson jumps in a Hilbert space. The results are obtained by using a fixed point theorem for condensing multivalued map due to Martelli.

Keywords Fixed point theorem · Multivalued map · Mild solution · Poisson jump

MSC 26A33, 34A08, 34A60, 47H10

1 Introduction

The theory and applications of fractional differential equations [FDEs] have notable contributions during the last few decades. Since the FDEs are the most powerful toll for describing the real-life phenomena more preciously, thus there is a rapid development in its applications [14]. In neurophysiology, the behavior of voltage potentials of spatially extended neurons has been described by stochastic differential equations [SDEs] driven by Poisson jump. SDEs driven by a Poisson process has applications in various fields such as storage systems, queuing systems, economic systems, and neurophysiology system. The study of SDE driven by a Poisson jump has considerable attentions (see [9, 12, 15]). Differential inclusions serve as an efficient tool in analysis of uncertain, nonlinear, and hybrid as well as switching and time-variant systems. Random differential and integral inclusions play an important role in characterizing many physical, biological, social, and engineering problems [3, 4]. The applicability of fractional differential inclusions [FDIs] in modeling of

P. Tamilalagan · P. Balasubramaniam (✉)
Department of Mathematics, Gandhigram Rural Institute—Deemed University, Gandhigram 624 302, Tamil Nadu, India
e-mail: balugru@gmail.com

P. Tamilalagan
e-mail: mathematicstamil@gmail.com

P.N. Agrawal et al. (eds.), *Mathematical Analysis and its Applications*, Springer Proceedings in Mathematics & Statistics 143,
DOI 10.1007/978-81-322-2485-3_39

various practical and engineering systems insists its necessity. Thus very few authors investigated the existence of solutions of FDIs [1, 5, 8]. However, to the best of authors' knowledge there is no work reported on semilinear fractional stochastic differential inclusions driven by Poisson jump.

In this manuscript, we study the existence of mild solution for the following semilinear fractional stochastic evolution inclusions driven by Poisson jumps described by

$$d\left[J_t^{1-\alpha}(x(t)-x(0))\right] \in [Ax(t) + F(x(\rho(t)))]dt + \int_Z L(t, x(t-), z)\widetilde{N}(dt, dz), t \in J := [0, b]$$
$$x(t) = \phi(t),\ t \in J_0 = [-r, 0],\ r > 0 \tag{1}$$

where $0 < \alpha < 1$, $J_t^{1-\alpha}$ is the $(1-\alpha)$ order Riemann–Liouville fractional integral operator, $A : D(A) \subset H \to H$ is the infinitesimal generator of analytic semigroup of bounded linear operators $\{T(t),\ t \geq 0\}$ on a separable Hilbert space H with inner product $\langle \cdot, \cdot \rangle$ and norm $\| \cdot \|$. $\rho : [0, \infty) \to [-r, \infty),\ r \geq 0$ be a suitable delay function, $x : [-r, b] \to H$ and $\phi : J_0 \to H$ is the cadlag function such that $\phi(t)$ is $\mathscr{F}_0$ measurable for all $t \in J_0$, $E\|\phi(0)\|^p < \infty$ and $\int_{-r}^{0} E\|\phi(s)\|^p ds < \infty,\ p \geq 2$. Let K be another separable Hilbert space with inner product $\langle \cdot, \cdot \rangle$ and norm $\| \cdot \|_K$. Let $\widetilde{N}(dt, dz)$ be a compensated Poisson random measure associated with the Poisson point process $k(\cdot)$. Let $L(K, H)$ be the space of all bounded linear operators from K into H with norm $\| \cdot \|$. Let $F : H \to 2^H$ and $L : J \times H \times (K - \{0\}) \to H$ are the appropriate functions to be defined later.

2 Preliminaries

In this section, we furnish some basic preliminaries, definitions, notations and lemmas, which are required to establish the main result. Let $(\Omega, \mathscr{F}, \{\mathscr{F}_t\}_{t \geq 0}, \mathbb{P})$ be a complete probability space with the filtration $\{\mathscr{F}_t\}_{t \geq 0}$, satisfying the usual conditions, that is, right continuous and $\mathscr{F}_0$ containing all P-null sets. Let $L_p(\Omega, \mathscr{F}, \mathbb{P}; H) \equiv L_p(\Omega, H)$ be the space of all p-integrable random variables with values in H, that are measurable with respect to $\{\mathscr{F}_t,\ t \in J\}$. Let $J_1 = [-r, b]$ and $C = C(J_1, H)$ denote the family of continuous H-valued stochastic processes $\{\xi(t) : t \in J_1\}$ which are $\mathscr{F}_t$-measurable and $\|\xi\| < \infty$, where

$$\|\xi\| = \|\xi\|_C = \sup_{t \in J_1} \left(E\|\xi(t)\|^p\right)^{\frac{1}{p}}.$$

Let $\mathscr{B}_\sigma(H)$ denotes the Borel σ-algebra of H. Let $k(t),\ t \geq 0$ be a stationary $\mathscr{F}_t$-adapted and K-valued Poisson point process. The counting measure N_k defined by

$$N_k((t_1, t_2] \times \Lambda)(\omega) := \sum_{t_1 < s \leq t_2} I_\Lambda(k(s)),$$

for any $\Lambda \in \mathscr{B}_\sigma(K)$ is called the Poisson random measure associated to the Poisson point process k. Then define the measure $\widetilde{N}$ by

$$\widetilde{N}(dt, dz) := N_k(dt, dz) - \pi(dt, dz),$$

where $\pi(dt, dz)$ is the compensator. We assume that k is σ-finite and stationary, there exists a characteristic measure λ such that $\pi(dt, dz) = dt\lambda(dz)$.

Definition 1 [14] The Riemann–Liouville fractional integral of order $\alpha > 0$ for the function $x : J \to H$ is defined by

$$J_t^\alpha x(t) = \frac{1}{\Gamma(\alpha)} \int_0^t (t-s)^{\alpha-1} x(s) ds$$

and the Laplace transform of the Riemann–Liouville's fractional integral is given by

$$L\{J_t^\alpha x(t)\} = \frac{1}{\lambda^\alpha} \hat{x}(\lambda), \text{ where } \hat{x}(\lambda) = \int_0^\infty e^{-\lambda t} x(t) dt, \; Re(\lambda) > w.$$

Definition 2 [10] The Riemann–Liouville fractional derivative of order $0 < \alpha < 1$ for the function $x : J \to H$ can be defined as

$$D_t^\alpha x(t) = \frac{d}{dt} J_t^{1-\alpha} x(t).$$

Definition 3 [11] The Caputo fractional derivative of order $0 < \alpha < 1$ for the function x can be defined in terms of Riemann–Liouville fractional derivative as follows:

$${}^c D_t^\alpha x(t) = D_t^\alpha (x(t) - x(0))$$

the Laplace transform of the Caputo fractional derivative is given by

$$L\{{}^c D_t^\alpha x(t)\} = \lambda^\alpha \hat{x}(\lambda) - \lambda^{\alpha-1} x(0).$$

Definition 4 [11] The Mainardi's function is defined by

$$M_\alpha(z) = \sum_{n=0}^{\infty} \frac{(-z)^n}{n! \Gamma(1 - \alpha n - \alpha)}, \quad 0 < \alpha < 1, \; z \in \mathbb{C},$$

it is clear that $\int_0^\infty M_\alpha(r) dr = 1, \; 0 < \alpha < 1$ on the otherhand, $M_\alpha(z)$ satisfies the following equality:

$$\int_0^\infty \frac{\alpha}{r^{\alpha+1}} M_\alpha\left(\frac{1}{r^\alpha}\right) e^{-\lambda r} dr = e^{-\lambda^\alpha}$$

$$\text{and} \int_0^\infty r^\delta M_\alpha(r) dr = \frac{\Gamma(\delta+1)}{\Gamma(\alpha\delta+1)}, \ \delta > -1, \ 0 < \alpha < 1.$$

Lemma 1 [4] *Let H be a Hilbert space and $\Phi : H \to BCC(H)$ be a u.s.c and condensing map. If the set*

$$U = \{x \in H : \hat{\lambda} x \in \Phi x \, for \, some \, \hat{\lambda} > 1\}$$

is bounded, then Φ has a fixed point.

The following lemma is crucial in the proof of our main result.

Lemma 2 [4] *Let J be a compact interval and H be a Hilbert space. Let F be a multivalued map which is measurable for each $u \in H$, upper semicontinuous with respect to u and for each fixed $u \in H$ the set $N_{F,u} = \{f \in L^p(H) : f(t) \in F(u)$ for a.e $t \in J$ is nonempty. Also let Π be a linear continuous mapping from $L^p(J, H)$ to $C(J, H)$, then the operator*

$$\Pi \circ N_F : C(J, H) \to BCC(C(J, H)), \ x \to (\Pi \circ N_F)(x) = \Pi(N_{F,x})$$

is a closed graph operator in $C(J, H) \times C(J, H)$.

Lemma 3 [2] *For any $p \geq 2$ there exists $c_p > 0$ such that*

$$E \sup_{0 \leq s \leq t} \left\| \int_0^t \int_Z H(s,z) \widetilde{N}(ds,dz) \right\|^p \leq c_p \left\{ E\left[\left(\int_0^t \int_Z \|H(s,z)\|^2 \lambda(dz) ds \right)^{\frac{p}{2}} \right] + E\left[\int_0^t \int_Z \|H(s,z)\|^p \lambda(dz) ds \right] \right\}.$$

Motivated by [10, 11], we present the following definition of mild solution for (1).

Definition 5 An H-valued stochastic process $x(t)$ satisfying $E\|\phi\|^p < \infty$ and $f \in L(H)$ is a selection of $F(x(\rho(t)))$ is called a mild solution of (1), if the following conditions are satisfied

(i) x(t) is $\mathscr{F}_t$-adapted and Cadlag,
(ii) $x(t)$ satisfies the following integral equations

$$x(t) = S_\alpha(t)\phi(0) + \int_0^t (t-s)^{\alpha-1} T_\alpha(t-s) f(s) ds + \int_0^t (t-s)^{\alpha-1} T_\alpha(t-s) \left[\int_Z L(s, x(s-), z) \widetilde{N}(ds,dz) \right]$$

where

$$S_\alpha(t)x = \int_0^\infty M_\alpha(r)T(t^\alpha r)xdr,\ t \geq 0,\ x \in H \tag{2}$$

$$\text{and} \qquad T_\alpha(t)x = \int_0^\infty \alpha r M_\alpha(r)T(t^\alpha r)xdr \tag{3}$$

The existence of solution for (1) is derived under the following assumptions:

H$_1$ The analytic semigroup $T(t)$ generated by A is compact for $t > 0$ and there exists $M > 0$ such that

$$\sup_{t\geq 0} \|T(t)\| \leq M,\ t \geq 0 \tag{4}$$

from (2) and (4) we have $\|S_\alpha(t)\| \leq M$ and from (3) and (4), we have $\|T_\alpha(t)\| \leq \frac{M}{\Gamma(\alpha)}$ (for details, see [10]).

H$_2$ $\rho : [0, \infty) \to [-r, \infty),\ r \geq 0$ is a continuous function such that $-r \leq \rho(t) \leq t$ for all $t \geq 0$.

H$_3$ $F : H \to BCC(H);\ u \to F(u)$ is measurable for each $u \in H$, u.s.c with respect to u and for each fixed $u \in H$ the set

$$N_{F,u} = \{f \in L^p(H) : f(t) \in F(u) \text{for a.e } t \in J\}$$

is nonempty.

H$_4$ $E\|F(u)\|^p = \sup\{E\|v\|^p : v \in F(u)\} \leq \eta(t)\psi(E\|u\|^p)$ for almost all $t \in J$ and $u \in H$, where $\eta \in L^p(J, \mathbb{R}_+)$ and $\psi : \mathbb{R}_+ \to (0, \infty)$ is continuous and increasing with

$$\frac{1}{r}\int_0^t (t-s)^{(\alpha-1)p}\eta(s)\psi(E\|x(\rho(s))\|^p)ds = \Lambda.$$

H$_5$ There exist positive constant M_L such that, for $x \in H$

$$\int_Z E\|L(t, x(t-), z)\|^p \lambda(dz) \leq M_L\|x(t)\|^p.$$

For more details on this section, the reader can refer [6, 7, 13].

3 Main Result

Theorem 1 *Assume that the hypotheses* **H$_1$–H$_5$** *are hold, then the initial value problem (1) has at least one mild solution on $J_1 = [-r, b]$, provide that*

$$3^{p-1}\left\{\frac{M^p b^{p-1}}{\Gamma^p(\alpha)}\Lambda + \frac{M^p c_p}{\Gamma^p(\alpha)}\left(\frac{M_L^{\frac{p}{2}} b^{\alpha p-\frac{p}{2}} + M_L b^{\alpha p-p+1}}{\alpha p - p + 1}\right)\right\} < 1. \tag{5}$$

Proof Consider the multivalued map $\Phi : C \to 2^C$, defined by

$$(\Phi x)(t) = \left\{ h \in C : \ h(t) = \begin{cases} \phi(t); \ t \in [-r, 0] \\ S_\alpha(t)\phi(0) + \int_0^t (t-s)^{\alpha-1} T_\alpha(t-s) f(s)ds, \\ + \int_0^t (t-s)^{\alpha-1} T_\alpha(t-s) \int_Z L(s, x(s-), z)\widetilde{N}(ds, dz), \ t \in J = [0, b] \end{cases} \right.$$

where $f \in N_{F,x} = \{f \in L^p(H) : f(t) \in F(x(\rho(s))) \text{ for a.e } t \in J\}$.

We shall prove that Φ is a completely continuous multivalued map u.s.c with convex closed values. The proof will be given in several steps.

Step 1 Φx is convex for each $x \in C$. Indeed, if h_1 and h_2 belong to Φx, then there exists $f_1, f_2 \in N_{F,x}$ such that for each $t \in J$, we have

$$h_i(t) = S_\alpha(t)\phi(0) + \int_0^t (t-s)^{\alpha-1} T_\alpha(t-s) f_i(s)ds$$
$$+ \int_0^t (t-s)^{\alpha-1} T_\alpha(t-s)\left[\int_Z L(s, x(s-), z)\widetilde{N}(ds, dz)\right], \ i = 1, 2.$$

Let $0 \le v \le 1$, then for each $t \in J$, we have

$$(vh_1 + (1-v)h_2)(t) = S_\alpha(t)\phi(0) + \int_0^t (t-s)^{\alpha-1} T_\alpha(t-s)(vf_1(s) + (1-v)f_2(s))ds$$
$$+ \int_0^t (t-s)^{\alpha-1} T_\alpha(t-s)\left[\int_Z L(s, x(s-), z)\widetilde{N}(ds, dz)\right]$$

since $N_{F,x}$ is convex, thus we have $(vh_1 + (1-v)h_2) \in \Phi x$.

Step 2 Φ maps bounded sets into bounded sets in C. Indeed it is enough to show that there exists a positive constant l such that for each $h \in \Phi x, \ x \in B_q = \{x \in C : \|x\|^p \le q\}$ one has $\|h\|^p \le l$. If $h \in \Phi x$, then there exists $f \in N_{F,x}$ such that for each $t \in J$, by $\mathbf{H_1}$–$\mathbf{H_5}$ and Lemma 3, we have

$$E\|h(t)\|^p \le 3^{p-1}\left\{E\|S_\alpha(t)\phi(0)\|^p + E\left\|\int_0^t (t-s)^{\alpha-1} T_\alpha(t-s) f(s)ds\right\|^p\right.$$
$$\left. + E\left\|\int_0^t (t-s)^{\alpha-1} T_\alpha(t-s)\left[\int_Z L(s, x(s-), z)\widetilde{N}(ds, dz)\right]\right\|^p\right\}$$
$$\le 3^{p-1}\left\{M^p E\|\phi(0)\|^p + \frac{M^p b^{p-1}}{\Gamma^p(\alpha)}\int_0^t (t-s)^{(\alpha-1)p} E\|f(s)\|^p ds\right.$$
$$+ \frac{M^p}{\Gamma^p(\alpha)} c_p E\left(\int_0^t \int_Z \left\|(t-s)^{\alpha-1} L(s, x(s-), z)\right\|^2 \lambda(dz)ds\right)^{\frac{p}{2}}$$
$$\left. + \frac{M^p}{\Gamma^p(\alpha)} c_p E\left(\int_0^t \int_Z \left\|(t-s)^{\alpha-1} L(s, x(s-), z)\right\|^p \lambda(dz)ds\right)\right\}$$

$$\leq 3^{p-1}\left\{M^pE\|\phi(0)\|^p + \frac{M^pb^{p-1}}{\Gamma^p(\alpha)}\int_0^t (t-s)^{(\alpha-1)p}\eta(s)\psi\left(E\|x(\rho(s))\|^p\right)ds\right.$$
$$\left.+\frac{M^pc_pM_L^{\frac{p}{2}}b^{\alpha p-\frac{p}{2}}q}{\Gamma^p(\alpha)(\alpha p-p+1)} + \frac{M^pc_pM_Lb^{\alpha p-p+1}q}{\Gamma^p(\alpha)(\alpha p-p+1)}\right\}$$
$$\leq 3^{p-1}\left\{M^pE\|\phi(0)\|^p + \frac{M^pb^{p-1}}{\Gamma^p(\alpha)}\sup_{t\in J}\psi(E\|x(t)\|^p)\int_0^t (t-s)^{(\alpha-1)p}\eta(s)ds\right.$$
$$\left.+\frac{M^pc_pq}{\Gamma^p(\alpha)}\left(\frac{M_L^{\frac{p}{2}}b^{\alpha p-\frac{p}{2}} + M_Lb^{\alpha p-p+1}}{\alpha p-p+1}\right)\right\}$$

then for each $h \in \Phi(B_q)$, we have

$$\|h\|^p \leq 3^{p-1}\left\{M^pE\|\phi(0)\|^p + \frac{M^pb^{p-1}}{\Gamma^p(\alpha)}\sup_{t\in J}\psi(E\|x(t)\|^p)\int_0^t (t-s)^{(\alpha-1)p}\eta(s)ds\right.$$
$$\left.+\frac{M^pc_pq}{\Gamma^p(\alpha)}\left(\frac{M_L^{\frac{p}{2}}b^{\alpha p-\frac{p}{2}} + M_Lb^{\alpha p-p+1}}{\alpha p-p+1}\right)\right\} := l.$$

Step 3 Φ maps bounded sets into equicontinuous sets of C. For each $x \in B_q$ and $h \in \Phi x$ there exists $f \in N_{F,x}$, we have

$$E\|h(t_2)-h(t_1)\|^p$$
$$= E\left\|S_\alpha(t_2)\phi(0) + \int_0^{t_2}(t_2-s)^{\alpha-1}T_\alpha(t_2-s)f(s)ds\right.$$
$$+\int_0^{t_2}(t_2-s)^{\alpha-1}T_\alpha(t_2-s)\left[\int_Z L(s,x(s-),z)\widetilde{N}(ds,dz)\right]$$
$$- S_\alpha(t_1)\phi(0) - \int_0^{t_1}(t_1-s)^{\alpha-1}T_\alpha(t_1-s)f(s)ds$$
$$\left.-\int_0^{t_1}(t_1-s)^{\alpha-1}T_\alpha(t_1-s)\left[\int_Z L(s,x(s-),z)\widetilde{N}(ds,dz)\right]\right\|^p$$
$$\leq 11^{p-1}\Big\{E\,\|[S_\alpha(t_2)-S_\alpha(t_1)]\phi(0)\|^p$$
$$+ E\left\|\int_0^{t_1-\varepsilon}(t_2-s)^{\alpha-1}[T_\alpha(t_2-s)-T_\alpha(t_1-s)]f(s)ds\right\|^p$$
$$+ E\left\|\int_{t_1-\varepsilon}^{t_1}[(t_2-s)^{\alpha-1}-(t_1-s)^{\alpha-1}]T_\alpha(t_1-s)f(s)ds\right\|^p$$
$$+ E\left\|\int_{t_1-\varepsilon}^{t_1}(t_2-s)^{\alpha-1}[T_\alpha(t_2-s)-T_\alpha(t_1-s)]f(s)ds\right\|^p$$
$$+ E\left\|\int_0^{t_1-\varepsilon}[(t_2-s)^{\alpha-1}-(t_1-s)^{\alpha-1}]T_\alpha(t_1-s)f(s)ds\right\|^p$$

$$
\begin{aligned}
&+ E\left\|\int_{t_1}^{t_2}(t_2-s)^{\alpha-1}T_\alpha(t_2-s)f(s)ds\right\|^p \\
&+ E\left\|\int_0^{t_1-\varepsilon}(t_2-s)^{\alpha-1}[T_\alpha(t_2-s)-T_\alpha(t_1-s)]\left[\int_Z L(s,x(s-),z)\widetilde{N}(ds,dz)\right]\right\|^p \\
&+ E\left\|\int_{t_1-\varepsilon}^{t_1}[(t_2-s)^{\alpha-1}-(t_1-s)^{\alpha-1}]T_\alpha(t_1-s)\left[\int_Z L(s,x(s-),z)\widetilde{N}(ds,dz)\right]\right\|^p \\
&+ E\left\|\int_{t_1-\varepsilon}^{t_1}(t_2-s)^{\alpha-1}[T_\alpha(t_2-s)-T_\alpha(t_1-s)]\left[\int_Z L(s,x(s-),z)\widetilde{N}(ds,dz)\right]\right\|^p \\
&+ E\left\|\int_0^{t_1-\varepsilon}[(t_2-s)^{\alpha-1}-(t_1-s)^{\alpha-1}]T_\alpha(t_1-s)\left[\int_Z L(s,x(s-),z)\widetilde{N}(ds,dz)\right]\right\|^p \\
&+ E\left\|\int_{t_1}^{t_2}(t_2-s)^{\alpha-1}T_\alpha(t_2-s)\left[\int_Z L(s,x(s-),z)\widetilde{N}(ds,dz)\right]\right\|^p\Bigg\}.
\end{aligned}
$$

As $t_2 \to t_1$ the right-hand side of the above inequality tends to zero, since the compactness of $T(t)$ for $t > 0$ implies the continuity in the uniform operator topology, the equicontinuity for the cases $t_1 < t_2 \le 0$ and $t_1 \le 0 \le t_2$ are obvious. As a consequence of the steps 2, 3 together with the Arzela Ascoli theorem it is concluded that $\Phi : C \to 2^C$ is a compact multivalued map and therefore a condensing map.

Step 4 Φ has a closed graph.

Let $x_n \to x_*,\ h_n \in \Phi x_n$ and $h_n \to h_*$. We shall prove that $h_* \in \Phi x_*,\ h_n \in \Phi x_n$ means that there exists $f_n \in N_{F,x_n}$ such that

$$
\begin{aligned}
h_n(t) = S_\alpha(t)\phi(0) &+ \int_0^t (t-s)^{\alpha-1}T_\alpha(t-s)f_n(s)ds \\
&+ \int_0^t (t-s)^{\alpha-1}T_\alpha(t-s)\left[\int_Z L(s,x_n(s-),z)\widetilde{N}(ds,dz)\right],\ t\in J.
\end{aligned}
$$

We must prove that there exists $f_* \in N_{F,x_*}$ such that

$$
\begin{aligned}
h_*(t) = S_\alpha(t)\phi(0) &+ \int_0^t (t-s)^{\alpha-1}T_\alpha(t-s)f_*(s)ds \\
&+ \int_0^t (t-s)^{\alpha-1}T_\alpha(t-s)\left[\int_Z L(s,x_*(s-),z)\widetilde{N}(ds,dz)\right],\ t\in J.
\end{aligned}
$$

Now, for every $t \in J$, L is continuous, we have

$$
\begin{aligned}
&\left\|\left(h_n(t) - S_\alpha(t)\phi(0) - \int_0^t (t-s)^{\alpha-1}T_\alpha(t-s)\left[\int_Z L(s,x_n(s-),z)\widetilde{N}(ds,dz)\right]\right)\right. \\
&\quad\left. - \left(h_*(t) - S_\alpha(t)\phi(0) - \int_0^t (t-s)^{\alpha-1}T_\alpha(t-s)\left[\int_Z L(s,x_*(s-),z)\widetilde{N}(ds,dz)\right]\right)\right\| \to 0
\end{aligned}
$$

as $n \to \infty$.

Consider the linear continuous operator $\Pi : L^p(H) \to C(J, H)$,

$$f \to \Pi(f)(t) = \int_0^t (t-s)^{\alpha-1} T_\alpha(t-s) f(s)ds$$

from Lemma 2, it follows that $\Pi \circ N_F$ is a closed graph operator. Moreover, we have

$$h_n(t) - S_\alpha(t)\phi(0) - \int_0^t (t-s)^{\alpha-1} T_\alpha(t-s) \int_Z L(s, x_n(s-), z)\widetilde{N}(ds, dz) \in \Pi(N_{F,x_n}).$$

Since $x_n \to x_*$, it follows from Lemma 2 that

$$\begin{aligned} h_*(t) - S_\alpha(t)\phi(0) - \int_0^t (t-s)^{\alpha-1} T_\alpha(t-s) \int_Z L(s, x_*(s-), z)\widetilde{N}(ds, dz) \\ = \int_0^t (t-s)^{\alpha-1} T_\alpha(t-s) f_*(s)ds \end{aligned}$$

for some $f_* \in N_{F,x_*}$. Therefore Φ is completely continuous multivalued map, u.s.c with convex closed values. In order to prove that Φ has a fixed point, we need one more step.

Step 5 Φ has a solution, if the set $U = \{x \in C : \hat{\lambda} x \in \Phi x, \text{ for some } \hat{\lambda} > 1\}$ is bounded, such that for $x \in U$, $E\|x(t)\|^p \leq r$. Assume that it is not true. Let $x \in U$ be a solution for $\hat{\lambda} x \in \Phi x$, for some $\hat{\lambda} > 1$ with $E\|x(t)\|^p > r$. Then there exists $f \in N_{F,x}$, we have

$$\begin{aligned} x(t) = \hat{\lambda}^{-1} S_\alpha(t)\phi(0) + \hat{\lambda}^{-1} \int_0^t (t-s)^{\alpha-1} T_\alpha(t-s) f(s)ds \\ + \hat{\lambda}^{-1} \int_0^t (t-s)^{\alpha-1} T_\alpha(t-s) \left[\int_Z L(s, x(s-), z)\widetilde{N}(ds, dz)\right], \quad t \in J. \end{aligned}$$

By $\mathbf{H_1}$–$\mathbf{H_2}$ and $\mathbf{H_4}$–$\mathbf{H_5}$, we have

$$\begin{aligned} E\|x(t)\|^p \leq 3^{p-1} \Bigg\{ E\left\|\hat{\lambda}^{-1} S_\alpha(t)\phi(0)\right\|^p + E\left\|\hat{\lambda}^{-1}\int_0^t (t-s)^{\alpha-1} T_\alpha(t-s) f(s)ds\right\|^p \\ + E\left\|\hat{\lambda}^{-1}\int_0^t (t-s)^{\alpha-1} T_\alpha(t-s)\left[\int_Z L(s, x(s-), z)\widetilde{N}(ds, dz)\right]\right\|^p \Bigg\} \\ r \leq 3^{p-1} \Bigg\{ M^p E\|\phi(0)\|^p + \frac{M^p b^{p-1}}{\Gamma^p(\alpha)} \int_0^t (t-s)^{(\alpha-1)p} \eta(s)\psi(E\|x(\rho(s))\|^p)ds \\ + \frac{M^p c_p r}{\Gamma^p(\alpha)} \left(\frac{M_L^{\frac{p}{2}} b^{\alpha p - \frac{p}{2}} + M_L b^{\alpha p - p + 1}}{\alpha p - p + 1} \right) \Bigg\} \end{aligned}$$

by diving both sides of the above inequality by r and taking $r \to \infty$, we obtain

$$1 \leq 3^{p-1} \left\{ \frac{M^p b^{p-1}}{\Gamma^p(\alpha)} \Lambda + \frac{M^p c_p}{\Gamma^p(\alpha)} \left(\frac{M_L^{\frac{p}{2}} b^{\alpha p - \frac{p}{2}} + M_L b^{\alpha p - p + 1}}{\alpha p - p + 1} \right) \right\}$$

which is contradiction to our assumption (5), thus Φ has a solution if U is bounded. This completes the proof.

4 Conclusion

In this manuscript, the Poisson jumps are incorporated with the stochastic differential inclusions lead to new systems in the world of fractional calculus. The existence of mild solutions for semilinear fractional stochastic evolution inclusions driven by Poisson jumps has been studied in a Hilbert space by means of fractional calculus and the fixed point theorem for condensing multivalued map due to Martelli.

Acknowledgments The work of authors are supported by Council of Scientific and Industrial Research, Extramural Research Division, Pusa, New Delhi, India under the grant No. 25(0217)/13/EMR-II.

References

1. Ait Dads, E., Benchohra, M., Hamani, S.: Impulsive fractional differential inclusions involving the Caputo fractional derivative. Fract. Calc. Appl. Anal. **12**, 15–38 (2009)
2. Applebaum, D.: Levy Processes and Stochastic Calculus. Cambridge University Press, Cambridge (2009)
3. Balasubramaniam, P., Ntouyas, S.K.: Controllability for neutral stochastic functional differential inclusions with infinite delay in abstract space. J. Math. Anal. Appl. **324**, 161–176 (2006)
4. Balasubramaniam, P., Ntouyas, S.K., Vinayagam, D.: Existence of solutions of semilinear stochastic delay evolution inclusions in a Hilbert space. J. Math. Anal. Appl. **305**, 438–451 (2005)
5. Chang, Y.K., Nieto, J.J.: Some new existence results for fractional differential inclusions with boundary conditions. Math. Comput. Model. **49**, 605–609 (2009)
6. Da Prato, G., Zabczyk, J.: Stochastic Equations in Infinite Dimensions. Cambridge University Press, Cambridge (1992)
7. Deimling, K.: Multivalued Differential Equations. De Gruyter, Berlin/New York (1992)
8. El-Sayed, A.M.A., Ibrahim, A.G.: Multivalued fractional differential equations. Appl. Math. Comput. **68**, 15–25 (1995)
9. Hausenblas, E.: SPDEs driven by Poisson random measure with non Lipschitz coefficients: existence results. Probab. Theory Relat. Fields **137**, 161–200 (2007)
10. Li, K., Peng, J.: Controllability of fractional neutral stochastic functional differential systems. Z. Angew. Math. Phys. **65**, 941–959 (2014)
11. Li, K., Peng, J., Gao, J.: Existence results for semilinear fractional differential equations via Kuratowski measure of noncompactness. Fract. Calc. Appl. Anal. **15**, 591–610 (2012)

12. Luo, J., Taniguchi, T.: The existence and uniqueness for non-Lipschitz stochastic neutral delay evolution equations driven by Poisson jumps. Stoch. Dyn. **9**, 135–152 (2009)
13. Pazy, A.: Semigroups of Linear Operators and Applications to Partial Differential Equations. Springer-Verlag, Berlin (1983)
14. Podlubny, I.: Fractional Differential Equations. Academic Press, San Diego (1998)
15. Taniguchi, T.: The existence and asymptotic behaviour of solutions to non-Lipschitz stochastic functional evolution equations driven by Poisson jumps. Stochastic **82**, 339–363 (2010)

A Numerical Investigation of Blood Flow in an Arterial Segment with Periodic Body Acceleration

Mamata Parida and Ameeya K. Nayak

Abstract A fluid convection flow driven by a periodic body acceleration with thermal stratification in an arterial cross section filled with an incompressible Newtonian fluid (blood) is studied. A two-dimensional computational visualization technique is used to study the steady flow behavior of the viscous electrically conducting fluid flow. The driving force is generated by putting an external magnetic field in the transverse direction of the flow. A numerical method based on the pressure correction iterative algorithm (SIMPLE) is adopted to compute the flow field and temperature along the arterial cross section. Variation over a wide range of parameters such as Prandtl number, Hartmann number, and Womersley number have been investigated for the flow and heat transfer characteristics.

Keywords Body acceleration · Magnetic field · Finite volume method

1 Introduction

In recent years, the study of magnetohydrodynamic (MHD) flow of blood has gained the attention of many researchers because of its wide range of physiological applications. Magnetohydrodynamic fluids are those physiological fluids which are electrically conducting. The hydrodynamic property of blood flow through arteries plays a vital role for understanding the function of the cardiovascular system under normal and diseased conditions. Several studies analytical as well as experimental have been carried out to analyze the flow of blood through arteries. It was observed by some investigators [2, 12, 15] that under certain conditions blood exhibit viscoelastic

M. Parida (✉)
Department of Mathematics, Utkal University, Bhubaneswar 751 004, India
e-mail: mamatakmru@gmail.com

A.K. Nayak
Department of Mathematics, Indian Institute of Technology Roorkee, Roorkee 247 667, India
e-mail: ameeyanayak@gmail.com

P.N. Agrawal et al. (eds.), *Mathematical Analysis and its Applications*, Springer Proceedings in Mathematics & Statistics 143,
DOI 10.1007/978-81-322-2485-3_40

behavior which may be due to the viscoelastic properties of the individual red cells and the internal structures formed by cellular interactions. Fung et al. [3] have studied blood flow mechanics in the arteries of different sizes. An experimental work was done by Taylor and Draney [14] for quantifying the blood flow velocity and pressure field in human artery. Blood as a dilute suspension of spherical particles which, though rigid, were free to move with the fluid and rotate under the influence of shearing force was studied by Jones [5]. A theoretical and experimental study was conducted by Berger et al. [1] to give an idea of the pressure drop and heat exchange in the fluid when it is subjected to move along a curved path. Several attempts have been made [8, 10, 16] to study the effect of magnetic field on the blood flow in the arteries in various physiological conditions. Heat transfer and fluid flow characteristics of blood in multistenosed arteries with the effect of magnetic field have been investigated in [13].

In the present paper, we have adopted a numerical method to study the characteristics of blood flow and heat transfer through a rectangular duct under the influence of periodic body acceleration and in the presence of transverse magnetic field since it has large-scale applications in biomedical devices [7]. Blood is considered to be Newtonian, viscous, incompressible, and electrically conducting fluid. Blood behaves as a non-Newtonian fluid in very narrow arteries, whereas its behavior is Newtonian in most of the arteries. The result of computation thus obtained for the physical quantities velocity, pressure, temperature, and stream function are presented graphically. The effects of magnetic field and body acceleration on axial blood flow, temperature, and pressure have been studied.

2 Problem Formulation and Numerical Method

2.1 Physical Configuration

We consider the mechanically driven flow of an incompressible electrically conducting Newtonian fluid within a long rectangular channel of length L, width w, and height h shown in Fig. 1a. The flow is assumed to be laminar and the binary fluid is assumed to be incompressible and the flow may be driven by stretching of the channel walls. Density variation of the fluid follows Boussinesq's assumption and changes with temperature only. Initially, the binary mixture is considered to be at rest with a uniform temperature T_∞ and constant density everywhere. An uniformly distributed

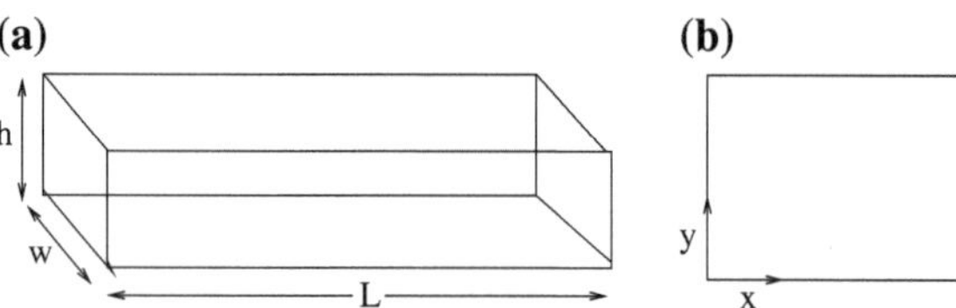

Fig. 1 Schematic of the flow configuration of the artery

external magnetic field of strength B_0 is applied along the transverse direction of the flow. The length, width, and height of the channel are along z, x and y directions, respectively. For computational purposes we have considered a channel cross section, where height is considered to be equal length as of width of the channel to study the flow properties and it is represented by a two-dimensional plane as shown in Fig. 1b.

2.2 Model Equations

As the fluid is assumed to be Newtonian and its density is supposed to be constant, except the gravitational force term in the Navier–Stokes equation, where it varies linearly with the local temperature fraction [4] and is given by,

$$\rho(T^\star) = \rho_0[1 - \beta_T(T^\star - T_\infty)]$$

where ρ_0 is the density of the undisturbed fluid and β_T is the volumetric coefficient of thermal expansion. Initially, the electrically conducting fluid is considered to be at rest. The flow is assumed to have periodic body acceleration given by

$$G^\star(t^\star) = a^\star \cos(\omega_b t^\star + \phi_g). \tag{1}$$

where $a^\star$, ω_b and ϕ_g denote the amplitude, frequency, and phase difference of body acceleration. We now introduce the nondimensional variables defined by

$$x = \frac{x^\star}{h}, y = \frac{y^\star}{h}, u = \frac{u^\star}{\omega h}, v = \frac{v^\star}{\omega h}, p = \frac{p^\star}{\mu\omega}, t = t^\star\omega, \theta = \frac{T^\star - T_w}{T_\infty - T_w} \tag{2}$$

Here ω is the frequency of pulse and T_w is the temperature of the wall. The governing Navier–Stokes equations and heat transport equations in nondimensional form with the Boussinesq-fluid assumption are given by

$$\frac{\partial u}{\partial x} + \frac{\partial v}{\partial y} = 0 \tag{3}$$

$$\frac{\partial u}{\partial t} + u\frac{\partial u}{\partial x} + v\frac{\partial u}{\partial y} = -\frac{1}{\alpha^2}\frac{\partial p}{\partial x} + \frac{1}{\alpha^2}\left(\frac{\partial^2 u}{\partial x^2} + \frac{\partial^2 u}{\partial y^2}\right) - \frac{H^2}{\alpha^2}u + \frac{1}{\alpha^2}G(t) \tag{4}$$

$$\frac{\partial v}{\partial t} + u\frac{\partial v}{\partial x} + v\frac{\partial v}{\partial y} = -\frac{1}{\alpha^2}\frac{\partial p}{\partial y} + \frac{1}{\alpha^2}\left(\frac{\partial^2 v}{\partial x^2} + \frac{\partial^2 v}{\partial y^2}\right) \tag{5}$$

$$\frac{\partial \theta}{\partial t} + u\frac{\partial \theta}{\partial x} + v\frac{\partial \theta}{\partial y} = \frac{1}{\alpha^2 Pr}\left(\frac{\partial^2 \theta}{\partial x^2} + \frac{\partial^2 \theta}{\partial y^2}\right) + \frac{H^2 Ec}{\alpha^2}u^2 \tag{6}$$

where $\alpha = h\sqrt{\frac{\omega\rho}{\mu}}$, is the Womersley number, μ is the coefficient of viscosity, $H = hB_0\sqrt{\frac{\sigma}{\mu}}$ is the Hartmann number, σ is the electrical conductivity, $Pr = \frac{\mu C_p}{\kappa_0}$ is the Prandtl number, C_p is the specific heat at constant pressure, κ_0 is the thermal conductivity, $Ec = \frac{\omega^2 h^2}{C_p(T_w - T_\infty)}$ is the Eckert number and $G(t) = a\ \cos(bt + \phi_g)$, $a = \frac{\rho h}{\omega\mu}a^\star$, $b = \frac{\omega_b}{\omega}$. Here in Eq. 4 the last two terms are due to the effect of applied magnetic field and the body acceleration. Initially ($t = 0$), the fluid is considered to be at rest and with an uniform temperature $\theta = 0$.

Boundary conditions: $t > 0$, $u = 0$, $v = 0$, $\frac{\partial\theta}{\partial x} = 0$.

On the sidewalls ($x = 0$; $x = 1$)$u = v = 0$; $\theta = 0$ on the lower lid ($y = 0$) $u = v = 0$; $\theta = 1$ on the upper lid ($y = 1$).

The local Nusselt number of the upper and lower plates of the channel is obtained by calculating the temperature gradient on the plates from the relation

$$Nu = -\frac{h}{T_w - T_\infty}\frac{\partial\theta}{\partial y}\,|_{y=0} \tag{7}$$

The stream function ψ has been computed from the velocity components by using the definition

$$u = \frac{\partial\psi}{\partial y} \qquad v = -\frac{\partial\psi}{\partial x} \tag{8}$$

2.3 Numerical Methods

In order to tackle the model nonlinear partial differential equations numerically, the method of Newton's linearization technique is applied, that is, when the values of the dependent variables at the nth iteration are known, the corresponding values of variables at the next iteration can be obtained by applying the Newton's linearization method as:

$$V_i^{n+1} = V_i^{n} + \Delta V_i^{n}$$

where V stands for u, v, and θ; ΔV_i^n represents the error at the nth iteration and i is the grid index. To find the numerical solution of the governing fluid flow equations together with the specified boundary conditions we opt a numerical method using control volume approach. This method involves integrating the continuity, momentum, and energy equations over a specified control volume on a staggered grid. In the staggered grid arrangement, the velocity components are stored at the midpoints of the cell faces to which they are normal and the physical quantities such as the pressure and temperature are placed at the cell center. The discretized form of the governing equations is obtained by integrating over each of the control volumes using the finite volume method. The u-momentum equation after integration over the u-control volume becomes

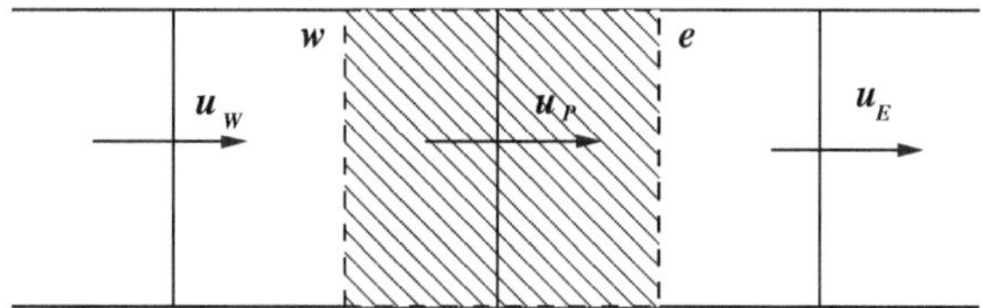

Fig. 2 Schematic diagram of u control volume for u [6, 11]

$$F_e u_e - F_w u_w + F_n u_n - F_s u_s = b \tag{9}$$

where F_e is the nonlinear coefficient of u_e and b contains the source terms and time-derivative terms. The convective term at any interface is estimated by a quadratic interpolation of u. For example, at the east face (Fig. 2) we have

$$u_e = \left(\frac{3}{8}u_E + \frac{3}{4}u_P - \frac{1}{8}u_W\right) \quad if \;\; F_e > 0 \tag{10}$$

$$u_e = \left(\frac{3}{4}u_E + \frac{3}{8}u_P - \frac{1}{8}u_{EE}\right) \quad if \;\; F_e < 0 \tag{11}$$

which can be summarized as

$$F_e u_e = \left(\frac{3}{8}u_E + \frac{3}{4}u_P - \frac{1}{8}u_W\right)[[F_e, 0]] - \left(\frac{3}{4}u_E + \frac{3}{8}u_P - \frac{1}{8}u_{EE}\right)[[-F_e, 0]] \tag{12}$$

The v-momentum equation is calculated in a similar manner. A third-order accurate QUICK (quadratic upstream interpolation for convective kinematics) scheme is employed to discretize the convective terms in the Navier–Stokes equations. A third-level implicit scheme is used for discretization of time derivatives. The pressure correction-based iterative algorithm SIMPLE (Semi-Implicit Method for Pressure Linked Equations) is used for solving the discretized equations. We have obtained a time-independent numerical solution which is convergent by advancing the flow field variables through a sequence of shorter time step $\Delta t = 0.001$. For the range of parameter values considered here, the flow field achieves a steady state after a transient state, and this steady state is independent of the initial conditions. In all the numerical computations that have been carried out and presented in the form of figures we have adopted the steady state condition considering time $t = 0.5$. The discretization of the governing Navier–Stokes equations and heat transfer equation results in a system of algebraic equations of the form

$$A^{\phi}\phi_{i-1,j}^{n+1} + B^{\phi}\phi_{i,j}^{n+1} + C^{\phi}\phi_{i+1,j}^{n+1} = D^{\phi} \tag{13}$$

where ϕ denotes u, v, and θ; A^{ϕ}, B^{ϕ} and C^{ϕ} gives the coefficient matrix and D^{ϕ} gives the pressure and source terms for ϕ at time level n and time derivative terms.

This system of algebraic equations can thus be written in matrix form with the coefficient matrix as a tridiagonal matrix. Due to coupling of energy equation with the momentum equations, the system of algebraic equations is solved through a block elimination method. Convergence criteria is employed of the form

$$|\phi_{i,j}{}^{n+1} - \phi_{i,j}{}^{n}| < \varepsilon \tag{14}$$

$$|1.0 - \frac{Nu_U}{Nu_L}| < \delta \tag{15}$$

Here i and j denote the cell indices, n is the time level, ϕ stands for u, v, or θ and Nu_U and Nu_L are the area averaged Nusselt number on the upper and lower lids, respectively, and the value of ε is considered to be 10^{-4} and that of δ is 10^{-3}.

3 Results and Discussion

We have investigated the blood flow phenomenon and heat transfer through an arterial segment under the effect of body acceleration as well as an external magnetic field. The physiological applicable data used for computation of numerical results are collected from the existing literatures [9] and are listed as: $\alpha = 3.0$, $H = 1.0$, $Pr = 21.0$, $b = 1.0$, $a = 1.0$, $\phi_g = 0.0$, $h = 1.0$, $Ec = 0.0002$, $T_\infty = 310.0\,\mathrm{K}$, $\rho = 1050.0\,\mathrm{kg/m^3}$, $\sigma = 0.8\,\mathrm{s/m}$. For computation purposes we have taken these values, however, any deviation from the listed values has been mentioned inside the figures. In order to validate the present mathematical model we have compared our results with that of Misra et al. [8] produced in Fig. 3. For the purpose of comparison, both the studies have been naturally brought to the same platform.

The distribution of axial velocity is presented in Fig. 4. Figure 4a depicts that the magnetic field parameter brings quantitative as well as qualitative changes in velocity profile. It can be observed that the velocity decreases as the magnetic strength parameter increases. Figure 4b includes the axial profile for various values of Womersley number. The variation of flow velocity can be determined from the inset figure. As we move toward the core region from the inflow region we find that the velocity decreases with increasing Womersley number. However, as the flow velocity attains a saturation point in the core of the channel the velocity increases as α increases. The nondimensional pressure distribution is presented in Fig. 5 for different values of Hartmann number and Womersley number. We observe that during the change of Hartmann number Fig. 5a the pressure variation is almost constant along the outflow region of the artery. However for $H = 1$, initially the pressure shows a negative variation as the reverse flow profile is observed. The pressure variation is found to be optimum in case of $\alpha = 2$. Figure 6 shows a variation in the dimensionless temperature along the flow axis. For different Prandtl numbers and Womersley numbers, we find that the temperature remains almost invariant close to the channel walls. Figure 6a presents the temperature variation for different values of Prandtl number along the cross section of an artery. For large values of Pr the effects of temperature

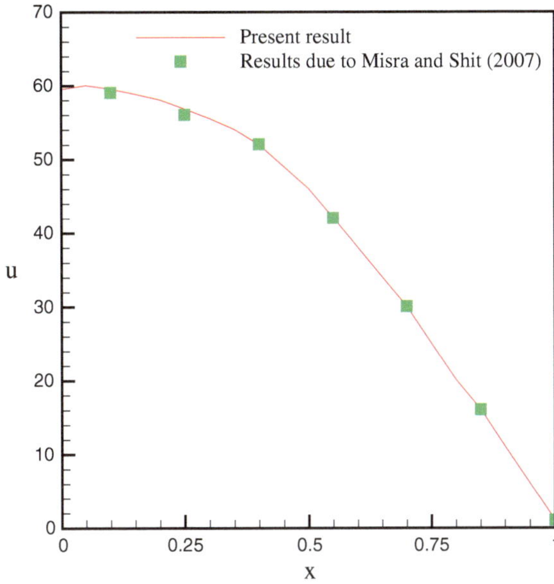

Fig. 3 Comparison of axial velocity profile for present solution in the absence of any magnetic field with that of Misra and Shit [8] when $\alpha = 4$ and $Re = 90$

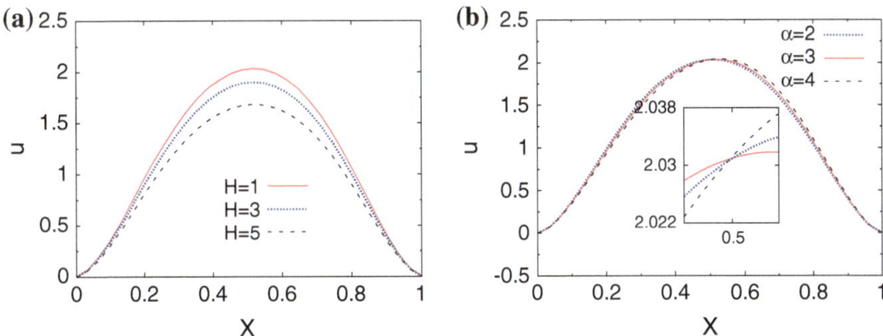

Fig. 4 Axial distribution of velocity profiles along the centerline symmetry, **a** for various values of H when $Pr = 21$ and $\alpha = 3$, **b** for various values of α when $Pr = 21$ and $H = 1$

variation decreases gradually. In Fig. 6b we find that when the Womersley number decreases, there is an increase in the variation of temperature profile. In Figs. 7, 8, and 9 we present the average temperature variation or rate of heat transfer (Nu) along the lower and upper wall of the artery. It is observed from Fig. 7 that the rate of heat transfer decreases with the increase of the Prandtl number Pr. The rate of heat transfer decreases with the increase of Hartmann number H as observed from Fig. 8. It is interesting to mention here that at the lower boundary the Nu rapidly decreases with the increase of H upto the core and beyond which, no significant change is noticed. Hence the rate of heat transfer can be increased whenever necessary by the

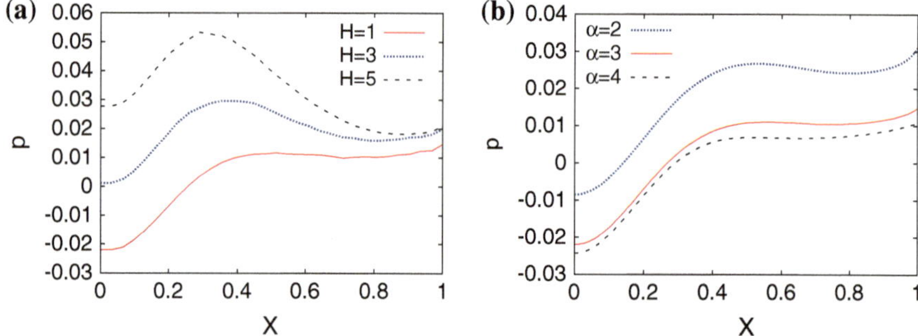

Fig. 5 Distribution of dimensionless pressure along x-axis **a** for various values of H when, $Pr = 21$ and $\alpha = 3$, and **b** for $Pr = 21$, $H = 1$, and $\alpha = 2, 3, 4$

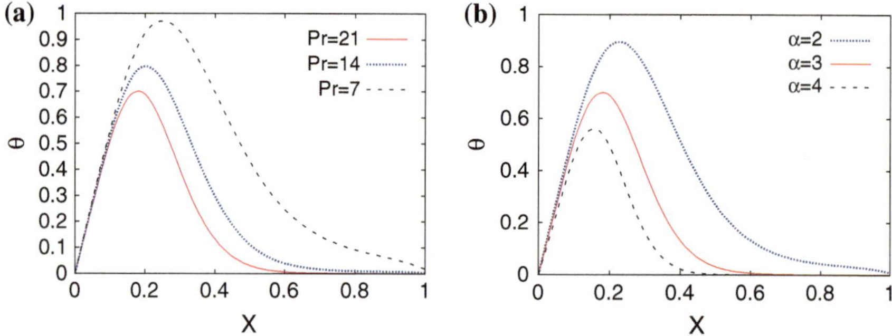

Fig. 6 Distribution of dimensionless temperature along x-axis **a** for various values of Pr when, $H = 1$ and $\alpha = 3$, and **b** for $Pr = 21$, $H = 1$, and $\alpha = 2, 3, 4$

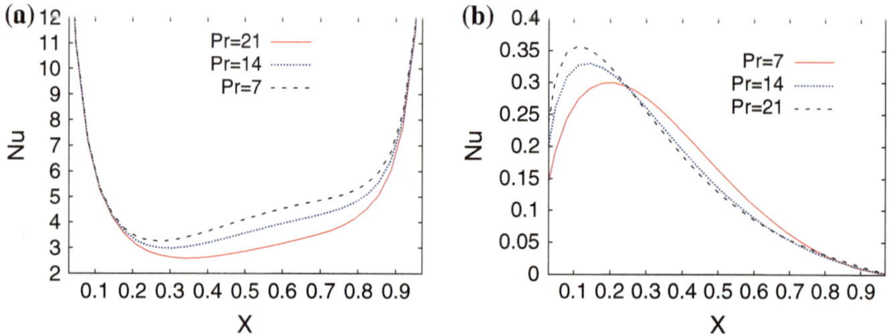

Fig. 7 The average heat transfer rate versus x-axis, where $H = 1$, $\alpha = 3$, and Pr is chosen to be, 7, 14, 21, **a** along lower boundary and **b** along upper boundary

application of the certain magnetic field strength. In Fig. 9a, b the rate of heat transfer for various values of Womersley numbers are presented for fixed values of Pr and H. The heat transfer rate is increasing with the increase of Womersley number at

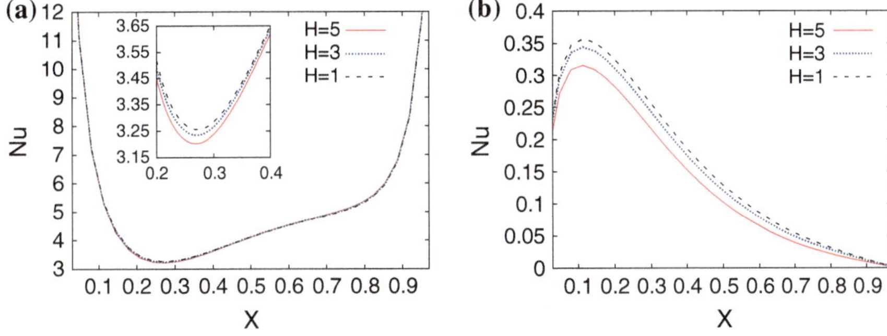

Fig. 8 The average heat transfer rate versus x-axis, for different values of H, where Pr is chosen to be =21 and α is 3, **a** along lower boundary and **b** along upper boundary

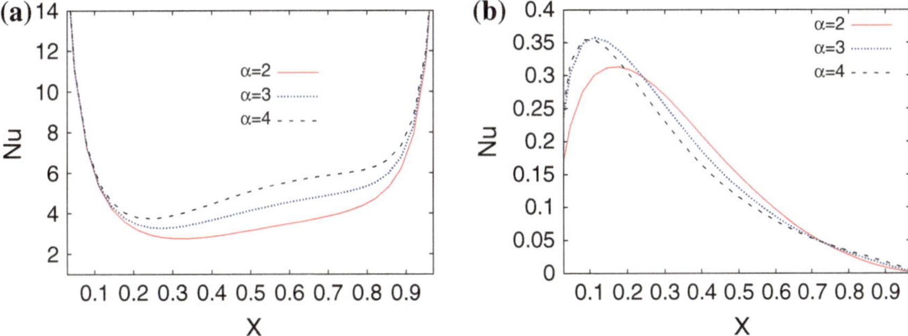

Fig. 9 The heat transfer rate along the arterial wall for different values of α, **a** along the lower boundary and **b** along the upper boundary, when $Pr = 21$ and $H = 1$

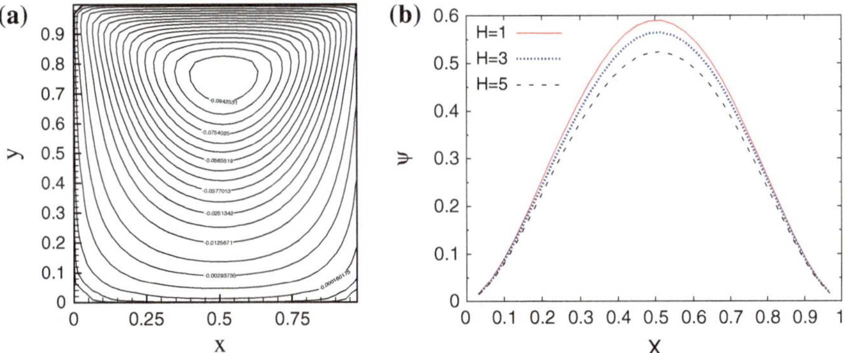

Fig. 10 **a** Contour plots, **b** surface plot of stream function along the cross section of the artery for $H = 1$, when $Pr = 21$ and $\alpha = 3$

the lower arterial wall, but the upper wall variation is large at the initial position but at the outflow region the variation is almost constant. In Fig. 10 we have presented the stream function variation for constant parametric values of Hartmann number, Womersley number, and Prandtl number. From Fig. 10 a we can observe that the stream function is symmetric about the midpoint of the channel where the maximum value of the function is attained.

4 Conclusion

A numerical approach for simulation of the blood flow in an artery under the combined effect of periodic acceleration and an external magnetic field has been studied. We have made an attempt to examine the effect of the Prandtl number, the Hartmann number, and the Womersley number on the flow and heat transport characteristic of blood. Some graphical presentations of the computed results have been performed. The study provides the fact that the axial velocity is largely influenced by the magnetic field parameter. This is due to the fact that when the biomagnetic fluid (blood) is subjected to a magnetic field, the action of magnetization introduces an orientation of the blood charged ions with the magnetic field. The Prandtl number and the Womersley number also play a major role on the velocity as well as on the pressure and temperature distribution.

References

1. Berger, S.A., Talbot, L., Yao, L.S.. Flow in curved pipes. Annu. Rev. Fluid Mech. **15**, 461–512 (1983)
2. Fukada, E., Kaibara, M.: Viscoelastic study of aggregation of red blood cells. Biorheology **17**, 17782 (1980)
3. Fung, Y.C., Zweifach, B.W.: Microcirculation, mechanics of blood flow in capillaries. Annu. Rev. Fluid Mech. **3**, 189–210 (1971)
4. Gebhart, B., Pera, L.: The nature of vertical natural convection flows resulting from the combined buoyancy effects of thermal and mass diffusion. Int. J. Heat Mass Transf. **14**, 2025–2050 (1971)
5. Jones, R.T.: Blood flow. Annu. Rev. Fluid Mech. **1**, 223–244 (1969)
6. Leonard, B.P.: A stable and accurate convective modeling procedure based on quadratic upstream interpolation. Comput. Meth. Appl. Mech. Eng. **19**, 59–98 (1979)
7. Lima, R., Imai, Y., Ishikawa, T., Takeda, M., Tanaka, S., Tsubota, K., Wada, S., Yamaguchi, T.: In vitro blood flow in a rectangular PDMS microchannel experimental observations using a confocal micro-PIV system. Biomed. Microdevices **10**(2), 153–167 (2008)
8. Misra, J.C., Shit, G.C.: Effect of magnetic field on blood flow through an artery: A numerical model. Comput. Technol. **12**, 3–16 (2007)
9. Misra, J.C., Chandra, S., Kundu, P.K., Shit, G.C.: Hydromagnetic flow and heat transfer of a second-grade viscoelastic fluid in a channel with oscillatory stretching walls. J. Eng. Math. **69**, 91–100 (2011)

10. Ogulu, A., Amos, E.: Modeling pulsatile blood flow within a homogeneous porous bed in the presence of a uniform magnetic field and time-dependent suction. Int. Commun. Heat Mass Transf. **34**, 989–995 (2007)
11. Shyy, W., Thakur, S., Wright, J.: Second-order upwind and central difference schemes for recirculating computation. AIAA J. **30**, 923–932 (1992)
12. Stoltz, J.F., Lucius, M.: Viscoelasticity and thixotropy of human blood. Biorheology. 18, 45373 (1981)
13. Tashtoush, B., Magableh, A.: Magnetic field effect on heat transfer and fluid flow characteristics of blood flow in multi-stenosis arteries. Heat Mass Transf. **44**, 297–304 (2008)
14. Taylor, C.A., Draney, M.T.: Experimental and computational methods in cardiovascular fluid mechanics. Annu. Rev. Fluid Mech. **36**, 197–231 (2004)
15. Thurston, G.B.: Viscoelasticity of human blood. Biophys. J. **12**, 120517 (1972)
16. Tzirtzilakis, E.E.: A mathematical model for blood flow in magnetic field. Phys. Fluids. **17**, 077103 (2005)

Improving R-Order Convergence of Derivative Free with Memory Method by Two Self-accelerator Parameters

Anuradha Singh and J.P. Jaiswal

Abstract The object of the present paper is to improve the R-order convergence of with memory method proposed by Eftekhari (Int J Differ Eqn 2014:6, 2014) [1]. To achieve this goal, one more iterative parameter is introduced, which is calculated with the help of Newton's interpolatory polynomial of degree five. It is shown that the R-order convergence of the proposed method is increased from 11.2915 to 13.4031 without any extra evaluation. Smooth as well as nonsmooth examples are presented to confirm theoretical result and superiority of the new scheme.

Keywords R-order convergence · Self accererating parameter · Computational efficiency

1 Introduction

Finding zeros of a scalar function f has importance among the most significant problems in not only the theory and practical of applied mathematics, but also of many branches of engineering sciences, physics, computer science, finance, to mention only some fields. These problems lead to a rich blend of mathematics, numerical analysis, and computational science. To solve these types of nonlinear equations, iterative methods such as Newton's method and its modification are usually used. During the last few years, multipoint methods have drawn the attention of many researchers. Multipoint iterative methods are defined as methods that require evaluation of functions and its derivatives at a number of values of the independent variable. The main goal and motivation in the construction of new methods is to achieve the

A. Singh · J.P. Jaiswal (✉)
Department of Mathematics, Maulana Azad National Institute of Technology, Bhopal 462051, Madhya Pradesh, India
e-mail: asstprofjpmanit@gmail.com

A. Singh
e-mail: singhanuradha87@gmail.com

P.N. Agrawal et al. (eds.), *Mathematical Analysis and its Applications*, Springer Proceedings in Mathematics & Statistics 143,
DOI 10.1007/978-81-322-2485-3_41

highest computational efficiency; in other words, it is desirable to attain as high as possible convergence order with a fixed number of function evaluations per iteration.

In our paper, the order of convergence and efficiency index have been improved in the neighborhood of a simple root. The efficiency index is defined [2] as $I = p^{\frac{1}{d}}$, where p is the order of convergence and d is the total number of function evaluations per iteration. By Kung and Traub conjecture [3], a multipoint method without memory performing $n+1$ function evaluations per iteration can have at most convergence order 2^n. The iterative methods that agree with this conjecture are known as so-called optimal methods. In [4], Petkovic et al. have presented a large collection of without memory multipoint methods for solving nonlinear equations. In the recent past, researchers have focused to improve the existing methods without additional evaluation of function and derivative in such a way it agrees with Kung and Traub conjecture. As a result, the methods give higher computational efficiency.

The main motive in constructing iterative algorithms for solving nonlinear equations is to achieve as high as possible convergence rate with a fixed number of function evaluations per iteration. To find this aim, the idea of with memory method, which use the information from current and previous iterations, is first initiated by Traub in 1964. The order of convergence of new multipoint methods with memory is greater than the order of convergence of the corresponding optimal multipoint methods without memory. Accelerated convergence is obtained by variation of self-accelerating parameters, which are recursively calculated as the iteration proceed using information from the current and previous iterations. The improved convergence rate, attained without extra evaluations, is a significant advantage of multipoint methods with memory. In this paper, we have presented modified and efficient version of existing with memory method.

Rest of the paper is organized as follows: Sect. 2 is devoted to the development and theoretical proof of the improved three-points with memory method. It is shown that the proposed scheme has greater convergence order without extra evaluation. As a result, it shows high computational efficiency. Finally, one smooth and nonsmooth examples are presented to justify the significance of the present work.

2 New Derivative-Free Iterative Method with Memory

In the convergence analysis of the new method, we employ the notation used in Traub's book [5]: if m_k and n_k are null sequences and $m_k/n_k \to C$, where C is a nonzero constant, we shall write $m_k = O(n_k)$ or $m_k \sim Cn_k$. We also use the concept of R-order of convergence introduced by Ortega and Rheinboldt [6]. Let x_k be a sequence of approximations generated by an iterative method (IM). If this sequence converges to a zero ξ of function f with the R-order $O_R((IM), \xi) \geq r$, we will write

$$e_{k+1} \sim D_{k,r} e_k^r,$$

where $D_{k,r}$ tends to the asymptotic error constant D_r of the iterative method (IM) when $k \to \infty$.

In [7] Thukral constructed the following derivative-free without memory method

$$\begin{aligned} w_k &= x_k + \beta f(x_k), \ k = 0, 1, 2, ..., \\ y_k &= x_k - \left(\frac{f(x_k)}{f[x_k, w_k]}\right), \\ z_k &= y_k - \left(\frac{f[x_k, w_k]}{f[w_k, y_k]}\right)\left(\frac{f(y_k)}{f[x_k, y_k]}\right), \\ x_{k+1} &= z_k - \left(1 - \frac{f(z_k)}{f(w_k)}\right)^{-1}\left(1 - \frac{f(y_k)^3}{f(w_k)^2 f(x_k)}\right)\left(\frac{f[x_k, y_k] f(z_k)}{f[y_k, z_k] f[x_k, z_k]}\right), \end{aligned} \tag{1}$$

where $\beta \in R^+$ and $f[.,.]$ denotes the usual divided difference. The author showed that it has optimal eighth-order of convergence. Very recently, Eftekhari [1] first replaced parameter β in the above method by iterative parameter β_k. This iterative parameter is also present in the coefficient of first term of the error expression. To achieve higher order convergence without extra evaluation, the author approximated the iterative parameter by Newton interpolatory polynomial of third and fourth degree, respectively. In fact the maximum R-order convergence of his proposed method is 11.2915. In this work, we discuss the modified version of the same with memory, which has more higher R-order of convergence as well as high computational efficiency. For this purpose, we introduce one more parameter and thus our proposed method is given by

$$\begin{aligned} w_k &= x_k + \beta f(x_k), \ k = 0, 1, 2, ..., \\ y_k &= x_k - \left(\frac{f(x_k)}{f[x_k, w_k] + \alpha f(w_k)}\right), \\ z_k &= y_k - \left(\frac{f[x_k, w_k]}{f[w_k, y_k]}\right)\left(\frac{f(y_k)}{f[x_k, y_k] + \alpha f(y_k)}\right), \\ x_{k+1} &= z_k - \left(1 - \frac{f(z_k)}{f(w_k)}\right)^{-1}\left(1 - \frac{f(y_k)^3}{f(w_k)^2 f(x_k)}\right)\left(\frac{f[x_k, y_k] f(z_k)}{f[y_k, z_k] f[x_k, z_k] + \alpha f(z_k)}\right), \end{aligned} \tag{2}$$

where $\alpha, \beta \in R^+$. The error expression of the above method is

$$e_{k+1} = M_{8,1}(1 + \beta c_1)^3(\alpha c_1 + c_2)^2 e_k^8 + O(e_k^9). \tag{3}$$

where $M_{8,1}$ is asymptotic constant, $c_i = \frac{f^{(i)}(\xi)}{i!}$ and ξ is the exact root. Since the above error equation contains both the parameters, which can be approximated in such a way that they increase the local convergence order. For this purpose, first we replace the parameters α and β by iterative parameters α_k and β_k, respectively and then approximation of these parameters are given by the following way

$$\beta_k = -\frac{1}{c_1} \approx -\frac{1}{\tilde{c_1}} = -\frac{1}{\tilde{N}_4'(x_k)},$$

$$\alpha_k = -\frac{c_2}{c_1} \approx -\frac{\widetilde{c_2}}{\widetilde{c_1}} = -\frac{\tilde{N}_5''(w_k)}{2\tilde{N}_5'(w_k)}, \tag{4}$$

where $\tilde{N}_4(t) = \tilde{N}_4(t; x_k, z_{k-1}, y_{k-1}, x_{k-1}, w_{k-1})$,
$\tilde{N}_5(t) = \tilde{N}_5(t; x_k, w_k, z_{k-1}, y_{k-1}, x_{k-1}, w_{k-1})$ are Newton's interpolatory polynomial of degree four and five, respectively. Before going to prove the main result, we state the following lemma which can be obtained using the error of Newton's interpolation, in the same manner as in [8]:

Lemma 1 *If* $\beta_k = -\frac{1}{\tilde{N}_4'(x_k)}$ *and* $\alpha_k = -\frac{\tilde{N}_5''(w_k)}{2\tilde{N}_5'(w_k)}$, *then the estimates*

$$(i)\ 1 + \beta_k c_1 \sim -\frac{c_5}{c_1} e_{k-1,z} e_{k-1,y} e_{k-1,w} e_{k-1},$$
$$(ii)\ \alpha_k c_1 + c_2 \sim c_6 e_{k-1,z} e_{k-1,y} e_{k-1,w} e_{k-1}.$$

The theoretical proof of the order of convergence of the proposed method is given by the following theorem:

Theorem 1 *If an initial approximation,* x_0 *is sufficiently close to a simple zero* ξ *of* $f(x) = 0$ *and the parameters* β_k *and* α_k *in the iterative scheme (2) is recursively calculated by the forms given in (4). Then the R-order of convergence of with memory scheme (2) with (4) is at least* $7 + \sqrt{41} = 13.4031$.

Proof First we assume that the R-orders of convergence of sequences x_k, w_k, y_k, z_k are at least r, l, m and n, respectively. Hence

$$e_{k+1} \sim D_{k,r} e_k^r \sim D_{k,r}(D_{k-1,r} e_{k-1}^r)^r \sim D_{k,r} D_{k-1,r}^r e_{k-1}^{r^2}. \tag{5}$$

and

$$e_{k,w} \sim D_{k,l} e_k^l \sim D_{k,l}(D_{k-1,r} e_{k-1}^r)^l \sim D_{k,l} D_{k-1,r}^l e_{k-1}^{rl}. \tag{6}$$

Similarly

$$e_{k,y} \sim D_{k,m} D_{k-1,r}^m e_{k-1}^{rm}, \tag{7}$$

$$e_{k,z} \sim D_{k,n} D_{k-1,r}^n e_{k-1}^{rn}. \tag{8}$$

By virtue of the above equations and Lemma 1, we have

$$1 + \beta_k c_1 \sim \left(-\frac{c_5}{c_1}\right)(D_{k-1,n})(D_{k-1,m})(D_{k-1,l}) e_{k-1}^{n+m+l+1}, \tag{9}$$

and

$$\alpha_k c_1 + c_2 \sim c_6(D_{k-1,n})(D_{k-1,m})(D_{k-1,l})e_{k-1}^{n+m+l+1}. \tag{10}$$

For the scheme (2), it can be derived that

$$e_{k,w} \sim (1 + \beta_k c_1)e_k, \tag{11}$$

$$e_{k,y} \sim M_{2,1}(1 + \beta_k c_1)(\alpha_k c_1 + c_2)e_k^2, \tag{12}$$

$$e_{k,z} \sim M_{4,1}(1 + \beta_k c_1)^2(\alpha_k c_1 + c_2)e_k^4, \tag{13}$$

and

$$e_{k+1} \sim M_{8,1}(1 + \beta_k c_1)^3(\alpha_k c_1 + c_2)^2 e_k^8, \tag{14}$$

where $M_{2,1}$, $M_{4,1}$ and $M_{8,1}$ are asymptotic constants. Using (9) in the Eq. (11) and then simplifying we obtain

$$e_{k,w} \sim \left(-\frac{c_5}{c_1}\right)(D_{k-1,n})(D_{k-1,m})(D_{k-1,l})(D_{k-1,r})e_{k-1}^{n+m+l+r+1}. \tag{15}$$

Similarly by virtue of (9) and (10), the Eqs. (12), (13) and (14), respectively, become

$$e_{k,y} \sim M_{2,1}\left(\frac{-c_5c_6}{c_1}\right)(D_{k-1,n}^2)(D_{k-1,m}^2)(D_{k-1,l}^2)(D_{k-1,r}^2)e_{k-1}^{2(n+l+m+1)+2r}, \tag{16}$$

$$e_{k,z} \sim M_{4,1}\left(\frac{c_5^2c_6}{c_1^2}\right)(D_{k-1,n}^3)(D_{k-1,m}^3)(D_{k-1,l}^3)(D_{k-1,r}^4)e_{k-1}^{3(n+l+m+1)+4r}, \tag{17}$$

and

$$e_{k+1} \sim M_{8,1}\left(\frac{-c_5^3c_6^2}{c_1^3}\right).(D_{k-1,n}^5)(D_{k-1,m}^5)(D_{k-1,l}^5)(D_{k-1,r}^8)e_{k-1}^{5(n+l+m+1)+8r}. \tag{18}$$

Now comparing the equal powers of e_{k-1} in Eqs. (6)–(15), (7)–(16), (8)–(17) and (5)–(18), we find the following system of nonlinear equations:

$$\begin{aligned} rl - r - (n + l + m + 1) &= 0, \\ rm - 2r - 2(n + l + m + 1) &= 0, \\ rn - 4r - 3(n + l + m + 1) &= 0, \\ r^2 - 8r - 5(n + l + m + 1) &= 0. \end{aligned}$$

Solving these equations, we get $l = \frac{1}{5}\left(4 + \sqrt{41}\right), m = \frac{2}{5}\left(4 + \sqrt{41}\right), n = \frac{1}{5}\left(17 + 3\sqrt{41}\right), r = 7 + \sqrt{41}$. And thus we proved the result.

Note 1: The efficiency index of the proposed method (2) with (4) is $(13.4031)^{1/4} = 1.9134$ which is more than $(11.2915)^{1/4} = 1.8831$ of method proposed by Eftekhari [1].

3 Numerical Results and Conclusions

In this section, the new method is applied to solve some nonlinear equations (smooth as well as nonsmooth) and compared with several with memory derivative-free methods. The absolute errors in the first three iterations are given in Tables 1 and 2, where the exact roots are computed with 1,000 significant digits. The computational order of convergence (COC) is defined by

$$COC - \frac{ln(|f(x_k)/f(x_{k-1})|)}{ln(|f(x_{k-1})/f(x_{k-2})|)}.$$

To test the performance of new method consider, the following two nonlinear functions (which are taken from [1, 9]):

$$\begin{aligned} 1. f_1(x) &= 10(x^4 + x), x < 0 \\ &= -10(x^3 + x), x \geq 0. \end{aligned}$$

$$2.\ f_2(x) = e^{x^2 + x\cos(x) - 1} + \sin(\pi x) + x\log(x\sin(x) + 1).$$

The effectiveness of the new scheme with memory method (2) with (4) (NWM) is confirmed by comparing this with some recently established with memory methods. Specifically, we consider the 12th-order method (31) (LWMa), (32) (LWMb), (33) (LWMc), and (34) (LWMd) introduced by Lotfi and Tavakoli in [10], 12th-order method (12)–(8) for ϕ_1 (EWMa), (12)–(8) for ϕ_2 (EWMb), and (12)–(8) for ϕ_3 (EWMc), introduced by Eftekhari in [1] and 14th-order method (22) (LSWMa) and (23) (LSWMb) introduced by Lotfi et al. in [11]. In the present scenario, high-order

Table 1 Numerical results for $f_1(x)$

Method	$\lvert x_1-\xi\rvert$	$\lvert x_2-\xi\rvert$	$\lvert x_3-\xi\rvert$	COC
	$x_0=-0.8$	$\gamma_0=0.01$	$\alpha_0=0.01$	$\xi=-1$
LWMa	0.18233e+1	0.99985e+0	1.00000e+0	2.2130
LWMb	0.86418e−1	0.52042e−9	0.45465e−109	11.824
LWMc	0.12019e−1	0.36933e−6	0.74211e−73	12.343
LWMd	0.36954e+0	0.36954e+0	0.36954e+0	1.0042
EWMa	0.57265e−1	0.10604e−9	0.32534e−106	11.117
EWMb	0.14034e+2	0.47011e+1	0.11965e+1	1.0065
EWMc	0.57265e−1	0.10604e−9	0.32534e−106	11.117
LSWMa	0.41716e−3	0.45375e−45	0.1.4833e−632	14.000
LSWMb	0.32191e−2	0.49928e−31	0.27724e−435	13.996
NWM	0.55140e−1	0.31086e−14	0.40744e−200	14.075

Table 2 Numerical results for $f_2(x)$

Method	$\lvert x_1-\xi\rvert$	$\lvert x_2-\xi\rvert$	$\lvert x_3-\xi\rvert$	COC
	$x_0=-0.8$	$\gamma_0=0.01$	$\alpha_0=0.01$	$\xi=-1$
LWMa	0.12710e−2	0.88275e−32	0.20559e−384	12.094
LWMb	0.32541e−2	0.12598e−29	056465e−356	11.906
LWMc	0.26347e−2	0.45753e−30	0.28758e−361	11.931
LWMd	0.19011e−2	0.1.3528e30	0.33386e−370	12.066
EWMa	0.25117e−2	0.18296e−26	0.93718e−294	11.073
EWMb	0.25118e−2	0.22661e−24	0.81372e−246	10.045
EWMc	0.25117e−2	0.18296e−26	0.93718e−294	11.073
LSWMa	0.28748e−2	0.46101e−36	0.58786e−519	14.290
LSWMb	0.15835e−2	0.86423e−39	0.72387e−532	13.598
NWM	0.24395e−2	0.77037e−33	0.32310e−464	14.143

methods are important because numerical applications use high precision in their computations; for this reason, numerical tests have been carried out using variable precision arithmetic in MATHEMATICA 8 with 1,000 significant digits.

The numerical results showed in Tables 1 and 2 are in concordance with the theory developed in this paper. From the results displayed in Tables 1 and 2, we can conclude that the order of convergence of the derivative-free method without memory can be made more higher by the method with memory by imposing one more parameter without any additional calculations and the computational efficiency of the with memory method is very high. The R-order of convergence is increased from 11.2915 to 13.4031 in accordance with the quality of the applied accelerating method given by (2) with (4). We can see that the self-accelerating parameters play a key role in increasing the order of convergence of the iterative method.

Acknowledgments The authors are grateful to editor and reviewers for their significant suggestions which improved of the quality of the paper.

References

1. Eftekhari, T.: On some iterative methods with memory and high efficiency index for solving nonlinear equations. Int. J. Differ. Eqn. **2014**(Article ID 495357), 6 (2014)
2. Ostrowski, A.M.: Solution of Equations and System of Equations. Prentice-Hall, Englewood Cliffs (1964)
3. Kung, H.T., Traub, J.F.: Optimal order of one point and multipoint iteration. J. Assoc. Comput. Math. **21**, 643–651 (1974)
4. Petkovic, M.S., Neta, B., Petkovic, L.D., Dzunic, J.: Multipoint Methods for Solving Nonlinear Equation. Academic Press, Elsevier (2012)
5. Traub, J.F.: Iterative Methods for Solution of Equations. Prentice Hall, New York (1964)
6. Ortega, J.M., Rheinboldt, W.C.: Iterative Solution of Nonlinear Equations in Several Variables. Academic Press, New York (1970)
7. Thukral, R.: A family of three-point derivative-free methods of eighth-order for solving nonlinear equations, J. Mod. Meth. Numer. Math. **3**(2), 11–21 (2012)
8. Dzunic, J.: On efficient two-parameter methods for solving nonlinear equations. Numer. Algorithm **63**, 549–569 (2013)
9. Hafiz, M.A., Bahgat, M.S.M.: Solving non-smooth equations using family of derivative-free optimal methods. J. Egypt. Math. Soc. **21**, 38–43 (2013)
10. Lotfi, T., Tavakoli, E.: On a new efficient Steffenssen-like iterative class by applying a suitable self-accelerator parameter. Sci. World J. **2014**(Article ID 769758), 9 (2014)
11. Lotfi, T., Soleymani, F., Shateyi, S., Assari, P., Haghani, F.K.: New mono- and bi-accelerator iterative methods with memory for nonlinear equations. Abstr. Appl. Anal. **2014**(Article ID 705674), 8 (2014)

Numerical Solutions of Differential Equations Using Modified B-spline Differential Quadrature Method

R.C. Mittal and Sumita Dahiya

Abstract In this article, a modified cubic B-spline differential quadrature method (MCB-DQM) is proposed to solve some of the basic differential equations. Here we have considered an ordinary differential equation of order two along with heat equation and one- and two-dimensional wave equations. A nonlinear ordinary differential equation of order two is also considered. The ordinary differential equation is reduced to a system of nonhomogeneous linear equations which is then solved by using the Gauss elimination method, whereas the heat equation and the one-dimensional and two-dimensional heat and wave equations are reduced to a system of ordinary differential equations. The system is then solved by the optimal four-stage three-order strong stability preserving time stepping Runge–Kutta (SSP-RK43) scheme. The reliability and efficiency of the method have been tested on six examples.

Keywords Ordinary differential equation · Heat equation · Wave equation cubic B-spline functions · Modified cubic B-spline quadrature method · System of ordinary differential equations · Gauss elimination method · Runge–Kutta fourth-order method

1 Introduction

To describe change, the most accurate way is to use differentials and derivatives, that is why differential equation arises in many different contexts. In this article, we have taken a second-order ordinary differential equation with boundary conditions of the form

$$\frac{d^2u}{dx^2} + A\frac{du}{dx} + Bu = f(x), \quad x \in [a, b] \tag{1}$$

R.C. Mittal · S. Dahiya (✉)
Department of Mathematics, IIT Roorkee, Roorkee 247667, Uttarakhand, India
e-mail: sumita.iitr@gmail.com

R.C. Mittal
e-mail: rcmmmfma@iitr.ernet.in

P.N. Agrawal et al. (eds.), *Mathematical Analysis and its Applications*,
Springer Proceedings in Mathematics & Statistics 143,
DOI 10.1007/978-81-322-2485-3_42

with the boundary conditions given as $u(a) = g_1(x) \quad and \quad u(b) = g_2(x)$ where $g_1(x)$, $g_2(x)$, A, and B are constants or functions of x.
Second, we have taken the wave equation. The wave is an important second-order partial differential equation which describes the nature of waves such as light waves, water waves, and sound waves. It arises in the fluid dynamics and electromagnetic fields. The problem of a vibrating string was earlier studied by Jean le Rond [1], d'Alembert [2], Leonhard Euler, Daniel Bernoulli, and Joseph-Louis Lagrange. The one-dimensional wave equation was discovered by d'Alembert in 1746, and then within 10 years Euler discovered the three-dimensional wave equation. Here we will restrict our results to one- and two-dimensional wave equations. The wave equation is a hyperbolic partial differential equation. It is time-dependent and concerns a time variable t, spatial variables $x_1, x_2, \ldots, x_n$, and a scalar function $u = u(x_1, x_2, \ldots, x_n; t)$ which can model the displacement of a wave. The wave equation is then given by

$$\frac{\partial^2 u}{\partial t^2} = c^2 \nabla^2 u \tag{2}$$

where ∇^2 is the spatial Laplacian and where c is a fixed constant. The equation alone does not specify a solution; a unique solution is obtained by setting a problem with initial conditions or boundary conditions. The one-dimensional wave equation is given by

$$\frac{\partial^2 u}{\partial t^2} = c^2 \frac{\partial^2 u}{\partial x^2} \tag{3}$$

and the two-dimensional wave equation is given by

$$\frac{\partial^2 u}{\partial t^2} = c^2 \left(\frac{\partial^2 u}{\partial x^2} + \frac{\partial^2 u}{\partial y^2} \right) \tag{4}$$

Third, we have considered the heat equation. The heat equation is a parabolic equation that describes the variation in temperature or distribution of heat in a given region over time. Generally in a coordinate system, heat equation is given by

$$\frac{\partial u}{\partial t} - \alpha \nabla^2 u = 0 \tag{5}$$

where α is a positive constant, and ∇ is the Laplacian operator. The heat equation is of fundamental importance in diverse scientific fields. In mathematics, it is the prototypical parabolic partial differential equation. In probability theory, the heat equation is connected with the study of Brownian motion via the Fokker–Planck equation. In financial mathematics, it is used to solve the Black–Scholes partial differential equation. The diffusion equation, a more general version of the heat equation, arises in connection with the study of chemical diffusion and other related processes. The heat equation is derived from the Fourier's Law and conservation of energy.

2 Description of Modified Cubic B-Spline Differential Quadrature Method (MCB-DQM)

Differential quadrature method [3] is a numerical technique to solve ordinary and partial differential equations. With this method, the spatial derivatives of unknown functions are approximated at any grid point using weighted sum of all the functional values at certain point in the whole computational domain. In two-dimensional DQM, first, we discretize the domain $D = \{(x, y) : a \leq x \leq b; c \leq y \leq d\}$ as $D^1 = \{(x_i, y_j), i = 1, 2, \ldots .N; j = 1, 2, \ldots, M\}$ by taking step length $\Delta x = x_i - x_{i+1}$ in x-axis direction and $\Delta y = y_j - y_{j-1}$ in y-axis direction. According to DQM, the approximation of the first-order partial derivative with respect to x of the dependent function $u(x, y, t)$, keeping y_j fixed, at point x_i is given as follows:

$$u_x(x_i, y_j, t) = \sum_{k=1}^{N} a_{ik}^{(1)} u(x_k, y_j, t), \qquad i = 1, 2, ..., N \tag{6}$$

Similarly, the approximation of the first-order partial derivative of the dependent function $u(x, y, t)$ with respect to y, keeping the point x_i fixed, at point y_j is given as follows:

$$u_y(x_i, y_j, t) = \sum_{k=1}^{N} \bar{a}_{jk}^{(1)} u(x_i, y_k, t), \qquad j = 1, 2, \ldots, M \tag{7}$$

where $a_{ij}^{(1)}$ and $\bar{a}_{jk}^{(1)}$ are unknown, representing the weighting coefficients of the first-order partial derivatives with respect to x and y. There are many approaches to calculate these weighting coefficients such as Shu's approach [4], Quan and Chang's approach [5, 6], and Bellman's approach [3]. In recent years, most of the differential quadrature method using various test functions such as Lagrange interpolation polynomial, Legendre polynomials, Lagrange interpolation cosine functions, spline functions, etc. are based on Shu's approach. Nowadays, most frequently used quadrature methods are based on sine–cosine expansion and Lagrange interpolation. Korkmaz and Dag [7, 8] have used cosine expansion-based differential quadrature method and sine differential quadrature method for many nonlinear partial differential equations. Mittal et al. [9–12] proposed polynomial-based differential quadrature method for numerical solutions of nonlinear partial differential equations. Here, in this article an approach based on modified cubic B-spline functions has been proposed to find the weighting coefficients of differential quadrature method. Computed results show that reported results are accurate.

2.1 Modified Cubic B-Spline Functions

In this method, a modification of cubic B-spline functions is used to find the weighting coefficients $a_{ij}^{(1)}$ and $\bar{a}_{jk}^{(1)}$. The cubic B-spline functions are defined as follows:

$$\varphi_m(x) = \frac{1}{h^3}\begin{cases} (x - x_{m-2})^3 & x \in [x_{m-2}, x_{m-1}) \\ (x - x_{m-2})^3 - 4(x - x_{m-1})^3 & x \in [x_{m-1}, x_m) \\ (x_{m+2} - x)^3 - 4(x_{m+1} - x)^3 & x \in [x_m, x_{m+1}) \\ (x_{m+2} - x)^3 & x \in [x_{m+1}, x_{m+2}) \\ 0 & \text{otherwise} \end{cases} \qquad m = 0, 1, \ldots, N+1 \tag{8}$$

where $\{\varphi_0(x), \varphi_1(x), \ldots, \varphi_N(x)\}$ forms a basis over the domain interval [a, b]. The values of cubic B-splines and their derivatives at the nodal points are given in Table 1. The modification in cubic B-spline functions is done in such a way, so that the resulting matrix becomes diagonally dominant. Modified cubic B-spline basis functions at the knots are defined as follows [13]:

$$\begin{aligned} \phi_1(x) &= \varphi_1(x) + 2\varphi_0(x), \\ \phi_2(x) &= \varphi_2(x) - \varphi_0(x), \\ \phi_l(x) &= \varphi_l(x), l = 3, 4, \ldots, N-2, \\ \phi_{N-1} &= \varphi_{N-1}(x) - \varphi_{N+1}(x) \\ \phi_N(x) &= \varphi_N(x) + 2\varphi_{N+1}(x) \end{aligned} \tag{9}$$

The function $\phi_l(x), l = 1, 2, \ldots, N$ again form a basis over the interval [a, b].

2.2 To Compute Weighting Coefficients

Keeping y-$axis$ fixed in Eq. (6), we find the weighting coefficients $a_{ik}^{(1)}$. Putting the functions $\phi_m(x),\ m = 1, 2, \ldots N$ in Eq. (6), we get

$$\phi_l^{'}(x_i, y_j) = \sum_{k=1}^{N} a_{ik}^{(1)}\phi_l(x_k, y_j), \quad j = 1, 2, \ldots, M \tag{10}$$

Table 1 Values of $\varphi_m(x)$ and its derivatives at the nodal points

	x_{m-2}	x_{m-1}	x_m	x_{m+1}	x_{m+2}
$\varphi_m(x)$	0	1	4	1	0
$\varphi_m^{'}(x)$	0	$3/h$	0	$-3/h$	0
$\varphi_m^{''}(x)$	0	$6/h^2$	$-12/h^2$	$6/h^2$	0

For any arbitrary choice of l, we get the following algebraic system of equations

$$\begin{bmatrix} \phi_{1,1} & \phi_{1,2} & & & & & \\ \phi_{2,1} & \phi_{2,2} & \phi_{2,3} & & & & \\ & \phi_{3,1} & \phi_{3,2} & \phi_{3,3} & & & \\ & \cdots & \cdots & \cdots & & & \\ & \cdots & & & \phi_{N-1,N-2} & \phi_{N-1,N-1} & \phi_{N-1,N} \\ & & & & \cdots & \phi_{N,N-1} & \phi_{N,N} \end{bmatrix} \begin{bmatrix} a_{i1}^{(1)} \\ a_{i2}^{(1)} \\ \vdots \\ \vdots \\ a_{iN-1}^{(1)} \\ a_{iN}^{(1)} \end{bmatrix} = \begin{bmatrix} \phi'_{1,i} \\ \phi'_{2,i} \\ \vdots \\ \vdots \\ \phi'_{N-1,i} \\ \phi'_{N,i} \end{bmatrix} \tag{11}$$

Here the Eq. (11) gives the systems of tridiagonal algebraic system of equations for each i. These tridiagonal system of equations can be easily solved by "Thomas Algorithm" providing the weighting coefficients of first-order derivatives $a_{ik}^{(1)}$. For $i = 1$, we have the following tridiagonal system of equations:

$$\begin{bmatrix} 6 & 1 & & & & \\ 0 & 4 & 1 & & & \\ & 1 & 4 & 1 & & \\ & \cdots & \cdots & \cdots & & \\ & \cdots & & & 1 & 4\,0 \\ & & & & \cdots & 1\,6 \end{bmatrix} \begin{bmatrix} a_{11}^{(1)} \\ a_{12}^{(1)} \\ \vdots \\ \vdots \\ a_{1N-1}^{(1)} \\ a_{1N}^{(1)} \end{bmatrix} = \begin{bmatrix} \frac{-6}{h} \\ \frac{6}{h} \\ \vdots \\ \vdots \\ 0 \\ 0 \end{bmatrix} \tag{12}$$

The solution of Eq. (12) by "Thomas algorithm" gives the weighting coefficients $a_{11}^{(1)}, a_{12}^{(1)}, \ldots, a_{1N}^{(1)}$. In the same way, we can also find out the weighting coefficients for $i = 2, 3, \ldots, N$. Using these coefficients we are able to find out the first-order partial derivatives. The higher order derivatives can be calculated by the following recurrence relation:

$$a_{ij}^{(r)} = r\left[a_{ij}^{(1)} a_{ii}^{(r-1)} - \frac{a_{ij}^{(r-1)}}{x_i - x_j}\right], \text{ for } i \neq j$$
$$i, j = 1, 2, \ldots, N; \qquad r = 2, 3, \ldots, N-1 \tag{13}$$

$$a_{ii}^{(r)} = -\sum_{j=1, j\neq i}^{N} a_{ij}^{(r)}, \qquad \text{for } i = j \tag{14}$$

Here $a_{ij}^{(r-1)}$ and $a_{ij}^{(r)}$ are the weighting coefficients of $(r-1)$th- and (r)th-order partial derivatives w.r.t. x.

We can find out the weighting coefficients $\bar{a}_{jk}^{(1)}$ of first-order partial derivatives with respect to y (keeping x-$axis$ fixed) in the same manner by putting modified cubic B-spline functions in Eq. (7). The second and higher order derivatives can be calculated in the same way by the recurrence formulae

$$\bar{a}_{ij}^{(r)} = r\left[\bar{a}_{ij}^{(1)}\bar{a}_{ii}^{(r-1)} - \frac{\bar{a}_{ij}^{(r-1)}}{y_i - y_j}\right], \text{ for } i \neq j$$
$$i, j = 1, 2, \ldots., N; \quad r = 2, 3, \ldots., N-1 \tag{15}$$

$$\bar{a}_{ii}^{(r)} = -\sum_{j=1, j\neq i}^{N} \bar{a}_{ij}^{(r)}, \quad \text{for } i = j \tag{16}$$

Here $\bar{a}_{ij}^{(r-1)}$ and $\bar{a}_{ij}^{(r)}$ are the weighting coefficients of $(r-1)$th- and (r)th-order partial derivatives w.r.t. y. In one-dimensional DQM, one can assume uniformly distributed N knots: $a = x_1 \leq x_2, \ldots, x_{N-1} \leq x_N = b$ such that $h = x_i - x_{i-1}$ on the real axis. By using modified cubic B-spline functions, the first-order derivative approximation is given as follows:

$$\phi_l^{'}(x_i) = \sum_{j=1}^{N} a_{ij}^{(1)}\phi_l(x_j), \quad \text{for } i = 1, 2, \ldots, N;\ k = 1, 2, \ldots, N \tag{17}$$

which provide the weighted coefficients a_{ij} for $i = 1, 2, \ldots, N$; $j = 1, 2, \ldots, N$. The second derivatives can be evaluated in the same way by the recurrence formulas

$$a_{ij}^{(2)} = 2a_{ij}^{(1)}\left[a_{ii}^{(1)} - \frac{1}{x_i - x_j}\right], \quad \text{for } i \neq j, \text{ and } a_{ii}^{(2)} = -\sum_{i=1, i\neq j}^{N} a_{ij}^{(1)} \tag{18}$$

3 Numerical Scheme Based on Modified Cubic B-Spline Differential Quadrature Method

Discretizing the spatial derivatives by applying the modified cubic B-spline differential quadrature method, Eq. (1), is reduced to the following system of linear equation

$$\sum_{k=1}^{N} a_{ik}^{(2)} u(x_k, t) + A\sum_{k=1}^{N} a_{ik}^{(1)} u(x_k, t) + Bu(x_k, t) = f(x_k), \tag{19}$$

This is a system of N equations in $N-2$ unknowns. After using the boundary conditions and doing certain manipulations we will solve a system of $N-2$ equations in $N-2$ unknowns. Gauss Elimination method will be used for this purpose.
For the one-dimensional wave equation, introducing an auxiliary function w, Eq. (3) reduces to a system of partial difference equations given as follows:

$$\begin{aligned}\frac{\partial w}{\partial t} &= \frac{\partial^2 u}{\partial x^2},\\ w &= \frac{\partial u}{\partial t}, \quad x \in [a,b],\ t \in (0,T]\end{aligned} \tag{20}$$

The initial value and boundary value conditions on w are given as

$$w(x,0) = w_0(x), \qquad x \in [a,b], \tag{21}$$

$$w(x,t) = g(x,t), \qquad x \in [a,b],\ t \in (0,T] \tag{22}$$

Discretizing the spatial derivatives by applying the modified cubic B-spline differential method, Eq. (3) reduces to

$$\frac{dw_{i,j}}{dt} = \sum_{k=1}^{N} a_{ik}^{(2)} u(x_k,t),$$

$$\frac{du_{ij}}{dt} = w(x_k,t) \tag{23}$$

with initial conditions

$$u(x_i,0) = u_0(x_i),\ w(x_i,0) = w_0(x_i) \tag{24}$$

and boundary conditions

$$u(x_i,t) = f(x_i),\ w(x_i,t) = g(x_i) \tag{25}$$

In the same way, an auxiliary function is introduced in the two-dimensional wave equation, and the Eq. (4) is reduced to

$$\begin{aligned}\frac{\partial w}{\partial t} &= \frac{\partial^2 u}{\partial x^2} + \frac{\partial^2 u}{\partial y^2},\\ w &= \frac{\partial u}{\partial t}, \quad (x,y) \in [a,b],\ t \in (0,T]\end{aligned} \tag{26}$$

The initial value and boundary value conditions on w are given as

$$w(x, y, 0) = w_0(x, y), \qquad (x, y) \in [a, b], \tag{27}$$

$$w(x, y, t) = g(x, y, t), \qquad (x, y) \in [a, b],\ t \in (0, T] \tag{28}$$

Discretizing the spatial derivatives by applying the modified cubic B-spline differential method, Eq. (3) reduces to

$$\frac{dw_{i,j}}{dt} = \sum_{k=1}^{N} a_{ik}^{(2)} u(x_k, y_j, t) + \sum_{k=1}^{N} \bar{a}_{jk}^{(2)} u(x_i, y_k, t),$$

$$\frac{du_{ij}}{dt} = w(x_i, y_j, t) \tag{29}$$

with initial conditions

$$u(x_i, y_j, 0) = u_0(x_i, y_j),\ w(x_i, y_j, 0) = w_0(x_i, y_j) \tag{30}$$

and boundary conditions

$$u(x_i, y_j, t) = f(x_i, y_j),\ w(x_i, y_j, t) = g(x_i, y_j) \tag{31}$$

For the heat equation given by Eq. (5), the same procedure is applied.
There are various methods to solve this system of ordinary differential equations. We used the optimal four-stage, order three strong stability preserving time stepping Runge–Kutta [SSP-RK43] scheme [14] to solve the system of ordinary differential equations. In this scheme the system in Eqs. (20) and (23) along with the initial and boundary conditions is integrated from time t_0 to $t_0 + \Delta t$ through the following operations:

$$\begin{aligned} u^{(1)} &= u^m + \frac{\Delta t}{2} L(u^m) \\ u^{(2)} &= u^{(1)} + \frac{\Delta t}{2} L(u^{(1)}) \\ u^{(3)} &= \frac{2}{3} u^m + \frac{u(2)}{3} + \frac{\Delta t}{6} L(u^{(2)}) \\ u^{(m+1)} &= u^{(3)} + \frac{\Delta t}{2} L(u^{(3)}) \end{aligned} \tag{32}$$

and consequently the solution u at a particular time level is completely known. Stability of the proposed system is guaranteed unconditionally as the eigenvalues of corresponding matrix will be real and negative.

Table 2 Comparison between exact values and calculated values with stepsize $h = \pi/40$

x	$MCB\text{-}DQM$	Exact
0.07	8.89041	8.89048
0.15	8.56967	8.56974
0.24	8.05311	8.05317
0.31	7.36956	7.36062
0.39	6.51566	6.51571
0.47	5.54507	5.54511
0.55	4.47771	4.47773
0.63	3.34387	3.34389
0.71	2.17444	2.17444

4 Numerical Experiments

In this section, the numerical solutions by the proposed method (MCB-DQM) are evaluated for some examples of hyperbolic problem.The computational work is done with the help of DEV C++. Existence of analytical solutions help to measure the accuracy of numerical methods. In the present study, the accuracy and efficiency of this method are measured for various numerical examples. The performance of the MCB-DQM method is measured by the maximum absolute error ϵ_k which is defined as $\epsilon_k = |u_{exact} - u_{mcb-dqm}|$.

Problem 1 Consider the following second-order one-dimensional ordinary differential equation

$$\frac{d^2u}{dx^2} + 4u = 8\sin(2x), \quad x \in (0, \pi/4) \tag{33}$$

with the boundary conditions given as

$$u(0) = 9 \quad \text{and} \quad u(\pi/4) = 1 \tag{34}$$

The exact solution is given by

$$u(x) = 9\cos(2x) + \sin(2x) - 2x\cos(2x)$$

We have compared the computed values by using differential quadrature method with modified B-splines with the exact results. The result in terms of absolute error is shown in Table 2.

Table 3 The absolute error at intermediate knots for time $t = 0.1, 0.5$ and 1

x	$t = 0.1$	$t = 0.5$	$t = 1$
0.1	6.14E-04	6.59E-04	4.86E-04
0.2	1.44E-04	4.37E-04	4.25E-04
0.3	4.28E-05	3.38E-04	4.22E-04
0.4	5.19E-05	2.53E-04	4.04E-04
0.5	2.90E-05	2.25E-04	3.99E-04
0.6	5.19E-05	2.52E-04	4.04E-04
0.7	4.28E-05	3.38E-04	4.22E-04
0.8	1.44E-04	4.37E-04	4.25E-04
0.9	6.14E-04	6.59E-04	4.86E-04

Problem 2 Consider the one-dimensional heat equation given by

$$\frac{\partial u}{\partial t} = \frac{1}{\pi^2}\left(\frac{\partial^2 u}{\partial x^2}\right), \quad x \in (0, 1) \tag{35}$$

with initial conditions

$$u(x, 0) = \sin(\pi x)$$

and boundary conditions

$$u(0, t) = 0, \ u(1, t) = 0$$

The exact solution is given by

$$u(x, t) = \sin(\pi x)\exp(-t), \ x \in [0, 1]$$

The calculated values are compared with the exact values for time $t = 0.1$, $t = 0.5$, and $t = 1$. The results are shown in Table 3.

Problem 3 Consider the two-dimensional heat equation given by

$$\frac{\partial u}{\partial t} = \frac{\partial^2 u}{\partial x^2} + \frac{\partial^2 u}{\partial y^2}, \ (x, y) \in (0, 1) \times (0, 1) \tag{36}$$

with initial conditions

$$u(x, y, 0) = \sin(\pi x)\sin(\pi y)$$

and boundary conditions

$$u(0, y, t) = 0, \ u(1, y, t) = 0$$

$$u(x, 0, t) = 0, \ u(x, 1, t) = 0 \tag{37}$$

Table 4 Comparison of exact values and computed values at intermediate points for time t = 0.1 and t = 1

x, y	$t = 0.1$		$t = 1$	
	MCB-DQM	Exact	MCB-DQM	Exact
0.1, 0.1	0.013378	0.013264	2.63842E-010	2.54963E-010
0.2, 0.2	0.048180	0.047992	9.50204E-010	9.22467E-010
0.3, 0.3	0.911731	0.090918	1.79811E-009	1.74755E-009
0.4, 0.4	0.125932	0.125646	2.48362E-009	2.41505E-009
0.5, 0.5	0.139208	0.138911	2.74544E-009	2.67001E-009
0.6, 0.6	0.125932	0.125646	2.48362E-009	2.41505E-009
0.7, 0.7	0.911731	0.090918	1.79811E-009	1.74755E-009
0.8, 0.8	0.048180	0.047992	9.50204E-010	9.22467E-010
0.9, 0.9	0.013378	0.013264	2.63842E-010	2.54963E-010

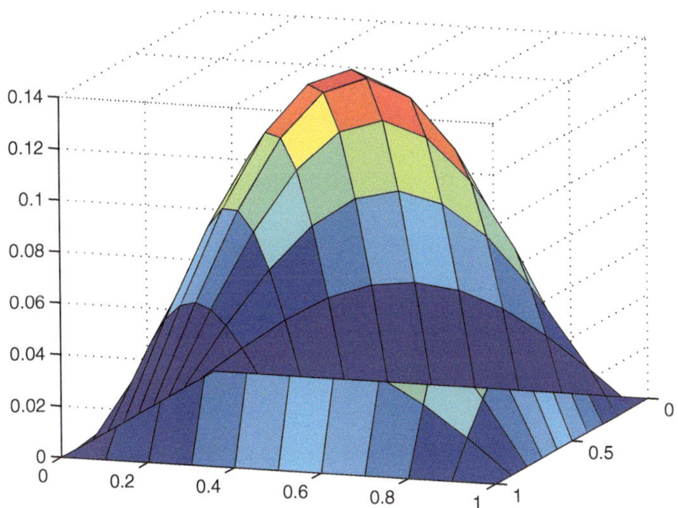

Fig. 1 Plots of numerical solution at time level $t = 0.1$ for Example 4

The exact solution is given by

$$u(x, y, t) = \sin(\pi x)\sin(\pi y)\exp(-2\pi^2 t), \ (x, y) \in [0, 1] \times [0, 1]$$

The calculated values are compared with the exact values for time $t = 0.1$ and $t = 1$. The results are shown in Table 4. The surface plots of numerical solutions at time level $t = 0.1$ is depicted in Fig. 1.

Problem 4 Consider the one-dimensional wave equation

Table 5 The absolute error at intermediate knots for time $t = 0.01, 0.1$ and 1

x	$t = 0.01$	$t = 0.1$	$t = 1$
0.1	2.0178E-013	4.7239E-009	6.8554E-007
0.2	3.4674E-013	3.1219E-011	1.3029E-006
0.3	4.7734E-013	4.2971E-011	1.7932E-006
0.4	5.6104E-013	5.0515E-011	2.1077E-006
0.5	5.8990E-013	5.3116E-011	2.2165E-006
0.6	5.6103E-013	5.0516E-011	2.1077E-006
0.7	4.7722E-013	4.2972E-011	1.7932E-006
0.8	3.4673E-013	2.9860E-011	1.3029E-006
0.9	2.0178E-013	4.7239E-009	6.8554E-007

$$\frac{\partial^2 u}{\partial t^2} = \frac{\partial^2 u}{\partial x^2}, \qquad x \in (0, 1) \tag{38}$$

along with the boundary conditions given by $u(0, t) = 0$ and $u(1, t) = 0$ and the initial conditions given by $u(x, 0) = 0$ and $\frac{\partial u}{\partial t}(x, 0) = \pi \sin(\pi x)$. The exact solution is given by $u(x, t) = \sin(\pi x) \sin(\pi t)$. The computed results are compared with the exact results for different values of time t, taking $\Delta t = 0.001$. The absolute error at intermediate knots for time $t = 0.01, 0.1$, and 1 is shown in Table 5.

Problem 5 Consider the two-dimensional wave equation given by

$$\frac{\partial^2 u}{\partial t^2} = \frac{1}{2}\left(\frac{\partial^2 u}{\partial x^2} + \frac{\partial^2 u}{\partial y^2}\right), \quad (x, y) \in (0, 1) \times (0, 1) \tag{39}$$

with initial conditions

$$u(x, y, 0) = 0$$

and boundary conditions

$$u(0, y, t) = 0, \;\; u(1, y, t) = 0$$

$$u(x, 0, t) = 0, \;\; u(x, 1, t) = 0 \tag{40}$$

The exact solution is given by

$$u(x, y, t) = \sin(\pi x) \sin(\pi y) \sin(\pi t), \;\; (x, y) \in [0, 1] \times [0, 1]$$

The calculated values are compared with the exact values for time $t = 0.1$ and $t = 1$. The results are shown in Table 6. The surface plots of the numerical solutions at time level $t = 1$ is depicted in Fig. 2.

Table 6 Comparison of exact values and computed values at intermediate points for time t = 0.1 and t = 1

x, y	$t = 0.1$		$t = 1$	
	Exact	MCB-DQM	Exact	MCB-DQM
0.1, 0.1	0.029508	0.029532	−0.029508	−0.029546
0.2, 0.2	0.106763	0.106740	−0.106763	−0.106311
0.3, 0.3	0.202254	0.202242	−0.202254	−0.200769
0.4, 0.4	0.279508	0.279484	−0.279508	−0.276865
0.5, 0.5	0.309017	0.308992	−0.309017	−0.305909
0.6, 0.6	0.279508	0.279484	−0.279508	−0.276865
0.7, 0.7	0.202254	0.202242	−0.202254	−0.200769
0.8, 0.8	0.106763	0.106741	−0.106763	−0.106311
0.9, 0.9	0.029508	0.029532	−0.029508	−0.029546

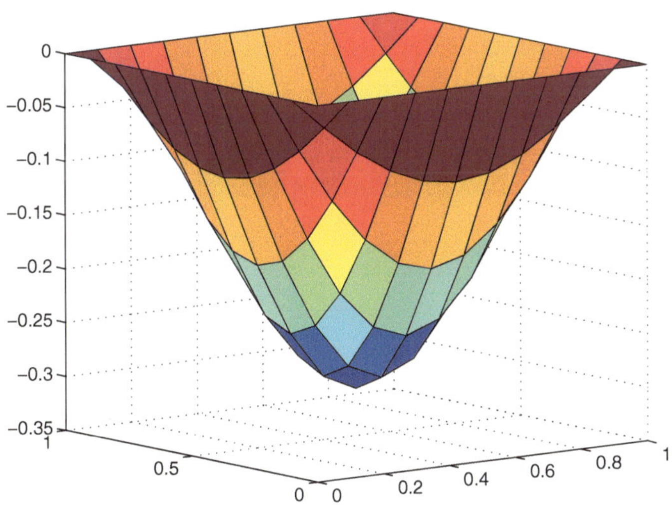

Fig. 2 Plots of numerical solution at time level $t = 1$ for Example 5

Problem 6 Consider the following nonlinear second-order one-dimensional partial differential equation

$$\frac{\partial u}{\partial t} = \frac{\partial^2 u}{\partial x^2} - u\frac{\partial u}{\partial x}, \quad x \in (0, 1) \tag{41}$$

with the boundary conditions given as

$$u(0) = 0 \quad \text{and} \quad u(1) = 0 \tag{42}$$

and initial condition given as

Table 7 Comparison between exact values and calculated values

x	t	*MCB-DQM*	Exact
0.25	0.4	0.317526	0.31752
	0.8	0.199558	0.19956
	1.0	0.165601	0.16560
	3.0	0.027761	0.02775
0.50	0.4	0.584541	0.58454
	0.8	0.367406	0.36740
	1.0	0.298352	0.29834
	3.0	0.041069	0.04106
0.75	0.4	0.645641	0.64562
	0.8	0.385369	0.38534
	1.0	0.295885	0.29586
	3.0	0.030443	0.03044

$$u(x,0) = 4x(1-x); \tag{43}$$

We have compared the computed values by using differential quadrature method with modified B-splines with the exact results. The description of the numerical solutions of this example for different values of t is shown in Table 7.

5 Conclusion

In this paper, we have proposed a modified cubic B-spline differential quadrature method (MCB-DQM) to solve second-order ordinary differential equation, wave equation, and heat equation. The numerical examples show that this scheme can produce highly accurate solutions. The main outcomes are as follows:

1. A technique based on modified cubic B-spline is proposed to find the weighting coefficients rather than using the traditional method of Lagrange interpolation [7].
2. Modifications in cubic B-spline are done in such a way that matrix size and complexity get reduced when applied with differential quadrature method.
3. The method is easy to implement and economical in data complexity, which results in less errors. We require less number of grid points, that means low memory storage, which can be counted as an advantage of (MCB-DQM). This method can be implemented easily to solve two-dimensional nonlinear partial differential equations.

References

1. Cannon, J.T., Dostrovsky, S.: The evolution of dynamics, vibration theory from 1687 to 1742. In: Studies in the History of Mathematics and Physical Sciences, vol. 6, pp. ix + 184. Springer, New York (1981). ISBN:0-3879-0626-6
2. Gerard, F.W.: The vibrating string controversy. Am. J. Phys. **55**(1), 33–37 (1987)
3. Bellman, R., Kashef, B.G., Casti, J.: Differential quadrature: a technique for the rapid solution of nonlinear partial differential equations. J. Comput. Phys **10**, 40–52 (1972)
4. Shu, C.: Differential Quadrature and its Application in Engineering. Athenaeum Press Ltd., Great Britain (2000)
5. Quan, J.R., Chang, C.T.: Newinsights in solving distributed system equations by the quadrature methods- I. Comput. Chem. Eng. **13**, 779–788 (1989)
6. Quan, J.R., Chang, C.T.: Newinsights in solving distributed system equations by the quadrature methods- II. Comput. Chem. Eng. **13**, 1017–1024 (1989)
7. Korkmaz, A., Dag, I.: Shock wave simulations using Sinc differential quadrature method. Eng. Comput. **28**(6), 654–674 (2011)
8. Korkmaz, A., Dag, I.: A differential quadrature algorithm for nonlinear Schrdinger equation. Nonlinear Dyn. **56**(12), 6983 (2009)
9. Mittal, R.C., Jiwari, R.: Numerical study of two-dimensional reactiondiffusion Brusselator system. Appl. Math. Comput. **217**(12), 5404–5415 (2011)
10. Mittal, R.C., Jiwari, R.: A differential quadrature method for solving Burgers-type equation. Int. J. Numer. Methods Heat Fluid Flow **22**(7), 880–895 (2012)
11. Jiwari, R., Pandit, S., Mittal, R.C.: Numerical simulation of two-dimensional sine-Gordon solitons by differential quadrature method. Comput. Phys. Commun. **183**(3), 600–616 (2012)
12. Jiwari, R., Pandit, S., Mittal, R.C.: A differential quadrature algorithm to solve the two dimensional linear hyperbolic telegraph equation with Dirichlet and Neumann boundary conditions. Appl. Math. Comput. **218**(13), 7279–7294 (2012)
13. Mittal, R.C., Jain, R.K.: Numerical solutions of nonlinear Burgers equation with modified cubic Bsplines collocation method. Appl. Math. Comput. **218**, 7839–7855 (2012)
14. Spiteri, J.R., Ruuth, S.J.: A new class of optimal high-order strongstability-preserving time-stepping schemes. SIAM J. Numer. Anal. **40**(2), 469–491 (2002)

Complete Controllability of a Delayed Semilinear Stochastic Control System

Urvashi Arora and N. Sukavanam

Abstract In this paper, complete controllability of a delayed semilinear stochastic system is considered under some basic and readily verified conditions. A fixed-point approach is employed for achieving the required result. At the end, an example is given to show the effectiveness of the result.

Keywords Complete controllability · Delayed system · Stochastic control system · Banach fixed point theorem

1 Introduction

Controllability concepts play a vital role in deterministic control theory. It is well known that controllability of deterministic equations is widely used in many fields of science and technology. Kalman [1] introduced the concept of controllability for finite-dimensional deterministic linear control systems. The basic concepts of control theory in finite- and infinite-dimensional spaces have been introduced in [2], and [3]. However, in many cases, some kind of randomness can appear in the problem, so that the system should be modeled by a stochastic form. Only few authors have studied the extensions of deterministic controllability concepts to stochastic control systems. Klamka et al. [4–7] studied the controllability of linear stochastic systems in finite-dimensional spaces with delay and without delay in control as well as in state. In [8–14], Mahmudov et al. established results for controllability of linear and semilinear stochastic systems in Hilbert Spaces. Shen and Sun [15] studied the controllability of stochastic first-order nonlinear systems with delay in control in finite-dimensional as well as in infinite-dimensional spaces. Sakthivel et al. [16] studied

U. Arora (✉) · N. Sukavanam
Department of Mathematics, Indian Institute of Technology Roorkee, Roorkee 247667, India
e-mail: urvashiaroraiitr@gmail.com

N. Sukavanam
e-mail: nsukavanam@gmail.com

P.N. Agrawal et al. (eds.), *Mathematical Analysis and its Applications*, Springer Proceedings in Mathematics & Statistics 143,
DOI 10.1007/978-81-322-2485-3_43

the controllability of nonlinear stochastic systems in finite-dimensional spaces using Banach fixed-point theorem. Throughout this paper, we use the following standard notations:

(a) $(\Omega, \Im, P)$: Let $\Im$ be the σ algebra on the nonempty set Ω and $P: \Im \to [0, 1]$ be the probability measure. Then the triple $(\Omega, \Im, P)$ is called a probability space.
(b) $\{\Im_t | t \in [0, T]\}$: the filtration generated by an n-dimensional Wiener process $\{\omega(s): 0 \leq s \leq t\}$ defined on the probability space.
(c) $L_2(\Omega, \Im_T, \mathbf{R}^n)$: the Hilbert space of all $\Im_T$-measurable square-integrable variables with values in $\mathbf{R}^n$.
(d) $L_2^{\Im}([0, T], \mathbf{R}^n)$: the Hilbert space of all square-integrable and $\Im_t$-measurable processes with values in $\mathbf{R}^n$.
(e) H_2: the Banach space of all square-integrable and $\Im_t$-adapted processes $\varphi(t)$ with norm

$$||\varphi||^2 = \sup_{t\in[0,T]} \mathbf{E}||\varphi(t)||^2$$

where $\mathbf{E}$ denotes the Expected value.
(f) $\mathbf{L}(X, Y)$: the space of all linear bounded operators from a Banach space X into a Banach space Y.
(g) $U_{ad} = L_2^{\Im}([0, T], \mathbf{R}^m)$: the set of admissible controls.

The problem of controllability of a linear stochastic system of the form

$$\begin{aligned} dx(t) &= [Ax(t) + Bu(t)]dt + \tilde{\sigma} d\omega(t) \quad t \in [0, b] \\ x(0) &= x_0 \end{aligned}$$

has been studied by various authors [7, 8, 14], where $\tilde{\sigma}: [0, b] \to \mathbf{R}^{n\times n}$.

Also, the complete controllability of linear stochastic system with state delay

$$dx(t) = [A_0 x(t) + A_1 x(t-h) + B_0 u(t)]dt + \sigma d\omega(t)$$

given the initial condition as a random function has been studied by Klamka [4].

In this paper, we examine the complete controllability of delayed semilinear stochastic control system given as:

$$\left.\begin{aligned} dx(t) &= [Ax(t) + Bu(t) + f(t, x(t-h))]dt + \sigma(t, x(t-h))d\omega(t), \quad t \in (0, T] \\ x(t) &= \psi(t) \text{for t} \in [-h, 0], \quad x(0) = x_0 = \psi(0)(say). \end{aligned}\right\} \quad (1)$$

where the state $x(t) \in \mathbf{R}^n$ and the control $u(t) \in \mathbf{R}^m$, A is an $n \times n$ constant matrix, B is an $n \times m$ constant matrix. $f: [0, T] \times \mathbf{R}^n \to \mathbf{R}^n$, $\sigma: [0, T] \times \mathbf{R}^n \to \mathbf{R}^{n\times n}$ are nonlinear functions, ω is an n-dimensional Wiener process and $h > 0$ is a constant point delay.

2 Preliminaries

It is well known that for given initial condition, any admissible control $u \in U_{ad}$, for $t \in [-h, T]$ and suitable nonlinear functions $f(t, x(t))$ and $\sigma(t, x(t))$, the mild solution of the semilinear stochastic differential state equation (1) can be represented as

$$x(t; x_0, u) = \begin{cases} \exp(At)x_0 + \int_0^t \exp(A(t-s))[Bu(s) + f(s, x(s-h))]ds \\ + \int_0^t \exp(A(t-s))\sigma(s, x(s-h))d\omega(s) \quad for \quad t > 0 \\ \psi(t) \quad for \quad t \in [-h, 0] \end{cases} \tag{2}$$

Let us introduce the following operators and sets. (see [16])
$L_T \in \mathbf{L}(U_{ad}, L_2(\Omega, \Im_T, \mathbf{R}^n))$ defined by

$$L_T u = \int_0^T \exp(A(T-s))Bu(s)ds$$

Then its adjoint operator $L_T^* : L_2(\Omega, \Im_T, \mathbf{R}^n) \to U_{ad}$ is given by

$$L_T^* z = B^* \exp(A^*(T-t))\mathbf{E}\{z|\Im_t\}$$

The set of all states reachable in time T from initial state $x(0) = x_0 \in L_2(\Omega, \Im_0, \mathbf{R}^n)$, using admissible controls is defined as

$$R_T(U_{ad}) = \{x(T; x_0, u) \in L_2(\Omega, \Im_T, \mathbf{R}^n) : u \in U_{ad}\}$$
$$where \quad x(T; x_0, u) = \exp(AT)x_0 + \int_0^T \exp(A(T-s))Bu(s)ds$$
$$+ \int_0^T \exp(A(T-s))(f(s, x(s-h))ds + \sigma(s, x(s-h))d\omega(s))$$

Let us introduce the linear controllability operator
$\Pi_0^T \in \mathbf{L}(L_2(\Omega, \Im_T, \mathbf{R}^n), L_2(\Omega, \Im_T, \mathbf{R}^n))$ as follows:

$$\Pi_0^T\{.\} = L_T(L_T)^*\{.\}$$
$$= \int_0^T \exp(A(T-t))BB^* \exp(A^*(T-t))\mathbf{E}\{.|\Im_t\}dt$$

The corresponding controllability operator for deterministic model is:

$$\Gamma_s^T = L_T(s)L_T^*(s)$$
$$= \int_s^T \exp(A(T-t))BB^* \exp(A^*(T-t))dt$$

Definition 1 The stochastic dynamic system (1) is said to be completely controllable on $[0, T]$ if

$$R_T(U_{ad}) = L_2(\Omega, \Im_T, \mathbf{R}^n)$$

i.e., all the points in $L_2(\Omega, \Im_T, \mathbf{R}^n)$ can be reached from the point x_0 in time T.

Lemma 1 *[17] Let $G: [0, T] \times \mathbf{R}^n \to \mathbf{R}^{n\times n}$ be a strongly measurable mapping such that $\int_0^T \mathbf{E}||G(t)||^p dt < \infty$. Then*

$$\mathbf{E}\left|\left| \int_0^t G(s)d\omega(s)\right|\right|^p \leq L_G \int_0^t \mathbf{E}||G(s)||^p ds, \tag{3}$$

for all $t \in [0, T]$ and $p \geq 2$, where L_G is the constant involving p and T.

Lemma 2 ***Schwartz inequality:*** *Let $\psi_1(x)$ and $\psi_2(x)$ be any two square-integrable real functions in $[a, b]$ then*

$$\left[\int_a^b \psi_1(x)\psi_2(x)dx\right]^2 \leq \int_a^b [\psi_1(x)]^2 dx \int_a^b [\psi_2(x)]^2 dx$$

3 Main Result

Lemma 3 *Assume that the operator (Π_0^T) is invertible. Then for arbitrary $x_T \in L_2(\Omega, \Im_T, \mathbf{R}^n)$, $f(.) \in L_2([0, T], \mathbf{R}^n)$,$\sigma(.) \in L_2([0, T], \mathbf{R}^{n\times n})$, the control defined as:*

$$u(t) = B^* \exp(A^*(T-t))\mathbf{E}\{(\Pi_0^T)^{-1} p(x)|\Im_t\} \tag{4}$$

where

$$p(x) = x_T - \exp(At)x_0 - \int_0^T \exp(A(T-s))f(s, x(s-h)ds$$
$$- \int_0^T \exp(A(T-s))\sigma(s, x(s-h))d\omega(s))$$

transfers the system (1) from $x_0 \in \mathbf{R}^n$ to the final state x_T at time T, provided the system (1) has a solution.

Proof By substituting (4) in (2), we can easily obtain the following

$$x(t;x_0,u) = \exp(At)x_0 + \int_0^t \exp(A(t-s))BB^* \exp(A^*(T-s))\mathbf{E}\{(\Pi_0^T)^{-1}p(x)|\Im_s\}ds + \int_0^t \exp(A(t-s))(f(s,x(s-h)) + \sigma(s,x(s-h))d\omega(s))$$

Hence, for a given final time t = T, we simply have the following equality:

$$x(T;x_0,u) = \exp(AT)x_0 + \int_0^T \exp(A(T-s))(BB^* \exp(A^*(T-s))\mathbf{E}\Big\{(\Pi_0^T)^{-1} \times \Big(x_T - \exp(AT)x_0 - \int_0^T \exp(A(T-s))(f(s,x(s-h))ds + \sigma(s,x(s-h))d\omega(s))\Big)\Big\}\Big|\Im_s ds + \int_0^T \exp(A(T-s))f(s,x(s-h))ds + \int_0^T \exp(A(T-s))\sigma(s,x(s-h))d\omega(s)$$

Thus, taking into account the form of the operator Π_0^T, we have

$$x(T;x_0,u) = \exp(AT)x_0 + (\Pi_0^T)(\Pi_0^T)^{-1}\Big(x_T - \exp(AT)x_0 - \int_0^T \exp(A(T-s))(f(s,x(s-h))ds + \sigma(s,x(s-h))d\omega(s))\Big) + \int_0^T \exp(A(T-s))(f(s,x(s-h))ds + \sigma(s,x(s-h))d\omega(s)) = x_T$$

Therefore, we see that the control $u(t)$ transfers the system (1) from the initial state $x_0 \in L_2(\Omega, \Im_T, \mathbf{R}^n)$ to the final state $x_T \in L_2(\Omega, \Im_T, \mathbf{R}^n)$ at time T.

Now we assume the following hypotheses:

(i) f and σ satisfy the Lipschitz condition with respect to x. i.e.,

$$||f(t,x_1) - f(t,x_2)||^2 \le L_1||x_1 - x_2||^2$$
$$||\sigma(t,x_1) - \sigma(t,x_2)||^2 \le L_2||x_1 - x_2||^2$$

(ii) (f,σ) is continuous on $[0,T] \times \mathbf{R}^n$ and satisfies $||f(t,x)||^2 + ||\sigma(t,x)||^2 \le L(||x||^2 + 1)$

(iii) The linear system corresponding to (1) is completely controllable.

Let us define the nonlinear operator **S**: $H_2 \to H_2$ for $t \in [-h, T]$ as follows:

$$(\mathbf{S}x)(t) = \begin{cases} \psi(t) \quad for \quad t \in [-h, 0] \\ \exp(At)x_0 + \int_0^t \exp(A(t-s))Bu(s)ds + \int_0^t \exp(A(t-s))f(s, x(s-h))ds \\ + \int_0^t \exp(A(t-s))\sigma(s, x(s-h))d\omega(s) \quad \text{for} \quad t \in [0, T] \end{cases}$$

From Lemma 3, the control $u(t)$ transfer the system (1) from the initial state x_0 to the final state x_T provided that the operator **S** has a fixed point. So, if the operator **S** has a fixed point then the system (1) is completely controllable.

Now for convenience, let us introduce the notation

$$l_1 = \max\{||\exp(At)||^2 : t \in [0, T]\}, \qquad l_2 = ||B||^2$$
$$l_3 = \mathbf{E}||x_T||^2, \qquad M = \max ||\Pi_0^T||^2$$

Lemma 4 *[12] For every $z \in L_2(\Omega, \Im_T, \mathbf{R}^n)$, there exists a process $\varphi(.) \in L_2([0, T], \mathbf{R}^{n \times n})$ such that*

$$z = \mathbf{E}z + \int_0^T \varphi(s)d\omega(s)$$
$$\Pi_0^T z = \Gamma_0^T \mathbf{E}z + \int_0^T \Gamma_s^T \varphi(s)d\omega(s)$$

Moreover,

$$\mathbf{E}||\Pi_0^T z||^2 \le M\mathbf{E}||\mathbf{E}\{z|\Im_T\}||^2$$
$$\le M\mathbf{E}||z||^2, \quad z \in L_2(\Omega, \Im_T, \mathbf{R}^n)$$

Note that if the assumption (iii) holds, then for some $\gamma > 0$

$$\mathbf{E}\langle \Pi_0^T z, z\rangle \ge \gamma \mathbf{E}||z||^2, \quad \text{for} \quad all \quad z \in L_2(\Omega, \Im_T, \mathbf{R}^n)$$

(see Mahmudov [8]) and consequently

$$\mathbf{E}||(\Pi_0^T)^{-1}||^2 \le \frac{1}{\gamma} = l_4$$

Theorem 1 *Assume that the conditions (i), (ii), and (iii) hold. In addition if the inequality*

$$4l_1(Ml_1l_4 + 1)(L_1T + L_2L_\sigma)T < 1 \tag{5}$$

holds, then the system (1) is completely controllable.

Proof As mentioned above, to prove the complete controllability it is enough to show that $\mathbf{S}$ has a fixed point in H_2. To do this, we use the contraction mapping principle. To apply the contraction mapping principle, first we show that $\mathbf{S}$ maps H_2 into itself. Now by Lemmas 1 and 2, we have

$$\begin{aligned}
\mathbf{E}||(\mathbf{S}x)(t)||^2 = \mathbf{E}\Bigg|\Bigg|\psi(t) + \exp(At)x_0 + \Pi_0^t\Big[\exp(A^*(T-t))(\Pi_0^T)^{-1} \times \\
\Big(x_T - \exp(AT)x_0 - \int_0^T \exp(A(T-s))f(s,x(s-h))ds \\
- \int_0^T \exp(A(T-s))\sigma(s,x(s-h))d\omega(s)\Big)\Big] \\
+ \int_0^t \exp(A(t-s))(f(s,x(s-h))ds + \sigma(s,x(s-h))d\omega(s))\Bigg|\Bigg|^2 \\
\leq 5||\psi||^2 + 5l_1||x_0||^2 + 5\mathbf{E}\Bigg|\Bigg|\Pi_0^t\Big[\exp(A^*(T-t))(\Pi_0^T)^{-1} \times \\
\Big(x_T - \exp(AT)x_0 - \int_0^T \exp(A(T-s))f(s,x(s-h))ds \\
- \int_0^T \exp(A(T-s))\sigma(s,x(s-h))d\omega(s)\Big)\Big]\Bigg|\Bigg|^2 \\
+ 5t\int_0^t ||\exp(A(t-s))||^2\mathbf{E}||f(s,x(s-h))||^2ds \\
+ 5L_\sigma\int_0^t ||\exp(A(t-s))||^2\mathbf{E}||\sigma(s,x(s-h))||^2ds \\
\leq 5||\psi||^2 + 5l_1||x_0||^2 + 20Ml_1l_4\Big(l_3 + l_1||x_0||^2 \\
+ Tl_1\int_0^T \mathbf{E}||f(s,x(s-h))||^2ds + l_1L_\sigma\int_0^T \mathbf{E}||\sigma(s,x(s-h))||^2ds\Big) \\
+ 5l_1\int_0^t (T\mathbf{E}||f(s,x(s-h))||^2 + L_\sigma\mathbf{E}||\sigma(s,x(s-h))||^2)ds \\
\leq B_1 + B_2\Big(\int_0^t (T\mathbf{E}||f(s,x(s-h))||^2 + L_\sigma\mathbf{E}||\sigma(s,x(s-h))||^2)ds\Big)
\end{aligned}$$

where $B_1 > 0$ and $B_2 > 0$ are suitable constants. It follows from the above and the condition (ii) that there exists $C_1 > 0$ such that

$$\begin{aligned}
\mathbf{E}||(Sx)(t)||^2 &\leq C_1\Big(1 + \int_0^T \mathbf{E}||x(r-h)||^2dr\Big) \\
&\leq C_1\Big(1 + T\sup_{-h\leq t\leq T}\mathbf{E}||x(t)||^2\Big)
\end{aligned}$$

for all $t \in [-h, T]$. Therefore, $\mathbf{S}$ maps H_2 into itself. Second, we show that $\mathbf{S}$ is a contraction mapping, indeed.

$$\begin{aligned}
\mathbf{E}\left\|(\mathbf{S}x)(t) - (\mathbf{S}y)(t)\right\|^2 &= \mathbf{E}\Bigg\|\Pi_0^t\Big[\exp(A^*(T-t))(\Pi_0^T)^{-1} \\
&\times\Big(\int_0^T \exp(A(T-s))(f(s, y(s-h)) - f(s, x(s-h)))ds \\
&+\int_0^T \exp(A(T-s))(\sigma(s, y(s-h)) - \sigma(s, x(s-h)))d\omega(s)\Big)\Big] \\
&+\int_0^t \exp(A(t-s))(f(s, x(s-h)) - f(s, y(s-h)))ds \\
&+\int_0^t \exp(A(t-s))(\sigma(s, y(s-h)) - \sigma(s, x(s-h)))d\omega(s)\Bigg\|^2 \\
&\le 4Ml_1^2 l_4\Big(T\int_0^T \mathbf{E}||f(s, x(s-h)) - f(s, y(s-h))||^2 ds \\
&+L_\sigma\int_0^T \mathbf{E}||\sigma(s, x(s-h)) - \sigma(s, y(s-h))||^2 ds\Big) \\
&+4l_1\Big(T\int_0^t \mathbf{E}||f(s, x(s-h)) - f(s, y(s-h))||^2 ds \\
&+L_\sigma\int_0^t \mathbf{E}||\sigma(s, x(s-h)) - \sigma(s, y(s-h))||^2 ds\Big) \\
&= 4Ml_1^2 l_4(L_1 T + L_2 L_\sigma)\int_0^T \mathbf{E}||x(s-h) - y(s-h)||^2 ds \\
&+4l_1(L_1 T + L_2 L_\sigma)\int_0^t \mathbf{E}||x(s-h) - y(s-h)||^2 ds \\
&\le 4l_1(Ml_1 l_4 + 1)(L_1 T + L_2 L_\sigma)\int_0^T \mathbf{E}||x(s-h) - y(s-h)||^2 ds
\end{aligned}$$

It results that

$$\sup_{t\in[-h,T]} \mathbf{E}||(\mathbf{S}x)(t) - (\mathbf{S}y)(t)||^2 \le 4l_1(Ml_1 l_4 + 1)(L_1 T + L_2 L_\sigma)T \sup_{t\in[-h,T]} \mathbf{E}||x(t) - y(t)||^2$$

Therefore, $\mathbf{S}$ is a contraction mapping if the inequality (5) holds. Then the mapping $\mathbf{S}$ has a unique fixed point $x(.)$ in H_2 which is the solution of the equation. Thus the system (1) is completely controllable. So, the theorem is proved.

Remark 1 If we consider the time-varying semilinear stochastic differential equation of the form

$$\left.\begin{aligned}
dx(t) &= [A(t)x(t) + B(t)u(t) + f(t, x(t-h)])dt + \sigma(t, x(t-h))d\omega(t), \\
&\quad t \in (0, T] \\
x(t) &= \psi(t), \text{ for t } \in [-h, 0], \quad x(0) = x_0.
\end{aligned}\right\} \quad (6)$$

where $A(t)$ and $B(t)$ are the matrices of $n \times n$ and $n \times m$, respectively, and f, σ are defined as previously. The solution of the above equation for $t > 0$ is

$$x(t; x_0, u) = \phi(t,0)x_0 + \int_0^t \phi(t,s)B(s)u(s)ds + \int_0^t \phi(t,s)f(s, x(s-h))ds$$
$$+ \int_0^t \phi(t,s)\sigma(s, x(s-h))d\omega(s) \quad (7)$$

If the functions f, σ satisfy Lipschitz conditions and linear growth conditions and the condition (iii) is also satisfied, then by suitably applying the above theorem, one can show that the system (6) is completely controllable.

4 Example

Consider a two-dimensional semilinear stochastic system with delay in state

$$dx(t) = [Ax(t) + Bu(t) + f(t, x(t-h))]dt + \sigma(t, x(t-h))d\omega(t), \quad t \in [0, T] \quad (8)$$

with initial condition $x_0 \in \mathbf{R}^2$ as a random function.
Here $\omega(t)$ is a two-dimensional Wiener process and

$$A = \begin{bmatrix} -1 & 1 \\ -1 & -1 \end{bmatrix}, \quad B = \begin{bmatrix} 1 & 0 \\ 0 & 1 \end{bmatrix}$$
$$f(t, x(t-h)) = \frac{1}{a}\begin{bmatrix} \sin x(t-h) \\ x(t-h) \end{bmatrix}, \quad \sigma(t, x(t-h)) = \frac{1}{b}\begin{bmatrix} x(t-h) & 0 \\ 0 & \cos x(t-h) \end{bmatrix}$$

Take the final point $x(T) \in \mathbf{R}^2$. For this system, the controllability matrix

$$\begin{aligned} \Gamma_0^T &= \int_0^T \exp(-At)BB^* \exp(-A^*t)dt \\ &= \frac{1}{2}(\exp(2b) - 1)\begin{bmatrix} 1 & 0 \\ 0 & 1 \end{bmatrix} \end{aligned}$$

is nonsingular if $T > 0$.
Moreover

$$||f(t, x(t-h) - f(t, y(t-h))||^2 \leq \frac{2}{a^2}||x(t-h) - y(t-h)||^2 \quad \text{and}$$
$$||\sigma(t, x(t-h) - \sigma(t, y(t-h))||^2 \leq \frac{2}{b^2}||x(t-h) - y(t-h)||^2$$

So, $L_1 = \frac{2}{a^2}$ and $L_2 = \frac{2}{b^2}$

Also, here $||A|| = 2$, $||B|| = \sqrt{2}$

We can see that the conditions of Theorem 1 are satisfied. So the system (8) is completely controllable.

Acknowledgments The authors express their sincere gratitude to the reviewers for their valuable suggestions for improving the paper. This research is supported by the Ministry of Human Resource and Development (MHRD) in the form of fellowship to the first author.

References

1. Kalman, R.E.: Controllability of linear systems. Contribut. Differ. Equ. **1**, 190–213 (1963)
2. Curtain, R.F., Zwart, H.: An introduction to infinite-dimensional linear systems theory. Texts in Applied Mathematics, p. 21. Springer, New York (1995)
3. Barnett, S.: Introduction to Mathematical Control Theory. Clarendon Press, Oxford (1975)
4. Klamka, J.: Stochastic controllability of linear systems with state delays. Int. J. Appl. Math. Comput. Sci. **17**(1), 5–13 (2007)
5. Klamka, J.: Stochastic controllability of linear systems with delay in control. Bullet. Polish Acad. Sci. **55**, 23–29 (2007)
6. Klamka, J., Socha, L.: Some remarks about stochastic controllability. IEEE Trans. Automatic Control AC **22**(5), 880–881 (1977)
7. Klamka, J., Socha, L.: Some remarks about stochastic controllability for delayed linear systems. Internat. J. Control **32**(3), 561–566 (1980)
8. Mahmudov, N.I.: Controllability of linear stochastic systems. IEEE Trans. Automat. Control **46**(5), 724–731 (2001)
9. Mahmudov, N.I.: Approximate controllability of semilinear deterministic and stochastic evolution equations in abstract spaces. SIAM J. Control Optim. **42**(5), 1604–1622 (2003)
10. Mahmudov, N.I., Zorlu, S.: Controllability of nonlinear stochastic systems. J. Control **76**, 95–104 (2003)
11. Mahmudov, N.I.: On controllability of semilinear stochastic systems in Hilbert spaces. IMA J. Math. Control Inform. **19**(4), 363–376 (2002)
12. Mahmudov, N.I.: Controllability of linear stochastic systems in Hilbert spaces. J. Math. Anal. Appl. **259**(1), 64–82 (2001)
13. Mahmudov, N.I., Semi, N.: Approximate controllability of semilinear control systems in Hilbert spaces. TWMS J. App. Eng. Math. **2**(1), 67–74 (2012)
14. Mahmudov, N.I., Denker, A.: On controllability of linear stochastic systems. Internat. J. Control **73**(2), 144–151 (2000)
15. Shen, L., Sun, J.: Relative controllability of stochastic nonlinear systems with delay in control. Nonlinear Anal. Real World Appl. **13**(6), 2880–2887 (2012)
16. Sakthivel, R., Kim, J.-H., Mahmudov, N.I.: On controllability of nonlinear stochastic systems. Rep. Math. Phys. **58**(3), 433–443 (2006)
17. Da Prato, G., Zabczyk, J.: Stochastic equations in infinite dimensions. Encyclopedia of Mathematics and its Applications. Cambridge University Press, Cambridge (1992)

On Mixed Integro-Differential Inequalities and Applications

S.D. Kendre and S.G. Latpate

Abstract In this paper, we establish mixed integro-differential inequalities which can be used as handy tools to study properties of solutions of certain mixed integro-differential and differential equations.

Keywords Integrodifferential inequality · Integral equations · Explicit bound

1 Introduction

Integral inequalities involving functions and their derivatives have played a significant role in the development of various branches of analysis. In the past few years with the development of the theory of nonlinear differential and integral equations, many authors have established several integral and integro-differential inequalities, see [1, 2, 4, 6–12, 14]. These inequalities play an important role in the study of some properties of differential, integral, and integro-differential equations. Existence of solutions of a certain mixed integral and integro-differential equations were studied in [3, 13] by M.B. Dhakne and H.L. Tidke.

In this paper, we establish mixed integro-differential inequalities which provide an explicit bound on unknown function. In particular, we extend the result established by B.G. Pachpatte in [12]. Some applications are also given to convey the importance of our results.

Before proceeding with the statement of our main result, we state some important integral inequalities that will be used in further discussion.

S.D. Kendre (✉)
Department of Mathematics, Savitribai Phule Pune University, Pune (MS) 411007, India
e-mail: sdkendre@yahoo.com

S.G. Latpate
Department of Mathematics, Nowrosjee Wadia College of Arts and Science, Pune 411001, India
e-mail: sglatpate@gmail.com

P.N. Agrawal et al. (eds.), *Mathematical Analysis and its Applications*,
Springer Proceedings in Mathematics & Statistics 143,
DOI 10.1007/978-81-322-2485-3_44

Lemma 1 (Fangcui Jiang and Fanwei Meng [5]). *Assume that* $a \geq 0$, $p \geq q \geq 0$, *and* $p \neq 0$, *then*

$$a^{\frac{q}{p}} \leq \frac{q}{p} k^{\frac{q-p}{p}} a + \frac{p-q}{p} k^{\frac{q}{p}}, \quad for\ any\ k > 0. \tag{1}$$

Theorem 1 (Pachpatte [12]) *Let* $u(t), a(t), b(t), c(t) \in C(I = [\alpha, \beta], R_+)$, $a(t)$ *be continuously differentiable on* I, $a'(t) \geq 0$ *and*

$$u(t) \leq a(t) + \int_\alpha^t b(s)u(s)ds + \int_\alpha^\beta c(s)u(s)ds, \quad t \in I. \tag{2}$$

If $p = \int_\alpha^\beta c(s) \exp\left(\int_\alpha^s b(\sigma)d\sigma\right) ds < 1$, *then*

$$u(t) \leq M \exp\left(\int_\alpha^t b(s)ds\right) + \int_\alpha^t a'(s) \exp\left(\int_\alpha^s b(\sigma)d\sigma\right) ds, \quad t \in I, \tag{3}$$

where

$$M = \frac{1}{1-p}\left[a(\alpha) + \int_\alpha^\beta c(s) \int_\alpha^s a'(\tau) \exp\left(\int_\tau^s b(\sigma)d\sigma\right) d\tau ds\right], \quad t \in I. \tag{4}$$

2 Main Results

In this section, we establish some nonlinear mixed integro-differential inequalities, which can be used in the analysis of various problems in the theory of nonlinear integrodifferential and Volterra-Fredholm integral equations.

Theorem 2 *Let* $u(t), u'(t), f(t), g(t), c(t), c'(t) \in C(I, R_+)$ *and* $u(\alpha) = 0$. *If*

$$[u'(t)]^p \leq c(t) + \int_\alpha^t f(s)u^q(s)ds + \int_\alpha^\beta g(s)[u'(s)]^p ds \tag{5}$$

and $\bar{Q}_5 = \int_\alpha^\beta g(s) \exp\left(\int_\alpha^s (\sigma-\alpha)^q n_1 f(\sigma)d\sigma\right) ds < 1$, *then*

$$\begin{aligned}
[u'(t)]^p \leq\ & \frac{c(\alpha) + \int_\alpha^\beta g(s)\left(\int_\alpha^s \left[c'(\tau) + (\tau-\alpha)^q n_2 f(\tau)\right] \exp\left(\int_\tau^s (\sigma-\alpha)^q n_1 f(\sigma)d\sigma\right) d\tau\right) ds}{1-\bar{Q}_5} \\
& \times \exp\left(\int_\alpha^t (s-\alpha)^q n_1 f(s)ds\right) \\
& + \int_\alpha^t \left[c'(s) + (s-\alpha)^q n_2 f(s)\right] \exp\left(\int_s^t (s-\alpha)^q n_1 f(s)ds\right) ds,
\end{aligned} \tag{6}$$

where $k > 0$, $n_1 = \frac{q}{p} k^{\frac{q-p}{p}}$, $n_2 = \frac{p-q}{p} k^{\frac{q}{p}}$ *and* p, q *are the same as defined in Lemma 1.*

Theorem 3 *Let* $u(t), u'(t), g(t), c(t), c'(t) \in C(I, R_+)$, $f(t,s)$, $f_t(t,s) \in C(D, R_+)$, $f(t,s)$ *be nondecreasing in* $t \in I$, *for each* $s \in I$ *and* $u(\alpha) = 0$. *If*

$$[u'(t)]^p \le c(t) + \int_\alpha^t f(t,s)u^q(s)ds + \int_\alpha^\beta g(s)[u'(s)]^p ds \tag{7}$$

and $\bar{Q}_6 = \int_\alpha^\beta g(s)\exp\left(\int_\alpha^s n_1 \bar{A}(\sigma)d\sigma\right)ds < 1$, *then*

$$[u'(t)]^p \le \frac{c(\alpha) + \int_\alpha^\beta g(s)\left(\int_\alpha^s \left[c'(\tau) + n_2\bar{A}(\tau)\right]\exp\left(\int_\tau^s n_1\bar{A}(\sigma)d\sigma\right)d\tau\right)ds}{1-\bar{Q}_6}$$
$$\times \exp\left(\int_\alpha^t n_1\bar{A}(s)ds\right) + \int_\alpha^t \left[c'(s) + n_2\bar{A}(s)\right]\exp\left(\int_s^t n_1\bar{A}(\sigma)d\sigma\right)ds, \tag{8}$$

where $\bar{A}(t) = (t-\alpha)^q f(t,t) + \int_\alpha^t (s-\alpha)^q f_t(t,s)ds$ *and* p, q, n_1, n_2 *are as same defined in Theorem 2.*

Theorem 4 *Let* $u(t), u'(t), c(t), c'(t) \in C(I, R_+)$, $f(t,s), g(t,s), f_t(t,s), g_t(t,s) \in C(D, R_+)$, $f(t,s), g(t,s)$ *be nondecreasing in* $t \in I$, *for each* $s \in I$ *and* $u(\alpha) = 0$. *If*

$$[u'(t)]^p \le c(t) + \int_\alpha^t f(t,s)u^q(s)ds + \int_\alpha^\beta g(t,s)[u'(s)]^p ds \tag{9}$$

and $\bar{Q}_7 = \int_\alpha^\beta g(\alpha,s)\exp\left(\int_\alpha^s n_1\bar{B}(\sigma)d\sigma\right)ds < 1$, *then*

$$[u'(t)]^p \le \frac{c(\alpha) + \int_\alpha^\beta g(\alpha,s)\left(\int_\alpha^s \left[c'(\tau) + n_2\bar{A}(\tau)\right]\exp\left(\int_\tau^s n_1\bar{B}(\sigma)d\sigma\right)d\tau\right)ds}{1-\bar{Q}_7}$$
$$\times \exp\left(\int_\alpha^t n_1\bar{B}(s)ds\right) + \int_\alpha^t \left[c'(s) + n_2\bar{A}(s)\right]\exp\left(\int_s^t n_1\bar{B}(\sigma)d\sigma\right)ds, \tag{10}$$

where p, q, n_1, n_2 *are the same as defined in Theorem 2,* $\bar{A}(t)$ *is the same as defined in Theorem 3 and* $\bar{B}(t) = (t-\alpha)^q f(t,t) + \int_\alpha^t (s-\alpha)^q f_t(t,s)ds + \frac{1}{n_1}\int_\alpha^\beta g_t(t,s)ds.$

2.1 Proofs of the Theorems 2–4

The proofs of the Theorems 2–4 resemble one another. Therefore, we give the detailed proof of Theorem 4 only and the proofs of Theorems 2 and 3, can be completed by closely looking at the proof of Theorem 4.

Proof Define a function $z(t)$ by the right-hand side of (9)

$$z(t) = c(t) + \int_{\alpha}^{t} f(t,s)u^{q}(s)ds + \int_{\alpha}^{\beta} g(t,s)[u'(s)]^{p}ds,$$

then $u(t) \leq \int_{\alpha}^{t} z^{\frac{1}{p}}(s)$, $z(t)$ is a nondecreasing,

$$z(\alpha) = c(\alpha) + \int_{\alpha}^{\beta} g(\alpha,s)[u'(s)]^{p}(s)ds \tag{11}$$

and

$$z'(t) \leq c'(t) + \left((t-\alpha)^{q} f(t,t) + \int_{\alpha}^{t}(s-\alpha)^{q} f_{t}(t,s)ds\right) z^{\frac{q}{p}}(t) + \left(\int_{\alpha}^{\beta} g_{t}(t,s)ds\right) z(t). \tag{12}$$

Applying Lemma 1 to (12), we have

$$\begin{aligned}
z'(t) &\leq c'(t) + \left((t-\alpha)^{q} f(t,t) + \int_{\alpha}^{t}(s-\alpha)^{q} f_{t}(t,s)ds\right) z^{\frac{q}{p}}(t) + \left(\int_{\alpha}^{\beta} g_{t}(t,s)ds\right) z(t)\\
&\leq c'(t) + \left((t-\alpha)^{q} f(t,t) + \int_{\alpha}^{t}(s-\alpha)^{q} f_{t}(t,s)ds\right)[n_1 z(t) + n_2] + \left(\int_{\alpha}^{\beta} g_{t}(t,s)ds\right) z(t)\\
&= c'(t) + n_1\left((t-\alpha)^{q} f(t,t) + \int_{\alpha}^{t}(s-\alpha)^{q} f_{t}(t,s)ds + \frac{1}{n_1}\int_{\alpha}^{\beta} g_{t}(t,s)ds\right) z(t)\\
&+ n_2\left((t-\alpha)^{q} f(t,t) + \int_{\alpha}^{t}(s-\alpha)^{q} f_{t}(t,s)ds\right)\\
&= c'(t) + n_1\bar{B}(t)z(t) + n_2\bar{A}(t),
\end{aligned}$$

or, equivalently

$$\left[\frac{z(t)}{\exp\left(\int_{\alpha}^{t} n_1\bar{B}(s)ds\right)}\right]' \leq [c'(t) + n_2\bar{A}(t)]\exp\left(-\int_{\alpha}^{t} n_1\bar{B}(s)ds\right). \tag{13}$$

By integrating (13), we get

$$\frac{z(t)}{\exp\left(\int_\alpha^t n_1\bar{B}(s)ds\right)} \leq z(\alpha) + \int_\alpha^t \left[c'(s) + n_2\bar{A}(s)\right]\exp\left(-\int_\alpha^s n_1\bar{B}(s)ds\right)ds,$$

i.e.

$$z(t) \leq z(\alpha)\exp\left(\int_\alpha^t n_1\bar{B}(s)ds\right) + \int_\alpha^t \left[c'(s) + n_2\bar{A}(s)\right]\exp\left(\int_s^t n_1\bar{B}(\sigma)d\sigma\right)ds. \tag{14}$$

As $[u'(t)]^p \leq z(t)$ from (14), we have

$$[u'(t)]^p \leq z(\alpha)\exp\left(\int_\alpha^t n_1\bar{B}(s)ds\right) + \int_\alpha^t \left[c'(s) + n_2\bar{A}(s)\right]\exp\left(\int_s^t n_1\bar{B}(\sigma)d\sigma\right)ds. \tag{15}$$

Now from (11) and (15), we have

$$z(\alpha) \leq c(\alpha) + z(\alpha)\int_\alpha^\beta g(\alpha, s)\exp\left(\int_\alpha^s n_1\bar{B}(\sigma)d\sigma\right)ds + \int_\alpha^\beta g(\alpha, s)\left(\int_\alpha^s \left[c'(\tau) + n_2\bar{A}(\tau)\right]\exp\left(\int_\tau^s n_1\bar{B}(\sigma)d\sigma\right)d\sigma\right)ds,$$

and hence

$$z(\alpha) \leq \frac{c(\alpha) + \int_\alpha^\beta g(\alpha, s)\left(\int_\alpha^s \left[c'(\tau) + n_2\bar{A}(\tau)\right]\exp\left(\int_\tau^s n_1\bar{B}(\sigma)d\sigma\right)d\tau\right)ds}{1 - \bar{Q}_7}. \tag{16}$$

The required inequality (10) follows, from inequalities (15) and (16). This completes the proof.

Theorem 5 *Let* $u(t), u'(t) \in C(I, R_+)$, $f(t, s), g(t, s), h_t(t, s) \in C(D, R_+)$, $u(\alpha) = 0$, $f(t, s), g(t, s)$ *be nondecreasing in* $t \in I$, *for each* $s \in I$ *and* $c \geq 0$, *be a constant. If*

$$[u'(t)]^p \leq c + \int_\alpha^t h(t, s)\left[u^q(s) + \int_\alpha^s f(s, \sigma)u^q(\sigma)d\sigma\right]ds + \int_\alpha^\beta g(t, s)[u'(s)]^p ds,$$

and

$$\bar{Q}_8 = \int_\alpha^\beta g(\alpha, s)\exp\left(\int_\alpha^s n_1\bar{B}_1(\sigma)d\sigma\right)ds < 1, \textit{then}$$

$$[u'(t)]^p \le \frac{c + \int_\alpha^\beta g(\alpha,s)\left(\int_\alpha^s n_2 \bar{A}_1(\tau)\exp\left(\int_\tau^s n_1 \bar{B}_1(\sigma)d\sigma\right)d\tau\right)ds}{1-\bar{Q}_8} \times \exp\left(\int_\alpha^t n_1 \bar{B}_1(\sigma)d\sigma\right) + \int_\alpha^t n_2 \bar{A}_1(s)\exp\left(\int_s^t n_1 \bar{B}_1(\sigma)d\sigma\right), \quad (17)$$

where

$$\bar{A}_1(t) = \left(\int_\alpha^t h_t(t,s)\left[(s-\alpha)^q + \int_\alpha^s (\sigma - \alpha^q) f(s,\sigma)d\sigma\right]ds + h(t,t)\left[(t-\alpha)^q + \int_\alpha^t (\sigma-\alpha)^q f(t,\sigma)d\sigma\right]ds\right),$$

$$\bar{B}_1(t) = \left(\int_\alpha^t h_t(t,s)\left[(s-\alpha)^q + \int_\alpha^s (\sigma - \alpha^q) f(s,\sigma)d\sigma\right]ds + h(t,t)\left[(t-\alpha)^q + \int_\alpha^t (\sigma-\alpha)^q f(t,\sigma)d\sigma\right]ds + \frac{1}{n_1}\int_\alpha^\beta g_t(t,s)ds\right)$$

and p, q, n_1, n_2 are the same as defined in Theorem 2.

Proof Let us define a function $z(t)$ by

$$z(t) = c + \int_\alpha^t h(t,s)\left[u^q(s) + \int_\alpha^s f(s,\sigma)u^q(\sigma)d\sigma\right]ds + \int_\alpha^\beta g(t,s)[u'(s)]^p ds,$$

then $u(t) \le \int_\alpha^t z^{\frac{1}{p}}(s)$,

$$z(\alpha) = c + \int_\alpha^\beta g(\alpha,s)[u'(s)]^p ds \quad (18)$$

and

$$z'(t) \le \left(\int_\alpha^t h_t(t,s)\left[(s-\alpha)^q + \int_\alpha^s (\sigma-\alpha^q) f(s,\sigma)d\sigma\right]ds + h(t,t)\left[(t-\alpha)^q + \int_\alpha^t (\sigma-\alpha)^q f(t,\sigma)d\sigma\right]\right) z^{\frac{q}{p}}(t) + \left(\int_\alpha^\beta g_t(t,s)ds\right)z(t). \quad (19)$$

Applying Lemma 1 to (19), we have

$$\begin{aligned}z'(t) &\le \left(\int_\alpha^t h_t(t,s)\left[(s-\alpha)^q + \int_\alpha^s (\sigma-\alpha^q) f(s,\sigma)d\sigma\right]ds \right. \\ &\quad \left. + h(t,t)\left[(t-\alpha)^q + \int_\alpha^t (\sigma-\alpha)^q f(t,\sigma)d\sigma\right]\right)[n_1 z(t) + n_2] + \left(\int_\alpha^\beta g_t(t,s)ds\right)z(t) \\ &\le n_1\left(\int_\alpha^t h_t(t,s)\left[(s-\alpha)^q + \int_\alpha^s (\sigma-\alpha^q) f(s,\sigma)d\sigma\right]ds\right.\end{aligned}$$

$$
\begin{aligned}
&+ h(t,t)\left[(t-\alpha)^q + \int_\alpha^t (\sigma-\alpha)^q f(t,\sigma)d\sigma\right] + \frac{1}{n_1}\int_\alpha^\beta g_t(t,s)ds\Bigg) z(t)\\
&+ n_2\left(\int_\alpha^t h_t(t,s)\left[(s-\alpha)^q + \int_\alpha^s (\sigma-\alpha^q) f(s,\sigma)d\sigma\right]ds\right.\\
&\left.+ h(t,t)\left[(t-\alpha)^q + \int_\alpha^t (\sigma-\alpha)^q f(t,\sigma)d\sigma\right]\right)\\
&\le n_1\bar{B}_1(t)z(t) + n_2\bar{A}_1(t),
\end{aligned}
$$

or, equivalently,

$$
\left[\frac{z(t)}{\exp\left(\int_\alpha^t n_1\bar{B}_1(\sigma)d\sigma\right)}\right]' \le n_2\bar{A}_1(t)\exp\left(-\int_\alpha^t n_1\bar{B}_1(\sigma)d\sigma\right). \quad (20)
$$

By integrating (20), we obtain an estimate

$$
z(t) \le z(\alpha)\exp\left(\int_\alpha^t n_1\bar{B}_1(\sigma)d\sigma\right) + \int_\alpha^t n_2\bar{A}_1(s)\exp\left(\int_s^t n_1\bar{B}_1(\sigma)d\sigma\right)ds. \quad (21)
$$

As $[u'(t)]^p \le z(t)$ from (21), we have

$$
[u'(t)]^p \le z(\alpha)\exp\left(\int_\alpha^t n_1\bar{B}_1(\sigma)d\sigma\right) + \int_\alpha^t n_2\bar{A}_1(s)\exp\left(\int_s^t n_1\bar{B}_1(\sigma)d\sigma\right). \quad (22)
$$

Now from Eq. (18) and (22), we have

$$
z(\alpha) \le \frac{c + \int_\alpha^\beta g(\alpha,s)\left(\int_\alpha^s n_2\bar{A}_1(\tau)\exp\left(\int_\tau^s n_1\bar{B}_1(\sigma)d\sigma\right)d\tau\right)ds}{1-\bar{Q}_8}. \quad (23)
$$

From (22) and (23), we get the required inequality (17) and hence the proof.

3 Applications

One of the main motivations for the study of different type inequalities given in the previous sections is to apply them as tools in the study of various classes of integral equations. In the following section we give application of some theorems of previous sections. In fact we discuss the boundedness behavior of solutions of a nonlinear mixed integral equations.

Example 1 Consider the following general mixed nonlinear integro-differential equation

$$
[y'(t)]^p = x(t) + \int_\alpha^t F\left(t,s,y^q(s)\right)ds + \int_\alpha^\beta G\left(t,s,[y'(s)]^p\right)ds, \quad (24)
$$

for $t \in I$, where $p \geq q \geq 0$ with $p \neq 0$, $y(t)$ is unknown function, $x \in C(I, R^n)$, $F, G \in C(D \times R^n, R^n)$, $I = [\alpha, \beta]$, $D = \left\{(t, s) \in I^2 : \alpha \leq s \leq t \leq \beta\right\}$, $R^n$is n dimensional Euclidean space with norm $|.|$.

We assume that every solution $y(t)$ of Eq. (24) exists on I and functions x, y, F, G involved in the Eq. (24) satisfy the following conditions:

$$|x(t)| \leq c(t), \tag{25}$$

$$|F(t, s, y^q)| \leq f(t, s)\, |y|^q\,, \tag{26}$$

$$|G(t, s, y^p)| \leq g(t, s) |y'|^p, \tag{27}$$

where c, f, g are the same as defined in Theorem 4. From Eqs. (24) and (25)–(27), we obtain

$$|y'(t)|^p \leq c(t) + \int_\alpha^t f(t, s)|y(s)|^q ds + \int_\alpha^\beta g(t, s)|y'(s)|^p ds. \tag{28}$$

Applying Theorem 4 to (28), we get the following explicit known bound:

$$|y'(t)|^p \leq \frac{c(\alpha) + \int_\alpha^\beta g(\alpha, s) \left(\int_\alpha^s \left[c'(\tau) + n_2 \bar{A}(\tau)\right] \exp\left(\int_\tau^s n_1 \bar{B}(\sigma) d\sigma\right) d\tau\right) ds}{1 - \bar{Q}_7} \times \exp\left(\int_\alpha^t n_1 \bar{B}(s) ds\right) + \int_\alpha^t \left[c'(s) + n_2 \bar{A}(s)\right] \exp\left(\int_s^t n_1 \bar{B}(\sigma) d\sigma\right) ds,$$

provided $\bar{Q}_7 < 1$, where $\bar{A}$, $\bar{B}$, $\bar{Q}_7, n_1, n_2$ are the same as defined in Theorem 4.

Example 2 We calculate the explicit bound on a solution of the following nonlinear integral equation:

$$[u'(t)]^3 = 4 + \int_{\frac{1}{3}}^t \frac{1}{s} u^2(s) ds + \int_{\frac{1}{3}}^{\frac{1}{2}} s[u'(s)]^3 ds, \tag{29}$$

where $u(t)$ is defined as in Theorem 2 and we assume that every solution $u(t)$ of (29) exists on $I = \left[\frac{1}{3}, \frac{1}{2}\right]$. Also, here

$$\bar{Q}_5 = \int_{\frac{1}{3}}^{\frac{1}{2}} s \exp\left(\frac{2}{3} \frac{1}{\sqrt[3]{k}} \int_0^s \sigma\, d\sigma\right) ds = \frac{3}{2} e^{\frac{1}{27 \sqrt[3]{k}}} \left(e^{\frac{5}{108 \sqrt[3]{k}}} - 1\right) \sqrt[3]{k} < 1, \text{ for } k > 0.$$

Hence, by Theorem 2 and Eq. (29), we get

$$u'(t) \leq \left[\frac{\exp\left(\frac{2}{3}\frac{1}{\sqrt[3]{k}}\int_{\frac{1}{3}}^{t} \sigma\, d\sigma\right)\left(\int_{\frac{1}{3}}^{\frac{1}{2}} \frac{1}{3} s \left(k^{2/3}\int_{\frac{1}{3}}^{s} \tau \exp\left(\frac{2}{3}\frac{1}{\sqrt[3]{k}}\int_{\tau}^{s} \sigma\, d\sigma\right) d\tau\right) ds + 4\right)}{1-\int_{\frac{1}{3}}^{\frac{1}{2}} s \exp\left(\frac{2}{3}\frac{1}{\sqrt[3]{k}}\int_{0}^{s}\sigma\, d\sigma\right) ds} + \frac{1}{3}k^{2/3}\left(\int_{\frac{1}{3}}^{t} s \exp\left(\frac{2}{3}\frac{1}{\sqrt[3]{k}}\int_{s}^{t}\sigma\, d\sigma\right) ds\right)\right]^{\frac{1}{3}}$$

$$= \left[\frac{1}{3}k^{2/3}\left(\frac{3}{2}\sqrt[3]{k}\, e^{\frac{t^2-\frac{1}{9}}{3\sqrt[3]{k}}} - \frac{3\sqrt[3]{k}}{2}\right) + \frac{\left(\frac{3}{4}\left(e^{\frac{5}{108\sqrt[3]{k}}}-1\right)k^{4/3} - \frac{5k}{144} + 4\right) e^{\frac{2\left(\frac{t^2}{2}-\frac{1}{18}\right)}{3\sqrt[3]{k}}}}{1-\frac{3}{2}e^{\frac{1}{27\sqrt[3]{k}}}\left(e^{\frac{5}{108\sqrt[3]{k}}}-1\right)\sqrt[3]{k}} \right]^{\frac{1}{3}}$$

By integrating the above inequality we get the desired bound for $u(t)$.

References

1. Abdeldaim, A., Yakout, M.: On some new inequalities of Gronwall-Bellman-Pachpatte type. Appl. Math. Comput. **217**, 7887–7899 (2011)
2. Bellman, R.: The stability of solutions of linear differential equations. Duke Math. J. **10**, 643–647 (1943)
3. Dhakne, M.B., Lamb, G.B.: On globle existance of abstrct nonlinear integrodifferential equations. Indian J. Pure Appl. Math. **33**(5), 665–676
4. El-Owaidy, H., Ragab, A., Abdeldaim, A.: On some new integral inequalities of Growall-Bellman type. Appl. Math. Comput. **106**, 289–303 (1999)
5. Jiang, F., Meng, F.: Explicit bounds on some new nonlinear integral inequality with delay. J. Comput. Appl. Math. **205**, 479–486 (2007)
6. Meng, F.W., Li, W.N.: On some new integral inequalitise and their applications. Appl. Math. Comput. **148**, 381–392 (2004)
7. Gronwall, H.T.: Note on the derivatives with respect to a parameter of the solutions of a system of differential equations. Ann. Math. **2**, 292–296 (1918–1919)
8. Li, L., Meng, F., He, L.: Some generalized integral inequalities and their applications. J. Math. Anal. Appl. **372**, 339–349 (2010)
9. Pachpatte, B.G.: A note on Gronwall type integral and integro differential inequalities. Tamkang J. Math. **8**, 53–59 (1977)
10. Pachpatte, B.G.: On some fundamental integro differential inequalities for differential equations. Chinese J. Math. **6**(1), 17–23 (1978)
11. Pachpatte, B.G.: Inequalities for Differential and Integral Equations. Academic Press, New York and London (1998)
12. Pachpatte, B.G.: Integral and finite difference inequalities and applications. In: North-Holland Mathematics Studies, vol. 205. (2006)
13. Tidke, H.L., Dhakne, M.B.: Existance of solutions for nolinear mixed type integrodifferential equation of second order. Surv. Math. Appl. **5**, 61–72 (2010)
14. Young-Ho, K.: Gronwall-Bellman and Pachpatte type integral inequalities with applications. Nonlinear Anal. **71**, e2641–e2656 (2009)

Buckling and Vibration of Functionally Graded Circular Plates Resting on Elastic Foundation

Neha Ahlawat and Roshan Lal

Abstract The axisymmetric vibrations of functionally graded circular plates subjected to uniform in-plane force resting on elastic foundation have been analysed on the basis of classical plate theory. The material properties, i.e. Young's modulus and density vary continuously through the thickness of the plate and obey a power law distribution of the volume fraction of the constituents. Differential transform method was employed to solve the differential equation governing the motion of simply supported plates. The effect of various plate parameters was studied on the first three modes of vibration. By allowing the frequency to approach zero, the critical buckling loads for the plate were computed. A comparison of results with those available in the literature has been presented.

Keywords Functionally graded circular plates · Buckling · Differential transform · Foundation · Axisymmetric vibrations

1 Introduction

Many researches dealing with vibration characteristics of functionally graded material (FGM) plates have appeared in the literature due to their wide application in nuclear energy reactors, solar energy generators, space shuttle, etc. FGMs are usually made from a mixture of ceramic and metal as they are able to withstand high temperature gradient environments while maintaining their structural integrity.

Numerous studies on the static/dynamic behaviour of FGM plates of various geometries have been made and reported in Refs. [1–6], to mention a few. Of these, Ref. [1] is an excellent review of the work upto 2012, on the deformation, stress, vibration and stability problems of FG plates. In Ref. [2], Feldman and Aboudi analysed the bifurcational buckling of FG rectangular plates employing a combination of micromechanical and structural approach. Abrate [3] studied the free vibrations, buckling

N. Ahlawat (✉) · R. Lal
Department of Mathematics, Indian Institute of Technology Roorkee, Roorkee 247667, India
e-mail: ahlawatneha@gmail.com

R. Lal
e-mail: rlatmfma@iitr.ac.in

P.N. Agrawal et al. (eds.), *Mathematical Analysis and its Applications*,
Springer Proceedings in Mathematics & Statistics 143,
DOI 10.1007/978-81-322-2485-3_45

and static deflections of FG plates. The technique of differential transform developed by Zhou [4] and Chen and Ho [5] for solving the initial value problems was employed by Malik and Dang [6] in studying the free vibration of continuous systems, particularly for thin beams. Later, this method was extended to study the free axisymmetric vibrations of isotropic/ FGM circular plates of uniform/non-uniform thickness with various boundary conditions/constraints in Refs. [7, 8]. Kumar and Lal [9] predicted the natural frequencies for axisymmetric vibrations of two-directional FG annular plates resting on Winkler foundation using differential quadrature method and Chebyshev collocation technique.

In the present study, the effect of in-plane force on the axisymmetric vibrations of FGM circular plates resting on elastic foundation using differential transformation method (DTM) has been studied. The material properties, i.e. Young's modulus and density are assumed to be graded in the thickness direction and these properties vary according to a power-law in terms of volume fractions of the constituents. The natural frequencies are obtained for simply supported boundary condition with different values of volume fraction index, in-plane force parameter and foundation parameter. A comparison of results is given.

2 Mathematical Formulation

Consider an FGM circular plate of radius a, thickness h, mass density ρ subjected to uniform in-plane tensile force N_0, resting on elastic foundation of modulus k_f and referred to a cylindrical polar coordinate system (R, θ, z), $z = 0$ being the middle plane of the plate. The top and bottom surfaces are $z = +h/2$ and $z = -h/2$, respectively. The line $R = 0$ is the axis of the plate. The equation of motion governing transverse axisymmetric vibration of the present model (Fig. 1) is given by [13]:

$$Dw_{,RRRR} + \frac{2}{R}Dw_{,RRR} - \frac{1}{R^2}\left[D + R^2N_0\right]w_{,RR} + \frac{1}{R^3}\left[D - R^2N_0\right]w_{,R} + k_f w + \rho h w_{,tt} = 0 \tag{1}$$

where w is the transverse deflection, D the flexural rigidity and ν the Poisson's ratio. Here, a comma followed by a suffix denotes the partial derivative with respect to that variable.

For a harmonic solution, the deflection w can be expressed as

$$w(R,t) = W(R)e^{i\omega t} \tag{2}$$

where ω is the radian frequency. Equation (1) reduces to

$$DW_{,RRRR} + \frac{2}{R}DW_{,RRR} - \frac{1}{R^2}DW_{,RR} + \frac{1}{R^3}DW_{,R} - N_0W_{RR} - \frac{N_0}{R}W_{,R} + k_f W - \rho h\omega^2 W = 0 \tag{3}$$

Assuming that the top and bottom surfaces of the plate are ceramic and metal-rich, respectively, for which the variations of the Young's modulus $E(z)$ and the density

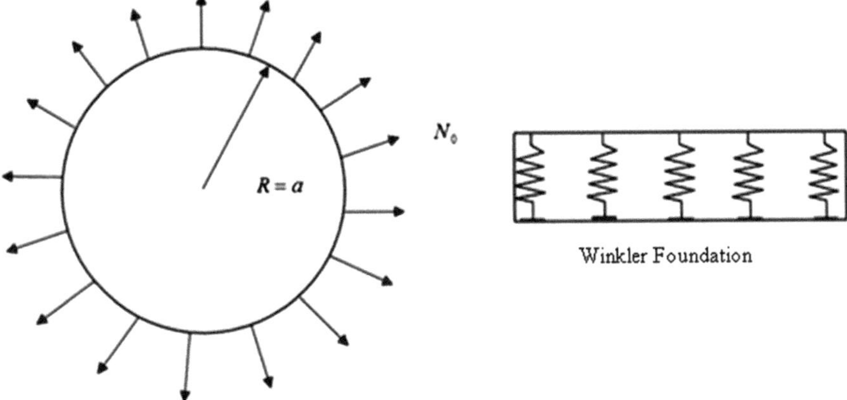

Fig. 1 Functionally graded circular plate under uniform tensile load N_0

$\rho(z)$ in the thickness direction are taken as

$$E(z) = (E_c - E_m)\left(\frac{z}{h} + \frac{1}{2}\right)^g + E_m \tag{4}$$

$$\rho(z) = (\rho_c - \rho_m)\left(\frac{z}{h} + \frac{1}{2}\right)^g + \rho_m \tag{5}$$

where E_c, ρ_c and E_m, ρ_m denote the Young's modulus and the density of ceramic and metal constituents, respectively, and g is the volume fraction index.

The flexural rigidity and mass density are given by

$$D = \frac{1}{1-\nu^2}\int_{-h/2}^{h/2} E(z)z^2 dz \tag{6}$$

$$\rho = \frac{1}{h}\int_{-h/2}^{h/2} \rho(z)dz \tag{7}$$

Substituting Eq. (4) and Eq. (5) into Eq. (6) and Eq. (7), we obtain

$$D = \frac{h^3}{1-\nu^2}\left[(E_c - E_m)\frac{g^2+g+2}{4(g+1)(g+2)(g+3)} + \frac{E_m}{12}\right] \tag{8}$$

$$\rho = \frac{\rho_c + \rho_m g}{g+1} \tag{9}$$

Using non-dimensional variables, $r = R/a, \quad f = W/a$, Eq. (3) now becomes

$$Df_{,rrrr} + \frac{2}{r}Df_{,rrr} - \frac{1}{r^2}Df_{,rr} + \frac{1}{r^3}Df_{,r} - N_0a^2 f_{,rr} - \frac{N_0}{r}a^2 f_{,r} + k_f f = \rho a^4 \Omega^2 fh \tag{10}$$

Substituting the values of D and ρ from Eq. (8) and Eq. (9), the Eq. (10) can be written as

$$r^3 Bf_{,rrrr} + 2r^2 Bf_{,rrr} - rBf_{,rr} + Bf_{,r} - Nr^3 f_{,rr} - Nr^2 f_{,r} + k_f r^3 f = r^3 \Omega^2 \left(\frac{\rho_c + \rho_m g}{\rho_c(g+1)}\right) f \tag{11}$$

where $D = D^* B,\ N = \frac{N_0}{D^*}a^2,\ \Omega^2 = \frac{\rho_c h a^4}{D^*}\omega^2,\ D^* = \frac{E_c h^3}{12(1-\nu^2)},\ K_f = \frac{k_f a^4}{D^*}$

$$B = \left[3\left(1 - \frac{E_m}{E_c}\right)\frac{g^2+g+2}{(g+1)(g+2)(g+3)} + \frac{E_m}{E_c}\right]$$

2.1 Boundary Condition

The relations which should be satisfied for a simply-supported plate at the edge are

$$f(1) = 0, \qquad M_r|_{r=1} = \left[-D\left\{\frac{d^2 f}{dr^2} + \nu\left(\frac{1}{r}\frac{df}{dr}\right)\right\}\right]_{r=1} = 0 \tag{12}$$

where M_r is the radial bending moment.

2.2 Regularity conditions

For the axisymmetric boundary conditions, the regularity conditions at the centre ($r = 0$) of the circular plate can be defined as

$$\frac{df}{dr}|_{r=0} = 0, \qquad Q_r|_{r=0} = \left(\frac{d^3 f}{dr^3} + \frac{1}{r}\frac{d^2 f}{dr^2} - \frac{1}{r^2}\frac{df}{dr}\right)_{r=0} = 0 \tag{13}$$

3 Method of Solution: Description of the Method

The differential transform of the kth derivative of $f(r)$ is given by

$$F_k = \frac{1}{k!}\left[\frac{d^k f(r)}{dr^k}\right]_{r=r_0} \tag{14}$$

where $f(r)$ is the original function and F_k is the transformed function.

The inverse transformation of the function F_k is defined as

$$f(r) = \sum_{k=0}^{\infty} (r - r_0)^k F_k$$

In actual applications, the function $f(r)$ is expressed by a finite series. So, the above expression may be written as

$$f(r) = \sum_{k=0}^{n} (r - r_0)^k F_k \tag{15}$$

The convergence of the natural frequencies decides the value of n. Some basic theorems which are frequently used in practical problems are given in Table 1.

3.1 Transformation of the Governing Differential Equation

Applying the transformation rules given in Table 1, the transformed form of the governing differential equation (11) around $r_0 = 0$ can be written as

$$B[(k^2 - 1)^2]F_{k+1} + K_f F_{k-3} - N(k-1)^2 F_{k-1} = \Omega^2 \left(\frac{\rho_c + \rho_m g}{\rho_c (g+1)} \right) F_{k-3} \tag{16}$$

3.2 Transformation of the Boundary/Regularity Conditions

By applying transformations rules given in Table 1, the Eq. (12) and Eq. (13) becomes

$$\sum_{k=0}^{n} F_k = 0, \quad \sum_{k=0}^{n} [k(k-1) + vk] F_k = 0 \tag{17}$$

$$F_1 = F_3 = F_5 = F_7 = \cdots = F_{4k+1} = F_{4k+3} = 0 \tag{18}$$

Table 1 Transformation rules for one-dimensional DTM

Original functions	Transformed functions
$f(r) = g(r) \pm h(r)$	$F_k = G_k \pm H_k$
$f(r) = \lambda g(r)$	$F_k = \lambda G_k$
$f(r) = g(r)h(r)$	$F_k = \sum_{l=0}^{k} G_l H_{k-l}$
$f(r) = \frac{d^n g(r)}{dr^n}$	$F_k = \frac{(k+n)!}{k!} G_{k+n}$
$f(r) = r^n$	$F_k = \delta(k-n) = \begin{cases} 1, & k = n \\ 0, & k \neq n \end{cases}$

4 Frequency Equations

The use of regularity condition (18) in the equation of motion, i.e. Eq. (16), gives

$$F_{2k} = \left[\frac{1}{16Bk^2(k-1)^2}\right] \cdot \left[4N(k-1)^2 F_{2k-2} + \left\{\Omega^2\left(\frac{\rho_c + \rho_m g}{\rho_c(g+1)}\right) - K_f\right\} F_{2k-4}\right] \quad (19)$$

Also, the use of regularity condition (18) in the relations (17) given for simply supported edge condition leads to

$$\sum_{k=0}^{n} F_{2k} = 0, \quad \sum_{k=0}^{n} [2k(2k-1) + 2\nu k] F_{2k} = 0 \quad (20)$$

Applying the boundary condition (20) on the resulting F_{2k} expressions, we get the following equations:

$$\Phi_{11}^{(m)}(\Omega) F_0 + \Phi_{12}^{(m)}(\Omega) F_2 = 0$$

$$\Phi_{21}^{(m)}(\Omega) F_0 + \Phi_{22}^{(m)}(\Omega) F_2 = 0 \quad (21)$$

where $\Phi_{11}^{(m)}$,$\Phi_{12}^{(m)}$,$\Phi_{21}^{(m)}$ and $\Phi_{21}^{(m)}$ are polynomials in Ω of degree m where $m = 2n$. Equation (21) can be expressed in matrix form as follows

$$\begin{bmatrix} \Phi_{11}^{(m)}(\Omega) & \Phi_{12}^{(m)}(\Omega) \\ \Phi_{21}^{(m)}(\Omega) & \Phi_{22}^{(m)}(\Omega) \end{bmatrix} \begin{Bmatrix} F_0 \\ F_2 \end{Bmatrix} = \begin{Bmatrix} 0 \\ 0 \end{Bmatrix} \quad (22)$$

For a non-trivial solution of Eq. (22), the frequency determinant must vanish and hence

$$\begin{vmatrix} \Phi_{11}^{(m)}(\Omega) & \Phi_{12}^{(m)}(\Omega) \\ \Phi_{21}^{(m)}(\Omega) & \Phi_{22}^{(m)}(\Omega) \end{vmatrix} = 0 \quad (23)$$

5 Numerical Results and Discussion

The frequency Eq. (23) provides the values of the frequency parameter Ω. The lowest three roots of this equation have been obtained using MATLAB. The values of elastic constants are

Table 2 Convergence study for first three modes of vibration for $N = 20$, $K_f = 100$, $g = 5$

No. of terms	First mode	Second mode	Third mode
10	17.2528	37.6839	68.9160
12	17.2528	37.6838	71.5054
14	17.2528	37.6838	71.6026
16	17.2528	37.6838	71.6037
18	17.2528	37.6838	71.6037

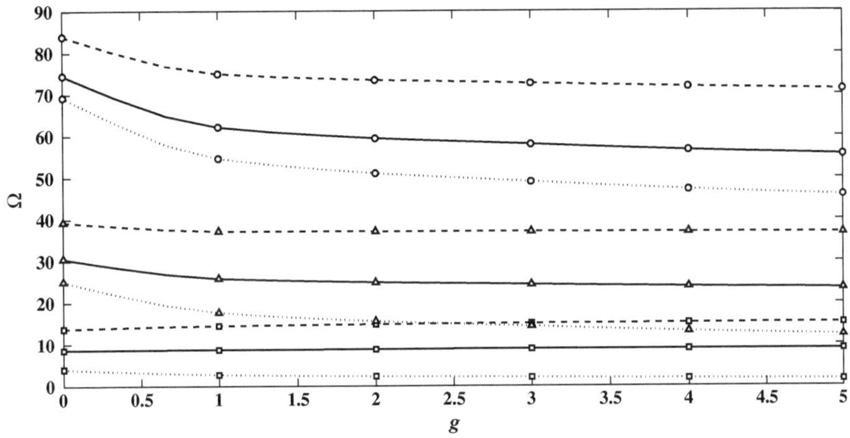

Fig. 2 Frequency parameter Ω for $K_f = 50$; _ _ _ _ _, $N = 20$; ________, $N = 0$; ...., $N = -10$, Box, firstmode; △, secondmode; ○, thirdmode

$$\text{Youngs modulus and density for aluminium: } E_m = 70\,Gpa,\ \rho_m = 2,702\,kg/m^3$$

$$\text{and for alumina: } E_c = 380\,Gpa,\ \rho_c = 3,800\,kg/m^3$$

The values of the volume fraction index g are taken as: 0, 1, 3, 5; the in-plane force parameter N : −20, −10, 0, 10, 20, foundation parameter K_f : 10, 50, 100 and $v = 0.3$. In order to choose an appropriate value of the number of terms 'n' in Eq. (20), a computer program has been developed and run for various values of g, N and K_f. The convergence of frequency parameter for the first three modes of vibration for a specified plate taking $g = 5$, $N = 20$ and $K_f = 100$ is shown in Table 2, as maximum deviations were observed for this data. The value of n has been fixed as 18. The values of frequency parameter Ω for different values of plate parameters has been given in Table 3. Figure 2 shows the behaviour of Ω verses g for $N = -10, 0, 20$ and $K_f = 50$ for all the three modes of vibration. The value of Ω

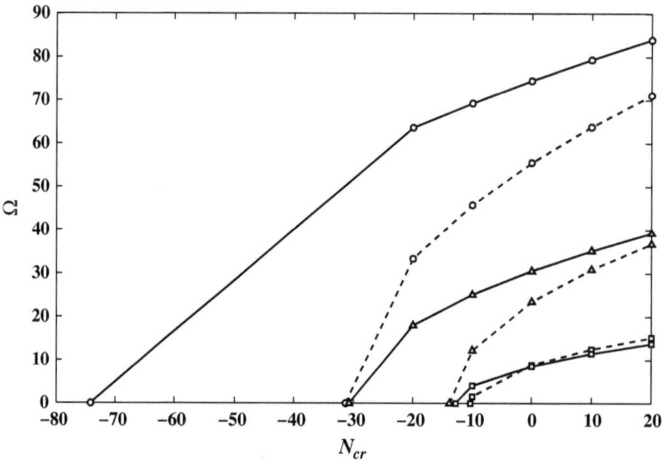

Fig. 3 Critical buckling loads N_{cr} for $K_f = 50$; _ _ _ _ _, $g = 5$; ________, $g = 0$; *Box*, first mode; △, second mode; ◯, third mode

increases with the increasing value of N, whatever be the value of g and decreases vice versa. The value of Ω is found to decrease with increasing value of g for both tensile and compressive values of N except in fundamental mode of vibration. In this case, Ω is found to increase continuously for tensile in-plane forces. The graph for critical buckling load for the three mode of vibration for $g = 0$ and 5 for the plate has been plotted in Fig. 3. It can be seen that the values of critical buckling loads N_{cr} for an isotropic plate ($g = 0$) are higher than that for an FGM plate ($g = 5$) for all the three modes of vibration. Figure 4 shows the effect of foundation parameter K_f on Ω for $g = 5$ and $g = 0$ for all the modes. It is clear that Ω increases with increasing value of K_f. The values of critical buckling load parameter N_{cr} in compression for different values of volume fraction index $g = 0$, 1, 3, 5 for $K_f = 50$ are reported in Table 4. The values of N_{cr} decrease with the increasing value of g. The results for N_{cr} for an isotropic plate have been compared in Table 5 with Gupta and Ansari [10] obtained by using Ritz method and Vol'mir [11] exact solutions. An excellent agreement among the results has been noticed.

The results for Ω have been compared in Table 6 with Leissa [12] obtained by series 151 solution and Wu et al. [13] obtained by generalized differential quadrature method. A close agreement of the results shows the versatility of the present technique.

Table 3 Frequency parameter Ω for FGM circular plate

g	N	$K_f = 10$			$K_f = 50$			$K_f = 100$		
↓	↓	I	II	III	I	II	III	I	II	III
0	-20	*	16.8557	63.3356	*	18.0032	63.6506	2.8229	19.3420	64.0422
	-10	*	24.2607	68.9946	4.0369	25.0716	69.2838	8.1423	26.0496	69.6437
	0	5.8614	29.8878	74.2235	8.6230	30.5496	74.4924	11.1515	31.3573	74.8273
	10	9.6085	34.6126	79.1076	11.5031	35.1857	79.3600	13.5027	35.8892	79.6744
	20	12.2575	38.7661	83.7072	13.7930	39.2786	83.9458	15.4999	39.9100	84.2431
1	-20	*	*	45.4645	*	*	45.9758	*	4.8035	46.6071
	-10	*	16.3378	54.2428	2.7105	17.7110	54.6720	8.1111	19.2905	55.2039
	0	5.3428	24.9600	61.7867	8.6776	25.8796	62.1639	11.5648	26.9852	62.6323
	10	9.8124	31.2926	68.5052	11.9598	32.0310	68.8456	14.1944	32.9306	69.2687
	20	12.8046	36.5442	74.6212	14.5160	37.1784	74.9339	16.4060	37.9562	75.3228
3	-20	*	*	37.3981	*	*	38.0747	*	*	38.9040
	-10	*	12.3957	48.5235	2.0263	14.3080	49.0469	8.2425	16.3876	49.6934
	0	5.2319	23.2909	57.5376	8.8566	24.3625	57.9797	11.9278	25.6392	58.5276
	10	10.0671	30.5213	65.3197	12.3456	31.3467	65.7094	14.7053	32.3489	66.1934
	20	13.2382	36.3408	72.2687	15.0438	37.0367	72.6212	17.0338	37.8887	73.0593
5	-20	*	*	32.6063	*	*	33.4045	*	*	34.3761
	-10	*	9.8802	45.2698	1.5612	12.2599	45.8480	8.2641	14.7025	46.5607
	0	5.1535	22.3306	55.0973	8.9020	23.4807	55.5734	12.0459	24.8435	56.1628
	10	10.1426	30.0014	63.4202	12.4723	30.8670	63.8342	14.8801	31.9160	64.3480
	20	13.3833	36.0767	70.7711	15.2250	36.7996	71.1424	17.2528	37.6838	71.6037

*denotes the fact that frequency parameter does not exist because of buckling

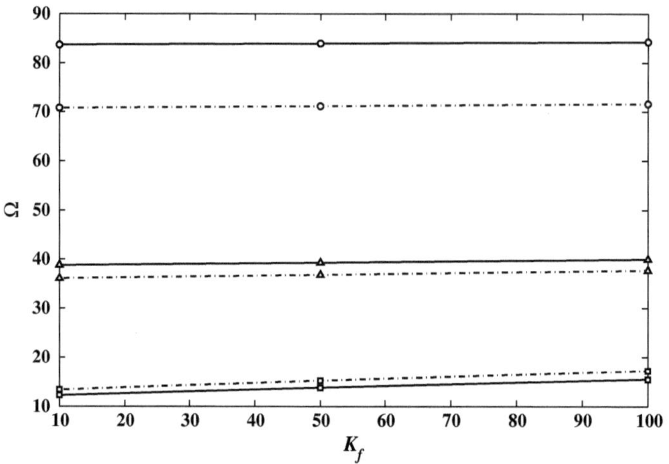

Fig. 4 Frequency parameter Ω for ________, $g = 5$ $N = 20$; _._._._._, $g = 0$ $N = 20$; and $K_f = 10, 50, 100$; *Box*, first mode; △, second mode; ○, third mode

Table 4 The critical buckling loads in compression for $K_f = 50$

Modes	$g = 0$	$g = 1$	$g = 3$	$g = 5$
I	12.8013	11.0752	10.5465	10.3115
II	30.7213	18.8872	15.3517	13.8384
III	74.1492	44.1788	35.1538	31.3409

Table 5 Comparison of critical buckling load parameter N_{cr} for isotopic plate ($g = 0$, $K_f = 0$)

Reference	First mode	Second mode	Third mode
Present	4.1978	29.0452	73.4768
Gupta and Ansari [10]	4.1978	29.0452	73.4768
Vol'mir [11]	4.1978	29.0452	73.4768

Table 6 Comparison of frequency parameter Ω for $N = 0$, $g = 0$, $K_f = 0$

Reference	First mode	Second mode	Third mode
Present	4.9351	29.7200	74.1561
Leissa [12]	4.977	29.76	74.20
Wu et al. [13]	4.935	29.72	74.156

6 Conclusions

Numerical results show that the values of the frequency parameter Ω for an isotropic plate are higher than that for FGM plate for the same set of values of other parameters. The value of the frequency parameter Ω increases with increasing value of in-plane force parameter, whatever be the value of g. The value of the frequency parameter increases in presence of the foundation, whatever be the value of g and N.

References

1. Jha, D.K., Kant, T., Singh, R.K.: A critical review of recent research on functionally graded plates. Compos. Struct. **96**, 833–849 (2013)
2. Feldman, E., Aboudi, J.: Buckling analysis of functionally graded plates subjected to uniaxial loading. Compos. Struct. **38**(1–4), 29–36 (1997)
3. Abrate, S.: Free vibration, buckling, and static deflections of functionally graded plates. Compos. Sci. Technol. **66**(14), 2383–2394 (2006)
4. Zhou, J.K.: Differential transformation and its application for electrical circuits. Huazhong University Press, Wuhan, China (1986)
5. Chen, C.K., Ho, S.H.: Solving partial differential equations by two dimensional differential transform. Appl. Math. Comput. **106**(2–3), 171–179 (1999)
6. Malik, M.: Huy Dang, H.: Vibration analysis of continuous systems by differential transformation. Appl. Math. Comput. **96**(1), 17–26 (1998)
7. Yalcin, H.S., Arikoglu, A., Ozkol, I.: Free vibration analysis of circular plates by differential transformation method. Appl. Math. Comput. **212**(2), 377–386 (2009)
8. Shariyat, M., Alipour, M.M.: A differential transform approach for modal analysis of variable thickness two-directional FGM circular plates on elastic foundations. Iran. J. Mech. Eng. **11**, 15–38 (2010)
9. Kumar, Y., Lal, R.: Prediction of frequencies of free axisymmetric vibration of two-directional functionally graded annular plates on Winkler foundation. Eur. J. Mech. A. Solids **42**, 219–228 (2013)
10. Gupta, U.S., Ansari, A.H.: Asymmetric vibrations and elastic stability of polar orthotropic circular plates of linearly varying profile. J. Sound Vib. **215**(2), 231–250 (1998)
11. Vol'mir, A.S.: Stability of elastic systems. FTD-MT-64-335, WP-FB, Ohio (1966)
12. Leissa, A.W.: Vibration of plates. NASA SP. 160, Washington (1969)
13. Wu, T.Y., Wang, Y.Y., Liu, G.R.: Free vibration analysis of circular plates using generalized differential quadrature rule. Comput. Meth. Appl. Mech. Eng. **191**, 5365–5380 (2002)

Discrete Fourier Transform and Extended Modified Hermite Polynomials

R.A. Malekar

Abstract Linear combinations of the solutions for modified second-order linear ordinary differential equations are related to eigenfunctions of discrete Fourier transform. This leads to in particular linear combinations of extended modified Hermite polynomials as eigenfunctions of discrete Fourier transform.

Keywords Discrete Fourier transform · Modified second order linear ordinary differential equation · Extended modified Hermite polynomials

1 Introduction

The Hemite functions are well-known eigenfunctions of the continuous Fourier transform. One of the problem associated with FT is to find a wider class of eigenfunctions. There has been work in this direction but the same has not been the case for the Discrete Fourier transform (DFT). The major work in this direction is based on the eigenfunctions of the continuous Fourier transform (FT) [1, 2]. The eigenfunctions of DFT are expressed in terms of derivatives of Jacobi theta functions [3]. Some of the well-known classical theta function identities are derived using the DFT [4, 5]. The recent work also uses the classical approach of constructing eigenfunctions of DFT in terms of Hermite and Gaussian functions, which are eigenfunctions of FT [6].

The eigenfunctions of DFT was generalized by Matveev [7] in terms of absolutely summable series. We extend the work of Matveev in this paper to generate eigenfunctions in terms of solutions to the modified second-order linear ordinary differential equations. This is illustrated in particular as a periodic extension of Hermite polynomials so that their linear combinations are eigenfunctions of DFT. This leads to another important class of eigenfunctions of DFT which comes from the solutions of modified second-order linear ordinary differential equation.

R.A. Malekar (✉)
National Defence Academy, Khadakwasala, Pune 411023, India
e-mail: rajanish1973@gmail.com

P.N. Agrawal et al. (eds.), *Mathematical Analysis and its Applications*,
Springer Proceedings in Mathematics & Statistics 143,
DOI 10.1007/978-81-322-2485-3_46

The well-known basic facts about DFT are discussed in Sect. 2. In Sect. 3 the linear combinations of solutions to modified second-order ODE are related as eigenfunctions of DFT. In Sect. 4 various properties of extended modified Hermite polynomials are discussed. In Sect. 5 linear combinations of extended modified Hermite polynomials are related to eigenfunctions of the DFT.

2 Preliminary Results

The matrix $\Phi(n)$ corresponding to the discrete Fourier transform of size n is given by

$$\Phi_{jk}(n) = \frac{1}{\sqrt{n}} q^{jk}, \qquad j, k = 0, \cdots, n-1, \qquad q = e^{\frac{2\pi i}{n}}. \tag{1}$$

Definition 1 For $f = (f_0, \cdots, f_{n-1})^t \in C^n$ we define the discrete Fourier transform $\tilde{f} \in C^n$ by $\tilde{f} = \Phi f = (\tilde{f}_0, \tilde{f}_1, \cdots, \tilde{f}_{n-1})$, where

$$\tilde{f}_k = \frac{1}{\sqrt{n}} \sum_{j=0}^{n-1} f_j \, e^{\frac{2\pi i jk}{n}}.$$

It is clear that $\Phi^4 = I$. Then any sequence (vector) see [7], f_j for $j = 0, 1, 2, \cdots, n-1$ generates eigenvector $v(k)$ of $\Phi : \Phi v = i^k v$ via formula

$$v_j(k) = f_j + (-1)^k \tilde{f}_j + (-1)^k f_{n-j} + (-i)^{3k} \tilde{f}_{-j}. \tag{2}$$

The multiplicities of the eigenvalues of the DFT $\Phi(n)$ are given by

$$\begin{array}{lllll}
n = 4m+2 \Rightarrow & m_1 = m, & m_2 = m+1, & m_3 = m, & m_0 = m+1. \\
n = 4m \quad\; \Rightarrow & m_1 = m, & m_2 = m, & m_3 = m-1, & m_0 = m+1. \\
n = 4m+1 \Rightarrow & m_1 = m, & m_2 = m, & m_3 = m, & m_0 = m+1. \\
n = 4m+3 \Rightarrow & m_1 = m+1, & m_2 = m+1, & m_3 = m, & m_0 = m+1
\end{array}$$

where m_k is the multiplicity of i^k see [8].

3 Modified Second-Order Linear ODE and DFT

The well-known special functions are solutions of second-order linear ODEs which are expressed as a power series. The fundamental existence and uniqueness theorem of ODE asserts that the linear second-order ODE of the form

$$\frac{d^2w}{dz^2} + P(z)\frac{dw}{dz} + Q(z)w(z) = 0 \tag{3}$$

with P(z), Q(z) are analytic and initial conditions $w(0) = c_0$ and $w^{'}(0) = c_1$ has a unique analytic solution in some region $|z| < R$ see [9].

Let

$$\sum_{m \geq 0} g_m = \sum_{m \geq 0} c_m z^m = w(z)$$

be a summable power series which is a solution of (3). In order to extend the summation over $\mathbb{Z}$, define

$$\begin{aligned} g_m &= 0 \qquad \text{if } m < 0, \\ &= c_m z^m \quad \text{if } m \geq 0. \end{aligned}$$

$$\sum_{m \in \mathbb{Z}} g_m = \sum_{m \geq 0} g_m = \sum_{m \geq 0} c_m z^m = w(z).$$

Let $q = e^{\frac{2\pi i}{n}}$. Define a continuous periodic function $w(z, t)$ of $t \in R$, and a continuous periodic sequence $\eta_{j+n} = \eta_j$ by

$$w(z, t) = \sum_{m \in \mathbb{Z}} g_m q^{mt} \quad \text{and} \quad \eta_j = \sum_{m \in Z} g_{nm+j}.$$

We have $w(z, t) = w(q^t z)$. Then by the change of variable in (3) the modified function $w(z, t)$ is the solution of the modified homogeneous linear differential equation given by

$$\frac{1}{q^{2t}}\frac{d^2w(z, t)}{dw^2} + P(q^t z)\frac{1}{q^t}\frac{dw(z, t)}{dz} + Q(q^t z)w(z, t) = 0. \tag{4}$$

We give the relationship between solutions of (4) and the eigenfunctions of DFT.

Theorem 1 *Let $w(z, t)$ be an analytic solution of modified linear second order homogeneous differential equation (4) with the initial conditions $w(0, t) = c_0$, $w'(0, t) = c_1$ and analytic coefficients. Then the vector $v(z, k)$ with the components $v_j(z, k)$ given by*

$$v_j(z, k) = \frac{1}{n}\sum_{l=0}^{n-1}\left(w(z, l) + (-1)^k w(z, -l)\right) q^{-jl} + \frac{(-i)^k}{\sqrt{n}}\left(w(z, j) + (-1)^k w(z, -j)\right) \tag{5}$$

is an eigenvector of DFT: $\Phi v = i^k v$.

Proof We note the following:

$$\sum_{j=0}^{n-1} q^{jl}\eta_j = w(z, l).$$

Applying the inversion formula we get

$$\sum_{m\geq 0} g_{nm+j} = \eta_j = \frac{1}{n}\sum_{l=0}^{n-1} w(z, l)q^{-jl}, \qquad j = 0, 1, 2\cdots, n - 1$$

Consider in (5) the following term:

$$\begin{aligned}
&\frac{(-i)^k}{\sqrt{n}}\Big(w(z, j) + (-1)^k w(z, -j)\Big) \\
&= \frac{(-i)^k}{\sqrt{n}} w(q^j z) + \frac{(-i)^{3k}}{\sqrt{n}} w(q^{-j} z) \\
&= \frac{(-i)^k}{\sqrt{n}} \sum_{m\geq 0} g_m q^{mj} + \frac{(-i)^{3k}}{\sqrt{n}} \sum_{m\geq 0} g_m q^{-mj} \\
&= \frac{(-i)^k}{\sqrt{n}} \sum_{r=0}^{n-1}\sum_{k\geq 0} g_{kn+r} q^{(kn+r)j} + \frac{(-i)^{3k}}{\sqrt{n}} \sum_{r=0}^{n-1}\sum_{k\geq 0} g_{kn+r} q^{-(kn+r)j} \\
&= \frac{(-i)^k}{\sqrt{n}} \sum_{r=0}^{n-1} \eta_r q^{rj} + \frac{(-i)^{3k}}{\sqrt{n}} \sum_{r=0}^{n-1} \eta_r q^{-rj}
\end{aligned}$$

Therefore (5) becomes

$$v_j(z, k) = \eta_j + (-1)^k \eta_{-j} + (-i)^k \tilde{\eta}_j + (-i)^{3k} \tilde{\eta}_{-j}. \tag{6}$$

From (2) it is clear that $v_j(z, k)$ is a component of the eigenvector for DFT. □

The particular cases of the differential equation (4) give rise to modified form of well-known special functions. The extended modified Hermite polynomials arise in this manner. We show in the next section that the linear combinations of the Hermite polynomials, the modified Hermite polynomials are eigenfunctions of DFT. These statements are verified in the following section. The properties of these functions are also studied.

4 Extended Modified Hermite Polynomials as Eigenfunctions of DFT

The classical orthogonal polynomials are solutions of second-order ODE. From Theorem 1 we have a linear combinations of all classical orthogonal polynomials such as Legendre, Hermite, Laguarre and Chebyshev in their modified form as an eigenfunctions of DFT. In this section we apply the result to the extended modified Hermite polynomials (EMHP) as eigenfunctions of DFT. The extended modified Hermite polynomials (EMHP) are defined by

$$H_n(z,t) = \sum_{r=0}^{[\frac{n}{2}]} \frac{(-1)^r\, n!}{r!\,(n-2r)!}(2z)^{n-2r} q^{rt} \quad \text{and} \quad t \in R, \qquad q = e^{\frac{2\pi i}{n}}$$

At $t = 0$ we get the classical Hermite polynomials. EMHP are related to Hermite polynomials by

$$H_n(z,t) = H_n\left(q^{-\frac{t}{2}} z\right) q^{\frac{nt}{2}}. \tag{7}$$

At $q^{-\frac{t}{2}} = i$, (7) reduces to modified Hermite polynomials given by

$$H_n(z,t) = H_n(iz) i^{-n}. \tag{8}$$

The modified polynomials of the type (8) are discussed in [10]. The paper [10] gives combinatorial interpretation of modified polynomials in terms of combinatorial probability defined on compound urn model. A real extension of classical Hermite polynomials in the modified form is discussed in [11]. They have been identified in the spectral approximation of boundary layer problems.

It is clear from (4), the functions $H_n(z,t)$ for a particular value of t is the solution of modified Hermite differential equation given by

$$q^t \frac{d^2y}{dz^2} - 2z\frac{dy}{dz} + n(n+1)y = 0. \tag{9}$$

The Rodrigues formula for EMHP in the modified form is given by

$$H_n(z,t) = (-1)^n\, q^{nt}\, e^{q^{-t}x^2} \frac{d^n}{dx^n} e^{-(q^{-t}x^2)}. \tag{10}$$

Simple calculations show that

$$H_0(z,t) = 1, \quad H_1(z,t) = 2z, \quad H_2(z,t) = 4z^2 - 2q^t,$$
$$H_3(z,t) = 8z^3 - 12zq^t, \quad H_4(z,t) = 16z^4 - 48q^t x^2 + 12q^{2t}.$$

In general, for all $n \geq 1$ the following relation holds (see [12]):

$$H_n(z,t) = 2zH_{n-1}(z,\breve{t}) - q^t H'_{n-1}(z,\breve{t}) \tag{11}$$

where $\breve{t} = \frac{(n-1)t}{n}$.

The following result follows from of Theorem 1:

Proposition 1 *The vector $H_n(z,k,t)$ with components*

$$v_j(z,k) = \frac{1}{n}\sum_{l=0}^{n-1}\Big(H_n(z,l) + (-1)^k H_n(z,-l)\Big)\, q^{-jl} + \frac{(-i)^k}{\sqrt{n}}\Big(H_n(z,j) + (-1)^k H_n(z,-j)\Big)$$

is an eigenvector of DFT $\Phi v = i^k v$. □

It is interesting to see the explicit form of eigenvectors of DFT $\Phi(n)$ in terms of EMHP for some particular values of n. For $n = 3$ eigenvector corresponding to eigenvalue +1 is given by

$$v(z,0) = \frac{1}{\sqrt{3}}\left[8z^3(\sqrt{3}+1) - 12z,\, 8z^3 - \left(\frac{\sqrt{3}-1}{2}\right)12z,\, 8z^3 - \left(\frac{\sqrt{3}-1}{2}\right)12z\right]^t.$$

The eigenvector corresponding to eigenvalue -1 is given by

$$v(z,2) = \frac{1}{\sqrt{3}}\left[8z^3(\sqrt{3}-1) + 12z,\, 8z^3 - \left(\frac{\sqrt{3}+1}{2}\right)3z,\, 8z^3 - \left(\frac{\sqrt{3}+1}{2}\right)12z\right]^t.$$

It is clear that all eigenvectors are real for real values of z though the polynomials used are with imaginary argument. It is possible to construct eigenvectors of DFT for further values of n applying the same techniques. As this construction is independent of the orthogonal polynomials involved, the same construction can be done for other orthogonal polynomials like the Laguerre and Chebyshev and Legender polynomials.

5 Conclusion

The construction of eigenvectors gives new class of eigenfunctions in terms of modified form of well-known special functions. This is an interesting example of the new way of generating the eigenvectors of DFT, which is now known in the literature. However, the EMHP are not orthogonal, they could be made orthogonal with suitable weight function. The further progress in this direction may be possible. The periodic extensions of these special functions should be of some interest in special function theory.

References

1. Grunbaum, F.A.: The eigenvectors of the discrete Fourier transform: a version of the Hermite functions. J. Math. Anal. Appl. **88**, 355–363 (1982)
2. Mehta, M.L.: Eigen values and eigenvectors of finite Fourier transform. J. Math. Phys. **28**(4), 781–785 (1987)
3. Ruzzi, M.: Jacobi theta functions and discrete Fourier transform. J. Math. Phys. **47**, 063507 (2006)
4. Malekar, R.A., Bhate, H.: Discrete Fourier transform and Jacobi theta function identities. J. Math. Phys. **51**, 023511 (2010)
5. Malekar, R.A., Bhate, H.: Discrete Fourier transform and Riemann identities for θ functions. Appl. Math. Lett. **25**, 1415–1419 (2012)
6. Hanna, M.T., et al.: Hermite Gausssian Like eigenvectors of the DFT Matrix based on the direct utilization of orthogonal projection matrices on its eigenspaces. IEEE Trans. Signal Process. **54**, (2006)
7. Matveev, V.B.: Intertwining relations between Fourier transform and discrete Fourier transform, the related functional identities and beyond. Inverse Prob. **17**, 633–657 (2001)
8. Auslander, L., Tolimieri, R.: Is computing with the finite Fourier transform pure or applied mathematics ? Bulletin American Math Soc **I** , 847–97 (1979)
9. Coddington, E.A., Levinson, N.: Theory of Ordinary Differential Equations. Tata McGraw-Hill Publishing Company Ltd, New Delhi (1992)
10. Charalambides, C.A.: Combinatorial probability interpretation of certain modified orthogonal polynomials. Eur. J. Comb. **29**, 1704–1716 (2008)
11. Adzic, N.: Modified Hermite polynomials in the spectral approximation for boundary layer problems. Bull. Austral. Math. Soc. 65L10 **45**, 267–276 (1992)
12. Folland, G.B.: Fourier Analysis and Its Applications. Wordsworth and Brooks/Cole Mathematics Series (1992)

A Numerical Simulation Based on Modified Keller Box Scheme for Fluid Flow: The Unsteady Viscous Burgers' Equation

B. Mayur Prakash, Ashish Awasthi and S. Jayaraj

Abstract In this paper the numerical solution of unsteady viscous Burgers' equation is presented. A combination of Modified Keller Box difference scheme and Hopf-Cole transformation is proposed to solve the Burgers' equation. The proposed scheme is an implicit scheme with second-order accuracy in space and time. Two test problems are considered to validate the proposed algorithm. Numerical results which are calculated for various values of kinematic viscosity and time steps are matching with the exact solution. It is also observed that, the proposed method yields satisfactory results for all the cases considered.

Keywords Burgers' equation · Hopf-Cole transformation · Finite difference method · Box method · Kinematic viscosity

1 Introduction

The Navier–Stokes equation is considered to be a cornerstone in fluid mechanics which, when expressed in its originality is a set of unsteady, nonlinear, second-order partial differential equations. Burgers' equation whose exact solution is well known, can be considered as a simplified form of the one-dimensional Navier–Stokes equation. Burgers' model is suitable for analysis in gas dynamics, shock wave theory,

B.M. Prakash (✉) · S. Jayraj
Department of Mechanical Engineering, National Institute of Technology Calicut, Kozhikode, India
e-mail: bonkile.mayur@gmail.com

S. Jayraj
e-mail: sjayaraj@nitc.ac.in

A. Awasthi
Department of Mathematics, National Institute of Technology Calicut, Kozhikode, India
e-mail: aawasthi@nitc.ac.in

P.N. Agrawal et al. (eds.), *Mathematical Analysis and its Applications*,
Springer Proceedings in Mathematics & Statistics 143,
DOI 10.1007/978-81-322-2485-3_47

cosmology, and traffic flow. The application of this model in various similar important fields, always require the solution of basic Burgers' equation.

In 1915, Harry Bateman (1882–1946) [1], an English mathematician, introduced the following Eq. (1) in his paper along with its initial (2) and boundary conditions (3):

$$\frac{\partial u}{\partial t} + u\frac{\partial u}{\partial x} = v\frac{\partial^2 u}{\partial x^2}, \quad 0 < x < L, \quad 0 < t < \tau \tag{1}$$

$$u(x, 0) = f(x), \quad 0 < x < L \tag{2}$$

$$u(0, t) = g_1(t), \quad u(L, t) = g_2(t), \quad 0 < t < \tau \tag{3}$$

where u, x, t and v are the velocity, spatial coordinate, time, and kinematic viscosity, respectively. The f, g_1, and g_2 are prescribed functions of variables depending upon the specific conditions for the problem to be solved. Later in 1948 Johannes Martinus Burgers (1895–1981) [4], a Dutch physicist, explained the mathematical modeling of turbulence with the help of Eq. (1). In order to honor the contributions of Burgers, this equation is well known as the "Burgers' equation." The simultaneous presence of nonlinear convective term ($u\frac{\partial u}{\partial x}$) and diffusive term ($v\frac{\partial^2 u}{\partial x^2}$) adds an additional feature to the Burgers' equation. When v approaches zero, Eq. (1) becomes inviscid Burgers' equation which is a model for nonlinear wave propagation. When u approaches zero, Eq. (1) becomes the heat equation.

Julian David Cole (1925–1999) [6] and E. Hopf (1902–1983) [8] independently introduced a transformation which converts Burgers' equation into linear heat equation and is solved exactly for an arbitrary initial condition. Hence, the transformation is famously known as Hopf-Cole transformation (4).

$$u(x, t) = -2v\frac{\theta_x}{\theta}, \tag{4}$$

where, θ satisfies the following heat equation:

$$\frac{\partial \theta}{\partial t} = v\frac{\partial^2 \theta}{\partial x^2} \tag{5}$$

Benton and Platman [2] were given 35 distinct analytical solutions of Burgers' equation with different initial conditions. Rodin [12] studied some approximate and exact solution of boundary value problem for Burgers' equation with the help of Hopf-Cole transformation. Kutluay et al. [11] used Hopf-cole transformation to convert Burgers' equation to heat equation. The transformed heat equation with the insulated boundary conditions was solved by explicit and exact-explicit finite difference method. Burns et al. [5] considered Burgers' equation with zero-Neumann boundary conditions to show that for moderate value of viscosity, numerical solution approaches nonconstant shock-type stationary solution. Based on Hopf-Cole linearization, Brander and Hedenfalk [3] solved Burgers' equation in one-space dimension for an arbitrary incident pulse of finite length. Restrictive Pade approximation classical implicit finite difference method was implemented by Gulsu [7], whose accuracy was demostrated by the two-test problem. In this paper, we consider the Modified Keller Box method [13]. Equation (1) is converted to linear heat equation (5) by the Hopf-Cole transformation (4), as explained by Kadalbajoo and Awasthi [9]. The present method has accuracy of second order in space and time. The accuracy and reliability of the present method is verified by performing several numerical experiments.

2 Difference Scheme

The solution domain of Eq. (5) is discretized with uniform mesh. The space interval [0,1] is divided into N equal subinterval. The time interval $[0, \tau]$ is divided into M equal subintervals. Assuming $\Delta x = 1/N$ as the mesh width in space and x_i is set as $x_i = i\Delta x$ for $i = 0, 1, ..., N$. Assuming $\Delta t = \tau/M$ as the mesh width in time and t^n is set as $t^n = n\Delta t$ for $n = 0, 1, ..., M$.

2.1 Keller Box Method

In Keller Box Method [10], second and higher derivatives of parabolic partial differential equation are replaced by first derivatives through the introduction of additional variables which result in a system of first-order equations. Equation (5) is written as a system of two first-order equations:

$$\frac{\partial \theta}{\partial x} = T \tag{6}$$

$$\frac{\partial \theta}{\partial t} = \nu \frac{\partial T}{\partial x} \tag{7}$$

We use only central differences about $(n + \frac{1}{2}, i - \frac{1}{2})$, making use of four points at the corners of a "box" (Fig. 1a). The resulting difference equations for Eqs. (6) and (7) are

$$\frac{\theta_i^{n+1} - \theta_{i-1}^{n+1}}{\Delta x} = T_{i-\frac{1}{2}}^{n+1} \tag{8}$$

$$\frac{\theta_{i-\frac{1}{2}}^{n+1} - \theta_{i-\frac{1}{2}}^{n}}{\Delta t} = \nu \left(\frac{T_i^{n+\frac{1}{2}} - T_{i-1}^{n+\frac{1}{2}}}{\Delta x} \right) \tag{9}$$

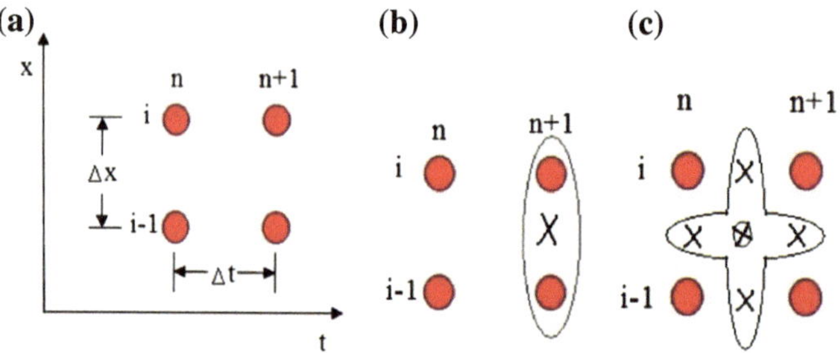

Fig. 1 **a** Grid for box scheme; **b** difference molecule for evaluation of $T^{n+1}_{i-\frac{1}{2}}$; **c** difference molecule for Eq. (9)

The discretized terms containing subscript or superscript $\frac{1}{2}$ in Eqs. (8) and (9) are defined as averages, for example,

$$\theta^{n+1}_{i-\frac{1}{2}} = \frac{\theta^{n+1}_{i} + \theta^{n+1}_{i-1}}{2} \tag{10}$$

$$T^{n+\frac{1}{2}}_{i} = \frac{T^{n}_{i} + T^{n+1}_{i}}{2} \tag{11}$$

Averaged expressions (10) and (11) are substituted into Eqs. (8) and (9). The resulting difference equations become

$$\frac{\theta^{n+1}_{i} - \theta^{n+1}_{i-1}}{\Delta x} = \frac{T^{n+1}_{i} + T^{n+1}_{i-1}}{2} \tag{12}$$

$$\frac{\theta^{n+1}_{i} + \theta^{n+1}_{i-1}}{\Delta t} = v\left(\frac{T^{n}_{i} - T^{n}_{i-1}}{\Delta x}\right) + \frac{\theta^{n}_{i} + \theta^{n}_{i-1}}{\Delta t} + v\left(\frac{T^{n+1}_{i} - T^{n+1}_{i-1}}{\Delta x}\right) \tag{13}$$

2.1.1 Modified Box Method

In Eqs. (12) and (13) T's can be expressed in terms of $\theta's$. Substituting Eq. (12) into Eq. (13), T^{n+1}_{i-1} is eliminated. Equation (12) is evaluated at time level n to eliminate T^{n}_{i-1}. Accordingly,

$$\frac{\theta^{n+1}_{i} + \theta^{n+1}_{i-1}}{\Delta t} = v\left(\frac{T^{n}_{i} + T^{n+1}_{i}}{\Delta x}\right) + \frac{\theta^{n}_{i} + \theta^{n}_{i-1}}{\Delta t} - v\left(\frac{\theta^{n+1}_{i} - \theta^{n+1}_{i-1}}{(\Delta x)^2} - \frac{T^{n+1}_{i}}{\Delta x} + 2\frac{\theta^{n}_{i} - \theta^{n}_{i-1}}{(\Delta x)^2} - \frac{T^{n}_{i}}{\Delta x}\right) \tag{14}$$

Equations (12) and (13) are rewritten with the i index advanced by 1.

$$\frac{\theta_{i+1}^{n+1} - \theta_i^{n+1}}{\Delta x} = \frac{T_{i+1}^{n+1} + T_i^{n+1}}{2} \tag{15}$$

$$\frac{\theta_{i+1}^{n+1} + \theta_i^{n+1}}{\Delta t} = v\left(\frac{T_{i+1}^{n} - T_i^{n}}{\Delta x}\right) + \frac{\theta_{i+1}^{n} + \theta_i^{n}}{\Delta t} + v\left(\frac{T_{i+1}^{n+1} - T_i^{n+1}}{\Delta x}\right) \tag{16}$$

To eliminate T_{i+1}^{n+1} and T_{i+1}^{n}, Eq. (15) is simply substituted into Eq. (16). The result is

$$\frac{\theta_{i+1}^{n+1} + \theta_i^{n+1}}{\Delta t} = 2v\left(\frac{(\theta_{i+1}^{n} - \theta_i^{n})}{(\Delta x)^2} - \frac{T_i^{n}}{\Delta x}\right) + \frac{\theta_{i+1}^{n} + \theta_i^{n}}{\Delta t} + 2v\left(\frac{(\theta_{i+1}^{n+1} - \theta_i^{n+1})}{(\Delta x)^2} - \frac{T_i^{n+1}}{\Delta x}\right) \tag{17}$$

Adding Eqs. (14) and (17) and after rearranging, the final expression is

$$\begin{aligned} &\left(1 - \frac{2v\Delta t}{(\Delta x)^2}\right)\theta_{i-1}^{n+1} + \left(2 + \frac{4v\Delta t}{(\Delta x)^2}\right)\theta_i^{n+1} + \left(1 - \frac{2v\Delta t}{(\Delta x)^2}\right)\theta_{i+1}^{n+1} \\ &= \left(1 + \frac{2v\Delta t}{(\Delta x)^2}\right)\theta_{i-1}^{n} + \left(2 - \frac{4v\Delta t}{(\Delta x)^2}\right)\theta_i^{n} + \left(1 + \frac{2v\Delta t}{(\Delta x)^2}\right)\theta_{i+1}^{n} \end{aligned} \tag{18}$$

Introducing $l = \dfrac{2v\Delta t}{(\Delta x)^2}$ in this equation the result is written in the tridiagonal format as

$$a_i\theta_{i-1}^{n+1} + b_i\theta_i^{n+1} + c_i\theta_{i+1}^{n+1} = d_i \tag{19}$$

where,

$$a_i = 1 - \frac{2v\Delta t}{(\Delta x)^2} = (1 - l) \tag{20}$$

$$b_i = 2 + \frac{4v\Delta t}{(\Delta x)^2} = 2(1 + l) \tag{21}$$

$$c_i = 1 - \frac{2v\Delta t}{(\Delta x)^2} = (1 - l) \tag{22}$$

$$\begin{aligned} d_i &= \left(1 + \frac{2v\Delta t}{(\Delta x)^2}\right)\theta_{i-1}^{n} + \left(2 - \frac{4v\Delta t}{(\Delta x)^2}\right)\theta_i^{n} + \left(1 + \frac{2v\Delta t}{(\Delta x)^2}\right)\theta_{i+1}^{n} \\ &= (1 + l)(\theta_{i-1}^{n} + \theta_{i+1}^{n}) + 2(1 - l)\theta_i^{n} \end{aligned} \tag{23}$$

After assembling the entire system of equations and applying boundary conditions, the general form is $F\theta = d$, where F and θ are the matrices of order N×N and N ×1 respectively, given by

$$F = \begin{pmatrix} 2(1+l) & 2(1-l) & & & & \\ (1-l) & 2(1+l) & (1-l) & & & \\ & \cdot & \cdot & \cdot & & \\ & & \cdot & \cdot & \cdot & \\ & & & \cdot & \cdot & \cdot \\ & & & (1-l) & 2(1+l) & (1-l) \\ & & & & 2(1-l) & 2(1+l) \end{pmatrix} \quad \theta = \begin{pmatrix} \theta_1^{n+1} \\ \theta_2^{n+1} \\ \cdot \\ \cdot \\ \cdot \\ \theta_{N-1}^{n+1} \\ \theta_N^{n+1} \end{pmatrix}$$

The approximate solution of Burgers' equation (1) in terms of the approximate solution of heat equation using the Hopf-Cole transformation (4) is given by

$$u_i^n = -\nu \left(\frac{\theta_{i+1}^n - \theta_{i-1}^n}{\Delta x \theta_i^n} \right) \tag{24}$$

The proposed scheme is unconditionally stable based on [14].

3 Numerical Experiments and Discussion

This section contain the results of two examples to validate the theoretical results obtained. Hopf-Cole transformation is used to find the exact solution of these examples. The fluid properties considered include lubricating oil at 40 °C ($\nu = 1\,\text{cm}^2/\text{s}$), saturated fatty acid methyl esters at 40 °C ($\nu = 0.1\,\text{cm}^2/\text{s}$), water at 20 °C ($\nu = 0.01\,\text{cm}^2/\text{s}$), water at 50 °C ($\nu = 0.005\,\text{cm}^2/\text{s}$) for numerical experiments.

Example 1 Burgers' equation (1) with initial condition and homogeneous boundary conditions

$$u(x,0) = sin(\pi x), \quad 0 < x < 1,$$

$$u(0,t) = u(1,t) = 0, \quad 0 \le t \le \tau$$

By using Hopf-Cole transformation (4) Eq. (1) is transformed to the linear heat equation (5) with initial condition (25) and boundary conditions (no heat transfer) (26)

$$\theta(x,0) = exp\left(-\frac{1}{2\pi\nu}[1 - cos(\pi x)]\right), \quad 0 < x < 1 \tag{25}$$

$$\theta_x(0,t) = \theta_x(1,t) = 0, \quad 0 \le t \le \tau. \tag{26}$$

Example 2 Burgers' equation (1) with the following initial condition and boundary conditions:

$$u(x,0) = 4x(1-x), \quad 0 < x < 1,$$

$$u(0,t) = 0 = u(1,t), \quad 0 \le t \le \tau.$$

Using Hopf-Cole transformation (4) Eq. (1) is transformed to linear heat equation (5) with initial condition and with boundary conditions (no heat transfer)

$$\theta(x,0) = exp\left(-\frac{1}{2v}\left[2x^2 - \frac{4x^3}{3}\right]\right), \quad 0 < x < 1 \tag{27}$$

$$\theta_x(0,t) = \theta_x(1,t) = 0, \quad 0 \le t \le \tau. \tag{28}$$

Exact solution was elaborated in [9] for both Examples (1) and (2).

Computed results are displayed in Tables 1, 2, 3 and 4 at different nodal points for diverse values of kinematic viscosity. It is found that the computed results show better agreement with the exact solution as the mesh size is refined. These results show that

Table 1 Comparison of the numerical solution with exact solution at different space points of Example 1 at $T = 0.1$ for $v = 1$ and $\Delta t = 0.0001$

x	N = 10	N = 20	N = 40	N = 80	Exact solution
0.1	0.10596	0.10871	0.10940	0.10958	0.10953
0.2	0.20298	0.20822	0.20954	0.20988	0.20979
0.3	0.28249	0.28973	0.29156	0.29202	0.29189
0.4	0.33683	0.34538	0.34754	0.34808	0.34792
0.5	0.35987	0.36891	0.37118	0.37175	0.37157
0.6	0.34787	0.35651	0.35868	0.35922	0.35904
0.7	0.30037	0.30775	0.30960	0.31006	0.30990
0.8	0.22087	0.22625	0.22760	0.22793	0.22781
0.9	0.11702	0.11986	0.12057	0.12075	0.12068

Table 2 Comparison of the numerical solution with the exact solution at different space points of Example 1 at $T = 0.01$ for $v = 10$ and $\Delta t = 0.00001$

x	N = 10	N = 20	N = 40	N = 80	Exact solution
0.1	0.11099	0.11378	0.11449	0.11466	0.11461
0.2	0.21127	0.21659	0.21793	0.21826	0.21816
0.3	0.29113	0.29845	0.30029	0.30075	0.30061
0.4	0.34274	0.35135	0.35352	0.35406	0.35389
0.5	0.36097	0.37002	0.37230	0.37287	0.37269
0.6	0.34386	0.35247	0.35464	0.35518	0.35501
0.7	0.29293	0.30026	0.30210	0.30257	0.30242
0.8	0.21307	0.21840	0.21974	0.22008	0.21997
0.9	0.11210	0.11490	0.11561	0.11578	0.11573

Table 3 Comparison of the numerical solution with the exact solution at different space points of Example 2 at $T = 0.1$ for $v = 1$ and $\Delta t = 0.0001$

x	N = 10	N = 20	N = 40	N = 80	Exact solution
0.1	0.10919	0.11203	0.11275	0.11293	0.11289
0.2	0.20921	0.21462	0.21600	0.21634	0.21625
0.3	0.29126	0.29873	0.30062	0.30109	0.30097
0.4	0.34742	0.35624	0.35846	0.35902	0.35886
0.5	0.37135	0.38067	0.38301	0.38360	0.38342
0.6	0.35914	0.36804	0.37028	0.37084	0.37066
0.7	0.31023	0.31784	0.31975	0.32022	0.32007
0.8	0.22821	0.23375	0.23514	0.23549	0.23537
0.9	0.12094	0.12387	0.12460	0.12478	0.12472

Table 4 Comparison of the numerical solution with the exact solution at different space points of Example 2 at $T = 0.01$ for $v = 10$ and $\Delta t = 0.00001$

x	N = 10	N = 20	N = 40	N = 80	Exact solution
0.1	0.11453	0.11742	0.11814	0.11833	0.11827
0.2	0.21802	0.22351	0.22489	0.22524	0.22513
0.3	0.30044	0.30799	0.30989	0.31037	0.31022
0.4	0.35372	0.36259	0.36483	0.36539	0.36521
0.5	0.37254	0.38187	0.38422	0.38481	0.38463
0.6	0.35491	0.36378	0.36602	0.36658	0.36640
0.7	0.30235	0.30991	0.31181	0.31229	0.31214
0.8	0.21993	0.22543	0.22681	0.22716	0.22705
0.9	0.11571	0.11860	0.11933	0.11951	0.11946

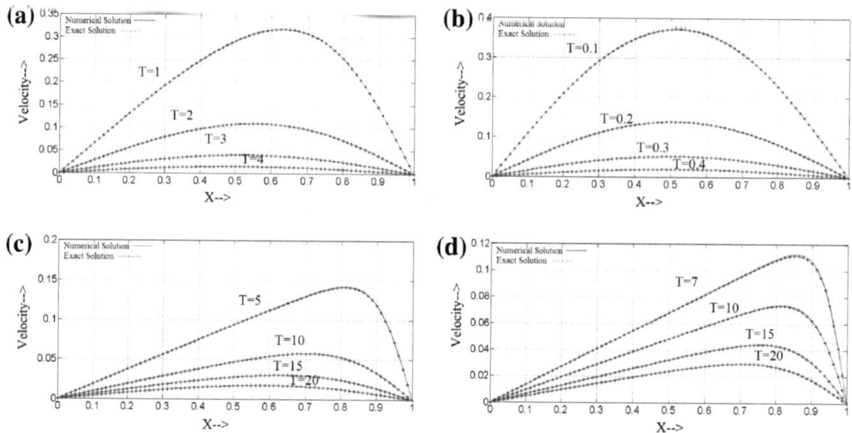

Fig. 2 Numerical solution of Example 1 at several times for N = 80 with distinguishable liquids having different values of v and Δt, **a** saturated fatty acid methyl esters at 40 °C with $v = 0.1$, $\Delta t = 0.0001$; **b** lubricating oil at 40 °C with $v = 1$, $\Delta t = 0.0001$; **c** water at 20 °C with $v = 0.01$, $\Delta t = 0.01$; **d** water at 50 °C with $v = 0.005$, $\Delta t = 0.01$

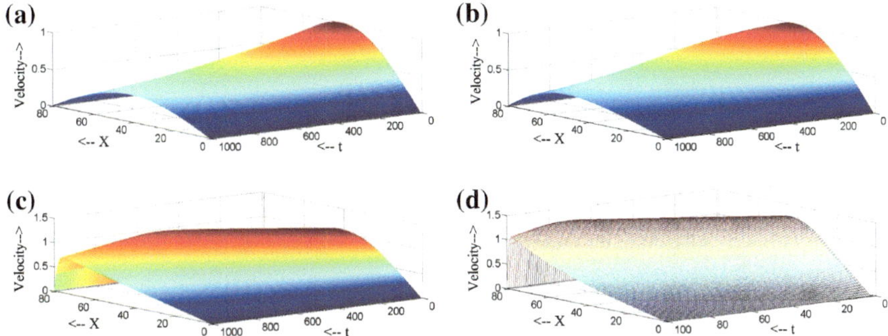

Fig. 3 Numerical solution profiles of Example 1 at $\Delta x = 0.0125$ for, **a** lubricating oil at 40 °C with $v = 1$, $\Delta t = 0.0001$; **b** Saturated fatty acid methyl esters at 40 °C with $v = 0.1$, $\Delta t = 0.001$; **c** Water at 20 °C with $v = 0.01$, $\Delta t = 0.001$; **d** Water at 50 °C with $v = 0.005$, $\Delta t = 0.01$

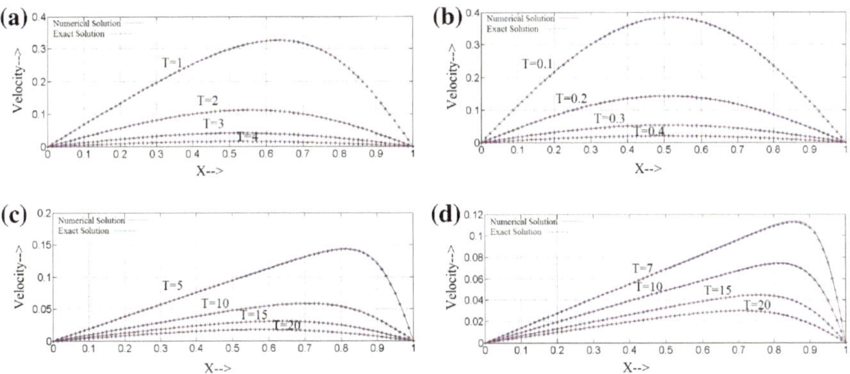

Fig. 4 Numerical solution of Example 2 at several times for N = 80 with distinguishable liquids having different values of v and Δt, **a** saturated fatty acid methyl esters at 40 °C with $v = 0.1$, $\Delta t = 0.0001$; **b** lubricating oil at 40 °C with $v = 1$, $\Delta t = 0.0001$; **c** water at 20 °C with $v = 0.01$, $\Delta t = 0.01$; **d** water at 50 °C with $v = 0.005$, $\Delta t = 0.01$

the scheme is consistent and accurate of order two in space and time. Figures 2 and 4 show the graphs for computed and exact solution at different times for various values of kinematic viscosity. From these graphs it is observed that the proposed method gives accurate results for any value of time step Δt. In order to show the physical behavior of the given problem, surface plots of the computed solutions are presented (Figs. 3 and 5) for different liquids with distinct values of kinematic viscosity. When the kinematic viscosity is very large and the fluid velocity is very slow, Reynold number (Re) becomes very much less compared to unity ($Re << 1$). This type of flow is known as creeping flow which occurs in the case of some paints, MEMS, viscous polymers, lubrication, etc. For the creeping flow this method is proved to give very accurate results. When $Re >> 1$, the inertia force is more dominating than

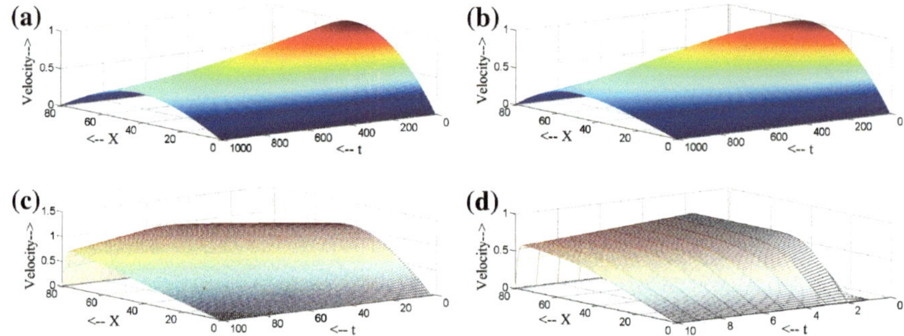

Fig. 5 Numerical solution profiles of Example 2 at $\Delta x = 0.0125$ for, **a** Lubricating oil at 40 °C with $\nu = 1$, $\Delta t = 0.0001$; **b** Saturated fatty acid methyl esters at 40 °C with $\nu = 0.1$, $\Delta t = 0.001$; **c** Water at 20 °C with $\nu = 0.01$, $\Delta t = 0.01$; **d** Water at 50 °C with $\nu = 0.005$, $\Delta t = 0.1$

viscous force and diffusion term $\left(\nu\frac{\partial^2 u}{\partial x^2}\right)$ tends to zero. Even if $Re >> 1$, it is possible that the inertia and viscous forces are of comparable magnitude, especially in the neighborhood of a solid boundary. This thin region is called the boundary layer and the proposed method works quite satisfactorily in this region as well.

4 Conclusions

The Modified Keller-Box difference scheme, which is a highly accurate and efficient method coupled with Hopf-Cole transformation, is used to study the properties of the solution of Burgers' equation for a wide range of kinematic viscosity. The present numerical experiments have confirmed that the present method is unconditionally stable. It is second order accurate in space and time. There is no requirement with respect to mesh size restriction. The results are in good agreement with the exact solution for modest values of the kinematic viscosity. The physical behavior of the solution is explained and it is concluded that the numerically calculated values are in close agreement with the exact solution.

Acknowledgments The authors thank the anonymous referees for their valuable time, effort, and extensive comments which helped to improve the quality of the paper.

References

1. Bateman, H.: Some recent researches on the motion of fluids. Mon. Weather Rev. **43**, 163–170 (1915)
2. Benton, E.R., Platzman, G.W.: A table of solutions of the one-dimensional Burgers' equations. Quart. Appl. Math. **30**, 195–212 (1972)
3. Brander, O., Hedenfalk, J.: A new formulation of the general solution to burgers equation. Wave Motion **28**, 319–332 (1998)

4. Burgers, J.: A Mathematical Model Illustrating the Theory of Turbulence. Advances in Applied Mechanics, pp. 171–199. Academic Press, New York (1948)
5. Burns, J., Balogh, A., Gilliam, D., Shubol, V.: Numerical stationary solutions for a viscous Burgers' equation. J. Math. Syst. Estim. Control **8**, 1–16 (1998)
6. Cole, J.: On a quaslinear parabolic equations occurring in aerodynamics. Quart. Appl. Math. **9**, 225–236 (1951)
7. Gulsu, M.: A finite difference approach for solution of Burgers' equation. Appl. Math. Comput. **175**, 1245–1255 (2006)
8. Hopf, E.: The partial differential equation $u_t + uu_x = \nu u_{xx}$. Comm. Pure Appl. Math. **3**, 201–230 (1950)
9. Kadalbajoo, M., Awasthi, A.: A numerical method based on crank-nicolson scheme for Burgers' equation. Appl. Math. Comput. **182**, 1430–1442 (2006)
10. Keller, H.B.: A new difference scheme for parabolic problems. In: Numerical Solutions of Partial Difference Equations, vol. 2. Academic Press, New York (1970)
11. Kutluay, S., Bahadir, A., Ozdes, A.: Numerical solution of one-dimensional burgers equation: explicit and exact-explicit finite difference methods. J. Comput. Appl. Math. **103**, 251–261 (1999)
12. Rodin, E.Y.: On some approximate and exact solutions of boundary value problems for Burgers equation. J. Math. Anal. Appl. **30**, 401–414 (1970)
13. Tannehill, J.C., Anderson, D.A., Pletcher, R.H.: Computational Fluid Mechanics and Heat Transfer. Taylor and Francis, Philadelphia (1997)
14. Vajiravelu, K., Prasad, K.V.: Keller-Box Method and Its Application. Walter de Gruyter and Co., Orlando (2014)

A Novel Approach to Surface Interpolation: Marriage of Coons Technique and Univariate Fractal Functions

A.K.B. Chand, P. Viswanathan and K.M. Reddy

Abstract The current article is intended to demonstrate that the theory of fractal functions when applied in conjunction with methods in the classical numerical analysis can supply new solution techniques that supplement and subsume the existing ones. To this end, in the first part of the paper, we review a $\mathscr{C}^1$-continuous rational cubic fractal interpolation function (FIF) introduced recently [Viswanathan and Chand, Elec. Trans. Numer. Anal. 41 (2014), pp. 420–442]. We carry out the convergence analysis of this univariate rational FIF and determine suitable values of the derivative parameters so that its global smoothness enhances to $\mathscr{C}^2$. In the subsequent part of the article, we apply Coons technique of transfinite interpolation in order to construct a new kind of $\mathscr{C}^1$-continuous bivariate fractal interpolation surface.

Keywords Fractal interpolation function · Convergence · Coons technique · Blending function · Fractal interpolation surface

1 Prologue

Fractal function proposed by Barnsley [1] is a powerful tool for the approximation of natural data sets and some complicated experimental variables. Fractal interpolation function (FIF) provides a totally new interpolation method which has proved to be advantageous over the traditional nonrecursive interpolation techniques. Traditional interpolation techniques such as polynomial, rational, trigonometric, and spline interpolation always render interpolating curves that are infinitely smooth except possibly at a finite number of points corresponding to the given interpolation data. FIF possesses the novelty of providing one of the very few methods of non-differentiable

A.K.B. Chand (✉) · P. Viswanathan · K.M. Reddy
Department of Mathematics, Indian Institute of Technology Madras, Chennai, India
e-mail: chand@iitm.ac.in

P. Viswanathan
e-mail: amritaviswa@gmail.com

K.M. Reddy
e-mail: mahipalnitw@gmail.com

P.N. Agrawal et al. (eds.), *Mathematical Analysis and its Applications*,
Springer Proceedings in Mathematics & Statistics 143,
DOI 10.1007/978-81-322-2485-3_48

interpolants. Nonsmooth version of FIF remains well-suited to approximate images of natural objects such as profiles of mountain ranges, tops of clouds, and stalactite hung roofs of caves, wherein regularity is almost missing [2]. Graph of the fractal interpolant possesses a non-integer dimension which can be used as a quantifier for the complexity of the underlying phenomenon.

Barnsley and Harrington [3] initiated the construction of smooth FIFs, and unfolded the striking relationship between fractal functions and splines. Smooth FIFs can be applied to generalize the classical interpolation techniques (see, for instance, [5–7, 10, 13]). Therefore, fractal interpolation offers the flexibility of choosing either a smooth or a nonsmooth approximant depending on the modeling problem at hand.

Apart from modeling a prescribed set of data with an interpolant having a suitable degree of smoothness, preserving fundamental shape properties such as positivity, monotonicity, and convexity inherent in the data is extremely important in many practical applications. There is a myriad of research articles in the field of classical approximation theory and numerical analysis targeting the problem of shape preserving interpolation, often directed at specific application domains. For a compendium of some of these, the reader may refer to the survey article [12]. However, the suitability of FIFs as a bona fide technique for shape preservation has not been fully exploited hitherto. Recently, the authors have initiated shape preserving fractal interpolation and approximation through their researches that appeared in [8–10, 15–17].

In this article, first we shall revisit (see Sect. 3) the $\mathscr{C}^1$-continuous rational cubic FIF with linear denominator introduced in [16]. Note that here we shall adopt a constructional approach and also that the FIF provides a fractal generalization of the rational cubic spline introduced in [14]. The convergence analysis of this cubic rational FIF is carried out in detail. Suitable values of derivatives are determined treating them as parameters so that global smoothness of the rational FIF enhances to $\mathscr{C}^2$. These findings when applied with suitable choice of shape parameters involved in the structure of the FIF provides an alternative to the standard moment construction of $\mathscr{C}^2$-cubic FIF discussed elaborately in [5].

In contrast to univariate fractal interpolants, the volume of literature available on bivariate fractal interpolants is less extensive, though its scope is wide. To build on the literature and to demonstrate that the cooperation of fractal theory with the tools in numerical analysis can produce new solution techniques, we apply the Coons' idea of transfinite interpolation [4] on these rational cubic FIFs. This culminates with a bivariate interpolant (surface) whose boundaries are cubic rational FIFs. Unlike the traditional interpolating surfaces, these fractal surfaces have partial derivatives that are smooth or nonsmooth depending on the choice of parameters, a fact that can find potential applications in various nonlinear and nonequilibrium phenomena.

2 Basics of Fractal Interpolation

This section targets to equip a novice reader with the preliminaries of fractal interpolation required for the current study. These materials are collected from well-known treatises [1–3].

We use the following notation throughout the article. For $m \in \mathbb{N}$, we denote by $\mathbb{N}_m$, the subset of $\mathbb{N}$ containing first m natural numbers.
For $n > 2$, let $x_1 < x_2 < \cdots < x_n$ be real numbers and $I = [x_1, x_n]$ be the closed interval that contains them. Let the prescribed set of interpolation data be $\{(x_i, y_i) \in I \times \mathbb{R} : i \in \mathbb{N}_n\}$. For $i \in \mathbb{N}_{n-1}$, set $I_i = [x_i, x_{i+1}]$ and let $L_i : I \to I_i$ be contraction homeomorphisms that obey the endpoint conditions

$$L_i(x_1) = x_i, \ L_i(x_n) = x_{i+1}. \tag{1}$$

Let $F_i : I \times \mathbb{R} \to \mathbb{R}$ be continuous maps satisfying

$$\left.\begin{array}{l} F_i(x_1, y_1) = y_i, \ F_i(x_n, y_n) = y_{i+1}, \ i \in \mathbb{N}_{n-1}, \\ |F_i(x, y) - F_i(x, y')| \le r_i |y - y'|, \ x \in I; \ y, y' \in \mathbb{R}; \ \text{for some } 0 \le r_i < 1. \end{array}\right\} \tag{2}$$

For $i \in \mathbb{N}_{n-1}$, define functions $\omega_i(x, y) = \big(L_i(x), F_i(x, y)\big)$. Consider the collection $\{I \times \mathbb{R}; \omega_i : i \in \mathbb{N}_{n-1}\}$ referred to as an iterated function system (IFS). The following is the most fundamental theorem in the field of fractal interpolation.

Theorem 1 ([1]) *The IFS $\{I \times \mathbb{R}; \omega_i : i \in \mathbb{N}_{n-1}\}$ defined above admits a unique attractor G, and G is the graph of a continuous function $f : I \to \mathbb{R}$ which obeys $f(x_i) = y_i$ for $i \in \mathbb{N}_n$.*

The function which has made its debut in the above theorem is termed a fractal interpolation function (FIF), and it satisfies the functional equation:

$$f(x) = F_i\big(L_i^{-1}(x), f \circ L_i^{-1}(x)\big), \ x \in I_i, \ i \in \mathbb{N}_{n-1}. \tag{3}$$

The most extensively studied FIFs in theory and applications hitherto are defined through the system of maps

$$L_i(x) = a_i x + b_i, \ F_i(x, y) = \alpha_i y + q_i(x), \tag{4}$$

where α_i are constants satisfying $0 \le |\alpha_i| < 1$ and q_i are continuous functions so that the "join-up conditions" in (2) imposed on the bivariate functions F_i are satisfied. The multiplier α_i is called a vertical scaling factor for the transformation ω_i and the vector $\boldsymbol{\alpha} = (\alpha_1, \alpha_2, \ldots, \alpha_{n-1}) \in (-1, 1)^{n-1}$ is called a scale vector of the IFS. The prescriptions in (1) uniquely determine the constants a_i and b_i appearing in the affine map L_i.

Let us recall that the function f determined by the IFS in (4), which takes the form

$$f(x) = \alpha_i f\big(L_i^{-1}(x)\big) + q_i\big(L_i^{-1}(x)\big), \tag{5}$$

is, in general, nonsmooth. The following theorem determines the conditions on α_i and functions q_i so that the FIF is $\mathcal{C}^p$-continuous.

Theorem 2 ([3]) *Let* $\{(x_i, y_i) : i \in \mathbb{N}_n\}$ *be a given data set with strictly increasing abscissae. Let* $L_i(x) = a_i x + b_i$ *satisfy (1) and* $F_i(x, y) = \alpha_i y + q_i(x)$ *obey (2) for* $i \in \mathbb{N}_{n-1}$. *Suppose that for some integer* $p \geq 0$, $|\alpha_i| < a_i^p$ *and* $q_i \in \mathscr{C}^p(I)$, $i \in \mathbb{N}_{n-1}$. *Let*

$$F_{i,k}(x, y) = \frac{\alpha_i y + q_i^{(k)}(x)}{a_i^k}, \quad y_{1,k} = \frac{q_1^{(k)}(x_1)}{a_1^k - \alpha_1}, \quad y_{n,k} = \frac{q_{n-1}^{(k)}(x_n)}{a_{n-1}^k - \alpha_{n-1}}, \quad k = 1, 2, \ldots, p.$$

If $F_{i-1,k}(x_n, y_{n,k}) = F_{i,k}(x_1, y_{1,k})$ *for* $i = 2, 3, \ldots, n-1$ *and* $k \in \mathbb{N}_p$, *then the IFS* $\{I \times \mathbb{R}; (L_i(x), F_i(x, y)) : i \in \mathbb{N}_{n-1}\}$ *determines a FIF* $f \in \mathscr{C}^p(I)$, *and* $f^{(k)}$ *is the FIF determined by* $\{I \times \mathbb{R}; (L_i(x), F_{i,k}(x, y)) : i \in \mathbb{N}_{n-1}\}$ *for* $k \in \mathbb{N}_p$.

3 Rational Cubic FIF with Linear Denominator Revisited

For $n \in \mathbb{N}$ with $n > 2$, let $\{(x_i, y_i) : i \in \mathbb{N}_n\}$ be a given set of interpolation data with strictly increasing abscissae. Let d_i be the derivative value (given or estimated by some standard methods) at the knot point x_i.

3.1 Construction of $\mathscr{C}^1$-Rational Cubic FIF

Consider the IFS in (4) where $q_i \in \mathscr{C}^1(I)$ is chosen such that

$$q_i(x) \equiv q_i^*(\theta) = \frac{P_i(\theta)}{Q_i(\theta)} = \frac{A_i(1-\theta)^3 + B_i\theta(1-\theta)^2 + C_i\theta^2(1-\theta) + D_i\theta^3}{(1-\theta)r_i + \theta t_i}, \tag{6}$$

where $\theta := \frac{x - x_1}{x_n - x_1}$ and $r_i > 0$, $t_i > 0$ are free shape parameters. The "join-up conditions" on the bivariate function F_i are equivalent to $q_i(x_1) = y_i - \alpha_i y_1$ and $q_i(x_N) = y_{i+1} - \alpha_i y_N$. This observation when coupled with (5) yields the conditions $f(x_i) = y_i$ and $f(x_{i+1}) = y_{i+1}$, which in turn determines the following coefficients:

$$A_i = r_i(y_i - \alpha_i y_1), \quad D_i = t_i(y_{i+1} - \alpha_i y_n). \tag{7}$$

In accordance with the principle of construction of a $\mathscr{C}^1$-spline FIF (see Theorem 2), we impose the condition $|\alpha_i| < a_i$ for $i \in \mathbb{N}_{n-1}$ on the scaling factors. Adhering to the notation in Theorem 2, let us denote $y_{1,1} = d_1$, $y_{n,1} = d_n$, $F_{i,1}(x_1, d_1) = d_i$, and $F_{i,1}(x_n, d_n) = d_{i+1}$ for $i \in \mathbb{N}_{n-1}$. Then, f is $\mathscr{C}^1$-continuous, the derivative f' is the fractal function corresponding to the IFS $\{I \times \mathbb{R}; (L_i(x), F_{i,1}(x, y)) : i \in \mathbb{N}_{n-1}\}$, and f' interpolates the data set $\{(x_i, d_i) : i \in \mathbb{N}_n\}$. Further, f' satisfies the functional equation:

$$f'\big(L_i(x)\big) = F_{i,1}\big(x, f'(x)\big) = \frac{\alpha_i f'(x) + q_i'(x)}{a_i}. \tag{8}$$

Imposing the conditions noted previously, namely $F_{i,1}(x_1, d_1) = d_i$, and $F_{i,1}(x_n, d_n) = d_{i+1}$ on (8), routine algebraic manipulations provide the following coefficients:

$$\begin{aligned} B_i &= (2r_i + t_i)y_i + r_i h_i d_i - \alpha_i[(2r_i + t_i)y_1 + r_i(x_n - x_1)d_1], \\ C_i &= (r_i + 2t_i)y_{i+1} - t_i h_i d_{i+1} - \alpha_i[(r_i + 2t_i)y_n - t_i(x_n - x_1)d_n], \end{aligned} \tag{9}$$

where $h_i = x_{i+1} - x_i$ is the length of the subinterval I_i. In view of (6), (7) and (9), the desired FIF with rational q_i takes the form

$$f\big(L_i(x)\big) = \alpha_i f(x) + \frac{P_i(x)}{Q_i(x)}, \tag{10}$$

$$\begin{aligned} P_i(x) \equiv P_i^*(\theta) = {} & r_i\{y_i - \alpha_i y_1\}(1-\theta)^3 + t_i\{y_{i+1} - \alpha_i y_n\}\theta^3 + \big\{(2r_i + t_i)y_i + r_i h_i d_i \\ & -\alpha_i[(2r_i + t_i)y_1 + r_i(x_n - x_1)d_1]\big\}\theta(1-\theta)^2 + \big\{(r_i + 2t_i)y_{i+1} \\ & -t_i h_i d_{i+1} - \alpha_i[(r_i + 2t_i)y_n - t_i(x_n - x_1)d_n]\big\}\theta^2(1-\theta), \\ Q_i(x) \equiv Q_i^*(\theta) = {} & (1-\theta)r_i + \theta t_i, \quad i \in \mathbb{N}_{n-1}, \ \theta = \frac{x - x_1}{x_n - x_1} \end{aligned}$$

The aforementioned rational FIF depends on the choice of scaling vector and may be denoted by f^{α} for clarity. The reader is invited to refer [16] for construction of the aforementioned rational cubic spline FIF using a different approach via $\boldsymbol{\alpha}$-fractal functions.

Remark 3 If the scaling factors $\alpha_i = 0$ for all $i \in \mathbb{N}_{n-1}$, then the rational cubic FIF f coincides with the $\mathscr{C}^1$-rational cubic spline $C \in \mathscr{C}^1(I)$ introduced in [14]. Let us note that C is defined in a piecewise manner such that for $x \in I_i$, $i \in \mathbb{N}_{n-1}$

$$\begin{aligned} C(x) = {} & \big[(1-\phi)r_i + \phi t_i\big]^{-1}\Big\{r_i y_i(1-\phi)^3 + [(2r_i + t_i)y_i + r_i h_i d_i]\phi(1-\phi)^2 \\ & + [(r_i + 2t_i)y_{i+1} - t_i h_i d_{i+1}]\phi^2(1-\phi) + t_i y_{i+1}\phi^3\Big\}, \end{aligned} \tag{11}$$

where $\phi = \frac{L_i^{-1}(x) - x_1}{x_n - x_1} = \frac{x - x_i}{h_i}$. If the shape parameters pertaining to each subinterval is taken according to $r_i = t_i$, then the rational cubic spline FIF reduces to the $\mathscr{C}^1$-cubic Hermite spline FIF studied in [10], which again with the zero scaling on each I_i reduces to the traditional nonrecursive $\mathscr{C}^1$-cubic Hermite interpolant.

3.2 Rational Cubic FIF with $\mathscr{C}^2$-Continuity

Assuming the derivative values d_i, $i \in \mathbb{N}_n$ to be free parameters, we derive suitable conditions on d_i so that the rational cubic spline $f \in \mathscr{C}^2(I)$. In view of principle of construction of a $\mathscr{C}^2$-FIF (cf. Theorem 2), we take the scaling factors such that $|\alpha_i| \le \kappa a_i^2 \; \forall\, i \in \mathbb{N}_{n-1}$, where $0 \le \kappa < 1$.

The functional equation of $f^{(2)}$ with $x = x_1$ and $x = x_n$ applied on the affine homeomorphisms L_i and L_{i-1} produce

$$\begin{aligned} f^{(2)}(x_i^+) = \frac{\alpha_i}{a_i^2} f^{(2)}(x_1) + \frac{1}{r_i^3 h_i}\{(2r_i^3 + 4r_i^2 t_i)[\Delta_i - \frac{\alpha_i}{h_i}(y_n - y_1)] \\ - (2r_i^3 + 2r_i^2 t_i)[d_i - \frac{\alpha_i}{h_i} d_1(x_n - x_1)] - 2r_i^2 t_i [d_{i+1} - \frac{\alpha_i}{h_i} d_n(x_n - x_1)]\}, \end{aligned} \tag{12}$$

$$\begin{aligned} f^{(2)}(x_i^-) = \frac{\alpha_{i-1}}{a_{i-1}^2} f^{(2)}(x_n) + \frac{1}{t_{i-1}^3 h_{i-1}}\{(2t_{i-1}^3 + 2r_{i-1} t_{i-1}^2)[d_i - \frac{\alpha_{i-1}}{h_{i-1}} d_n(x_n - x_1)] \\ + 2r_{i-1} t_{i-1}^2 [d_{i-1} - \frac{\alpha_{i-1}}{h_{i-1}} d_1(x_n - x_1)] - (2t_{i-1}^3 + 4r_{i-1} t_{i-1}^2)[\Delta_{i-1} - \frac{\alpha_{i-1}}{h_{i-1}}(y_n - y_1)]\}, \end{aligned} \tag{13}$$

where $\Delta_i := \frac{y_{i+1} - y_i}{h_i}$ denotes the secant slope. Continuity of $f^{(2)}$ at the interior points x_i, $i = 2, 3, \ldots, n-1$ demands $f^{(2)}(x_i^+) = f^{(2)}(x_i^-)$, which in view of (12)–(13) yields

$$\begin{aligned} &- \alpha_i a_{i-1}^2 r_i t_{i-1} h_i h_{i-1} f^{(2)}(x_1) + 2r_{i-1} a_i^2 a_{i-1}^2 r_i h_i d_{i-1} + 2a_i^2 a_{i-1}^2 t_{i-1} h_{i-1} t_i d_{i+1} \\ &- [2a_i^2 a_{i-1}^2 t_{i-1} h_{i-1} \frac{\alpha_i}{h_i}(x_n - x_1)(r_i + t_i) + 2a_i^2 a_{i-1}^2 r_i r_{i-1} h_i \frac{\alpha_{i-1}}{h_{i-1}}(x_n - x_1)] d_1 \\ &+ [2a_i^2 a_{i-1}^2 t_{i-1} h_{i-1}(r_i + t_i) + 2a_i^2 a_{i-1}^2 r_i h_i (r_{i-1} + t_{i-1})] d_i \\ &- [2a_i^2 a_{i-1}^2 t_{i-1} t_i h_{i-1} \frac{\alpha_i}{h_i}(x_n - x_1) + 2a_i^2 a_{i-1}^2 r_i h_i \frac{\alpha_{i-1}}{h_{i-1}}(x_n - x_1)(t_{i-1} + r_{i-1})] d_n \\ &+ \alpha_{i-1} a_i^2 r_i t_{i-1} h_i h_{i-1} f^{(2)}(x_n) \\ &= 2a_i^2 a_{i-1}^2 t_{i-1} h_{i-1}[\Delta_i - \frac{\alpha_i}{h_i}(y_n - y_1)](r_i + 2t_i) + 2a_i^2 a_{i-1}^2 r_i h_i \times \\ &\quad [\Delta_{i-1} - \frac{\alpha_{i-1}}{h_{i-1}}(y_n - y_1)](2r_{i-1} + t_{i-1}), \;\; i = 2, 3, \ldots, n-1. \end{aligned} \tag{14}$$

Next, from the functional equation for $f^{(2)}$ we get

$$\begin{aligned} r_1 h_1 (a_1^2 - \alpha_1) f^{(2)}(x_1) + 2a_1^2[1 - \frac{\alpha_1}{h_1}(x_n - x_1)](r_1 + t_1) d_1 + 2t_1 a_1^2 d_2 \\ - 2t_1 a_1^2 (x_n - x_1) \frac{\alpha_1}{h_1} d_n = 2[\Delta_1 - \frac{\alpha_1}{h_1}(y_n - y_1)] a_1^2 (r_1 + 2t_1). \end{aligned} \tag{15}$$

Similarly,

$$\begin{aligned} &- t_{n-1} h_{n-1} (a_{n-1}^2 - \alpha_{n-1}) f^{(2)}(x_n) - 2r_{n-1} a_{n-1}^2 \frac{\alpha_{n-1}}{h_{n-1}}(x_n - x_1) d_1 \\ &+ 2r_{n-1} a_{n-1}^2 d_{n-1} + 2a_{n-1}^2 [1 - \frac{\alpha_{n-1}}{h_{n-1}}(x_n - x_1)](r_{n-1} + t_{n-1}) d_n \\ &= 2[\Delta_{n-1} - \frac{\alpha_{n-1}}{h_{n-1}}(y_n - y_1)] a_{n-1}^2 (t_{n-1} + 2r_{n-1}). \end{aligned} \tag{16}$$

Now, for $i = 2, 3, \ldots, n-1$, let us introduce the following notation:

$$\lambda_i = \frac{r_{i-1}r_i h_i}{t_{i-1}t_i h_{i-1} + r_{i-1}r_i h_i}, \quad \mu_i = 1 - \lambda_i, \quad A_i = -\frac{\alpha_i r_i \mu_i h_i}{2a_i^2 t_i},$$

$$B_i = \frac{\alpha_{i-1} r_i h_i \mu_i}{2a_{i-1}^2 t_i}, \; A_i^* = -\frac{\alpha_i \mu_i (x_n - x_1)(r_i + t_i)}{h_i t_i} - \frac{\alpha_{i-1}\lambda_i (x_n - x_1)}{h_{i-1}},$$

$$C_i = \frac{\mu_i (r_i + t_i)}{t_i} + \frac{\lambda_i (r_{i-1} + t_{i-1})}{r_{i-1}},$$

$$B_i^* = -\frac{\alpha_i \mu_i (x_n - x_1)}{h_i} - \frac{\alpha_{i-1}\lambda_i (r_{i-1} + t_{i-1})(x_n - x_1)}{h_{i-1} r_{i-1}},$$

$$\beta_i = \frac{\mu_i (r_i + 2t_i)[\Delta_i - \frac{\alpha_i}{h_i}(y_n - y_1)]}{t_i} + \frac{\lambda_i (2r_{i-1} + t_{i-1})[\Delta_{i-1} - \frac{\alpha_{i-1}}{h_{i-1}}(y_n - y_1)]}{r_{i-1}}.$$

Using the above notation, the continuity condition (14) can be reformulated as

$$A_i^* d_1 + A_i f^{(2)}(x_1) + \lambda_i d_{i-1} + C_i d_i + \mu_i d_{i+1} + B_i f^{(2)}(x_n) + B_i^* d_n = \beta_i. \tag{17}$$

Letting

$$A_1^* = 2a_1^2[1 - \frac{\alpha_1}{h_1}(x_n - x_1)](r_1 + t_1), \; A_1 = r_1 h_1 (a_1^2 - \alpha_1), \; \mu_1 = 2a_1^2 t_1, \; \lambda_n = 2r_{n-1}a_{n-1}^2,$$

$$B_1^* = -\mu_1 \frac{\alpha_1}{h_1}(x_n - x_1), \; \beta_1 = 2a_1^2 (r_1 + 2t_1)\left[\Delta_1 - \frac{\alpha_1}{h_1}(y_n - y_1)\right], \; A_n^* = -\lambda_n \frac{\alpha_{n-1}}{h_{n-1}}(x_n - x_1)$$

$$B_n = -t_{n-1}h_{n-1}(a_{n-1}^2 - \alpha_{n-1}), \; B_n^* = 2a_{n-1}^2 (r_{n-1} + t_{n-1})\left[1 - \frac{\alpha_{n-1}}{h_{n-1}}(x_n - x_1)\right],$$

$$\beta_n = 2a_{n-1}^2 (t_{n-1} + 2r_{n-1})\left[\Delta_{n-1} - \frac{\alpha_{n-1}}{h_{n-1}}(y_n - y_1)\right],$$

(15)–(16) can be rewritten as

$$\left.\begin{aligned} A_1^* d_1 + A_1 f^{(2)}(x_1) + \mu_1 d_2 + B_1^* d_n &= \beta_1, \\ A_n^* d_1 + \lambda_n d_{n-1} + B_n f^{(2)}(x_n) + B_n^* d_n &= \beta_n. \end{aligned}\right\} \tag{18}$$

The matrix form that defines Eqs. (17) and (18) is

$$\begin{bmatrix} A_1^* & A_1 & \mu_1 & \cdots & 0 & 0 & B_1^* \\ A_2^* + \lambda_2 & A_2 & C_2 & \cdots & 0 & B_2 & B_2^* \\ A_3^* & A_3 & \lambda_3 & \cdots & 0 & B_3 & B_3^* \\ \vdots & \vdots & \vdots & \vdots & \vdots & \vdots & \vdots \\ A_{n-2}^* & A_{n-2} & 0 & \cdots & \mu_{n-2} & B_{n-2} & B_{n-2}^* \\ A_{n-1}^* & A_{n-1} & 0 & \cdots & C_{n-1} & B_{n-1} & B_{n-1}^* + \mu_{n-1} \\ A_n^* & 0 & 0 & \cdots & \lambda_n & B_n & B_n^* \end{bmatrix} \begin{bmatrix} d_1 \\ f^{(2)}(x_1) \\ d_2 \\ \vdots \\ d_{n-1} \\ f^{(2)}(x_n) \\ d_n \end{bmatrix} = \begin{bmatrix} \beta_1 \\ \beta_2 \\ \beta_3 \\ \vdots \\ \beta_{n-2} \\ \beta_{n-1} \\ \beta_n \end{bmatrix}.$$

Hence, the linear system governing the $\mathscr{C}^2$-continuity of the rational cubic FIF consists of a coefficient matrix of order $N \times (N+2)$. By prescribing suitable boundary conditions similar to that of the classical rational cubic spline [14], the rectangular system reduces to a square system.

If we take zero scaling factor in each interval, then the continuity equation (17) reduces to

$$\lambda_i d_{i-1} + C_i d_i + \mu_i d_{i+1} = \frac{t_{i-1}h_{i-1}(r_i + 2t_i)\Delta_i + r_i h_i(2r_{i-1} + t_{i-1})\Delta_{i-1}}{t_{i-1}t_i h_{i-1} + r_{i-1}r_i h_i},$$

which coincides with the linear tridiagonal system corresponding to the $\mathscr{C}^2$-continuity of the classical rational cubic spline C (see Eq. (4.1) in [14] or Eq. (2) in [11]). The feasibility of the above construction depends on the existence and uniqueness of the solution of the linear system corresponding to a fixed choice of the scaling factors and the shape parameters, which follows from the existence and uniqueness of the fixed point of Read-Bajraktarévic operator (see [5]). Using suitable boundary conditions, and solving the corresponding square system of equations, the values $d_i, i \in \mathbb{N}_n$ are determined.

Remark 4 For the shape parameters satisfying $r_i = t_i \,\forall\, i \in \mathbb{N}_{n-1}$, our fractal rational cubic FIF reduces to the $\mathscr{C}^1$-cubic Hermite FIF. Consequently, the above elements of $\mathscr{C}^2$ theory renders, in particular, the construction of a $\mathscr{C}^2$-cubic spline FIF through slopes d_i. For the construction of $\mathscr{C}^2$-cubic spline FIF through moments, the reader is invited to refer [5].

4 Convergence Analysis

The goal in the current section is to establish theorem that can eventually handle the convergence of the developed rational cubic FIF f under the assumption that the original function $\Phi \in \mathscr{C}^1(I)$. Convergence result for the rational cubic spline FIF f is derived using the convergence result for its classical nonrecursive counterpart C (cf. (12)) and an upper bound for the uniform distance between f and C. The following lemma establishes an upper bound for the uniform distance between the classical rational cubic spline $C = f^{\mathbf{0}}$ and the perturbed fractal function $f = f^{\boldsymbol{\alpha}}$. Proof follows on lines similar to that of Lemma 4.1 (cf. [5]), and is hence omitted.

Lemma 5 *Let f and C, respectively, be the rational cubic spline FIF and the classical rational cubic spline interpolant for the data $\{(x_i, y_i) : i \in \mathbb{N}_n\}$. Let the rational functions $q_i(\alpha_i, x) = \frac{P_i(\alpha_i,x)}{Q_i(x)}$ involved in the FIF f satisfy $\left|\frac{\partial q_i}{\partial \alpha_i}(\xi_i, x)\right| \le D_0 \,\forall\, (\xi_i, x) \in (-a_i, a_i) \times I_i$ and $i \in 1, 2, \ldots, n-1$. Then with* $|\boldsymbol{\alpha}|_\infty := \max\{|\alpha_i| : i \in \mathbb{N}_{n-1}\}$,

$$\|f - C\|_\infty \le \frac{|\boldsymbol{\alpha}|_\infty}{1 - |\boldsymbol{\alpha}|_\infty}(\|C\|_\infty + D_0).$$

With this preparation, an upper bound for the L^∞-norm of the error in approximating the data generating function $\Phi \in \mathscr{C}^1(I, \mathbb{R})$ with the rational cubic spline FIF f (cf. (10)) is deduced in the following theorem. For simplicity of presentation, we introduce the following notation: $|y|_\infty = \max\{|y_i| : 1 \le i \le n\}$, $|d|_\infty = \max\{|d_i| : 1 \le i \le n\}$, $|I| = x_n - x_1$, $h = \max\{h_i : i \in \mathbb{N}_{n-1}\}$, $\overline{\gamma} = \max\{\gamma_i : i \in \mathbb{N}_{n-1}\}$, and $\underline{\delta} = \min\{\delta_i : i \in \mathbb{N}_{n-1}\}$. We shall assume that the shape parameters are selected so as to prevent the situation: quantities $\overline{\gamma}$ and $\underline{\delta}$ approach to zero as the norm of the partition approaches zero.

Theorem 6 *Let $\Phi \in \mathscr{C}^1(I, \mathbb{R})$ be the function generating the data $\{(x_i, y_i) : i \in \mathbb{N}_n\}$ and let f be the corresponding rational cubic spline FIF defined through the functional equation (10). Then*

$$\|\Phi - f\|_\infty \le \frac{\overline{\gamma}}{4\underline{\delta}}(4\omega(f; h) + h|d|_\infty) + \frac{|\boldsymbol{\alpha}|_\infty}{1 - |\boldsymbol{\alpha}|_\infty}\Big\{|y|_\infty + \max\{|y_1|, |y_n|\} + \frac{1}{4}\big(h|d|_\infty + |I| \max\{|d_1|, |d_n|\}\big)\Big\}.$$

In particular, f converges to Φ uniformly, as the norm of the partition approaches zero.

Proof For the data generating function Φ, let C and f, respectively, be the classical rational cubic spline and the rational cubic spline FIF with respect to the data $\{(x_i, y_i) : i \in \mathbb{N}_n\}$. We will use the following triangle inequality for our calculations:

$$\|\Phi - f\|_\infty \le \|\Phi - C\|_\infty + \|C - f\|_\infty. \tag{19}$$

By Lemma 5,

$$\|f - C\|_\infty \le \frac{|\boldsymbol{\alpha}|_\infty}{1 - |\boldsymbol{\alpha}|_\infty}\Big(\|C\|_\infty + D_0\Big) \tag{20}$$

With a careful observation of the expressions of functions C and q_i, and with rigorous calculations, the following bounds for $\|C\|_\infty$ and D_0 can be obtained:

$$\|C\|_\infty \le |y|_\infty + \frac{h}{4}|d|_\infty. \tag{21}$$

On similar lines D_0 can be selected as a constant such that

$$D_0 \ge \max\{|y_1|, |y_n|\} + \frac{1}{4}|I| \max\{|d_1|, |d_n|\}. \tag{22}$$

Substituting the bounds (21)–(22) in (20) we see that

$$\|f - C\|_\infty \le \frac{|\boldsymbol{\alpha}|_\infty}{1 - |\boldsymbol{\alpha}|_\infty}\Big\{|y|_\infty + \max\{|y_1|, |y_n|\} + \frac{1}{4}(h|d|_\infty + |I| \max\{|d_1|, |d_n|\})\Big\}. \tag{23}$$

Next we shall turn our attention to the first summand appearing in (19), which corresponds to the error bound for the classical rational cubic spline C (cf. (11)).

$$C(x) - \Phi(x) = \frac{1}{Q_i(\theta)}\Big\{(1-\theta)^3 r_i\big(y_i - \Phi(x)\big) + \theta(1-\theta)^2\big[(2r_i + t_i)(y_i - \Phi(x)) + r_i h_i d_i\big]$$
$$+ \theta^2(1-\theta)\big[(r_i + 2t_i)(y_{i+1} - \Phi(x)) - t_i h_i d_{i+1}\big] + \theta^3 t_i\big(y_{i+1} - \Phi(x)\big)\Big\}.$$

Thus, we infer

$$|C(x) - \Phi(x)| \le \frac{\gamma_i}{\delta_i}\Big\{\omega(\Phi; h_i) + \frac{h_i}{4}\max\{|d_i|, |d_{i+1}|\}\Big\},$$

which implies

$$\|\Phi - C\|_\infty \le \frac{\overline{\gamma}}{4\underline{\delta}}(4\omega(\Phi; h) + h|d|_\infty). \tag{24}$$

Now substituting (23) and (24) in (19), we obtain

$$\|\Phi - f\|_\infty \le \frac{\overline{\gamma}}{4\underline{\delta}}(4\omega(\Phi; h) + h|d|_\infty) + \frac{|\boldsymbol{\alpha}|_\infty}{1 - |\boldsymbol{\alpha}|_\infty}\Big\{|y|_\infty + \max\{|y_1|, |y_n|\}$$
$$+ \frac{1}{4}(h|d|_\infty + |I|\max\{|d_1|, |d_n|\})\Big\}.$$

Recalling the fact that the modulus of continuity $\omega(\Phi; h)$ of a continuous function Φ on a closed interval I approaches zero as $h \to 0$ and noting that $|\boldsymbol{\alpha}|_\infty < \frac{h}{|I|}$, we can deduce the uniform convergence of f to Φ. This finishes the proof. □

5 Construction of Bicubic Partially Blended Rational FIS

Let $\Delta = \{(x_i, y_j, z_{i,j}) : i \in \mathbb{N}_M, j \in \mathbb{N}_N\}$ be a set of bivariate interpolation data, where $x_1 < x_2 < \cdots < x_i < x_{i+1} < \cdots < x_M$ and $y_1 < y_2 < \cdots < y_j < y_{j+1} < \cdots < y_N$. Let $R_{i,j} := I_i \times J_j = [x_i, x_{i+1}] \times [y_j, y_{j+1}]$; $i \in \mathbb{N}_{M-1}, j \in \mathbb{N}_{N-1}$, be the generic subrectangular region and let us put $R := I \times J = [x_1, x_M] \times [y_1, y_N]$. Following the one-dimensional scheme, let us denote by $h_i = x_{i+1} - x_i$, $h_j^* = y_{j+1} - y_j$, $i \in \mathbb{N}_{M-1}$, $j \in \mathbb{N}_{N-1}$. Let $z_{i,j}^x$ and $z_{i,j}^y$ be the x-partials and y-partials at the point (x_i, y_j). In this section, we wish to construct a $\mathscr{C}^1$-continuous bivariate function $\Psi : R \to \mathbb{R}$ such that $\Psi(x_i, y_j) = z_{i,j}$, $\frac{\partial \Psi}{\partial x}(x_i, y_j) = z_{i,j}^x$, and $\frac{\partial \Psi}{\partial y}(x_i, y_j) = z_{i,j}^y$ for $i \in \mathbb{N}_m$ and $j \in \mathbb{N}_n$. This is achieved by blending the univariate rational cubic FIFs using the partially bicubic Coons technique [4].

For each $i \in \mathbb{N}_M$, let us consider the univariate data set $\Delta_{x_i} := \{(x_i, y_j, z_{i,j}, z_{i,j}^y) : j \in \mathbb{N}_N\}$, which is the data associated with the i-th grid line parallel to y-axis. Consider the affine maps $L_j^* : [y_1, y_N] \to [y_j, y_{j+1}]$ defined by $L_j^*(y) = c_j y + d_j$ satisfying $L_j^*(y_1) = y_j$ and $L_j^*(y_N) = y_{j+1}$, $j \in \mathbb{N}_{N-1}$. For a fixed $i \in \mathbb{N}_M$, let $\alpha_{i,j}^*$

be the scaling factors along the vertical grid line $x = x_i$ such that $|\alpha^*_{i,j}| < c_j < 1$ and let the shape parameters be selected so as to satisfy $r^*_{i,j} > 0$ and $t^*_{i,j} > 0$ for all $j \in \mathbb{N}_{N-1}$. Following the functional equation for the univariate rational cubic FIF given in (10), we infer that the rational cubic FIF S^* corresponding to Δ_{x_i}, $i \in \mathbb{N}_M$ enjoys the functional equation:

$$S^*(x_i, y) = \alpha^*_{i,j} S^*\big(x_i, L_j^{*-1}(y)\big) + \frac{P^*_{i,j}(\phi)}{Q^*_{i,j}(\phi)}, \tag{25}$$

where

$$\begin{aligned} P^*_{i,j}(\phi) = {} & r^*_{i,j}(z_{i,j} - \alpha^*_{i,j} z_{i,1})(1-\phi)^3 + t^*_{i,j}(z_{i,j+1} - \alpha^*_{i,j} z_{i,N})\phi^3 + \Big\{(2r^*_{i,j} + t^*_{i,j})z_{i,j} \\ & + r^*_{i,j} h^*_j z^y_{i,j} - \alpha^*_{i,j}\big[(2r^*_{i,j} + t^*_{i,j})z_{i,1} + r^*_{i,j}(y_N - y_1)z^y_{i,1}\big]\Big\}(1-\phi)^2\phi + \Big\{(2t^*_{i,j} + r^*_{i,j})z_{i,j+1} \\ & - t^*_{i,j} h^*_j z^y_{i,j+1} - \alpha^*_{i,j}\big[(2t^*_{i,j} + r^*_{i,j})z_{i,N} - t^*_{i,j}(y_N - y_1)z^y_{i,N}\big]\Big\}(1-\phi)\phi^2, \\ Q^*_{i,j}(\phi) = {} & (1-\phi)r^*_{i,j} + \phi t^*_{i,j}, \\ \text{and } \phi = {} & \frac{L_j^{*-1}(y) - y_1}{y_N - y_1} = \frac{y - y_j}{h^*_j}, \; y \in J_j, \; j \in \mathbb{N}_{N-1}. \end{aligned}$$

Next, let us consider the univariate data sets obtained by taking sections of R with the lines $y = y_j$, $j \in \mathbb{N}_N$, namely $\Delta_{y_j} := \{(x_i, y_j, z_{i,j}, z^x_{i,j}) : i \in \mathbb{N}_M\}$. Let $L_i : [x_1, x_M] \to [x_i, x_{i+1}]$ be affine maps $L_i(x) = a_i x + b_i$ satisfying $L_i(x_1) = x_i$ and $L_i(x_M) = x_{i+1}$, $i \in \mathbb{N}_{M-1}$. For each fixed $j \in \mathbb{N}_N$, let the scaling factors obey $|\alpha_{i,j}| < a_i < 1$ and the shape parameters fulfill $r_{i,j} > 0$ and $t_{i,j} > 0$ for $i \in \mathbb{N}_{M-1}$. Modifying and adapting the expression for the univariate rational cubic FIF given in the preceding section, we obtain the fractal function S interpolating Δ_{y_j} for $j \in \mathbb{N}_N$:

$$S(x, y_j) = \alpha_{i,j} S\big(L_i^{-1}(x), y_j\big) + \frac{P_{i,j}(\theta)}{Q_{i,j}(\theta)}, \tag{26}$$

wherein

$$\begin{aligned} P_{i,j}(\theta) = {} & r_{i,j}(z_{i,j} - \alpha_{i,j} z_{1,j})(1-\theta)^3 + t_{i,j}(z_{i+1,j} - \alpha_{i,j} z_{M,j})\theta^3 + \Big\{(2r_{i,j} + t_{i,j})z_{i,j} + r_{i,j} h_i z^x_{i,j} \\ & - \alpha_{i,j}\big[(2r_{i,j} + t_{i,j})z_{1,j} + r_{i,j}(x_M - x_1)z^x_{1,j}\big]\Big\}(1-\theta)^2\theta + \Big\{(r_{i,j} + 2t_{i,j})z_{i+1,j} \\ & - t_{i,j} h_i z^x_{i+1,j} - \alpha_{i,j}\big[(r_{i,j} + 2t_{i,j})z_{M,j} - t_{i,j}(x_M - x_1)z^x_{M,j}\big]\Big\}(1-\theta)\theta^2, \\ Q_{i,j}(\theta) = {} & (1-\theta)r_{i,j} + \theta t_{i,j}, \\ \text{and } \theta = {} & \frac{L_i^{-1}(x) - x_1}{x_M - x_1} = \frac{x - x_i}{h_i}, \; x \in I_i, \; i \in \mathbb{N}_{M-1}. \end{aligned}$$

In what follows, we shall blend these univariate FIFs given in (25)–(26) using well-known bicubic partially blended Coons patch to obtain the desired surface. Consider the network of FIFs $S(x, y_j)$, $S(x, y_{j+1})$, $S^*(x_i, y)$, and $S^*(x_{i+1}, y)$ for $i \in \mathbb{N}_{M-1}$ and $j \in \mathbb{N}_{N-1}$. Consider the cubic Hermite functions $b^i_{0,3}(x) = (1-\theta)^2(1+2\theta)$,

$b^i_{3,3}(x) = \theta^2(3-2\theta)$, $b^j_{0,3}(y) = (1-\phi)^2(1+2\phi)$, and $b^j_{3,3}(y) = \phi^2(3-2\phi)$. These functions are called the blending functions, because their effect is to blend together four separate boundary curves to provide a single well-defined surface. On each individual patch $R_{i,j} = I_i \times J_j, i \in \mathbb{N}_{M-1}$ and $j \in \mathbb{N}_{N-1}$, we define the surface

$$\begin{aligned}\Psi(x,y) =& \left[b^i_{0,3}(x)\ b^i_{3,3}(x)\right]\begin{bmatrix} S^*(x_i,y) \\ S^*(x_{i+1},y)\end{bmatrix} + \left[b^j_{0,3}(y)\ b^j_{3,3}(y)\right]\begin{bmatrix} S(x,y_j) \\ S(x,y_{j+1})\end{bmatrix} \\ &- \left[b^i_{0,3}(x)\ b^i_{3,3}(x)\right]\begin{bmatrix} z_{i,j} & z_{i,j+1} \\ z_{i+1,j} & z_{i+1,j+1}\end{bmatrix}\begin{bmatrix} b^j_{0,3}(y) \\ b^j_{3,3}(y)\end{bmatrix}, \\ :=& \Psi_1(x,y) + \Psi_2(x,y) - \Psi_3(x,y).\end{aligned} \tag{27}$$

It is plain to see that Ψ_1 provides a surface which is cubic blending of two boundaries $S^*(x_i, y)$ and $S^*(x_{i+1}, y)$ and Ψ_2 gives a surface which is cubic blending of the remaining opposite pair of edge curves. In this process, the corners will be added twice and Ψ_3 provides the "correction" so that the successive substitution of $x = x_i$, $x = x_{i+1}$, $y = y_j$, and $y = y_{j+1}$ quickly confirms that the surface patch has the four original curves as its boundaries. The word bicubic partially blended is chosen to suggest that only two of the cubic Hermite basis functions are used for the blending process, and the blended surface is a fractal surface in the sense that the boundaries consist of FIFs.

The following theorem is a direct consequence of the properties of the univariate FIFs forming the boundaries of Ψ and the blending functions.

Theorem 7 *The bicubic partially blended rational fractal surface Ψ (cf. (27)) satisfies the interpolation conditions $\Psi(x_i, y_j) = z_{i,j}$, $\frac{\partial\Psi}{\partial x}(x_i, y_j) = z^x_{i,j}$ and $\frac{\partial\Psi}{\partial y}(x_i, y_j) = z^y_{i,j}$, for $i \in \mathbb{N}_M$, $j \in \mathbb{N}_N$, and possesses $\mathscr{C}^1$-continuity.*

6 Numerical Examples

The aim of this section is to illustrate the rational cubic fractal interpolation surface (FIS) with some examples. Consider the positive surface interpolation data (see Table 1) with 16 points taken at random. Let us note that in Table 1, the 1st, 2nd,

Table 1 Interpolation data for positive blending rational cubic FISs

$\downarrow x/y \rightarrow$	0.5	1.5	2.5	4
1	(1.1, 0.1, 4.3)	(3.7, 0.7, 2.1)	(219.1, 0.6, 5.8)	(812.1, 0.11, 4)
2	(213.6, 1.7, 3)	(493.2, 3.8, 2.9)	(11.2, 2.1, 9.1)	(819.9, 1, 0.1)
3	(736.2, 2.8, 3)	(312.9, 1.55, 11.1)	(18.2, 1, 12.8)	(27.1, 2, 3.2)
4	(4, 3.7, 2.7)	(7.9, 8.99, 3.9)	(9.2, 3.22, 8.5)	(11.88, 1.7, 9.22)

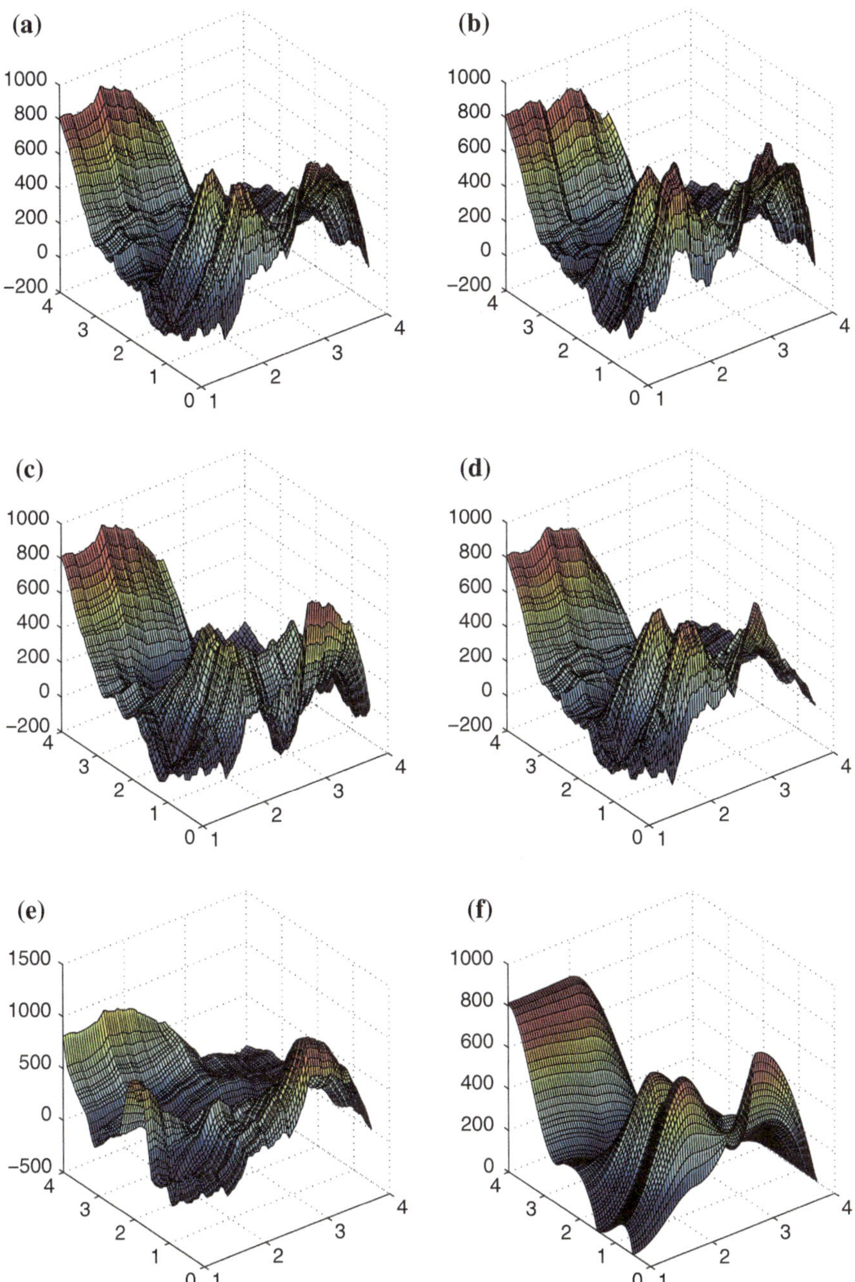

Fig. 1 Partially bicubically blended rational FIS. **a** Rational cubic FIS. **b** Effect of change in α in Fig. 1a. **c** Effect of change in α^* in Fig. 1a. **d** Effect of change in **r** and **t** in Fig. 1a. **e** Effect of change in $\mathbf{r}^*$ and $\mathbf{t}^*$ in Fig. 1a. **f** Classical rational cubic FIS

and 3rd components of (., ., .) represent the function value, the first order partial derivatives in x-direction, and y-direction at (x_i, y_j), where i, j are in $\{1, 2, 3, 4\}$.

For simplicity of presentation, let us represent the parameters, namely the (horizontal) scaling factors $|\alpha_{i,j}| < a_i < 1$ and the (horizontal) shape parameters $r_{i,j} > 0$, $t_{i,j} > 0$ with the aid of matrices of order $(M-1) \times N$ as follows: $\boldsymbol{\alpha} = [\alpha_{i,j}]$, $r = [r_{i,j}]$, and $t = [t_{i,j}]$. Similarly, the parameters along the vertical grid lines may be represented using matrices of order $M \times (N-1)$: $\boldsymbol{\alpha}^* = [\alpha^*_{i,j}]$, $r^* = [r^*_{i,j}]$, and $t^* = [t^*_{i,j}]$. We shall refer to the parameter matrices $\boldsymbol{\alpha}, \boldsymbol{\alpha}^*$ as scaling matrices and r, t, r^*, t^* as shape matrices.

The details of the scaling and the shape parameters used in the construction of Fig. 1a–f are provided in Tables 2 and 3. By using the matrices of the scaling and the shape parameters (see Tables 2 and 3), a rational cubic FIS is generated in Fig. 1a, which is taken as a reference surface. We construct a blending cubic rational FIS in Fig. 1b by changing the scaling matrix $\boldsymbol{\alpha}$ with respect to the IFS matrices of Fig. 1a. The effects of perturbed $\boldsymbol{\alpha}^*$ in Fig. 1a are captured in Fig. 1c. Similarly, by changing the shape matrices r, t, r^* and t^* we obtain Fig. 1d, e. We retrieve the classical rational cubic surface plotted in Fig. 1f by taking $\boldsymbol{\alpha} = [0]_{3\times4}$ and $\boldsymbol{\alpha}^* = [0]_{4\times3}$. We selected the parameters at random except for the mild conditions imposed in the construction of the cubic spline FIS. Among various values of the scaling factors and the shape parameters satisfying the conditions imposed for $\mathscr{C}^1$-continuity, we can choose the values so that the constructed interpolant satisfies certain additional conditions, for instance, preserving shape inherent in the data set. This is reserved for a future work.

Let us conclude the paper with some remarks. By using the convergence of the univariate FIFs forming the boundaries, it can be shown that the fractal surface Ψ converges to the original function Φ generating the bivariate data. It is shown that by appropriate choices of derivative parameters, the global continuity of the univariate rational cubic spline can be made $\mathscr{C}^2$. By using bicubically blended Coons patch (instead of partially bicubically blended scheme) on a network of $\mathscr{C}^2$ boundary curves, we will be able to obtain a $\mathscr{C}^2$-continuous surface. From the shape preserving properties of the networks of boundary curves that follow from [16], the shape preserving properties of transfinite interpolating surface Ψ can be deduced. These topics will be the focal point of the articles that follow.

Table 2 Scaling matrices in the construction of blending rational cubic FISs in Fig. 1a–f

Scaling matrices in x-direction	Figure	Scaling matrices in y-direction	Figure
$\boldsymbol{\alpha} = \begin{bmatrix} -0.31 & -0.31 & -0.3 & -0.2 \\ -0.3 & -0.29 & -0.28 & -0.25 \\ 0.19 & 0.29 & 0.31 & 0.30 \end{bmatrix}$	1a, c–e, f	$\boldsymbol{\alpha}^* = \begin{bmatrix} -0.28 & -0.27 & -0.28 \\ 0.26 & 0.25 & 0.24 \\ 0.27 & 0.28 & 0.29 \\ -0.27 & -0.28 & -0.28 \end{bmatrix}$	1a, b, d, e
$\boldsymbol{\alpha} = \begin{bmatrix} 0.3100 & 0.3100 & 0.3300 & 0.2000 \\ -0.3010 & -0.3290 & -0.3280 & -0.3250 \\ 0.3190 & 0.3285 & 0.3090 & 0.3000 \end{bmatrix}$	1b	$\boldsymbol{\alpha}^* = \begin{bmatrix} 0.28 & 0.27 & 0.279 \\ -0.26 & -0.25 & -0.24 \\ 0.27 & -0.28 & 0.285 \\ 0.2690 & 0.279 & 0.28 \end{bmatrix}$	1c

Table 3 Scaling matrices in the construction of blending rational cubic FISs in Fig. 1a–f

Matrices of shape parameters in x-direction	Figure	Matrices of shape parameters in y-direction	Figure
$r = \begin{bmatrix} 2 & 8.5 & 13.7 & 56 \\ 40 & 23 & 98 & 45 \\ 75 & 35.5 & 17 & 8 \end{bmatrix}$	1a–c, e, f	$r^* = \begin{bmatrix} 3.5 & 20 & 11 \\ 4.5 & 10 & 23 \\ 8.9 & 40 & 54 \\ 27 & 30 & 19 \end{bmatrix}$	1a–d, f
$r = \begin{bmatrix} 40 & 2 & 98 & 4 \\ 75 & 35 & 17 & 8 \\ 2 & 34 & 14 & 16 \end{bmatrix}$	1d	$r^* = \begin{bmatrix} 4 & 10 & 23 \\ 8 & 40 & 54 \\ 3 & 20 & 10 \\ 27 & 3 & 19 \end{bmatrix}$	1e
$t = \begin{bmatrix} 1 & 1 & 1 & 1 \\ 1 & 1 & 1 & 1 \\ 1 & 1 & 1 & 1 \end{bmatrix}$	1a–c, e, f	$t^* = \begin{bmatrix} 10 & 10 & 10 \\ 10 & 10 & 10 \\ 10 & 10 & 10 \\ 10 & 10 & 10 \end{bmatrix}$	1a–d, f
$t = \begin{bmatrix} 1 & 1 & 1 & 1 \\ 2 & 2 & 2 & 2 \\ 1 & 1 & 1 & 1 \end{bmatrix}$	1d	$t^* = \begin{bmatrix} 1 & 1 & 1 \\ 1 & 1 & 1 \\ 1 & 1 & 1 \\ 1 & 1 & 1 \end{bmatrix}$	1e

Acknowledgments The authors are thankful to the Department of Science & Technology, India for the SERC DST Project No. SR/S4/MS: 694/10.

References

1. Barnsley, M.F.: Fractal functions and interpolations. Constr. Approx. **2**, 303–329 (1986)
2. Barnsley, M.F.: Fractals Everywhere. Academic Press, Orlando (1988)
3. Barnsley, M.F., Harrington, A.N.: The calculus of fractal interpolation functions. J. Approx. Theory **57**(1), 14–34 (1989)
4. Böhm, W.: A survey of curve and surface methods in CAGD. Comput. Aided Geom. Des. **1**, 1–60 (1984)
5. Chand, A.K.B., Kapoor, G.P.: Generalized cubic spline fractal interpolation functions. SIAM J. Numer. Anal. **44**(2), 655–676 (2006)
6. Chand, A.K.B., Navascués, M.A.: Natural bicubic spline fractal interpolation. Nonlinear Anal. TMA **69**, 3679–3691 (2008)
7. Chand, A.K.B., Navascués, M.A.: Generalized Hermite fractal interpolation. Rev. R. Acad. Cienc. Zaragoza **64**(2), 107–120 (2009)
8. Chand, A.K.B., Vijender, N.: Monotonicity preserving rational quadratic fractal interpolation functions. Adv. Num. Anal. **2014**(Art ID 504825), 1–17 (2014)
9. Chand, A.K.B., Vijender, N., Agarwal, R.P.: Rational iterated function system for positive/monotonic shape preservation. Adv. Differ. Equ. **2014**(30), 1–19 (2014)
10. Chand, A.K.B., Viswanathan, P.: A constructive approach to cubic Hermite Fractal Interpolation Function and its constrained aspects. BIT Numer. Math. **53**, 841–865 (2013)

11. Duan, Q., Chen, T.Z., Djidjeli, K., Price, W.G., Twizell, E.H.: Convexity control and approximation properties of interpolating curves. Korean J. Comput. Appl. Math. **7**(2), 97–405 (2000)
12. Greiner, H.: A survey on univariate data interpolation and approximation by splines of given shape. Math. Comput. Model. **15**(10), 97–108 (1991)
13. Navascués, M.A.: Fractal polynomial interpolation. Z. Anal. Anwend. **24**(2), 401–418 (2005)
14. Shrivastava, M., Joseph, J.: $\mathscr{C}^2$-rational cubic spline involving tension parameters. Proc. Indian Acad. Sci. (Math. Sci.) **110**(3), 305–314 (2000)
15. Viswanathan, P., Chand, A.K.B.: A C^1 rational cubic fractal interpolation function: convergence and associated parameter identification problem. Acta Appl. Math. **136**(1), 19–41 (2015)
16. Viswanathan, P., Chand, A.K.B.: α-fractal rational splines for constrained interpolation. Electron. Tran. Numer. Anal. **41**, 420–442 (2014)
17. Viswanathan, P., Chand, A.K.B., Navascués, M.A.: Fractal perturbation preserving fundamental shapes: bounds on the scale factors. J. Math. Anal. Appl. **419**(2), 804–817 (2014)

Constrained 2D Data Interpolation Using Rational Cubic Fractal Functions

A.K.B. Chand and K.R. Tyada

Abstract In this paper, we construct the $\mathscr{C}^1$-rational cubic fractal interpolation function (RCFIF) and its application in preserving the constrained nature of a given data set. The $\mathscr{C}^1$-RCFIF is the fractal design of the traditional rational cubic interpolant of the form $\frac{p_i(\theta)}{q_i(\theta)}$, where $p_i(\theta)$ and $q_i(\theta)$ are the cubic polynomials with three tension parameters. We derive the uniform error bound between the RCFIF with the original function in $\mathscr{C}^3[x_1, x_n]$. When the data set is constrained between two piecewise straight lines, we deduce the sufficient conditions on the parameters of the RCFIF so that it lies between those two lines. Numerical examples are given to support that our method is interactive and smooth.

Keywords Fractal interpolation · Rational function · Rational fractal interpolation · Convergence · Constrained data · Positivity.

1 Introduction

In the literature a wide range of development on shape preserving classical spline interpolation techniques have been discussed, see [1–3, 13–18]. Although the classical splines interpolate the data smoothly, certain derivatives of the classical interpolants are either piecewise smooth or globally smooth in nature. Therefore, the classical interpolants are not suitable to approximate functions that have irregular nature or fractality in their first-order derivatives, see for instance [9–11, 19, 20]. On the other hand, fractal interpolation is an ideal tool in such scenario as well. Since the classical polynomial spline interpolant representation available in the literature is unique for given data, and it simply depends on the data points, it is difficult to

A.K.B. Chand · K.R. Tyada (✉)
Department of Mathematics, Indian Institute of Technology Madras, Chennai 600036, India
e-mail: kurmaths86@gmail.com

A.K.B. Chand
e-mail: chand@iitm.ac.in

P.N. Agrawal et al. (eds.), *Mathematical Analysis and its Applications*, Springer Proceedings in Mathematics & Statistics 143,
DOI 10.1007/978-81-322-2485-3_49

preserve all the hidden shape properties of the given data, for example, data between two piecewise straight lines, positivity, monotonicity, or convexity. For this reason a user needs an interactive shape preserving smooth curve representation of interpolation data. In these cases, the rational interpolation functions with shape parameters are preferred over various shape preserving interpolating techniques, see [15].

Fractal interpolation is a modern and advance technique to analyze various scientific data. Barnsley [4] coined the term fractal interpolation function which was constructed based on the theory of iterated functions system (IFS). An IFS ensures an attractor which is the graph of a continuous function that interpolates the given data points. FIFs are the fixed points of Read-Bajraktarević operator [4], which is defined on suitable function spaces. Barnsley and Harrington [5] introduced the construction of k- times differentiable polynomial spline FIF with a fixed type of boundary conditions. The polynomial spline FIFs with general body conditions were studied recently in [6–8]. Dalla and Drakopoulos [12] introduced polar fractal interpolation functions and developed the range restriction concept for a FIF. A specific feature of spline FIF is that its certain derivative can be used to capture the irregularity associated with the original function from where the interpolation data is obtained.

The use of fractal splines for constrained curve interpolation has been extensively investigated in the literature, see [1, 3, 14, 18], and references therein. Abbas [1] constructed a $\mathscr{C}^1$-piecewise rational cubic function to preserve the shape of constrained 2D and 3D data. Awang [3] developed a $\mathscr{C}^2$-rational cubic function to 2D constrained data interpolation. Duan [14] constructed a kind of rational spline based on the function values to constrain the interpolating curve to lies between two piecewise straight lines. Hussain and collaborators [16–18] used different types of $\mathscr{C}^1$-piecewise rational cubic functions to preserve the shape of various constrained data.

The shape preservation of scientific data through different types of smooth rational FIFs are studied recursively in [8–11, 19, 20]. Inspired by the work of Duan in constrained interpolation, we have constructed the smooth RCFIF so that it can be used for shape preservation. In particular, when the interpolation data set lies in between the two given piecewise straight lines, the IFS parameters of the proposed RCFIF are restricted so that it lies between these straight lines.

The paper is organized as follows: In Sect. 2, the general theory of FIF for a given data set is reviewed. The construction of $\mathscr{C}^1$ RCFIFs passing through a set of data points is discussed in Sect. 3. In Sect. 4, we have deduced the error estimation of the RCFIF with an original function for convergence results. In Sect. 5, the range of scaling factors and shape parameters are restricted according to the sufficient conditions so that the developed RCFIF lies between two piecewise straight lines followed by conclusions in Sect. 6.

2 Basics of FIF Theory

Let $\mathscr{P} : \{x_1, x_2, \dots, x_n\}$ be a partition of the real compact interval $I = [x_1, x_n]$, where $x_1 < x_2 < \cdots < x_n$. Denote $\Lambda := \{1, 2, \dots, n-1\}$ and $\Lambda^* := \{1, 2, \dots, n\}$.

Let a set of data points $\{(x_j, f_j) \in I \times K : j \in \Lambda^*\}$ be given, where K is a compact set in $\mathbb{R}$. Let $I_i = [x_i, x_{i+1}]$ and $L_i : I \to I_i,\ i \in \Lambda$ be contractive homeomorphisms such that $L_i(x_1) = x_i,\ L_i(x_n) = x_{i+1}$ for $i \in \Lambda$, and

$$|L_i(x) - L_i(x^*)| \le l_i\ |x - x^*| \quad \forall\, x, x^* \in I,\ 0 < l_i < 1. \tag{1}$$

Let $C = I \times K$, and consider $n-1$ continuous mappings $F_i : C \to K$ satisfying $F_i(x_1, f_1) = f_i,\ \ F_i(x_n, f_n) = f_{i+1},\ \ i \in \Lambda$,

$$|F_i(x, f) - F_i(x, \tilde{f})| \le |\lambda_i||f - \tilde{f}|\ \forall\ x \in I,\quad \forall\ f, \tilde{f} \in K \text{ for } 0 \le |\lambda_i| < 1. \tag{2}$$

Now, define functions $\omega_i : C \to I_i \times K$ such that $\omega_i(x, f) = (L_i(x), F_i(x, f))$ $\forall\, i \in \Lambda$.

Proposition 1 (Barnsley and Harrington [4]) *The IFS $\{C;\ \omega_i, i \in \Lambda\}$ defined above admits a unique attractor G such that G is the graph of a continuous function $g : I \to K$ which obeys $g(x_j) = f_j$ for $j \in \Lambda^*$.*

The above function g is called a FIF associated with the IFS $\{I \times K; \omega_i(x, f) = (L_i(x), F_i(x, f)),\ i \in \Lambda\}$. The functional representation of g follows from the fixed point of the Read-Bajraktarević operator T [4]. The FIF g satisfies the following functional equation:

$$Tg(x) \equiv F_i(L_i^{-1}(x), g \circ L_i^{-1}(x)) = g(x),\quad x \in I_i,\quad i \in \Lambda. \tag{3}$$

The following IFS has been studied extensively in the literature of FIF theory:

$$\{C; \omega_i(x, f) = (L_i(x), F_i(x, f)),\ i \in \Lambda\}, \tag{4}$$

where $L_i(x) = a_i x + b_i,\ \ F_i(x, f) = \lambda_i f + M_i(x)$ with $M_i : I \to \mathbb{R}$ are suitable continuous functions such that (2) is satisfied. The multiplier λ_i is called a scaling factor of the transformation ω_i, and $\lambda = (\lambda_1, \lambda_2, \ldots, \lambda_{n-1})$ is the scale vector associated with the IFS (4). The scaling factors give an additional degree of freedom to FIFs over its counterparts in classical interpolation and allow us to modify their shape preserving properties. The existence of a spline FIF is given by Barnsley and Harrington [5] based on the calculus of fractal functions, and that result has been extended for the existence of rational spline FIF in the following theorem [11]:

Theorem 1 *Let $\{(x_j, f_j) : j \in \Lambda^*\}$ be the given data set such that $x_1 < x_2 < \cdots < x_n$. Suppose that $L_i(x) = a_i x + b_i$, where $a_i = \frac{x_{i+1}-x_i}{x_n-x_1},\ \ b_i = \frac{x_n x_i - x_1 x_{i+1}}{x_n - x_1}$ and $F_i(x, f) = \alpha_i f + M_i(x),\ \ M_i(x) = \frac{p_i(x)}{q_i(x)},\ p_i(x)$ and $q_i(x)$ are chosen polynomials of degree r and s, respectively, and $q_i(x) \neq 0\ \forall\ x \in [x_1, x_n]$ for $i \in \Lambda$. Suppose for some integer $p \ge 0,\ |\alpha_i| < a_i^p, i \in \Lambda$. Let $F_{i,m}(x, f) = \frac{\lambda_i f + M_i^{(m)}(x)}{a_i^m},\ \ f_{1,m} = \frac{M_1^{(m)}(x_1)}{a_1^m - \lambda_1},\ f_{n,m} = \frac{M_{n-1}^{(m)}(x_n)}{a_{n-1}^m - \lambda_{n-1}},\ \ m = 1, 2, ..., p$, where $M_i^{(m)}(x)$ represents the mth derivative of $M_i(x)$ with respect to x. If $F_{i,m}(x_n, f_{n,m}) =$*

$F_{i+1,m}(x_1, f_{1,m})$, $i = 1, 2, \ldots, n-2$, $m = 1, 2, \ldots, p$, *then the IFS* $\{I \times K; \omega_i(x, f) = (L_i(x), F_i(x, f)), i \in \Lambda\}$ *determines a rational FIF* $\Phi \in \mathscr{C}^p[x_1, x_n]$ *such that* $\Phi(L_i(x)) = \lambda_i \Phi(x) + M_i(x)$, *and* $\Phi^{(m)}$ *is the FIF determined by* $\{I \times K; \omega_{i,m}(x, f) = (L_i(x), F_{i,m}(x, f)),\ i = 1, \ldots, n-1\}$ *for* $m = 1, 2, \ldots, p$.

3 $\mathscr{C}^1$-Rational Cubic Fractal Interpolation Function

In this section, we construct the RCFIF with three shape parameters in each subinterval with the help of Theorem 1. Let $\{(x_j, f_j),\ j \in \Lambda^*\}$ be a given set of interpolation data for an original function Ψ such that $x_1 < x_2 < \cdots < x_n$. Consider the IFS $\{I \times K; \omega_i(x, f) = (L_i(x), F_i(x, f)),\ i \in \Lambda\}$, where $L_i(x) = a_i x + b_i$ and $F_i(x, f) = \lambda_i f(x) + M_i(x)$, $M_i(x) = \frac{p_i(x)}{q_i(x)}$, where $p_i(x)$ and $q_i(x)$ are cubic polynomials, $q_i(x) \neq 0\ \forall\ x \in [x_1, x_n]$, and $|\lambda_i| < a_i$, $i \in \Lambda$. Let $F_i^{(1)}(x, d) = \frac{\lambda_i d + M_i^{(1)}(x)}{a_i}$, where $M_i^{(1)}(x)$ is the first order derivative of $M_i(x)$, $x \in [x_1, x_n]$. $F_i(x, f)$ satisfying the following join-up conditions:

$$F_i(x_1, f_1) = f_i,\ F_i(x_n, f_n) = f_{i+1},\ F_i^{(1)}(x_1, d_1) = d_i,\ F_i^{(1)}(x_n, d_n) = d_{i+1}, \quad (5)$$

where d_i denote the first-order derivative of Ψ with respect to x at knot x_i. The attractor of the above IFS will be the graph of a $\mathscr{C}^1$-rational cubic FIF. From (3) one can observe that our FIF can be written as

$$\Phi(L_i(x)) = \lambda_i \Phi(x) + M_i(x) = \lambda_i \Phi(x) + \frac{p_i(\theta)}{q_i(\theta)}, \quad (6)$$

where

$$p_i(\theta) = (1-\theta)^3 U_i + \theta(1-\theta)^2 V_i + \theta^2(1-\theta) W_i + \theta^3 X_i,$$
$$q_i(\theta) = (1-\theta)^3 \alpha_i + \theta(1-\theta)\gamma_i + \theta^3 \beta_i,$$

$\theta = \frac{x - x_1}{l}$, $l = x_n - x_1$, $x \in I$, and α_i, β_i and γ_i are positive real shape parameters. To ensure that the rational cubic FIF is $\mathscr{C}^1$-continuous, the following interpolation conditions are imposed:

$$\Phi(L_i(x_1)) = f_i,\ \Phi(L_i(x_n)) = f_{i+1},\ \Phi'(L_i(x_1)) = d_i,\ \Phi'(L_i(x_n)) = d_{i+1}. \quad (7)$$

From (6) and (7), it is clear that at $x = x_1$ we get

$$\Phi(L_i(x_1)) = f_i \Rightarrow f_i = \lambda_i f_1 + \frac{U_i}{\alpha_i} \Rightarrow U_i = \alpha_i(f_i - \lambda_i f_1).$$

Similarly, at $x = x_n$ we obtain

$$\Phi(L_i(x_n)) = f_{i+1} \Rightarrow f_{i+1} = \lambda_i f_n + \frac{X_i}{\beta_i} \Rightarrow X_i = \beta_i(f_{i+1} - \lambda_i f_n).$$

Again from (6) and (7), at $x = x_1$ we observe that

$$\Phi'(L_i(x_1)) = d_i \Rightarrow a_i d_i = \lambda_i d_1 + \frac{V_i - \gamma_i(f_i - \lambda_i f_1)}{\ell \alpha_i}$$
$$\Rightarrow V_i = \gamma_i(f_i - \lambda_i f_1) + \ell \alpha_i(a_i d_i - \lambda_i d_1).$$

Similarly, at $x = x_n$ we notice that

$$\Phi'(L_i(x_n)) = d_{i+1} \Rightarrow a_i d_{i+1} = \lambda_i d_n + \frac{\gamma_i(f_{i+1} - \lambda_i f_n) - W_i}{\ell \beta_i}$$
$$\Rightarrow W_i = \gamma_i(f_{i+1} - \lambda_i f_n) - \ell \beta_i(a_i d_{i+1} - \lambda_i d_n).$$

Now substituting U_i, V_i, W_i and X_i in (6), we get the required $\mathscr{C}^1$-RCFIF with the numerator,

$$p_i(\theta) = \alpha_i(f_i - \lambda_i f_1)(1-\theta)^3 + \{\gamma_i(f_i - \lambda_i f_1) + \ell\alpha_i(a_i d_i - \lambda_i d_1)\}\theta(1-\theta)^2$$
$$+ \{\gamma_i(f_{i+1} - \lambda_i f_n) - \ell\beta_i(a_i d_{i+1} - \lambda_i d_n)\}\theta^2(1-\theta) + \beta_i(f_{i+1} - \lambda_i f_n)\theta^3.$$

In most applications, the derivatives $d_j (j \in \Lambda^*)$ are not given, and hence must be calculated either from the given data or by some numerical methods. In this paper we have calculated d_j, $j \in \Lambda^*$ from the given data using the arithmetic mean method.

Remark 1 If $\lambda_i = 0$ for all $i \in \Lambda$, the RCFIF Φ becomes the classical rational cubic interpolation function $S(x)$(say) that is defined in [2] on each subinterval $[x_i, x_{i+1}]$ as

$$S(x) = \frac{p_i(z)}{q_i(z)}, \quad x \in [x_i, x_{i+1}], \tag{8}$$

where $z = \frac{x - x_i}{h_i}$, $h_i = x_{i+1} - x_i$,
$p_i(z) = \alpha_i f_i(1-z)^3 + (\gamma_i f_i + h_i\alpha_i d_i)z(1-z)^2 + (\gamma_i f_{i+1} - h_i\beta_i d_{i+1})z^2(1-z) + \beta_i f_{i+1} z^3$, and $q_i(z) = \alpha_i(1-z)^3 + \gamma_i z(1-z) + \beta_i z^3$.

4 Convergence Analysis

Due to the implicit expression of the RCFIF Φ, it is not easy to compute the uniform error bound $\|\Phi - \Psi\|_\infty$ by using any standard numerical analysis techniques. Hence we derive an upper bound of the error by using the classical counterpart S of Φ with the help of

$$\|\Phi - \Psi\|_\infty \le \|\Phi - S\|_\infty + \|S - \Psi\|_\infty, \tag{9}$$

where S is given by (8).

Now the error estimation between the original function Ψ and the classical rational cubic function S in an arbitrary subinterval $I_i = [x_i, x_{i+1}]$ can be found by using the Peano-Kernel theorem and the details are given in [2].

Proposition 2 *The error between the classical rational cubic function defined in (8) and the original function* $\Psi \in C^3[x_1, x_n]$ *is*

$$|\Psi(x) - S(x)| \leq \|\Psi^{(3)}\| h_i^3 c_i, \; x \in [x_i, x_{i+1}], \tag{10}$$

$c_i = \max_{0 \leq z \leq 1} \Theta(\alpha_i, \beta_i, z)$,

$$\Theta(\alpha_i, \beta_i, z) = \begin{cases} \max \Theta_1(\alpha_i, \beta_i, z) & \text{for } 0 \leq z \leq z^*, \\ \max \Theta_2(\alpha_i, \beta_i, z) & \text{for } z^* \leq z \leq 1, \end{cases}$$

where $z^* = 1 - \frac{\beta_i}{(\gamma_i - \beta_i)}$ *and* $\Theta_1(\alpha_i, \beta_i, \gamma_i, z)$, *and* $\Theta_2(\alpha_i, \beta_i, \gamma_i, z)$ *are obtained from the proof of Theorem 3.1 in [2].*

Theorem 2 *Let* Φ *be the* $\mathscr{C}^1$ *continuous RCFIF and* $\Psi \in C^3[x_1, x_n]$ *is the data generating function with respect to the given data* $\{(x_j, f_j), j \in \Lambda^*\}$. *Let* d_j *be the bounded first order derivative at the knot* x_j, $j \in \Lambda^*$. *Let* $|\lambda|_\infty = \max\{|\lambda_i|, i \in \Lambda\}$ *and the shape parameters* $\alpha_i > 0$, $\beta_i > 0$ *and* $\gamma_i > \max\{\alpha_i, \beta_i\}$ *for* $i \in \Lambda$. *Then*

$$\|\Psi - \Phi\|_\infty \leq \|\Psi^{(3)}\|_\infty h^3 c + \frac{|\lambda|_\infty}{1 - |\lambda|_\infty}(E(h) + E^*(h)), \tag{11}$$

where $h = \max_{1 \leq i \leq n-1}\{h_i\}$, $E(h) = \|\Psi\|_\infty + 2hE_1$, $E^*(h) = F + 2hE_2$ *with* $E_1 = \max_{1<j<n}\{|d_j|\}$, $F = \max\{|f_1|, |f_n|\}$, $E_2 = \max\{|d_1|, |d_n|\}$, *and* $c = \max_{1 \leq i \leq n-1}\{c_i\}$, c_i *is defined as in Proposition 2.*

Proof Consider the space $\mathscr{F}^* = \{g \in C^1(I, \mathbb{R}) \mid g(x_1) = f_1, g(x_n) = f_n, g'(x_1) = d_1, g'(x_n) = d_n\}$. From (2) and (6), the Read-Bajraktarević operator $T_\lambda^* : \mathscr{F}^* \to \mathscr{F}^*$ for the RCFIF can be written as

$$T_\lambda^* g(x) = \lambda_i g\left(L_i^{-1}(x)\right) + \frac{p_i\left(L_i^{-1}(x), \lambda_i\right)}{q_i\left(L_i^{-1}(x)\right)}, \; x \in I_i, \; i \in \Lambda. \tag{12}$$

Note that Φ is the fixed point of T_λ^* with $\lambda \neq 0$ and S is the fixed point of T_0^*. Since T_λ^* is a contractive operator with the contraction factor $|\lambda|_\infty$, we have

$$\|T_\lambda^* \Phi - T_\lambda^* S\|_\infty \leq |\lambda|_\infty \|\Phi - S\|_\infty. \tag{13}$$

From (12), we have

$$|T_\lambda^* S(x) - T_0^* S(x)| \le |\lambda|_\infty \left(\|S\|_\infty + \left| \frac{\partial \left\{ \frac{p_i\left(L_i^{-1}(x),\tau_i\right)}{q_i\left(L_i^{-1}(x)\right)} \right\}}{\partial \lambda_i} \right| \right), \ |\tau_i| \in (0, \lambda_i), \tag{14}$$

where the mean value theorem for function of several variables is used in this calculation. Now we wish to find out the error bounds of the terms on the right sides of (14). From the classical rational cubic trigonometric function (8), it is easy to observe that

$$S(x) = \sigma_1(\alpha_i, \beta_i, \gamma_i, z) f_i + \sigma_2(\alpha_i, \beta_i, \gamma_i, z) f_{i+1} + \sigma_3(\alpha_i, \beta_i, \gamma_i, z) d_i - \sigma_4(\alpha_i, \beta_i, \gamma_i, z) d_{i+1}, \tag{15}$$

where $\sigma_1(\alpha_i, \beta_i, \gamma_i, z) = \frac{1}{q_i(z)} \left\{\alpha_i(1-z)^3 + \gamma_i z(1-z)^2\right\} \ge 0$,
$\sigma_2(\alpha_i, \beta_i, \gamma_i, z) = \frac{1}{q_i(z)} \left\{\gamma_i z^2(1-z) + \beta_i z^3\right\} \ge 0$,
$\sigma_3(\alpha_i, \beta_i, \gamma_i, z) = \frac{h_i}{q_i(z)} \left\{\alpha_i z(1-z)^2\right\} \ge 0$,
$\sigma_4(\alpha_i, \beta_i, \gamma_i, z) = \frac{h_i}{q_i(z)} \left\{\beta_i z^2(1-z)\right\} \ge 0$.
It is easy to verify that $\sigma_1(\alpha_i, \beta_i, \gamma_i, z) + \sigma_2(\alpha_i, \beta_i, \gamma_i, z) = 1$.

Also, for $\alpha_i > 0$, $\beta_i > 0$, $\gamma_i > 0$ and choosing $\gamma_i > \max\{\alpha_i, \beta_i\}$ we obtain the following inequality:

$$\begin{aligned} \sigma_3(\alpha_i, \beta_i, \gamma_i, z) + \sigma_4(\alpha_i, \beta_i, \gamma_i, z) &= \frac{h_i}{q_i(z)} \left\{\alpha_i z(1-z)^2 + \beta_i z^2(1-z)\right\} \\ &\le h_i \left\{ \frac{\alpha_i}{\gamma_i} + \frac{\beta_i}{\gamma_i} \right\} \le 2h_i. \end{aligned}$$

Thus, $|S(x)| \le \max_{j=i,i+1}\{|f_j|\} + 2h_i \max_{j=i,i+1}\{|d_j|\}$. Since the above estimation is true for $i \in \Lambda$, we get the following estimation:

$$\|S\|_\infty \le E(h) := \|\Psi\|_\infty + 2hE_1, \tag{16}$$

Since $q_i(x)$ is independent of λ_i, from the first term in the right side of (14),

$$\begin{aligned} \frac{\partial \left\{ \frac{p_i(L_i^{-1}(x),\tau_i)}{q_i(L_i^{-1}(x))} \right\}}{\partial \lambda_i} &= \sigma_1(\alpha_i, \beta_i, \gamma_i, z) f_1 + \sigma_2(\alpha_i, \beta_i, \gamma_i, z) f_n + \sigma_3(\alpha_i, \beta_i, \gamma_i, z) d_1 \\ &\quad - \sigma_4(\alpha_i, \beta_i, \gamma_i, z) d_n. \end{aligned}$$

Now by applying a similar argument, the following estimate can be obtained:

$$\left| \frac{\partial \left\{ \frac{p_i(L_i^{-1}(x),\tau_i)}{q_i(L_i^{-1}(x))} \right\}}{\partial \lambda_i} \right| \le E^*(h) := F + 2hE_2. \tag{17}$$

Substituting (16) and (17) into (14), we have

$$|T^*_\lambda S(x) - T^*_0 S(x)| \leq |\lambda|_\infty (E(h) + E^*(h)), \ \ x \in [x_1, x_n].$$

Consequently, we obtain

$$\|T^*_\lambda S - T^*_0 S\|_\infty \leq |\lambda|_\infty (E(h) + E^*(h)). \tag{18}$$

Using (13) and (18) in
$\|\Phi - S\|_\infty = \|T^*_\lambda \Phi - T^*_0 S\|_\infty \leq \|T^*_\lambda \Phi - T^*_\lambda S\|_\infty + \|T^*_\lambda S - T^*_0 S\|_\infty$,
we have the following estimation:

$$\|\Phi - S\|_\infty \leq \frac{|\lambda|_\infty (E(h) + E^*(h))}{1 - |\lambda|_\infty}. \tag{19}$$

From Proposition 2, we have $|\Psi(x) - S(x)| \leq |\Psi^{(3)}| h_i^3 c_i$ which gives that

$$\|\Psi - S\|_\infty \leq \|\Psi^{(3)}\|_\infty h^3 c, \tag{20}$$

where h and c are defined in the statement of theorem. Substituting (19) and (20) in (9), we obtain the desired upper bound in (11). □

Convergence Result: Assume that $\max_{1 \leq j \leq n}\{|d_j|\}$ is bounded for every partition of the domain I. Since $|\lambda_i| < a_i, i \in \Lambda \Rightarrow |\lambda|_\infty < \frac{h}{\ell}$, and Theorem 2 proves that the RCFIF Φ converges uniformly to the original function Ψ as $h \to 0$. Additionally, if $|\lambda_i| < a_i^3 = \frac{h_i^3}{\ell^3}$ for $i \in \Lambda$, then $\|\Psi - \Phi\|_\infty = O(h^3)$ as $h \to 0$.

5 Constrained $\mathscr{C}^1$-RCFIF

In this section, we discuss on the construction of a constrained RCFIFs whose graph lie in between two piecewise straight lines 'L^u' and 'L^b' when the given interpolation data are distributed between 'L^u' and 'L^b'. In general, a RCFIF may not lie in between 'L^u' and 'L^b' with an arbitrary choice of IFS parameters. In order to avoid this circumstance, it is required to deduce sufficient data dependent restrictions on the scaling factor λ_i and on the shape parameters α_i, β_i and γ_i so that the RCFIF preserves the shape of the constrained data.

Suppose that the line 'L^u' is defined piecewise over $[x_i, x_{i+1}]$ such that $L^u(x_j) = f^u_j \ \forall j \in \Lambda^*$. Similarly, '$L^b$' is defined piecewise over $[x_i, x_{i+1}]$ such that $L^b(x_j) = f^b_j \ \forall j \in \Lambda^*$. The IFSs for '$L^u$' and '$L^b$' over I are given by $\{\mathbb{R}; (L_i(x), F^u_i(x)), i \in \Lambda\}$, $\{\mathbb{R}; (L_i(x), F^b_i(x)), i \in \Lambda\}$ where $F^b_i(x) = (1-\theta)\mu_i + \theta\eta_i$, and $F^u_i(x) = (1-\theta)\mu^*_i + \theta\eta^*_i$ where $\theta = \frac{x - x_1}{x_n - x_1}$ with $\mu_i = m_i x_i + c_i, \eta_i = m_i x_{i+1} + c_i$ and $\mu^*_i = m^*_i x_i + c^*_i, \eta^*_i = m^*_i x_{i+1} + c^*_i, \ i \in \Lambda$.

Theorem 3 *Let Φ be the RCFIF (6) defined over the interval $[x_1, x_n]$ with respect to the given data $\{(x_j, y_j), j \in \Lambda^*\}$. Further, assume that the data points lie above the straight line 'L^u' and below the straight line 'L^b'. Then the RCFIF Φ lies in between the straight line 'L^u' and 'L^b' if the following conditions are satisfied for all $i \in \Lambda$:*

(i) Select the scaling factors as

$$0 < \lambda_i < \min\{\lambda_i^u, \lambda_i^b\}, \tag{21}$$

(ii) Select the shape parameters as

$$\alpha_i > 0,\ \beta_i > 0 \text{ and } \gamma_i > \max\{\gamma_i^u, \gamma_i^b\}, \tag{22}$$

where λ_i^u, γ_i^u, λ_i^b and γ_i^b are defined in (29)–(32) respectively.

Proof Let $\{(x_j, f_j) : j \in \Lambda^*\}$ be the given set of data points lying in between the straight lines L^u and L^b. Then,

$$m_i x_j + c_i = f_j^b < f_j < f_j^u = m_i^* x_j + c_i^* \ \forall\, i \in \Lambda\ j \in \Lambda^*.$$

Since 'L^u' and 'L^b' are FIFs associated with the IFSs $\{\mathbb{R}; (L_i(x), F_i^u(x)),\ i \in \Lambda\}$ and $\{\mathbb{R}; (L_i(x), F_i^b(x)),\ i \in \Lambda\}$ respectively, then the functional equations of 'L^u' and 'L^b' are

$$\begin{aligned} L^b(L_i(x)) &= m_i L_i(x) + c_i = \mu_i(1-\theta) + \eta_i\theta = r_i(\theta) \text{ (say)},\\ L^u(L_i(x)) &= m_i^* L_i(x) + c_i^* = \mu_i^*(1-\theta) + \eta_i^*\theta = r_i^*(\theta) \text{ (say)}, \end{aligned} \tag{23}$$

where $L_i(x) = a_i x + b_i$ with $a_i = \frac{x_{i+1}-x_i}{x_n-x_1}$ and $b_i = \frac{x_n x_i - x_1 x_{i+1}}{x_n - x_1}$, $\theta = \frac{x-x_1}{\ell}$, $\ell = x_n - x_1$.

Note that at $x = x_1$, $\mu_i = m_i x_i + c_i$, $\mu_i^* = m_i^* x_i + c_i^*$ and at $x = x_n$, $\eta_i = m_i x_{i+1} + c_i$, $\eta_i^* = m_i^* x_{i+1} + c_i^*$. Thus the curve will lie in between the straight lines 'L^u' and 'L^b' if the $\mathscr{C}^1$-RCFIF Φ satisfies the following conditions:

$$L^b(L_i(x)) < \Phi(L_i(x)) < L^u(L_i(x))\ \forall\, x \in [x_1, x_n],\ i \in \Lambda. \tag{24}$$

Let $\theta_j = \frac{x_j - x_1}{x_n - x_1}$ and let $r_i^j = r_i(\theta_j)$, $r_i^{*j} = r_i^*(\theta_j)$. Assume that $\lambda_i \in [0, a_i), i \in \Lambda$ as $\Phi \in \mathscr{C}^1[x_1, x_n]$. In order to the RCFIF Φ lies between the piecewise straight lines 'L^u' and 'L^b', it is clear from (24) that for the next generation of interpolation points should satisfy the following inequalities:

$$\begin{aligned} &r_i(\theta_j) < \Phi(L_i(x_j)) < r_i^*(\theta_j) \Rightarrow r_i^j < \Phi(L_i(x_j)) < r_i^{*j},\\ &\Rightarrow \lambda_i r_i^j + \frac{p_i(\theta_j)}{q_i(\theta_j)} < \lambda_i f_j + \frac{p_i(\theta_j)}{q_i(\theta_j)} < \lambda_i r_i^{*j} + \frac{p_i(\theta_j)}{q_i(\theta_j)}. \end{aligned} \tag{25}$$

For the validity of $r_i^j < \lambda_i f_j + \frac{p_i(\theta_j)}{q_i(\theta_j)} < r_i^{*j}$, we need to impose the following conditions from (25):

$$r_i^j < \lambda_i r_i^j + \frac{p_i(\theta_j)}{q_i(\theta_j)}, \text{ and } \lambda_i r_i^{*j} + \frac{p_i(\theta_j)}{q_i(\theta_j)} < r_i^{*j}.$$

Therefore, the RCFIF lies in between the straight lines 'L^u' and 'L^b' if

$$\Omega_{1,i}(\theta_j) := (\lambda_i - 1)r_i^j + \frac{p_i(\theta_j)}{q_i(\theta_j)} \geq 0 \;\forall\, \theta \in [0,1],\; i \in \Lambda\; j \in \Lambda^*, \tag{26}$$

$$\Omega_{2,i}(\theta_j) := (\lambda_i - 1)r_i^{*j} + \frac{p_i(\theta_j)}{q_i(\theta_j)} \leq 0 \;\forall\, \theta \in [0,1], \text{ for every } i \in \Lambda\; j \in \Lambda^*. \tag{27}$$

After some algebraic simplifications, $\Omega_{1,i}(\theta)$ is reformulated as

$$\Omega_{1,i}(\theta_j) = \frac{p_i^*(\theta_j)}{q_i(\theta_j)} > 0, \tag{28}$$

where $p_i^*(\theta_j) = (1-\theta_j)^3 U_i^* + \theta_j(1-\theta_j)^2 V_i^* + \theta_j^2(1-\theta_j) W_i^* + \theta_j^3 X_i^*$, with $U_i^* = U_i + \alpha_i(\lambda_i - 1)r_i^j$, $V_i^* = V_i + \gamma_i(\lambda_i - 1)r_i^j$, $W_i^* = W_i + \gamma_i(\lambda_i - 1)r_i^j$, and $X_i^* = X_i + \beta_i(\lambda_i - 1)r_i^j$.

It is clear that the shape parameters $\alpha_i > 0$, $\beta_i > 0$ and $\gamma_i > 0$ guarantee that the denominator in (28) is positive. Thus the RCFIF preserves the constrained aspect of the constrained data if (28) holds for all $i \in \Lambda$ i.e., if the numerator $p_i^*(\theta_j)$ is positive. Therefore, $p_i^*(\theta_j) > 0$ if each U_i^*, V_i^*, W_i^* and X_i^* are positive.

Since $\alpha_i > 0$ and $U_i^* = U_i + \alpha_i(\lambda_i - 1)r_i^j = \alpha_i(f_i - \lambda_i f_1 + (\lambda_i - 1)r_i^j),\; j \in \Lambda^*$, the choice of

$$\lambda_i < \Xi_i := \min\left\{\frac{f_i - r_i^j}{f_1 - r_i^j} : j \in \Lambda^*\right\} \text{ yields } U_i^* > 0.$$

Similarly, since $\beta_i > 0$ and $X_i^* = X_i + \beta_i(\lambda_i - 1)r_i^j = \beta_i(f_{i+1} - \lambda_i f_n + (\lambda_i - 1)r_i^j),\; j \in \Lambda^*$, the selection of

$$\lambda_i < \Im_i := \min\left\{\frac{f_{i+1} - r_i^j}{f_n - r_i^j} : j \in \Lambda^*\right\} \text{ ensures } X_i^* > 0.$$

Consider $V_i^* = V_i + \gamma_i(\lambda_i - 1)r_i^j = \gamma_i(f_i - \lambda_i f_1 + (\lambda_i - 1)r_i^j) + \ell\alpha_i(a_i d_i - \lambda_i d_1)$. Then for $a_i d_i - \lambda_i d_1 > 0$, arbitrary $\alpha_i > 0$ and $\gamma_i > 0$ provide $V_i^* > 0$. Otherwise for $\alpha_i > 0$, the choice of

$$\gamma_i > \Upsilon_i := \max\left\{\frac{-\ell\alpha_i(a_i d_i - \lambda_i d_1)}{f_i - \lambda_i f_1 + (\lambda_i - 1)r_i^j} : j \in \Lambda^*\right\} \text{ results } V_i^* > 0.$$

Similarly consider $W_i^* = W_i + \gamma_i(\lambda_i - 1)r_i^j = \gamma_i\left(f_{i+1} - \lambda_i f_n + (\lambda_i - 1)r_i^j\right) - \ell\beta_i(a_i d_{i+1} - \lambda_i d_n)$. Then for $(a_i d_{i+1} - \lambda_i d_n) < 0$, arbitrary $\beta_i > 0$ and $\gamma_i > 0$ provide $X_i^* > 0$. Otherwise for $\beta_i > 0$, the selection of

$$\gamma_i > \aleph_i := \max\left\{\frac{\ell\beta_i(a_i d_{i+1} - \lambda_i d_n)}{f_{i+1} - \lambda_i f_n + (\lambda_i - 1)r_i^j} : j \in \Lambda^*\right\} \text{ produces } W_i^* > 0.$$

Hence $\Omega_{1,i}(\theta_j) > 0 \; \forall \, i \in \Lambda, \; j \in \Lambda^*$ when

(i) the scaling factors are chosen as

$$\lambda_i < \lambda_i^u := \min\{a_i, \Xi_i, \Im_i\} \tag{29}$$

(ii) the shape parameters are chosen as $\alpha_i > 0$, $\beta_i > 0$ and

$$\gamma_i > \gamma_i^u := \max\{0, \Upsilon_i, \aleph_i\}. \tag{30}$$

Using the similar argument as above, we deduce that $\Omega_{2,i}(\theta_j) < 0 \; \forall \, \theta \in [0, 1]$, $i \in \Lambda \; j \in \Lambda^*$, i.e., the RCFIF Φ lies below the straight line 'L^u' when

(i) the scaling factors are selected as

$$\lambda_i < \lambda_i^b := \min\{\Xi_i^*, \Im_i^*\} \tag{31}$$

(ii) the shape parameters are selected as $\alpha_i > 0$, $\beta_i > 0$ and

$$\gamma_i > \gamma_i^b := \max\{\Upsilon_i^*, \aleph_i^*\}, \tag{32}$$

where $\Xi_i^* := \min\left\{\frac{r_i^{*j} - f_i}{r_i^{*j} - f_1} : j \in \Lambda^*\right\}$, $\Im_i^* := \min\left\{\frac{r_i^{*j} - f_{i+1}}{r_i^{*j} - f_n} : j \in \Lambda^*\right\}$, $\Upsilon_i^* = \max\left\{\frac{-\ell\alpha_i(a_i d_i - \lambda_i d_1)}{f_i - \lambda_i f_1 + (\lambda_i - 1)r_i^{*j}} : j \in \Lambda^*\right\}$ and $\aleph_i^* = \max\left\{\frac{\ell\beta_i(a_i d_{i+1} - \lambda_i d_n)}{f_{i+1} - \lambda_i f_n + (\lambda_i - 1)r_i^{*j}} : j \in \Lambda^*\right\}$.

Thus the RCFIF preserves the constraining nature of the data and lies between the straight lines if the IFS parameters are selected according to (21) and (22). □

Remark 2 It is clear that positivity preserving interpolation is a special case of the above developed constrained interpolation setting. By considering $r_i^j = 0$ in (26) and $r_i^{*j} = \infty$ in (27) for $i \in \Lambda, \; j \in \Lambda^*$, the RCFIF (6) preserves the positivity feature of the given data with respect to the restricted IFS parameters calculated from the Theorem 3. Since $r_i^{*j} = \infty$, there is no need of captivating the RCFIF from above by a piecewise straight line.

5.1 Numerical Example

We present a numerical example to illustrate the construction of the $\mathscr{C}^1$-RCFIFs and related constrained interpolation problem discussed in the previous section. For this we considered the interpolating data set $\{(0, 1), (0.5, 0.2), (1, 0.8), (1.5, 0.4)\}$ which is constrained in between the two straight lines taken from [14]:

$$L^u = \begin{cases} -\frac{8}{5}x + 1.07, & 0 \le x \le 0.5, \\ \frac{6}{5}x - 0.33, & 0.5 \le x \le 1.0, \\ -\frac{4}{5}x + 1.67, & 1.0 \le x \le 1.5, \end{cases}, \quad L^b = \begin{cases} -\frac{8}{5}x + 0.93, & 0 \le x \le 0.5, \\ \frac{6}{5}x - 0.47, & 0.5 \le x \le 1.0, \\ -\frac{4}{5}x + 1.53, & 1.0 \le x \le 1.5. \end{cases} \tag{33}$$

The derivative values at the knots are calculated using the arithmetic mean method. The constrained $\mathscr{C}^1$-RCFIF are generated iteratively using the IFS parameters given in Table 1. For simplicity, we have fixed two of the shape parameters $\alpha_i = 1$ and $\beta_i = 1,\ i = 1, 2, 3$. For arbitrary choice of rational IFS parameters, the RCFIF Φ_1 may not preserve the constrained nature of the given data, see for instance Fig. 1a. So by using Theorem 3, we have calculated the restrictions on the IFS parameters that satisfy the constrained inequalities (21) and (22), so that the RCFIF (6) must be $\mathscr{C}^1$-continuous in [0, 1.5] and bounded between the upper straight line 'L^u' and the lower straight line 'L^b'. The choice of scaling factors and shape parameters as per Theorem 3 are shown in Table 1. Figure 1b is generated as the graph of such RCFIF Φ_2 which preserves the constrained nature of given data for a particular restricted IFS parameters. The constrained RCFIF Φ_3 in Fig. 1c is generated with a perturbation in λ_2, and it has major effects in second subinterval, while the changes in third subinterval are also noticeable in comparison with Φ_2. This illustrates the global effect of scaling parameter in a rational FIF. These effects are distributed according

Table 1 Scaling factors and shape parameters used in the RCFIFs

Figure	Scaling factors (λ)	Shape parameters (γ)
1a	$\lambda = (-0.1, 0.1323, 0.1324)$	$\gamma = (8.4148, 19.1938, 12.0329)$
1b	$\lambda = (0.0123, 0.1123, 0.1323)$	$\gamma = (20, 100, 50)$
1c	$\lambda = (0.0123, 0.01, 0.1323)$	$\gamma = (20, 100, 50)$
1d	$\lambda = (0.0123, 0.01, 0.01)$	$\gamma = (20, 100, 50)$
1e	$\lambda = (0.0123, 0.1123, 0.01)$	$\gamma = (50, 100, 50)$
1f	$\lambda = (0, 0, 0)$	$\gamma = (50, 100, 50)$

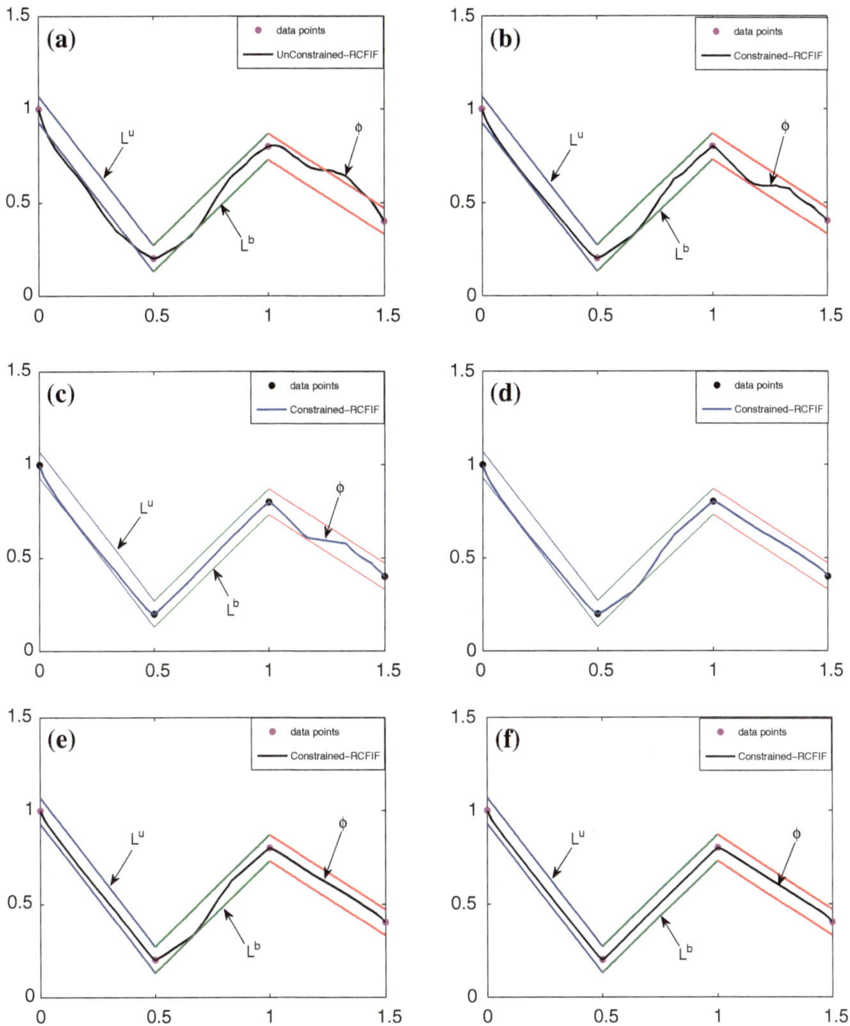

Fig. 1 Rational cubic fractal interpolation functions. **a** Unconstrained RCFIF Φ_1. **b** Constrained RCFIF Φ_2. **c** Constrained RCFIF Φ_3, effects of λ_2. **d** Constrained RCFIF Φ_4, effects of $\lambda_3(\lambda_2)$. **e** Constrained RCFIF Φ_5, effects of λ_3, γ_1. **f** Constrained classical rational spline S

to the code space related with map L_2 in the given domain. Next, we modify only λ_3 with respect to IFS parameters of Φ_3 to generate Φ_4. The perturbation effects of scaling parameter(s) on the shape of Φ_4 are worth to be noted in comparison with the shape of Φ_3 (Φ_2). By changing the scaling factor λ_3 and shape parameter γ_1, we have constructed the constrained RCFIF Φ_5 with pleasing effects in $[x_1, x_2]$ and in $[x_3, x_4]$ whose graph is shown in Fig. 1e. We observe that in $[x_1, x_2]$ and $[x_3, x_4]$, the RCFIF looks like converging to straight lines. The effect of fractality or irregularity in some portions of RCFIF can be restricted by setting the associated scaling factors to zero therein. Finally by setting all the scaling factors to zero, we have generated the graph of classical rational cubic interpolant S in Fig. 1f.

6 Conclusions

In this paper we have constructed $\mathscr{C}^1$-RCFIF to preserve the constrained aspect of given data. The RCFIF reduces to the traditional rational cubic interpolant by setting all scaling factors to zero. The developed RCFIF converges uniformly to the data generating original function as $h \to 0$, and additionally if $|\lambda_i| < a_i^3$, then the order of convergence is $O(h^3)$. We have developed the sufficient data-dependent conditions on the rational IFS parameters to preserve the shape of the given data in such away that the RCFIF lies between two piecewise straight lines. The effects of the rational IFS parameters on the shape of the RCFIFs are illustrated. The developed RCFIF can be used for the visualization of data with/without slopes at the knots. Applications of the proposed RCFIF in geometric modeling problems are under investigation.

Acknowledgments The first author is thankful to the Department of Science & Technology, India for the SERC DST Project No. SR/S4/MS: 694/10.

References

1. Abbas, M., Jamal, E., Ali, J.M.: Shape preserving constrained data visualization using spline functions. Int. J. Appl. Math. Stat. **29**(5), 34–50 (2012)
2. Abbas., M, Majid., A.A, Awang., M.N.H, Ali, J.M.: Local convexity shape-preserving data visualization by spline function. ISRN Math. Anal. **2012**(Article ID: 174048), 1–14 (2012)
3. Awang, M.N.H., Abbas. M., Majid, A.A., Ali, J.M.: Data visualization for constrained data using $\mathscr{C}^2$ rational cubic spline. In: Proceedings of the World Congress on Engineering and Computer Science, vol. 1 (2013)
4. Barnsley, M.F.: Fractal functions and interpolation. Constr. Approx. **2**, 303–329 (1986)
5. Barnsley, M.F., Harrington, A.N.: The calculus of fractal interpolation functions. J. Approx. Theory **57**(1), 14–34 (1989)
6. Chand, A.K.B., Kapoor, G.P.: Generalized cubic spline fractal interpolation functions. SIAM J. Numer. Anal. **44**(2), 655–676 (2006)
7. Chand, A.K.B., Navascués, M.A.: Generalized Hermite fractal interpolation. Rev. R. Acad. Cienc. Zaragoza **64**(2), 107–120 (2009)
8. Chand, A.K.B., Viswanathan, P.: Cubic Hermite and cubic spline fractal interpolation functions. AIP Conf. Proc. **1479**, 1467–1470 (2012)
9. Chand, A.K.B., Viswanathan, P.: A constructive approach to cubic Hermite fractal interpolation function and its constrained aspects. BIT Numer. Math. **53**(4), 841–865 (2013)
10. Chand, A.K.B., Vijender, N., Agarwal, R.P.: Rational iterated function system for positive/monotonic shape preservation. Adv. Differ. Eqn. **2014**(30), 1–19 (2014)
11. Chand, A.K.B., Vijender, N., Navascués, M.A.: Shape preservation of scientific data through rational fractal splines. Calcolo **51**(2), 329–362 (2014)

12. Dalla, L., Drakapoulos, V.: On the parameter identification problem in the plane and polar fractal interpolation functions. J. Approx. Theory **101**, 289–302 (1999)
13. Duan, Q., Xu, G., Liu, A., Wang, X., Cheng, F.: Constrained interpolation using rational cubic spline with linear denominators. Korean J. Comput. Appl. Math. **6**, 203–215 (1999)
14. Duan, Q., Wang, L., Twizell, E.H.: A new $\mathscr{C}^2$ rational interpolation based on function values and constrained control of the interpolant curves. J. Appl. Math. Comput. **61**, 311–322 (2005)
15. Gregory, J.A., Delbourgo, R.: Shape preserving piecewise rational interpolation. SIAM J. Stat. Comput. **6**(4), 967–976 (1985)
16. Hussian, M.Z., Sarfraz, M., Hussain, M.: Scientific data visualization with shape preserving $\mathscr{C}^1$-rational cubic interpolation. Eur. J. Pure Appl. Math. **3**(2), 194–212 (2010)
17. Sarfraz, M., Hussain, M.Z.: Data visualization using rational spline interpolation. J. Comp. Appl. Math **189**, 513–525 (2006)
18. Tahira, S.S., Sarfraz, M., Hussain., M.Z.: Shape preserving constrained data visualization using rational functions. J. Prime Res. Math. **7**, 35–51 (2011)
19. Viswanathan, P., Chand, A.K.B.: A $\mathscr{C}^1$-rational cubic fractal interpolation function: convergence and associated parameter identification problem. Acta. Appl. Math. **136**(1), 19–41 (2015)
20. Viswanathan, P., Chand, A.K.B., Navascués, M.A.: Fractal perturbation preserving fundamental shapes: Bounds on the scale factors. J. Math. Anal. Appl. **419**(2), 804–817 (2014)

Transverse Vibrations of Nonhomogeneous Rectangular Kirchhoff Plates of Variable Thickness

Roshan Lal and Renu Saini

Abstract In this paper the free transverse vibrations of nonhomogeneous rectangular Kirchhoff plates of linearly varying thickness along one direction have been studied using generalized differential quadrature method. The nonhomogeneity of the plate material is assumed to arise due to linear variations in Young's modulus and density of the plate material with the in-plane coordinates. Numerical results have been computed when the plate is clamped at all the four edges. The effect of various plate parameters such as nonhomogeneity parameters, density parameters, thickness parameter, and aspect ratio on the frequencies has been investigated for the first two modes of vibration. Three-dimensional mode shapes for a specified plate have been plotted. A comparison of results with those available in the literature has been presented.

Keywords Rectangular · Nonhomogeneity · GDQ · Variable thickness

1 Introduction

Nowadays, the study of nonhomogeneous materials is of great interest due to their wide applications in various fields of engineering such as aerospace, mechanical, nuclear, marine, structural engineering, etc. Plywood, timber, fiber-reinforced plastic, etc., are good examples of nonhomogeneous materials. The mechanical properties of such materials display spatial variations. In this regard, some of the high-strength light-weight nonhomogeneous/composite materials fabricated by mixing two or more materials, e.g., carbon fiber and epoxies are being used for aerospace applications and in high performance sporting goods. The nonhomogeneity of a structure is characterized by a number of factors governing its structural features. For plate-type structure these features are geometrical imperfections, including foreign

R. Lal · R. Saini (✉)
Department of Mathematics, Indian Institute of Technology Roorkee,
Roorkee 247667, India
e-mail: renusaini189@gmail.com

R. Lal
e-mail: rlatmfma@iitr.ac.in

P.N. Agrawal et al. (eds.), *Mathematical Analysis and its Applications*,
Springer Proceedings in Mathematics & Statistics 143,
DOI 10.1007/978-81-322-2485-3_50

materials and reinforcements of various types [1–3]. Sometimes, plate-type structural elements have to work under high temperature environment which causes nonhomogeneity in the material, particularly in aerospace industry, modern missile technology, and microelectronics. Further, these plates with appropriate thickness variation have significantly greater efficiency for vibration compared to plates of uniform thickness and also provide the advantage of material saving and hence cost requirement. Thus their design requires an accurate analysis for their vibration characteristic. In particular, rectangular plates are key components in ocean structures and in the aerospace industry. In this regard, numerous studies dealing with the vibration of rectangular plates of nonuniform thickness have been carried out and reported in references [4–7], to mention a few. Various models for nonhomogeneity of the plate material have been proposed in the literature [8–12]. In these papers, it is considered that nonhomogeneity of the plate material arises due to change in only one space variable except in references [11, 12]. The present study analyzes the effect of nonhomogeneity on the free transverse vibration of thin rectangular plates of varying thickness employing generalized differential quadrature (GDQ) method when the plate is clamped at all the four edges. Nonhomogeneity of the plate material is assumed to arise due to linear variation in Young's modulus and density of the plate material with the in-plane coordinates while the thickness is varying linearly along one direction. The effect of various parameters on the natural frequencies has been investigated for the first two modes of vibration. A comparison of results has been presented.

2 Mathematical Formulation

Configuration of a nonhomogeneous isotropic rectangular plate of length a, breadth b, thickness h, and density ρ is shown in Fig. 1. The x- and y-axes are taken along the edges of the plate, the axis of z is perpendicular to the xy-plane. The middle surface being $z = 0$ and origin is at the one of the corners of the plate. The differential equation governing the transverse vibration of such plates [12], is given by

$$\nabla^2(D\nabla^2 w)-(1-\nu)\left[\frac{\partial^2 D}{\partial x^2}\frac{\partial^2 w}{\partial y^2}-2\left(\frac{\partial^2 D}{\partial x\partial y}\right)\left(\frac{\partial^2 w}{\partial x\partial y}\right)+\frac{\partial^2 D}{\partial y^2}\frac{\partial^2 w}{\partial x^2}\right]-\rho h\frac{\partial^2 w}{\partial t^2}=0 \quad (1)$$

where ∇^2 is Laplacian operator, $D = Eh^3/12(1-\nu^2)$ is the flexural rigidity, $w(x, y, t)$ is the transverse displacement E is the Young's modulus, ν is the Poisson ratio, and t is the time.

For a harmonic solution, the displacement w is assumed to be

$$w(x, y, t) = \bar{w}(x, y, t)e^{i\omega t} \quad (2)$$

where ω is the circular frequency in radians.

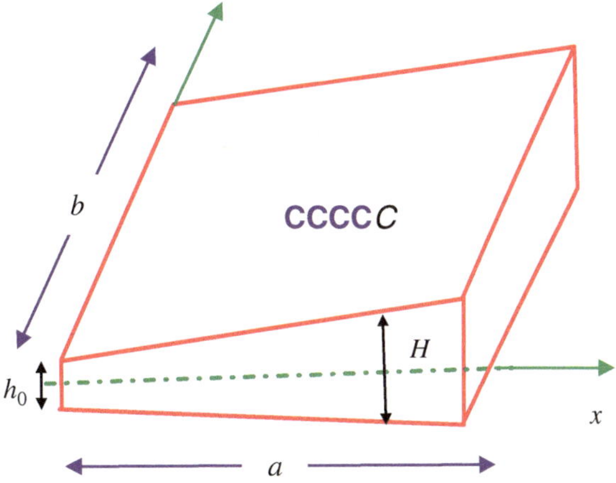

Fig. 1 Geometry of the rectangular plate

Using Eq. (2), Eq. (1) reduces to

$$D\left(\frac{\partial^4 \bar{w}}{\partial x^4}+\frac{\partial^4 \bar{w}}{\partial y^4}\right)+2D\frac{\partial^4 \bar{w}}{\partial x^2 \partial y^2}+2\frac{\partial D}{\partial x}\left(\frac{\partial^3 \bar{w}}{\partial x^3}+\frac{\partial^3 \bar{w}}{\partial x \partial y^2}\right)+2\frac{\partial D}{\partial y}\left(\frac{\partial^3 \bar{w}}{\partial y^3}+\frac{\partial^3 \bar{w}}{\partial x^2 \partial y}\right)$$
$$+\frac{\partial^2 \bar{w}}{\partial x^2}\left(\frac{\partial^2 D}{\partial x^2}+\nu\frac{\partial^2 D}{\partial y^2}\right)+\frac{\partial^2 \bar{w}}{\partial y^2}\left(\nu\frac{\partial^2 D}{\partial x^2}+\frac{\partial^2 D}{\partial y^2}\right)$$
$$+2(1-\nu)\frac{\partial^2 D}{\partial x \partial y}\frac{\partial^2 \bar{W}}{\partial x \partial y}-\rho h \omega^2 \bar{w}=0 \tag{3}$$

Taking the following nondimensional variables $X = x/a$, $Y = y/b$, $H = h/a$, $W = \bar{w}/a$ and assuming that Young's modulus and density of the plate material vary with the in-plane coordinates by the relations

$$\begin{aligned} E(X,Y) &= E_0(1+\alpha_1 X+\alpha_2 Y) \\ \rho(X,Y) &= \rho_0(1+\beta_1 X+\beta_2 Y) \end{aligned} \tag{4}$$

and thickness of the plate varies linearly in X-direction, given by

$$H(X) = h_0(1+\gamma X) \tag{5}$$

where E_0, ρ_0 and h_0 are the Young's modulus, density, and thickness of the plate at $X = 0$, $Y = 0$, γ is the thickness parameter, α_1, α_2 are the nonhomogeneity parameters, and β_1, β_2 are the density parameters, respectively.
Equation (3) now reduces to

$$A_0\left(\frac{\partial^4 W}{\partial X^4}+2\lambda^2\frac{\partial^4 W}{\partial X^2\partial Y^2}+\lambda^4\frac{\partial^4 W}{\partial Y^4}\right)+A_1\left(\frac{\partial^3 W}{\partial X^3}+\lambda^2\frac{\partial^3 W}{\partial X\partial Y^2}\right)$$
$$+A_2\left(\lambda^4\frac{\partial^3 W}{\partial Y^3}+\lambda^2\frac{\partial^3 W}{\partial X^2\partial Y}\right)+A_3\left(\frac{\partial^2 W}{\partial X^2}+\lambda^2\frac{\partial^2 W}{\partial Y^2}\right)+A_4\frac{\partial^2 W}{\partial X\partial Y}-A_5 W=0 \tag{6}$$

where $A_0=(1+\alpha_1 X+\alpha_2 Y)(1+\gamma X)^2$
$A_1=(6\gamma(1+\alpha_1 X+\alpha_2 Y)+2\alpha_1(1+\gamma X))(1+\gamma X)$
$A_2=2\alpha_2(1+\gamma X)^2$, $A_3=6\gamma^2(1+\alpha_1 X+\alpha_2 Y)+6\alpha_1\gamma(1+\gamma X)$
$A_4=6(1-\nu)\lambda^2\gamma\alpha_2(1+\gamma X)$, $A_5=\Omega^2(1+\beta_1 X+\beta_2 Y)$
$\lambda=a/b,\ \Omega^2=12\rho(1-\nu^2)\omega^2/aE_0h_0^2$

Equation (6) is a fourth-order partial differential equation of variable coefficients with respect to X and Y. The clamped boundary condition (i.e., CCCC) has been considered in the present paper and the relation that should satisfied at clamped edges given by:

$$W=\frac{dW}{dX}=0,\ W=\frac{dW}{dY}=0,\ \text{at } X=0 \text{ or } X=1,\ \text{and } Y=0 \text{ or } Y=1,\ \text{respectively.}$$

3 Method of Solution: Generalized Differential Quadrature (GDQ) Method

The computational domain of a rectangular plate is $0\le X\le 1, 0\le Y\le 1$. If N and M are the number of grid points in X and Y directions, respectively, then the total number of function values in the whole domain is $N\times M$. According to GDQ method [13], the nth and mth order derivatives of $W(X, Y)$ with respect to X, Y and its mixed derivative with respect to X and Y are given by

$$\frac{\partial^n W(X,Y)}{\partial X^n}=\sum_{l=1}^{N}a_{ij}^{(n)}W(X_l,Y_j)$$
$$\frac{\partial^m W(X,Y)}{\partial Y^m}=\sum_{l=1}^{M}b_{ij}^{(m)}W(X_i,Y_l) \tag{7}$$
$$\frac{\partial^{m+n} W(X,Y)}{\partial X^n\partial Y^m}=\sum_{l_1=1}^{N}\sum_{l_2=1}^{M}a_{il_1}^{(n)}b_{jl_2}^{(m)}W(X_{l_1},Y_{l_2})$$

$i=1,2,\ldots,N;\ j=1,2,\ldots,M;\ n=1,2,\ldots,N-1;\ m=1,2,\ldots,M-1$

where $a_{ij}^{(n)}$ and $b_{ij}^{(m)}$ are the weighting coefficients associated with nth and mth order derivatives with respect to $X$and Y respectively. The weighting coefficient of first-order derivative are determined as

$$a_{ij}^{(1)} = \begin{cases} \frac{P^{(1)}(X_i)}{(X_i - X_j)P^{(1)}(X_j)}, & j \neq i \\ -\sum_{j=1, j\neq i}^{N} a_{ij}^{(1)}, & j = i, \end{cases} \tag{8}$$

for $i, j = 1, 2, \ldots, N$

where $P^{(1)}(X_i) = \prod_{j=1, j\neq i}^{N} (X_i - X_j)$
Similarly, for the second and higher order derivatives the recurrence relationships are obtained as follows:

$$a_{ij}^{(n)} = \begin{cases} n\left(a_{ii}^{(n-1}a_{ij}^{(1)} - \frac{a_{ij}^{(n-1)}}{(X_i - X_j)}\right), & j \neq i \\ -\sum_{j=1, j\neq i}^{N} a_{ij}^{(n)}, & j = i, \end{cases} \tag{9}$$

for $i, j = 1, 2, \ldots, N,\ n = 2, 3, \ldots, N-1$.
The corresponding coefficients $b_{ij}^{(m)}$ associated with derivatives with respect to Y required can be similarly determined [13].
Discretizing Eq. (6) at the internal grid points (X_i, Y_j), with $3 \leq i \leq N-2$ and $3 \leq j \leq M-2$, it reduces to

$$\begin{aligned} A_0(i,j)&\left(\sum_{l=1}^{N} a_{il}^{(4)} W_{l,j} + 2\lambda^2 \sum_{l_1=1}^{N}\sum_{l_2=1}^{M} a_{il_1}^{(2)} b_{jl_2}^{(2)} W_{l_1,l_2} + \lambda^4 \sum_{l=1}^{M} b_{ij}^{(m)} W_{i,l}\right) \\ &+ A_1(i,j)\left(\sum_{l=1}^{N} a_{il}^{(3)} W_{l,j} + \lambda^2 \sum_{l_1=1}^{N}\sum_{l_2=1}^{M} a_{il_1}^{(1)} b_{jl_2}^{(2)} W_{l_1,l_2}\right) \\ &+ A_2(i,j)\left(\lambda^4 \sum_{l=1}^{M} b_{ij}^{(3)} W_{i,l} + \lambda^2 \sum_{l_2=1}^{M} a_{il_1}^{(2)} b_{jl_2}^{(1)} W_{l_1,l_2}\right) \\ + A_3(i,j)&\left(\sum_{l=1}^{N} a_{il}^{(2)} W_{l,j} + \nu\lambda^2 \sum_{l=1}^{M} b_{ij}^{(2)} W_{i,l}\right) + A_4(i,j) \sum_{l_1=1}^{N}\sum_{l_2=1}^{M} a_{il_1}^{(1)} b_{jl_2}^{(1)} W_{l_1,l_2} \\ &- A_5(i,j) W_{i,j} = 0 \end{aligned} \tag{10}$$

Similarly, the boundary conditions can be non-dimensionalized and then discretized using GDQ. Here, the grid points chosen for collocation are the zeroes of shifted Chebyshev polynomials and are given by

$$X_{i+1} = \frac{1}{2}\left[1 + \cos\left(\frac{2i-1}{N-2}\frac{\pi}{2}\right)\right], Y_{i+1} = \frac{1}{2}\left[1 + \cos\left(\frac{2j-1}{M-2}\frac{\pi}{2}\right)\right] \tag{11}$$

$$i = 1, 2, \ldots, N-2\ j = 1, 2, \ldots, M-2$$

4 Numerical Results and Discussions

Equation (10) together with boundary condition form a eigenvalue problem [13], which has been solve numerically using GDQ. The values of various plate parameters are taken as follows: Nonhomogeneity parameters $\alpha_1, \alpha_2 = (-0.5(0.1)0.5)$, density parameters $\beta_1, \beta_2 = (-0.5(0.1)0.5)$, thickness parameter $\gamma = (-0.5(0.1)0.5)$, aspect ratio $a/b = (0.25(0.25)2.0)$ and Poisson ratio $\nu = 0.3$.
The values of grid points N and M have been fixed as 15, since further increase in the values of grid points, frequency parameter remain constant at the fourth place of decimals. The convergence of frequency parameter Ω for a particular set $\alpha_1 = \alpha_2 = \beta_1 = \beta_2 = \gamma = 0.5, a/b = 1$ is shown in Table 1.

Figure 2 shows the behavior of frequency parameter Ω with nonhomogeneity parameter α_1 for $\alpha_2 = \pm 0.5, \gamma = 0.5, \beta_1 = \pm 0.5, \beta_2 = 0.5$ and $a/b = 1$ for the first two modes of vibration. It is observed that the value of frequency parameter Ω increases with increasing values of nonhomogeneity parameter α_1. Further, it is increases with increasing values of α_2 while it decreases with increasing values of β_1 keeping all other parameters fixed.

The effect of thickness parameter on the frequency parameter Ω for $\alpha_1 = 0.5, \alpha_2 = \pm 0.5, \beta_1 = \beta_2 = 0.5$ and $a/b = 1$ for the first two mode of vibration has

Table 1 Convergence study for frequency parameter Ω for the first three modes of vibration

No. of terms	I mode	II mode	III mode
$N = M = 8$	44.5957	89.7378	90.5676
$N = M = 10$	44.5942	90.5409	90.9843
$N = M = 12$	44.5940	90.5478	90.9795
$N = M = 13$	44.5940	90.5477	90.9795
$N = M = 14$	44.5940	90.5477	90.9795
$N = M = 15$	44.5940	90.5477	90.9795

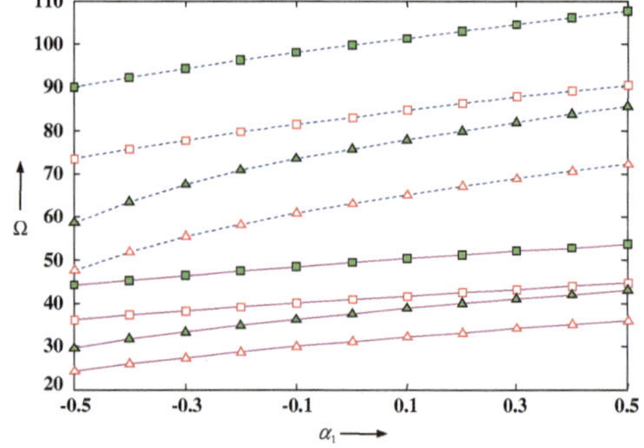

Fig. 2 Frequency parameter Ω for CCCC square plate for $\beta_2 = \gamma = 0.5$. I mode:—;II mode: ....; □, $\alpha_2 = 0.5$, $\beta_1 = 0.5$; ■, $\alpha_2 = 0.5$ $\beta_1 = -0.5$;Δ, $\alpha_2 = -0.5$, $\beta_1 = 0.5$; ▲, $\alpha_2 = -0.5$, $\beta_1 = -0.5$

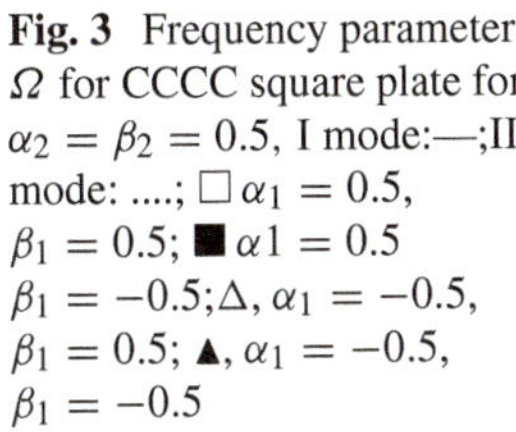

Fig. 3 Frequency parameter Ω for CCCC square plate for $\alpha_2 = \beta_2 = 0.5$, I mode:—;II mode: ....; □ $\alpha_1 = 0.5$, $\beta_1 = 0.5$; ■ $\alpha 1 = 0.5$ $\beta_1 = -0.5$;Δ, $\alpha_1 = -0.5$, $\beta_1 = 0.5$; ▲, $\alpha_1 = -0.5$, $\beta_1 = -0.5$

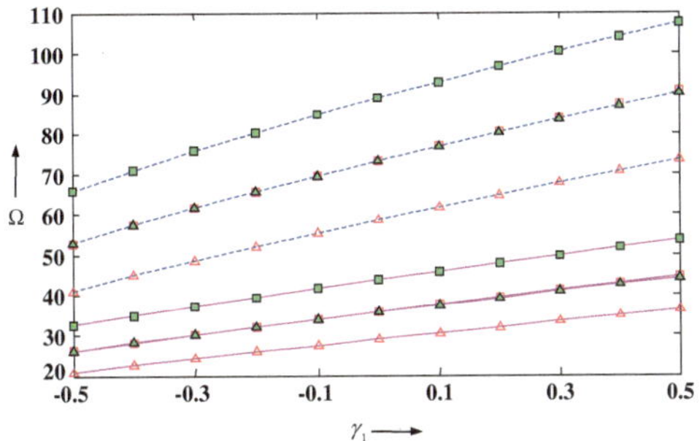

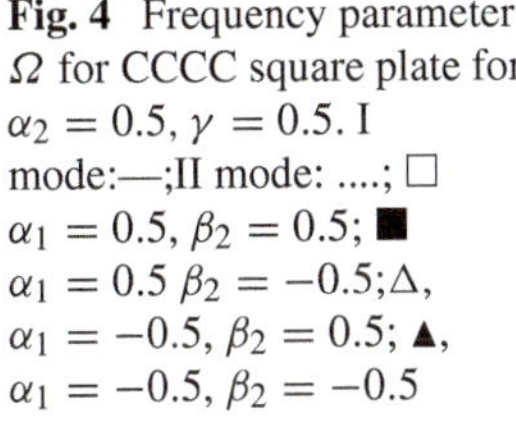

Fig. 4 Frequency parameter Ω for CCCC square plate for $\alpha_2 = 0.5, \gamma = 0.5$. I mode:—;II mode: ....; □ $\alpha_1 = 0.5, \beta_2 = 0.5$; ■ $\alpha_1 = 0.5$ $\beta_2 = -0.5$;Δ, $\alpha_1 = -0.5, \beta_2 = 0.5$; ▲, $\alpha_1 = -0.5, \beta_2 = -0.5$

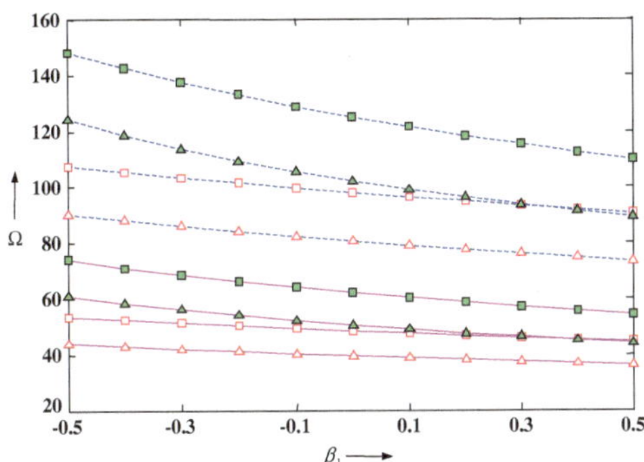

been shown in Fig. 3. It is seen that the frequency parameter Ω increases with the increasing values of whatever be the value of other plate parameters.

The value of frequency parameter Ω increases with increasing values of α_1, while it decreases with increasing values of β_1. The rate of increase of frequency parameter Ω with γ is higher for second mode than that of first mode.

Figure 4 depicts the behavior of the frequency parameter Ω with the density parameter β_1 for $\alpha_1 = \pm 0.5, \beta_2 = \pm 0.5, \alpha_2 = 0.5, \gamma = 0.5$ and $a/b = 1$ for the first two modes of vibration. It is found that the frequency parameter Ω decreases with increasing values of density parameter β_1. The value of Ω increases with increasing values of α_1. The rate of decrease of Ω with β_1 increases with increasing values of α_1, while it is decreases with increasing values of β_1. This rate is higher in the second mode compared to the first mode.

Figure 5 illustrates the behavior of frequency parameter Ω with increasing values of aspect ratio a/b for $\alpha_1 = \beta_1 = \pm 0.5, \alpha_2 = \beta_2 = 0.5$ and $\gamma = 0.5$ for the first two modes of vibration. It is clear that the frequency parameter Ω increases with increasing values of a/b. The rate of increase of Ω with a/b increases with increase

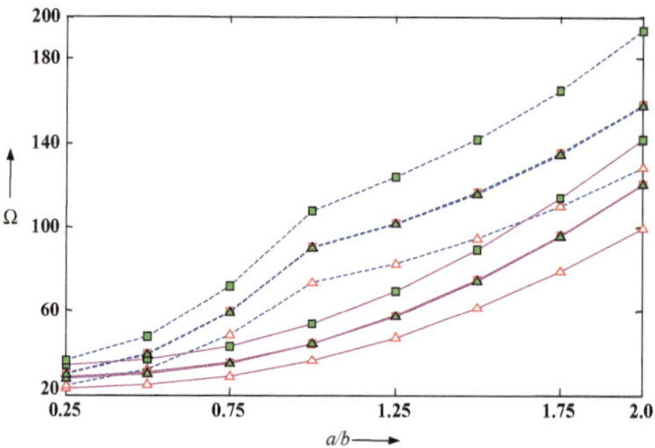

Fig. 5 Frequency parameter Ω for CCCC square plate for $\alpha_2 = \beta_2 = \gamma = 0.5$. I mode:—;II mode: ....; □, $\alpha_1 = 0.5, \beta_1 = 0.5$; ■ $\alpha_1 = 0.5\ \beta_1 = -0.5$;Δ, $\alpha_1 = -0.5, \beta_1 = 0.5$; ▲, $\alpha_1 = -0.5, \beta_1 = -0.5$

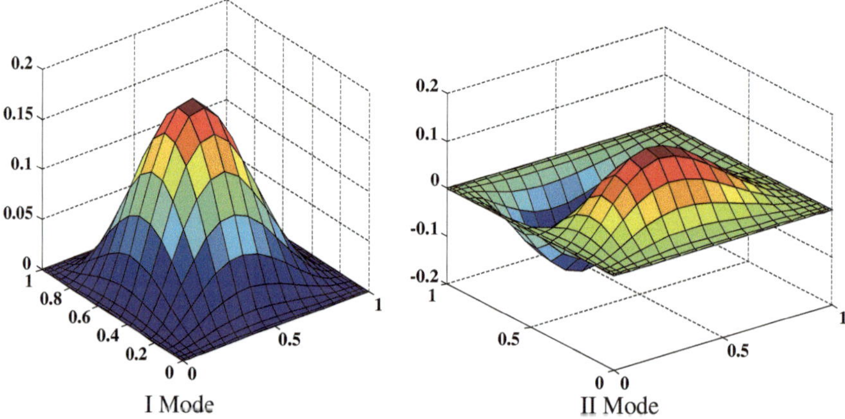

Fig. 6 Three dimensional mode shapes of CCCC square plate; for $\alpha_1 = \alpha_2 = \beta_1 = \beta_2 = \gamma = 0.5$

in the number of modes. This rate of increase is much higher for $a/b > 1$ compared to $a/b < 1$.

Three-dimensional mode shapes for a specified plate have been shown in Fig. 6 for the first three modes of vibration.

A comparison of frequency parameter for homogeneous square plate with the results available in the literature and obtained by other methods is presented in Table 2. A close agreement of results is obtained.

Table 2 Comparison of frequency parameter Ω for homogeneous ($\alpha_1 = \alpha_2 = \beta_1 = \beta_2 = 0$) plate

References	γ	I mode	II mode	III mode
[14]	0.0	35.992	73.413	73.413
[15]	0.0	35.986	73.395	73.395
[16]	0.0	35.986	73.395	73.395
[17]	0.0	35.99	73.41	–
[18]	0.0	35.99	73.419	73.419
[19]	0.0	35.45	72.03	72.03
[20]	0.0	35.989	73.399	73.399
Present	0.0	35.9852	73.3938	73.3938
[16]	0.2	39.5097	80.5201	80.5857
[20]	0.2	39.513	80.525	80.591
Present	0.2	39.5094	80.5184	80.5842
[15]	0.4	42.9088	87.2835	87.5259
[20]	0.4	42.913	87.901	87.901
Present	0.4	42.9084	87.2822	87.5233
[20]	0.5	44.574	90.564	90.926
Present	0.5	44.5694	90.5543	90.9158
[20]	−0.4	28.377	57.530	57.890
Present	−0.4	28.3737	57.5225	57.8837

5 Conclusion

The effect of nonhomogeneity and thickness variation on the vibration characteristics of isotropic rectangular plates has been studied on the basis of classical plate theory using generalized differential quadrature (GDQ) method. The thickness of the plate is taken linear along one direction. The nonhomogeneity of the plate material is assumed to arise due to the linear variations in Young's modulus and density of the plate material with both the in-plane coordinates. It is found that the values of frequency parameter Ω increases with increasing values of nonhomogeneity parameters α_1 and α_2 aspect ratio, while it decreases with increasing values of density parameters β_1 and β_2 keeping other plate parameters fixed. The frequency parameter Ω also increases with increasing values of thickness parameter γ. The present analysis will be helpful to design engineers dealing with nonhomogeneity in obtaining the desired frequency by taking one or more plate parameters considered here.

Acknowledgments The authors wish to express their sincere thanks to the learned reviewers for their constructive comments in improving the paper. One of the authors, Renu Saini, is thankful to the Ministry of Human Resources and Development (MHRD), India for the financial support to carry out this research work.

References

1. Jones, R.M.: Mechanics of Composite Materials, 2nd edn. Taylor and Francis, Philadelphia (1999)
2. Hudramovich, V.S.: Features of nonlinear deformation and critical states of shell systems with geometrical imperfections. Int. Appl. Mech. **43**(12), 1323–1355 (2006)
3. Hudramovich, V.S.: Contact mechanics of shell structures under local loading. Int. Appl. Mech. **45**(7), 708–729 (2009)
4. Grigorenko, A.Y., Tregubenko, T.V.: Numerical and experimental analysis of natural vibrations of rectangular plates with variable thickness. Int. Appl. Mech. **36**(2), 131–134 (2000)
5. Kang, S.W., Kim, S.H.: Vibration analysis of simply supported rectangular plates with unidirectionally, arbitrarily varying thickness. J. Sound Vib. **312**, 551–562 (2008)
6. Civalek, O.: Fundamental frequency of isotropic and orthotropic rectangular plates with linearly varying thickness by discrete singular convolution method. Appl. Math. Mod. **33**, 3825–3835 (2009)
7. Manna, M.C.: Free vibration of tapered isotropic rectangular plates. J. Vib. Control **8**(1), 76–91 (2011)
8. Lal, R., Dhanpati: Transverse vibrations of non-homogeneous orthotropic rectangular plates of variable thickness: a spline technique. J. Sound Vib. **306**, 203–214 (2007)
9. Lal, R., Dhanpati: Quintic splines in the study of buckling and vibration of non-homogeneous orthotropic rectangular plates with variable thickness. Int. J. Appl. Math. Mech. **3**, 18–35 (2007)
10. Lal, R., Saini, R.: Buckling and vibration of non-homogeneous rectangular plates subjected to linearly varying in-plane force. Shock Vib. **20**, 879–894 (2013)
11. Lal, R., Kumar, Y., Gupta, U.S.: Transverse vibrations of nonhomogeneous rectangular plates of uniform thickness using boundary characteristic orthogonal polynomials. Int. J. Appl. Math. Mech. **6**, 93–109 (2010)
12. Saini, R., Lal, R.: Vibration analysis of non-homogeneous rectangular plates using GDQ. Int. J. Adv. Sci. Tech. ISSN **2348–5426**, 166–172 (2014)
13. Shu, C.: Differential Quadrature and Its Application in Engineering. Springer-Verlag, London (2000)
14. Leissa, A.W.: Free vibrations of rectangular plates. J. Sound Vib. **31**, 257–293 (1973)
15. Bhat, R.B.: Natural frequencies of rectangular plates using orthogonal polynomials in the Rayleigh-Ritz method. J. Sound Vib. **102**, 493–499 (1985)
16. Bhat, R.B., Laura, P.A.A., Gutierrez, R.G., Cortinez, V.H.: Numerical experiment on the determination of natural frequencies of transverse vibrations of rectangular plates of non-uniform thickness. J. Sound Vib. **139**, 205–219 (1990)
17. Liew, K.M., Lam, K.Y., Chow, S.T.: Free vibration analysis of rectangular plate using orthogonal plate function. Compost. Struct. **34**, 75–89 (1990)
18. Bardell, N.S.: Free vibration analysis of a flat plate using the hierarchical finite element method. J. Sound Vib. **151**, 263–289 (1991)
19. Kerboua, Y., Lakis, A.A., Thomas, M., Marcouiller, L.: Hybrid method for vibration analysis of rectangular plates. Nucl. Eng. Des. **237**, 791–801 (2007)
20. Singh, B., Saxena, V.: Transverse vibration of a rectangular plate with bidirectional thickness variation. J. Sound Vib. **198**, 51–65 (1996)

Compartmental Disease Models with Heterogeneous Populations: A Survey

R.N. Mohapatra, Donald Porchia and Zhisheng Shuai

Abstract Compartmental models for infectious disease transmission among heterogeneous host populations are surveyed. Mathematical methodologies for analyzing heterogeneous disease models are reviewed. Specifically, three methods are provided to establish the global stability of the endemic equilibrium for a multigroup SIS model. The survey is concluded with several open problems.

Keywords Infectious disease · Mathematical model · Global stability · Lyapunov function · Comparison metho

1 Introduction

Heterogeneity is one of the most important characteristics in modeling infectious diseases due to the variability of both the pathogen and the host population. Ignoring these heterogeneous structures in the disease transmission can fail in prediction of disease outbreak and misevaluate disease control and intervention strategies. For example, in [1] Hethcote and Yorke point out that 60 % of gonorrhea infections in the United States was caused by less than 2 % of the human population, showing that heterogeneity in individual behavior significantly affects the disease dynamics. Superspreading and the effect of individual variation have been highlighted in recent disease outbreaks such as the 2002–2004 severe acute respiratory syndrome (SARS) outbreak [2, 3]. Heterogeneity in the host population can be the result from factors such as age, gender, and genetic heterogeneity of host individuals, and geographical locations such as individuals living in the same city, community, or country. It may

R.N. Mohapatra · D. Porchia · Z. Shuai (✉)
Department of Mathematics, University of Central Florida, Orlando, FL 32816, USA
e-mail: Zhisheng.Shuai@ucf.edu

R.N. Mohapatra
e-mail: Ram.Mohapatra@ucf.edu

D. Porchia
e-mail: Donald.Porchia@ucf.edu

P.N. Agrawal et al. (eds.), *Mathematical Analysis and its Applications*,
Springer Proceedings in Mathematics & Statistics 143,
DOI 10.1007/978-81-322-2485-3_51

also be from disease-dependent host parameters such as susceptibility to the disease, disease transmission rate, or recovery rate.

Multigroup models have been widely used in the literature to capture and model the effect of host heterogeneity on disease transmission. The pioneering multigroup model of SIS type has been proposed by Lajmanovich and Yorke [4] in 1976, describing the transmission of gonorrhea among heterogeneous host populations that have different levels of sexual activities. They successfully establish the global dynamics of the multigroup SIS model by constructing suitable Lyapunov functions. Specifically, the global stability of the endemic equilibrium is proved by analyzing the intersection of invariant sets in which Lyapunov functions do not vary along solutions of the model. Their results in [4] have been widely cited and used to establish the global dynamics for other multigroup models; however, the technique of using multiple Lyapunov functions to show the global stability, which is reviewed in Sect. 3.3, has seen less application in the literature.

Following the pioneering work [4], studies on various multigroup models have been conducted; see, e.g., [5–15] and the references therein. However, due to the large scale and complexity of multigroup models, progress in the mathematical analysis of their global dynamics has been slow. In particular, the questions of uniqueness and global stability of the endemic equilibrium, when the basic reproduction number $\mathscr{R}_0$ is greater than 1, remain largely open. Hethcote [9] established the global stability of the endemic equilibrium for multigroup SIR models without vital dynamics. Beretta and Capasso [6] derived sufficient conditions for global stability with constant group population sizes. Thieme [14] proved global stability of the endemic equilibrium of multigroup SEIRS models under certain restrictions. For a class of SIRS models with constant group sizes, Lin and So [12] proved that the endemic equilibrium is globally asymptotically stable if the cross-group contact rates are small or if the recovery rates in each group are small. On the other hand, results in the opposite direction also exist in the literature. For a multigroup SIR model with proportionate incidence, uniqueness of endemic equilibria may not hold [11, 15] when $\mathscr{R}_0$ is greater than 1.

In 2006, Guo et al. [16] established the complete global dynamics for a class of multigroup SIR models with varying group sizes. In particular, it was proved that when $\mathscr{R}_0$ is greater than 1, the endemic equilibrium of the model is unique and globally asymptotically stable. This completely resolved the open question on the uniqueness and global stability of the endemic equilibrium. The proof relies on the use of a class of global Lyapunov functions and Kirchhoff's Matrix Tree Theorem (see Theorem A in Apendix). Lyapunov functions of this type have previously been used in the ecological literature (e.g., see [17–19]) and was rediscovered for several classes of epidemic models (see [20–22]). The key to the analysis was in using graph theory to provide a complete description of the complicated patterns exhibited in the derivative of the Lyapunov function. This graph-theoretic approach to the construction of Lyapunov functions has been further developed in the series of papers [23–25], and has now become a standard tool for analyzing heterogeneous disease models. For example, the graph-theoretic approach has been applied to heterogeneous models incorporating nonlinear incidence functions in [24, 26, 27], discrete and distributed time delays for disease latency in [28–31], disease progression/relapse in [32–34], vaccine strategies in [35–37], and random fluctuations in [38–40]. The approach has also been applied to heterogeneous models for specific diseases, such as, cholera in [41, 42] and dengue in [43, 44].

Besides Lyapunov function methods, other methods such as comparison arguments [45] and the theory of monotone dynamical systems [46] have also been applied to establish global dynamics for heterogeneous models. In Sect. 2, we revisit the classic multigroup SIR model [4], and in Sect. 3 review mathematical methods that can be applied to the model analysis. Specifically, we provide three methods to prove the global stability of the endemic equilibrium when $\mathscr{R}_0 > 1$: the original proof by Lajmanovich and Yorke in [4], the graph-theoretic approach as shown in [25, Sec. 6], and the comparison method in [45]. We conclude with discussions and several open problems in Sect. 4.

2 Revisit the Multigroup SIS Model

In this section, we revisit the classical multigroup SIS model [4] and summarize by results in the model analysis.

The total host population is divided into n host groups; the population size for each group k is denoted as N_k. The host group N_k is further categorized into two compartments: the compartment containing susceptible individuals and the compartment containing infectious individuals, with population sizes S_k and I_k, respectively. Thus, $N_k = S_k + I_k$. The new infections in host group k are due to within-group transmission $\beta_{kk} S_k I_k$ and all between-group transmissions (cross infection) $\beta_{kj} S_k I_j$ for $j \neq k$. Thus the multigroup SIS model, which was first proposed in [4] for the transmission of gonorrhea in a heterogeneous host population, takes the following form:

$$\begin{aligned} S_k' &= \Lambda_k - \sum_{j=1}^{n} \beta_{kj} S_k I_j - \mu_k S_k + \gamma_k I_k, \\ I_k' &= \sum_{j=1}^{n} \beta_{kj} S_k I_j - (\mu_k + \gamma_k) I_k, \qquad k = 1, 2, \ldots, n. \end{aligned} \tag{2.1}$$

Here Λ_k represents the new input (e.g., birth, migration) into group k, μ_k represents the mortality rate in group k, γ_k represents the recovery rate of infectious individuals in group k, and β_{kj} represents the transmission coefficients between susceptible individuals in group k and infectious individuals in group j. Parameters γ_k, β_{kj} are assumed to be nonnegative and Λ_k, μ_k positive. In addition, the contact matrix $[\beta_{kj}]$ is assumed to be irreducible; biologically, this is the same as assuming that any two groups have a direct or indirect transmission route.

Adding the two equations in (2.1) gives $N_k' = \Lambda_k - \mu_k N_k$, and thus $\lim_{t\to\infty} N_k(t) = \Lambda_k/\mu_k$. By the theory of asymptotically autonomous systems [47, 48], it is sufficient to study the long time behaviors of the reduced system

$$I_k' = \sum_{j=1}^{n} \beta_{kj}\Big(\frac{\Lambda_k}{\mu_k} - I_k\Big)I_j - (\mu_k + \gamma_k)I_k, \quad k = 1, 2, \ldots, n. \tag{2.2}$$

The multigroup model (2.1) (or the reduced model (2.2)) also includes as a special case other models in the literature. For example, a network-based SIS model in [49, 50] takes the following form:

$$\rho_k' = -\rho_k + \lambda k(1 - \rho_k)\frac{\sum_{j=1}^{n} jP(j)\rho_j}{\sum_{j=1}^{n} jP(j)}, \quad k = 1, 2, \ldots, n, \tag{2.3}$$

where ρ_k represents the relative density of infected nodes with degree k (i.e., the probability that a node with k links is infected), $P(k) \geq 0$ represents the density of nodes with degree k, λ represents the effective disease spreading rate, and n is the maximum degree of all nodes. Model (2.2) is equivalent to (2.3) if we set

$$\beta_{kj} = \frac{\lambda kjP(j)}{\sum_{l=1}^{n} lP(l)}, \quad \Lambda_k = \mu_k, \quad \gamma_k = 1 - \mu_k. \tag{2.4}$$

Now we start to analyze model (2.1) within a feasible region.

Lemma 1 *The feasible region* $\Gamma = \left\{(S_1, I_1, \ldots, S_n, I_n) \in \mathbb{R}_+^{2n} \mid S_k + I_k = \frac{\Lambda_k}{\mu_k}\right\}$ *is positively invariant with respect to* (2.1).

Model (2.1) always admits a disease-free equilibrium $P_0 = (S_1^0, 0, \ldots, S_n^0, 0)$ in Γ with $S_k^0 = \frac{\Lambda_k}{\mu_k}$. There might be an endemic equilibrium $P^* = (S_1^*, I_1^*, \ldots, S_n^*, I_n^*)$ with $S_k^*, I_k^* > 0$, which lies in $\mathrm{int}(\Gamma)$, the interior of Γ.

Following the next generation matrix approach [51, 52], define two $n \times n$ matrices $F = [\beta_{kj} S_k^0]$ and $V = \mathrm{diag}\{\mu_1 + \gamma_1, \ldots, \mu_n + \gamma_n\}$, representing the disease incidence matrix and disease transfer matrix, respectively. Then the basic reproduction number is defined as the spectral radius of the next generation matrix FV^{-1}; that is

$$\mathscr{R}_0 = \rho\left(FV^{-1}\right) = \rho\left(\left[\frac{\beta_{kj} S_k^0}{\mu_j + \gamma_j}\right]\right). \tag{2.5}$$

For the network SIS model (2.3), using (2.4), the disease-free equilibrium and the basic reproduction number can be evaluated as $P_0 = (1, 0, \ldots, 1, 0)$ and $\mathscr{R}_0 = \rho([\beta_{kj}]) = \frac{\lambda}{\sum_{l=1}^{n} lP(l)}\rho([kjP(j)])$, respectively. Since the matrix $[kjP(j)]$ has rank 1, the spectral radius can be calculated as $\sum_{l=1}^{n} l^2 P(l)$. Therefore, the basic reproduction number for (2.3) becomes $\mathscr{R}_0 = \lambda \frac{<k^2>}{<k>}$ with $<k^j> = \sum_{l=1}^{n} l^j P(l)$, agreeing with the threshold value in [49, 50].

The basic reproduction number $\mathscr{R}_0$ completely determines the disease dynamics of (2.1) as shown in the following sharp threshold result.

Theorem 1 *Assume that contact matrix* $[\beta_{kj}]$ *is irreducible.*

(a) The disease-free equilibrium P_0 *is globally asymptotically stable in* Γ *if* $\mathscr{R}_0 \leq 1$ *and becomes unstable if* $\mathscr{R}_0 > 1$.
(b) If $\mathscr{R}_0 > 1$, *then model* (2.1) *is uniformly persistent, and there exists a unique endemic equilibrium* P^* *that is globally asymptotically stable in* $\mathrm{int}(\Gamma)$.

3 Stability of the Multigroup SIS Model

In the following we provide the proof for the sharp threshold result (Theorem 1) in Sect. 2. In particular, three different methods are provided for the global stability of the endemic equilibrium.

3.1 Global Stability of the Disease-Free Equilibrium

In this subsection, we prove the global stability of the disease-free equilibrium when $\mathscr{R}_0 \leq 1$. Let $x = (I_1, \ldots, I_n)^T$ denote the disease compartment in (2.1) and F, V defined as above. It follows from Lemma 1 that $x' \leq (F - V)x$. Following the matrix-theoretic method in [25, Sect. 2], a Lyapunov function L can be constructed for (2.1); that is, $L = w^T V^{-1} x$, where w^T is a left eigenvector of the nonnegative matrix $V^{-1}F$ corresponding to the eigenvalue $\rho(V^{-1}F) = \rho(FV^{-1}) = \mathscr{R}_0$. Differentiating L along (2.1) gives

$$L' = w^T V^{-1} x' \leq w^T V^{-1}(F - V)x = (\mathscr{R}_0 - 1) w^T x. \tag{3.1}$$

Hence, $L' \leq 0$ if $\mathscr{R}_0 \leq 1$. Furthermore, $L' = 0$ implies that $S_k = S_k^0$ for all k. By the first equation of (2.1), it follows that $I_k = 0$ for all k. Hence the largest invariant set where $L' = 0$ is the singleton $\{P_0\}$. By LaSalle's invariance principle [53], P_0 is globally asymptotically stable in Γ when $\mathscr{R}_0 \leq 1$.

If $\mathscr{R}_0 > 1$, then by (3.1), $L' = (\mathscr{R}_0 - 1) w^T x > 0$ when $S_k = S_k^0$ and $I_k > 0$, for all k. Continuity thus shows that $L' > 0$ in a neighborhood of P_0. Solutions in the positive cone sufficiently close to P_0 move away from P_0, implying that P_0 is unstable. Using a uniform persistence result from [54] and an argument as in the proof of Proposition 3.3 of [55], it can be shown that, when $\mathscr{R}_0 > 1$, instability of P_0 implies uniform persistence of (2.1). Uniform persistence and the positive invariance of the compact set Γ imply the existence of an endemic equilibrium of (2.1) (see Theorem D.3 in [56] or Theorem 2.8.6 in [57]).

3.2 Global Stability of the Endemic Equilibrium: Lyapunov Function

In this subsection we apply the graph-theoretic method to construct a Lyapunov function to prove the global stability of the endemic equilibrium P^* when $\mathscr{R}_0 > 1$.

Let $D_k = I_k - I_k^* - I_k^* \ln \frac{I_k}{I_k^*}$. Differentiating and using the equilibrium equation gives

$$\begin{aligned}
D_k' =& \frac{I_k - I_k^*}{I_k}\left(\sum_{j=1}^n \beta_{kj}\left(\frac{\Lambda_k}{\mu_k} - I_k\right) I_j - \sum_{j=1}^n \beta_{kj}\left(\frac{\Lambda_k}{\mu_k} - I_k^*\right) I_j^* \frac{I_k}{I_k^*}\right) \\
=& -\sum_{j=1}^n \beta_{kj} I_j \frac{(I_k - I_k^*)^2}{I_k} + \sum_{j=1}^n \beta_{kj}\left(\frac{\Lambda_k}{\mu_k} - I_k^*\right) I_j^* \left(1 - \frac{I_k}{I_k^*} + \frac{I_j}{I_j^*} - \frac{I_k^* I_j}{I_k I_j^*}\right) \\
\leq& \sum_{j=1}^n \beta_{kj}\left(\frac{\Lambda_k}{\mu_k} - I_k^*\right) I_j^* \left(1 - \frac{I_k}{I_k^*} + \frac{I_j}{I_j^*} - \frac{I_k^* I_j}{I_k I_j^*}\right) \\
\leq& \sum_{j=1}^n \beta_{kj}\left(\frac{\Lambda_k}{\mu_k} - I_k^*\right) I_j^* \left(\frac{I_j}{I_j^*} - \ln\frac{I_j}{I_j^*} - \frac{I_k}{I_k^*} + \ln\frac{I_k}{I_k^*}\right) := \sum_{j=1}^n a_{kj} G_{kj},
\end{aligned}$$

with $a_{kj} = \beta_{kj}\left(\frac{\Lambda_k}{\mu_k} - I_k^*\right) I_j^* \geq 0$ and $G_{kj} = \frac{I_j}{I_j^*} - \ln\frac{I_j}{I_j^*} - \frac{I_k}{I_k^*} + \ln\frac{I_k}{I_k^*}$. The second inequality follows from the fact that $1 - x \leq -\ln x$ for all $x > 0$. Let $H_k = \frac{I_k}{I_k^*} - \ln\frac{I_k}{I_k^*}$, then $G_{kj} = H_j - H_k$. A weighted digraph $\mathscr{G}$ can be constructed to associate with the weight matrix $A = [a_{kj}]$; see Appendix for more details. Notice that along any directed cycle $\mathscr{C}$ of $(\mathscr{G}, \Lambda)$,

$$\sum_{(s,r)\in\mathscr{E}(\mathscr{C})} G_{rs} = \sum_{(s,r)\in\mathscr{E}(\mathscr{C})} (H_s - H_r) = 0.$$

Here $\mathscr{E}(\mathscr{C})$ denotes the arc set of $\mathscr{C}$. Since all assumptions of Theorem B in Appendix hold, let c_i be as given in Theorem A in Appendix, then by Theorem B, $D = \sum_{k=1}^n c_k D_k$ is a Lyapunov function for (2.2). Using this Lyapunov function, the irreducibility of $[\beta_{ij}]$, and LaSalle's Invariance Principle, it can be proved that $\{(I_1^*, \ldots, I_n^*)\}$ is the largest invariant set for (2.2), and thus if $\mathscr{R}_0 > 1$, then the positive equilibrium $(I_1^*, \ldots, I_n^*)$ is globally asymptotically stable for (2.2). As a consequence, P^* is globally asymptotically stable and thus unique in int(Γ) for (2.1).

3.3 Global Stability of the Endemic Equilibrium: Multiple Lyapunov Functions

In [4] Lajmanovich and Yorke construct two Lyapunov functions for (2.1), and are able to prove the global stability of the endemic equilibrium by analyzing the intersection of invariant sets where Lyapunov functions do not vary along its solutions. We summarize their proof in this subsection.

Let $D_1 = \max\{M-1, 0\}$ and $D_2 = \max\{1-m, 0\}$, where $M = \max_k \left\{\frac{I_k}{I_k^*}\right\}$ and $m = \min_k\{\frac{I_k}{I_k^*}\}$. Without loss of generality, assume $M = \frac{I_1}{I_1^*}$ and $m = \frac{I_2}{I_2^*}$ in some time interval $\mathscr{I}$. Thus, if $M \geq 1$ in $\mathscr{I}$, then $I_1 \geq I_1^*$, and by (2.2)

$$\begin{aligned} D_1' = M' &= \sum_{j=1}^{n} \beta_{1j}\Big(\frac{\Lambda_1}{\mu_1} - I_1\Big)\frac{I_j}{I_1^*} - (\mu_1+\gamma_1)\frac{I_1}{I_1^*} \\ &\leq \sum_{j=1}^{n} \beta_{1j}\Big(\frac{\Lambda_1}{\mu_1} - I_1^*\Big)\frac{I_j}{I_1^*} - (\mu_1+\gamma_1)\frac{I_1}{I_1^*} \\ &\leq \sum_{j=1}^{n} \beta_{1j}\Big(\frac{\Lambda_1}{\mu_1} - I_1^*\Big)\frac{I_j^* I_1}{(I_1^*)^2} - (\mu_1+\gamma_1)\frac{I_1}{I_1^*} = 0. \end{aligned}$$

The second inequality follows from the fact that $\frac{I_1}{I_1^*} \geq \frac{I_j}{I_j^*}$ for all j and the last equality follows from the equilibrium equation. Similarly, if $m \leq 1$ in $\mathscr{I}$, then

$$\begin{aligned} D_2' = -m' &= \sum_{j=1}^{n} \beta_{2j}\Big(I_2 - \frac{\Lambda_2}{\mu_2}\Big)\frac{I_j}{I_2^*} + (\mu_2+\gamma_2)\frac{I_2}{I_2^*} \\ &\leq \sum_{j=1}^{n} \beta_{2j}\Big(I_2^* - \frac{\Lambda_2}{\mu_2}\Big)\frac{I_j}{I_2^*} + (\mu_2+\gamma_2)\frac{I_2}{I_2^*} \\ &\leq \sum_{j=1}^{n} \beta_{2j}\Big(I_2^* - \frac{\Lambda_2}{\mu_2}\Big)\frac{I_j^* I_2}{(I_2^*)^2} + (\mu_2+\gamma_2)\frac{I_2}{I_2^*} = 0. \end{aligned}$$

The invariant sets where $D_1' = 0$ and $D_2' = 0$ are $E_1 = \{(I_1, \ldots, I_n) \in \mathbb{R}_+^n \mid I_k \leq I_k^*, k = 1, \ldots, n\}$ and $E_2 = \{(I_1, \ldots, I_n) \in \mathbb{R}_+^n \mid I_k \geq I_k^*, k = 1, \ldots, n\}$, respectively. Notice that $E_1 \cap E_2 = \{(I_1^*, \ldots, I_n^*)\}$. Therefore, by LaSalle's Invariance Principle, $(I_1^*, \ldots, I_n^*)$ is globally asymptotically stable in $\mathrm{int}(\Gamma)$ if $\mathscr{R}_0 > 1$.

3.4 Global Stability of the Endemic Equilibrium: Comparison Method

In this subsection we provide another proof for the global stability of the endemic equilibrium for (2.1), which follows the comparison argument for the network SIS model (2.3) used in [45] and the fact that (2.2) is a cooperative system [46]. In the following we show that $\liminf_{t\to\infty} I_k(t) = \limsup_{t\to\infty} I_k(t)$ for all k. Specifically, we construct two sequences: one is a decreasing sequence bounded from below (upper solutions) and the other an increasing sequence bounded from above (lower solutions).

Let $u_k^{(1)} = \frac{\Lambda_k}{\mu_k}$ for all k, and by Lemma 1, $I_k(t) \le u_k^{(1)}$ for all $t \ge 0$. Define the upper solution sequence

$$u_k^{(m+1)} = \frac{\sum_{j=1}^n \beta_{kj} \frac{\Lambda_k}{\mu_k} u_j^{(m)}}{\mu_k + \gamma_k + \sum_{j=1}^n \beta_{kj} u_j^{(m)}}, \quad 1 \le k \le n, \quad m = 1, 2, \ldots \tag{3.2}$$

It can be shown that $I_k(t) \le u_k^{(m)}$ for all $m = 2, 3, \ldots$ and $t \ge 0$ when applying the reduction argument to differential inequalities

$$I_k' \le \sum_{j=1}^n \beta_{kj} \left(\frac{\Lambda_k}{\mu_k} - I_k \right) u_j^{(m)} - (\mu_k + \gamma_k) I_k = \sum_{j=1}^n \beta_{kj} \frac{\Lambda_k}{\mu_k} u_j^{(m)} - I_k \left(\mu_k + \gamma_k + \sum_{j=1}^n \beta_{kj} u_j^{(m)} \right).$$

In addition, it follows from (3.2) that $u_k^{(2)} < \frac{\Lambda_k}{\mu_k} = u_k^{(1)}$, and induction argument shows that for $m = 2, 3, \ldots$

$$u_k^{(m+1)} = \frac{\sum_{j=1}^n \beta_{kj} \frac{\Lambda_k}{\mu_k} u_j^{(m)}}{\mu_k + \gamma_k + \sum_{j=1}^n \beta_{kj} u_j^{(m)}} < \frac{\sum_{j=1}^n \beta_{kj} \frac{\Lambda_k}{\mu_k} u_j^{(m-1)}}{\mu_k + \gamma_k + \sum_{j=1}^n \beta_{kj} u_j^{(m-1)}} = u_k^{(m)}.$$

Hence, the sequence $\{u_k^{(m)}\}$ is decreasing so its limit exists. Denote $u_k = \lim_{m\to\infty} u_k^{(m)}$. It follows from taking the limit on both sides of (3.2) that

$$\sum_{j=1}^n \beta_{kj} \left(\frac{\Lambda_k}{\mu_k} - u_k \right) u_j - (\mu_k + \gamma_k) u_k = 0, \tag{3.3}$$

implying that $(u_1, \ldots u_n)$ is the equilibrium of (2.2).

By Theorem 1, model (2.1) is uniformly persistent when $\mathscr{R}_0 > 1$; thus there exists $0 < \varepsilon \ll 1$ such that $\liminf_{t\to\infty} I_k(t) \ge \varepsilon$ for all k. We keep ε small enough such that

$$\varepsilon < \min_k \left\{ \frac{(\mathscr{R}_0 - 1)(\mu_k + \gamma_k)}{\sum_{j=1}^n \beta_{kj}} \right\}. \tag{3.4}$$

Since matrices $\left[\frac{\beta_{kj}S_k^0}{\mu_j+\gamma_j}\right]$ and $\left[\frac{\beta_{kj}S_k^0}{\mu_k+\gamma_k}\right]$ are similar, it follows that $\rho\left(\left[\frac{\beta_{kj}S_k^0}{\mu_k+\gamma_k}\right]\right) = \mathscr{R}_0 > 1$. Let $(\eta_1,\ldots,\eta_n)^T$ with $0 < \eta_k < \varepsilon$, $1 \le k \le n$, be the Perron eigenvector for $\left[\frac{\beta_{kj}S_k^0}{\mu_k+\gamma_k}\right]$ corresponding to the eigenvalue $\mathscr{R}_0$; that is,

$$\sum_{j=1}^{n} \frac{\beta_{kj}S_k^0}{\mu_k+\gamma_k}\eta_j = \mathscr{R}_0\eta_k. \tag{3.5}$$

Now consider the lower solutions $\left\{l_k^{(m)}\right\}$ with $l_k^{(1)} = \eta_k$ and

$$l_k^{(m+1)} = \frac{\sum_{j=1}^n \beta_{kj}\frac{\Lambda_k}{\mu_k} l_j^{(m)}}{\mu_k+\gamma_k+\sum_{j=1}^n \beta_{kj} l_j^{(m)}}, \quad 1 \le k \le n, \quad m = 1, 2, \ldots \tag{3.6}$$

Choose T large enough such that $I_k(t) \ge \varepsilon$ for all $t > T$ and $k = 1, \ldots, n$. It follows from (3.4) and (3.5) that

$$l_k^{(2)} = \frac{\sum_{j=1}^n \beta_{kj}\frac{\Lambda_k}{\mu_k}\eta_j}{\mu_k+\gamma_k+\sum_{j=1}^n \beta_{kj}\eta_j} > \frac{\sum_{j=1}^n \beta_{kj}S_k^0\eta_j}{\mathscr{R}_0(\mu_k+\gamma_k)} = \eta_k = l_k^{(1)}.$$

Reduction can further be used to show that $I_k(t) \ge l_k^{(m)}$ for all $m = 2, 3, \ldots$ and $t \ge T$, and that $\{l_k^{(m)}\}$ is increasing so its limit exists. Denote $l_k = \lim_{m\to\infty} l_k^{(m)}$, and $l_k > 0$ satisfies the same equation as u_k in (3.3). Therefore, $(l_1, \ldots, l_n)$ is the nontrivial equilibrium of (2.2).

Given the uniqueness of the endemic equilibrium of (2.2) (see Sects. 3.2 or 3.3), the inequalities $l_k \le I_k(t) \le u_k$ imply that all solutions of (2.2) approach to the unique endemic equilibrium.

4 Discussions and Open Problems

The classic multigroup SIR model (2.1), which was first proposed in [4], is revisited in Sect. 2. The methods that can be applied to analyze this kind of heterogeneous disease models are reviewed in Sect. 3. The graph-theoretic method in Sect. 3.2 can be used to guide the construction of Lyapunov functions for heterogeneous models, which requires individual Lyapunov functions for homogeneous models as building blocks. The method using multiple Lyapunov functions as shown in [4] (see Sect. 3.3) requires strong techniques in combining the invariant sets derived from each Lyapunov functions, which makes it hard to apply for other heterogeneous models. Comparison method as in [45] (also see Sect. 3.4) requires certain monotone properties on the flow generated by the model. Due to these limitations, the global stability of the endemic equilibrium for several multigroup infectious disease models remains open. For example, in model (2.1), each group population size N_k either is a constant or approaches a constant, which allows the reduction to model (2.2) consisting of only infectious compartments. If disease-induced mortality is introduced to each group (i.e., adding a term $\alpha_k I_k$ at the end of the equation for I_k in (2.1)), then such a reduction does not hold any more. It remains open whether the endemic equilibrium is unique and whether it is globally asymptotically stable when $\mathscr{R}_0 > 1$. The global dynamics of the SEIRS model has recently been resolved by Cheng and Yang [58], based on previous results using the theory of compound differential equations in [59], but no complete studies on global dynamics of the multigroup SEIRS model have been done yet.

Multigroup models can also be used to model spatial heterogeneity, in which between-group transmissions are due to pathogen and/or host movements; see, for example, [60, 61]. In [61], the multigroup model is also called the Lagrangian model, in comparison with Eulerian (multipatch) models that explicitly incorporate pathogen and/or host movements. Demographic and movement data can be used to derive the contact matrix $[\beta_{kj}]$ for this situation; see, for example, the gravity model used in [62] to derive between-group transmission for the cholera outbreak in Haiti.

Acknowledgments The authors would like to acknowledge the financial support from the College of Sciences and the Department of Mathematics at the University of Central Florida.

Appendix: Graph-Theoretic Method to the Construction of Lyapunov Functions

Given a weighted digraph $\mathscr{G}$ with n vertices, define the $n \times n$ weight matrix $A = [a_{ij}]$ with entry $a_{ij} > 0$ equal to the weight of arc (j, i) if it exists, and 0 otherwise. We denote such a weighted digraph by $(\mathscr{G}, A)$. A digraph $\mathscr{G}$ is strongly connected if, for any pair of distinct vertices i, j, there exists a directed path from i to j (and also from j to i). A weighted digraph $(\mathscr{G}, A)$ is strongly connected if and only if the weight matrix A is irreducible [63]. The Laplacian matrix $L = [\ell_{ij}]$ of $(\mathscr{G}, A)$ is defined as

$$\ell_{ij} = \begin{cases} -a_{ij} & \text{for } i \neq j, \\ \sum_{k \neq i} a_{ik} & \text{for } i = j. \end{cases} \tag{1}$$

The following result gives a graph-theoretic description of the cofactors of the diagonal entries of L. We refer the reader to [64] for its proof.

Theorem A (Kirchhoff's Matrix Tree Theorem) *Assume $n \geq 2$ and let c_i be the cofactor of ℓ_{ii} in L. Then*

$$c_i = \sum_{\mathscr{T} \in \mathbb{T}_i} w(\mathscr{T}), \qquad i = 1, 2, \ldots, n, \tag{2}$$

where $\mathbb{T}_i$ is the set of all spanning trees $\mathscr{T}$ of $(\mathscr{G}, A)$ that are rooted at vertex i, and $w(\mathscr{T})$ is the weight of $\mathscr{T}$. If $(\mathscr{G}, A)$ is strongly connected, then $c_i > 0$ for $1 \le i \le n$.

Let U be an open set in $\mathbb{R}^m$. Consider a differential equation system

$$z_k' = f_k(z_1, z_2, \ldots, z_m), \qquad k = 1, 2, \ldots, m, \tag{3}$$

with $z = (z_1, z_2, \ldots, z_m) \in U$. The following result can be used to construct Lyapunov functions for (3), and its proof can be found in [24, Sect. 3] or [25, Sect. 3].

Theorem B *Suppose that the following assumptions are satisfied.*

(1) There exist functions $D_i : U \to \mathbb{R}$, $G_{ij} : U \to \mathbb{R}$, and constants $a_{ij} \ge 0$ such that, for every $1 \le i \le n$, $D_i' = D_i'|_{(3)} \le \sum_{j=1}^n a_{ij} G_{ij}(z)$ for $z \in U$.
(2) For $A = [a_{ij}]$, each directed cycle $\mathscr{C}$ of $(\mathscr{G}, A)$ has $\sum_{(s,r) \in \mathscr{E}(\mathscr{C})} G_{rs}(z) \le 0$ for $z \in U$, where $\mathscr{E}(\mathscr{C})$ denotes the arc set of the directed cycle $\mathscr{C}$.

Then, the function $D(z) = \sum_{i=1}^n c_i D_i(z)$, with constants $c_i \ge 0$ as given in Theorem A, satisfies $D' = D'|_{(3)} \le 0$; that is, D is a Lyapunov function for (3).

References

1. Hethcote, H.W., Yorke, J.A.: Gonorrhea Transmission Dynamics and Control. Lecture Notes in Biomath, vol. 56. Springer, Berlin (1984)
2. Galvani, A.P., May, R.M.: Dimensions of superspreading. Nature **438**, 293–295 (2005)
3. Lloyd-Smith, J.O., Schreiber, S.J., Kopp, P.E., Getz, W.M.: Superspreading and the effect of individuals variation on disease emergence. Nature **438**, 355–359 (2005)
4. Lajmanovich, A., Yorke, J.A.: A deterministic model for gonorrhea in a nonhomogeneous population. Math. Biosci. **28**, 221–236 (1976)
5. Aronsson, G., Mellander, I.: A deterministic model in biomathematics, asymptotic behavior and threshold conditions. Math. Biosci. **49**, 207–222 (1980)
6. Beretta, E., Capasso, V.: Global stability results for a multigroup SIR epidemic model. In: Hallam, T.G., Gross, L.J., Levin, S.A. (eds.) Mathematical Ecology, pp. 317–342. World Scientific, Singapore (1986)
7. Hethcote, H.W.: Mathematical models for the spread of infectious diseases. In: Ludwig, D., Cooke, K.L. (eds.) Epidemiology, pp. 122–131. SIAM, Philadelphia (1975)
8. Hethcote, H.W.: Qualitative analyses of communicable disease models. Math. Biosci. **28**, 335–356 (1976)
9. Hethcote, H.W.: An immunization model for a heterogeneous population. Theor. Popu. Biol. **14**, 338–349 (1978)
10. Hethcote, H.W., Thieme, H.R.: Stability of the endemic equilibrium in epidemic models with subpopulations. Math. Biosci. **75**, 205–227 (1985)
11. Huang, W., Cooke, K.L., Castillo-Chavez, C.: Stability and bifurcation for a multiple-group model for the dynamics of HIV/AIDS transmission. SIAM J. Appl. Math. **52**, 835–854 (1992)
12. Lin, X., So, J.W.H.: Global stability of the endemic equilibrium and uniform persistence in epidemic models with subpopulations. J. Austral. Math. Soc. Ser. B **34**, 282–295 (1993)

13. Rass, L., Radcliffe, J.: Global asymptotic convergence results for multitype models. Int. J. Appl. Math. Comput. Sci. **10**, 63–79 (2000)
14. Thieme, H.R.: Local stability in epidemic models for heterogeneous populations. In: Capasso, V., Grosso, E., Paveri-Fontana, S.L. (eds.) Mathematics in Biology and Medicine. Lecture Notes in Biomath, vol. 57, pp. 185–189. Springer, Berlin (1985)
15. Thieme, H.R.: Mathematics in Population Biology. Princeton University Press, Princeton (2003)
16. Guo, H., Li, M.Y., Shuai, Z.: Global stability of the endemic equilibrium of multigroup SIR epidemic models. Can. Appl. Math. Q. **14**, 259–284 (2006)
17. Freedman, H.I., So, J.W.H.: Global stability and persistence of simple food chains. Math. Biosci. **76**, 69–86 (1985)
18. Goh, B.S.: Global stability in many-species systems. Am. Nat. **111**, 135–143 (1977)
19. Hsu, S.B.: On global stability of a predator-prey system. Math. Biosci. **39**, 1–10 (1978)
20. Guo, H., Li, M.Y.: Global dynamics of a staged progression model for infectious diseases. Math. Biosci. Eng. **3**, 513–525 (2006)
21. Korobeinikov, A., Maini, P.K. A Lyapunov function and global properties for SIR and SEIR epidemiological models with nonlinear incidence, Math. Biosci. Eng. **1**, 57–60 (2004)
22. Korobeinikov, A., Wake, G.C.: Lyapunov functions and global stability for SIR, SIRS, and SIS epidemiological models. Appl. Math. Lett. **15**, 955–960 (2002)
23. Guo, H., Li, M.Y., Shuai, Z.: A graph-theoretic approach to the method of global Lyapunov functions. Proc. Am. Math. Soc. **136**, 2793–2802 (2008)
24. Li, M.Y., Shuai, Z.: Global-stability problems for coupled systems of differential equations on networks. J. Differ. Eqn. **248**, 1–20 (2010)
25. Shuai, Z., van den Driessche, P.: Global stability of infectious disease models using Lyapunov functions. SIAM J. Appl. Math. **73**, 1513–1532 (2013)
26. Sun, R.: Global stability of the endemic equilibrium of multigroup SIR models with nonlinear incidence. Comput. Math. Appl. **60**, 2286–2291 (2010)
27. Sun, R., Shi, J.: Global stability of multigroup epidemic model with group mixing and nonlinear incidence rates. Appl. Math. Comput. **218**, 280–286 (2011)
28. Li, M.Y., Shuai, Z., Wang, C.: Global stability of multi-group epidemic models with distributed delays. J. Math. Anal. Appl. **361**, 38–47 (2010)
29. Shu, H., Fan, D., Wei, J.: Global stability of multi-group SEIR epidemic models with distributed delays and nonlinear transmission. Nonlinear Anal. Real World Appl. **13**, 1581–1592 (2012)
30. Shuai, Z., van den Driessche, P.: Impact of heterogeneity on the dynamics of an SEIR epidemic model. Math. Biosci. Eng. **9**, 393–411 (2012)
31. Yuan, Z., Wang, L.: Global stability of epidemiological models with group mixing and nonlinear incidence rates. Nonlinear Anal. Real World Appl. **11**, 995–1004 (2010)
32. Guo, H., Li, M.Y., Shuai, Z.: Global dynamics of a general class of multi-stage models for infectious diseases. SIAM J. Appl. Math. **72**, 261–279 (2012)
33. Wang, J., Pang, J., Liu, X.: Modelling diseases with relapse and nonlinear incidence of infection: a multi-group epidemic model. J. Biol. Dyn. **8**, 99–116 (2014)
34. Yuan, Z., Zou, X.: Global threshold property in an epidemic model for disease with latency spreading in a heterogeneous host population. Nonlinear Anal. Real World Appl. **11**, 3479–3490 (2010)
35. Ding, D., Ding, X.: Global stability of multi-group vaccination epidemic models with delays. Nonlinear Anal. Real World Appl. **12**, 1991–1997 (2011)
36. Kuniya, T.: Global stability of a multi-group SVIR epidemic model. Nonlinear Anal. Real World Appl. **14**, 1135–1143 (2013)
37. Wang, J., Takeuchi, Y., Liu, S.: A multi-group SVEIR epidemic model with distributed delay and vaccination. Int. J. Biomath. **5**(3), 235–258 (2012)
38. Ji, C., Jiang, D., Shi, N.: Multigroup SIR epidemic model with stochastic perturbation. Phys. A **390**, 1747–1762 (2011)
39. Ji, C., Jiang, D., Yang, Q., Shi, N.: Dynamics of a multigroup SIR epidemic model with stochastic perturbation. Automatica **48**, 121–131 (2012)

40. Yang, Q., Mao, X.: Extinction and recurrence of multi-group SEIR epidemic models with stochastic perturbations. Nonlinear Anal. Real World Appl. **14**, 1434–1456 (2013)
41. Eisenberg, M.C., Shuai, Z., Tien, J.H., van den Driessche, P.: A cholera model in a patchy environment with water and human movement. Math. Biosci. **246**, 105–112 (2013)
42. Shuai, Z., van den Driessche, P.: Global dynamics of cholera models with differential infectivity. Math. Biosci. **234**, 118–126 (2011)
43. Ding, D., Wang, X. Ding, X.: Global stability of multigroup dengue disease transmission model, J. Appl. Math. **2012**, no. 342472
44. Huang, G., Wang, J., Zu, J.: Global dynamics of multi-group dengue disease model with latency distributions, Math. Meth. Appl. Sci. (2014) (in press)
45. Wang, L., Dai, G.-Z.: Global stability of virus spreading in complex heterogeneous networks. SIAM J. Appl. Math. **68**, 1495–1502 (2008)
46. Smith, H.L.: Monotone Dynamical Systems. An Introduction to the Theory of Competitive and Cooperative Systems. American Mathematical Society, Providence (1995)
47. Markus, L.: Asymptotically autonomous differential systems. In: Lefschetz, S. (ed.) Contributions to the Theory of Nonlinear Oscillations III. Annals of Mathematics Studies, vol. 36, pp. 17–29. Princeton University Press, Princeton (1956)
48. Mischaikow, K., Smith, H., Thieme, H.R.: Asymptotically autonomous semiflows: chain recurrence and Lyapunov functions. Trans. Am. Math. Soc. **347**, 1669–1685 (1995)
49. Pastor-Satorras, R., Vespignani, A.: Epidemic spreading in scale free networks. Phys. Rev. Lett. **86**, 3200–3203 (2001)
50. Pastor-Satorras, R., Vespignani, A.: Epidemic dynamics in finite size scale-free networks. Phys. Rev. E **65**, 035108 (2002)
51. Diekmann, O., Heesterbeek, J.A.P., Metz, J.A.J.: On the definition and the computation of the basic reproduction ratio R_0 in models for infectious diseases in heterogeneous populations. J. Math. Biol. **28**, 365–382 (1990)
52. van den Driessche, P., Watmough, J.: Reproduction numbers and sub-threshold endemic equilibria for compartments models of disease transmission. Math. Biosci. **180**, 29–48 (2002)
53. LaSalle, J.P.: The Stability of Dynamical Systems, Regional Conference Series in Applied Mathematics. SIAM, Philadelphia (1976)
54. Freedman, H.I., Ruan, S., Tang, M.: Uniform persistence and flows near a closed positively invariant set. J. Dyn. Differ. Eqn. **6**, 583–600 (1994)
55. Li, M.Y., Graef, J.R., Wang, L., Karsai, J.: Global dynamics of a SEIR model with varying total population size. Math. Biosci. **160**, 191–213 (1999)
56. Smith, H.L., Waltman, P.: The Theory of the Chemostat: Dynamics of Microbial Competition. Cambridge University Press, Cambridge (1995)
57. Bhatia, N.P., Szegö, G.P.: Dynamical Systems: Stability Theory and Applications. Lecture Notes in Mathematics, vol. 35. Springer, Berlin (1967)
58. Cheng, Y., Yang, X.: On the global stability of SEIRS models in epidemiology. Can. Appl. Math. Q. **20**, 115–133 (2012)
59. Li, M.Y., Muldowney, J.S., van den Driessche, P.: Global stability of the SEIRS model in epidemiology. Can. Appl. Math. Q. **7**, 409–425 (1999)
60. Lloyd, A.L., May, R.M.: Spatial heterogeneity in epidemic models. J. Theor. Biol. **179**, 1–11 (1996)
61. Cosner, C., Beier, J.C., Cantrell, R.S., Impoinvil, D., Kapitanski, L., Potts, M.D., Troyo, A., Ruan, S.: The effects of human movement on the persistence of vector-borne diseases. J. Theory Biol. **258**, 550–560 (2009)
62. Tuite, A.R., Tien, J.H., Eisenberg, M.C., Earn, D.J.D., Ma, J., Fisman, D.N.: Cholera epidemic in Haiti, 2010: using a transmission model to explain spatial spread of disease and identify optimal control interventions. Ann. Intern. Med. **154**, 593–601 (2011)
63. Berman, A., Plemmons, R.J.: Nonnegative Matrices in the Mathematical Sciences. Academic Press, New York (1979)
64. Moon, J.W.: Counting Labelled Trees. William Clowes and Sons, London (1970)

Radially Symmetric Vibrations of Exponentially Tapered Clamped Circular Sandwich Plate Using Harmonic Differential Quadrature Method

Rashmi Rani and Roshan Lal

Abstract In the present paper, axisymmetric vibrations of a circular sandwich plate with relatively stiff core of exponentially varying thickness have been investigated. The face sheets are treated as membranes of constant thickness and the core is assumed to be solid as well as moderately thick. The equations of motion have been derived using Hamilton's energy principle. The frequency equation for clamped boundary condition is obtained by employing harmonic differential quadrature method. The lowest three roots of this equation have been reported as the frequencies for the first three modes of vibration. The effect of various plate parameters on the natural frequencies has been studied. Three-dimensional mode shapes for a specified sandwich plate have been illustrated. A comparison of the results with published work has been made.

Keywords Sandwich plate · HDQ method · Mode shapes

1 Introduction

A sandwich essentially consists of two thin faces sandwiching a light core between them. It is one of the most useful forms of composite structures which has wide applications in aerospace and many other industries. Sandwich construction provides several key benefits over the conventional structures, such as very high bending stiffness, low weight, cost effectiveness, durability together with a very high stiffness-to-weight ratio and high bending strength. Due to these extra ordinary features, sandwich plates are used for both interior and exterior components of aircraft (e.g., overhead bins, floor panels, radome, aerodynamic fairings), space vehicles, trains, ships, boats, cargo containers, and in residential construction [1, 2]. In many practical situations, particularly in the design of aerospace vehicles such as wings, control

R. Rani (✉) · R. Lal
Indian Institute of Technology Roorkee, Roorkee, India
e-mail: rani.msrashmi9@gmail.com

R. Lal
e-mail: rlatmfma@iitr.ac.in

P.N. Agrawal et al. (eds.), *Mathematical Analysis and its Applications*,
Springer Proceedings in Mathematics & Statistics 143,
DOI 10.1007/978-81-322-2485-3_52

surfaces (e.g., ailerons, elevators, and rudders), and rotor blades of helicopter, it becomes essential to use tapered sandwich construction for grater structural and aerodynamic efficiency. It has necessitated to study the dynamic behavior of sandwich plates of nonuniform thickness with a fair amount of accuracy. Until now, various theories for multilayered structures, particularly for composite and sandwich plates, have been developed and given in references [3, 4], to mention a few. The work up to 2008 has been reported by Carrera and Brichetto in their excellent survey article [5] on numerical assessment of classical and refined theories for the analysis of sandwich plates. Very recently, solution of the static buckling for a uniformly compressed rectangular sandwich plate having two parallel edges simply supported using the generalized Galerkin method has been given by Lopatin and Morzov [6]. Khalili et al. [7] used finite element procedure based on second-order Lagrangian elements and Galerkin-type formulation for the analysis of rectangular multilayered and sandwich plates. In the recent time, harmonic differential quadrature method has emerged as a powerful technique to solve a variety of problems in engineering and physical sciences and gives highly accurate solution with minimal computational effort [8].

In the present work, harmonic differential quadrature (HDQ) method has been employed to study the axisymmetric behavior of a circular sandwich plate with isotropic core of exponentially varying thickness in radial direction using a refined theory. The face sheets are isotropic and treated as membranes of constant thickness. The effect of transverse shear deformation and rotatory inertia is retained in the core. The frequency equation for clamped boundary condition has been obtained. The lowest three roots of these frequency equations have been reported as the natural frequencies for the first three modes of vibration. Effect of various plate parameters has been studied on natural frequencies for clamped boundary condition. Comparison of results with those available in the literature has been presented.

2 Mathematical Formulation

Consider a circular sandwich plate of radius a and thickness $2(h_c + h_f)$ referred to cylindrical polar coordinate (r, θ, z) being the middle surface of the plate and also the plane of symmetry. The line $r = 0$ is the axis of the plate. A cross-sectional view of the plate with exponentially varying core thickness $h_c(r)$, the facing thickness $h_f(\ll h_c)$, and facing slope ϕ is shown in Fig. 1. Any location in the lower or upper facing is identified by its r-coordinate or by its ϕ-coordinate, where $\phi = \phi(r)$. The variables ϕ and r are connected by the relations

$$r = R_\theta \sin\phi \quad and \quad dr = R_\phi \cos\phi d\phi \tag{1}$$

where R_ϕ is the radius of curvature of the core-facing interface and R_θ is the length of the normal between any point on the core-facing interface and the axis of sandwich plate. The thickness variation of the core with radial distance is given by:

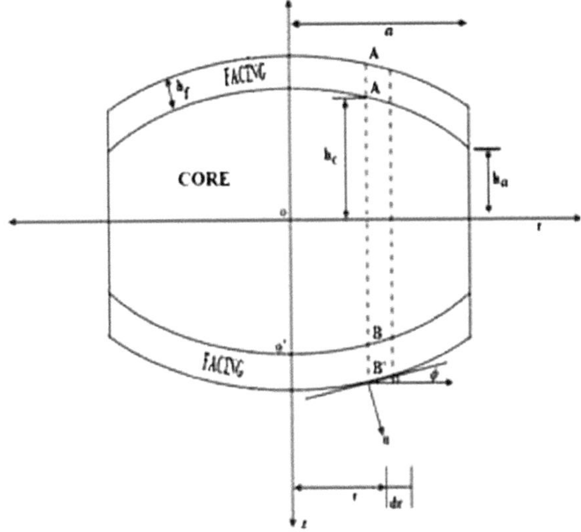

Fig. 1 Cross-section of circular sandwich plate with core of exponentially varying thickness, i.e., $h_c = h_o e^{\alpha x}$

$$\frac{dh_c}{d} r = -\tan\phi, \quad h_c = h_o e^{\alpha x}$$

and

$$\frac{d\phi}{dr} = -\cos^2\phi \frac{d^2 h_c}{dr^2}$$

where α is taper parameter and h_o is the thickness of the core at the center of the plate. The differential equations governing the axisymmetric vibration of such plate [9], are given by

$$U_o \frac{d^2\psi}{dx^2} + U_1 \frac{d\psi}{dx} + (U_2 - \Omega^2 P_2) + U_3 \frac{d^2 W}{dx^2} + U_4 \frac{dW}{dx} = 0, \tag{2}$$

$$U_5 \frac{d^2\psi}{dx^2} + U_6 \frac{d\psi}{dx} + U_7\psi + U_8 \frac{d^2 W}{dx^2} + U_9 \frac{dW}{dx} - \Omega^2 P_{10} W = 0, \tag{3}$$

where

$$U_o = R_c H_c^2 x + 3R_c H_c \frac{dH_c}{dx} x^2 + 3R_f H_f x \cos^3\phi \left(H_c + 2\frac{dH_c}{dx}\right)$$
$$- 9R_f H_f H_c x^2 \cos^2\phi \sin\phi \frac{d\phi}{dx}, \quad U_3 = -3R_f H_f x^2 \sin\phi \cos^2\phi,$$

$$U_2 = 3R_c v_c H_c \frac{dH_c}{dx} x - R_c H_c^2 + 3R_f H_f \left(x \frac{dH_c}{dx} \cos^3\phi - H_c \sec\phi \right) - 3k_s x^2$$
$$- 3R_f H_f \frac{d^2 H_c}{dx^2} \cos^3 \phi x^2 - 3R_f H_f H_c v_f \sin\phi \frac{d\phi}{dx} x - 9R_f H_f H_c x^2 \cos^2\phi \sin\phi \frac{dH_c}{dx} \frac{d\phi}{dx},$$

$$U_4 = 3R_f H_f x \sin\phi (v_f - \cos^2\phi) - 3k_s x^2 - 3R_f H_f x^2 \cos\phi (\cos^2\phi - 2\sin^2\phi) \frac{d\phi}{dx}$$
$$U_5 = R_f H_f H_c x \sin\phi \cos^2\phi, \quad P_2 = -x^2 H_c (H_c + 3H_f R_\rho \sec\phi),$$
$$U_6 = k_s H_c x R_f H_f x \sin\phi (2x \frac{dH_c}{dx} \cos^2\phi + v_f H_c + H_c \cos^2\phi$$
$$+ R_f H_f H_c x \cos\phi (\cos^2\phi - 2\sin^2\phi) \frac{d\phi}{dx},$$
$$U_7 = R_f H_f \sin\phi \left(\sin\phi\cos\phi - v_f \frac{dH_c}{dx} \right) + k_s \left(H_c + x \frac{dH_c}{dx} \right)$$
$$+ R_f H_f (x \sin\phi (2\cos^2\phi - \sin^2\phi) - v_f H_c \cos\phi) \frac{d\phi}{dx},$$
$$U_o = (R_c H_c^2 + 3R_f H_f H_c \cos^3\phi) x^2,$$

$$P_{10} = -x(H_c + H_f R_\rho \sec\phi), \quad U_8 = x(k_s H_c + R_f H_f \cos\phi \sin^2\phi),$$

$$\Omega^2 = \frac{\rho_c a^2 \Omega^2}{\mu_c} \quad and \quad R_\rho = \frac{\rho_f}{\rho_c}$$

The solution of Eqs. (2) and (3) together with regularity condition $\Psi = Q_r = 0$ [10] at the center $x = 0$ and clamped boundary condition at the edge $x = 1$ gives rise to a two point boundary value problem with variable coefficients whose closed form solution is not possible. Keeping this in view, an approximate solution is obtained by employing HDQ method.

3 Method of Solution

Let x_i, $1 = 1, 2, ..., m$ be the m grid points in the applicability range [0, 1] of the plate. According to HDQ method [8], the nth-order derivatives of $W(x)$ and $\Psi(x)$ with respect to x at the ith-point x_i are given by

$$(W_x^n, \ \Psi_x^n) = \sum_{j=1}^{m} C_{ij}^{(n)} (W(x_j)), \ \ (\Psi(x_j)), \quad i = 1, 2, ..., m \tag{4}$$

where C_{ij}^n are the weighting coefficients and the first-order weighting coefficients, i.e., C_{ij}^n for $n = 1$ have been given as follows:

$$C_{ij}^{(1)} = \frac{(\pi/2)M^1(x_i)}{M^1(x_j)\sin[(x_i - x_j)/2]}$$

where

$$M^1(x_i) = \prod_{\binom{j=1}{j\neq i}}^{m} \sin\left(\frac{x_i - x_j}{2}\pi\right), \quad i, j = 1, 2, ..., m \quad but, \quad j \neq i \tag{5}$$

and the second-order coefficients are generated from the recurrence relation

$$C_{ij}^{(2)} = C_{ij}^{(1)}\left(C_{ij}^{(1)} - \pi ctg\left(\frac{x_i - x_j}{2}\right)\pi\right), \quad i, \quad j = 1, 2, ..., m, \quad but \quad j \neq i, \tag{6}$$

with

$$C_{ii}^{(n)} = -\sum_{\binom{j=1}{j\neq i}}^{m} C_{ij}^{(n)}, \quad i = 1, 2, ..., m \quad and \quad n = 1 \quad or \quad 2. \tag{7}$$

Now, discretizing Eqs. (2) and (3) at the grid points $x = x_i, \quad i = 2, 3, ..., (m-1)$, and substituting the values of first two derivatives of W and Ψ from Eq. (4), we get

$$\sum_{j=1}^{m}(U_{0,i}C_{i,j}^{(2)} + U_{1,i}C_{i,j}^{(1)})\Psi_j + \sum_{j=1}^{m}(U_{3,i}C_{i,j}^{(2)} + U_{4,i}C_{i,j}^{(1)})W_j + (U_{2,i} - \Omega^2 P_{2,i})\Psi_i = 0, \tag{8}$$

$$\sum_{j=1}^{m}(U_{5,i}C_{i,j}^{(2)} + U_{6,i}C_{i,j}^{(1)})\Psi_j + \sum_{j=1}^{m}(U_{8,i}C_{i,j}^{(2)} + U_{9,i}C_{i,j}^{(1)})W_j + U_{7,i}\Psi_i - \Omega^2 P_{10,i}W_i = 0, \tag{9}$$

The satisfaction of Eqs. (8) and (9) at $(m-2)$ internal grid points $x_i, \quad i = 2, ...,$ (m-1) together with the regularity condition: $\Psi = Q_r = 0$ (Wu et al. [10]), at the center of the plate provides a set of $(2m-2)$ equations in terms of unknowns $W_j = W(x_j)$ and $\Psi_j = \Psi(x_j), \quad j = 1, 2, ..., m$. The resulting system of equations can be written in matrix form as

$$[U][C] = [0], \tag{10}$$

where U and C are the matrices of orders $(2m-2)\times 2m$ and $(2m \times 1)$, respectively. The above $(m-2)$ internal grid points chosen for collocation are the zeros of shifted Chebyshev polynomial of order $(m-2)$ with orthogonality range (0, 1), given by

$$x_{k+1} = \frac{1}{2}\left[1+\cos\left(\frac{2k-1}{m-2}\frac{\pi}{2}\right)\right].$$

4 Boundary Condition and Frequency Equation

By satisfying the relation: (i) $\Psi = W = 0$: for clamped edge (C-plate), a set of two homogeneous equations is obtained. These equations together with the field Eq. (10) give a complete set of $2m$ equations in terms of $2m$ unknowns which can be written as

$$\begin{bmatrix} U \\ U^C \end{bmatrix}[C] = [0], \tag{11}$$

where U^C is a matrix of order $2 \times 2m$. For a nontrivial solution of Eq. (11), the frequency determinant must vanish and hence

$$\begin{vmatrix} U \\ U^C \end{vmatrix} = 0. \tag{12}$$

5 Numerical Results and Discussion

The frequency Eq. (12) provides the values of the frequency parameter Ω and solved using MATLAB for various values of plate parameters. The numerical values of the lowest three roots have been reported as the first three natural frequencies to investigate the influence of the taper parameters, core thickness at the center, and face thickness for clamped boundary condition. In the work reported here, the values of various plate parameters are taken as follows:
$\alpha = -0.5(0.1)0.5$, $H_o = 0.05(0.05)0.30$, $H_f = 0.0025(0.0025)0.02$.

The material for the core and the facings are taken to be PVC (Poly vinyl chloride) and aluminum, respectively, for which the various constants are, $R_c = 2.85$, $R_f = 1232.21$, $R_\rho = 20.76$, $\nu_c = 0.3$ and $\nu_f = 0.3$ and $k_s = 1$ from Ref. [8].

To choose the appropriate number of grid points m, a computer program used to evaluate the frequencies, was run for $m = 5(1)20$ for different sets of plate parameters for clamped boundary condition. The numerical values showed a consistent improvement with the increase in the number of grid points m. In all the computations, the number of grid points has been taken as 12, since further increase in m does not improve the result even at the fourth place of decimal. The convergence of

Table 1 Convergence study for the first three frequencies

Boundary condition		Clamped plate	
m/mode	I	II	III
10	3.0664	4.6264	6.8181
11	3.0664	4.6264	6.8170
12	3.0664	4.6264	6.8171
13	3.0664	4.6264	6.8171
14	3.0664	4.6264	6.8171
15	3.0664	4.6264	6.8171
16	3.0664	4.6264	6.8171
17	3.0664	4.6264	6.8171
18	3.0664	4.6264	6.8171
19	3.0664	4.6264	6.8171
20	3.0664	4.6264	6.8171

Table 2 Comparison of results for isotropic circular plate

Boundary condition		Clamped plate	
Method/mode	I	II	III
Exact [11]	10.2158	39.7711	89.1041
Finite element [12]	10.2159	39.7766	89.1708
Rayleigh-Ritz [13]	10.2160	39.7710	89.1030
Receptance [14]	10.2160	39.7710	89.1041
Ritz method [15]	10.2158	39.7711	–
symplectic method [16]	10.2160	39.7710	–
Present	10.2158	39.7711	89.1041

frequency parameter Ω for a particular set $\alpha = 0.5$, $H_o = 0.3$, $H_f = 0.02$ is shown in Table 1.

A comparison of results for specified uniform isotropic circular plate ($\alpha = 0$, $H_f = 0$, $\nu_c = 0.3$, $H_o = 0.001$, $k_s = 1$) with exact solution [11] and obtained by Finite element method [12], Rayleigh–Ritz method [13], receptance method [14], Ritz method [15], symplectic method [16] has been presented in Table 2. An excellent agreement of the frequencies shows the versatility of the present method.

The numerical results are given in Figs. 2, 3 and 4. Figure 2 depicts the effect of taper parameter α on the frequency parameter Ω for $H_f = 0.005$ and two values of $H_o = 0.1, 0.2$, for all the three modes of vibration for clamped plate. It is observed that frequency parameter Ω increases with the increasing values of taper parameter α. The rate of increase of frequency parameter Ω with taper parameter α decreases

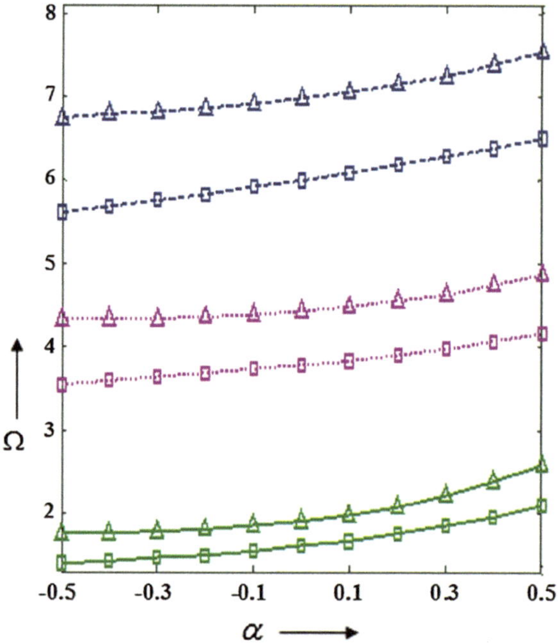

Fig. 2 Taper parameter α versus frequency parameter Ω for $H_f = 0.005$; $H_o = 0.1$; □, $H_o = 0.2$; Δ. First mode:—— ; Second mode:·········; Third mode:– – –

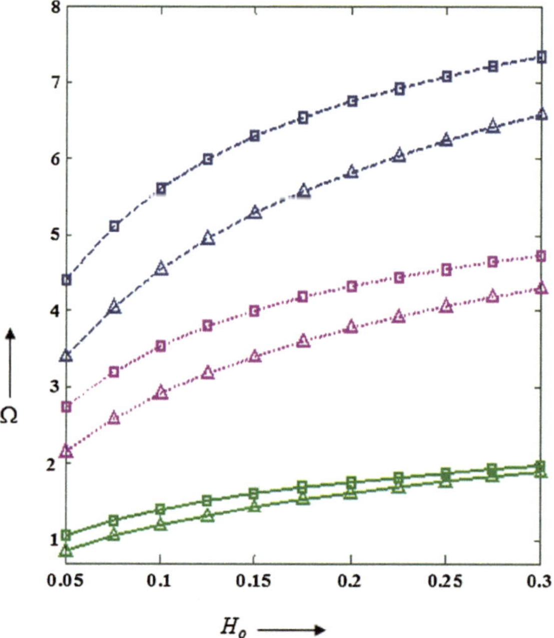

Fig. 3 Core thickness at the center H_o versus frequency parameter Ω for $\alpha = 0.5$: $H_f = 0.005$; □, $H_f = 0.010$; Δ. First mode:—— ; Second mode:·········; Third mode:– – –

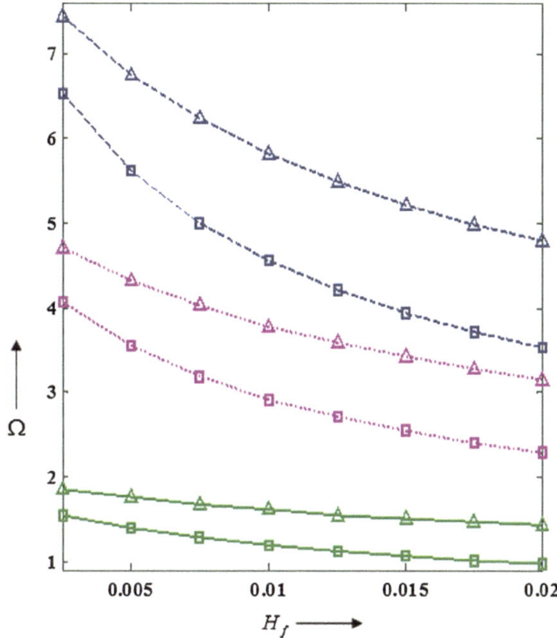

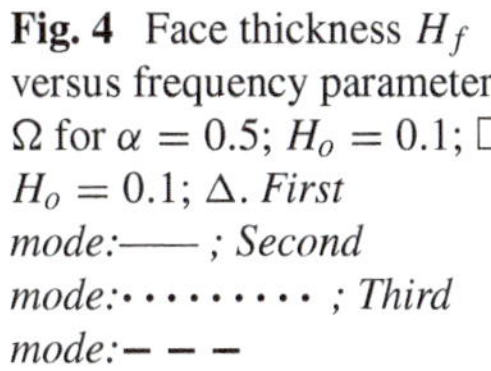
Fig. 4 Face thickness H_f versus frequency parameter Ω for $\alpha = 0.5$; $H_o = 0.1$; □, $H_o = 0.1$; Δ. *First mode:*—— ; *Second mode:*········· ; *Third mode:*– – –

with the increase in the number of modes. The effect of H_o is more pronounced for $\alpha = -0.5$ as compared to $\alpha = 0.5$.

Figure 3 demonstrates the effect of core thickness at the center H_o on the frequency parameter Ω for $\alpha = 0.5$ and two different values of face thickness $H_f = 0.005, 0.01$, for all the three modes for clamped plate. It is observed that the frequency parameter Ω increases with the increasing values of H_o for all the three modes. This effect is more pronounced for $H_o = 0.05$ as compared to $H_o = 0.3$. Further, the rate of increase in the values of Ω is comparatively higher for $H_o \leq 0.175$ as compared to $H_o > 0.175$. This rate increases with the increase in number of modes with the increasing values of H_f.

Figure 4 depicts the behavior of facing thickness H_f on the frequency parameter Ω for $\alpha = 0.5$ and two different values of $H_o = 0.1, 0.2$, for all the three modes of vibration for clamped plate. It can be seen that the frequency parameter Ω decreases with the increasing values of face thickness H_f for all the three modes. This effect is more pronounced for $H_f = 0.02$ as compared to $H_f = 0.0025$ for the increasing values of H_o from 0.1 to 0.2. This effect increases with the increase in number of modes. The rate of decrease in the values of frequency parameter Ω with increasing values of H_f is found to increase with increasing number of modes. Three-dimensional mode shapes for specified clamped circular sandwich plate taking $H_o = 0.1$, $H_f = 0.005$, and $\alpha = 0.5$ have been plotted and shown in Fig. 5.

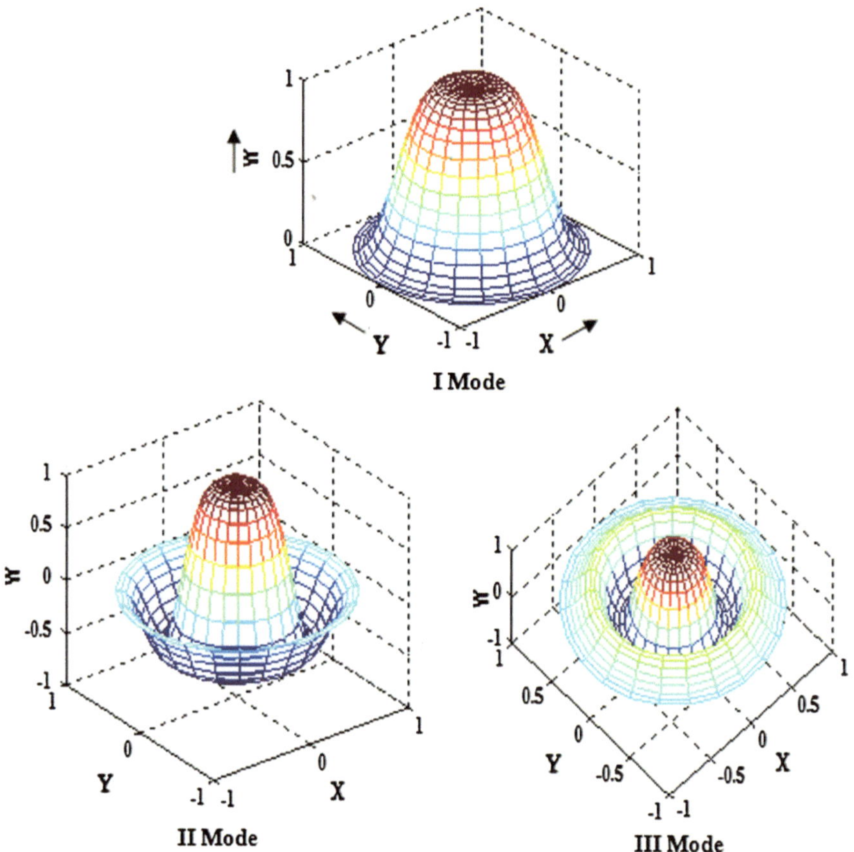

Fig. 5 First three mode shapes for clamped circular sandwich plate

6 Conclusion

The radially symmetric vibration of clamped circular sandwich plates with core of exponentially varying thickness has been analyzed employing HDQ method. The effect of transverse shear deformation and rotatory inertia has been retained in the core and the face sheets are treated as membrane of constant thickness. It is observed that the frequency parameter Ω,
(i) increases with the increasing values of taper parameter α and core thickness at the center H_o,
(ii) decreases with the increasing values of face thickness parameter H_f.

The present analysis will be of great help to the design engineers dealing with sandwich structures in obtaining the desired frequency by varying one or more plate parameters involved in the present model.

Acknowledgments One of the authors, Rashmi Rani is thankful to Ministry of Human Resources and Development (MHRD), India for the financial support to carry out this research work.

References

1. Vel, S.S., Caccese, V., Zhao, H.: Elastic coupling effects in tapered sandwich panels with laminated anisotropic composite facings. J. Compos. Mater. **39**, 2161–2183 (2005)
2. Mustapha, S., Ye, L.: Damage identification and assessment in tapered sandwich structures using guided waves. Key Eng. Mater. **558**, 25–38 (2013)
3. Altenbach, H.: Theories for laminated and sandwich plates. Mech. Compos. Mater. **34**, 333–348 (1998)
4. Carrera, E., Ciuffreda, A.: Bending of composites and sandwich plates subjected to localized lateral loadings: a comparison of various theories. Compos. Struct. **68**, 185–202 (2005)
5. Carrera, E., Brischetto, S.: A survey with numerical assessment of classical and refined theories for the analysis of sandwich plates. Appl. Mech. Rev. **62**, 010803–010819 (2009)
6. Lopatin, A.V., Morozov, E.V.: Buckling of a uniformly compressed rectangular SSCF composite sandwich plate. Compos. Struct. **105**, 108–115 (2013)
7. Khalili, S.M.R., Shariyat, M., Rajabi, I.: A finite element based global local theory for static analysis of rectangular sandwich and laminated composite plates. Compos. Struct. **107**, 177–189 (2014)
8. Civalek, O., Catal, H.H.: Linear static and vibration analysis of circular and annular plates by the harmonic differential quadrature (HDQ) method. Eng. Arch. Fac. Osmangazi University, **XVII**(1), (2003)
9. Lal, R., Rani, R.: Mode shapes and frequencies of radially symmetric vibrations of annular sandwich plates of variable thickness. Acta Mech. **225**, 1565–1580 (2014)
10. Wu, T.Y., Wang, Y.Y., Liu, G.R.: Free vibration analysis of circular plates using generalized differential quadrature rule. Comput. Methods Appl. Mech. Eng. **191**, 5365–5380 (2002)
11. Leissa, A.W.: Vibration of Plates. U.S. Government Printing Office (NASA SP 160), Washington, DC (1969)
12. Pardoen, G.C.: Axisymmetric vibration and stability of circular plate. Comput. struct. **9**, 89–95 (1978)
13. Kim C.S., Dickinson S.M.: On the free, transverse vibration of annular and circular, thin, sectorial plates subject to certain complicating effects. J. Sound Vib. **134**, 407–421 (1989)
14. Azimi, S.: Free vibration of circular plates with elastic edge supports using the receptance method. J. Sound Vib. **120**(1), 19–35 (1998)
15. Ansari A.H.: Vibration of plates of variable thickness. Ph.D. Thesis, University of Roorkee, India (2000)
16. Zhou Z.H., Wong K.W., Xu X.S., Leung, A.Y.T.: Natural vibration of circular and annular thin plates by Hamiltonian approach. J. Sound Vib. **330**, 1005–1017 (2011)

Hybrid Projective Synchronization of Fractional-Order Neural Networks with Time Delays

G. Velmurugan and R. Rakkiyappan

Abstract In this paper, the problem of hybrid projective synchronization of fractional-order neural networks with time delay is extensively investigated. The fractional-order neural networks with hub structure and time delay is considered. By using stability theorem of linear fractional order systems with multiple time delays and constructing an appropriate linear feedback control, some new sufficient conditions are derived to ensure projective synchronization of fractional-order neural networks with time delays. It means that the response system can be synchronized with the drive system based on the choice of a scaling matrix. Finally, a numerical example is provided to demonstrate the effectiveness of our theoretical results.

Keywords Hybrid projective synchronization · Fractional-order · Neural networks · Time delays

1 Introduction

Fractional calculus has become an active area of research in recent years due to their widespread applications in various fields of science and engineering, such as dielectric polarization, viscoelasticity, heat conduction, biology, etc. [1–3]. Generally, most of the real-world problems are modeled by fractional-order dynamical systems rather than integer-order ones. That is fractional-order systems provide more accurate result than the integer-order systems. In [4], the authors pointed out that fractional derivatives provide an excellent tool for the description of memory and hereditary properties

G. Velmurugan (✉) · R. Rakkiyappan
Department of Mathematics, Bharathiar University, Coimbatore 641046, Tamil Nadu, India
e-mail: gvmuruga@gmail.com

R. Rakkiyappan
e-mail: rakkigru@gmail.com

P.N. Agrawal et al. (eds.), *Mathematical Analysis and its Applications*, Springer Proceedings in Mathematics & Statistics 143,
DOI 10.1007/978-81-322-2485-3_53

of various processes. Recently, many of the researchers have focused their interest and attention to analysis the fractional-order dynamical systems and many good results have been reported in the existing literature [5–7].

In the past few decades, the dynamical analysis of neural networks has received increasing interest and hot topic of research because of their potential applications in numerous fields, such as pattern recognition, associative memory and combinatorial optimization, etc. [8, 9]. In fact, fractional-order systems have infinite memory. According to that feature, the incorporation of a memory term into a neural network model is an extremely important improvement. Therefore, it is necessary to investigate the dynamical analysis of fractional-order neural networks (FNNs). As we know that, time delay is an unavoidable factor in the practical applications. It follows that, many authors extensively analysis the FNNs with time delays and some remarkable results have been proposed in the literature [10–14].

On the other hand, synchronization of fractional-order chaotic systems and FNNs have received much attention in the area of nonlinear science and it has been applied many fields such as image processing, secure communication and ecological system. In [15], the authors have been introduced synchronization of chaotic systems. In the literature, there are many types of synchronization have been exposed and investigated, such as complete synchronization, anti-phase synchronization, projective synchronization, etc. [15–19]. There are several methods have been provided for the synchronization based on linear feedback control, adaptive control, sliding mode control, etc. Moreover, complete synchronization and anti-phase synchronization are the special case of projective synchronization. However, projective synchronization was first introduced in [20] and it has been providing faster communication with its proportional feature. This feature can be used to extend binary digital to M-nary digital communication for achieving fast communication in [21]. Thus, the analysis of projective synchronization is very important in both theoretical and application point of view. Recently, several important results have been derived for projective synchronization in the literature [22–26]. In [22], the authors extensively studied the modified projective synchronization of time-delayed fractional-order chaotic systems. Some new sufficient conditions were obtained to realize projective synchronization of FNNs with open loop control and adaptive control in [23]. To the best of our knowledge, there are few results of the projective synchronization of FNNs.

Motivated by the above discussion, the problem of hybrid projective synchronization of FNNs with time delay is studied in this paper. Some new sufficient conditions are derived to ensure the hybrid projective synchronization of FNNs with time delays by using linear feedback control. Here, we use the Adams-Bashforh-moulton predictor-corrector method [27] to solve FNNs by numerically.

This paper organized as follows. In Sect. 2, some basic definitions of fractional calculus are given. Some new sufficient criteria for hybrid projective synchronization of FNNs with time delays are obtained in Sect. 3. In Sect. 4, a numerical example is provided to show the effectiveness of our main results. The conclusion of this paper is given in Sect. 5.

2 Preliminaries

In this section, we provide some basic definitions of fractional calculus. Throughout this paper, we use the Caputo fractional-order derivative.

Definition 1 [1] The fractional integral of order α for a function g is defined as

$$I^{\alpha} g(t) = \frac{1}{\Gamma(\alpha)} \int_0^t (t-\tau)^{\alpha-1} g(\tau) d\tau, \tag{1}$$

where $t \geq t_0$ and $\alpha > 0$, $\Gamma(\cdot)$ is the gamma function defined as $\Gamma(\alpha) = \int_0^{\infty} t^{\alpha-1} e^{-t} dt$.

Definition 2 [1] The Caputo fractional derivative of order α for a function $g(t)$ is

$$_C D_t^{\alpha} g(t) = \frac{1}{\Gamma(n-\alpha)} \int_0^t \frac{g^{(n)}(\tau)}{(t-\tau)^{\alpha-n+1}} d\tau, \tag{2}$$

where $t > 0$ and n is a positive integer such that $n - 1 < \alpha < n \in Z^+$.

Definition 3 [14] The Laplace transform of the Caputo fractional-order derivatives is

$$L\{_C D_t^{\alpha} g(t); s\} = s^{\alpha} G(s) - \sum_{k=0}^{n-1} s^{\alpha-k-1} g^{(k)}(0), \ n-1 < \alpha \leq n,$$

where $G(s)$ is the Laplace transform of $g(t)$, $g^k(0) = 0, \ k = 1, 2, \cdots, n$, are the initial conditions.

In this paper, consider the FNNs with time delays as drive system is described as

$$D^{\alpha} x_i(t) = -c_i x_i(t) + \sum_{j=1}^{n} a_{ij} f_j(x_j(t)) + \sum_{j=1}^{n} b_{ij} f_j(x_j(t-\tau)), \ i = 1, 2, \cdots, n, \tag{3}$$

where n corresponds to the number of units. $0 < \alpha < 1$, $x_i(t)$ is the state vector of the ith unit. $f_j(\cdot)$ denotes the nonlinear activation function. a_{ij} and b_{ij} denotes the connection weight matrices without delay and with delay. $c_i > 0$ is the self-feedback connection weight matrix and τ is the constant time delay. Equation (3) can be rewritten as in the vector form as follows

$$D^{\alpha} x(t) = -Cx(t) + \widehat{A} f(x(t)) + \widehat{B} f(x(t-\tau)), \tag{4}$$

where $x(t) = (x_1(t), x_2(t), \cdots, x_n(t))^T \in \mathscr{R}^n$, $C = \text{diag}(c_1, c_2, \cdots, c_n) \in \mathscr{R}^{n \times n}$, $\widehat{A} = (a_{ij})_{n \times n} \in \mathscr{R}^{n \times n}$, $\widehat{B} = (b_{ij})_{n \times n} \in \mathscr{R}^{n \times n}$, $f(x(t)) = (f_1(x_1(t)), f_2(x_2(t)),$

$\cdots, f_n(x_n(t)))^T$ and $f(x(t-\tau)) = (f_1(x_1(t-\tau)), f_2(x_2(t-\tau)), \cdots, f_n(x_n(t-\tau)))^T$. System (4) can be linearize as follows

$$D^\alpha x(t) = -Cx(t) + \widehat{A}Jx(t) + \bar{x}(t-\tau), \tag{5}$$

where J is the Jacobian matrix of $f(x(t))$ and $\bar{x}(t-\tau) = (\sum_{j=1}^{n} b_{1j}q_{1j}x_j(t-\tau), \cdots, \sum_{j=1}^{n} b_{nj}q_{nj}x_j(t-\tau))^T$ is the linearization vector of $\widehat{B}f(x(t-\tau))$ at the equilibrium point. Also, denote $\bar{A} = \widehat{A}J$ and $\bar{B} = (b_{ij}q_{ij})_{n\times n}$, then (5) can be rewritten as

$$D^\alpha x(t) = -Cx(t) + \bar{A}x(t) + \bar{B}x(t-\tau). \tag{6}$$

The corresponding response system is given by

$$D^\alpha y(t) = -Cy(t) + \bar{A}y(t) + \bar{B}y(t-\tau) + \psi(t), \tag{7}$$

where $\psi(t) = (\psi_1(t), \psi_2(t), \cdots, \psi_n(t))^T$ is the control input. Here, consider the linear feedback control to realize synchronization between the derive system (6) and response system (7). The controller $\psi(t)$ is defined as

$$\psi(t) = \Gamma(y(t) - \beta x(t)), \tag{8}$$

where $\Gamma = \text{diag}(\gamma_1, \gamma_2, \cdots, \gamma_n) \in \mathscr{R}^{n\times n}$ is a feedback gain matrix. The initial conditions associated with the system (6) and (7) is $x(t) = x(0)$ and $y(t) = y(0)$, $t \in [-\tau, 0]$.

Definition 4 If there exists a real scaling matrix $\beta \in \mathscr{R}^{n\times n}$, such that for any two solutions $x(t)$ and $y(t)$ of system (6) and system (7) with different initial values denoted by $x(0)$ and $y(0)$, one has

$$\lim_{t\to\infty} \|y(t) - \beta x(t)\| = 0, \tag{9}$$

then, drive system (6) and response system (7) are said to be globally hybrid projectively synchronized.

3 Main Results

In this section, some new sufficient criteria has been derived to ensure that system (6) and (7) is projectively synchronized under linear feedback control.

Let us define $w(t) = y(t) - \beta x(t)$ be the synchronization errors. From (6) and (7), the error system as follows

$$D^\alpha w(t) = -Cw(t) + \bar{A}w(t) + \bar{B}w(t-\tau) + \Gamma w(t), \tag{10}$$

with initial conditions $w(t) = \delta(0)$, $t \in [-\tau, 0]$.

Taking the Laplace transform, we have

$$\begin{aligned}
s^{\alpha}E_1(s)-s^{\alpha-1}\delta_1(0) = {} & (-c_1+\gamma_1+\bar{a}_{11})E_1(s)+\bar{a}_{12}E_2(s)+\cdots+\bar{a}_{1n}E_n(s) \\
& +\bar{b}_{11}e^{-s\tau}\left(E_1(s)+\int_{-\tau}^{0}e^{-st}\delta_1(t)dt\right) \\
& +\bar{b}_{12}e^{-s\tau}\left(E_2(s)+\int_{-\tau}^{0}e^{-st}\delta_2(t)dt\right) \\
& +\cdots+\bar{b}_{1n}e^{-s\tau}\left(E_n(s)+\int_{-\tau}^{0}e^{-st}\delta_n(t)dt\right), \\
s^{\alpha}E_1(s)-s^{\alpha-1}\delta_1(0) = {} & (-c_1+\gamma_1+\bar{a}_{11}+\bar{b}_{11}e^{-s\tau})E_1(s)+(\bar{a}_{12}+\bar{b}_{12}e^{-s\tau})E_2(s)+\cdots \\
& +(\bar{a}_{1n}+\bar{b}_{1n}e^{-s\tau})E_n(s)+\bar{b}_{11}e^{-s\tau}\int_{-\tau}^{0}e^{-st}\delta_1(t)dt \\
& +\bar{b}_{12}e^{-s\tau}\int_{-\tau}^{0}e^{-st}\delta_2(t)dt+\cdots+\bar{b}_{1n}e^{-s\tau}\int_{-\tau}^{0}e^{-st}\delta_n(t)dt, \\
s^{\alpha}E_2(s)-s^{\alpha-1}\delta_2(0) = {} & (-c_2+\gamma_2+\bar{a}_{22}+\bar{b}_{22}e^{-s\tau})E_2(s)+(\bar{a}_{21}+\bar{b}_{21}e^{-s\tau})E_1(s)+\cdots \\
& +(\bar{a}_{2n}+\bar{b}_{2n}e^{-s\tau})E_n(s)+\bar{b}_{21}e^{-s\tau}\int_{-\tau}^{0}e^{-st}\delta_1(t)dt \\
& +\bar{b}_{22}e^{-s\tau}\int_{-\tau}^{0}e^{-st}\delta_2(t)dt+\cdots+\bar{b}_{2n}e^{-s\tau}\int_{-\tau}^{0}e^{-st}\delta_n(t)dt, \\
\cdots \\
\cdots \\
s^{\alpha}E_n(s)-s^{\alpha-1}\delta_n(0) = {} & (-c_n+\gamma_n+\bar{a}_{nn}+\bar{b}_{nn}e^{-s\tau})E_n(s)+(\bar{a}_{n1}+\bar{b}_{n1}e^{-s\tau})E_1(s)+\cdots \\
& +(\bar{a}_{n,n-1}+\bar{b}_{n,n-1}e^{-s\tau})E_{n-1}(s)+\bar{b}_{n1}e^{-s\tau}\int_{-\tau}^{0}e^{-st}\delta_1(t)dt \\
& +\bar{b}_{n2}e^{-s\tau}\int_{-\tau}^{0}e^{-st}\delta_2(t)dt+\cdots+\bar{b}_{nn}e^{-s\tau}\int_{-\tau}^{0}e^{-st}\delta_n(t)dt, \quad (11)
\end{aligned}$$

where $E(s)$ is the Laplace transform of $w(t)$ with $E(s) = L(w(t))$. The vector form of (11) is

$$\Delta(s)\cdot E(s) = k(s), \tag{12}$$

where $\Delta(s)$ is the characteristic matrix of system (11) and $k(s)$ is the remainder nonlinear part of system (11), such as

$$\Delta(s) = \begin{pmatrix} s^{\alpha}+d_1-\bar{b}_{11}e^{-s\tau} & -\bar{a}_{12}-\bar{b}_{12}e^{-s\tau} & \cdots & -\bar{a}_{1n}-\bar{b}_{1n}e^{-s\tau} \\ -\bar{a}_{21}-\bar{b}_{21}e^{-s\tau} & s^{\alpha}+d_2-\bar{b}_{22}e^{-s\tau} & \cdots & -\bar{a}_{2n}-\bar{b}_{2n}e^{-s\tau} \\ \cdots & \cdots & \cdots & \cdots \\ -\bar{a}_{n1}-\bar{b}_{n1}e^{-s\tau} & -\bar{a}_{n2}-\bar{b}_{n2}e^{-s\tau} & \cdots & s^{\alpha}+d_n-\bar{b}_{nn}e^{-s\tau} \end{pmatrix} \tag{13}$$

with $d_i = c_i - \gamma_i - \bar{a}_{ii},\ (i = 1, 2, \cdots, n)$ and

$$\begin{aligned}
k_1(s) = &\ s^{\alpha-1}\delta_1(0) + \bar{b}_{11}e^{-s\tau}\int_{-\tau}^{0} e^{-st}\delta_1(t)dt + \bar{b}_{12}e^{-s\tau}\int_{-\tau}^{0} e^{-st}\delta_2(t)dt + \cdots \\
&+ \bar{b}_{1n}e^{-s\tau}\int_{-\tau}^{0} e^{-st}\delta_n(t)dt, \\
k_2(s) = &\ s^{\alpha-1}\delta_2(0) + \bar{b}_{21}e^{-s\tau}\int_{-\tau}^{0} e^{-st}\delta_1(t)dt + \bar{b}_{22}e^{-s\tau}\int_{-\tau}^{0} e^{-st}\delta_2(t)dt + \cdots \\
&+ \bar{b}_{2n}e^{-s\tau}\int_{-\tau}^{0} e^{-st}\delta_n(t)dt, \\
\cdots \\
k_n(s) = &\ s^{\alpha-1}\delta_n(0) + \bar{b}_{n1}e^{-s\tau}\int_{-\tau}^{0} e^{-st}\delta_1(t)dt + \bar{b}_{n2}e^{-s\tau}\int_{-\tau}^{0} e^{-st}\delta_2(t)dt + \cdots \\
&+ \bar{b}_{nn}e^{-s\tau}\int_{-\tau}^{0} e^{-st}\delta_n(t)dt.
\end{aligned}$$

Now, $\tau = 0$, the systems (10) becomes

$$D^{\alpha}w(t) = (-C + \Gamma + \bar{A} + \bar{B})w(t) = \mathscr{M}w(t), \tag{14}$$

where,

$$\mathscr{M} = \begin{pmatrix} -c_1 + \gamma_1 + \bar{a}_{11} + \bar{b}_{11} & \bar{a}_{12} + \bar{b}_{12} & \cdots & \bar{a}_{1n} + \bar{b}_{1n} \\ \bar{a}_{21} + \bar{b}_{21} & -c_2 + \gamma_2 + \bar{a}_{22} + \bar{b}_{22} & \cdots & \bar{a}_{2n} + \bar{b}_{2n} \\ \cdots & \cdots & \cdots & \cdots \\ \bar{a}_{n1} + \bar{b}_{n1} & \bar{a}_{n2} + \bar{b}_{n2} & \cdots & -c_n + \gamma_n + \bar{a}_{nn} + \bar{b}_{nn} \end{pmatrix}. \tag{15}$$

Suppose $C = 0$, $\bar{A} = 0$ and $\Gamma = 0$, then (10) becomes the system in [6] and we have the following results.

Theorem 1 ([6]) *If all the roots of the characteristic equation* $det(\Delta(s)) = 0$ *have negative real parts, then the zero solution of (10) is Lyapunov asymptotically stable.*

Theorem 2 ([6]) *If* $\alpha \in (0, 1),\ C = 0,\ \bar{A} = 0,\ \Gamma = 0$, *all the eigenvalues of* $\mathscr{M}$ *satisfy* $|arg(\lambda)| > \frac{\alpha\pi}{2}$ *and the characteristic equation* $det(\Delta(s)) = 0$ *has no pure imaginary roots for* $\tau > 0$, *then the zero solution of (10) is Lyapunov asymptotically stable.*

If $\alpha \in (0, 1),\ C \neq 0,\ \bar{A} \neq 0,\ \Gamma \neq 0$, then Theorem 2 is not valid to study the stability of (10). Therefore, we consider the following Theorem.

Theorem 3 ([14]) *If $\alpha \in (0, 1)$, all the eigenvalues of $\mathcal{M}$ satisfy $|arg(\lambda)| > \frac{\pi}{2}$ and the characteristic equation $det(\Delta(s)) = 0$ has no pure imaginary roots for $\tau > 0$, then the zero solution of (10) is Lyapunov asymptotically stable.*

3.1 The FNNs with Hub Structure and Time Delays

Hub structure is a common feature of neural networks, which is used to understand the mechanism of complex recurrent networks. Now, consider the FNNs with hub structure and time delays

$$\begin{cases} D^{\alpha}x_1(t) = -c_1x_1(t) + \sum_{j=1}^{n} a_{1j}f_j(x_j(t)) + b_1f_1(x_1(t-\tau)), \\ D^{\alpha}x_i(t) = -c_ix_i(t) + a_{i1}f_1(x_1(t)) + a_{ii}f_i(x_i(t)) + bf_i(x_i(t-\tau)),\ i = 2, \cdots, n, \end{cases} \tag{16}$$

where $c_i > 0$, the first neuron is the center of the hub and all the other $i-1$ neurons are connected directly only to the central neuron and to themselves.

The response system defined as

$$\begin{cases} D^{\alpha}y_1(t) = -c_1y_1(t) + \sum_{j=1}^{n} a_{1j}f_j(y_j(t)) + b_1f_1(y_1(t-\tau)) + \psi_1(t), \\ D^{\alpha}y_i(t) = -c_iy_i(t) + a_{i1}f_1(y_1(t)) + a_{ii}f_i(y_i(t)) + bf_i(y_i(t-\tau)) + \psi_i(t),\ i = 2, \cdots, n. \end{cases} \tag{17}$$

The linear forms of equations (16) and (17) are

$$D^{\alpha}x(t) = -Cx(t) + \bar{A}x(t) + \bar{B}x(t-\tau), \tag{18}$$

$$D^{\alpha}y(t) = -Cy(t) + \bar{A}y(t) + \bar{B}y(t-\tau) + \psi(t). \tag{19}$$

From (18) and (19), the error system as

$$D^{\alpha}w(t) = -Cw(t) + \bar{A}w(t) + \bar{B}w(t-\tau) + \Gamma w(t), \tag{20}$$

where $C = \text{diag}(c_1, \cdots, c_n)$, $\bar{A} = \begin{pmatrix} \bar{a}_{11} & \bar{a}_{12} & \bar{a}_{13} & \cdots & \bar{a}_{1n} \\ \bar{a}_{21} & \bar{a}_{22} & 0 & \cdots & 0 \\ \cdots & \cdots & \cdots & \cdots & \cdots \\ \bar{a}_{n1} & 0 & 0 & \cdots & \bar{a}_{nn} \end{pmatrix}$, $\bar{B} = \begin{pmatrix} \bar{b}_1 & 0 & 0 & \cdots & 0 \\ 0 & \bar{b} & 0 & \cdots & 0 \\ \cdots & \cdots & \cdots & \cdots & \cdots \\ 0 & 0 & 0 & \cdots & \bar{b} \end{pmatrix}$.

Theorem 4 *When $\alpha \in (0, 1)$, $\bar{c}_2 - \bar{b} > 0$, $\bar{c}_1 + \bar{c}_2 - \bar{b}_1 - \bar{b} > 0$, $(\bar{c}_1 - \bar{b}_1)(\bar{c}_2 - \bar{b}) - \Phi > 0$, where $\bar{c}_1 = c_1 - \gamma_1 - \bar{a}_{11}$, $\bar{c}_2 = c_i - \gamma_i - \bar{a}_{ii}$, $(i = 2, \cdots, n)$, $\Phi = \sum_{i=2}^{n}(\bar{a}_{i1}\bar{a}_{1i})$.*

(i) if $\Phi = 0$ and $\bar{b}^2 - \bar{c}_2^2 \sin^2 \frac{\alpha\pi}{2} < 0$ and $\bar{b}_1^2 - \bar{c}_1^2 \sin^2 \frac{\alpha\pi}{2} < 0$, then the zero solution of (20) is Lyapunov asymptotically stable;

(ii) if $\Phi \neq 0$ and $\bar{b}^2 - \bar{c}_2^2 \sin^2 \frac{\alpha\pi}{2} < 0$, then the zero solution of (20) is Lyapunov asymptotically stable.

Proof Taking the Laplace transform of equations (20) and using the same method of finding for $\Delta(s)$ in the preliminaries, we have

$$\Delta(s) = \begin{pmatrix} s^{\alpha}+\bar{c}_1-\bar{b}_1e^{-s\tau} & -\bar{a}_{12} & -\bar{a}_{13} & \cdots & -\bar{a}_{1n} \\ -\bar{a}_{21} & s^{\alpha}+\bar{c}_2-\bar{b}e^{-s\tau} & 0 & \cdots & 0 \\ -\bar{a}_{31} & 0 & s^{\alpha}+\bar{c}_2-\bar{b}e^{-s\tau} & \cdots & 0 \\ \cdots & \cdots & \cdots & \cdots & \cdots \\ -\bar{a}_{n1} & 0 & 0 & \cdots & s^{\alpha}+\bar{c}_2-\bar{b}e^{-s\tau} \end{pmatrix}. \quad (21)$$

Hence, $\Delta(s)$ is $n \times n$ matrix ($n \geq 3$ in hub structure). Generally, characteristic equation $\det(\Delta(s)) = 0$ satisfies

$$\begin{aligned} \det(\Delta(s)) = & \left(s^{\alpha}+\bar{c}_2-\bar{b}e^{-s\tau}\right)^{n-2} \\ & \times \left((s^{\alpha}+\bar{c}_1-\bar{b}_1e^{-s\tau})(s^{\alpha}+\bar{c}_2-\bar{b}e^{-s\tau})-\Phi\right)=0. \end{aligned} \quad (22)$$

From (22), if $\Phi = 0$ then $\left(s^{\alpha}+\bar{c}_2-\bar{b}e^{-s\tau}\right)=0$ or $\left(s^{\alpha}+\bar{c}_1-\bar{b}_1e^{-s\tau}\right)=0$, where $\Phi = \sum_{i=2}^{n} \bar{a}_{i1}\bar{a}_{1i}$.

Next prove that $\det(\Delta(s)) = 0$ has no pure imaginary roots for any $\tau > 0$.

Suppose that there exists $s = \eta i = |\eta|(\cos\frac{\pi}{2} + i\sin(\pm\frac{\pi}{2})$, that is a pure imaginary root of $s^{\alpha}+\bar{c}_2-\bar{b}e^{-s\tau} = 0$, where η is a real number. If $\eta > 0$, $s = \eta i = |\eta|(\cos\frac{\pi}{2} + i\sin(\frac{\pi}{2}))$ and if $\eta < 0$, $s = \eta i = |\eta|(\cos\frac{\pi}{2} - i\sin(\frac{\pi}{2}))$. Substituting $s = \eta i = |\eta|(\cos\frac{\pi}{2} + i\sin(\pm\frac{\pi}{2}))$ into $s^{\alpha}+\bar{c}_2-\bar{b}e^{-s\tau} = 0$ which gives

$$|\eta|^{\alpha}\left(\cos\frac{\alpha\pi}{2} + i\sin(\pm\frac{\alpha\pi}{2})\right) + \bar{c}_2 - \bar{b}(\cos\eta\tau - i\sin\eta\tau) = 0. \quad (23)$$

From (23), we separate the real and imaginary parts

$$|\eta|^{\alpha}\cos\frac{\alpha\pi}{2} + \bar{c}_2 = \bar{b}\cos\eta\tau \quad (24)$$

and

$$|\eta|^{\alpha}\sin(\pm\frac{\alpha\pi}{2}) = -\bar{b}\sin\eta\tau. \quad (25)$$

Squaring and adding equations (24) and (25), one can obtain

$$|\eta|^{2\alpha} + 2|\eta|^{\alpha}\left(\bar{c}_2\cos\frac{\alpha\pi}{2}\right) + \bar{c}_2^2 - \bar{b}^2 = 0. \quad (26)$$

Obviously, when $0 < \alpha < 1$ and $\bar{b}^2 - \bar{c}_2^2\sin^2\frac{\alpha\pi}{2} < 0$, the above equation (26) has no real solutions, i.e., $\det(\Delta(s)) = 0$ has no pure imaginary roots for any $\tau > 0$. Similarly, if the $s^{\alpha}+\bar{c}_1-\bar{b}_1e^{-s\tau} = 0$, we have $\bar{b}_1^2 - \bar{c}_1^2\sin^2\frac{\alpha\pi}{2} < 0$. If $\Phi \neq 0$ then from (22), we have $s^{\alpha}+\bar{c}_2-\bar{b}e^{-s\tau} = 0$ and $(s^{\alpha}+\bar{c}_1-\bar{b}_1e^{-s\tau})(s^{\alpha}+\bar{c}_2-\bar{b}e^{-s\tau})-\Phi \neq 0$. Clearly, we also derived $\bar{b}^2 - \bar{c}_2^2\sin^2\frac{\alpha\pi}{2} < 0$. Hence, the conditions (i) and (ii) of Theorem 4 is easily derived from the above results.

Moreover, prove that all the eigenvalues of $\mathscr{M}$ satisfy $|\arg(\lambda)| > \frac{\pi}{2}$. The coefficient matrix $\mathscr{M}$ of system (20) satisfies

$$\mathscr{M} = \begin{pmatrix} -\bar{c}_1 + \bar{b}_1 & \bar{a}_{12} & \bar{a}_{13} \cdots & \bar{a}_{1n} \\ \bar{a}_{21} & -\bar{c}_2 + \bar{b} & 0 \quad \cdots & 0 \\ \cdots & \cdots & \cdots \cdots & \cdots \\ \bar{a}_{n1} & 0 & 0 \quad \cdots & -\bar{c}_2 + \bar{b} \end{pmatrix}.$$

Choose $\bar{c}_2 - \bar{b} > 0$, $\bar{c}_1 + \bar{c}_2 - \bar{b}_1 - \bar{b} > 0$, $(\bar{c}_1 - \bar{b}_1)(\bar{c}_2 - \bar{b}) - \Phi > 0$, the eigenvalues of $\mathscr{M}$ have negative real parts, i.e., all the eigenvalues of $\mathscr{M}$ satisfy $|\arg(\lambda)| > \frac{\pi}{2}$. Thus, the proof of the Theorem 4 is completed. □

4 Numerical Example

In this section, a numerical example is given to show the effectiveness of our results.

Example 1 Consider the FNNs with hub structure and time delays drive system (16) and response system (17) with the following parameter values, such as $\alpha = 0.95$, $\tau = 0.03$, $c_1 = 2$, $c_2 = 10$, $c_3 = 1$, $c_4 = 2$, $a_{11} = 2$, $a_{22} = 1$, $a_{33} = 2$, $a_{44} = 5$, $a_{12} = -2$, $a_{21} = 0$, $a_{13} = 3$, $a_{31} = 0$, $a_{14} = 1$, $a_{41} = 0$, $b_1 = -5$, $b = -12$, $x(0) = (-0.5, 3.5, -0.3, 0.1)^T$ and $y(0) = (0.6, -2.0, 0.5, -0.3)^T$. The condition (i) of Theorem 4 is satisfied for given parameter values. Thus, hybrid projective synchronization between drive system (16) and response system (17) can be achieved with $\Gamma = \text{diag}(-4, -20, -10, -8)$. The convergence behavior of the error system (20) and the state trajectories of system (16) and (17) are shown in Fig. 1 with the scaling matrix $\beta = \text{diag}(2.5, 2.5, 2.5, 2.5)$.

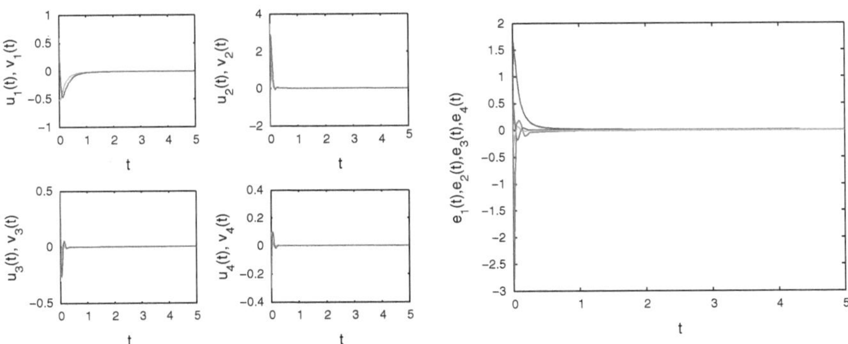

Fig. 1 Time responses and state trajectories of FNNs (16) and (17). The error state curves of the hybrid projective synchronization between (16) and (17) with $\alpha = 0.95$, $\Gamma = \text{diag}(-4, -20, -10, -8)$, $\tau = 0.03$ and scaling matrix $\beta = \text{diag}(2.5, 2.5, 2.5, 2.5)$

5 Conclusion

In this paper, the hybrid projective synchronization of fractional-order neural networks with time delays have been extensively studied. First, we considered the fractional-order neural networks with hub structure and time delays. Some new sufficient conditions for hybrid projective synchronization of fractional-order neural networks with hub structure and time delays have been derived by using linear feedback control, stability theorem of linear fractional order systems with multiple time delays and appropriate scaling matrix. In addition, the scaling matrix $\beta = I\,(\beta = -I)$ where I is the Identity matrix, then hybrid projective synchronization becomes complete synchronization (anti-phase synchronization) of the considered network. Finally, a numerical example is given to show the effectiveness of our main results.

Acknowledgments This work was supported by NBHM research project No. 2/48(7)/2012/NBHM(R.P.)/R and D-II/12669.

References

1. Podlubny, I.: Fractional Differential Equations. Academic Press, New York (1999)
2. Koeller, R.C.: Application of fractional calculus to the theory of viscoelasticity. J. Appl. Mech. **51**, 294–298 (1984)
3. Heaviside, O.: Electromagnetic Theory. Chelsea, New York (1971)
4. Petras, I.: A note on the fractional-order cellular neural networks. In: International Joint Conference on Neural Networks, pp. 1021–1024 (2006)
5. Li, Y., Chen, Y., Podlubny, I.: Mittag-Leffler stability of fractional order nonlinear dynamic systems. Automatica **45**, 1965–1969 (2009)
6. Deng, W., Li, C.: Lü, J.: Stability analysis of linear fractional differential system with multiple time delays. Nonlinear Dynam. **48**, 409–416 (2007)
7. Shen, J., Lam, J.: Non-existence of finite-time stable equilibria in fractional-order nonlinear systems. Automatica **50**, 547–551 (2014)
8. Carpenter, G.A.: Neural network models for pattern recognition and associative memory. Neural Netw. **2**, 243–257 (1989)
9. Cochocki, A., Rolf, U.: Neural Networks for Optimization and Signal Processing, 1st edn. Wiley, New York
10. Lundstrom, B., Higgs, M., Spain, W., Fairhall, A.: Fractional differentiation by neocortical pyramidal neurons. Nature Neurosci. **11**, 1335–1342 (2008)
11. Boroomand, A., Menhaj, M.: Fractional-order Hopfield neural networks. Lect. Notes Comput. Sci. **5506**, 883–890 (2009)
12. Chen, L., Chai, Y., Wu, R., Ma, T., Zhai, H.: Dynamic analysis fo a class of fractional-order neural networks with delay. Neurocomputing **111**, 190–194 (2013)
13. Chen, J., Zeng, Z., Jiang, P.: Global Mittag-Leffler stability and synchronization of memristor-based fractional-order neural networks. Neural Netw. **51**, 1–8 (2014)
14. Wang, H., Yu, Y., Wen, G.: Stability analysis of fractional-order Hopfield neural networks with time delays. Neural Netw. **55**, 98–109 (2014)
15. Perora, L.M., Carroll, T.L.: Synchronization in chaotic systems. Phys. Rev. Lett. **64**, 821–824 (1990)
16. Zhu, H., He, Z.S., Zhou, S.B.: Lag synchronization of the fractional-order system via nonlinear observer. Int. J. Mod. Phys. B **25**, 3951–3964 (2011)

17. Taghvafard, H., Erjaee, G.H.: Phase and anti-phase synchronization of fractional order chaotic systems via active control. Commun. Nonlinear Sci. Numer. Simul. **16**, 4079–4088 (2011)
18. Wang, X.Y., He, Y.J.: Projective synchronization of fractional order chaotic system based on linear separation. Phys. Lett. A **372**, 435–441 (2008)
19. Kuntanapreeda, S.: Robust synchronization of fractional-order unified chaotic systems via linear control. Comput. Math. Appl. **63**, 183–190 (2012)
20. Mainieri, R., Rehacek, J.: Projective synchronization in three-dimensional chaotic systems. Phys. Rev. Lett. **82**, 3024–3045 (1999)
21. Chee, C.Y., Xu, D.: Chaos-based M-nary digital communication technique using controller projective synchronization. IEE Proc. G (Circuits, Devices and Systems) **153**, 357–360 (2006)
22. Wang, S., Yu, Y., Wen, G.: Hybrid projective synchronization of time-delayed fractional-order chaotic systems. Nonlinear Anal. Hybrid Syst. **11**, 129–138 (2014)
23. Yu, J., Hu, C., Jiang, H., Fan, X.: Projective synchronization for fractional neural networks. Neural Netw. **49**, 87–95 (2014)
24. Wang, S., Yu, Y.G., Diao, M.: Hybrid projective synchronization of chaotic fractional order systems with different dimensions. Phys. A **389**, 4981–4988 (2010)
25. Zhou, P., Zhu, W.: Function projective synchronization for fractional-order chaotic systems. Nonlinear Anal. Real World Appl. **12**, 811–16 (2011)
26. Wang, X.Y., Zhang, X.P., Ma, C.: Modified projective synchronization of fractional-order chaotic systems via active sliding mode control. Nonlinear Dynam. **69**, 511–17 (2012)
27. Bhalekar, S., Daftardar-Gejji, V.: A predictor-corrector scheme for solving nonlinear delay differential equations of fractional order. J. Fract. Calc. Appl. **1**, 1–8 (2011)

Approximations of Solutions of a Class of Neutral Differential Equations with a Deviated Argument

Pradeep Kumar, Dwijendra N. Pandey and D. Bahuguna

Abstract In this paper, we study the approximations of solutions to a class of nonlinear neutral differential equations with a deviated argument in a Hilbert space. We consider an associated integral equation corresponding to the given problem and a sequence of approximate integral equations. We establish the existence and uniqueness of solutions to every approximate integral equation using the fixed point theory. Then, we prove the convergence of the solutions of the approximate integral equations to the solution of the associated integral equation. Next, we consider the Faedo–Galerkin approximations of solutions and prove some convergence results.

Keywords Abstract differential equation with a deviated argument · Banach fixed point theorem · Analytic semigroup · Faedo–Galerkin approximations

MSC: 34K30 · 34G20 · 47H06

1 Introduction

In the present study, we are concerned with the approximations of solutions to the following class of neutral differential equation with a deviated argument in a separable Hilbert space $(H, \|.\|, (., .))$:

P. Kumar (✉) · D. Bahuguna
Department of Mathematics and Statistics, Indian Institute of Technology Kanpur, Kanpur 208016, India
e-mail: prdipk@gmail.com

D. Bahuguna
e-mail: dwij.iitk@gmail.com

D.N. Pandey
Department of Mathematics, Indian Institute of Technology Roorkee, Roorkee 247667, India
e-mail: dhiren@iitk.ac.in

P.N. Agrawal et al. (eds.), *Mathematical Analysis and its Applications*, Springer Proceedings in Mathematics & Statistics 143,
DOI 10.1007/978-81-322-2485-3_54

$$\frac{d}{dt}[u(t)+g(t,u(t))]+A[u(t)+g(t,u(t))]=f(t,u(t),u(h(u(t),t))),$$
$$u(0)=u_0, 0<t\leq T<\infty. \quad (1)$$

where $A : D(A) \subset H \to H$ is a close, densely defined, positive definite, self-adjoint linear operator. Functions $f,\ g$, and h are suitably defined satisfying certain conditions to be stated later.

Haloi et al. [1] studied the following neutral differential equation with a deviated argument

$$\frac{d}{dt}[u(t)+g(t,u(a(t)))]+A(t)[u(t)+g(t,u(a(t)))]=f(t,u(t),u([h(u(t),t)])),\quad t>0,$$
$$u(0)=u_0. \quad (2)$$

$A(t)$, for each $t \geq 0$, generates an analytic semigroup of bounded linear operators on X. The nonlinear functions $f,\ g$, and h satisfy suitable growth conditions in their arguments and $a : [0,T] \to [0,T]$ satisfies the delay property. In this paper, we study Eq. (2) for an autonomous case, i.e., $A(t) = A$.

Ezzinibi et al. [2] have established the existence and stability of solution of the following nonlinear partial neutral functional differential equations with infinite delay:

$$\frac{d}{dt}[u(t)-g(t,u_t)]=A[u(t)-g(t,u_t)]+f(t,u_t),\quad t\geq 0,$$
$$u_0=\varphi\in C_0,$$

where the operator A is the Hille–Yosida operator not necessarily densely defined on the Banach space B. The functions g and f are continuous from $[0,\infty)\times C_0$ into B.

In the present work we are interested in the Faedo–Galerkin approximations of solutions to (1). This technique basically uses the idea of finite-dimensional projections of solutions which gives rise to a sequence of approximate solutions. These approximate solutions are then required to be shown to converge to the solution of the problem under consideration. Initially, the Faedo–Galerkin approximations of solutions to the particular case of (1) where $g, h \equiv 0$ and $f(t,u) = M(u)$ has been considered by Milleta [3]. For a nice introduction and related study of various problems in this direction, we refer to (see [4–10]) and reference cited therein.

For the earlier works on existence and uniqueness of solutions to the differential equations with deviating arguments, we refer to (see [1, 11–15]) and references cited therein.

The plan of the paper is as follows: In the second section, we provide some of the notations, notions, and results required for later sections. In the third section, we consider an integral equation associated with (1) and then consider a sequence of approximate integral equations and establish the existence and uniqueness of a solution to each of the approximate integral equation. Also, we prove the convergence of the solutions of the approximate integral equations and show that the limiting function satisfies the associated integral equation. In the fourth section, we consider the Faedo–Galerkin approximations of solutions and prove some convergence results for such approximations. In the last section, we have given an example to show some of the applications of the results obtained in the earlier sections.

2 Preliminaries and Assumptions

In this section, we shall provide the assumptions, notations, notions, and related results needed for the subsequent sections. We assume that the operator A satisfies the following:

(H1) A is a closed, positive definite, self-adjoint, linear operator from the domain $D(A) \subset H$ of A into H such that $D(A)$ is dense in H. Also we assume that A has the pure point spectrum

$$0 < \lambda_0 \leq \lambda_1 \leq \lambda_2 \leq \cdots, \lambda_m \leq \cdots,$$

where $\lambda_m \to \infty$ as $m \to \infty$ and a corresponding complete orthonormal system of eigenfunctions $\{u_i\}$, i.e., $Au_i = \lambda_i u_i$ and $(u_i, u_j) = \delta_{ij}$, where $\delta_{ij} = 1$ if $i = j$ and zero otherwise.

These assumptions on A guarantee that $-A$ generates an analytic semigroup, denoted by $S(t), t \geq 0$.

We mention some notions and preliminaries essential for our purpose. It is well known that there exist constants $\tilde{M} \geq 1$ and $\omega \geq 0$ such that

$$\|S(t)\| \leq \tilde{M} e^{\omega t}, \quad t \geq 0.$$

Without loss of generality we may assume that $\|S(t)\|$ is uniformly bounded by M, i.e., $\|S(t)\| \leq M$ and $0 \in \rho(-A)$. In this case it is possible to define the fractional power A^α for $0 \leq \alpha \leq 1$ as closed linear operator with domain $D(A^\alpha) \subseteq H$ (cf. Pazy [16], pp. 69–75 and p. 195). Furthermore, $D(A^\alpha)$ is dense in H and the expression

$$\|x\|_\alpha = \|A^\alpha x\|,$$

defines a norm on $D(A^\alpha)$. Henceforth, we represent by H_α the space $D(A^\alpha)$ endowed with the norm $\|\cdot\|_\alpha$. Also, for each $\alpha > 0$, we define $H_{-\alpha} = (H_\alpha)^*$, the dual space of H_α is a Banach space endowed with the norm $\|x\|_{-\alpha} = \|A^{-\alpha} x\|$.

Lemma 1 *[16] Suppose that $-A$ is the infinitesimal generator of an analytic semigroup $S(t)$, $t \geq 0$ with $\|S(t)\| \leq M$ for $t \geq 0$ and $0 \in \rho(-A)$. Then we have the following properties:*

(i) H_α is a Banach space for $0 \leq \alpha \leq 1$.
(ii) For $0 < \delta \leq \alpha < 1$, the embedding $H_\alpha \hookrightarrow H_\delta$ is continuous.

(iii) A^{α} *commutes with* $S(t)$ *and there exists a constant* $C_{\alpha} > 0$ *depending on* $0 \leq \alpha \leq 1$ *such that*

$$\|A^{\alpha} S(t)\| \leq C_{\alpha} t^{-\alpha}, \quad t > 0.$$

It can be seen easily that $\mathscr{C}_t^{\alpha} = C([0, t]; H_{\alpha})$, for all $t \in [0, T]$, is a Banach space endowed with the supremum norm,

$$\|\psi\|_{t,\alpha} := \sup_{0 \leq \eta \leq t} \|\psi(\eta)\|_{\alpha}, \quad \psi \in \mathscr{C}_t^{\alpha}.$$

We set,

$$\mathscr{C}_T^{\alpha-1} = C([0, T]; H_{\alpha-1}) = \{y \in \mathscr{C}_T^{\alpha} : \|y(t) - y(s)\|_{\alpha-1} \leq L|t - s|, \forall\ t, s \in [0, T]\},$$

where L is a suitable positive constant to be specified later.

We assume the following conditions:

(H2): Let $U_1 \subset$ Dom(f) is an open subset of $\mathbb{R}_+ \times H_{\alpha} \times H_{\alpha-1}$ and for each $(t, u, v) \in U_1$ there is a neighborhood $V_1 \subset U_1$ of (t, u, v). The nonlinear map $f : \mathbb{R}_+ \times H_{\alpha} \times H_{\alpha-1} \to H$ satisfies the following condition:

$$\|f(t, x, \psi) - f(s, y, \tilde{\psi})\| \leq L_f[|t - s|^{\theta_1} + \|x - y\|_{\alpha} + \|\psi - \tilde{\psi}\|_{\alpha-1}],$$

where $0 < \theta_1 \leq 1$, $0 \leq \alpha < 1$, $L_f > 0$ is a constant, $(t, x, \psi), (s, y, \tilde{\psi}) \in V_1$.

(H3): Let $U_2 \subset$ Dom(h) is an open subset of $H_{\alpha} \times \mathbb{R}_+$ and for each $(x, t) \in U_2$ there is a neighborhood $V_2 \subset U_2$ of (x, t). The map $h : H_{\alpha} \times \mathbb{R}_+ \to \mathbb{R}_+$ satisfies the following condition:

$$|h(x, t) - h(y, s)| \leq L_h[\|x - y\|_{\alpha} + |t - s|^{\theta_2}],$$

where $0 < \theta_2 \leq 1$, $0 \leq \alpha < 1$, $L_h > 0$ is a constant, $(x, t), \ (y, s) \in V_2$ and $h(., 0) = 0$.

(H4): Let $U_3 \subset$ Dom(g) is an open subset of $[0, T] \times H_{\alpha-1}$ and for each $(t, x) \in U_3$ there is a neighborhood $V_3 \subset U_3$ of (x, t). There exist positive constants $0 < \alpha < \beta < 1$, such that the function $A^{\beta} g$ is continuous for $(t, u) \in [0, T_0] \times H_{\alpha-1}$ such that

$$\|A^{\beta} g(t, x) - A^{\beta} g(s, y)\| \leq L_g\{|t - s| + \|x - y\|_{\alpha-1}\}, \text{ and}$$
$$4L_g\|A^{\alpha-\beta-1}\| = \eta < 1$$

where $L_g, \eta > 0$ are positive constants and $(x, t), \ (y, s) \in V_3$.

3 Approximate Solutions and Convergence

The existence of a solution to (1) is closely related to the following integral equation (3):

Definition 1 A continuous function $u : [0, T] \to H$ is said to be a mild solution of Eq. (1) if u is the solution of the following integral equation:

$$u(t) = S(t)[u(0) + g(0, u_0)] - g(t, u(t)) + \int_0^t S(t-s)f(s, u(s), u(h(u(s), s)))ds, \ t \in [0, T] \tag{3}$$

and satisfies the initial condition $u(0) = u_0$.

Definition 2 By a solution of the problem (1), we mean a function $u : [0, T] \to H$ satisfying the following three conditions:

(i) $u(.) + g(., u(.)) \in \mathscr{C}_T^{\alpha-1} \cap C^1((0, T), H) \cap C([0, T], H)$,
(ii) $u(t) \in D(A)$, and $(t, u(t), u(h(u(t), t))) \in U_1$,
(iii) $\frac{d}{dt}[u(t) + g(t, u(t))] + A[u(t) + g(t, u(t))] = f(t, u(t), u(h(u(t), t)))$, for all $t \in (0, T]$,
(iv) $u(0) = u_0$.

Let $H_n \subseteq H$ denote the finite-dimensional subspace spanned by $\{u_0, u_1, \ldots, u_n\}$ and let $P^n : H \longrightarrow H_n$ be the corresponding projection operator for $n = 0, 1, 2, \ldots$. We define

$$g_n : \mathbb{R}_+ \times H \longrightarrow H \ \text{ as } \ g_n(t, u(t)) = g(t, P^n u(t)). \tag{4}$$

Also, we define

$$f_n : \mathbb{R}_+ \times H \times H \longrightarrow H$$

given by

$$f_n(s, u(s), u(h(u(s), s))) = f(s, P^n u(s), P^n u[h_n(u(s), s)]). \tag{5}$$

For a fixed $R > 0$, we choose $0 < T_0 = T_0(\alpha, \beta, u_0) \leq T$ such that

$$C_\alpha L_f[2 + LL_h]\frac{T_0^{1-\alpha}}{1-\alpha} \leq 1 - \eta \tag{6}$$

where $\eta = 4L_g\|A^{\alpha-\beta-1}\| < 1$,

$$T_0 \leq \left(\frac{R}{4}(1-\alpha)(C_\alpha[2 + LL_h]L_f)^{-1}\right)^{\frac{1}{1-\alpha}},$$

and satisfying the following

$$\|(S(t) - I)A^\alpha[u_0 + g_n(0, u_0)]\| + \|A^{\alpha-\beta}\|L_g[T_0 + R] \leq \frac{R}{2}, \tag{7}$$

for all $t \in [0, T_0]$.

$$C_\alpha N \frac{T_0^{1-\alpha}}{1-\alpha} \leq \frac{R}{2}. \tag{8}$$

For more details of choosing such a T_0, we refer Theorem 2.2 of [15].

We set

$$\mathscr{W} = \{u \in \mathscr{C}_{T_0}^{\alpha} \cap \mathscr{C}_{T_0}^{\alpha-1} : u(0) = u_0, \quad \|u - u_0\|_{T_0,\alpha} \leq R\}.$$

Clearly, $\mathscr{W}$ is a closed and bounded subset of $\mathscr{C}_{T_0}^{\alpha-1}$.

We define a map $\mathscr{F}_n : \mathscr{W} \to \mathscr{W}$ given by

$$\begin{aligned}(\mathscr{F}_n u)(t) = \; & S(t)[u_0 + g_n(0, u_0)] - g_n(t, u(t)) + \int_0^t S(t-s) \\ & \times f_n(s, u(s), u(h(u(s), s)))ds, t \in [0, T_0].\end{aligned}$$

Theorem 1 *Let us assume that the assumptions* (H1)–(H4) *are satisfied and* $u_0 \in D(A^\alpha)$ *for* $0 \leq \alpha < 1$. *Then there exists a unique* $u_n \in \mathscr{C}_{T_0}^{\alpha-1} \cap \mathscr{C}_{T_0}^{\alpha}$ *such that* $\mathscr{F}_n u_n = u_n$ *for each* $n = 0, 1, 2, \ldots$, *i.e.,* u_n *satisfies the approximate integral equation*

$$\begin{aligned}u_n(t) = \; & S(t)[u_0 + g_n(0, u_0)] - g_n(t, u(t)) \\ & + \int_0^t S(t-s) f_n(s, u(s), u(h(u(s), s)))ds, \quad t \in [0, T_0].\end{aligned}$$

Proof In order to prove this theorem first we need to show that $\mathscr{F}_n u \in \mathscr{C}_{T_0}^{\alpha-1}$ for any $u \in \mathscr{C}_{T_0}^{\alpha-1}$. Clearly, $\mathscr{F}_n : \mathscr{C}_{T_0}^{\alpha} \to \mathscr{C}_{T_0}^{\alpha}$.

If $u \in \mathscr{C}_{T_0}^{\alpha-1}$, $\quad T_0 > t_2 > t_1 > 0$, and $0 \leq \alpha < 1$, then we get

$$\begin{aligned}&\|(\mathscr{F}_n u)(t_2) - (\mathscr{F}_n u)(t_1)\|_{\alpha-1} \\ &\quad \leq \|(S(t_2) - S(t_1))(u_0 + g_n(0, u_0))\|_{\alpha-1} \\ &\qquad + \|A^{\alpha-1-\beta}\| \|A^\beta g_n(t_2, u(t_2)) - A^\beta g_n(t_1, u(t_1))\| \\ &\qquad + \int_0^{t_1} \|(S(t_2 - s) - S(t_1 - s))A^{\alpha-1}\| \|f_n(s, u(s), u(h(u(s), s)))\| ds \\ &\qquad + \int_{t_1}^{t_2} \|S(t_2 - s)A^{\alpha-1}\| \|f_n(s, u(s), u(h(u(s), s)))\| ds.\end{aligned} \tag{9}$$

The continuity of $u \in \mathscr{C}_{T_0}^{\alpha}$ and the assumptions (H2)–(H3) will imply that the function $f(s, u(s), u(h(u(s), s)))$ is continuous on $[0, T_0]$. Therefore, we can easily show that there exists a positive constant N such that

$$\|f(s, u(s), u(h(u(s), s)))\| \leq N = L_f[T_0^{\theta_1} + R(1 + LL_h) + LL_h T_0^{\theta_2}] + N_0,$$

where $N_0 = \|f(0, u_0, u_0)\|$.

Similarly, we have

$$\|A^{\beta} g(t, u(t))\| \leq L_g[T_0 + R] + \|g(0, u_0)\|_{\beta} = N_1.$$

For the first part of right-hand side of (9), we have,

$$\begin{aligned}
&\|(S(t_2) - S(t_1))(u_0 + g_n(0, u_0))\|_{\alpha-1} \\
&\qquad \leq \int_{t_1}^{t_2} \|A^{\alpha-1} S'(s)(u_0 + g_n(0, u_0))\| ds \\
&\qquad = \int_{t_1}^{t_2} \|A^{\alpha} S(s)(u_0 + g_n(0, u_0))\| ds \\
&\qquad \leq \int_{t_1}^{t_2} \|S(s)\| [\|u_0\|_{\alpha} + \|A^{\alpha-\beta}\| \|g_n(0, u_0)\|_{\beta}] ds \\
&\qquad \leq C_1(t_2 - t_1),
\end{aligned} \tag{10}$$

where $C_1 = [\|u_0\|_{\alpha} + \|A^{\alpha-\beta}\| \|g_n(0, u_0)\|_{\beta}] M$.

For the second part of right-hand side of (9), we can see that

$$\begin{aligned}
&\|A^{\alpha-\beta-1}\| \|A^{\beta} g_n(t_2, u(t_2)) - A^{\beta} g_n(t_1, u(t_1))\| \\
&\leq \|A^{\alpha-\beta-1}\| L_g[(t_2 - t_1) + \|u(t_2) - u(t_1)\|_{\alpha-1}] \\
&\leq \|A^{\alpha-\beta-1}\| [L_g(1 + L)](t_2 - t_1) \\
&\leq C_2(t_2 - t_1).
\end{aligned} \tag{11}$$

where $C_2 = \|A^{\alpha-\beta-1}\| [L_g(1 + L)]$. To handle the third part of the right-hand side of (9), observe that,

$$\begin{aligned}
\|(S(t_2 - s) - S(t_1 - s))\|_{\alpha-1} &\leq \int_0^{t_2-t_1} \|A^{\alpha-1} S'(l) S(t_1 - s)\| dl \\
&\leq \int_0^{t_2-t_1} \|S(l) A^{\alpha} S(t_1 - s)\| dl \\
&\leq M C_{\alpha}(t_2 - t_1)(t_1 - s)^{-\alpha}.
\end{aligned} \tag{12}$$

Now we use the inequality (12) to get the bound for third part as given below,

$$\int_0^{t_1} \|(S(t_2 - s) - S(t_1 - s))A^{\alpha-1}\| \|f_n(s, u(s), u(h(u(s), s)))\| ds \\ \leq C_3(t_2 - t_1), \tag{13}$$

where $C_3 = NMC_\alpha \frac{T_0^{1-\alpha}}{1-\alpha}$.

For the bound for fourth part, we have,

$$\int_{t_1}^{t_2} \|S(t_2 - s)A^{\alpha-1}\| \|f_n(s, u(s), u(h(u(s), s)))\| ds \leq C_4(t_2 - t_1), \tag{14}$$

where $C_4 = \|A^{\alpha-1}\| MN$.

We use the inequalities (10), (11), (13)–(14) in inequality (9) to get the following inequality:

$$\|(\mathscr{F}_n u)(t_2) - (\mathscr{F}_n u)(t_1)\|_{\alpha-1} \leq L|t_2 - t_1|,$$

where, $L = \max\{C_i,\ i = 1, 2, \ldots 4\}$. Hence, $\mathscr{F}_n : \mathscr{C}_{T_0}^{\alpha-1} \to \mathscr{C}_{T_0}^{\alpha-1}$ follows.

Our next task is to show that $\mathscr{F}_n : \mathscr{W} \to \mathscr{W}$. Now, for $t \in [0, T_0]$ and $u \in \mathscr{W}$, we have

$$\begin{aligned}
&\|(\mathscr{F}_n u)(t) - u_0\|_\alpha \\
&\leq \|(S(t) - I)A^\alpha[u_0 + g_n(0, u_0)]\| \\
&\quad + \|A^{\alpha-\beta}\| \|A^\beta g_n(s, u(s))) - A^\beta g_n(0, u(0))\| \\
&\quad + \int_0^t \|S(t - s)A^\alpha\| \|f_n(s, u(s), u(h(u(s), s))]) \| ds \\
&\leq \|(S(t) - I)A^\alpha[u_0 + g_n(0, u_0)]\| + \|A^{\alpha-\beta}\| L_g[T_0 + R] \\
&\quad + C_\alpha N \frac{T_0^{1-\alpha}}{1 - \alpha}.
\end{aligned}$$

Hence, from inequalities (7) and (8), we get

$$\|\mathscr{F}_n u - u_0\|_{T_0,\alpha} \leq R.$$

Therefore, $\mathscr{F}_n : \mathscr{W} \to \mathscr{W}$.

Now, if $t \in [0, T_0]$ and $u, v \in \mathscr{W}$, then

$$\begin{aligned}
&\|(\mathscr{F}_n u)(t) - (\mathscr{F}_n v)(t)\|_\alpha \\
&\le \|A^{\alpha-\beta}\| \|A^\beta g_n(t, u(s)) - A^\beta g_n(t, v(s))\| \\
&\quad + \int_0^t \|S(t-s)A^\alpha\| \| f_n(s, u(s), u(h[u(s), s])) \\
&\qquad\qquad - f_n(s, v(s), v(h[v(s), s])))\| ds. \qquad (15)
\end{aligned}$$

We have the following inequalities,

$$\|A^\beta g_n(s, u(s))) - A^\beta g_n(t, v(t))\| \le L_g \|A^{-1}\| \|u - v\|_{T_0,\alpha}, \qquad (16)$$

$$\begin{aligned}
&\| f_n(s, u(s), u(h(u(s), s))) - f_n(s, v(s), v[h(v(s), s)])\| \\
&\qquad \le L_f[2 + LL_h]\|u - v\|_{T_0,\alpha}. \qquad (17)
\end{aligned}$$

We use the inequalities (16) and (17) in the inequality (15) and get

$$\begin{aligned}
\|(\mathscr{F}_n u)(t) - (\mathscr{F}_n v)(t)\|_\alpha \le &\left[\left(L_g\|A^{\alpha-\beta-1}\| + C_{1+\alpha-\beta} L_g \frac{T_0^{\beta-\alpha}}{\beta-\alpha}\right)\right. \\
&\left.+ C_\alpha L_f[2 + LL_h]\frac{T_0^{1-\alpha}}{1-\alpha}\right)\Bigg] \|u - v\|_{T_0,\alpha}.
\end{aligned}$$

Hence from inequality (6), we get the following inequality given below

$$\|\mathscr{F}_n u - \mathscr{F}_n v\|_{T_0,\alpha} < \|u - v\|_{T_0,\alpha}.$$

Therefore, the map $\mathscr{F}_n$ has a unique fixed point $u_n \in \mathscr{W}$ which is given by,

$$\begin{aligned}
u_n(t) = &S(t)[u_0 + g_n(0, u_0)] - g_n(t, u(t)) \\
&+ \int_0^t S(t-s) f_n(s, u(s), u(h(u(s), s)))ds \quad t \in [0, T_0]. \qquad (18)
\end{aligned}$$

Hence, the mild solution u_n of Eq. (1) is given by the Eq. (18) and belongs to $\mathscr{W}$, hence, the theorem is proved.

Lemma 2 *Let* (H1)–(H4) *hold. If* $u_0 \in D(A^\alpha)$ *then* $u_n(t) \in D(A^\vartheta)$, *for all* $t \in (0, T]$ *where* $0 \le \vartheta \le \beta < 1$. *Furthermore, if* $u_0 \in D(A)$ *then* $u_n(t) \in D(A^\vartheta)$, *for all* $t \in [0, T]$ *where* $0 \le \vartheta \le \beta < 1$.

Proof From Theorem 1, we have the existence of a unique $u_n \in \mathscr{C}^{\alpha}_{T_0} \cap \mathscr{C}^{\alpha-1}_{T_0}$ satisfying (18). Part (a) of Theorem 2.6.13 in Pazy [16] implies that for $t > 0$ and $0 \leq \vartheta < 1$, $S(t) : H \to D(A^{\vartheta})$ and for $0 \leq \vartheta \leq \beta < 1$, $D(A^{\beta}) \subseteq D(A^{\vartheta})$. (H2)–(H4) implies that the map $t \mapsto A^{\beta} g(t, u_n(t))$ is Hölder continuous on $[0, T]$ with the exponent $\rho = \min\{\gamma, \vartheta\}$ since the Hölder continuity of u_n can be easily established using the similar arguments from (9) to (13). Also from Theorem 1.2.4 in Pazy [16], we have $S(t)x \in D(A)$ if $x \in D(A)$. The required result follows from these facts and the fact that $D(A) \subseteq D(A^{\vartheta})$ for $0 \leq \vartheta \leq 1$.

Lemma 3 *Let (H1) and (H2) hold. If $u_0 \in D(A^{\alpha})$ and $t_0 \in (0, T_0]$ then*

$$\|u_n(t)\|_{\vartheta} \leq U_{t_0}, \quad \alpha < \vartheta < \beta, \quad t \in [t_0, T_0], \quad n = 1, 2, \ldots,$$

for some constant U_{t_0}, dependent of t_0 and

$$\|u_n(t)\|_{\vartheta} \leq U_0, \quad 0 < \vartheta \leq \alpha, \quad t \in (0, T_0], \quad n = 1, 2, \ldots,$$

for some constant U_0. Moreover, if $u_0 \in D(A)$, then there exists a constant U_0, such that

$$\|u_n(t)\|_{\vartheta} \leq U_0, \quad 0 < \vartheta < \beta, \quad t \in [0, T_0], \quad n = 1, 2, \ldots.$$

Proof First, we assume that $u_0 \in D(A^{\alpha})$. Applying A^{ϑ} on both the sides of (18) and using (iii) of Lemma 1, for $t \in [t_0, T]$ and $\alpha < \vartheta < \beta$, we have

$$\begin{aligned}
\|u_n(t)\|_{\vartheta} \leq & \|A^{\vartheta} S(t)(u_0 + g_n(0, u_0)\| + \|A^{\vartheta-\beta}\| \, \|A^{\beta} g_n(t, u_n(t))\| \\
& + \int_0^t \|S(t-s)A^{\vartheta}\| \, \|f_n(s, u_n(s), u_n[h(u_n(s), s)])\| ds \\
\leq & C_{\vartheta} t_0^{-\vartheta}(\|u_0\| + \|g_n(0, u_0\|) + \|A^{\vartheta-\beta}\| N_1 \\
& + C_{\vartheta} N \frac{T^{1-\vartheta}}{1-\vartheta} \leq U_{t_0}.
\end{aligned}$$

Again, for $t \in (0, T_0]$ and $0 < \vartheta \leq \alpha$, $u_0 \in D(A^{\vartheta})$ and

$$\begin{aligned}
\|u_n(t)\|_{\vartheta} \leq & M(\|A^{\vartheta} u_0\| + \|g_n(0, \tilde{u_0}\|_{\vartheta}) + \|A^{\vartheta-\beta}\| N_1 \\
& + C_{\vartheta} N \frac{T^{1-\vartheta}}{1-\vartheta} \leq U_0.
\end{aligned}$$

Furthermore, If $u_0 \in D(A)$ then $u_0 \in D(A^{\vartheta})$ for $0 < \vartheta \leq 1$ and we can easily get the required estimate. This completes the proof of the proposition.

4 Convergence of Solutions

In this section, we establish the convergence of the solution $u_n \in X_\alpha(T)$ of the approximate integral equation (18) to a unique solution u of (3).

Theorem 2 *Let* (H1)–(H4) *hold. If* $u_0 \in D(A^\alpha)$, *then for any* $t_0 \in (0, T_0]$,

$$\lim_{m\to\infty} \sup_{\{n\geq m,\ t_0\leq t\leq T\}} \|u_n(t) - u_m(t)\|_\alpha = 0.$$

Proof Let $0 < \alpha < \vartheta < \beta$. For $n \geq m$, we have

$$\begin{aligned}
&\|f_n(t, u_n(t), u_n[h(u_n(t), t)]) - f_m(t, u_m(t), u_m[h(u_m(t), t)])\| \\
&\quad \leq \|f_n(t, u_n(t), u_n[h(u_n(t), t)]) - f_n(t, u_m(t), u_m[h(u_m(t), t)])\| \\
&\qquad + \|f_n(t, u_m(t), u_m[h(u_m(t), t)]) - f_m(t, u_m(t), u_m[h(u_m(t), t)])\| \\
&\quad \leq L_f(2 + LL_h)\|u_n(t) - u_m(t)\|_\alpha + L_f[\|(P^n - P^m)u_m(t)\|_\alpha \\
&\qquad + \|A^{-1}\|\|(P^n - P^m)u_m(h(u_m(t), t))\|_\alpha].
\end{aligned}$$

Also,

$$\|(P^n - P^m)u_m(t)\|_\alpha \leq \|A^{\alpha-\vartheta}(P^n - P^m)A^\vartheta u_m(t)\| \leq \frac{1}{\lambda_m^{\vartheta-\alpha}}\|A^\vartheta u_m(t)\|.$$

Thus, we have

$$\begin{aligned}
&\|f_n(t, u_n(t), u_n[h(u_n(t), t)]) - f_m(t, u_m(t), u_m[h(u_m(t), t)])\| \\
&\leq L_f(2 + LL_h)\|u_n(t) - u_m(t)\|_\alpha + L_f\Big[\frac{1}{\lambda_m^{\vartheta-\alpha}}\|A^\vartheta u_m(t)\| \\
&\qquad + \frac{\|A^{-1}\|}{\lambda_m^{\vartheta-\alpha}}\|A^\vartheta u_m(h(u_m(t), t))\|\Big].
\end{aligned}$$

Similarly,

$$\begin{aligned}
&\|A^\beta g_n(t, u_n(t)) - A^\beta g_m(t, u_m(t))\| \\
&\leq \|A^\beta g_n(t, u_n(t)) - A^\beta g_n(t, u_m(t))\| + \|A^\beta g_n(t, u_m(t)) - A^\beta g_m(t, u_m(t))\| \\
&\leq L_g\|A^{-1}\|\left[\|u_n(t) - u_m(t)\|_\alpha + \frac{1}{\lambda_m^{\vartheta-\alpha}}\|A^\vartheta u_m(t)\|\right].
\end{aligned}$$

Now, for $0 < t_0' < t_0$, we may write

$$
\begin{aligned}
&\|u_n(t) - u_m(t)\|_\alpha \\
&\leq \|S(t)A^\alpha(g_n(0, u_0 - g_m(0, u_0)\| + \|A^{\alpha-\beta}\| \, \|A^\beta g_n(t, u_n(t)) - A^\beta g_m(t, u_m(t))\| \\
&\quad + \Big(\int_0^{t_0'} + \int_{t_0'}^{t}\Big)\|A^\alpha S(t-s)\| \, \|f_n(s, u_n(s), u_n[h(u_n(s), s)]) \\
&- f_m(s, u_m(s), u_m[h(u_m(s), s)])\|ds.
\end{aligned}
$$

We estimate the first term as

$$
\begin{aligned}
\|S(t)A^\alpha(g_n(0, u_0 - g_m(0, u_0)\| &\leq M\|A^{\alpha-\beta}\| \, \|A^\beta g(0, P^n u_0) - A^\beta g(0, P^m u_0)\| \\
&\leq M\|A^{\alpha-\beta-1}\|L_g\|(P^n - P^m)A^\alpha u_0\|.
\end{aligned}
$$

The first and the third integrals are estimated as

$$
\begin{aligned}
\int_0^{t_0'} \|A^\alpha S(t-s)\| \, \|f_n(s, u_n(s), u_n[h(u_n(s), s)]) - f_m(s, u_m(s), u_m[h(u_m(s), s)])\|ds \\
\leq 2C_\alpha N(t_0 - t_0')^{-\alpha} t_0'.
\end{aligned}
$$

For the second and the fourth integrals, we have

$$
\begin{aligned}
&\int_{t_0'}^{t} \|A^\alpha S(t-s)\| \, \|f_n(s, u_n(s), u_n[h(u_n(s), s)]) - f_m(s, u_m(s), u_m[h(u_m(s), s)])\|ds \\
&\leq C_\alpha L_f \int_{t_0'}^{t} (t-s)^{-\alpha}\Big[(2 + LL_h)\|u_n(s) - u_m(s)\|_\alpha + \frac{1}{\lambda_m^{\vartheta-\alpha}}\|A^\vartheta u_m(s) \\
&\quad + \frac{\|A^{-1}\|}{\lambda_m^{\vartheta-\alpha}}\|A^\vartheta u_m(h(u_m(t), t))\|\Big]ds \\
&\leq C_\alpha L_f\Big((1 + \|A^{-1}\|)\frac{U_{t_0'} {T_0}^{1-\alpha}}{\lambda_m^{\vartheta-\alpha}(1-\alpha)} + (2 + LL_h)\int_{t_0'}^{t}(t-s)^{-\alpha}\|u_n(s) - u_m(s)\|_\alpha ds\Big).
\end{aligned}
$$

Therefore,

$$
\begin{aligned}
\|u_n(t) - u_m(t)\|_\alpha \leq & M\|A^{\alpha-\beta-1}\|L_g\|(P^n - P^m)A^\alpha u_0\| \\
& + \|A^{\alpha-\beta-1}\|L_g\left(\|u_n(t) - u_m(t)\|_\alpha + \frac{U_{t_0'}}{\lambda_m^{\vartheta-\alpha}}\right) \\
& + 2\left(\frac{C_\alpha N}{(t_0 - t_0')^\alpha}\right)t_0' + D_\alpha\frac{U_{t_0'}}{\lambda_m^{\vartheta-\alpha}} \\
& + \int_{t_0'}^{t}\left(\frac{C_\alpha L_f(2 + LL_h)}{(t-s)^\alpha}\right)\|u_n(s) - u_m(s)\|_\alpha ds,
\end{aligned}
$$

where

$$D_\alpha = C_\alpha L_f(1 + \|A^{-1}\|)\frac{{T_0}^{1-\alpha}}{1-\alpha}.$$

Since $\|A^{\alpha-\beta-1}\|L_g < 1$, we have

$$\begin{aligned}\|u_n(t) - u_m(t)\|_\alpha \le & \frac{1}{(1 - \|A^{\alpha-\beta-1}\|L_g)}\Big\{M\|(P^n - P^m)A^\alpha u_0\| + \|A^{\alpha-\beta-1}\|L_g\frac{U_{t_0'}}{\lambda_m^{\vartheta-\alpha}} \\ & + 2\left(\frac{C_\alpha N}{(t_0 - t_0')^\alpha}\right)t_0' + D_\alpha\frac{U_{t_0'}}{\lambda_m^{\vartheta-\alpha}} \\ & + \int_{t_0'}^t\left(\frac{C_\alpha L_f(2 + LL_h)}{(t-s)^\alpha}\right)\|u_n(s) - u_m(s)\|_\alpha ds.\end{aligned}$$

Lemma 5.6.7 in [16] implies that there exists a constant C such that

$$\begin{aligned}&\|u_n(t) - u_m(t)\|_\alpha \\ &\le \frac{1}{(1 - \|A^{\alpha-\beta-1}\|L_g)}\Big\{M\|(P^n - P^m)A^\alpha u_0\| + (\|A^{\alpha-\beta-1}\|L_g + D_\alpha)\frac{U_{t_0'}}{\lambda_m^{\vartheta-\alpha}} \\ &+ 2\left(\frac{C_\alpha N}{(t_0 - t_0')^\alpha}\right)t_0'\Big\}C.\end{aligned}$$

Taking supremum over $[t_0, T]$ and letting $m \to \infty$, we obtain

$$\begin{aligned}&\lim_{m\to\infty}\sup_{\{n\ge m, t\in[t_0,T]\}}\|u_n(t) - u_m(t)\|_\alpha \\ &\le \frac{2}{(1 - \|A^{\alpha-\beta-1}\|L_g)}\left(\frac{C_\alpha N}{(t_0 - t_0')^\alpha}\right)C.\end{aligned}$$

As t_0' is arbitrary, the right-hand side may be made as small as desired by taking t_0' sufficiently small. This completes the proof of the proposition.

Corollary 1 *If* $u_0 \in D(A)$ *then*

$$\lim_{m\to\infty}\sup_{\{n\ge m,\ 0\le t\le T\}}\|u_n(t) - u_m(t)\|_\alpha = 0.$$

With the help of Theorems 1 and 2, the convergence of the solutions $u_n(t)$ of the approximate integral equations (18) follows from the next result.

Theorem 3 *Let* (H1)–(H4) *hold and let* $u_0 \in D(A^\alpha)$. *Then there exists a unique function* $u_n \in \mathscr{W}$

$$u_n(t) = S(t)[u(0) + g_n(0, u_0)] - g_n(t, u(t)) \\ + \int_0^t S(t-s) f_n(s, u_n(s), u_n[h_n(u_n(s), s)])ds, \ t \in [0, T_0]$$

and $u \in \mathscr{W}$

$$u(t) = S(t)[u(0) + g(0, u_0)] - g(t, u(t)) \\ + \int_0^t S(t-s) f(s, u(s), u(h(u(s), s)))ds, \ t \in [0, T_0]$$

such that $u_n \to u$ *as* $n \to \infty$ *in* $\mathscr{W}$ *and* u *satisfies* (3) *on* $[0, T_0]$.

5 Faedo–Galerkin Approximations

In this section, we will discuss the Faedo–Galerkin Approximations of solutions and prove the convergence results for such approximations.

For any $0 < t < T_0$, we have a unique $u \in \mathscr{W}$ satisfying the integral equation

$$u(t) = S(t)[u(0) + g(0, u_0)] - g(t, u(t)) \\ + \int_0^t S(t-s) f(s, u(s), u(h(u(s), s)))ds.$$

Then it has the representation

$$u(t) = \sum_{i=0}^{\infty} \alpha_i(t)\phi_i, \quad \alpha_i(t) = (u(t), \phi_i), \quad i = 0, 1, \ldots \tag{19}$$

where ϕ_i' s are defined in $(H1)$.

Also, we have a unique solution $u_n \in H_\alpha(T_0)$ from the approximate integral equations

$$u_n(t) = S(t)[u(0) + g_n(0, u_0)] - g_n(t, u_n(t)) \\ + \int_0^t S(t-s) f_n(s, u_n(s), u_n(h(u_n(s), s)))ds. \tag{20}$$

Let $P^n u_n(t) = \hat{u}_n(t)$ is the orthogonal projection of (20) on the first n elements of $\{\phi_i\}$ satisfying the following equations:

$$P^n u_n(t) = S(t)[P^n u(0) + P^n g_n(0, u_0)] - P^n g_n(t, u_n(t)) \\ + \int_0^t S(t-s) P^n f_n(s, u_n(s), u_n(h(u_n(s), s)))ds,$$

by using (4) and (5), we get the following

$$\hat{u}_n(t) = S(t)[P^n u(0) + P^n g(0, u_0)] - P^n g(t, \hat{u}_n(t)) + \int_0^t S(t-s)P^n f(s, \hat{u}_n(s), \hat{u}_n(h(\hat{u}_n(s), s)))d\theta ds. \quad (21)$$

The solution $\hat{u}_n$ of (21) has the following representation:

$$\hat{u}_n(t) = \sum_{i=0}^{n} \alpha_i^n(t)\phi_i, \quad \alpha_i^n(t) = (\hat{u}_n(t), \phi_i), \quad i = 0, 1, \ldots \quad (22)$$

Then we get a system of equations from (21) and (22)

$$\begin{aligned}&\frac{d^\beta}{dt^\beta}\left[\alpha_i^n(t) + H_i^n(t, \alpha_0^n, \alpha_1^n, \ldots, \alpha_n^n)\right] + \lambda_i[\alpha_i^n(t) + H_i^n(t, \alpha_0^n, \alpha_1^n, \ldots, \alpha_n^n)]\\ &\quad = F_i^n(t, \alpha_0^n, \alpha_1^n, \cdots, \alpha_n^n, \tau_0^n, \tau_1^n, \cdots, \tau_n^n)\\ &\alpha_i^n(0) = u_i,\end{aligned}$$

where

$$\begin{aligned}F_i^n &= \left(f\left(t, \sum_{i=0}^{n} \alpha_i^n \phi_i, \sum_{i=0}^{n} \tau_i^n \phi_i\right), \phi_i\right),\\ H_i^n &= \left(g\left(t, \sum_{i=0}^{n} \alpha_i^n \phi_i\right), \phi_i\right),\\ \tau_i^n &= \alpha_i^n\left(h\left(\alpha_0^n, \alpha_1^n, \cdots, \alpha_n^n, t\right)\right)\end{aligned}$$

and $u_i = (u_0, \phi_i)$ for $i = 1, 2, \cdots, n$.

Convergence of $\alpha_i^n(t) \to \alpha_i(t)$ follows from following theorem and the fact that:

$$A^\alpha[u(t) - \hat{u}_n(t)] = A^\alpha\left[\sum_{i=0}^{\infty}(\alpha_i(t) - \alpha_i^n(t))\phi_i\right] = \sum_{i=0}^{\infty} \lambda_i^\alpha(\alpha_i(t) - \alpha_i^n(t))\phi_i.$$

Thus, we have

$$\|A^\alpha[u(t) - \hat{u}_n(t)]\|^2 \geq \sum_{i=0}^{n} \lambda_i^{2\alpha}(\alpha_i(t) - \alpha_i^n(t))^2.$$

Theorem 4 *Let (H1) and (H2) hold. Then we have the following:*

(a) If $u_0 \in D(A^\alpha)$*, then for any* $0 < t_0 \leq T_0$*,*

$$\lim_{n\to\infty} \sup_{t_0\leq t\leq T_0} \left[\sum_{i=0}^{n} \lambda_i^{2\alpha}(\alpha_i(t) - \alpha_i^n(t))^2\right] = 0.$$

(b) If $u_0 \in D(A)$*, then*

$$\lim_{n\to\infty} \sup_{0\leq t\leq T_0} \left[\sum_{i=0}^{n} \lambda_i^{2\alpha}(\alpha_i(t) - \alpha_i^n(t))^2\right] = 0.$$

The assertion of this theorem follows from the facts mentioned above and the following result:

Proposition 1 *Let (H1) and (H2) hold and let* T *be any number such that* $0 < T < t_{\max}$*, then we have the following:*

(a) If $u_0 \in D(A^\alpha)$*, then for any* $0 < t_0 \leq T_0$*,*

$$\lim_{n\to\infty} \sup_{\{n\geq m, t_0\leq t\leq T_0\}} \|A^\alpha[\hat{u}_n(t) - \hat{u}_m(t)]\| = 0.$$

(b) If $u_0 \in D(A)$*, then*

$$\lim_{n\to\infty} \sup_{\{n\geq m, 0\leq t\leq T_0\}} \|A^\alpha[\hat{u}_n(t) - \hat{u}_m(t)]\| = 0.$$

Proof For $n \geq m$, we have

$$\begin{aligned}\|A^\alpha[\hat{u}_n(t) - \hat{u}_m(t)]\| &= \|A^\alpha[P^n u_n(t) - P^m u_m(t)]\| \\ &\leq \|P^n[u_n(t) - u_m(t)]\|_\alpha + \|(P^n - P^m)u_m\|_\alpha \\ &\leq \|u_n(t) - u_m(t)\|_\alpha + \frac{1}{\lambda_m^{\vartheta-\alpha}}\|A^\vartheta u_m\|.\end{aligned}$$

If $u_0 \in D(A^\alpha)$ then the result in (a) follows from Proposition 2. If $u_0 \in D(A)$, (b) follows from Corollary 1.

Theorem 5 *Let* (H1)–(H4) *hold and let* $u_0 \in D(A^\alpha)$*. Then there exists a unique function* $\hat{u}_n \in \mathscr{W}$

$$\begin{aligned}\hat{u}_n(t) &= S(t)[u(0) + g_n(0, u_0)] - g_n(t, \hat{u}_n(t)) \\ &\quad + \int_0^t S(t-s) f_n(s, \hat{u}_n(s), \hat{u}_n[h_n(\hat{u}_n(s), s)])ds, \; t \in [0, T_0]\end{aligned}$$

and $u \in \mathscr{W}$

$$
\begin{aligned}
u(t) &= S(t)[u(0) + g(0, u_0)] - g(t, u(t)) \\
&\quad + \int_0^t S(t-s) f(s, u(s), u(h(u(s), s))) ds, \ t \in [0, T_0]
\end{aligned}
$$

such that $\hat{u}_n \to u$ *as* $n \to \infty$ *in* $\mathscr{W}$ *and* u *satisfies (3) on* $[0, T_0]$.

6 Examples

Let $X = L^2(0, 1)$. We consider the following partial differential equations with a deviated argument,

$$
\begin{cases}
\partial_t [w(t, x) + \partial_x f_1(t, w(t, x))] - \partial_x^2 [w(t, x) + \partial_x f_1(t, w(t, x))] \\
\quad = f_2(x, w(t, x)), + f_3(t, x, w(t, x)), \quad x \in (0, 1), \ t > 0, \\
\quad w(t, 0) = w(t, 1) = 0, \ t \in [0, T], \ 0 < T < \infty, \\
\quad w(0, x) = u_0, \ x \in (0, 1),
\end{cases}
\tag{23}
$$

where

$$
f_2(x, w(t, x)) = \int_0^x K(x, s) w(s, h(t)(a_1|w(s, t)| + b_1|w_s(s, t)|)) ds.
$$

The function $f_3 : \mathbb{R}_+ \times [0, 1] \times \mathbb{R} \to \mathbb{R}$ is measurable in x, locally Hölder continuous in t, locally Lipschitz continuous in u and uniformly in x. Further, we assume that $a_1, b_1 \geq 0$, $(a_1, b_1) \neq (0, 0)$, $h : \mathbb{R}_+ \to \mathbb{R}_+$ is locally Hölder continuous in t with $h(0) = 0$ and $K : [0, 1] \times [0, 1] \to \mathbb{R}$.

We define an operator A as follows:

$$
Au = -u'' \quad \text{with} \quad u \in D(A) = \{u \in H_0^1(0, 1) \cap H^2(0, 1) : \ u'' \in X\}. \tag{24}
$$

Here clearly the operator A is self-adjoint with compact resolvent and is the infinitesimal generator of an analytic semigroup $S(t)$. Now we take $\alpha = 1/2$, $D(A^{1/2}) = H_0^1(0, 1)$ is the Banach space endowed with the norm,

$$
\|x\|_{1/2} := \|A^{1/2} x\|, \quad x \in D(A^{1/2})
$$

and we denote this space by $X_{1/2}$. Also, for $t \in [0, T]$, we denote

$$
C_t^{1/2} = C([0, t]; D(A^{1/2})),
$$

endowed with the sup norm

$$\|\psi\|_{t,1/2} := \sup_{0\leq\eta\leq t} \|\psi(\eta)\|_{\alpha}, \quad \psi \in \mathscr{C}_t^{1/2}.$$

We observe some properties of the operators A and $A^{1/2}$ defined by (24). For $u \in D(A)$ and $\lambda \in \mathbb{R}$, with $Au = -u'' = \lambda u$, we have $\langle Au, u\rangle = \langle \lambda u, u\rangle$; that is,

$$\langle -u'', u\rangle = |u'|_{L^2}^2 = \lambda|u|_{L^2}^2$$

so $\lambda > 0$. A solution u of $Au = \lambda u$ is of the form

$$u(x) = C\cos(\sqrt{\lambda}x) + D\sin(\sqrt{\lambda}x)$$

and the conditions $u(0) = u(1) = 0$ imply that $C = 0$ and $\lambda = \lambda_n = n^2\pi^2$, $n \in \mathbb{N}$. Thus, for each $n \in \mathbb{N}$, the corresponding solution is given by

$$u_n(x) = D\sin(\sqrt{\lambda_n}x).$$

We have $\langle u_n, u_m\rangle = 0$ for $n \neq m$ and $\langle u_n, u_n\rangle = 1$ and hence $D = \sqrt{2}$. For $u \in D(A)$, there exists a sequence of real numbers $\{\alpha_n\}$ such that

$$u(x) = \sum_{n\in\mathbb{N}} \alpha_n u_n(x), \quad \sum_{n\in\mathbb{N}} (\alpha_n)^2 < +\infty \quad \text{and} \quad \sum_{n\in\mathbb{N}} (\lambda_n)^2(\alpha_n)^2 < +\infty.$$

We have

$$A^{1/2}u(x) = \sum_{n\in\mathbb{N}} \sqrt{\lambda_n}\, \alpha_n\, u_n(x)$$

with $u \in D(A^{1/2})$; that is, $\sum_{n\in\mathbb{N}} \lambda_n(\alpha_n)^2 < +\infty$. $X_{-\frac{1}{2}} = H^1(0,1)$ is a Sobolev space of negative index with the equivalent norm $\|.\|_{-\frac{1}{2}} = \sum_{n=1}^{\infty} |\langle ., u_n\rangle|^2$. For more details on the Sobolev space of negative index, we refer to Gal [15].

The Eq. (23) can be reformulated as the following abstract equation in $X = L^2(0,1)$:

$$\frac{d}{dt}[u(t) + g(t, u(t))] + A[u(t) + g(t, u(t))] = f(t, u(t), u(h(u(t), t))) \quad t > 0,$$
$$u(0) = u_0, \tag{25}$$

where $u(t) = w(t, .)$ that is $u(t)(x) = w(t, x)$, $x \in (0, 1)$. The function $g : \mathbb{R}_+ \times X_{1/2} \to X$, such that $g(t, u(t))(x) = \partial_x f_1(t, w(t, x))$ and the operator A is same as in Eq. (24).

The function $f : \mathbb{R}_+ \times X_{1/2} \times X_{-1/2} \to X$, is given by

$$f(t, \psi, \xi)(x) = f_2(x, \xi) + f_3(t, x, \psi), \tag{26}$$

where $f_2 : [0, 1] \times X \to H_0^1(0, 1)$ is given by

$$f_2(t, \xi) = \int_0^x K(x, y)\xi(y)dy, \tag{27}$$

and $f_3 : \mathbb{R} \times [0, 1] \times H^2(0, 1) \to H_0^1(0, 1)$ satisfies the following:

$$\|f_3(t, x, \psi)\| \le Q(x, t)(1 + \|\psi\|_{H^2(0,1)}) \tag{28}$$

with $Q(., t) \in X$ and Q is continuous in its second argument. We can easily verify that the function f satisfied the assumptions (H1)–(H4). For more details see [15].

Acknowledgments We highly appreciate the valuable comments and suggestions of the referees on our manuscript, which helped to considerably improve the quality of the manuscript. The first author would like to thank Prof. Shobha Madan, HOD, Mathematics and Statistics, IIT Kanpur for providing financial support to attend the conference. Also, the first and third authors would like to acknowledge the financial aid from the Department of Science and Technology, New Delhi, under its research project SR/S4/MS:796/12.

References

1. Haloi, R., Pandey, D.N., Bahuguna, D.: Existence of solutions to a non-autonomous abstract neutral differential equation with deviated argument. J. Nonl. Evol. Equ. Appl. **5**, 75–90 (2011)
2. Adimy, M., Bouzahir, H., Ezzinbi, K.: Existence and stability for some partial neutral functional differential equations with infinite delay. J. Math. Anal. Appl. **294**, 438–461 (2004)
3. Miletta, P.D.: Approximation of solutions to evolution equations. Math. Meth. Appl. Sci. **17**, 753–763 (1994)
4. Bazley, N.W.: Approximation of wave equations with reproducing nonlinearities. Nonlinear Anal. **3**, 539–546 (1979)
5. Bazley, N.W.: Approximation of operators with reproducing nonlinearities. Manuscripta Math. **18**, 353–369 (1976)
6. Bahuguna, D., Muslim, M.: A study of nonlocal history-valued retarded differential equations using analytic semigroups. Nonlinear Dyn. Syst. Theory **6**, 63–75 (2006)
7. Bahuguna, D., Shukla, R.: Approximations of solutions to nonlinear sobolev type evolution equation. Elec. J. Diff. Equ. **31**, 1–16 (2003)
8. Bahuguna, D., Srivastava, S.K., Singh, S.: Approximations of solutions to semilinear integrodifferential equations. Numer. Funct. Anal. Optim. **22**, 487–504 (2001)
9. Bahuguna, D., Muslim, M.: Approximation of solutions to retarded differential equations with applications to population dynamics. J. Appl. Math. Stoch. Anal. **1**, 1–11 (2005)
10. Muslim, M.: Approximation of solutions to History-valued neutral functional differential equations. Comput. Math. Appl. **51**, 537–550 (2006)
11. El'sgol'ts, L.E., Norkin, S.B.: Introduction to the Theory of Differential Equations with Deviating Arguments. Academic Press (1973)

12. Driver, R.D.: A two-body problem of classical electrodynamics: the one-dimensional case. Ann. Phys. **21**, 122–142 (1963)
13. Oberg, R.J.: On the local existence of solutions of certain functional-differential equations. Proc. Amer. Math. Soc. **20**, 295–302 (1969)
14. Ezzinbi, K., Xianlong, Fu, Hilal, K.: Existence and regularity in the α-norm for some neutral partial differential equations with nonlocal conditions. Nonlinear Anal. **67**, 1613–1622 (2007)
15. Gal, C.G.: Nonlinear abstract differential equations with deviated argument. J. Math. Anal. Appl. **333**, 177–189 (2007)
16. Pazy, A.: Semigroups of Linear Operators and Applications to Partial Differential Equations. Springer (1983)

Approximation of Solutions to Fractional Integro-Differential Equations with Finite Delay

Renu Chaudhary and Dwijendra N. Pandey

Abstract In the present work, we study the approximation of solutions to fractional integro-differential equations with finite delay in an arbitrary separable Hilbert space H. We consider an associated sequence of approximate integral equations. We establish the existence and uniqueness of the solutions to each approximate integral equation using the fixed-point arguments. Then we prove the convergence of the solutions of the approximate integral equations to the solution of the given fractional integro-differential equation. Finally, we consider the Faedo–Galerkin approximations of the solutions and prove some convergence results.

Keywords Analytic semigroup · Banach fixed point theorem · Faedo–Galerkin approximations · Fractional integro-differential equation · Mild solution

1 Introduction

The investigation of fractional differential equations, that is, calculus of derivatives of any arbitrary real or complex order, has gained importance and popularity during the past three decades. In various problems of physics, mechanics and engineering, fractional differential equations have been proved to be a valuable tool in the modelling of many phenomena. These differential equations are also very important to describe the memory and hereditary properties of various materials and phenomenon. More details on the theory and its applications we refer to Kilbas and Trujillo [1], Lakshmikantham [2], Miller and Ross [3], Podlubny [4], and Kilbas and Samko [5].

R. Chaudhary (✉) · D.N. Pandey
Department of Mathematics, Indian Institute of Technology Roorkee, Roorkee 247667, Uttarakhand, India
e-mail: rrenu94@gmail.com

D.N. Pandey
e-mail: dwij.iitk@gmail.com

P.N. Agrawal et al. (eds.), *Mathematical Analysis and its Applications*, Springer Proceedings in Mathematics & Statistics 143, DOI 10.1007/978-81-322-2485-3_55

In the present work we are concerned with the approximation of solution to following fractional integro-differential equation with finite delay in a separable Hilbert space $(H, \|\cdot\|, (,))$

$$\mathbf{D}_t^q[u(t) + f(t, u_t)] + Au(t) = g(t, u_t) + K(u)(t), \qquad t \in (0, T], \tag{1}$$

$$u(t) = \phi(t), \qquad t \in [-h, 0],\ 0 < h < \infty, \tag{2}$$

where K is the nonlinear Volterra operator $K(u)(t) = \int_0^t M(t-s)k(u)(s)ds$, $0 < T < \infty$, $\mathbf{D}_t^q$ is Caputo fractional derivative, where $0 < q < 1$. $A : D(A) \subset H \to H$ is closed, positive definite and self-adjoint linear operator with densely defined domain $D(A)$ which is the infinitesimal generator of an analytic semigroup of bounded linear operator on H. The nonlinear operator k is defined on $D(A^\alpha)$ for some α. The map M is a real-valued continuous function defined on R_+. The nonlinear functions f and g are defined from $[0, T] \times C_0$ into H and $\phi \in C_0$. Here, $C_0 := C([-h, 0]; H)$ be the Banach space of all continuous functions from $[-h, 0]$ into H endowed with the supremum norm $\|u\|_0 := \sup_{-h \le s \le 0} \|u(s)\|$, $u \in C_0$. Further, for $t \in [0, T]$, $C_t := C([-h, t]; H)$ be the Banach space of all continuous functions from $[-h, t]$ into H endowed with supremum norm $\|u\|_t := \sup_{-h \le s \le t} \|u(s)\|$, $u \in C_t$. For $u \in C_T$ and $0 \le t \le T$, $u_t \in C_0$ be the function defined by $u_t(\theta) = u(t + \theta)$ for $\theta \in [-h, 0]$.

Initial studies concerning existence, uniqueness and finite time blow-up of solutions for the following equation:

$$u'(t) + Au(t) = M(u(t)), \quad u(0) = \phi, \tag{3}$$

have been considered by Murakami [6] and Segal [7].

Although Faedo–Galerkin method is useful for convergence yet the convergence behaviour in many cases is not known. Bazely in [8] and [9] showed the uniform convergence of the approximations to solutions of the nonlinear wave equation

$$u''(t) + Au(t) + M(u(t)) = 0, \quad u(0) = \phi, \quad u'(0) = \psi, \tag{4}$$

on any closed subinterval $[0, T]$ of existence of the solution. Miletta [10] has established the convergence of Faedo–Galerkin approximate solutions to (3).

In [11] Bahuguna and Srivastava extended the results of Miletta [10] and considered the Faedo–Galerkin approximations of the solutions to following functional integro-differential equation:

$$\frac{du(t)}{dt} = -Au(t) + M(u(t)) + \int_0^t g(t-s)k(u(s))ds, \tag{5}$$

in a separable real Hilbert space H.

The case of (1) in which $q = 1$, $K = 0$ and $u_t(\theta) = u(t)$, i.e. $\theta = 0$ have been studied by Bahuguna [12], the case in which $u_t(\theta) = u(t)$, i.e. $\theta = 0$ have been studied by Chaddha [13] and the case in which $q = 1$ and $K = 0$ have been studied by Dubey [14]. Thus this paper is the generalization of [12–14]. For more details to the existence of an approximate solution to different differential equations see [15, 16].

The manuscript is organized as follows: In Sect. 2, we recall some necessary preliminaries, lemmas, theorems and assumptions. In Sect. 3, we show the existence and uniqueness of approximate solutions. In Sect. 4, the convergence of approximate integral equation to the associated integral equation is established. In Sect. 5, we consider the Faedo–Galerkin approximate solutions and prove the convergence of such approximations.

2 Preliminaries and Assumptions

In this section, we have some basic definitions, assumptions and properties of fractional calculus which will be used further in this paper.

Definition 1 The Riemann–Liouville integral of order q with the lower limit zero for a function $f \in L^1((0, T); H)$ is defined by

$$I_t^q f(t) = \frac{1}{\Gamma(q)} \int_0^t (t-s)^{q-1} f(s) ds, \quad t > 0, \quad q > 0.$$

Here q is called the order of fractional integration.

Definition 2 The Riemann–Liouville fractional derivative of order q for a function $f \in L^1((0, T); H)$ is defined by

$$D_t^q f(t) = \frac{d^m}{dt^m}[\frac{1}{\Gamma(m-q)} \int_0^t \frac{f(s)}{(t-s)^{q+1-m}} ds],$$

where $m - 1 < q < m$ and $m \in \mathbb{N}$.

Definition 3 The Caputo fractional derivative of order q for a function $f \in C^{m-1}((0, T); H) \cap L^1((0, T); H)$ is defined by

$$\mathbf{D}_t^q f(t) = \frac{1}{\Gamma(m-q)} \int_0^t (t-s)^{m-q-1} f^m(s) ds,$$

where $m - 1 < q < m$ and $m \in \mathbb{N}$.

We shall use the following assumptions:

(H1) Since A is a closed, positive definite, self-adjoint linear operator from the domain $D(A) \subset H$ of A into H such that $D(A)$ is dense in H. Therefore, we can assume that A has the pure point spectrum $0 < \lambda_0 \leq \lambda_1 \leq \lambda_2 \leq \cdots \leq \lambda_m \leq \cdots$ with $\lambda_m \to 0$ and a corresponding complete orthonormal system of eigenfunctions $\{\phi_j\}$, i.e. $A\phi_j = \lambda_j \phi_j$ and $<\phi_i, \phi_j>= \delta_{ij}$, where $\delta_{ij} = \begin{cases} 1, & \text{if } i = j; \\ 0, & \text{otherwise.} \end{cases}$

These assumptions on A guarantee that $-A$ is the infinitesimal generator of an analytic semigroup $Q(t)$ then for $c > 0$, large enough, $-(A + cI)$ is invertible and generates a bounded analytic semigroup. This allows us to reduce the general case in which $-A$ is the infinitesimal generator of an analytic semigroup to the case in which the semigroup is bounded and the generator is invertible.

Hence without loss of generality we can assume that $\|Q(t)\| \leq C$ for $t > 0$ and $0 \in \rho(-A)$, where $\rho(-A)$ is the resolvent set of $-A$, i.e. we can define the positive fractional power A^α as closed linear operator with domain $D(A^\alpha)$ for $\alpha \in (0, 1]$. Moreover, $D(A^\alpha)$ is dense in H with the norm $\|u\|_\alpha = \|A^\alpha u\|$.

Here, we signify the space $D(A^\alpha)$ by H_α endowed with the $\alpha - norm(\|\cdot\|_\alpha)$. Also, we have that $H_k \hookrightarrow H_\alpha$ for $0 < \alpha < 1$ and therefore, the embedding is continuous. For study on the fractional powers of closed linear operators, we refer to book by Pazy [17].

(H2) The function $\phi(t) \in D(A)$, for all $t \in [-h, 0]$ and ϕ is locally *Hölder* continuous on $[-h, 0]$ and define

$$\widetilde{\phi}(t) = \begin{cases} \phi(t), & t \in [-h, 0]; \\ \phi(0), & t \in [0, T]. \end{cases}$$

(H3) For $0 < \alpha < 1$, the nonlinear map g defined from $[0, T] \times C_0^\alpha$ into H is continuous and there exists a non-decreasing function G_R from $[0, \infty)$ *into* $[0, \infty)$ depending on $R > 0$ such that

$$\|g(t, u_1) - g(t, u_2)\| \leq G_R(t)\|u_1 - u_2\|_{0,\alpha}$$

and

$$\|g(t, u)\| \leq G_R(t),$$

for (t, u_1), (t, u_2) and (t, u) in $[0, T] \times B_R(C_0^\alpha, \widetilde{\phi})$, where $B_R(Z, z_0) = \{z \in Z| \|z - z_0\|_Z \leq R\}$ for any Banach space Z with its norm $\|.\|_Z$.

(H4) There exist positive constants $0 < \alpha < \beta < 1$ such that the function $A^\beta f$ defined from $[0, T] \times C_0^\alpha$ into C_0^β is continuous and there exist constants L and $0 < \gamma \leq 1$ such that

$$\|A^\beta f(t, u_1) - A^\beta f(s, u_2)\| \leq L_f\{| t - s |^\gamma + \|u_1 - u_2\|_{0,\alpha}\}$$

and $2L_f\|A^{\alpha-\beta}\| < 1$, for all (t, u_1), (s, u_2) in $[0, T] \times B_R(C_0^\alpha, \widetilde{\phi})$.

(H5) For $0 < \alpha < 1$, the nonlinear map k defined on $D(A^\alpha)$ into H is continuous and there exists constant $C_k > 0$ such that

$$\|k(u_1)(t) - k(u_2)(t)\| \leq C_k(t)\|A^\alpha(u_1 - u_2)\|$$

and

$$\|k(u)(t)\| \leq C_k(t),$$

for $u_1,\ u_2 \in D(A^\alpha)$.

Definition 4 A continuous function $u : [-h, T] \to H$ is said to be a mild solution for the problem (1)–(2) if $u(\cdot)$ satisfies the following integral equation:

$$u(t) = \begin{cases} \phi(t), & t \in [-h, 0]; \\ S_q(t)[\phi(0) + f(0, \widetilde{\phi})] - f(t, u_t) & \\ + \int_0^t (t-s)^{q-1} Q_q(t-s) A f(s, u_s) ds & \\ + \int_0^t (t-s)^{q-1} Q_q(t-s)[g(s, u_s) + K(u)(s)] ds, & t \in [0, T]. \end{cases} \tag{6}$$

The operators $S_q(t)$ and $Q_q(t)$ are defined as follows:

$$S_q(t) = \int_0^\infty \varsigma_q(\xi) Q(t^q \xi) d\xi,$$

$$Q_q(t) = q \int_0^\infty \xi \varsigma_q(\xi) Q(t^q \xi) d\xi,$$

where $\varsigma_q(\xi) = \frac{1}{q}\xi^{1-\frac{1}{q}} \times \psi_q(\xi^{-\frac{1}{q}})$ is a probability density function defined on $(0, \infty)$, i.e. $\varsigma_q(\xi) \geq 0$, $\int_0^\infty \varsigma_q(\xi) d\xi = 1$ and $\psi_q(\xi) = \frac{1}{\pi} \sum_{n=1}^{\infty} (-1)^{n-1} \xi^{-nq-1} \frac{\Gamma(nq+1)}{n!} \sin(n\pi q), 0 < \xi < \infty$. For further details on mild solution see [18].

Lemma 1 (Pazy [17]) *Let $-A$ be the infinitesimal generator of an analytic semigroup $\{Q(t) : t \geq 0\}$ such that $\| Q(t) \| \leq C$, for $t \geq 0$ and $0 \in \rho(-A)$. Then,*

1. *For $0 < \alpha \leq 1$, H_α is a Hilbert space.*
2. *The operator $A^\alpha Q(t)$ is bounded for every $t > 0$ and*

$$\|AQ(t)\| \leq Ct^{-1},$$
$$\|A^\alpha Q(t)\| \leq C_\alpha t^{-\alpha}.$$

Lemma 2 (Zhou [19]) *The operators $\{S_q(t), t \geq 0\}$ and $\{Q_q(t), t \geq 0\}$ are bounded linear operators such that*

(i) $\|S_q(t)z\| \leq C\|z\|$, $\|Q_q(t)z\| \leq \frac{qC}{\Gamma(1+q)}\|z\|$ *and* $\|A^\alpha Q_q(t)z\| \leq \frac{qC_\alpha \Gamma(2-\alpha)t^{-q\alpha}}{\Gamma(1+q(1-\alpha))}\|z\|$, *for any* $z \in H$.

(ii) The families $\{S_q(t) : t \geq 0\}$ *and* $\{Q_q(t) : t \geq 0\}$ *are strongly continuous.*
(iii) If $Q(t)$ *is compact, then* $S_q(t)$ *and* $Q_q(t)$ *are compact operators for any* $t > 0$.

3 Approximate Solutions and Convergence

In this section, we show the existence of a mild solution of the Eqs. (1)–(2) on $[0, T]$ for some T, $0 < T < \infty$. Let H_n denote the finite-dimensional subspace of the Hilbert space H spanned by $\{u_0, u_1, \cdots, u_n\}$ and let $P^n : H \to H_n$ for $n = 1, 2, \cdots,$ be the corresponding projection operators.

Let $0 < T_0 < T$ be an arbitrarily fixed constant and let

$$A_1 = \max_{0 \leq t \leq T} \|A^\beta f(t, \widetilde{\phi})\|,$$

$$M_0 = \sup_{0 \leq t \leq T_0} \|M(t)\|.$$

For each n, we define $f_n : [0, T] \times C_0^\alpha \to C_0^\beta$ and $g_n : [0, T] \times C_0^\alpha \to H$, respectively, by $f_n(t, u_t) = f(t, P^n u_t)$, $g_n(t, u_t) = g(t, P^n u_t)$ nd $k : D(A^\alpha) \to H$ and K by

$$k_n(u) = k(P^n u),$$
$$K_n(u)(t) = \int_0^t M(t-s) k_n(u(s)) ds.$$

We choose $T_0, 0 < T_0 \leq T$ sufficiently small such that

$$\|(S_q(t) - I)A^\alpha(\phi(0) + f_n(0, \widetilde{\phi})\| \leq (1-\mu)\frac{R}{3},$$

$$\|A^{\alpha-\beta}\| L_f T_0^\gamma + C_{1+\alpha-\beta} \frac{\Gamma(1-(\alpha-\beta))}{\Gamma(1+q(\beta-\alpha))} (L_f R + A_1) \frac{T_0^{q(\beta-\alpha)}}{(\beta-\alpha)}$$
$$+ \frac{C_\alpha \Gamma(2-\alpha)}{\Gamma(1+q(1-\alpha))} C(T) \frac{T_0^{q(1-\alpha)}}{(1-\alpha)} < (1-\mu)\frac{R}{6},$$
$$\frac{C_{1+\alpha-\beta}\Gamma(1-(\alpha-\beta))}{\Gamma(1+q(\beta-\alpha))} L_f \frac{T_0^{q(\beta-\alpha)}}{(\beta-\alpha)} + \frac{C_\alpha \Gamma(2-\alpha)}{\Gamma(1+q(1-\alpha))} C(T) \frac{T_0^{q(1-\alpha)}}{1-\alpha} < (1-\mu),$$

where $C(T) = G_R(T) + T_0 M_0 C_k(T)$, $R = \sqrt{2(R^2 + 3\|\widetilde{\phi}\|_{0,\alpha^2})}$, C_α, $C_{1+\alpha-\beta}$ and $\mu = \|A^{\alpha-\beta}\| L_f < 1$, are constants.

Let $A^\alpha : C_t^\alpha \to C_t$ be defined as $(A^\alpha \psi)(t) = A^\alpha(\psi(t))$ and $(P^n u_t)(s) = P^n(u(t+s))$, for all $s \in [-h, 0]$, $t \in [0, T_0]$. Now, we consider

$$B_R = B_R(C_{T_0}^\alpha, \widetilde{\phi}) = \{u \in C_{T_0}^\alpha : \|u - \widetilde{\phi}\|_\alpha \leq R\}$$

and define the operator Q_n on B_R as follows:

$$(Q_n u)(t) = \begin{cases} \widetilde{\phi}(t), & t \in [-h, 0]; \\ S_q(t)[\widetilde{\phi}(0) + f_n(0, \widetilde{\phi})] - f_n(t, u_t) \\ + \int_0^t (t-s)^{q-1} Q_q(t-s) A f_n(s, u_s) ds \\ + \int_0^t (t-s)^{q-1} Q_q(t-s)[g_n(s, u_s) + K(u)(s)] ds, & t \in [0, T_0]. \end{cases} \tag{7}$$

Theorem 1 *Suppose that conditions* $(H1)$–$(H4)$ *are fulfilled and* $\phi(t) \in D(A^\alpha)$, *for all* $t \in [-h, 0]$. *Then, there exists a unique fixed point* $u_n \in B_R$ *of the map* Q_n, *i.e.* $Q_n u_n = u_n$ *for each* $n = 0, 1, 2, \cdots$; *that is,* u_n *satisfies the approximate integral equation*

$$u_n(t) = \begin{cases} \widetilde{\phi}(t), & t \in [-h, 0]; \\ S_q(t)[\widetilde{\phi}(0) + f_n(0, \widetilde{\phi})] - f_n(t, (u_n)_t) + \\ \int_0^t (t-s)^{q-1} Q_q(t-s) A f_n(s, (u_n)_s) ds \\ + \int_0^t (t-s)^{q-1} Q_q(t-s)[g_n(s, (u_n)_s) + K_n(u_n)(s)] ds, & t \in [0, T_0]. \end{cases} \tag{8}$$

Proof To prove the theorem, we first show that map $t \to (Q_n u)(t)$ is continuous from $[-h, T_0]$ into $D(A^\alpha)$ with respect to norm $\| \cdot \|_\alpha$. Thus, for any $u \in B_R(C_T^\alpha, \widetilde{\phi})$ and $t \in [-h, 0]$, we have

$$(Q_n u)(t+h) - (Q_n u)(t) = \widetilde{\phi}(t+h) - \widetilde{\phi}(t),$$

for sufficiently small $h > 0$. Further, for $t \in (0, T_0]$ and sufficiently small $h > 0$, we have

$$A^\alpha[(Q_n u)(t+h) - (Q_n u)(t)] =$$

$$\begin{aligned} & [S_q(t+h) - S_q(t)] A^\alpha (\widetilde{\phi}(0) + f_n(0, \widetilde{\phi}) - A^{\alpha-\beta}[A^\beta f_n(t+h, u_{t+h}) - A^\beta f_n(t, u_t)] \\ & + \int_0^t [(t+h-s)^{q-1} - (t-s)^{q-1}] A^{1+\alpha-\beta} Q_q(t+h-s) A^\beta f_n(s, u_s) ds \\ & + \int_0^t (t-s)^{q-1} A^{1+\alpha-\beta} [Q_q(t+h-s) - Q_q(t-s)] A^\beta f_n(s, u_s) ds \\ & + \int_t^{t+h} (t+h-s)^{q-1} A^{1+\alpha-\beta} Q_q(t+h-s) A^\beta f_n(s, u_s) ds \\ & + \int_0^t [(t+h-s)^{q-1} - (t-s)^{q-1}] A^\alpha Q_q(t+h-s)[g_n(s, u_s) + K_n(u_n)(s)] ds \\ & + \int_0^t (t-s)^{q-1} A^\alpha [Q_q(t+h-s) - Q_q(t-s)][g_n(s, u_s) + K_n(u_n)(s)] ds \\ & + \int_t^{t+h} (t+h-s)^{q-1} A^\alpha Q_q(t+h-s)[g_n(s, u_s) + K_n(u_n)(s)] ds \\ = & \sum_{i=1}^{8} J_i. \end{aligned} \tag{9}$$

For $z \in H$, we have

$$[Q(t+h)^q\xi) - Q(t^q\xi)]z = \int_t^{t+h} \frac{d}{d\tau} Q(\tau^q\xi)zd\tau = \int_t^{t+h} q\xi(\tau)^{q-1} AQ(\tau^q\xi)zd\tau.$$

Here

$$\begin{aligned}
\|J_1\| &\le \int_0^\infty \varsigma_q(\xi)\|Q((t+h)^q\xi) - Q(t^q\xi)A^\alpha(\widetilde{\phi}(0) + f_n(0,\widetilde{\phi})\|d\xi \\
&\le \int_0^\infty \varsigma_q(\xi) \int_t^{t+h} q\xi\tau^{q-1}\|A^\alpha Q(\tau^q\xi)\|d\tau\|A(\widetilde{\phi}(0) + f_n(0,\widetilde{\phi})\|d\xi \\
&\le qC_\alpha \int_0^\infty \xi^{1-\alpha}\varsigma_q(\xi) \int_t^{t+h} \tau^{q(1-\alpha)-1}\|A(\widetilde{\phi}(0) + f_n(0,\widetilde{\phi})\|d\tau d\xi
\end{aligned}$$

$$\begin{aligned}
&\le \frac{\Gamma(1-q)}{\Gamma(1+q(1-\alpha))} C_\alpha\|A(\widetilde{\phi}(0) + f_n(0,\widetilde{\phi}))\|[(t+h)^{q(1-\alpha)} - t^{q(1-\alpha)}] \\
&\le \frac{qC_\alpha\Gamma(2-\alpha)}{\Gamma(1+q(1-\alpha))}\|A(\widetilde{\phi}(0) + f_n(0,\widetilde{\phi}))\|h[t+\theta(h)]^{q(1-\alpha)-1} \\
&\le \frac{qC_\alpha\|A(\widetilde{\phi}(0) + f_n(0,\widetilde{\phi}))\|\Gamma(2-\alpha)}{\Gamma(1+q(1-\alpha))}\theta^{q(1-\alpha)-1}(h)^{q(1-\alpha)}, \qquad (10)
\end{aligned}$$

where $0 < \theta < 1$ [see [18, 20].
For J_2 we have

$$\begin{aligned}
\|J_2\| &\le \|A^{\alpha-\beta}\|\|A^\beta f_n(t+h, u_{t+h}) - A^\beta f_n(t, u_t)\| \\
&\le L_f\|A^{\alpha-\beta}\|[| h |^\gamma + \|P^n u_{t+h} - P^n u_t\|_\alpha] \\
&\le L_f\|A^{\alpha-\beta}\|[| h |^\gamma + \|u_{t+h} - u_t\|_\alpha] \qquad (11)
\end{aligned}$$

and

$$\begin{aligned}
\|A^\beta f_n(t,u_t)\| &\le \|A^\beta f_n(t,u_t) - A^\beta f_n(t,\widetilde{\phi}(0))\| + \|A^\beta f_n(t,\widetilde{\phi}(0))\| \\
&\le L_f\|P^n u_t - \widetilde{\phi}(0)\|_\alpha + A_1 \\
&\le L_f R + A_1. \qquad (12)
\end{aligned}$$

Also,

$$\begin{aligned}
\|J_3\| &\le \frac{qC_{1+\alpha-\beta}\Gamma(1-(\alpha-\beta))}{\Gamma(1+q(\beta-\alpha))}(L_f R + A_1)\int_0^t (t-s)^{m_1-1}\Big[(t+h-s)^{-m_1k_1} - (t-s)^{-m_1k_1}\Big]ds \\
&\le \frac{qC_{1+\alpha-\beta}\Gamma(1-(\alpha-\beta))}{\Gamma(1+q(\beta-\alpha))}(L_f R + A_1)k_1\psi^{k_1-1}(1-d)^{-m_1(1-k_1)-1}h^{m_1(1-k_1)}, \qquad (13)
\end{aligned}$$

where $m_1 = 1 - q(1-\alpha-\beta)$, $k_1 = \frac{(1-q)}{1-q(1+\alpha-\beta)}$, $d = (1 - (\frac{k_1}{m_1})^{1/k_1 m_1})$ and $0 < \psi \le 1$ (see [16, 20]).

$$\begin{aligned}\|J_4\| &\le \frac{qC_{1+\alpha-\beta}\Gamma(1-(\alpha-\beta))}{\Gamma(1+q(\beta-\alpha))}(L_f R + A_1)\int_0^t (t-s)^{q-1} \\ &\quad \times[(t-s)^{-q(1+\alpha-\beta)} - (t+h-s)^{-q(1+\alpha-\beta)}]ds \\ &\le \frac{qC_{1+\alpha-\beta}\Gamma(1-(\alpha-\beta))}{\Gamma(1+q(\beta-\alpha))}\frac{(L_f R + A_1)}{1+\alpha-\beta}d_2^{\alpha-\beta}(1-d_3)^{-q(\beta-\alpha)-1}h^{q(\beta-\alpha)},\end{aligned} \tag{14}$$

where $0 < d_2 \le 1$ and $d_3 = (1-(1+\alpha-\beta/q))^{1/q(1+\alpha-\beta)}$.

$$\begin{aligned}\|J_5\| &\le \frac{qC_{1+\alpha-\beta}\Gamma(1-(\alpha-\beta))}{\Gamma(1+q(\beta-\alpha))}(L_f R + A_1)\int_t^{t+h} (t+h-s)^{q(\beta-\alpha)-1}ds \\ &\le \frac{C_{1+\alpha-\beta}\Gamma(1-(\alpha-\beta))}{\Gamma(1+q(\beta-\alpha))}(L_f R + A_1)\frac{h^{q(\beta-\alpha)}}{(\beta-\alpha)}.\end{aligned} \tag{15}$$

Similarly,

$$\begin{aligned}\|J_6\| &\le \frac{qC_\alpha\Gamma(2-\alpha)}{\Gamma(1+q(1-\alpha))}C(T)\int_0^t (t-s)^{m_2-1}[(t+h-s)^{-m_1k_2} - (t-s)^{-m_2k_2}]ds \\ &\le \frac{qC_\alpha\Gamma(2-\alpha)}{\Gamma(1+q(1-\alpha))}C(T)k_2\psi_1^{k_2-1}(1-d_4)^{-m_2(1-k_2)-1}h^{m_2(1-k_2)},\end{aligned} \tag{16}$$

where $m_2 = 1 - q\alpha$, $k_2 = \frac{1-q}{1-q\alpha}$, $0 < \theta_1 \le 1$ and $d_4 = (1 - \left(\frac{k_2}{m_2}\right)^{1/k_2m_2})$.

$$\|J_7\| \le \frac{qC_\alpha\Gamma(2-\alpha)}{\Gamma(1+q(1-\alpha))}C(T)\theta_2^{\alpha-1}(1-d_5)^{-q(1-\alpha)-1}(h)^{q(1-\alpha)}, \tag{17}$$

where $d_5 = (1-(\alpha/q)^{1/q\alpha})$ and $0 < \theta_2 \le 1$.

$$\|J_8\| \le \frac{C_\alpha\Gamma(2-\alpha)}{\Gamma(1+q(1-\alpha))}C(T)\frac{(h)^{q(1-\alpha)}}{(1-\alpha)}. \tag{18}$$

from (9)–(18) we have the map Q_n which is $H\ddot{o}lder$ continuous from $[-h, T_0]$ into $D(A^\alpha)$ with $\alpha-$norm.

Now we show that $Q_n(u) \in B_R(C_T^\alpha, \widetilde{\phi})$ for any $u \in B_R(C_T^\alpha, \widetilde{\phi})$. Thus, for any $t \in [-h, 0]$,

$$\|(Q_n u)(t) - \phi(t)\|_\alpha = 0.$$

Further, for $t \in (0, T_0]$

$$
\begin{aligned}
\|(Q_n u)(t) - \widetilde{\phi}\|_\alpha &\le \|(S_q(t) - I)A^\alpha(\widetilde{\phi}(0) + f_n(0, \widetilde{\phi}))\| \\
&\quad + \|A^{\alpha-\beta}\|\|A^\beta f_n(0, \widetilde{\phi}) - A^\beta f_n(t, u_t)\| \\
&\quad + \int_0^t (t-s)^{q-1}\|A^{1+\alpha-\beta}Q_q(t-s)\|\|A^\beta f_n(s, u_s)\|ds \\
&\quad + \int_0^t (t-s)^{q-1}\|A^\alpha Q_q(t-s)\|[\|g_n(s, u_s)\| + \|K(u)(s)\|]ds \\
&\le (1-\mu)\frac{R}{3} + \|A^{\alpha-\beta}\|L_f[T_0^\gamma + \|u_t - \phi\|_\alpha] \\
&\quad + \frac{C_{1+\alpha-\beta}\Gamma(1-(\alpha-\beta))}{\Gamma(1+q(1-\alpha))}(L_f R + A_1)\frac{T_0^{q(\beta-\alpha)}}{(\beta-\alpha)} \\
&\quad + \frac{C_\alpha \Gamma(2-\alpha)}{\Gamma(1+q(1-\alpha))}C(T)\frac{T_0^{q(1-\alpha)}}{(1-\alpha)} \\
&\le (1-\mu)\frac{R}{3} + (1-\mu)\frac{R}{6} + \mu R \le R.
\end{aligned}
$$

Thus

$$\|Q_n u - \widetilde{\phi}\|_{T_0,\alpha} \le R.$$

Hence Q_n maps $B_R(C^\alpha_{T_0}, \widetilde{\phi})$ into $B_R(C^\alpha_{T_0}, \widetilde{\phi})$. Now we show that Q_n is a strict contraction on $B_R(C^\alpha_{T_0}, \widetilde{\phi})$. For $u, v \in B_R(C^\alpha_{T_0}, \widetilde{\phi})$ and $t \in [-h, 0]$, we have

$$\|(Q_n u)(t) - (Q_n v)(t)\|_\alpha = 0.$$

Again, if $t \in (0, T_0]$ and $u, v \in B_R(C^\alpha_{T_0}, \widetilde{\phi})$, then we have

$$
\begin{aligned}
\|(Q_n u)(t) - (Q_n v)(t)\|_\alpha &\le \|A^{\alpha-\beta}\|\|A^\beta f_n(t, u_t) - A^\beta f_n(t, v_t)\|_\alpha \\
&\quad + \int_0^t (t-s)^{q-1}\|A^{1+\alpha-\beta}Q_q(t-s)\|\|A^\beta f_n(s, u_s) - A^\beta f_n(s, v_s)\|ds \\
&\quad + \int_0^t (t-s)^{q-1}\|A^\alpha Q_q(t-s)\|[\|g_n(s, u_s) - g_n(s, v_s)\| \\
&\quad + \|K_n(u)(s) - K_n(v)(s)\|]ds. \qquad (19)
\end{aligned}
$$

Now,

$$\|A^\beta f_n(t, u_t) - A^\beta f_n(t, v_t)\| \le L_f\|u - v\|_{T_0,\alpha}, \qquad (20)$$

$$\|g_n(s, u_s) - g_n(s, v_s)\| \le G_R(T)\|u - v\|_{T_0,\alpha} \qquad (21)$$

and

$$\|K_n(u)(s) - K_n(v)(s)\|_\alpha \le M_0 \int_0^s \|k_n(u)(\tau) - k_n(v)(\tau)\|_\alpha d\tau$$
$$\le M_0 T_0 C_k(T)\|u - v\|_{T_0,\alpha}. \tag{22}$$

Using (20), (21) and (22) in (19), we get

$$\|(Q_n u)(t) - (Q_n v)(t)\|_\alpha \le [\|A^{\alpha-\beta}\|L_f + \frac{C_{1+\alpha-\beta}\Gamma(1-(\alpha-\beta))}{\Gamma(1+q(\beta-\alpha))}L_f\frac{T_0^{q(\beta-\alpha)}}{(\beta-\alpha)}$$
$$+ \frac{C_\alpha \Gamma(2-\alpha)}{\Gamma(1+q(1-\alpha))}C(T)\frac{T_0^{q(1-\alpha)}}{(1-\alpha)}] \times \|u - v\|_{T_0,\alpha}.$$

Taking supremum on t over $[-h, T_0]$, we get

$$\|(Q_n u)(t) - (Q_n v)(t)\|_{T_0,\alpha} \le \left[\|A^{\alpha-\beta}\|L_f + \frac{C_{1+\alpha-\beta\Gamma(1-(\alpha-\beta))}}{\Gamma(1+q(\beta-\alpha))}L_f\frac{T_0^{q(\beta-\alpha)}}{(\beta-\alpha)}\right.$$
$$\left. + \frac{C_\alpha \Gamma(2-\alpha)}{\Gamma(1+q(1-\alpha))}C(T)\frac{T_0^{q(1-\alpha)}}{(1-\alpha)}\right] \times \|u - v\|_{T_0,\alpha}.$$

The above estimate implies that Q_n is a strict contraction on $B_R(C^\alpha_{T_0}, \widetilde{\phi})$. Hence, there exists a unique $u_n \in B_R(C^\alpha_{T_0}, \widetilde{\phi})$ such that $Q_n u_n = u_n$. Clearly, u_n satisfies (8) on $[0, T_0]$.

Lemma 3 *Let the hypotheses* $(H1)$*–*$(H4)$ *hold. If* $\phi(t) \in D(A)$ *for* $t \in [-h, 0]$*, then* $u_n(t) \in D(A^\eta)$*, for all* $t \in [-h, T_0]$ *with* $0 < \eta \le \beta < 1$.

Proof For all $t \in [-h, 0]$ it is obvious. For $t \in (0, t_0]$, from Theorem (1) we have the existence of a unique $u_n \in B_R(C^\alpha_{T_0}, \widetilde{\phi})$ satisfying (8). Theorem (1.2.4) in Pazy [17] implies that for $t > 0$ and $0 \le \eta < 1$, $\tau(t) : H \to D(A^\eta)$ and for $0 \le \eta \le \beta < 1$, $D(A^\beta) \subseteq D(A^\eta)$. (H4) implies that the map $t \to A^\beta f(t, (u_n)_t)$ is *Hölder* continuous on $[0, T_0]$. Similarly, *Hölder* continuity of u_n can be easily established using (9)–(18). It follows that (see Theorem 4.3.2 in [17])

$$\int_0^t (t-s)^{q-1} Q_q(t-s) A^\beta f_n(s, (u_n)_s) ds \in D(A).$$

Also, from Theorem (1.2.4) in Pazy [17], we have $\tau(t)u \in D(A)$ if $u \in D(A)$. The required result follows from these facts and the fact that $D(A) \subseteq D(A^\eta)$ for $0 \le \eta \le 1$.

Proposition 1 *Let the hypotheses* $(H1)$*–*$(H4)$ *hold. If* $\phi(t) \in D(A)$ *for* $t \in [-h, 0]$*, then there exists a constant* U_0 *such that*

$$\|u_n(t)\|_\mu \le U_0, \ \ 0 \le \mu < \beta < 1, \ \ t \in [-h, T_0], \ \ n = 1, 2, ...$$

Proof If $\phi(t) \in D(A)$, for all $t \in [-h, 0]$, hence, $\phi(t) \in D(A^\beta)$, for all $t \in [-h, 0]$ and $0 \le \mu \le \beta < 1$. Now applying A^μ on both the sides of (8), then for all $t \in [-h, 0]$, we have

$$\|u_n(t)\|_\mu = \|\phi(t)\|_\mu \le \|\phi\|_{0,\mu} = U_0.$$

Now for $t \in (0, T_0]$, we have

$$\begin{aligned} \|A^\mu u_n(t)\| \le\ & \|A^\mu S_q(t)(\widetilde{\phi}(0) + f_n(0, \widetilde{\phi}))\| + \|A^{\mu-\beta}\|\|A^\beta f_n(t, (u_n)_t)\| \\ & + \int_0^t (t-s)^{q-1}\|A^{1+\mu-\beta} Q_q(t-s)\|\|A^\beta f_n(s, (u_n)_s)\| ds \\ & + \int_0^t (t-s)^{q-1}\|A^\mu Q_q(t-s)\|[\|g_n(s, (u_n)_s)\| + \|K_n(u_n)(s)\| ds. \end{aligned} \tag{23}$$

Thus we have the required result for $t \in [-h, T_0]$.

4 Convergence of Solutions

In this section, we establish the convergence of the solution $u_n \in C^\alpha_{T_0}$ of the approximate integral equation (8) to a unique solution $u(\cdot) \in C^\alpha_{T_0}$ of (6) on $[-h, T_0]$.

Theorem 2 *Let the hypotheses* $(H1)$–$(H4)$ *hold. If* $\phi(t) \in D(A^\alpha)$ *for* $t \in [-h, 0]$, *then*

$$\lim_{k\to\infty} \sup_{\{n\ge k, t_0 \le t \le T_0\}} \|u_n(t) - u_k(t)\|_\alpha = 0, \tag{24}$$

for all $t_0 \in (0, T_0]$.

Proof Let $n \ge k \ge n_0$, where n_0 is large enough and $n, k, n_0 \in N$. For $-h \le t_0 \le 0$, we have

$$\|A^\alpha[u_n(t) - u_k(t)]\| = \|\phi(t) - \phi(t)\|_\alpha = 0.$$

For $t_0 \in (0, T_0]$ and $0 < \alpha < \eta < \beta < 1$,

$$\begin{aligned} \|g_n(t, (u_n)_t) - g_k(t, (u_k)_t)\| &\le \|g_n(t, (u_n)_t - g_n(t, (u_k)_t)\| + \|g_n(t, (u_k)_t) - g_k(t, (u_k)_t)\| \\ &\le G_R(T)[\|(u_n)_t - (u_k)_t\|_\alpha + \|(P^n - P^k)(u_k)_t\|_\alpha]. \end{aligned}$$

Also,

$$\begin{aligned}\|(P^n - P^k)(u_k)_t\|_\alpha &\le \|A^{\alpha-\mu}(P^n - P^k)A^\mu(u_k)_t\| \\ &\le \frac{1}{\lambda_k^{\mu-\alpha}}\|(u_k)_t\|_\mu.\end{aligned}$$

Thus

$$\begin{aligned}\|g_n(t,(u_n)_t) - g_k(t,(u_k)_t)\| &\le G_R(T)[\|(u_n)_t - (u_k)_t\|_\alpha + \frac{1}{\lambda_k^{\mu-\alpha}}\|(u_k)_t\|_\mu] \\ &\le G_R(T)[\|u_n - u_k\|_{t,\alpha} + \frac{U_0}{\lambda_k^{\mu-\alpha}}].\end{aligned}$$

Similarly,

$$\|A^\beta f_n(t,(u_n)_t) - A^\beta f_k(t,(u_k)_t)\| \le L_f\left[\|u_n - u_k\|_{t,\alpha} + \frac{U_0}{\lambda_k^{\mu-\alpha}}\right]$$

and

$$\|k_n(u_n)(t) - k_k(u_k)(t)\| \le C_k(T)\left[\|u_n(t) - u_k(t)\|_\alpha + \frac{1}{\lambda_k^{\mu-\alpha}}\|u_k(t)\|_\mu\right].$$

Therefore,

$$\begin{aligned}\|K_n(u_n)(t) - K_k(u_k)(t)\| &\le M_0\int_0^t \|k_n(u_n)(\tau) - k_k(u_k)(\tau)\|d\tau \\ &\le T_0M_0C_k(T)[\|u_n - u_k\|_{t,\alpha} + \frac{U_0}{\lambda_k^{\mu-\alpha}}].\end{aligned}$$

We choose t_0' such that $0 < t_0' < t < T_0$, we have

$$\begin{aligned}\|u_n(t) - u_k(t)\|_\alpha \le\ & \|S_q(t)A^\alpha(f_n(0,\widetilde{\phi}) - f_k(0,\widetilde{\phi}))\| \\ &+ \|A^{\alpha-\beta}\|\|A^\beta f_n(t,(u_n)_t) - A^\beta f_k(t,(u_k)_t)\| \\ &+ \left(\int_0^{t_0'} + \int_{t_0'}^t\right)(t-s)^{q-1}\|A^{1+\alpha-\beta}Q_q(t-s)\|\|A^\beta f_n(s,(u_n)_s) \\ &- A^\beta f_k(s,(u_k)_s)\|ds \\ &+ \left(\int_0^{t_0'} + \int_{t_0'}^t\right)(t-s)^{q-1}\|A^\alpha Q_q(t-s)\|\|g_n(s,(u_n)_s) - g_k(s,(u_k)_s)\|ds \\ &+ \left(\int_0^{t_0'} + \int_{t_0'}^t\right)(t-s)^{q-1}\|A^\alpha Q_q(t-s)\|\|K_n(u_n)(s) - K_k(u_k)(s)\|ds.\end{aligned} \tag{25}$$

The first term of above inequality can be evaluated as

$$\begin{aligned}\|S_q(t)A^\alpha(f_n(0,\widetilde{\phi})-f_k(0,\widetilde{\phi})\| &\le C\|A^{\alpha-\beta}\|\|A^\beta f_n(0,\widetilde{\phi})-A^\beta f_k(0,\widetilde{\phi})\|\\ &\le C\|A^{\alpha-\beta}\|L_f\|(P^n-P^k)\widetilde{\phi}\|_{0,\alpha}.\end{aligned}$$

First, third and fifth integral of the inequality (25) can be evaluated as

$$\begin{aligned}&\int_0^{t_0'}(t-s)^{q-1}\|A^{1+\alpha-\beta}Q_q(t-s)[A^\beta f_n(s,(u_n)_s)-A^\beta f_k(s,(u_k)_s)]\|ds\\ &\quad\le \frac{2qC_{1+\alpha-\beta}\Gamma(1-(\alpha-\beta))}{\Gamma(1+q(\beta-\alpha))}(L_fR+A_1)(T_0-t_0')^{q(\beta-\alpha)-1}t_0'.\\ &\int_0^{t_0'}(t-s)^{q-1}\|A^\alpha Q_q(t-s)\|[\|g_n(s,(u_n)_s)-g_k(s,(u_k)_s)\|]ds\\ &\quad\le \frac{2qC_\alpha\Gamma(2-\alpha)}{\Gamma(1+q(1-\alpha))}G_R(T)(T_0-t_0')^{q(1-\alpha)-1}t_0'.\\ &\int_0^{t_0'}(t-s)^{q-1}\|A^\alpha Q_q(t-s)\|[\|K_n(u_n)(s)-k_k(u_k)(s)\|]ds\\ &\quad\le \frac{2qC_\alpha\Gamma(2-\alpha)}{\Gamma(1+q(1-\alpha))}T_0M_0C_k(T)(T_0-t_0')^{q(1-\alpha)-1}t_0'.\end{aligned}$$

We estimate the second, forth and sixth integral of the inequality (25) as

$$\begin{aligned}&\int_{t_0'}^{t}(t-s)^{q-1}\|A^{1+\alpha-\beta}Q_q(t-s)\|[\|A^\beta f_n(s,(u_n)_s)-A^\beta f_k(s,(u_k)_s)\|]ds\\ &\le \frac{qC_{1+\alpha-\beta}\Gamma(1-(\alpha-\beta))}{\Gamma(1+q(\beta-\alpha))}L_f\int_{t_0'}^{t}(t-s)^{q(\beta-\alpha)-1}[\|(u_n)_s-(u_k)_s\|_\alpha+\frac{1}{\lambda_k^{\mu-\alpha}}\|(u_k)_s\|_\mu]ds\\ &\le \frac{qC_{1+\alpha-\beta}\Gamma(1-(\alpha-\beta))}{\Gamma(1+q(\beta-\alpha))}L_f[\frac{U_0T_0^{q(\beta-\alpha)}}{q(\beta-\alpha)\lambda_k^{\mu-\alpha}}+\int_{t_0'}^{t}(t-s)^{q(\beta-\alpha)-1}\|u_n-u_k\|_{s,\alpha}ds].\end{aligned}$$

$$\begin{aligned}&\int_{t_0'}^{t}(t-s)^{q-1}\|A^\alpha Q_q(t-s)\|\|g_n(s,(u_n)_s)-g_k(s,(u_k)_s)\|ds\\ &\le \frac{qC_\alpha\Gamma(2-\alpha)}{\Gamma(1+q(1-\alpha))}G_R(T)\int_{t_0'}^{t}(t-s)^{q(1-\alpha)-1}[\|(u_n)_s-(u_k)_s\|+\frac{1}{\lambda_k^{\mu-\alpha}}\|(u_k)_s\|_\mu ds]\\ &\le \frac{qC_\alpha\Gamma(2-\alpha)}{\Gamma(1+q(1-\alpha))}G_R(T)[\frac{U_0T_0^{q(1-\alpha)}}{q(1-\alpha)\lambda_k^{\mu-\alpha}}+\int_{t_0'}^{t}(t-s)^{q(1-\alpha)-1}\|u_n-u_k\|_{s,\alpha}ds].\end{aligned}$$

$$\int_{t_0'}^{t}(t-s)^{q-1}\|A^{\alpha}Q_q(t-s)\|[\|K_n(u_n)(s)-K_k(u_k)(s)\|]ds$$
$$\leq \frac{qC_\alpha\Gamma(2-\alpha)}{\Gamma(1+q(1-\alpha))}T_0M_0C_k(T)\int_{t_0'}^{t}(t-s)^{q(1-\alpha)-1}[\|(u_n)(s)-(u_k)(s)\|_\alpha+\frac{1}{\lambda_k^{\mu-\alpha}}\|(u_k)_s\|_\mu]ds$$
$$\leq \frac{qC_\alpha\Gamma(2-\alpha)}{\Gamma(1+q(1-\alpha))}T_0M_0C_k(T)[\frac{U_0T_0^{q(1-\alpha)}}{q(1-\alpha)\lambda_k^{\mu-\alpha}}+\int_{t_0'}^{t}(t-s)^{q(1-\alpha)-1}\|u_n-u_k\|_{s,\alpha}ds].$$

Thus,

$$\begin{aligned}\|u_n(t)-u_k(t)\|_\alpha \leq\ & C\|A^{\alpha-\beta}\|L_f\|(P^n-P^k)\widetilde{\phi}\|_{0,\alpha}+\|A^{\alpha-\beta}\|L_f\left[\|u_n-u_k\|_{t,\alpha}+\frac{U_0}{\lambda_k^{\mu-\alpha}}\right]\\ &+2\Big[\frac{qC_{1+\alpha-\beta}\Gamma(1-(\alpha-\beta))}{(T_0-t_0')^{-q(\beta-\alpha)+1}\Gamma(1+q(\beta-\alpha))}(L_fR+A_1)\\ &+\frac{qC_\alpha\Gamma(2-\alpha)}{(T_0-t_0')^{-q(1-\alpha)+1}\Gamma(1+q(1-\alpha))}C(T)]t_0'+K_{\alpha,\beta}\frac{U_0}{\lambda_k^{\mu-\alpha}}\\ &+\int_{t_0'}^{t}[\frac{qC_{1+\alpha-\beta}\Gamma(1-(\alpha-\beta))}{\Gamma(1+q(\beta-\alpha))}L_f(t-s)^{q(\beta-\alpha)-1}\\ &+\frac{qC_\alpha\Gamma(2-\alpha)}{\Gamma(1+q(1-\alpha))}C(T)(t-s)^{q(1-\alpha)-1}\Big]\|u_n-u_k\|_{s,\alpha}ds,\end{aligned}\tag{26}$$

where

$$K_{\alpha,\beta}=\frac{qC_{1+\alpha-\beta}\Gamma(1-(\alpha-\beta))}{\Gamma(1+q(\beta-\alpha))}L_f\frac{T_0^{q(\beta-\alpha)}}{q(\beta-\alpha)}+\frac{qC_\alpha\Gamma(2-\alpha)}{\Gamma(1+q(1-\alpha))}C(T)\frac{T_0^{q(1-\alpha)}}{q(1-\alpha)}.$$

and further, we have

$$\begin{aligned}\|u_n(t)-u_k(t)\|_\alpha \leq\ & B_1+\|A^{\alpha-\beta}\|L_f(\|u_n-u_k\|_{t,\alpha})+B_2t_0'+\frac{B_3}{\lambda_k^{\mu-\alpha}}\\ &+B_4\int_{t_0'}^{t}(t-s)^{q(1-\alpha)-1}\|u_n-u_k\|_{s,\alpha}ds,\end{aligned}\tag{27}$$

where

$$\begin{aligned}B_1 &= C\|A^{\alpha-\beta}\|L_f\|(P^n-P^k)\widetilde{\phi}\|_{0,\alpha},\\ B_2 &= 2\Big[\frac{qC_{1+\alpha-\beta}\Gamma(1-(\alpha-\beta))}{(T_0-t_0')^{-q(\beta-\alpha)+1}\Gamma(1+q(\beta-\alpha))}(L_fR+A_1)\\ &\quad+\frac{qC_\alpha\Gamma(2-\alpha)}{(T_0-t_0')^{-q(1-\alpha)+1}\Gamma(1+q(1-\alpha))}C(T)\Big],\\ B_3 &= \big[\|A^{\alpha-\beta}\|L_f+K_{\alpha,\beta}\big]U_0,\end{aligned}$$

$$B_4 = \frac{qC_{1+\alpha-\beta}\Gamma(1-(\alpha-\beta))}{\Gamma(1+q(\beta-\alpha))}L_f + \frac{qC_\alpha\Gamma(2-\alpha)}{\Gamma(1+q(1-\alpha))}C(T).$$

Replace t by $t+\theta$ in above inequality where $t_0'-t \le \theta \le 0$, we get

$$\begin{aligned}\|u_n(t+\theta)-u_k(t+\theta)\|_\alpha \le & B_1 + \|A^{\alpha-\beta}\|L_f(\|u_n-u_k\|_{t,\alpha}) + B_2 t_0' + \frac{B_3}{\lambda_k^{\mu-\alpha}} \\ & + B_4\int_{t_0'}^{t+\theta}(t+\theta-s)^{q(1-\alpha)-1}\|u_n-u_k\|_{s,\alpha}ds.\end{aligned}$$

Now put $s-\theta=\gamma$ in the integral term of above inequality and we get

$$\begin{aligned}\|u_n(t+\theta)-u_k(t+\theta)\|_\alpha \le & B_1 + \|A^{\alpha-\beta}\|L_f(\|u_n-u_k\|_{t,\alpha}) + B_2 t_0' + \frac{B_3}{\lambda_k^{\mu-\alpha}} \\ & + B_4\int_{t_0'-\theta}^{t}(t-\gamma)^{q(1-\alpha)-1}\|u_n-u_k\|_{\gamma+\theta,\alpha}d\gamma \\ \le & B_1 + \|A^{\alpha-\beta}\|L_f(\|u_n-u_k\|_{t,\alpha}) + B_2 t_0' + \frac{B_3}{\lambda_k^{\mu-\alpha}} \\ & + B_4\int_{t_0'-\theta}^{t}(t-\gamma)^{q(1-\alpha)-1}\|u_n-u_k\|_{\gamma,\alpha}d\gamma \\ \le & B_1 + \|A^{\alpha-\beta}\|L_f(\|u_n-u_k\|_{t,\alpha}) + B_2 t_0' + \frac{B_3}{\lambda_k^{\mu-\alpha}} \\ & + B_4\int_{t_0'}^{t}(t-\gamma)^{q(1-\alpha)-1}\|u_n-u_k\|_{\gamma,\alpha}d\gamma.\end{aligned}$$

Thus,

$$\begin{aligned}\sup_{t_0'-t\le\theta\le 0}\|u_n(t+\theta)-u_k(t+\theta)\|_\alpha \le & B_1 + \|A^{\alpha-\beta}\|L_f(\|u_n-u_k\|_{t,\alpha}) + B_2 t_0' + \frac{B_3}{\lambda_k^{\mu-\alpha}} \\ & + B_4\int_{t_0'}^{t}(t-\gamma)^{q(1-\alpha)-1}\|u_n-u_k\|_{\gamma,\alpha}d\gamma. \quad (28)\end{aligned}$$

Since $u_n(t+\theta)=\phi(t+\theta)$ for $t+\theta\le 0$ and for all $n\ge n_0$, thus, we have

$$\begin{aligned}\sup_{-h-t\le\theta\le 0}\|u_n(t+\theta)-u_k(t+\theta)\|_\alpha \le & \sup_{0\le\theta+t\le t_0'}\|u_n(t+\theta)-u_k(t+\theta)\|_\alpha \\ & + \sup_{t_0'-t\le\theta\le 0}\|u_n(t+\theta)-u_k(t+\theta)\|_\alpha.\end{aligned} \quad (29)$$

For $0<t\le t_0'$, we have from (27)

$$\|u_n(t) - u_k(t)\|_\alpha \leq B_1 + \|A^{\alpha-\beta}\| L_f(\|u_n - u_k\|_{t,\alpha}) + B_2 t_0' + \frac{B_5}{\lambda_k^{\mu-\alpha}}, \tag{30}$$

where $B_5 = \|A^{\alpha-\beta}\| L U_0$. Using inequalities (28) and (30) in (29), we get

$$\begin{aligned}\sup_{-h\leq t+\theta\leq t} \|u_n(t+\theta) - u_k(t+\theta)\|_\alpha \leq & 2B_1 + 2\|A^{\alpha-\beta}\| L_f(\|u_n - u_k\|_{t,\alpha}) + 2B_2 t_0' \\ & + \frac{B_3 + B_5}{\lambda_k^{\mu-\alpha}} + B_4 \int_{t_0'}^{t} (t-\gamma)^{q(1-\alpha)-1} \|u_n - u_k\|_{\gamma,\alpha} d\gamma.\end{aligned}$$

Since we have $2L_f\|A^{\alpha-\beta}\| < 1$. We have

$$\begin{aligned}\|u_n(t) - u_k(t)\|_\alpha \leq & \frac{1}{(1 - 2L_f\|A^{\alpha-\beta}\|)} [2B_1 + 2B_2 t_0' + \frac{B_3 + B_5}{\lambda_k^{\mu-\alpha}} \\ & + B_4 \int_{t_0'}^{t} (t-\gamma)^{q(1-\alpha)-1} \|u_n - u_k\|_{\gamma,\alpha}] d\gamma.\end{aligned}$$

Lemma (5.6.7) in [17] implies that there exists a constant M such that

$$\|u_n(t) - u_k(t)\|_\alpha \leq \frac{1}{(1 - 2L_f\|A^{\alpha-\beta}\|)} \left[2B_1 + 2B_2 t_0' + \frac{B_3 + B_5}{\lambda_k^{\mu-\alpha}} \right] M.$$

Letting $k \to \infty$ and since t_0' is arbitrary, the right-hand side may be made as small as desired by taking t_0' sufficiently small, which gives the required result.

Theorem 3 *Let the hypotheses (H1)–(H4) hold and let $\phi(t) \in D(A^\alpha)$ for $t \in [-h, 0]$. Then there exists a unique function $u \in C_{T_0}^\alpha$ such that $u_n \to u$ as $n \to \infty$ in $C_{T_0}^\alpha$ and u satisfies (6) on $[-h, T_0]$.*

Proof Let $\phi(t) \in D(A^\alpha)$ for $t \in [-h, 0]$. since for $0 < t \leq T_0$, $A^\alpha u_n(t)$ converges to $A^\alpha u(t)$ as $n \to \infty$ and $u_n(t) = u(t) = \phi(t)$, for all n and $t \in [-h, 0]$, we have, for $-h = t = T$, $A^\alpha u_n(t)$ converges to $A^\alpha u(t)$ in H as $n \to \infty$. Since $u_n \in B_R(C_{T_0}^\alpha, \widetilde{\phi})$, it follows that $u \in B_R(C_{T_0}^\alpha, \widetilde{\phi})$ and for any $0 < t_0 \leq T$,

$$\lim_{n\to\infty} \sup_{t_0\leq t\leq T_0} \|u_n(t) - u(t)\|_\alpha = 0.$$

Also, we have

$$\sup_{t_0\leq t\leq T_0} \|g_n(t, (u_n)_t) - g(t, u_t)\| \leq G_R(T)(\|u_n - u\|_{T_0,\alpha} + \|(P^n - I)u\|_{T_0,\alpha}) \to 0,$$

$$\sup_{t_0\leq t\leq T_0} \|A^\beta f_n(t, (u_n)_t) - A^\beta f(t, u_t)\| \leq L_f(\|u_n - u\|_{T_0,\alpha} + \|(P^n - I)u\|_{T_0,\alpha}) \to 0,$$

$$\sup_{t_0 \le t \le T_0} \|k_n(u_n)(s) - k_k(u_k)(s)\| \le C_k(T)(\|u_n - u\|_{T,\alpha} + \|(P^n - I)u\|_{T_0}) \to 0.$$

As $n \to \infty$. For $t_0 \in (0, t_0)$, the integral equation (8) can be written as

$$\begin{aligned}(u_n)(t) = {} & S_q(t)[\widetilde{\phi}(0) + f_n(0, \widetilde{\phi})] - f_n(t, (u_n)_t) \\ & + \left(\int_0^{t_0} + \int_{t_0}^{t}\right)(t-s)^{q-1} Q_q(t-s) A f_n(s, (u_n)_s) ds \\ & + \left(\int_0^{t_0} + \int_{t_0}^{t}\right)(t-s)^{q-1} Q_q(t-s) g_n(s, (u_n)_s) ds \\ & + \left(\int_0^{t_0} + \int_{t_0}^{t}\right)(t-s)^{q-1} Q_q(t-s) K_n(u_n)(s) ds.\end{aligned}$$

First, third and fifth integral can be evaluated as

$$\begin{aligned}\left\|\int_0^{t_0}(t-s)^{q-1} Q_q(t-s) A f_n(s, (u_n)_s) ds\right\| &\le \int_0^{t_0}(t-s)^{q-1}\|A^{1-\beta} Q_q(t-s) A^{\beta} f_n(s, (u_n)_s)\| ds \\ &\le \frac{qC_{1-\beta}\Gamma(1+\beta)}{\Gamma(1+q\beta)}(L_f R + A_1) T_0^{q\beta-1} t_0. \\ \left\|\int_0^{t_0}(t-s)^{q-1} Q_q(t-s) g_n(s, (u_n)_s) ds\right\| &\le \frac{C}{\Gamma(q)} G_R(T) T_0^{q-1} t_0. \\ \left\|\int_0^{t_0}(t-s)^{q-1} Q_q(t-s) K_n(u_n)(s) ds\right\| &\le \frac{C}{\Gamma(q)} T_0 M_0 C_k(t) T_0^{q-1} t_0.\end{aligned}$$

Thus, we have

$$\begin{aligned}&\|u_n(t) - S_q(t)[\widetilde{\phi}(0) + f_n(0, \widetilde{\phi})] + f_n(t, (u_n)_t) - \int_{t_0}^{t}(t-s)^{q-1} Q_q(t-s) A f_n(s, (u_n)_s) ds \\ &\quad - \int_{t_0}^{t}(t-s)^{q-1} Q_q(t-s)[g_n(s, (u_n)_s) + K_n(u_n)(s)] ds\| \\ &\quad \le \left[\frac{qC_{1-\beta}\Gamma(1+\beta)}{\Gamma(1+q\beta)}(L_f(R) + A_1) T_0^{q\beta-1} + \frac{C}{\Gamma(q)} C(T) T_0^{q-1}\right] t_0.\end{aligned}$$

Let $n \to \infty$, we get

$$\begin{aligned}&\|u(t) - S_q(t)[\phi(0) + f(0, \phi)] + f(t, u_t) - \int_{t_0}^{t}(t-s)^{q-1} Q_q(t-s) A f(s, u_s) ds \\ &\quad - \int_{t_0}^{t}(t-s)^{q-1} Q_q(t-s) g(s, u_s) ds - \int_{t_0}^{t}(t-s)^{q-1} Q_q(t-s) K(u)(s) ds\| \\ &\quad \le \left[\frac{qC_{1-\beta}\Gamma(1+\beta)}{\Gamma(1+q\beta)}(L_f(R) + A_1) T_0^{q\beta-1} + \frac{C}{\Gamma(q)} C(T) T_0^{q-1}\right] t_0.\end{aligned}$$

Since $t_0 \in (0, T_0]$ is arbitrary, hence we conclude that u is the solution of the integral equation (6).

Now, we show the uniqueness of the solutions to (6). Let u_1, u_2 be two solutions on the interval $[-h, T_0]$. For $t \in [-h, 0]$ uniqueness is obvious and for $t \in (0, T_0]$, we have

$$\begin{aligned}
\|u_1(t) - u_2(t)\|_\alpha &\leq \|A^{\alpha-\beta}\| \|A^\beta f(t, (u_1)_t) - A^\beta f(t, (u_2)_t)\| \\
&\quad + \int_0^t (t-s)^{q-1} \|A^{1+\alpha-\beta} Q_q(t-s)[A^\beta f(s, (u_1)_s) - A^\beta f(s, (u_2)_s)]\| ds \\
&\quad + \int_0^t (t-s)^{q-1} \|A^\alpha Q_q(t-s)[g(s, (u_1)_s) - g(s, (u_2)_s)]\| ds \\
&\quad + \int_0^t (t-s)^{q-1} \|A^\alpha Q_q(t-s)[K(u_1)(s) - K(u_2)(s)]\| ds \\
&\leq \|A^{\alpha-\beta}\| L_f \|(u_1)_t - (u_2)_t\|_\alpha + \frac{q C_{1+\alpha-\beta} \Gamma(1-(\alpha-\beta))}{\Gamma(1+q(\beta-\alpha))} L_f \\
&\quad \times \int_0^t (t-s)^{q(\beta-\alpha)-1} \|(u_1)_s - (u_2)_s\|_\alpha ds \\
&\quad + \frac{q C_\alpha \Gamma(2-\alpha)}{\Gamma(1+q(1-\alpha))} G_R(t) \int_0^t (t-s)^{q(1-\alpha)-1} \|(u_1)_s - (u_2)_s\|_\alpha ds \\
&\quad + \frac{q C_\alpha \Gamma(2-\alpha)}{\Gamma(1+q(1-\alpha))} T_0 M_0 C_k(t) \int_0^t (t-s)^{q(1-\alpha)-1} \|u_1(s) - u_2(s)\|_\alpha ds \\
&\leq \|A^{\alpha-\beta}\| L_f \sup_{-h \leq \theta \leq 0} \|u_1(t+\theta) - u_2(t+\theta)\|_\alpha \\
&\quad + C' \int_0^t (t-s)^{q(1-\alpha)-1} \sup_{-h \leq \theta \leq 0} \|u_1(s+\theta) - u_2(s+\theta)\|_\alpha ds \\
&\quad + \frac{q C_\alpha \Gamma(2-\alpha)}{\Gamma(1+q(1-\alpha))} T_0 M_0 C_k(t) \int_0^t (t-s)^{q(1-\alpha)-1} \|u_1(s) - u_2(s)\|_\alpha ds,
\end{aligned}$$

where $C' = \frac{q C_{1+\alpha-\beta} \Gamma(1-(\alpha-\beta))}{\Gamma(1+q(\beta-\alpha))} L_f + \frac{q C_\alpha \Gamma(2-\alpha)}{\Gamma(1+q(1-\alpha))} G_R(t)$.

Let $\bar{\theta} \in [-t, 0]$ and $t \in [0, T_0]$ and suppose $T_0 \leq h$, hence, we have $0 \leq t \leq h$. For $t \leq -\bar{\theta}$, we have $u_1(t+\bar{\theta}) = u_2(t+\bar{\theta})$. For $t \geq -\bar{\theta}$, we have

$$\begin{aligned}
\|u_1(t+\bar{\theta}) - u_2(t+\bar{\theta})\|_\alpha &\leq \|A^{\alpha-\beta}\| L_f \sup_{-h \leq \theta \leq 0} \|u_1(t+\bar{\theta}+\theta) - u_2(t+\bar{\theta}+\theta)\|_\alpha \\
&\quad + C' \int_0^{t+\bar{\theta}} (t+\bar{\theta}-s)^{q(1-\alpha)-1} \sup_{-h \leq \theta \leq 0} \|u_1(s+\theta) - u_2(s+\theta)\|_\alpha ds \\
&\quad + \frac{q C_\alpha \Gamma(2-\alpha)}{\Gamma(1+q(1-\alpha))} T_0 M_0 C_k(t) \int_0^t (t-s)^{q(1-\alpha)-1} \|u_1(s) - u_2(s)\|_\alpha ds.
\end{aligned}$$

Substitute $s = \eta + \bar{\theta}$, we get

$$\begin{aligned}\|u_1(t+\bar\theta)-u_2(t+\bar\theta)\|_\alpha \le{}& \|A^{\alpha-\beta}\|L_f \sup_{-h\le\theta\le 0}\|u_1(t+\bar\theta+\theta)-u_2(t+\bar\theta+\theta)\|_\alpha\\ &+C'\int_{-\bar\theta}^{t}(t-\eta)^{q(1-\alpha)-1}\sup_{-h\le\theta\le 0}\|u_1(\eta+\bar\theta+\theta)-u_2(\eta+\bar\theta+\theta)\|_\alpha d\eta\\ &+\frac{qC_\alpha\Gamma(2-\alpha)}{\Gamma(1+q(1-\alpha))}T_0M_0C_k(t)\int_0^t(t-s)^{q(1-\alpha)-1}\|u_1(s)-u_2(s)\|_\alpha ds.\end{aligned}$$

Let $\theta=\gamma-\bar\theta$, then we get

$$\begin{aligned}\|u_1(t+\bar\theta)-u_2(t+\bar\theta)\|_\alpha \le{}& \|A^{\alpha-\beta}\|L_f \sup_{-h+\bar\theta\le\gamma\le 0}\|u_1(t+\gamma)-u_2(t+\gamma)\|_\alpha\\ &+C'\int_{-\bar\theta}^{t}(t-\eta)^{q(1-\alpha)-1}\sup_{-h+\bar\theta\le\gamma\le 0}\|u_1(\eta+\gamma)-u_2(\eta+\gamma)\|_\alpha d\eta\\ &+\frac{qC_\alpha\Gamma(2-\alpha)}{\Gamma(1+q(1-\alpha))}T_0M_0C_k(t)\int_0^t(t-s)^{q(1-\alpha)-1}\|u_1(s)-u_2(s)\|_\alpha ds.\end{aligned}$$

Since $u_1(\eta+\gamma)=u_2(\eta+\gamma)$ on $[-h+\bar\theta,-h]$ then

$$\begin{aligned}\|(u_1)_t(\bar\theta)-(u_2)_t(\bar\theta)\|_\alpha \le{}& \|A^{\alpha-\beta}\|L_f \sup_{-h\le\gamma\le 0}\|(u_1)_t(\gamma)-(u_2)_t(\gamma)\|_\alpha\\ &+C'\int_{-\bar\theta}^{t}(t-\eta)^{q(1-\alpha)-1}\sup_{-h\le\gamma\le 0}\|(u_1)_\eta(\gamma)-(u_2)_\eta(\gamma)\|_\alpha d\eta\\ &+\frac{qC_\alpha\Gamma(2-\alpha)}{\Gamma(1+q(1-\alpha))}T_0M_0C_k(t)\int_0^t(t-s)^{q(1-\alpha)-1}\|u_1(s)-u_2(s)\|_\alpha ds\\ \le{}& \|A^{\alpha-\beta}\|L_f \sup_{-h\le\gamma\le 0}\|(u_1)_t(\gamma)-(u_2)_t(\gamma)\|_\alpha\\ &+C'\int_{0}^{t}(t-\eta)^{q(1-\alpha)-1}\sup_{-h\le\gamma\le 0}\|(u_1)_\eta(\gamma)-(u_2)_\eta(\gamma)\|_\alpha d\eta\\ &+\frac{qC_\alpha\Gamma(2-\alpha)}{\Gamma(1+q(1-\alpha))}T_0M_0C_k(t)\int_0^t(t-s)^{q(1-\alpha)-1}\|u_1(s)-u_2(s)\|_\alpha ds.\end{aligned}$$

Taking supremum on $\bar\theta$ over $[-h,0]$, we get

$$\begin{aligned}\|(u_1)_t-(u_2)_t\|_{0,\alpha} \le{}& \|A^{\alpha-\beta}\|L_f\|(u_1)_t-(u_2)_t\|_{0,\alpha}+C'\int_0^t(t-\eta)^{q(1-\alpha)-1}\|(u_1)_\eta-(u_2)_\eta\|_{0,\alpha}d\eta\\ &+\frac{qC_\alpha\Gamma(2-\alpha)}{\Gamma(1+q(1-\alpha))}T_0M_0C_k(t)\int_0^t(t-s)^{q(1-\alpha)-1}\|u_1(s)-u_2(s)\|_\alpha ds.\end{aligned}$$

Since $\|A^{\alpha-\beta}\|L_f<1$, we have

$$\|(u_1)_t-(u_2)_t\|_{0,\alpha} \le \frac{C''}{1-\|A^{\alpha-\beta}\|L_f}\int_0^t(t-\eta)^{q(1-\alpha)-1}\|(u_1)_\eta-(u_2)_\eta\|_{0,\alpha}d\eta. \tag{31}$$

where $C'' = C' + \frac{qC_\alpha \Gamma(2-\alpha)}{\Gamma(1+q(1-\alpha))} T_0 M_0 C_k(t)$. Using Lemma (5.6.7) in Pazy [17], we get

$$\|(u_1)_t - (u_2)_t\|_{0,\alpha} = 0,$$

for all $t \in [0, T_0]$. and from the fact that

$$\|u_1(t) - u_2(t)\| \leq \frac{1}{\lambda_0^\alpha} \|(u_1)_t - (u_2)_t\|_\alpha$$

it implies that $u_1 = u_2$ on $[0, T_0]$.

5 Faedo–Galerkin Approximations

In this section, we discuss the Faedo–Galerkin Approximation of solution and prove the convergence result for such approximation.

For any $0 < T_0 < T$, we have a unique $u \in C_{T_0}^\alpha$ satisfying the following integral equation:

$$u(t) = \begin{cases} \phi(t), & t \in [-h, 0]; \\ S_q(t)[\phi(0) + f(0, \widetilde{\phi})] - f(t, u_t) + \int_0^t (t-s)^{q-1} Q_q(t-s) A f(s, u_s) ds & \\ + \int_0^t (t-s)^{q-1} Q_q(t-s)[g(s, u_s) + K(u)(s)] ds, & t \in [0, T_0]. \end{cases} \tag{32}$$

Also, we have a unique solution $u_n \in C_{T_0}^\alpha$ of the approximate integral equation

$$u_n(t) = \begin{cases} \widetilde{\phi}(t), & t \in [-h, 0]; \\ S_q(t)[\phi(0) + f_n(0, \widetilde{\phi})] - f_n(t, (u_n)_t) + \int_0^t (t-s)^{q-1} Q_q(t-s) & \\ A f_n(s, (u_n)_s) ds + \int_0^t (t-s)^{q-1} Q_q(t-s)[g_n(s, (u_n)_s) + K_n(u_n)(s)] ds, & t \in (0, T_0]. \end{cases} \tag{33}$$

If we project (33) onto H_n, we get the Faedo–Galerkin approximation $\widehat{u}_n(t) = P^n u_n(t)$ satisfying

$$\widehat{u}_n(t) = P^n u_n(t) = \begin{cases} P^n \widetilde{\phi}(t), & t \in [-h, 0]; \\ S_q(t)[P^n \phi(0) + P^n f(0, P^n \phi)] - P^n f(t, P^n(\widehat{u}_n)_t) & \\ + \int_0^t (t-s)^{q-1} A Q_q(t-s) P^n f(s, P^n(\widehat{u}_n)_t) ds & \\ + \int_0^t (t-s)^{q-1} Q_q(t-s)[P^n g(s, P^n(\widehat{u}_n)_t) + P^n K(\widehat{u}_n)(s)] ds, & t \in [0, T_0]. \end{cases} \tag{34}$$

The solutions u of (32) and $\widehat{u}_n$ of (34), have the representation

$$u(t) = \Sigma_{i=0}^\infty \alpha_i(t) u_i, \;\; \alpha_i(t) = (u(t), u_i), i = 0, 1, \cdots; \tag{35}$$

$$\widehat{u}_n(t) = \Sigma_{i=0}^{\infty}\alpha_i^n(t)u_i, \ \ \alpha_i^n(t) = (\widehat{u}_n(t), u_i), i = 0, 1, \cdots \tag{36}$$

Using (36) in (34), we get the following system of first-order functional differential equations

$$\frac{d^q}{dt^q}(\alpha_i^n(t) + F_i^n(t, (\alpha_0^n)_t, \cdots, (\alpha_i^n)_t)) + \lambda_i\alpha_i^n(t) = G_i^n(t, (\alpha_0^n)_t, ..., (\alpha_n^n)_t) + \int_0^t M(t-s)k_i^n(\alpha_0^n(s), \cdots, \alpha_n^n(s))ds, \tag{37}$$

$$\alpha_i^n(0) = \phi_i,$$

here

$$F_i^n(t, (\alpha_0^n)_t, \cdots, (\alpha_i^n)_t)) = \left(f(t, \sum_{i=0}^{n}(\alpha_i^n)_t u_i), u_i \right),$$

$$G_i^n(t, (\alpha_0^n)_t, \cdots, (\alpha_n^n)t) = \left(g(t, \sum_{i=0}^{n}(\alpha_i^n)_t u_i), u_i \right),$$

$$k_i^n(\alpha_0^n(t), \cdots, \alpha_n^n(t)) = \left(k(\sum_{i=0}^{n}(\alpha_i^n)(t)u_i), u_i \right)$$

and $\phi_i = (\phi, u_i)$ for $i = 1, 2, \ldots n$.

The system (37) determines the $\alpha_i^n(t)'s$. Now, we shall prove the convergence of $\alpha_i^n(t) \to \alpha_i(t)$. We have

$$A^\alpha[u(t) - \widehat{u}(t)] = A^\alpha\left[\sum_{i=0}^{\infty}(\alpha_i(t) - \alpha_i^n(t))u_i\right] = \sum_{i=0}^{\infty}\lambda_i^\alpha(\alpha_i(t) - \alpha_i^n(t))u_i.$$

Therefore, we have

$$\|A^\alpha[u(t) - \widehat{u}(t)]\|^2 \geq \sum_{i=0}^{\infty}\lambda_i^{2\alpha}(\alpha_i(t) - \alpha_i^n(t))^2.$$

Theorem 4 *Let the hypotheses $(H1)$–$(H4)$ hold. If $\phi(0) \in D(A^\alpha)$ for $t \in [-h, 0]$, then*

$$\lim_{n\to\infty}\sup_{t_0\leq t\leq T_0}\left[\sum_{i=0}^{\infty}\lambda_i^{2\alpha}(\alpha_i(t) - \alpha_i^n(t))^2\right] = 0.$$

The statement of this theorem follows from the facts mentioned above and the following result:

Proposition 2 *Let the hypotheses* $(H1)$*–*$(H4)$ *hold. If* $\phi(0) \in D(A^{\alpha})$ *for* $t \in [-h, 0]$*, then*

$$\lim_{n\to\infty} \sup_{n\geq m, t_0\leq t\leq T_0} \|A^{\alpha}\,[\widehat{u}_n(t) - \widehat{u}_m(t)]\,\| = 0.$$

Proof For $n = m$, we have

$$\begin{aligned}\|A^{\alpha}[\widehat{u}_n(t) - \widehat{u}_m(t)]\| &= \|A^{\alpha}[P^n u_n(t) - P^m u_m(t)]\| \\ &\leq \|P^n[u_n(t) - u_m(t)]\|_{\alpha} + \|(P^n - p^m)u_m\|_{\alpha} \\ &\leq \|u_n(t) - u_m(t)\|_{\alpha} + \frac{1}{\lambda_m^{\mu-\alpha}}\|A^{\mu}u_m\|.\end{aligned}$$

If $\phi(t) \in D(A)$ for $t \in [-h, 0]$, then the result follows from Theorem (4.1).

References

1. Kilbas, A.A., Srivastava, H.M., Trujillo, J.J.: Theory and applications of fractional differential equations. North-Holland Mathematics Studies. Elsevier, Amsterdam (2006)
2. Lakshmikantham, V.: Theory of fractional functional differential equations. Nonlinear Anal. **69**(10), 3337–3343 (2008)
3. Miller, K.S., Ross, B.: An introduction to the fractional calculus and fractional differential equations. A Wiley-Interscience Publication, Wiley, New York (1993)
4. Podlubny, I.: Fractional differential equations. Mathematics in Science and Engineering. Academic Press, San Diego (1999)
5. Samko, S.G., Kilbas, A.A., Marichev, O.I.: Fractional Integrals and Derivatives, Translated from the 1987 Russian Original. Gordon and Breach, Yverdon (1993)
6. Murakami, H.: On non-linear ordinary and evolution equations. Funkcial. Ekvac. **9**, 151–162 (1966)
7. Segal, I.: Non-linear semi-groups. Ann. Math. **78**(2), 339–364 (1963)
8. Bazley, N.W.: Approximation of wave equations with reproducing nonlinearities. Nonlinear Anal. **3**(4), 539–546 (1979)
9. Bazley, N.W.: Global convergence of Faedo-Galerkin approximations to nonlinear wave equations. Nonlinear Anal. **4**(3), 503–507 (1980)
10. Miletta, P.D.: Approximation of solutions to evolution equations. Math. Methods Appl. Sci. **17**(10), 753–763 (1994)
11. Bahuguna, D., Srivastava, S.K.: Approximation of solutions to evolution integrodifferential equations. J. Appl. Math. Stochastic Anal. **9**(3), 315–322 (1996)
12. Bahuguna, D., Shukla, R.: Approximations of solutions to nonlinear Sobolev type evolution equations. Electron. J. Differ. Equ. **2003**(1), 16pp. (electronic)
13. Chaddha, A., Pandey, D.N.: Approximations of solutions for a Sobolev type fractional order differential equation. Nonlinear Dyn. Syst. Theory **14**(1), 11–29 (2014)
14. Dubey, R.S.: Approximations of solutions to abstract neutral functional differential equation. Numer. Funct. Anal. Optim. **32**(3), 286–308 (2011)
15. Muslim, M., Bahuguna, D.: Existence of solutions to neutral differential equations with deviated argument, Electron. J. Qual. Theory Differ. Equ. **2008**(27), 12 pp. (2008)

16. Muslim, M., Nandakumaran, A.K.: Existence and approximations of solutions to some fractional order functional integral equations. J. Integral Equations Appl. **22**(1), 95–114 (2010)
17. Pazy, A.: Semigroups of linear operators and applications to partial differential equations. Applied Mathematical Sciences. Springer, New York (1983)
18. El-Borai, M.M.: Some probability densities and fundamental solutions of fractional evolution equations. Chaos Solitons Fractals **14**(3), 433–440 (2002)
19. Zhou, Y., Jiao, F.: Existence of mild solutions for fractional neutral evolution equations. Comput. Math. Appl. **59**(3), 1063–1077 (2010)
20. Abujabal, H.A.S., El-Borai, M.M.: On the Cauchy problem for some abstract nonlinear differential equations. Korean J. Comput. Appl. Math. **3**(2), 279–290 (1996)

An Efficient Hybrid Approach for Simulating MHD Nanofluid Flow over a Permeable Stretching Sheet

Rama Bhargava, Mania Goyal and Pratibha

Abstract The problem of magnetohydrodynamics boundary layer flow and heat transfer on a permeable stretching surface in a nanofluid under the effect of heat generation and partial slip is simulated using numeric symbolic approach. The Brownian motion and thermophoresis effects are also considered. The boundary layer equations governed by the PDEs are transformed into a set of ODEs with the help of transformations. The differential equations are solved by variational finite element method as well as hybrid approach. The results obtained by the two approaches match well. The effects of different controlling parameters on the flow field and heat transfer characteristics are examined. The comparison confirms excellent agreement. The efficiency of the hybrid approach is demonstrated through a table. The present study is of great interest in coating and suspensions, cooling of metallic plate, oils and grease, paper production, coal water or coal-oil slurries, heat exchangers technology, materials processing exploiting.

Keywords Nanofluids · MHD · Stretching sheet · FEM · Hybrid approach

1 Introduction

The complicated mathematical models arising in problems of fluid flow and heat transfer has forced for finding numerical solutions using grid-based methods, e.g., finite element methods, finite volume method, etc. However, due to the basic problem of meshing and remeshing, these methods consume a lot of computational time.

R. Bhargava · M. Goyal (✉) · Pratibha
Department of Mathematics, Indian Institute of Technology Roorkee, Roorkee 247667, India
e-mail: goyalmania87@gmail.com

R. Bhargava
e-mail: rbharfma@iitr.ac.in

Pratibha
e-mail: pratifma@iitr.ac.in

P.N. Agrawal et al. (eds.), *Mathematical Analysis and its Applications*, Springer Proceedings in Mathematics & Statistics 143, DOI 10.1007/978-81-322-2485-3_56

Further, the occurrence of many integrals by default in the elements as well in the global matrix lead to many inaccuracies. Presently, symbolic computation is vastly applied in finite element analysis to solve system of equations, to derive stiffness matrices, which provide closed-form expressions for integration and can deliver a drastic savings of CPU time. Symbolic computation provide closed-form expressions for integration. Kaminski [1] developed the semi-analytical probabilistic version of the finite element method (FEM) related to the homogenization problem. The hybrid computational implementation of the system MAPLE with homogenization-oriented FEM code MCCEFF was invented to provide probabilistic FEM analysis.

Alns and Mardal [2] generated a low-level C++ code based on symbolic expressions to accomplish a high degree of abstraction in the problem definition while surpassing the run-time efficiency of traditional hand written C++ codes for FEM.

Integrations were performed analytically by Griffith et al. [3] in closed form with the help of computer algebra software for 3-, 6-, 10- and 15-noded triangles. The analytical routines ran significantly faster. Videla et al. [4] generated exact expressions for the stiffness matrix of an 8-node plane elastic finite element using computer algebra software. The reduction in CPU time was over 50 %. Eyheramendy and Saad [5] recently explored a track in which the symbolic framework is integrated into the classic numerical one which provides a dynamic binding of symbolically created formula.

It was observed that a hybrid approach, combining FEM with symbolic computations has a lot of advantages namely:

1. It increases accuracy.
2. It saves a lot of computational time.

Convective heat transfer can be enhanced by changing the flow geometry, boundary conditions, or by enhancing thermal conductivity of the fluid. Researchers have also tried to increase the thermal conductivity of base fluids by suspending micro- or larger-sized solid particles in fluids, since the thermal conductivity of solid is typically higher than that of liquid. Modern nanotechnology provides new opportunities to process and produce materials with average crystallite sizes below 50 nm. Fluids with nanoparticles suspended in them are called nanofluids, a term first proposed by Choi [6]. Choi et al. [7], and Masuda et al. [8] have shown that a very small amount of nanoparticles (usually less than 5 %), when dispersed uniformly and suspended stably in base fluids, can provide dramatic improvements in the thermal conductivity and in the heat transfer coefficient of the base fluid. Nanofluid is a suspension of nanoparticles in the base fluid. A comprehensive survey of convective transport in nanofluids was made by Buongiorno [9]. Khan and Pop [10] have used the model of Kuznetsov and Nield [11], Makinde and Aziz [12], Khan and Pop [10].

It is now a well-accepted fact that many fluids of industrial and geophysical importance are non-Newtonian. Due to much attention in many industrial applications, the research on boundary layer behaviour of a viscoelastic fluid over a continuously stretching surface keeps going. McCormack and Crane [13] have provided comprehensive discussion on boundary layer flow caused by stretching of an elastic flat sheet moving in its own plane with a velocity varying linearly with distance. Several researchers, viz., Gupta and Gupta [14], Dutta et al. [15], Chen and Char [16] extended the work of Crane [13] by including the effects of heat and mass transfer under different situations. Later, Rajagopal et al. [17] and Chang [18] presented an analysis on flow of viscoelastic fluid over a stretching sheet. The above sources all utilize the no-slip condition. Wang [19] discussed the partial slip effects on the planar stretching flow.

A study of utilizing heat source or sink in moving fluids assumes a greater significance in all situations that deal with exothermic or endothermic chemical reaction and those concerned with dissociating fluids. Sparrow and Cess [20] investigated the steady stagnation point flow and heat transfer in the presence of temperature-dependent heat absorption. Later, Azim et al. [21] discussed the effect of viscous Joule heating on MHD-conjugate heat transfer for a vertical flat plate in the presence of heat generation. One of the latest works is the study of the heat transfer characteristic in the mixed convection flow of a nanofluid along a vertical plate with heat source/sink, studied by Rana and Bhargava [22].

In real situations in nanofluids, the base fluid does not satisfy the properties of Newtonian fluids, hence it is more justified to consider them as viscoelastic fluids. In this paper, the base fluid is taken as second grade fluid. To the best of our knowledge, no studies have so far investigated to analyze the partial slip effect on the boundary layer flow of viscoelastic nanofluid over a permeable stretching sheet under the effect of MHD and heat generation. The objective of the present paper is to extend the work of Noghrehabadi [23] by taking base fluid as second grade fluid with a new approach. The hybrid technique is used to simulate the heat and mass transfer characteristics of the flow. Therefore, working with symbolic-numeric environment, we solve our models using FEM, working with closed-form analytical expressions, directly transforming complex analytical expressions into numerical tools and drastic savings of CPU time.

The results so obtained are compared with those of FEM and shown in Table III which shows a drastic saving in computational time. The effects of flow controlling parameters on the fluid velocity, temperature, nanoparticle concentration are shown. The heat transfer rate and the nanoparticle volume fraction rate have been demonstrated graphically and discussed. Our aim is to reduce, substantially, the CPU time, which is a subject of concern when dealing with large FEM meshes. The stiffness terms of elements are computed analytically through symbolic computation which will provide closed-form expressions for integration and has provided a drastic savings of CPU time. The analytical integration also ensures accurate results even for distorted elements.

2 Mathematical Formulation

Consider two-dimensional, steady, incompressible, laminar flow of non-Newtonian nanofluid past a stretching sheet in a quiescent fluid. The velocity of the stretching sheet is $u_w = U = cx$. The x-axis is taken along the plate in the vertically upward direction and the y-axis is taken normal to the plate. A transverse magnetic field of strength B_o is applied parallel to the y-axis. The surface of plate is maintained at uniform temperature and concentration, T_w and C_w, respectively, and these values are assumed to be greater than the ambient temperature and concentration, T_∞ and C_∞, respectively. Moreover, it is assumed that both the fluid phase and nanoparticles are in thermal equilibrium state. The thermophysical properties of the nanofluid are assumed to be constant. The pressure gradient and external forces are neglected. The governing equations are:

$$\frac{\partial u}{\partial x} + \frac{\partial v}{\partial y} = 0 \tag{1}$$

$$u\frac{\partial u}{\partial x} + v\frac{\partial u}{\partial y} = \nu\frac{\partial^2 u}{\partial y^2} - \frac{\sigma}{\rho_f}B_o^2 u + \frac{\alpha_1}{\rho_f}\left[\frac{\partial u}{\partial x}\frac{\partial^2 u}{\partial y^2} + u\frac{\partial^3 u}{\partial x \partial^2} + \frac{\partial u}{\partial y}\frac{\partial^2 v}{\partial y^2} + v\frac{\partial^3 u}{\partial y^3}\right] \tag{2}$$

$$u\frac{\partial T}{\partial x} + v\frac{\partial T}{\partial y} = \alpha_m\frac{\partial^2 T}{\partial y^2} + \frac{Q_o}{(T - T_\infty)_f} + \tau\left[D_B\frac{\partial C}{\partial y}\frac{\partial T}{\partial y} + \frac{D_T}{T_\infty}\left(\frac{\partial C}{\partial y}\right)^2\right] \tag{3}$$

$$u\frac{\partial C}{\partial x} + v\frac{\partial C}{\partial y} = D_B\frac{\partial^2 C}{\partial y^2} + \frac{D_T}{T_\infty}\frac{\partial^2 T}{\partial y^2} \tag{4}$$

The boundary conditions for the velocity, temperature, and concentration fields are given as follows:

$$u = U + \kappa\nu\frac{\partial u}{\partial y}, v = v_w, T = T_w, C = C_w \text{ at } y = 0 \tag{5}$$

$$u = 0, T = T_\infty, C = C_\infty \text{ as } y \to \infty \tag{6}$$

where u and v are the velocity components along the x and y-directions, respectively, p is the pressure, ρ_f is the density of base fluid, ρ_p is the nanoparticle density, μ is the absolute viscosity of the base fluid, ν is the kinematic viscosity of the base fluid, σ is the electrical conductivity of the base fluid, α_1 is the material fluid parameter, T is the fluid temperature, α_m is the thermal diffusivity, $\tau = (\rho C)_p/(\rho C)_f$ is the ratio of effective heat capacity of the nanoparticle material to heat capacity of the fluid, C is the nanoparticle volume fraction, D_B and D_T are the Brownian diffusion coefficient and the thermophoresis diffusion coefficient, T_∞ is the free stream temperature, C_p is the specific heat at constant pressure, and g, k are the acceleration due to gravity, the thermal conductivity of the fluid respectively.

To transform the governing equations into a set of similarity equations, the following dimensionless parameters are introduced:

$$\eta = \sqrt{\frac{c}{\nu}}y, u = cxf'(\eta), v = -\sqrt{c\nu}f(\eta), \theta(\eta) = \frac{T - T_\infty}{T_w - T_\infty}, \phi(\eta) = \frac{C - C_\infty}{C_w - C_\infty}. \tag{7}$$

The transformed momentum, energy, and concentration equations together with the boundary conditions given by (1)–(4), (5), (6) can be written as

$$f''' + ff'' - f'^2 - Mf' - \alpha\left(f''^2 - 2f'f''' + ff^{iv}\right) = 0 \tag{8}$$

$$\frac{1}{Pr}\theta'' + f\theta' + Nb\theta'\phi' + Nt\theta'^2 + Q\theta = 0 \tag{9}$$

$$\phi'' + Lef\phi' + \frac{Nt}{Nb}\theta'' = 0 \tag{10}$$

The transformed boundary conditions are

$$f(0) = s,\ f'(0) = 1 + Kf''(0), \theta(0) = 1, \phi(0) = 1 \text{ at } \eta = 0 \tag{11}$$

$$f'(\infty) \to 0, \theta(\infty) \to 0, \phi(\infty) \to 0 \text{ as } \eta \to \infty \tag{12}$$

where primes denote differentiation with respect to η and the seven parameters appearing in Eqs. (8–10) are defined as follows:

$$Pr = \frac{\nu}{\alpha_m}, Le = \frac{\nu}{D_B}, M = \frac{\sigma}{\rho_f}\frac{B_o^2}{c}, \alpha = \frac{\alpha_1 c}{\mu}, Q = \frac{Q_o}{c\rho C_f}$$
$$Nb = \frac{(\rho C)_p D_B(C_w - C_\infty)}{(\rho C)_f \nu}, Nt = \frac{(\rho C)_f D_T(T_w - T_\infty)}{(\rho C)_f T_\infty \nu} \tag{13}$$

In Eq. (13), Pr, Le, M, α, Q, Nb, and Nt denote the Prandtl number, the Lewis number, the magnetic field strength parameter, the viscoelastic parameter, the heat source/sink parameter, the Brownian motion parameter, and the thermophoresis parameter, respectively.

The physical quantities of interest are the local heat flux Nu and the local mass diffusion flux Sh from the vertical moving plate, which are defined as

$$Nu = \frac{xq_w}{k(T_w - T_\infty)}, Sh = \frac{xh_w}{D_B(C_w - C_\infty)} \tag{14}$$

where τ_w is the wall skin friction, q_w is the surface heat flux, and h_w is the wall mass flux given by

$$q_w = -k\left(\frac{\partial T}{\partial y}\right)_y = 0, h_w = -D_B\left(\frac{\partial C}{\partial y}\right)_y = 0 \tag{15}$$

Using (7) in (14), one can obtain

$$Re_x^{-1/2}Nu_x = -\theta'(0) = Nur, Re_x^{-1/2}Sh_x = -\phi'(0) = Shr, \tag{16}$$

where $Re_x = u_w(x)x/\nu$ is the local Reynolds number based on the stretching velocity $u_w(x)$. Kuznetsov and Nield [11] referred $Re_x^{-1/2}Nu_x$ and $Re_x^{-1/2}Sh_x$ as the reduced Nusselt number $Nur = -\theta'(0)$ and reduced Sherwood number $Shr = -\phi'(0)$, respectively.

3 Method of Solution

The problem has been undertaken by FEM and hybrid approach. The Finite Element Method (FEM) is a numerical and computer-based technique of solving a variety of practical engineering problems that arise in different fields. The method essentially consists of assuming the piecewise continuous function for the solution and obtaining the parameters of the functions in a manner that reduces the error in the solution. The steps involved in the finite element analysis are as follows:

- Discretization of the domain into set of finite elements.
- Weighted integral formulation of the differential equation.
- Defining an approximate solution over the element.
- Substitution of the approximate solution and the generation of the element equations.
- Assembly of the stiffness matrices for each element.
- Imposition of the boundary conditions.
- Solution of assembled equations.

The entire flow domain is divided into 10,000 quadratic elements of equal size. Each element is three-noded and therefore the whole domain contains 20001 nodes. A system of equations have been obtained which is solved numerically. The code of the algorithm has been executed in MATLAB. Excellent convergence was achieved for all the results.

4 Results and Discussion

Nonlinear ordinary differential equations (8)–(10) together with the boundary conditions (11) and (12) are solved numerically using FEM and numeric symbolic approach. The numerical computations have been carried out for different values of the parameters involved. The aim of the present study is to examine the variations of different quantities of parameters as given $0 \leq K \leq 10, 0 \leq \alpha \leq 10, 0 \leq Pr \leq 70, 0.1 \leq Nt \leq 0.5, 0.1 \leq Nb \leq 0.5$, and $5 \leq Le \leq 30$ and to show the efficiency of FEM with symbolic approach. The computational work is carried out by taking size of the element $\Delta\eta = 0.0001$. It is observed that, if the number of elements is increased or the size of the element is decreased in the same domain, even then the accuracy is not affected.

Figure 1 demonstrates that the effect of increasing value of slip parameter K is to shift the streamlines toward stretching boundary and thereby reduce thickness of the momentum boundary layer. Therefore, the effect of slip parameter K is seen to decrease the boundary layer velocity while the temperature and concentration are increased with increase in the slip parameter. Figures 4, 5 and 6 show the effect of viscoelastic parameter α on the evolution of fluid motion and subsequent on the

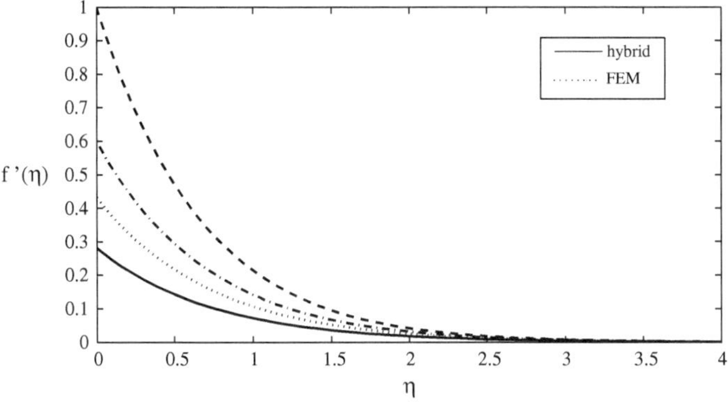

Fig. 1 Effect of K on velocity profile for $Pr = Le = M = 1.0$, $Nb = Nt = 0.5$, $Q = 0.1$, $\alpha = 0.5$, $s = 0.5$

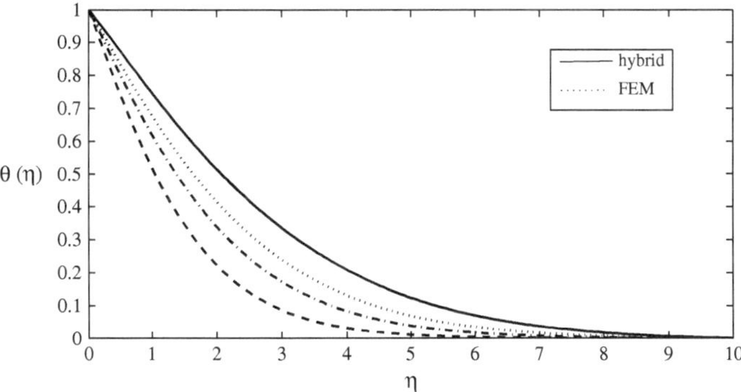

Fig. 2 Effect of K on temperature profile for $Pr = Le = M = 1.0$, $Nb = Nt = 0.5$, $Q = 0.1$, $\alpha = 0.5$, $s = 0.5$

distribution of heat and mass across the sheet as time evolves. From this plot it is evident that increasing values of viscoelastic parameter α opposes the motion of the liquid close to the stretching sheet and assists the motion of the liquid far away from the stretching sheet. Increasing values of second-grade parameter enables the liquid to flow at a faster rate due to which there is decline in the heat transfer. This is responsible for the increase in momentum boundary layer, whereas the thermal and concentration boundary layers reduce when the viscoelastic effects intensify (Figs. 2 and 3).

The variations in velocity field, temperature distribution, and nanoparticle concentration profile for various values of M are presented in Figs. 7, 8 and 9. It is clear from these figures that the velocity decreases, whereas the temperature and concentration increase with the increase in the magnetic field parameter. The hydromagnetic force in Eq. (7) is a linear Lorentzian body force which acts transverse to the direction of application, i.e., in the negative x-direction, parallel to the plate surface. It is

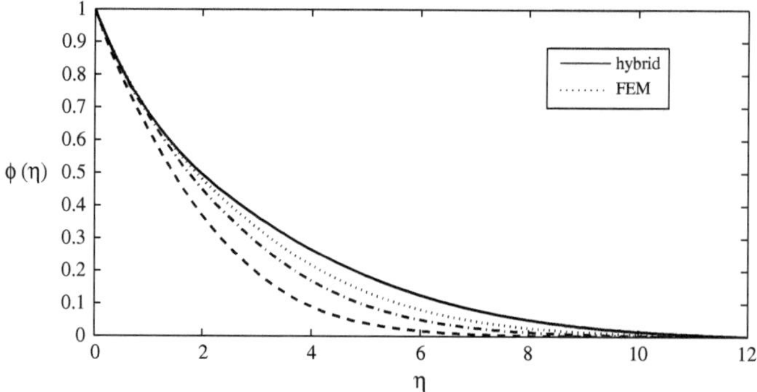

Fig. 3 Effect of K on nanoparticle concentration profile for $Pr = Le = M = 1.0, Nb = Nt = 0.5, Q = 0.1, \alpha = 0.5, s = 0.5$

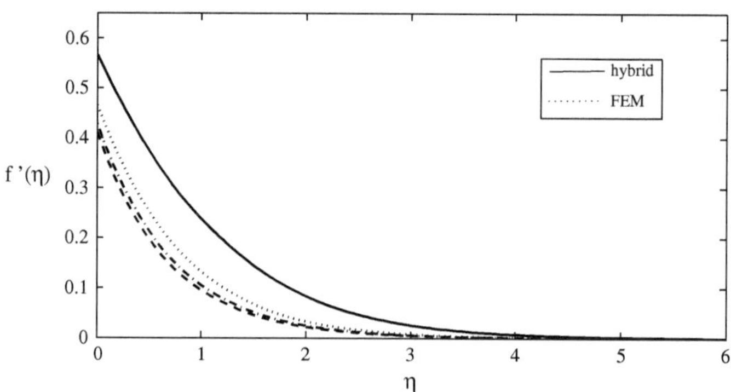

Fig. 4 Effect of α on velocity profile for $Pr = Le = M = K = 1.0, Nb = Nt = 0.5, Q = 0.1, s = 0.5$

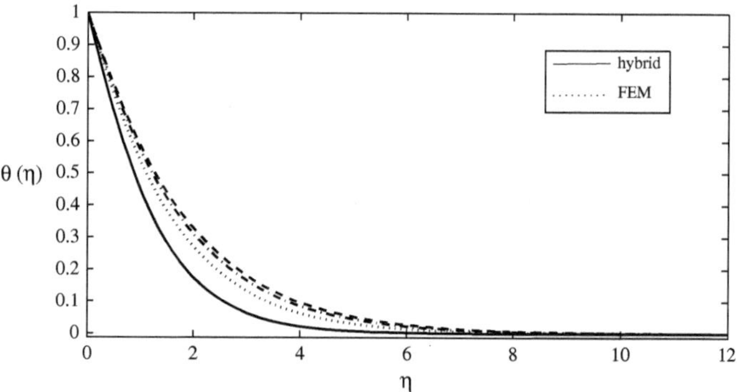

Fig. 5 Effect of α on temperature profile for $Pr = Le = M = K = 1.0, Nb = Nt = 0.5, Q = 0.1, s = 0.5$

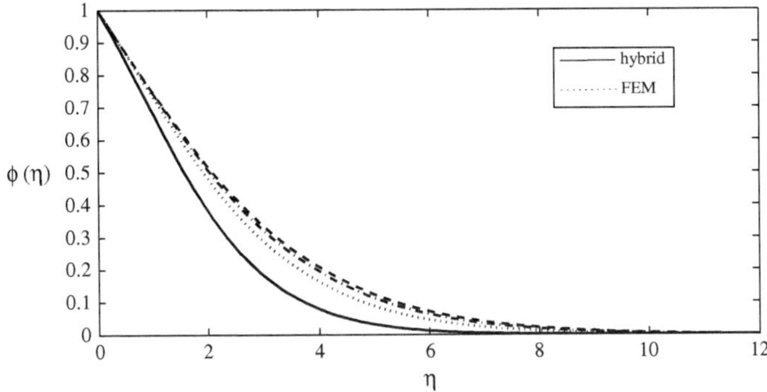

Fig. 6 Effect of α on nanoparticle concentration profile for $Pr = Le = M = K = 1.0, Nb = Nt = 0.5, Q = 0.1, s = 0.5$

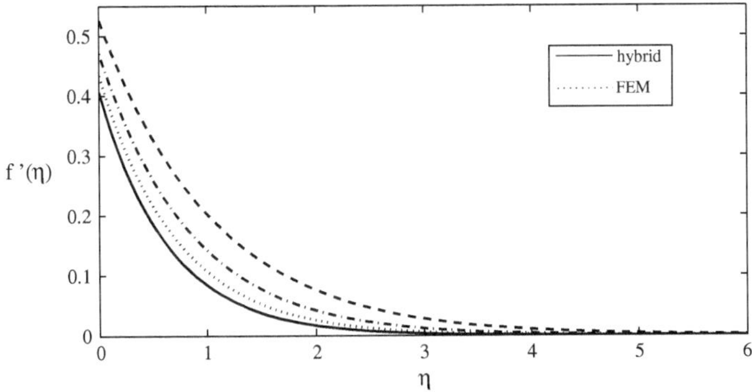

Fig. 7 Effect of M on velocity profile for $Pr = Le = K = 1.0, Nb = Nt = 0.5, Q = 0.1, s = 0.5, \alpha = 0.5$

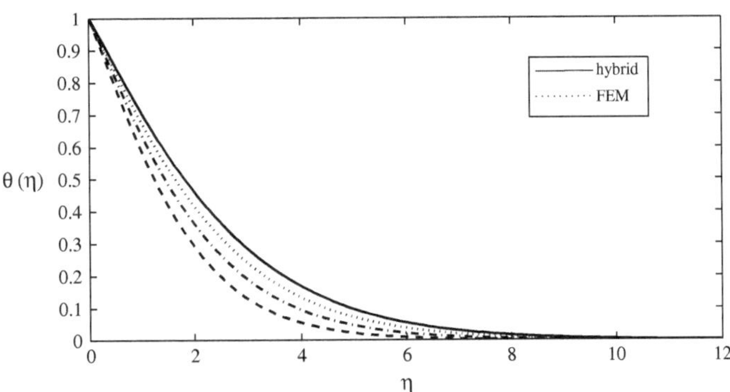

Fig. 8 Effect of M on temperature profile for $Pr = Le = K = 1.0, Nb = Nt = 0.5, Q = 0.1, s = 0.5, \alpha = 0.5$

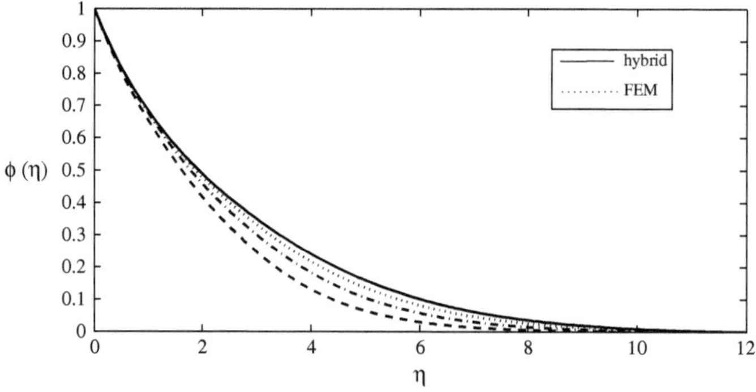

Fig. 9 Effect of M on nanoparticle concentration profile for $Pr = Le = K = 1.0, Nb = Nt = 0.5, Q = 0.1, s = 0.5, \alpha = 0.5$

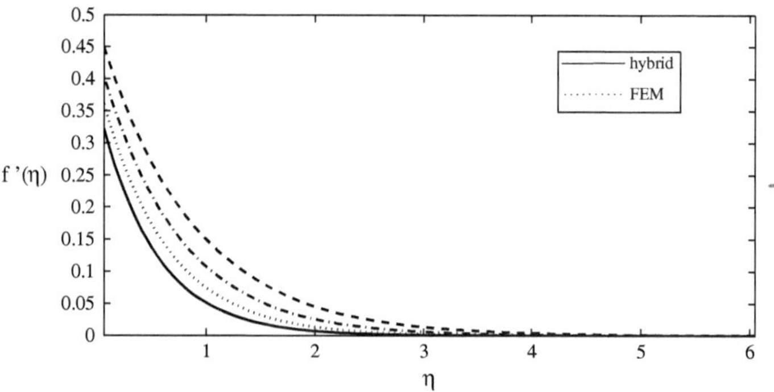

Fig. 10 Effect of s on velocity profile for $Pr = Le = K = M = 1.0, Nb = Nt = 0.5, Q = 0.1, \alpha = 0.5$

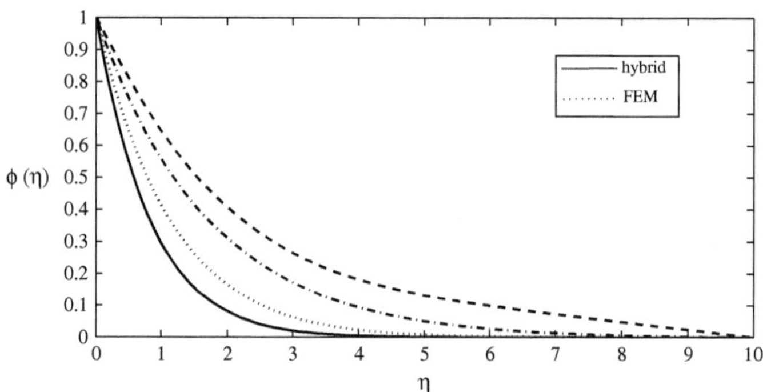

Fig. 11 Effect of s on nanoparticle concentration profile for $Pr = Le = K = M = 1.0, Nb = Nt = 0.5, Q = 0.1, \alpha = 0.5$

Table 1 Comparison of results for Nur when $\alpha = K = M = Q = s = 0$, $Nb = Nt = 0$

Pr	Wang [24]	Gorla and Sidawi [25]	Present results
0.07	0.0656	0.0656	0.0695
0.20	0.1691	0.1691	0.1694
0.70	0.4539	0.5349	0.4541
2.00	0.9114	0.9114	0.9120
7.00	1.8954	1.8905	1.8954
20.0	3.3539	3.3539	3.3539
70.0	6.4622	6.4622	6.4623

Table 2 Variation of Nur and Shr with Nb, Nt and Le when $s = 0.5$, $K = 1.0$, $Pr = 1.0$, $Q = 0.1$, $M = 1.0$, $\alpha = 0.5$

Nb	Nt	$Le = 1$		$Le = 10$	
		Nur	Shr	Nur	Shr
0.1	0.1	0.50278	0.25287	0.48513	4.6238
	0.2	0.48406	−0.15312	0.45966	4.3467
	0.3	0.46576	−0.52569	0.43552	4.1034
	0.4	0.44788	−0.86599	0.41262	3.8909
	0.5	0.43041	−1.1751	0.39089	3.7066
0.3	0.1	0.44173	0.44173	0.41204	4.8534
	0.2	0.42444	0.40848	0.39010	4.7790
	0.3	0.40757	0.34848	0.36947	4.7140
	0.4	0.39110	0.24442	0.34984	4.6575
	0.5	0.37502	0.15688	0.33120	4.6089
0.5	0.1	0.38546	0.62732	0.34938	4.8975
	0.2	0.36959	0.56696	0.33066	4.8621
	0.3	0.35412	0.51221	0.31292	4.8313
	0.4	0.33903	0.46290	0.29611	4.8048
	0.5	0.32430	0.41882	0.28015	4.7821

directly proportional to the applied magnetic field, B_0. This force inhibits momentum development and decelerates the flow. The supplementary work done in dragging the conducting nanofluid against the action of the magnetic field, B_0, is manifested as thermal energy. This heats the conducting nanofluid and elevates temperatures. The warming of the boundary layer therefore also aids in nanoparticle diffusion which causes a rise in nanoparticle volume fraction, ϕ.

Figures 10 and 11 depict the effects of suction parameter S on velocity and concentration profile. It is noticed that both momentum and concentration boundary layer thickness decrease with the increase in suction parameter.

In the present study, the local rate of heat transfer Nur and local rate of mass transfer at the sheet Shr, defined in Eq. (16), are the important characteristics. The numerical values of Nur and Shr are exhibited in Tables 1 and 2. Table 1 shows

Table 3 Comparison of CPU time in the two approaches (in seconds)

Number of elements	Usual FEM		Hybrid approach	
	Element matrix	Assembled matrix	Element matrix	Assembled matrix
1	0.07	0.15	0.03	0.08
10	1.31	1.63	0.67	0.88
20	2.15	4.15	1.05	2.02
50	8.89	9.83	4.48	4.87
100	17.82	22.98	8.89	11.45
200	46.85	58.72	24.02	29.58
500	120.08	141.12	61.54	71.32
1000	425.05	503.39	211.43	212.18

that the excellent correlation between the current FEM computations and the earlier results of Wang [24] and Gorla and Sidawi [25].

5 Conclusions

The problem of MHD boundary-layer flow of a viscoelastic nanofluid past a stretching sheet has been taken with numeric symbolic approach to exhibit the effect of partial slip (i.e., Navier's condition) and heat source/sink on the fluid flow and heat transfer characteristics. The result can be summarized as follows:

1. With the increase in the second grade parameter α, the velocity and the momentum boundary layer thickness increases, however, the temperature and nanoparticles concentration decrease.
2. There is a decrease in the velocity, but temperature and concentration are found to increase with increase in velocity slip parameter K.
3. Magnetic field decelerates the flow,as expected and enhances temperatures and nanoparticle volume fraction (concentration) distributions in the boundary layer.
4. With increase in the slip parameter K, heat transfer rate and mass transfer rate decrease which can be used for controlling the heat flow e.g., in reactors, etc.
5. With increase of thermophoretic number Nt, the effect of velocity slip parameter K on reduced Nusselt number Nur and reduced Sherwood number Shr increases and decreases respectively.
6. The reduced Nusselt number and reduced Sherwood number show the effectiveness of heat flow.
7. The results obtained by FEM and Hybrid approach tally nicely as shown in graphs and a drastic saving in computational time as shown in Table 3. Obviously it will be quite effective in more complicated problems. The CPU time which is a

subject of concern when dealing with large FEM meshes, in real-life problems is expected to curtail drastically which will be of great use.

Acknowledgments Authors 1 and 3 are thankful to SERB, DST for providing financial support for completing this paper under the project sponsored by them.

References

1. Kaminski, M.: On semi-analytical probabilistic finite element method for homogenization of the periodic fiber-reinforced composites. Int. J. Numer. Meth. Eng. **86**, 1144–1162 (2011)
2. Alns, M.S., Mardal, K.: On the efficiency of symbolic computations combined with code generation for finite element methods. ACM Trans. Math. Softw. **37**, 1–26 (2010)
3. Griffiths, D., Huang, J., Schiermeyer, R.P.: Elastic stiffness of straight-sided triangular finite elements by analytical and numerical integration. Commun. Numer. Meth. Eng. **25**, 247–262 (2009)
4. Videla, L., Baloa, T., Griffiths, D., Cerrolaza, M.: Exact integration of an 8-node plane elastic finite element by symbolic computation. Numer. Meth. Part. Diff. Eqns. **24**, 249–261 (2008)
5. Eyhermandy, D., Saad, R.: A dynamic approach for automating finite element code development. In: 11th World Congress on Computational Mechanics (WCCM XI), July 20–25, 2014. Spain, Barcelona (2014)
6. Choi, S.: Enhancing thermal conductivity of fluids with nanoparticles in developments and applications of Non-Newtonian flows. ASME FED 231/MD **66**, 99–105 (1995)
7. Choi, S., Zhang, Z., Yu, W., Lockwood, F., Grulke, E.: Anomalously thermal conductivity enhancement in nanotube suspensions. Appl. Phys. Lett. **79**, 2252–2254 (2001)
8. Masuda, H., Ebata, A., Teramae, K., Hishinuma, N.: Alteration of thermal conductivity and viscosity of liquid by dispersing ultra-fine particles. Netsu Bussei **7**, 227–33 (1993)
9. Buongiorno, J.: Convective transport in nanofluids. ASME J. Heat Transf. **128**, 240–250 (2006)
10. Khan, W., Pop, I.: Boundary-layer flow of a nanofluid past a stretching sheet. Int. J. Heat Mass Transf. **53**, 2477–2483 (2010)
11. Kuznetsov, A., Nield, D.: Natural convective boundary-layer flow of a nanofluid past a vertical plate. Int. J. Therm. Sci. **49**, 243–247 (2010)
12. Makinde, O., Aziz, A.: Boundary layer flow of a nanofluid past a stretching sheet with a convective boundary condition. Int. J. Therm. Sci. **50**, 1326–1332 (2011)
13. McCormack, P., Crane, L.: Physical Fluid Dynamics. Academic Press, New York (1973)
14. Gupta, P., Gupta, A.: Heat and mass transfer on a stretching sheet with suction or blowing. Can. J. Chem. Eng. **55**, 744–746 (1977)
15. Dutta, B., Roy, P., Gupta, A.: Temperature field in the flow over a stretching sheet with uniform heat flux. Int. Commun. Heat Mass Transf. **12**, 89–94 (1985)
16. Chen, C., Char, M.: Heat transfer of a continuous stretching surface with suction or blowing. J. Math Anal. Appl. **135**, 568–580 (1988)
17. Rajagopal, K., Na, T., Gupta, A.: Flow of a viscoelastic fluid over a stretching sheet. Rheol. Acta **23**, 213–221 (1984)
18. Chang, W.: The non-uniqueness of the flow of a viscoelastic fluid over a stretching sheet. Q. Appl. Math. **47**, 365–366 (1989)
19. Wang, C.: Flow due to a stretching boundary with partial slip-an exact solution of the Navier Stokes equations. Chem. Eng. Sci. **57**, 3745–3747 (2002)
20. Sparrow, E., Cess, R.: Temperature dependent heat sources or sinks in a stagnation point flow. Appl. Sci. Res. **10**, 185–197 (1961)
21. Azim, M., Mamun, A., Rahman, M.: Viscous joule heating MHD-conjugate heat transfer for a vertical flat plate in the presence of heat generation. Int. Commun. Heat Mass Transfer **37**, 666–74 (2010)

22. Rana, P., Bhargava, R.: Numerical study of heat transfer enhancement in mixed convection flow along a vertical plate with heat source/sink utilizing nanofluids. Commun. Nonlinear Sci. Numer. Simulat. **16**, 4318–4334 (2011)
23. Noghrehabadi, A., Pourrajab, R., Ghalambaz, M.: Effect of partial slip boundary condition on the flow and heat transfer of nanofluids past stretching sheet prescribed constant wall temperature. Int. J. Therm. Sci. **54**, 253–261 (2012)
24. Wang, C.: Free convection on a vertical stretching surface. J. Appl. Math. Mech. **69**, 418–4334 (1989)
25. Gorla, R., Sidawi, I.: Free convection on a vertical stretching surface with suction and blowing. Appl. Sci. Res. **52**, 247–257 (1994)

Some Advanced Finite Element Techniques for the Problems of Mechanics: A Review

Rama Bhargava and Rangoli Goyal

Abstract Reduced efficiency and difficult complex domain discretization for grid-based numerical methods, e.g. FEM/FVM have led to the formulation of advanced finite element techniques. Propagation of fire, detection of cracks in ice or bones, bursting of stars, etc. are some such domains where the grid-based methods fail. The present paper contains a review of some advanced techniques—αFEM and MeshFree Methods, along with the details about their implementation and development.

Keywords Meshfree Methods · Alpha FEM · EFGM · MLPG · N-SFEM · Parallelization

1 Introduction

A mathematical model is the best approximation to the physical world. Such models are assembled based on guaranteed conservation principles and/or empirical observations. Numerical methods are essential for the effective simulation of the mathematical model as the underlying partial differential equations (PDE) usually have to be approximated. Many numerical methods have been developed to achieve this task. The grid-based methods introduce a finite number of nodes and can be based on the principles of weighted residual methods. The three classical choices for grid-based methods are the finite difference method (FDM), the finite element method (FEM) and the finite volume method (FVM).

The FDM is the oldest among the grid-based methods. It is based upon the application of a local Taylor expansion to approximate the differential equations. It uses a topologically quadrangular network of lines to build the mesh for discretization of the

R. Bhargava · R. Goyal (✉)
Department of Mathematics, Indian Institute of Technology Roorkee,
Roorkee 247667, India
e-mail: goyaldma@iitr.ac.in

R. Bhargava
e-mail: rbharfma@iitr.ac.in

P.N. Agrawal et al. (eds.), *Mathematical Analysis and its Applications*,
Springer Proceedings in Mathematics & Statistics 143,
DOI 10.1007/978-81-322-2485-3_57

PDE. This method fails when we have complex geometries in multiple dimensions. This drawback encouraged the use of an integral form of the PDEs and consequently the development of the finite element and finite volume techniques. FEM and FVM involve subdividing the domain into a large number of finite elements/control volumes and then solving the governing equations of fluid flow. A system of algebraic equations is formed and solved using various iterative methods. The numerical methods differ in the definition and derivation of the algebraic equations.

2 Failure of Grid-Based Methods

The Finite Element Method (FEM) [1] and Finite Volume Method (FVM) [2] may be the most well-known members of these thoroughly developed mesh-based methods. The grid-based methods need a priori definition of the connectivity of the nodes, i.e. they rely on a mesh. Element-dependent solution is obtained with mesh-based methods, i.e. shape quality of elements and element density affect the solution. Distorted or low-quality meshes lead to higher errors and demand re-meshing, a time and human effort consuming task, which is not guaranteed to be feasible in finite time for complex three-dimensional geometries. Also, they are not well suited to treat problems with moving discontinuities or the ones that do not align with element edges. Re-meshing is difficult for three dimensions and requires projection of quantities between consecutive meshes with significant degradation of accuracy.

In contrast, a comparably new class of numerical methods has been developed which approximates partial differential equations only based on a set of nodes without the need for an additional mesh. The present paper is devoted to a review of a few advanced finite element techniques, which are highly efficient for solving the mathematical models, in particular the problem of mechanics.

3 MeshFree Methods

3.1 Introduction

Definition 1 MeshFree method is a method used to establish system algebraic equations for the whole problem domain without the use of a predefined mesh for the domain discretization.

MeshFree methods (MFs) [3, 4] use a set of nodes distributed within the problem domain as well as sets of nodes distributed on the boundaries of the domain to represent (not discretize) the problem domain and its boundaries. These sets of distributed nodes are called as field nodes. These nodes do not form a mesh, implying that no priori information about the relationship between the nodes (for the interpolation or approximation of the unknown functions of field variables) is required.

3.2 Features of MeshFree Methods

We list some of the important features of Meshfree Methods (MFs) while comparing them with the analogous properties of mesh-based methods.

- **Absence of Mesh**
  In MFs the connectivity of nodes is determined at run-time. There is no sensitivity regarding the mesh alignment. h-adaptivity is simpler as nodes have to be added, and their connectivity will be determined at run-time. p-adaptivity is also simpler as compared to mesh-based methods.
  No re-meshing during the solution for the problems with large deformations of the domain or moving discontinuities is required.
- **Continuity of shape functions**
  The shape functions for MFs can be constructed to have any desired order of continuity. In problems where the discontinuities can be physically justified such as cracks, different material properties, etc., the continuity of shape functions and derivatives is not desirable.
- **Convergence**
  Numerical experiments suggest that the convergence results for MFs are significantly better than the results obtained by mesh-based shape functions [5]. However, theory fails to predict this for higher order of convergence.
- **Computational effort**
  MFs are often substantially more time-consuming than their mesh-based counterparts. The shape functions for MFs are complex in nature unlike the polynomial type shape functions for mesh-based methods. Thus, more number of integration points are required for sufficiently accurate evaluation of the integrals of weak form. The resulting global system of equations has in general a larger bandwidth for MFs than for comparable mesh-based methods [6]. However, the assembly of the resulting matrix is simpler.
- **Essential boundary conditions**
  The shape functions of most of the MFs do not satisfy the Kronecker delta property, i.e. $\phi_i(x_j) = \delta_{ij}$. Thus, the imposition of essential boundary conditions requires attention and may lower the convergence of the method.

3.3 Comparison—FEM and MeshFree Methods

	FEM	MeshFree methods
Element Mesh	Yes	No
Shape function formulation	Based on predefined elements	Based on local support domains
Stiffness matrix	Symmetric	May or may not be depending on technique used
Imposition of essential boundary condition	Easy and standard	Exceptional treatments may be required depending on method used
Computation speed	Fast	Slower compared to FEM depending on method used
Accuracy	Accurate compared to FDM	Usually more accurate than FEM
Adaptive analysis	Difficult for 3D cases and complex geometries	Easier
Stage of development	Well developed	Young, with challenging issues
Availability of commercial software packages	Many	Few

3.4 Solution Procedure for MeshFree Methods

We now list the steps in a Meshfree Method [3].

1. Domain Representation:

In the MFs, the problem domain and its boundary are modelled and represented by using sets of nodes distributed in the problem domain and on its boundary. These nodes carry the values of the field variables in MFs formulation, and are thus called field nodes. The density of the nodes depends on desired accuracy and existing resources. The nodal distribution is usually not uniformly relatively fine (coarse) discretization in regions where a high (low) gradient of strains and/or stresses is expected. The density can be automatically controlled using adaptive algorithms in the code.

2. Function interpolation/approximation:

Since there is no mesh in MFs, the field variable (e.g. a component of the displacement) u at any point at $x = (x, y, z)$ within the problem domain is interpolated using function values at field nodes within a small local support domain of the point at x, i.e. $u(x) = \Sigma_{i=1}^{n} \phi_i(x)u_i$, where n is the number of nodes which are included in the local support domain of x, u_i is the nodal field variable at ith node and $\phi_i(x)$ is the shape function at ith node determined using the nodes included in support domain of x.

A local support domain of a point x governs the number of nodes to be used to support or approximate the function value at x. It can have different shapes and dimensions, usually circular or rectangular. For different points of interest x, there can be differently dimensionalized and shaped support domains.

3. Formation of system equations

The discrete equations of MFs can be framed using the shape functions and strong- or weak-form system equations. The partial differential equation for the mathematical model is said to be the strong form of system equation. The weak form places a weaker consistency on the approximate function and is achieved by introducing an integral operation to the system equation based on a mathematical or physical principle. These equations are often written in nodal matrix form and are assembled into the global system matrices for the entire problem domain. The discretized system equations of MFs are similar to those of FEM in terms of bandwidth. They can be symmetric or asymmetric depending on the method used.

4. Solve the global MFs equations

Solving for symmetric (or asymmetric) system of equations using either the standard linear algebraic equation solvers (for static problems), such as the Gauss elimination method, LU decomposition method, etc. or the standard eigenvalue equation solvers (for free vibration and buckling problems), such as Jacobis method, Givens method, Inverse iteration, etc.

3.5 Meshfree Interpolation/Approximation Techniques

A good shape function should satisfy certain following basic requirements [3].

- Ideally, the shape function should possess the Kronecker delta function property making it easier for imposition of essential boundary conditions.
- The algorithm must be stable and computationally efficient.
- The construction of shape functions should satisfy a certain order of consistency to ensure the convergence of numerical results.
- Smaller and compact domain for field variable approximation so that the banded system matrix can be handled with good computational efficiency.

3.6 Categories of Meshfree Methods

The idea of meshfree analysis dates back from 1977, with Monaghan and Gingold [7] developing a Lagrangian method based on the Kernel Estimates method to model astrophysics problems such as exploding stars and dust clouds that had no boundaries. This method, named Smoothed Particle Hydrodynamics (SPH), is a particle method based on the idea of replacing the fluid by a set of moving particles and transforming the governing partial differential equations into the kernel estimates integrals.

The first meshfree method based on the Galerkin technique was only introduced over a decade after Monaghan and Gingold first published the SPH method. The Diffuse Element Method (DEM) was introduced by Nayroles and Touzot in 1991. Many authors believe that it was only after the DEM that the idea of a meshfree technique appealed to the research community. The idea behind the DEM was to replace the FEM interpolation within an element by the Moving Least Square (MLS) local interpolation.

The three categories of MFs are given in the following table.

Categories	MeshFree approximation techniques
Integral representation	Smoothed Particle Hydrodynamics (SPH) Reproducing Kernel Particle Method (RKPM)
Series representation	Moving Least Squares (MLS) Point Interpolation Methods (PIM, RPIM) Partition of Unity (PU) methods
Differential representation	General Finite difference method (GFDM)

- In the **integral representation method**, the function is characterized using its information in a local domain (smoothing domain or influence domain) with the help of a weighted integral operation. The consistency is achieved by selecting an appropriate weight function. It is often used in the smoothed particle hydrodynamics (SPH).
- The **series representation methods** have an extensive history. They are well established in FEM and are now used in Meshfree methods based on arbitrarily distributed nodes. The consistency is ensured by ensuring the completeness of the basis functions. The moving least square (MLS) approximation is the most widely used method. The point interpolation method (PIM) using radial basis function (or RPIM) is also often used.
- The **differential representation method** has also been established and in use for a long time in the finite difference method (FDM). The finite difference approximation is not globally compatible, and the consistency is ensured by the theory of Taylor series. Differential representation methods are generally used for establishing system equations based on strong-form formulations, such as FDM and the general finite difference method (GFDM).

3.7 Smoothed Particle Hydrodynamics Approach

The SPH method was introduced in 1977 by Lucy. The name SPH comes from the smoothing character of the particles point properties to the kernel function, therefore leading to a continuous field. Similar to finite element formulation, $u^h(x)$ can be written as $u^h(x) = \Sigma\phi_i(x)u_i$, where, $\phi_i(x)$ are the SPH shape functions given by $\phi_i(x) = W(x - x_i)\Delta V_i$, where ΔV_i represents the volume of node i. The SPH shape functions satisfy neither the Kronecker Delta property nor Partition of Unity.

3.8 Reproducing Kernel Particle Method (RKPM)

The reproducing kernel particle method (RKPM) [8] is an advancement of the continuous SPH approximation near the boundaries. For increasing the order of completeness of the approximation, a correction function $C(x, \xi)$ is introduced into the approximation:

$$u^h(x) = \int_{\Omega} u(\xi)W(x - \xi, h)C(x, \xi)d\xi$$

where $C(x, \xi)$ is the correction function. An example of the correction function in one dimension is $C(x, \xi) = c_1(x) + c_2(x)(\xi - x)$, where, $c_1(x)$ and $c_2(x)$ are coefficients which are evaluated by enforcing the corrected kernel to reproduce the function.

A remarkable point to be observed is that if we choose: $\Delta V_i = 1$, the RKPM and MLS are the same.

3.9 Moving Least Squares Approximation

This method was presented by Shepard in late 1960s for assembling smooth approximations to definite cloud of points. The MLS approximation [3, 9] has two main features that make it popular:

(1) The approximated field function is continuous and smooth in the complete problem domain.
(2) It is capable of constructing an approximation with the preferred order of consistency.

The MLS approximation of field variable $u(x)$ is defined as

$$u^h(x) = \Sigma_{j=1}^{m} p_j(x)a_j(x) = p^T(x)a(x)$$

where $p(x)$ is the basis function of coordinates, $x^T = [x, y]$ for 2D problem, and m is number of basis functions. The basis function is constructed using monomials from the Pascal triangle to guarantee minimum completeness. $a(x)$ is a vector of coefficients, which can be obtained by minimizing the following weighted discrete L_2 norm.

$$J = \Sigma_{i=1}^{n} W(x - x_i)[p^t(x_i)a(x) - u_i]^2$$

where n is the number of nodes in support domain of x and u_i is the nodal parameter of u at $x = x_i$. The number of nodes, n used in MLS approximation is typically much greater than number of coefficients, m because of which the approximated function, u^h does not pass through the nodal values.

The stationarity of J with respect to $a(x)$ gives

$$\frac{\partial J}{\partial a} = 0$$

which gives set of linear equations $A(x)a(x) = B(x)U_s$, where, U_s is the vector that has nodal parameters of field function for all nodes in support domain and $A(x)$ is the weighted moment matrix. The above equation is solved for $a(x) = A^{-1}(x)B(x)U_s$ in the approximation of field variable $u^h(x)$. The shape function for the ith node is defined by

$$\phi_i(x) = \Sigma_{i=1}^{m} p_j(x)(A^{-1}(x)B(x))_{ji} = p^T(x)(A^{-1}B)_i$$

The consistency of MLS approximation depends on the complete order of the monomial taken in the polynomial basis. If the complete order of monomial is k, then the shape function will possess C^k consistency.

3.10 Point Interpolation Method

In this method, approximation is acquired by allowing the interpolation function to pass through the function values at each scattered node within the defined domain of support. The formulation of PIM starts with following finite series representation

$$u^h(x, x_Q) = \Sigma_{i=1}^{n} B_i(x)a_i(x_Q)$$

where $B_i(x)$ are basis functions defined in coordinate space $x^T = [x, y, z]$, n is the number of nodes in support domain of x_Q and $a_i(x_Q)$ is the coefficient for basis function $B_i(x)$.

Two types of PIM have been developed, one that uses polynomial functions and the other that uses radial basis functions.

3.11 Various MeshFree Methods

Due to brevity of paper, we will be discussing only a few methods. There are various other methods such as, Diffuse Element Method, Least-squares Meshfree Method, Local Boundary Integral Equation, Natural Element Method, Hybrid meshfree methods [10], etc.

3.12 Element-Free Galerkin Method

It mainly deals with FEM shape function being replaced by a nodal-based approximation. Moving least square (MLS) approximation is used for the construction of the shape function in EFGM [11, 12]. For the calculation of system matrices, the cells of the background mesh are used for integration and hence, it is not truly a meshless method. The typical solution procedure using EFGM are:

- Nodal discretization of the solution area.
- Construct the approximation function using Moving Least Squares Approximation.
- Build the equivalent form of the physical problems (PDEs) using the Galerkin Weak form.
- Substitute the approximation and their derivatives into the equivalent form, and construct the solution matrix.
- Enforcement of essential boundary conditions using Penalty method. In penalty method, we introduce a penalty factor to correct the difference between the variable of MLS approximation and the given variable on the essential boundary.
- Solve the solution matrix to obtain the results.

3.13 Meshless Local Petrov–Galerkin

The main idea of MLPG [13, 14] is that the implementation of the integral form of the weighted residual method is restricted to a very small local subdomain of a node. This means that weak form is fulfilled at each node in problem domain in local integral sense and thus, it is more stable. MLS shape functions are used. For building the equivalent form of the physical problems, the Petrov–Galerkin form is used which gives us the freedom to choose weight and trail functions individualistically. There are two major drawbacks of MLPG method.

1. Asymmetry of matrices because of Petrov–Galerkin formulation.
2. For domains intersecting with the boundary of problem domain, the local integration is tricky.

3.14 Node-Based Smooth FEM (N-SFEM)

The smoothed finite element procedures were developed by joining Meshfree methods and Finite Element Method. It gives upper bound solutions to problems of solid mechanics. In this numerical procedure, we do not construct the shape functions explicitly [15]. We perform point interpolation using nodes within the element that cloud the point of interest. The procedure for NS-FEM [16, 17] briefly follows:

1. Discretize the domain into a set of elements to find the node coordinates and connectivity between the elements.
2. Compute the area of cell Ω_k associated with node k, find neighbouring cells for each node.
3. For all the nodes, using the node connectivity, evaluate the stiffness matrix and force vector associated with the cell of node. Assemble to form the system stiffness matrix and force vector.
4. Implement essential boundary conditions.
5. Solve system equations to find displacements and hence, value of stress and strain at nodes of interest.

4 αFEM

The αFEM [17–19] is a novel advancement of FEM in which the gradients of strains are scaled by a factor $\alpha \in [0, 1]$. The idea is to find a combined model of the standard FEM and NS-FEM which makes the finest use of the upper bound property of the N-SFEM and the lower bound property of the standard FEM. The procedure is similar to standard FEM except that the stiffness matrix and the Jacobi matrix are substituted by scaled strain matrix and equivalent scaled Jacobi matrix, respectively [20]. For overestimation problems, exact α approach is used and for underestimation problems, zero α approach is used.

The area of triangular element (say V_e) is divided into four parts with scale factor α. Three quadrilaterals at corners are scaled down by a factor of $(1-\alpha^2)$ with an equal area of $\frac{(1-\alpha^2)V_e}{3}$. The remaining Y-shaped part in the middle has area $\alpha^2 V_e$. The Y-shaped area is evaluated using FEM while the three quadrilaterals at corners are evaluated using N-SFEM [19].

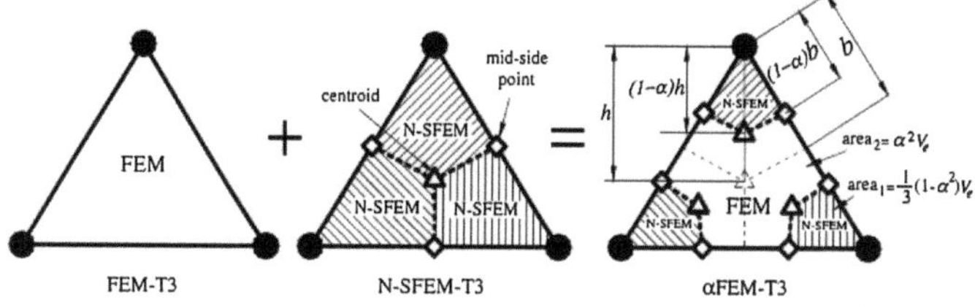

Hence,

$$K_{IJ}^{\alpha FEM} = \Sigma_{k=1}^{n} K_{IJ(k)}^{N-SFEM} + \Sigma_{k=1}^{e} K_{IJ(k)}^{FEM}$$

where n is the total number of nodes and e is the total number of elements in problem domain.

For 2D problems using triangular elements

$$\Sigma_{k=1}^{e} K_{IJ(k)}^{FEM} = B_I^T M B_J \alpha^2 V_e$$

$$\Sigma_{k=1}^{e} K_{IJ(k)}^{N-SFEM} = (1 - \alpha^2) B_I^T M B_J V^k$$

where B_i is the strain gradient matrix of the ith element around node k, M is matrix of material constants, V_e is the volume of element e and V^k is volume of cell k.

The result of αFEM is a continuous function of α from results of N-SFEM to that of standard FEM. According to numerical experiments on different linear problems, it is recommended to use directly $\alpha = 0.6$ for 2D problems and $\alpha = 0.7$ for 3D problems. This method is variationally consistent and has the same computational complexity as that of FEM.

5 Hybrid Meshfree Methods

Attempts have also been made to couple Meshfree methods that are formulated using MLS shape functions and Meshfree methods that are formulated using Point interpolation method (PIM) shape functions or Finite element (FE) shape functions [10]. The aim is to simplify the procedure of imposing the boundary conditions. Some examples of hybrid Meshfree methods are: SPH coupling with FEM, EFG coupling with BEM, MLPG coupling with BEM or FEM, etc.

6 Parallelization of EFGM

Although, EFGM is very efficient for solving problems of irregular geometry or phase transition problem, the major disadvantage of EFGM is the increased computational cost. The additional computational cost of EFGM is from several sources listed as follows:

1. The need to identify the number of nodes in support domain of a quadrature point.
2. The relative complexity of the shape functions, which increase the cost of evaluating them and their derivatives.
3. The additional expense of dealing with essential boundary conditions. To overcome this difficulty of high computational time, parallel implementation of EFGM can be done. In EFGM, the process of forming stiffness matrices at each quadrature point is totally independent. Therefore, in a parallel computing set-

up, stiffness matrices can be computed at different work stations and finally assembled at the master server.

7 Related Problems

Fluid flow and thermal effects during phase change (melting and solidification) are of great interest in a number of manufacturing processes, where a solid material is formed by the freezing of a liquid. Its main characteristic is that a moving interface separates two phases with different physical properties [21]. Temperature differences in the melt give rise to buoyancy forces that produce significant convective flow. Due to the problem complexity, direct application of numerical methods (FEM) to the problems of phase change is not an easy task.

As the nature of surface influences the rate of heat transfer, the analysis of natural convection heat transfer with different types of heat transfer surfaces [22, 23] (e.g. wavy surfaces or rough surfaces) becomes important, e.g. the transfer of heat generated due to friction in car tyres. Numerical simulation (FEM and BEM) of fluid flow problems with complex geometries is a challenge as they require re-meshing at each stage of simulation.

Therefore, to tackle the above mechanics problems, Meshfree techniques are used.

Li et al. [18] have utilized αFEM for simulation of a phase transition problem during the cryosurgery treatment of tumour tissue, in which transition occurs between the healthy blood tissues and tumour tissues. In their study, a comparison has been shown between obtained FEM and αFEM results, which illustrate that αFEM results obtained with 291 nodes are in well agreement with FEM results obtained with 12876 nodes. Li et al. [24] applied αFEM to analyze 2D underwater exterior scattering problems in the unbounded domain.

8 Conclusions

In this paper, an overview of Meshfree methods and FEM has been presented. We have discussed properties and advantages of Meshfree methods compared to standard finite elements. The MFs themselves have been explained in detail, taking into account the different viewpoints and origins of each method. We have concentrated more on pointing out important characteristic features rather than on explaining how the method functions. Since standard Meshfree methods do not fulfil the KroneckerDelta property, essential boundary conditions cannot be imposed as straightforwardly as in finite element methods. We have summarized on how to incorporate essential boundary conditions using the penalty method.

From the literature review, it is observed that the results obtained with αFEM are much more accurate and can be used to improve solutions for non-linear problems of large deformation.

We hope this paper to be a helpful tool for the reader's successful work with advanced finite element techniques.

Acknowledgments The second author would like to thank Department of Science and Technology, Government of India, for its financial support through the award of a research grant.

References

1. Reddy, J.N.: An Introduction to the Finite Element Method. McGraw-Hill Inc., New York (2005)
2. Causona, D.M., Mingham, C.G., Qia, L.: Introductory Finite Volume Methods for PDEs. Ventus publishing (2011)
3. Liu, G.R.: Mesh Free Methods Moving beyond the finite element Method. CRC Press, USA (2003)
4. Babuska, I., Banerjee, U., Osborn, J.E.: Meshless and generalized finite element methods: a survey of some major results. Meshfree Methods for Partial Differential Equations, vol. 26. Springer Verlag, Berlin (2002)
5. Li, S., Liu, W.K.: Meshfree and particle methods and their applications. Appl. Mech. Rev. **55**, 1–34 (2002)
6. Fries, T., Matthies, H.: Classification and Overview of Meshfree Methods. Scientific Computing, Germany (2003)
7. Gingold, R.A., Monaghan, J.J.: Smoothed particle hydrodynamics: theory and application to non-spherical stars. Mon. Not. R. Astron. Soc. **181**, 375389 (1977)
8. Chen, J.S., Yoon, S., Wang, H.P., Liu, W.K.: An improved reproducing kernel particle method for nearly incompressible finite elasticity. Comp. Methods Appl. Mech. Eng. **181**, 117–145 (2000)
9. Breitkopf, P., Touzot, G., Villon, P.: Consistency approach and diffuse derivation in element free methods based on moving least squares approximation. Comput. Assist. Mech. Eng. Sci. **5**, 479501 (1998)
10. Pian, T.H.H., Wu, C.C.: Hybrid and Incompatible Finite Element Methods. CRC Press, Boca Raton (2006)
11. Chung, H.J., Belytschko, T.: An error estimate in the EFG method. Comput. Mech. **21**, 91–100 (1998)
12. Beissel, S., Belytschko, T.: Nodal integration of the element-free Galerkin method. Comp. Methods Appl. Mech. Eng. **139**, 49–74 (1996)
13. Atluri, S.N., Zhu, T.: A new meshless local PetrovGalerkin (MLPG) approach in computational mechanics. Comput. Mech. **22**, 117127 (1998)
14. Gu, T.Y., Liu, G.R.: A meshless local PetrovGalerkin (MLPG) method for free and forced vibration analyses for solids. Comput. Mech. **27**(3), 188198 (2001)
15. Dai, K.Y., Liu, G.R., Nguyen, T.T.: An n-sided polygonal smoothed finite element method (nSFEM) for solid mechanics. Finite Elem. Anal. Des. **43**, 847860 (2007)
16. Liu, G.R., Dai, K.Y., Nguyen, T.T.: A smoothed finite element method for mechanics problems. Comput. Mech. **39**, 859877 (2007)
17. Liu, G.R., Trung, N.T.: Smoothed Finite Element Methods. CRC Press, USA (2010)
18. Li, E., Liu, G.R., Tan, V., He, Z.C.: An efficient algorithm for phase change problem in tumor treatment using α-FEM. Int. J. Therm. Sci. **49**, 1954–1967 (2010)
19. Liu, G.R., Nguyen-Thoi, T., Lam, K.Y.: A novel alpha finite element method (αFEM) for exact solution to mechanics problems using triangular and tetrahedral elements. Comput. Methods Appl. Mech. Eng. **197**, 38833897 (2008)

20. Trung, N.T., Hung, N.X.: About applying directly the alpha finite element method for solid mechanics using triangular and tetrahedral elements. Vietnam J. Mech. VAST **32**(4), 235–246 (2010)
21. Belhamadia, Y., Kane, A.S., Fortin, A.: An enhanced mathematical model for phase change problems with natural convection. Int. J. Numer. Anal. Model. **3**(2), 192–206 (2012)
22. Nikfar, M., Mahmoodi, M.: Meshless local PetrovGalerkin analysis of free convection of nano fluid in a cavity with wavy side walls. Eng. Anal. Boundary Elem. **36**, 433445 (2012)
23. Das, P.K., Mahmud, S.: Numerical investigation of natural convection inside a wavy enclosure. Int. J. Therm. Sci. **42**, 397406 (2003)
24. Li, W., Li, Y.F., Chai, Y.B.: Calculation of underwater acoustic scattering problems in unbounded domain using the alpha finite element method, ICCM2014. Cambridge, England

Duality for a Nondifferentiable Multiobjective Second-Order Fractional Programming Problem Involving (F, α, ρ, d)—V-type-I Functions

Ramu Dubey and S.K. Gupta

Abstract In this paper, a nondifferentiable multiobjective fractional programming problem in which a support function appears in the numerator and denominator of each objective function as well as in each constraint function. We propose Schaible type dual for a nondifferentiable second-order fractional programming problem. Next, we prove appropriate duality results using second-order (F, α, ρ, d)—V-type-I convexity assumptions.

Keywords Fractional programming problem · Support function · Duality results · (F, α, ρ, d)—V-type-I functions

1 Introduction

Mangasarian [1] introduced the concept of second-order duality for nonlinear programming problems. By introducing an additional vector $p \in R^n$, he formulated the second-order dual and established duality theorems under convexity assumptions. One significant practical use of duality is that it provides bounds for the value of the objective function. Second-order duality may provide tighter bounds than first-order duality because there are more parameters involved in it. In the dual formulation of Mangasarian the same vector p appears at three places. In [2], Hanson introduced the dual by replacing the same vector p by three different vectors p, q, and $r \in R^n$ thus providing applicability to a wider class of functions that may give still tighter bounds than Mangasarian's dual. The author also established weak and strong duality theorems under second-order type-I assumptions.

Mond [3] introduced second-order convex functions and established second-order duality results. Bector and Chandra [4] formulated second-order Mond-Weir type

R. Dubey (✉) · S.K. Gupta
Department of Mathematics, Indian Institute of Technology Roorkee, Roorkee 247667, India
e-mail: rdubeyjiya@gmail.com

S.K. Gupta
e-mail: skgiitr@gmail.com

P.N. Agrawal et al. (eds.), *Mathematical Analysis and its Applications*, Springer Proceedings in Mathematics & Statistics 143,
DOI 10.1007/978-81-322-2485-3_58

dual for a nondifferentiable fractional program and established duality results using the concept of second-order pseudoconvexity and quasiconvexity. Jeyakumar [5] and Yang [6] also discussed second-order dual formulation under ρ-convexity and its generalizations. Suneja et al. [7] presented a pair of Mond-Weir type multiobjective second-order symmetric dual programs and obtained duality results using second-order η-convex function. Preda [8] defined (F, ρ)-convex functions which generalize the definition of second-order (F, ρ)-convex functions given in Aghezzaf [9] and Zhang and Mond [10]. Ahmad and Husain [11] extended the concept to second-order (F, α, ρ, d)-convex functions and obtained duality results for a second-order Mond-Weir type multiobjective dual. Recently, Hachimi and Aghezzaf [12] extended the notion of (F, α, ρ, d)-type I functions to second-order generalized (F, α, ρ, d)-type I functions and established mixed duality results.

In this paper, we have formulated a dual model for a nondifferentiable multiobjective second-order fractional programming problem. In this dual, we have generalized the models already existing in the literature replacing the same vector $p \in R^n$ involved in the objective and constraint functions with five different vectors p, q, r, s and $t \in R^n$ and thus provide applicability to a wider class of functions and may give more tighter bounds than Mangasarian dual. Using $(F, \alpha, \rho, d) - V$-type-I functions, duality theorems are derived for a Schaible type dual program.

2 Notations and Preliminaries

Definition 1 [13] Let C be a compact convex set in R^n. The support function of C is defined by

$$S(x|C) = \max\{x^T y : y \in C\}.$$

A support function, being convex and everywhere finite, has a subdifferential, that is, there exists a $z \in R^n$ such that

$$S(y|C) \geq S(x|C) + z^T(y - x), \quad \forall x \in C.$$

The subdifferential of $S(x|C)$ is given by

$$\partial S(x|C) = \{z \in C : z^T x = S(x|C)\}.$$

For a convex set $D \subset R^n$, the normal cone to D at a point $x \in D$ is defined by

$$N_D(x) = \{y \in R^n : y^T(z - x) \leq 0, \quad \forall z \in D\}.$$

When C is a compact convex set, $y \in N_C(x)$ if and only if $S(y|C) = x^T y$, or equivalently, $x \in \partial S(y|C)$.
Throughout the paper, we use the index sets $K = \{1, 2, \ldots, k\}$ and $M = \{1, 2, \ldots, m\}$.

Consider the following multiobjective programming problem:

$$\begin{aligned}\textbf{(P)}\ \ \text{Minimize} \quad & f(x) = (f_1(x), f_2(x), \ldots, f_k(x)) \\ \text{subject to} \quad & X^0 = \{x \in X : h_j(x) \le 0, j \in M\},\end{aligned}$$

where $X \subset R^n$ is open and $f_i : X \to R$ and $h_j : X \to R,\ i \in K, j \in M$ are twice differentiable functions on X.

Definition 2 [14] A solution $x^0 \in X^0$ is said to be an efficient (or Pareto optimal) solution of (P) if there exists no $x \in X^0$ such that for some $i \in K,\ \ f_i(x) < f_i(x^0)$ and $f_j(x) \le f_j(x^0),\ \ \forall j \in K$.

Definition 3 [12] A functional $F : X \times X \times R^n \to R$ is said to be sublinear with respect to the third variable if for all $(x, u) \in X \times X$,
(i) $F_{x,u}(a_1 + a_2) \le F_{x,u}(a_1) + F_{x,u}(a_2)$, for all $a_1, a_2 \in R^n$,
(ii) $F_{x,u}(\alpha a) = \alpha F_{x,u}(a)$, for all $\alpha \in R_+$ and $a \in R^n$.

Definition 4 [15] The functions (f, h) are said to be second-order $(F, \alpha, \rho, d)-V$-type-I at $u \in X$ if there exist vectors $\alpha = (\alpha_1^1, \alpha_2^1, \ldots, \alpha_k^1,\ \alpha_1^2, \alpha_2^2, \ldots, \alpha_m^2)$, $\rho = (\rho_1^1, \rho_2^1, \ldots, \rho_k^1,\ \rho_1^2, \rho_2^2, \ldots, \rho_m^2)$ and $d \in R$ where $\alpha_i^1,\ \alpha_j^2 : X \times X \to R_+ \setminus \{0\}$, $\rho_i^1,\ \rho_j^2 \in R$ for all $i \in K,\ j \in M$ and $d : X \times X \to R$ such that for each $x \in X^0$ and $p, q, r, s, t \in R^n$ and for all $i \in K,\ j \in M$, we have

$$f_i(x) - f_i(u) \ge F_{x,u}\Big(\alpha_i^1(x, u)\big(\nabla f_i(u) + \nabla^2 f_i(u)p\big)\Big) - \frac{1}{2}q^T \nabla^2 f_i(u) r + \rho_i^1 d^2(x, u)$$

and

$$-h_j(u) \ge F_{x,u}\Big(\alpha_j^2(x, u)\big(\nabla h_j(u) + \nabla^2 h_j(u)p\big)\Big) - \frac{1}{2}s^T \nabla^2 h_j(u) t + \rho_j^2 d^2(x, u).$$

Remark 1 If $\alpha_i^1 = \alpha^1(x, y),\ \ \alpha_j^2 = \alpha^2(x, y)$ and $p = q = r = s = t$, then the Definition 4 reduces to Hachimi and Aghezzaf [12].

Definition 5 [15] The functions (f, h) are said to be second-order semistrictly $(F, \alpha, \rho, d) - V$-type-I at $u \in X$ if there exist vectors $\alpha = (\alpha_1^1, \alpha_2^1, \ldots, \alpha_k^1,\ \alpha_1^2, \alpha_2^2, \ldots, \alpha_m^2)$, $\rho = (\rho_1^1, \rho_2^1, \ldots, \rho_k^1,\ \rho_1^2, \rho_2^2, \ldots, \rho_m^2)$ and $d \in R$ where $\alpha_i^1,\ \alpha_j^2 : X \times X \to R_+ \setminus \{0\}$, $\rho_i^1,\ \rho_j^2 \in R$ for all $i \in K,\ j \in M$ and $d : X \times X \to R$ such that for each $x \in X^0$ and $p, q, r, s, t \in R^n$ and for all $i \in K,\ j \in M$, we have

$$f_i(x) - f_i(u) > F_{x,u}\Big(\alpha_i^1(x, u)\big(\nabla f_i(u) + \nabla^2 f_i(u)p\big)\Big) - \frac{1}{2}q^T \nabla^2 f_i(u) r + \rho_i^1 d^2(x, u)$$

and

$$-h_j(u) \geq F_{x,u}\left(\alpha_j^2(x,u)\big(\nabla h_j(u) + \nabla^2 h_j(u)p\big)\right) - \frac{1}{2}r^T \nabla^2 h_j(u)t + \rho_j^2 d^2(x,u).$$

Consider the following multiobjective non-differentiable programming problem:

$$\textbf{(PP)} \quad \text{Minimize } G(x) = \left(\frac{f_1(x) + S(x|C_1)}{g_1(x) - S(x|D_1)}, \frac{f_2(x) + S(x|C_2)}{g_2(x) - S(x|D_2)}, \ldots, \frac{f_k(x) + S(x|C_k)}{g_k(x) - S(x|D_k)}\right)$$
$$\text{subject to } \; x \in X^0 = \{x \in X : h_j(x) + S(x|E_j) \leq 0, \; j \in M\},$$

where $x \in X \subset R^n$, $f_i, g_i : X \to R \; (i \in K)$ and $h_j : X \to R \; (j \in M)$ are continuously differentiable functions.

$f_i(.) + S(.|C_i) \geq 0$ and $g_i(.) - S(.|D_i) > 0$; C_i, D_i and E_j are compact convex sets in R^n and $S(x|C_i)$, $S(x|D_i)$ and $S(x|E_j)$ denote the support functions of compact convex sets, C_i, D_i and E_j for all $i \in K$, $j \in M$, respectively.

Theorem 1 (Karush-Kuhn-Tucker type Necessary Condition) [16] *Assume that $\bar{x}$ is an efficient solution of (PP) at which the Kuhn-Tucker constraint qualification is satisfied. Then there exist $0 < \bar{\lambda} \in R^k$, $0 \leq \bar{y}_j \in R^m$, $\bar{z}_i \in R^n$, $\bar{v}_i \in R^n$ and $\bar{w}_j \in R^n$, $i \in K$, $j \in M$ such that*

$$\sum_{i=1}^{k} \bar{\lambda}_i \nabla\left(\frac{f_i(\bar{x}) + \bar{x}^T \bar{z}_i}{g_i(\bar{x}) - \bar{x}^T \bar{v}_i}\right) + \sum_{j=1}^{m} \bar{y}_j \nabla(h_j(\bar{x}) + \bar{x}^T \bar{w}_j) = 0,$$

$$\sum_{j=1}^{m} \bar{y}_j (h_j(\bar{x}) + \bar{x}^T \bar{w}_j) = 0,$$

$$\bar{x}^T \bar{z}_i = S(\bar{x}|C_i),$$

$$\bar{x}^T \bar{v}_i = S(\bar{x}|D_i),$$

$$\bar{x}^T \bar{w}_j = S(\bar{x}|E_j),$$

$$\bar{z}_i \in C_i, \; \bar{v}_i \in D_i, \; \bar{w}_j \in E_j, \; i \in K, \; j \in M.$$

3 Duality Model

Consider the following Schaible type dual (DP) of (PP):

(DP) maximize $(\beta_1, \beta_2, \ldots, \beta_k)$

subject to

$$\sum_{i=1}^{k} \lambda_i \nabla \Big[(f_i(u) + u^T z_i) - \beta_i (g_i(u) - u^T v_i) \Big] + \sum_{j=1}^{m} y_j \nabla (h_j(u) + u^T w_j)$$
$$+ \sum_{i=1}^{k} \lambda_i (\nabla^2 f_i(u) - \beta_i \nabla^2 g_i(u)) p + \sum_{j=1}^{m} y_j \nabla^2 h_j(u) p = 0, \quad (1)$$

$$(f_i(u) + u^T z_i) - \beta_i (g_i(u) - u^T v_i) \geq 0, \ \ i \in K, \quad (2)$$

$$\sum_{i=1}^{k} \lambda_i q^T (\nabla^2 f_i(u) - \beta_i \nabla^2 g_i(u)) r \leq 0, \quad (3)$$

$$\sum_{j=1}^{m} y_j (h_j(u) + u^T w_j - \frac{1}{2} s^T \nabla^2 h_j(u) t) \geq 0, \quad (4)$$

$$z_i \in C_i, \ \ v_i \in D_i, \ \ w_j \in E_j, \ \ i \in K, \ \ j \in M, \quad (5)$$

$$y_j \geq 0, \ \beta_i \geq 0, \ \lambda_i > 0, \ i \in K, \ j \in M. \quad (6)$$

We now discuss the weak duality, strong, and strict converse duality results for the pair (PP) and (DP).

Theorem 2 (Weak Duality Theorem) *Let x be a feasible solution for (PP) and $(u, \beta, z, v, y, \lambda, w, p, q, r, s, t)$ be feasible for (DP). Suppose that:*

(i) For any $i \in k$, $j \in M$, $\big(f_i(.) + (.)^T z_i - \beta_i (g_i(.) - (.)^T v_i),\ h_j(.) + (.)^T w_j\big)$ is second-order
$(F, \alpha, \rho, d) - V$-type-I at u,

(ii) $\alpha_i^1(x, u) = \alpha_j^2(x, u) = \alpha(x, u)$, $\ \forall\, i \in K$ and $j \in M$,

(iii) $\sum_{i=1}^{k} \lambda_i \rho_i^1 + \sum_{j=1}^{m} y_j \rho_j^2 \geq 0$.

Then, the following cannot hold:

$$\frac{f_i(x) + S(x|C_i)}{g_i(x) - S(x|D_i)} \leq \beta_i, \ \textit{for all } i \in K \tag{7}$$

and

$$\frac{f_r(x) + S(x|C_r)}{g_r(x) - S(x|D_r)} < \beta_r, \ \textit{for some } \ r \in K. \tag{8}$$

Proof Let (7) and (8) hold, then using $\lambda_i > 0$, $x^T z_i \leq S(x|C_i)$, $x^T v_i \leq S(x|D_i)$, we have

$$\sum_{i=1}^{k} \lambda_i (f_i(x) + x^T z_i - \beta_i (g_i(u) - x^T v_i)) < 0. \tag{9}$$

Since $(f_i(.) + (.)^T z_i - \beta_i(g_i(.) - (.)^T v_i), \ h_j(.) + (.)^T w_j)$ is second-order $(F, \alpha, \rho, d) - V$-type- I at u, therefore for $i \in K$ and $j \in M$, we have

$$\begin{aligned}
&(f_i(x) + x^T z_i - \beta_i(g_i(x) - x^T v_i)) - (f_i(u) + u^T z_i - \beta_i(g_i(u) - u^T v_i)) \\
&\geq F_{x,u}\Big[\alpha_i^1(x, u)\{\nabla(f_i(u) + u^T z_i - \beta_i(g_i(u) - u^T v_i)) \\
&\quad + (\nabla^2 f_i(u) - \beta_i \nabla^2 g_i(u))p\}\Big] \\
&\quad - \frac{1}{2} q^T (\nabla^2 f_i(u) - \beta_i \nabla^2 g_i(u)) r + \rho_i^1 d^2(x, u)
\end{aligned} \tag{10}$$

and

$$\begin{aligned}
-(h_j(u) + u^T w_j) \geq F_{x,u}\Big[\alpha_j^2(x, u)\Big\{\nabla(h_j(u) + u^T w_j) + \nabla^2 h_j(u) p\Big\}\Big] \\
- \frac{1}{2} s^T \nabla^2 h_j(u) t + \rho_j^2 d^2(x, u).
\end{aligned} \tag{11}$$

Using (2) and multiplying (10) by λ_i and (11) by y_j, summing over $i \in K$ and $j \in M$, we get

$$\sum_{i=1}^{k}\lambda_i\Big[f_i(x)+x^Tz_i-\beta_i(g_i(x)-x^Tv_i)\Big]-\sum_{j=1}^{m}y_j\left(h_j(u)+u^Tw_j\right)$$
$$\geq F_{x,u}\Bigg[\sum_{i=1}^{k}\lambda_i\alpha_i^1(x,u)\nabla(f_i(u)+u^Tz_i-\beta_i(g_i(u)-u^Tv_i))$$
$$+\sum_{j=1}^{m}\alpha_j^2(x,u)y_j\nabla(h_j(u)+u^Tw_j)$$
$$+\sum_{i=1}^{k}\alpha_i^1(x,u)\lambda_i(\nabla^2 f_i(u)-\beta_i\nabla^2 g_i(u))p+\sum_{j=1}^{m}\alpha_j^2(x,u)y_j\nabla^2 h_j(u)p\Bigg]$$
$$-\frac{1}{2}\sum_{i=1}^{k}\lambda_i q^T(\nabla^2 f_i(u)-\beta_i\nabla^2 g_i(u))r$$
$$-\frac{1}{2}\sum_{j=1}^{m}y_j s^T\nabla^2 h_j(u)t+\left(\sum_{i=1}^{k}\lambda_i\rho_i^1+\sum_{j=1}^{m}y_j\rho_j^2\right)d^2(x,u). \tag{12}$$

Further, using $\alpha_i^1(x,u)=\alpha_j^2(x,u)=\alpha(x,u)$ and feasibility conditions (3), (4) in (12), we get

$$\sum_{i=1}^{k}\lambda_i\left[f_i(x)+x^Tz_i-\beta_i(g_i(x)-x^Tv_i)\right]$$
$$\geq F_{x,u}\Bigg[\alpha(x,u)\Bigg(\sum_{i=1}^{k}\lambda_i\nabla(f_i(u)+u^Tz_i-\beta_i\left(g_i(u)-u^Tv_i\right)\Bigg)$$
$$+\sum_{j=1}^{m}y_j\nabla(h_j(u)+u^Tw_j)$$
$$+\sum_{i=1}^{k}\lambda_i\left(\nabla^2 f_i(u)-\beta_i\nabla^2 g_i(u)\right)p+\sum_{j=1}^{m}y_j\nabla^2 h_j(u)p\Bigg]$$
$$+\sum_{i=1}^{k}\left(\lambda_i\rho_i^1+\sum_{j=1}^{m}y_j\rho_j^2\right)d^2(x,u). \tag{13}$$

Using hypothesis (iii), we have

$$\sum_{i=1}^{k}\lambda_i\Big[f_i(x)+x^Tz_i-\beta_i(g_i(x)-x^Tv_i)\Big]$$

$$\geq F_{x,u}\Big[\alpha(x,u)\Big(\sum_{i=1}^{k}\lambda_i\nabla(f_i(u)+u^Tz_i-\beta_i(g_i(u)$$

$$-u^Tv_i)))+\sum_{j=1}^{m}y_j\nabla(h_j(u)+u^Tw_j)$$

$$+\sum_{i=1}^{k}\lambda_i\nabla^2 f_i(u)-\beta_i\nabla^2 g_i(u)p+\sum_{j=1}^{m}y_j\nabla^2 h_j(u)p\Big]. \tag{14}$$

Using feasibility condition (1) and the result $F_{x,u}(0)=0$, we get

$$\sum_{i=1}^{k}\lambda_i\Big[f_i(x)+x^Tz_i-\beta_i(g_i(x)-x^Tv_i)\Big]\geq 0,$$

which contradicts (9) and hence the result.

Theorem 3 (Strong Duality Theorem) *If $\bar{u}$ is an efficient solution of (PP) and let the Kuhn-Tucker constraint qualification be satisfied. Then there exist $\bar{\lambda}\in R^k$, $\bar{y}\in R^m$, $\bar{z_i}\in R^n$, $\bar{v_i}\in R^n$ and $\bar{w_j}\in R^n$, $i\in K$, $j\in M$, such that $(\bar{u},\bar{\beta},\bar{z},\bar{v},\bar{y},\bar{\lambda},\bar{w},\bar{p}=0,\bar{q}=0,\bar{r},\bar{s}=0,\bar{t})$ is a feasible solution of (DP) and the objective function values of (PP)and (DP) are equal. Furthermore, if the conditions of Theorem 1 hold for all feasible solutions of (PP) and (DP), then $(\bar{u},\bar{\beta},\bar{z},\bar{v},\bar{y},\bar{\lambda},\bar{w},\bar{p}=0,\bar{q}=0,\bar{r},\bar{s}=0,\bar{t})$ is an efficient solution of (DP).*

Proof Since $\bar{u}$ is an efficient solution of (PP) at which the Kuhn-Tucker constraint qualification is satisfied, so by the Karush-Kuhn-Tucker type necessary conditions (Theorem 1), there exist $\bar{\mu}\in R^k,\bar{y}\in R^m,\bar{z_i}\in R^n,\bar{v_i}\in R^n$ and $\bar{w_j}\in R^n$, $i\in K, j\in M$, such that

$$\sum_{i=1}^{k}\bar{\mu_i}\nabla\left(\frac{f_i(\bar{u})+\bar{u}^T\bar{z_i}}{g_i(\bar{u})-\bar{u}^T\bar{v_i}}\right)+\sum_{j=1}^{m}\bar{y_j}\nabla(h_j(\bar{u})+\bar{u}^T\bar{w_j})=0, \tag{15}$$

$$\sum_{j=1}^{m}\bar{y_j}(h_j(\bar{u})+\bar{u}^T\bar{w_j})=0, \tag{16}$$

$$\bar{u}^T\bar{z_i}=S(\bar{u}|C_i), \tag{17}$$

$$\bar{u}^T \bar{v_i} = S(\bar{u}|D_i), \tag{18}$$

$$\bar{u}^T \bar{w_j} = S(\bar{u}|E_j), \tag{19}$$

$$\bar{z_i} \in C_i, \quad \bar{v_i} \in D_i, \quad \bar{w_j} \in E_j, \tag{20}$$

$$\bar{\mu_i} > 0, \quad \bar{y_j} \geq 0. \tag{21}$$

Equation (15) can be written as

$$\sum_{i=1}^{k} \frac{\bar{\mu_i}}{g_i(\bar{u}) - \bar{u}^T \bar{v_i}} \left(\nabla(f_i(\bar{u}) + \bar{u}^T \bar{z_i}) - \frac{f_i(\bar{u}) + \bar{u}^T \bar{z_i}}{g_i(\bar{u}) - \bar{u}^T \bar{v_i}} \nabla(g_i(\bar{u}) - \bar{u}^T \bar{v_i}) \right) + \sum_{j=1}^{m} \bar{y_j} \nabla(h_j(\bar{u}) + \bar{u}^T \bar{w_j}) = 0. \tag{22}$$

Setting $\bar{\lambda_i} = \dfrac{\bar{\mu_i}}{g_i(\bar{u}) - \bar{u}^T \bar{v_i}}$ and $\bar{\beta_i} = \dfrac{f_i(\bar{u}) + \bar{u}^T \bar{z_i}}{g_i(\bar{u}) - \bar{u}^T \bar{v_i}}, \quad i \in K$, we get

$$\sum_{i=1}^{k} \bar{\lambda_i} \nabla \left(f_i(\bar{u}) + \bar{u}^T \bar{z_i} - \bar{\beta_i}(g_i(\bar{u}) - \bar{u}^T \bar{v_i}) \right) + \sum_{j=1}^{m} \bar{y_j} \nabla(h_j(\bar{u}) + \bar{u}^T \bar{w_j}) = 0, \tag{23}$$

$$f_i(\bar{u}) + \bar{u}^T \bar{z_i} - \bar{\beta_i}(g_i(\bar{u}) - \bar{u}^T \bar{v_i}) = 0, \tag{24}$$

$$\bar{\lambda_i} > 0, \ \bar{\beta_i} \geq 0. \tag{25}$$

Thus, $(\bar{u}, \bar{\beta}, \bar{z}, \bar{v}, \bar{y}, \bar{\lambda}, \bar{w}, \bar{p} = 0, \ \bar{q} = 0, \ \bar{r}, \ \bar{s} = 0, \ \bar{t})$ is feasible for (DP) and the objective function values of (PP) and (DP) are equal.

We now show that $(\bar{u}, \bar{\beta}, \bar{z}, \bar{v}, \bar{y}, \bar{\lambda}, \bar{w}, \bar{p} = 0, \ \bar{q} = 0, \ \bar{r}, \ \bar{s} = 0, \ \bar{t})$ is an efficient solution of (DP). If not, then there exists a feasible solution $(u', \beta', z', v', y', w', \bar{\lambda}, p' = 0 = q' = 0, \ r' = s' = 0, t')$ of (DP) such that

$$\frac{f_i(\bar{u}) + \bar{u}^T \bar{z_i}}{g_i(\bar{u}) - \bar{u}^T \bar{v_i}} \leq \frac{f_i(u') + u'^T z_i'}{g_i(u') - u'^T v_i'}, \quad \forall i \in K$$

and

$$\frac{f_r(\bar{u}) + \bar{u}^T \bar{z}_r}{g_r(\bar{u}) - \bar{u}^T \bar{v}_r} < \frac{f_r(u') + u'^T z_r'}{g_r(u') - u'^T v_r'}, \quad \text{for some } r \in K.$$

This contradicts the Theorem 2. Hence, $(\bar{u}, \bar{\beta}, \bar{z}, \bar{v}, \bar{y}, \bar{\lambda}, \bar{w}, \bar{p} = 0, \bar{q} = 0, \bar{r}, \bar{s} = 0, \bar{t})$ is an efficient solution of (DP).

Theorem 4 (Strict Converse Duality Theorem) *Let $\bar{x}$ be a feasible solution for (PP) and $(\bar{u}, \bar{\beta}, \bar{z}, \bar{v}, \bar{y}, \bar{\lambda}, \bar{w}, \bar{p}, \bar{q}, \bar{r}, \bar{s}, \bar{t})$ be feasible for (DP). Suppose that*

(i) $\sum_{i=1}^{k} \bar{\lambda}_i \left[f_i(\bar{x}) + \bar{x}^T \bar{z}_i - \bar{\beta}_i (g_i(\bar{x}) - \bar{x}^T \bar{v}_i) \right] \leq 0,$

(ii) For any $i \in K$, $j \in M$, $\Big(f_i(.) + (.)^T \bar{z}_i - \bar{\beta}_i (g_i(.) - (.)^T \bar{v}_i),\ h_j(.) + (.)^T w_j \Big)$ *is second-order semi-strictly $(F, \alpha, \rho, d) - V$-type-I at $\bar{u}$,*

(iii) $\alpha_i^1(\bar{x}, \bar{u}) = \alpha_j^2(\bar{x}, \bar{u}) = \alpha(\bar{x}, \bar{u})$, *for all $i \in K$ and $j \in M$,*

(iv) $\sum_{i=1}^{k} \bar{\lambda}_i \rho_i^1 + \sum_{j=1}^{m} \bar{y}_j \rho_j^2 \geq 0.$

Then, $\bar{x} = \bar{u}$.

Proof We suppose that $\bar{x} \neq \bar{u}$ and exhibit a contradiction. Since $(\bar{u}, \bar{\beta}, \bar{z}, \bar{v}, \bar{y}, \bar{\lambda}, \bar{w}, \bar{p}, \bar{q}, \bar{r}, \bar{s}, \bar{t})$ is feasible solution for (DP), then by the dual constraint (1), we have

$$F_{\bar{x},\bar{u}}\Bigg[\sum_{i=1}^{k} \bar{\lambda}_i \nabla \Big\{ (f_i(\bar{u}) + \bar{u}^T \bar{z}_i) - \bar{\beta}_i (g_i(\bar{u}) - \bar{u}^T \bar{v}_i) \Big\} + \sum_{j=1}^{m} \bar{y}_j \nabla (h_j(\bar{u}) + \bar{u}^T \bar{w}_j)$$
$$+ \sum_{i=1}^{k} \bar{\lambda}_i (\nabla^2 f_i(\bar{u}) - \bar{\beta}_i \nabla^2 g_i(\bar{u})) \bar{p} + \sum_{j=1}^{m} \bar{y}_j \nabla^2 h_j(\bar{u}) \bar{p} \Bigg] = 0. \tag{26}$$

By hypothesis (ii), we get

$$(f_i(\bar{x}) + \bar{x}^T \bar{z}_i - \bar{\beta}_i (g_i(\bar{x}) - \bar{x}^T v_i)) - (f_i(\bar{u}) + \bar{u}^T \bar{z}_i - \bar{\beta}_i (g_i(\bar{u}) - \bar{u}^T \bar{v}_i))$$
$$> F_{\bar{x},\bar{u}}\Big[\alpha_i^1(\bar{x}, \bar{u}) \Big\{ \nabla (f_i(\bar{u}) + \bar{u}^T \bar{z}_i - \bar{\beta}_i (g_i(\bar{u}) - \bar{u}^T \bar{v}_i)) + (\nabla^2 f_i(\bar{u}) - \bar{\beta}_i \nabla^2 g_i(\bar{u})) \bar{p} \Big\} \Big]$$
$$- \frac{1}{2} \bar{q}^T (\nabla^2 f_i(\bar{u}) - \bar{\beta}_i \nabla^2 g_i(\bar{u})) \bar{r} + \rho_i^1 d^2(\bar{x}, \bar{u}) \tag{27}$$

and

$$-(h_j(\bar{u}) + \bar{u}^T \bar{w}_j) \geq F_{\bar{x},\bar{u}}\Big[\alpha_j^2(\bar{x}, \bar{u}) [\nabla (h_j(\bar{u}) + \bar{u}^T \bar{w}_j) + \nabla^2 h_j(\bar{u}) \bar{p}] \Big]$$
$$- \frac{1}{2} \bar{s}^T \nabla^2 h_j(\bar{u}) \bar{t} + \rho_j^2 d^2(\bar{x}, \bar{u}). \tag{28}$$

Using (2) and multiplying (27) by $\bar{\lambda}_i$ and (28) by $\bar{y}_j$, summing over $i \in K$ and $j \in M$, we have

$$\sum_{i=1}^{k}\bar{\lambda}_i[f_i(\bar{x}) + \bar{x}^T\bar{z}_i - \bar{\beta}_i(g_i(\bar{x}) - \bar{x}^T\bar{v}_i)] - \sum_{j=1}^{m}\bar{y}_j\left(h_j(\bar{u}) + \bar{u}^T\bar{w}_j\right)$$

$$> F_{\bar{x},\bar{u}}\left[\sum_{i=1}^{k}\bar{\lambda}_i\alpha_i^1(\bar{x},\bar{u})\nabla(f_i(\bar{u}) + \bar{u}^T\bar{z}_i - \bar{\beta}_i(g_i(\bar{u}) + \bar{u}^T\bar{v}_i))\right.$$

$$+ \sum_{j=1}^{m}\alpha_j^2(\bar{x},\bar{u})\bar{y}_j\nabla(h_j(\bar{u}) + \bar{u}^T\bar{w}_j) + \sum_{i=1}^{k}\alpha_i^1(\bar{x},\bar{u})\bar{\lambda}_i(\nabla^2 f_i(\bar{u})$$

$$\left.- \bar{\beta}_i\nabla^2 g_i(\bar{u}))\bar{p} + \sum_{j=1}^{m}\alpha_j^2(\bar{x},\bar{u})\bar{y}_j\nabla^2 h_j(\bar{u})\bar{p}\right] - \frac{1}{2}\sum_{i=1}^{k}\bar{\lambda}_i\bar{q}^T(\nabla^2 f_i(\bar{u}) - \bar{\beta}_i\nabla^2 g_i(\bar{u}))\bar{r}$$

$$- \frac{1}{2}\sum_{j=1}^{m} y_j\bar{s}^T\nabla^2 h_j(\bar{u})\bar{t} + \left(\sum_{i=1}^{k}\bar{\lambda}_i\rho_i^1 + \sum_{j=1}^{m} y_j\rho_j^2\right)d^2(\bar{x},\bar{u}). \tag{29}$$

Finally, using $\alpha_i^1(\bar{x},\bar{u}) = \alpha_j^2(\bar{x},\bar{u}) = \alpha(\bar{x},\bar{u})$ and feasibility conditions (3) and (4) in (29), we get

$$\sum_{i=1}^{k}\bar{\lambda}_i[f_i(\bar{x}) + \bar{x}^T\bar{z}_i - \bar{\beta}_i(g_i(\bar{x}) - \bar{x}^T\bar{v}_i)] > F_{\bar{x},\bar{u}}\left[\alpha(\bar{x},\bar{u})\left(\sum_{i=1}^{k}\bar{\lambda}_i\nabla(f_i(\bar{u})\right.\right.$$

$$+ \bar{u}^T\bar{z}_i - \bar{\beta}_i(g_i(\bar{u}) - \bar{u}^T\bar{v}_i)) + \sum_{j=1}^{m}\bar{y}_j\nabla(h_j(\bar{u})$$

$$\left.\left.+ \bar{u}^T\bar{w}_j) + \sum_{i=1}^{k}\bar{\lambda}_i(\nabla^2 f_i(\bar{u}) - \bar{\beta}_i\nabla^2 g_i(\bar{u}))\bar{p} + \sum_{j=1}^{m}\bar{y}_j\nabla^2 h_j(\bar{u})\bar{p}\right)\right]$$

$$+ \sum_{i=1}^{k}\left(\bar{\lambda}_i\rho_i^1 + \sum_{j=1}^{m}\bar{y}_j\rho_j^2\right)d^2(\bar{x},\bar{u}). \tag{30}$$

Using hypothesis (iii), we get

$$\sum_{i=1}^{k}\bar{\lambda}_i[f_i(\bar{x}) + \bar{x}^T\bar{z}_i - \bar{\beta}_i(g_i(\bar{x}) - \bar{x}^T\bar{v}_i)]$$

$$> F_{\bar{x},\bar{u}}\left[\alpha(\bar{x},\bar{u})\left(\sum_{i=1}^{k}\bar{\lambda}_i\nabla(f_i(\bar{u}) + \bar{u}^T\bar{z}_i - \bar{\beta}_i(g_i(\bar{u}) - \bar{u}^T\bar{v}_i))\right)\right.$$

$$+ \sum_{j=1}^{m} \bar{y}_j \nabla(h_j(\bar{u}) + \bar{u}^T \bar{w}_j)$$

$$+ \sum_{i=1}^{k} \bar{\lambda}_i \left((\nabla^2 f_i(\bar{u}) - \bar{\beta}_i \nabla^2 g_i(\bar{u}))\bar{p} + \sum_{j=1}^{m} \bar{y}_j \nabla^2 h_j(\bar{u}) \bar{p} \right) \Bigg].$$

Using Eq. (26), the above inequality implies

$$\sum_{i=1}^{k} \bar{\lambda}_i [f_i(\bar{x}) + \bar{x}^T \bar{z}_i - \bar{\beta}_i (g_i(\bar{x}) - \bar{x}^T \bar{v}_i)] > 0,$$

which contradicts hypothesis (i). Hence, $\bar{x} = \bar{u}$.

4 Special Cases

(i) For $p = 0,\ q = 0,\ s = 0$ and $k = 1$, then second-order dual (DP) becomes a first-order problem, given by Husain and Jabeen [17].
(ii) For $p = 0,\ q = 0,\ s = 0,\ C_i = \{0\},\ D_i = \{0\},\ i \in K$ and $E_j = \{0\}$, $j \in M$, then (PP) reduces to the problem studied in Egudo [14].
(iii) If $k = 1,\ C_i = \{0\},\ D_i = \{0\},\ i \in K$ and $E_j = \{0\},\ j \in M$, then (PP) becomes the problem considered by Mond and Weir [18].
(iv) If $D_i = \{0\},\ i \in K$, then (PP) and (DP) reduced to the problems considered in Gulati and Geeta [19].

Acknowledgments The authors express their sincere gratitude to the reviewer for her/his valuable suggestions for improving the presentation of the paper. This research is supported by "Ministry of Human Resource and Development," New Delhi (India) in the form of fellowship to the first author.

References

1. Mangasarian, O.L.: Second-and higher-order duality in nonlinear programming. J. Math. Anal. Appl. **51**, 607–620 (1975)
2. Hanson, M.A.: Second-order invexity and duality in mathematical programming. Opsearch **30**, 313–320 (1993)
3. Mond, B.: Second-order duality for nonlinear programs. Opsearch **11**, 90–99 (1974)
4. Bector, C.R., Chandra, S.: First and second order duality for a class of nondifferentiable fractional programming problems. J. Inf. Optim. Sci. **7**, 335–348 (1986)
5. Jeyakumar, V.: First and second order fractional programming duality in nonlinear programming. J. Math. Anal. Appl. **51**, 607–620 (1975)
6. Yang, X.Q.: Second-order global optimality conditions for convex composite optimization. Math. Program. **81**, 327–347 (1998)

7. Suneja, S.K., Lalitha, C.S., Khurana, S.: Second order symmetric duality in multiobjective programming. European J. Oper. Res. **144**, 492–500 (2003)
8. Preda, V.: On efficiency and duality for multiobjective programs. J. Math. Anal. Appl. **166**, 265–377 (1992)
9. Aghezzaf, B.: Second-order mixed type duality in multiobjective programming problems. J. Math. Anal. Appl. **285**, 97–106 (2003)
10. Zhang, J., Mond, B.: Second-order duality for multiobjective nonlinear programming generalized convexity. In: Glover, B.M., Craven, B.D., Ralph, D. (eds.) Proceedings of the Optimization Miniconference III, pp. 79–95. University of Ballarat, Ballarat (1997)
11. Ahmad, I., Husain, Z.: Second-order (F, α, ρ, d)-convexity and duality in multiobjective programming. Inform. Sci. **176**, 3094–3103 (2006)
12. Hachimi, M., Aghezzaf, B.: Second-order duality in multiobjective programming involving generalized type I functions. J. Math. Anal. Appl. **296**, 382–392 (2004)
13. Gupta, S.K., Kailey, N., Sharma, M.K.: Multiobjective second-order nondifferentiable symmetric duality involving (F, α, ρ, d)-convex function. J. Appl. Math. Inform. **28**, 1395–1408 (2010)
14. Egudo, R.R.: Multiobjective fractional duality. Bull. Austral. Math. Soc. **37**, 367–378 (1988)
15. Jayswal, A., Kumar, D., Kumar, R.: Second order duality for nondifferentiable multiobjective programming problem involving (F, α, ρ, d)-V-type I functions. Optim. Lett. **4**, 211–226 (2010)
16. Miettinen, K.M.: Nonlinear Multiobjective Optimization. Kluwer Academic, Bosten, M. A. (1999)
17. Husain, I., Jabeen, Z.: On fractional programming containing support functions. J. Appl. Math. Comput. **18**, 361–376 (2005)
18. Mond, B., Weir, T.: Duality for fractional programming with generalized convexity conditions. J. Inform. Optim. Sci. **3**, 105–124 (1982)
19. Gulati, T.R., Geeta, M.: Duality in nondifferentiable multiobjective fractional programming with generalized invexity. J. Math. Anal. Appl. **35**, 103–118 (2011)

Consistency and Unconditional Stability of a Positive Upwind Scheme for One-Dimensional Species Transport Equation

S. Prabhakaran and L. Jones Tarcius Doss

Abstract A positivity preserving upwind scheme for one-dimensional species transport equation is discussed in this article. It is proved that the proposed numerical scheme is unconditionally stable. Consistency of the scheme is also discussed in detail. It is shown that the local truncation error is consistent with the advection-diffusion-reaction equation when $\Delta t \to 0$ and inconsistent when $\Delta x \to 0$. Hence, the numerical approximation converges to exact solution only when $\Delta t \to 0$.

Keywords Positivity preserving · Stability · Consistency · Advection · Difussion · Reaction

1 Introduction

The problem related to one-dimensional unsteady state species transport is given as follows: Find $u(x, t)$ satisfying the governing equation

$$R\frac{\partial u}{\partial t} - D\frac{\partial^2 u}{\partial x^2} + v\frac{\partial u}{\partial x} = -ku \qquad 0 < x < \infty, \quad t > 0, \tag{1}$$

subject to the boundary conditions:

$$u(0, t) = u_0 \quad t > 0 \tag{2}$$

$$\lim_{x\to\infty} u(x, t) = 0 \quad t > 0 \tag{3}$$

and initial condition:

S. Prabhakaran · L.J.T. Doss (✉)
College of Engineering Guindy, Anna University, Chennai 600 025, India
e-mail: jones@annauniv.edu

S. Prabhakaran
e-mail: karan.shivam@gmail.com

P.N. Agrawal et al. (eds.), *Mathematical Analysis and its Applications*,
Springer Proceedings in Mathematics & Statistics 143,
DOI 10.1007/978-81-322-2485-3_59

$$u(x, 0) = 0, \quad 0 < x < \infty, \tag{4}$$

where, u is the property being transported; v the prescribed transport velocity; D the diffusion coefficient; R the retardation factor; k the first order reaction coefficient.

The above equation is used to model the chemical species transportation with first-order reaction with the assumption that the degradation reaction occurs in the liquid phase. This problem is useful to find movement of solute particles with the ground-water flow beneath the earth surface. These types of advection-diffusion-reaction equations are also used to model air pollution, exponential traveling wave, bacterial growth, tumor growth, colonization of Europe by oaks, adsorption of contaminants, etc.

Analytical solution to this problem has been discussed by researchers Cho [2] and Clement et al. [3], etc. Analytical solution to (1) with conditions (2), (3) and (4) is given by (see [3]):

$$u(x, t) = \frac{u_0}{2} exp\left(\frac{vx}{2D}\right)\left[exp\left(-\frac{mx}{2D}\right) erfc\left(\frac{Rx - mt}{\sqrt{4DRt}}\right) + exp\left(\frac{mx}{2D}\right) erfc\left(\frac{Rx + mt}{\sqrt{4DRt}}\right)\right], \tag{5}$$

where $m = \sqrt{v^2 + 4kD}$.

Positivity preserving finite difference schemes are discussed by Chen-Charpentier [1] and Karahan [5]. In this article, we have discussed stability and consistency of positivity preserving scheme through upwind finite volume formulation. Illustrative figures are given to show the variation in consistency.

2 Mathematical Description

The vector form of (1) is given by

$$R\frac{\partial u}{\partial t} - \overline{\nabla} \cdot (D\overline{\nabla} u) + \overline{v} \cdot \overline{\nabla}(u) = -ku. \tag{6}$$

The numerical scheme is derived on the following control volume (CV) (Fig. 1)

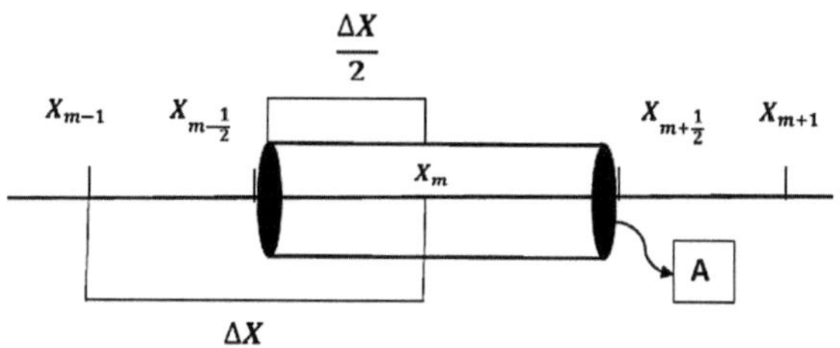

Fig. 1 Control volume

In this article, the species transport is considered in the x direction only. Hence, we assume that the other two dimensions y and z are infinitesimal. Further, $m+1, m, m-1$ are nodal indices and $m-\frac{1}{2}$ and $m+\frac{1}{2}$ are face indices of control volume. Let $\Delta x > 0$ be the spatial discretization length with nodal points $x_0 = 0$ and $x_m = x_0 + m\Delta x$. Further, let $x_{m+\frac{1}{2}} = \frac{x_m + x_{m+1}}{2}$ and $x_{m-\frac{1}{2}} = \frac{x_m + x_{m-1}}{2}$. Integrating the governing Eq. (6) over the local control volume CV in the time interval $(t, t+\Delta t)$, we obtain (Versteg and Malalasekara [7])

$$R\int_t^{t+\Delta t}\int_{CV}\frac{\partial u}{\partial t}dVdt = \int_t^{t+\Delta t}\int_{CV}\overline{\nabla}\cdot(D\overline{\nabla}u)dVdt - \int_t^{t+\Delta t}\int_{CV}\overline{v}\cdot\overline{\nabla}udVdt - \int_t^{t+\Delta t}\int_{CV}kudVdt.$$

Then, the volume integral is converted to a integral over a boundary surface by applying Gauss divergence theorem and a suitable numerical approximation is used at the boundary surface.

$$R\int_{CV}(u^{n+1}-u^n)dV = \int_t^{t+\Delta t}\int_S \overrightarrow{n}\cdot D\overline{\nabla}udSdt - \int_t^{t+\Delta t}\int_S \overrightarrow{n}\cdot(u\overline{v})dSdt - \int_t^{t+\Delta t}\int_{CV}kudVdt$$

where $\overrightarrow{n}$ is the unit normal to the surface S (Here, the surface is cross sectional area A). The one-dimensional formulation of above is given by

$$R(u_m^{n+1}-u_m^n)\int_{CV}dV = \int_t^{t+\Delta t}\left[\left(DA\frac{\partial u}{\partial x}\right)_{m+\frac{1}{2}} - \left(DA\frac{\partial u}{\partial x}\right)_{m-\frac{1}{2}}\right]dt - \int_t^{t+\Delta t}\left[(Auv)_{m+\frac{1}{2}} - (Auv)_{m-\frac{1}{2}}\right]dt - \int_t^{t+\Delta t}u_m\int_{CV}kdVdt$$

where u_m is the average over the control volume. Let us assume that the area of cross section A is uniform and velocity v is constant. Using the explicit weighed average over the time interval, we have that

$$R(U_m^{n+1}-U_m^n)A\Delta x = DA\left[\left(\frac{\partial u}{\partial x}\right)^n_{m+\frac{1}{2}} - \left(\frac{\partial u}{\partial x}\right)^n_{m-\frac{1}{2}}\right]\Delta t - vA\left[u^n_{m+\frac{1}{2}} - u^n_{m-\frac{1}{2}}\right]\Delta t - kU_m^nA\Delta x\Delta t.$$

where U_m^n be the approximation of $u(x, t)$ at the nodal point (x_m, t_n). Using the central difference approximation for the derivative term, we obtain

$$(U_m^{n+1} - U_m^n) = \frac{D\Delta t}{R\Delta x^2}\left[U_{m-1}^n - 2U_m^n + U_{m+1}^n\right] - \frac{v\Delta t}{R\Delta x}\left[u_{m+\frac{1}{2}}^n - u_{m-\frac{1}{2}}^n\right] - \frac{kU_m^n\Delta t}{R}.$$

The upwind approximation in a positive direction to advection term in above is given by the following:

$$u_{m+\frac{1}{2}}^n = U_m^n \qquad u_{m-\frac{1}{2}}^n = U_{m-1}^n.$$

Therefore, we have

$$(U_m^{n+1} - U_m^n) = \frac{D\Delta t}{R\Delta x^2}\left[U_{m-1}^n - 2U_m^n + U_{m+1}^n\right] - \frac{v\Delta t}{R\Delta x}\left[U_m^n - U_{m-1}^n\right] - \frac{kU_m^n\Delta t}{R}.$$

Let us replace U_m^n with U_m^{n+1} on the right-hand side of the above equation to get positivity preserving numerical scheme [1, 5].

$$\left[1 + \frac{2D\Delta t}{R\Delta x^2} + \frac{v\Delta t}{R\Delta x} + \frac{k\Delta t}{R}\right]U_m^{n+1} = \left[\frac{D\Delta t}{R\Delta x^2} + \frac{v\Delta t}{R\Delta x}\right]U_{m-1}^n + U_m^n + \left[\frac{D\Delta t}{R\Delta x^2}\right]U_{m+1}^n. \tag{7}$$

The above can be written as

$$U_m^{n+1} = aU_{m-1}^n + bU_m^n + cU_{m+1}^n \tag{8}$$

where

$$a = \frac{\frac{D\Delta t}{R\Delta x^2} + \frac{v\Delta t}{R\Delta x}}{1 + \frac{2D\Delta t}{R\Delta x^2} + \frac{v\Delta t}{R\Delta x} + \frac{k\Delta t}{R}} \qquad b = \frac{1}{1 + \frac{2D\Delta t}{R\Delta x^2} + \frac{v\Delta t}{R\Delta x} + \frac{k\Delta t}{R}}$$

$$c = \frac{\frac{D\Delta t}{R\Delta x^2}}{1 + \frac{2D\Delta t}{R\Delta x^2} + \frac{v\Delta t}{R\Delta x} + \frac{k\Delta t}{R}}.$$

Here, all the coefficients D, R, v and k are positive. Also, the mesh lengths Δt and Δx are positive. Therefore a, b, c are all positive and hence positivity is preserved in this upwind scheme.

3 Stability

The general form of an explicit finite difference numerical scheme is given as

$$U_m^{n+1} = aU_{m-1}^n + bU_m^n + cU_{m+1}^n. \tag{9}$$

Let $U_m^n = B\xi^n e^{im\theta}$. Then using von Neumann stability analysis in (9), we obtain

$$\xi = ae^{-i\theta} + b + ce^{i\theta} = b + (a + c)(\cos\theta) + i(c - a)\sin\theta.$$

The amplification factor ξ should meet the condition $|\xi| \leq 1$ in order to get a stable scheme which is equivalently $|\xi|^2 \leq 1$ (Smith [6]). Therefore we get

$$\begin{aligned}
& b^2 + (a + c)^2\cos^2\theta + 2b(a + c)\cos\theta + (c - a)^2\sin^2\theta \leq 1 \\
\Leftrightarrow\ & (a^2 + c^2)(\cos^2\theta + \sin^2\theta) + b^2 + 2ac(\cos^2\theta - \sin^2\theta) + 2b(a + c)\cos\theta \leq 1 \\
\Leftrightarrow\ & a^2 + b^2 + c^2 + 2ac(\cos^2\theta - \sin^2\theta) + 2b(a + c)\cos\theta \leq 1 \\
\Leftrightarrow\ & (a + b + c)^2 - 2b(a + c)(1 - \cos\theta) - 2ac(1 - \cos 2\theta) \leq 1 \\
\Leftrightarrow\ & (a + b + c)^2 \leq 1 + 4b(a + c)\sin^2\frac{\theta}{2} + 4ac\sin^2\theta \\
\Leftrightarrow\ & (a + b + c)^2 \leq 1 + 4b(a + c)\sin^2\frac{\theta}{2} + 16ac\sin^2\frac{\theta}{2}\cos^2\frac{\theta}{2} \\
\Leftrightarrow\ & (a + b + c)^2 + 16ac\sin^4\frac{\theta}{2} \leq 1 + 4b(a + c)\sin^2\frac{\theta}{2} + 16ac\sin^2\frac{\theta}{2}.
\end{aligned}$$

Maximizing the trigonometric functions in the above inequality with respect to their argument θ, we obtain

$$(a + b + c)^2 \leq 1 + 4b(a + c). \tag{10}$$

We now establish the stability condition for the generalized explicit scheme (9) as follows: If the coefficients a, b, and c of (9) satisfy the following conditions:

(i) $a \geq 0, b \geq 0$, and $c \geq 0$
(ii) $(a + b + c)^2 \leq 1 + 4b(a + c)$

then the scheme (9) is stable. This is evident from previous derivation. The scheme (8) is stable only when it satisfies the above-mentioned stability criteria. Clearly, the coefficients a, b, and c from (8) are positive. Further, substituting these coefficients in (10), we have that

$$\left(\frac{1 + \frac{2D\Delta t}{R\Delta x^2} + \frac{v\Delta t}{R\Delta x}}{1 + \frac{2D\Delta t}{R\Delta x^2} + \frac{v\Delta t}{R\Delta x} + \frac{k\Delta t}{R}}\right)^2 \leq 1 + 4\frac{\frac{2D\Delta t}{R\Delta x^2} + \frac{v\Delta t}{R\Delta x}}{\left(1 + \frac{2D\Delta t}{R\Delta x^2} + \frac{v\Delta t}{R\Delta x} + \frac{k\Delta t}{R}\right)^2}$$

$$\left(1 + \frac{2D\Delta t}{R\Delta x^2} + \frac{v\Delta t}{R\Delta x}\right)^2 \leq \left(1 + \frac{2D\Delta t}{R\Delta x^2} + \frac{v\Delta t}{R\Delta x} + \frac{k\Delta t}{R}\right)^2 + 4\left(\frac{2D\Delta t}{R\Delta x^2} + \frac{v\Delta t}{R\Delta x}\right)$$

Let $\alpha = \frac{2D\Delta t}{R\Delta x^2}$, $\beta = \frac{v\Delta t}{R\Delta x}$ and $\gamma = \frac{k\Delta t}{R}$, then we get

$$(1 + \alpha + \beta)^2 \leq (1 + \alpha + \beta + \gamma)^2 + 4(\alpha + \beta)$$

$$4\alpha + 4\beta + 2\gamma + 2\alpha\gamma + 2\beta\gamma \geq 0.$$

The constants α, β are γ all positive, because the coefficients D, R, v, and k and also the mesh lengths Δt and Δx are positive. Therefore, the above inequality holds for any choice of parameters and mesh lengths. Hence, the scheme (8) is unconditionally stable.

4 Truncation Error and Consistency

The truncation error $T_{m,n}$ for the explicit scheme at interior nodal point (x_m, t_n) is defined by Smith [6]

$$T_{m,n} = \frac{1}{\Delta t}\left[u(x_m, t_{n+1}) - U_m^{n+1}\right]$$

where $u(x_m, t_{n+1})$ and U_m^{n+1} are the values of exact and numerical solution at (x_m, t_{n+1}) respectively. Let $U_m^{n+1} = aU_{m-1}^n + bU_m^n + cU_{m+1}^n$, we have that

$$\Delta t T_{m,n} = u(x_m, t_{n+1}) - aU_{m-1}^n - bU_m^n - cU_{m+1}^n.$$

Following the usual procedure of obtaining the truncation error, we replace numerical solution by exact solution

$$\begin{aligned}\Delta t T_{m,n} &= u(x_m, t_{n+1}) - au(x_{m-1}, t_n) - bu(x_m, t_n) - cu(x_{m+1}, t_n)\\ &= u(x_m, t_n + \Delta t) - au(x_m - \Delta x, t_n) - bu(x_m, t_n) - cu(x_m + \Delta x, t_n).\end{aligned}$$

Expanding using Taylor series, we have that

$$\begin{aligned}\Delta t T_{m,n} &= \left\{\left[u + \Delta t\frac{\partial u}{\partial t} + \frac{\Delta t^2}{2}\frac{\partial^2 u}{\partial t^2} + \cdots\right] - a\left[u - \Delta x\frac{\partial u}{\partial x} + \frac{\Delta x^2}{2}\frac{\partial^2 u}{\partial x^2} + \cdots\right]\right.\\ &\quad \left. - bu - c\left[u + \Delta x\frac{\partial u}{\partial x} + \frac{\Delta x^2}{2}\frac{\partial^2 u}{\partial x^2} + \cdots\right]\right\}_{(x_m,t_n)}\\ &= \left\{\Delta t\frac{\partial u}{\partial t} + \frac{\Delta t^2}{2}\frac{\partial^2 u}{\partial t^2} - (a + b + c - 1)u + (a - c)\Delta x\frac{\partial u}{\partial x}\right.\\ &\quad \left. - (a + c)\frac{\Delta x^2}{2}\frac{\partial^2 u}{\partial x^2} + \cdots\right\}_{(x_m,t_n)}\end{aligned}$$

$$= \Delta t \left\{ \frac{\partial u}{\partial t} + \frac{\Delta t}{2} \frac{\partial^2 u}{\partial t^2} + \left(\frac{1}{1 + \frac{2D\Delta t}{R\Delta x^2} + \frac{v\Delta t}{R\Delta x} + \frac{k\Delta t}{R}} \right) \left(-\frac{ku}{R} + \frac{v}{R} \frac{\partial u}{\partial x} - \frac{D}{R} \frac{\partial^2 u}{\partial x^2} + \frac{v\Delta x}{2R} \frac{\partial^2 u}{\partial x^2} + \cdots \right) \right\}_{(x_m, t_n)} \quad \text{using } a, b \text{ and } c.$$

$$T_{m,n} = \left\{ \frac{\partial u}{\partial t} + \frac{\Delta t}{2} \frac{\partial^2 u}{\partial t^2} + \left(\frac{1}{1 + \frac{2D\Delta t}{R\Delta x^2} + \frac{v\Delta t}{R\Delta x} + \frac{k\Delta t}{R}} \right) \left(-\frac{ku}{R} + \frac{v}{R} \frac{\partial u}{\partial x} - \frac{D}{R} \frac{\partial^2 u}{\partial x^2} + \frac{v\Delta x}{2R} \frac{\partial^2 u}{\partial x^2} + \cdots \right) \right\}_{(x_m, t_n)} \quad (11)$$

It is clear that the local truncation error is not consistent with the partial differential equation (1). Suppose that $\Delta t \rightarrow 0$ then the truncation error is consistent with (1) and the order of truncation error is Δx. Also note that the error is inconsistent when $\Delta x \rightarrow 0$.

5 Results and Discussion

An artificial boundary is fixed at the farther end and allowed to approach infinity to handle boundary condition (3) in the numerical computation. It is easy to fix the boundary since the initial condition (4) is zero initial condition (Table 1).

The conclusion from Sect. 3 is that the proposed scheme is unconditionally stable. Also from Sect. 4, it is shown that the scheme is consistent only when $\Delta t \rightarrow 0$ and inconsistent when $\Delta x \rightarrow 0$. This claim is also supported by the numerical computation which is shown in Figs. 2 and 3. From Fig. 2 and Table 2, it is evident that the numerical approximation is consistent with exact solution when $\Delta t \rightarrow 0$. Hence, the positivity preserving unconditionally stable upwind scheme converges to exact solution only when $\Delta t \rightarrow 0$. Also, it can be concluded from Fig. 3 that the numerical scheme is inconsistent when $\Delta x \rightarrow 0$.

Table 1 Parameters used for computation

Parameter	Value
u_0	1 mg/l
k	$0.01\,\mathrm{h}^{-1}$
R	2
v	$1\,\mathrm{cm\,h}^{-1}$
T	50 h
D	0.18

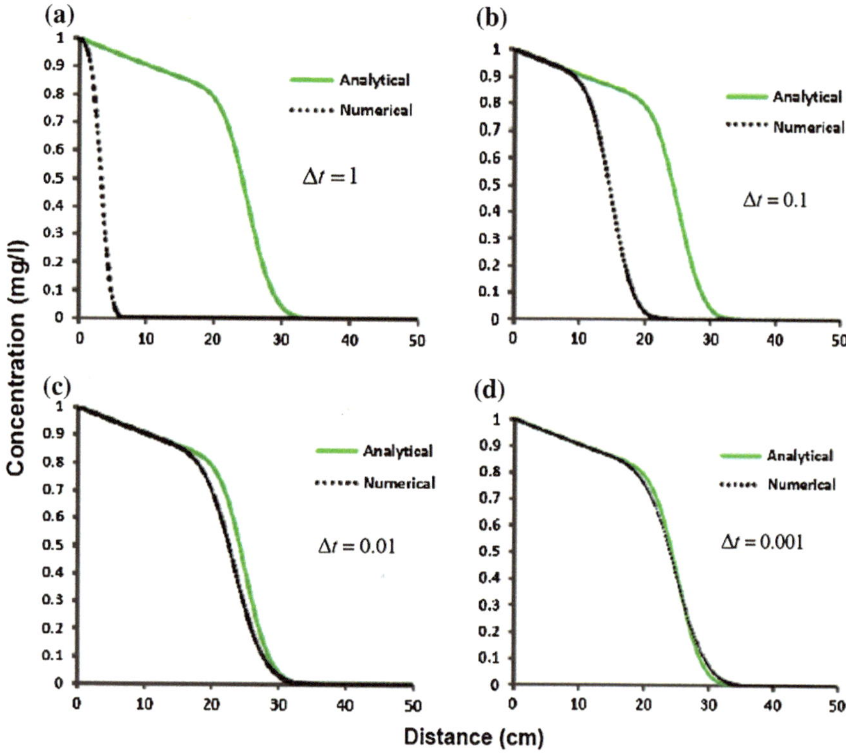

Fig. 2 Convergence of numerical solution when $\Delta t \to 0$

Table 2 L^1 error when $\Delta x = 0.2$ and Δt varies

Δt	L^1 error
1	95
0.1	41.9
0.01	5.96
0.001	2.22

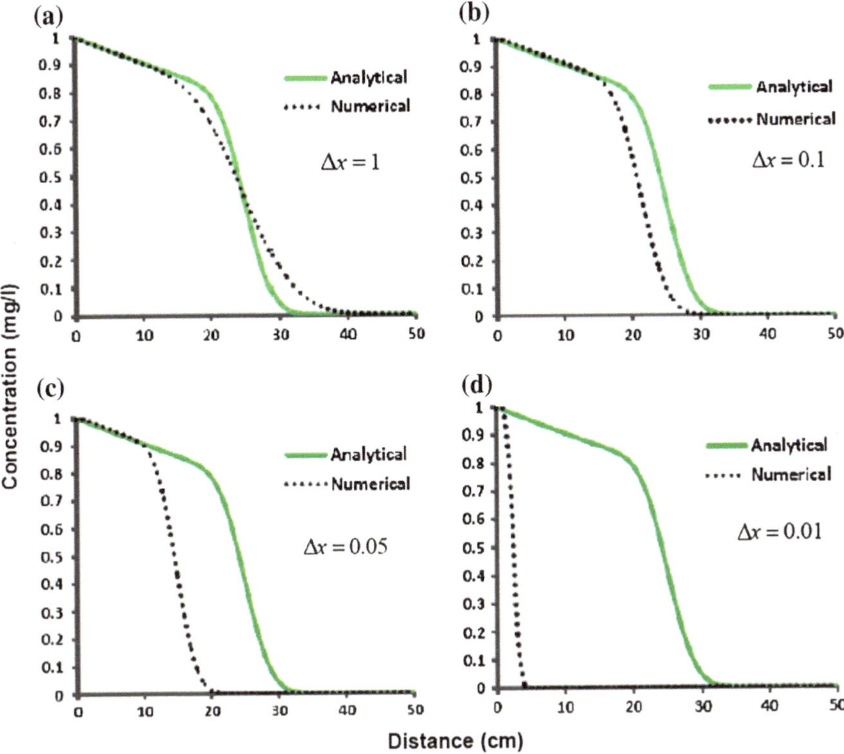

Fig. 3 Divergence of numerical solution when $\Delta x \rightarrow 0$

Acknowledgments The authors would like to thank DST-SERB for providing financial support through the project SR/S4/MS:775/12 to carry out this research work.

References

1. Chen-Charpentier, B.M., Kojouharov, H.V.: An unconditionally positivity preserving scheme for advection diffusion reaction equations. Math. Comput. Model. **57**(910), 21772185 (2013)
2. Cho, C.M.: Convective transport of ammonium with nitrification in soil. Can. J. Soil Sci. **51**(3), 339350 (1971)
3. Clement, T.P., Quezada, C.R., Lee, K.-K.: Generalised solution to multi-dimensional multi species transport equations coupled with a first-order reaction network involving distinct retardation factors. Adv. Water Resour. **27**, 507–520 (2004)
4. Eymard, R., Gallout, T., Herbin, R.: Handbook of Numerical Analysis. Addison-Wesley, New York (2000)
5. Karahan, H.: Unconditional stable explicit finite difference technique for the advection diffusion equation using spreadsheets. Adv. Eng. Soft. **38**(2), 80–86 (2007)
6. Smith, G.D.: Numerical Solutions of Partial Differential Equations: Finite Difference Methods. Oxford University Press, Oxford (1985)
7. Versteg, H.K., Malalasekara, W.: An Introduction to Computational Fluid Dynamics: A Finite Volume Method. Addison-Wesley, New York (1996)

Numerical Solution for a Coupled System of Singularly Perturbed Initial Value Problems with Discontinuous Source Term

S. Chandra Sekhara Rao and Sheetal Chawla

Abstract In this work, we study a numerical method for a coupled system of singularly perturbed initial value problems having discontinuous source term. The leading term of each equation is multiplied by a distinct small positive parameter, due to which the overlapping initial and interior layers are generated in the solution. The problem is discretized using backward Euler difference scheme which involves an appropriate piecewise-uniform variant of Shishkin mesh that is fitted to both the initial and interior layers. The method is proved to be uniformly almost first-order accurate with respect to all the parameters. Numerical results are presented in support of the theory.

Keywords Singular perturbation · Initial layer · Interior layer · Coupled system · Discontinuous source term · Uniformly convergent · Shishkin mesh

1 Introduction

Consider a coupled system of singularly perturbed initial value problem with discontinuous source term on the unit interval $\Omega = (0, 1)$, and assume a single discontinuity in the source term at a point $d \in \Omega$. Let $\Omega_1 = (0, d)$ and $\Omega_2 = (d, 1)$ and the jump at d in any function is given as $[\omega](d) = \omega(d+) - \omega(d-)$. The corresponding initial value problem is to find $u_1, \ldots, u_m \in C^0(\overline{\Omega}) \cap C^1(\Omega_1 \cup \Omega_2)$, such that

$$\boldsymbol{Lu} := \boldsymbol{Eu}' + \boldsymbol{Au} = \boldsymbol{f} \quad \text{in } \Omega_1 \cup \Omega_2, \tag{1}$$

S. Chandra Sekhara Rao (✉) · S. Chawla
Department of Mathematics, Indian Institute of Technology Delhi,
Hauz Khas 110016, New Delhi, India
e-mail: scsr@maths.iitd.ac.in

S. Chawla
e-mail: chawlaasheetal@gmail.com

P.N. Agrawal et al. (eds.), *Mathematical Analysis and its Applications*,
Springer Proceedings in Mathematics & Statistics 143,
DOI 10.1007/978-81-322-2485-3_60

with initial conditions

$$\boldsymbol{u}(0) = \boldsymbol{p}, \tag{2}$$

where $\boldsymbol{E} = diag(\varepsilon_1, \ldots, \varepsilon_m)$ with small parameters $\varepsilon_1, \ldots, \varepsilon_m$ are such that $0 < \varepsilon_1 \leq \cdots \leq \varepsilon_m \leq 1$,

$$\boldsymbol{A}(x) = (a_{ij}(x))_{m\times m} \quad \text{and} \quad \boldsymbol{f}(x) = (f_j(x))_{m\times 1} \tag{3}$$

are given. We assume that the coupling matrix satisfies the following positivity conditions:

$$a_{ii}(x) > \sum_{j\neq i, j=1}^{m} |a_{ij}(x)|, \text{ for } 1 \leq i \leq m, \text{ and } a_{ij}(x) \leq 0 \text{ for } i \neq j, \tag{4}$$

and for some constant α, we have

$$0 < \alpha < \min_{x\in\overline{\Omega}, 1\leq i\leq m} \sum_{j=1}^{m} (a_{ij}(x)) \tag{5}$$

The source terms $f_1(x), \ldots, f_m(x)$ are sufficiently smooth on $\overline{\Omega} \setminus \{d\}$. The solution components $u_1, \ldots, u_m$ of the problem (1) and (2) have overlapping initial layers at $x = 0$ and have overlapping interior layers to the right side of point of discontinuity at $x = d$.

Shishkin [8] laid down the framework for singularly perturbed reaction-diffusion problems with discontinuous coefficients. Dunne and Riordan [2] discussed the singularly perturbed initial value problem with discontinuous coefficients in scalar case. Parameter-robust numerical methods for systems of singularly perturbed differential equations were analyzed in [7]. A parameter-uniform numerical method was constructed in [3] for a system of singularly perturbed initial value problem where all of the singular perturbation parameters are equal. In this case all the solution components have an initial layer of same width, due to which the analysis is simpler. The case where the small parameter is associated with only one equation was considered in [4]. The most difficult and general case is that each component of the solution has its own initial layer that overlaps and interacts with others and this was considered in [1, 5, 9] . In [9], first order (up to logarithmic factor) uniformly convergent numerical method was developed. A hybrid finite difference scheme on a piecewise-uniform Shishkin mesh was considered in [1] and the scheme was almost second-order accurate, uniformly in both small parameters. In all these works the source term is smooth. In the present work, the source term is discontinuous, due to which each solution component has an initial layer as well as interior layer.

This paper is organized as follows. Section 2, presents the properties of the exact solution. The mesh and the scheme that approximate singularly perturbed initial value problem with discontinuous source term are constructed in Sect. 3. In Sect. 4, it is proved that the numerical approximation computed by finite difference method is almost first-order accurate in the maximum norm, uniformly with respect to the parameters $\varepsilon_1, \ldots, \varepsilon_m$. In Sect. 5, the results of numerical experiments are presented for validation of the theoretical results.

Notations We use C to denote a generic positive constant and $\mathbf{C} = (C, C, \ldots, C)^T$ to denote a generic positive constant vector which are independent of the perturbation parameters and the discretization parameter N, but may not be the same at each occurrence. Define $\boldsymbol{v} \leq \boldsymbol{w}$ if $v_i \leq w_i$, for $1 \leq i \leq m$. We consider the maximum norm and denote it by $\| \, . \, \|_S$, where S is a closed and bounded subset of $\overline{\Omega}$. We define $\| v \|_S = \max_{x \in S} |v(x)|$ and $\| \boldsymbol{v} \|_S = \max\{\| v_1 \|_S, \| v_2 \|_S, \ldots, \| v_m \|_S\}$.

2 Properties of the Exact Solution

Theorem 1 *The problem (1) and (2) has a solution* $\boldsymbol{u} = (u_1, \ldots, u_m)^T$ *with* $u_1, \ldots, u_m \in C(\Omega) \bigcap C^1(\Omega_1 \bigcup \Omega_2)$.

Proof The result can be proved by following the similar technique considered in [6].

Theorem 2 *Suppose that a function* $\boldsymbol{u} \in C^1(\Omega_1 \cup \Omega_2)^m$ *satisfies* $\boldsymbol{u}(0) \geq \boldsymbol{0}, \boldsymbol{L}\boldsymbol{u}(x) \geq \boldsymbol{0}$ *for all* $x \in \Omega_1 \cup \Omega_2$. *Then* $\boldsymbol{u}(x) \geq \boldsymbol{0}$, $x \in \overline{\Omega}$.

Proof Let $u_i(p_i) = \min_{x \in \overline{\Omega}}\{u_i(x)\}$, for $1 \leq i \leq m$. Assume without loss of generality $u_1(p_1) \leq u_i(p_i)$, for $2 \leq i \leq m$. If $u_1(p_1) \geq 0$, then there is nothing to prove. Suppose that $u_1(p_1) < 0$, then the proof is completed by showing that this leads to contradiction. Note that $p_1 \neq \{0\}$, so either $p_1 \in \Omega_1 \cup \Omega_2$ or $p_1 = d$.

In the first case

$$
\begin{aligned}
(\boldsymbol{L}\boldsymbol{u})_1(p_1) &= \varepsilon_1 u_1'(p_1) + \sum_{j=1}^{m} a_{1j}(p_1) u_j(p_1) \\
&= \varepsilon_1 u_1'(p_1) + \sum_{j=1}^{m} a_{1j}(p_1) u_j(p_1) + \sum_{j=2}^{m} a_{1j}(p_1) u_1(p_1) - \sum_{j=2}^{m} a_{1j}(p_1) u_1(p_1) < 0.
\end{aligned}
$$

In the second case, since $\boldsymbol{u} \in C(\Omega)^m$ and $u_1(d) < 0$, there exists a neighborhood $N_h = (d - h, d)$ such that $u_1(x) < 0$ for all $x \in N_h$. Now choose a point $x_1 \neq d, x_1 \in N_h$ such that $u_1(x_1) > u_1(d)$. It follows from the mean value theorem that, for some $x_2 \in N_h$, $u_1'(x_2) = \frac{u_1(d) - u_1(x_1)}{d - x_1} < 0$, since $x_2 \in N_h$.

Thus $(\boldsymbol{L}\boldsymbol{u})_1(x_2) = \varepsilon_1 u_1'(x_2) + \sum_{j=1}^{m} a_{1j}(x_2) u_j(x_2) < 0$.

Lemma 1 *Let $\boldsymbol{u}$ be the solution of (1) and (2). Then,*

$$\|\boldsymbol{u}\|_{\overline{\Omega}} \le \max\left\{\|\boldsymbol{u}(0)\|, \frac{1}{\alpha}\|\boldsymbol{L}\boldsymbol{u}\|_{\Omega_1\cup\Omega_2}\right\}.$$

An immediate consequence of this result is that the solution $\boldsymbol{u}$ is unique.

To derive sharper bounds on the derivatives of the solution, the solution is decomposed into a sum, composed of a regular component $\boldsymbol{v}$ and a singular component $\boldsymbol{w}$. That is, $\boldsymbol{u} = \boldsymbol{v} + \boldsymbol{w}$. The regular component $\boldsymbol{v}$, is defined as the solution of the following problem:

$$\boldsymbol{L}\boldsymbol{v}(x) = \boldsymbol{f}(x),\ x \in \Omega_1 \cup \Omega_2,\ \boldsymbol{v}(0) = \boldsymbol{A}^{-1}(0)\boldsymbol{f}(0),\ \boldsymbol{v}(d+) = \boldsymbol{A}^{-1}(d+)\boldsymbol{f}(d+). \quad (6)$$

The singular component $\boldsymbol{w}$, is defined as the solution of the following problem:

$$\boldsymbol{L}\boldsymbol{w}(x) = \boldsymbol{0},\ x \in \Omega_1 \cup \Omega_2,\ \boldsymbol{w}(0) = \boldsymbol{u}(0) - \boldsymbol{v}(0),\ [\boldsymbol{w}](d) = -[\boldsymbol{v}](d). \quad (7)$$

Theorem 3 *Let $\boldsymbol{A}(x)$ satisfy (4) and (5). Then the components v_i, $1 \le i \le m$ of the regular component $\boldsymbol{v}$ and its derivatives satisfy the bounds for all $x \in \Omega_1 \cup \Omega_2$, and $k = 0, 1, 2$,*

$$\|\boldsymbol{v}^{(k)}\|_{\Omega_1\cup\Omega_2} \le C \quad for \quad k = 0, 1 \qquad with$$
$$\|v_i''\|_{\Omega_1\cup\Omega_2} \le C\, \varepsilon_i^{-1}.$$

Consider the following layer functions:

$$B_{\varepsilon_{l_i}}(x) = e^{-\alpha x/\varepsilon_i},$$
$$B_{\varepsilon_{r_i}}(x) = e^{-\alpha(x-d)/\varepsilon_i}.$$

Lemma 2 *Let $\boldsymbol{A}(x)$ satisfy (4) and (5). Then the components w_i, $1 \le i \le m$ of the singular component $\boldsymbol{w}$ and its derivatives satisfy the bounds for all $x \in \Omega_1 \cup \Omega_2$,*

$$|w_i(x)| \le \begin{cases} C\mathrm{B}_{\varepsilon_{l_m}}(x), & x \in \Omega_1 \\ C\mathrm{B}_{\varepsilon_{r_m}}(x), & x \in \Omega_2, \end{cases} \qquad |w_i'(x)| \le \begin{cases} C\displaystyle\sum_{q=i}^{m} \frac{\mathrm{B}_{\varepsilon_{l_q}}(x)}{\varepsilon_q}, & x \in \Omega_1 \\ C\displaystyle\sum_{q=i}^{m} \frac{\mathrm{B}_{\varepsilon_{r_q}}(x)}{\varepsilon_q}, & x \in \Omega_2, \end{cases}$$

$$|\varepsilon_i w_i''(x)| \le \begin{cases} C\displaystyle\sum_{q=1}^{m} \frac{\mathrm{B}_{\varepsilon_{l_q}}(x)}{\varepsilon_q}, & x \in \Omega_1 \\ C\displaystyle\sum_{q=1}^{m} \frac{\mathrm{B}_{\varepsilon_{r_q}}(x)}{\varepsilon_q}, & x \in \Omega_2. \end{cases}$$

Proof We have $\boldsymbol{u} = \boldsymbol{v}+\boldsymbol{w}$ and by Lemma 1 $|\boldsymbol{w}(0)| \leq \boldsymbol{C}$ and $|\boldsymbol{w}(d+)| \leq \boldsymbol{C}$. Define the barrier function $\Upsilon := CB_{\varepsilon_{lm}}(x)\boldsymbol{e}$, with C chosen sufficiently large such that $\Upsilon \geq |\boldsymbol{w}|$ at $x = 0, d+$,

$$\boldsymbol{L}\Upsilon = CB_{\varepsilon_{lm}}\left(\sum_{j=1}^{m} a_{1j} - \frac{\varepsilon_1}{\varepsilon_m}\alpha, \ldots, \sum_{j=1}^{m} a_{mj} - \frac{\varepsilon_m}{\varepsilon_m}\alpha\right) \geq \boldsymbol{0} = |\boldsymbol{L}\boldsymbol{w}|_{\Omega_1}.$$

Using continuous maximum principle, we get the required bound on $\boldsymbol{w}$. Now to bound first-order derivative of w_i, consider $\varepsilon_i w_i' + \sum_{j=1}^{m} a_{ij}w_j = 0$, together with the bound on $\boldsymbol{w}$. This implies that

$$|w_i'| \leq \begin{cases} C\varepsilon_i^{-1}B_{\varepsilon_{lm}}(x), \ x \in \Omega_1, \\ C\varepsilon_i^{-1}B_{\varepsilon_{rm}}(x), \ x \in \Omega_2. \end{cases}$$

Now to find the sharper bound consider the system of $m-1$ equations

$$\hat{\boldsymbol{E}}\hat{\boldsymbol{w}}' + \hat{\boldsymbol{A}}\hat{\boldsymbol{w}} = \boldsymbol{g},$$

where $\hat{\boldsymbol{E}}, \hat{\boldsymbol{A}}$ is the matrix obtained by deleting the last row and column from $\boldsymbol{E}, \boldsymbol{A}$, respectively, and the components of $\boldsymbol{g}$ are $g_i = -a_{im}w_m$, for $1 \leq i \leq m-1$. Using the bounds derived earlier and the decomposition of $\hat{\boldsymbol{w}} = \boldsymbol{q}+\boldsymbol{r}$, into regular and singular component we get the required result. Now to bound second-order derivatives, differentiate $\varepsilon_i w_i' + \sum_{j=1}^{m} a_{ij}w_j = 0$ once and using the estimates of w_i', we get the required bounds on singular component $\boldsymbol{w}$ and its derivatives.

Lemma 3 *For all i, j such that $1 \leq i \leq j \leq m$, there exists a unique point $x_{i,j} \in (0, d)$ such that $\varepsilon_i^{-1}\mathrm{B}_{\varepsilon_{l_i}}(x_{i,j}) = \varepsilon_j^{-1}\mathrm{B}_{\varepsilon_{l_j}}(x_{i,j})$. Also, $\varepsilon_i^{-1}\mathrm{B}_{\varepsilon_{r_i}}(d + x_{i,j}) = \varepsilon_j^{-1}\mathrm{B}_{\varepsilon_{r_j}}(d + x_{i,j})$. On $[0, x_{i,j})$ we have $\varepsilon_i^{-1}\mathrm{B}_{\varepsilon_{l_i}}(x) > \varepsilon_j^{-1}\mathrm{B}_{\varepsilon_{l_j}}(x)$ and on $(x_{i,j}, d)$ we have $\varepsilon_i^{-1}\mathrm{B}_{\varepsilon_{l_i}}(x) < \varepsilon_j^{-1}\mathrm{B}_{\varepsilon_{l_j}}(x)$. Similarly on $(d, d+x_{i,j})$ we have $\varepsilon_i^{-1}\mathrm{B}_{\varepsilon_{r_i}}(x) > \varepsilon_j^{-1}\mathrm{B}_{\varepsilon_{r_j}}(x)$ and on $(d + x_{i,j}, 1]$ we have $\varepsilon_i^{-1}\mathrm{B}_{\varepsilon_{r_i}}(x) < \varepsilon_j^{-1}\mathrm{B}_{\varepsilon_{r_j}}(x)$.*

For the analysis of the convergence, a more precise decomposition of the components of the singular component $\boldsymbol{w}$ is required.

Theorem 4 *The singular component $\boldsymbol{w}$ can be decomposed in this way as follows, for $1 \leq i \leq m$:*

$$w_i(x) = \sum_{q=1}^{m} w_{i,\varepsilon_q}(x)$$

where

$$|w'_{i,\varepsilon_q}(x)| \leq \begin{cases} C\dfrac{B_{\varepsilon_{l_q}}(x)}{\varepsilon_q}, & x \in \Omega_1 \\ C\dfrac{B_{\varepsilon_{r_q}}(x)}{\varepsilon_q}, & x \in \Omega_2, \end{cases} \qquad |\varepsilon_i w''_{i,\varepsilon_q}(x)| \leq \begin{cases} C\dfrac{B_{\varepsilon_{l_q}}(x)}{\varepsilon_q}, & x \in \Omega_1 \\ C\dfrac{B_{\varepsilon_{r_q}}(x)}{\varepsilon_q}, & x \in \Omega_2. \end{cases}$$

Proof Define a function w_{i,ε_1} as follows:

$$w_{i,\varepsilon_1}(x) = w_i(x) - \sum_{q=2}^{m} w_{i,\varepsilon_q}(x)$$

and for $1 < q \leq m$, we have

$$w_{i,\varepsilon_q} = \begin{cases} \Sigma_{k=0}^{2} \frac{[(x-x_{q-1,q})^k]}{k!} w_i^{(k)}(x_{q-1,q}), & x \in [0, x_{q-1,q}), \\ w_i(x) - \displaystyle\sum_{r=q+1}^{m} w_{i,\varepsilon_r}(x), & x \in [x_{q-1,q}, d), \\ \Sigma_{k=0}^{2} \frac{[(x-d-x_{q-1,q})^k]}{k!} w_1^{(k)}(d + x_{q-1,q}), & x \in (d, d + x_{q-1,q}), \\ w_i(x) - \displaystyle\sum_{r=q+1}^{m} w_{i,\varepsilon_r}(x), & x \in [d + x_{q-1,q}, 1]. \end{cases}$$

Now we establish the bounds on the second derivative.
For $x \in [x_{m-1,m}, d) \cup [d + x_{m-1,m}, 1]$.

$$|\varepsilon_i w''_{i,\varepsilon_m}(x)| = |\varepsilon_i w''_i(x)| \leq C\sum_{q=1}^{m} \frac{B_{\varepsilon_{l_q}}(x)}{\varepsilon_q} \leq C\frac{B_{\varepsilon_{l_m}}(x)}{\varepsilon_m}.$$

For $x \in [0, x_{m-1,m}) \cup (d, d + x_{m-1,m})$.

$$|\varepsilon_i w''_{i,\varepsilon_m}(x)| = |\varepsilon_i w''_i(x_{m-1,m})| \leq C\sum_{q=1}^{m} \frac{B_{\varepsilon_{l_q}}(x_{m-1,m})}{\varepsilon_q} \leq C\frac{B_{\varepsilon_{l_m}}(x_{m-1,m})}{\varepsilon_m} \leq C\frac{B_{\varepsilon_{l_m}}(x)}{\varepsilon_m}.$$

Now, for each $m - 1 \leq q \leq 2$, it follows that
For $x \in [x_{q-1,q}, d) \cup [d + x_{q-1,q}, 1]$, $w''_{i,\varepsilon_q}(x) = 0$.
For $x \in [0, x_{q-1,q}) \cup (d, d + x_{q-1,q}]$.

$$|\varepsilon_i w''_{i,\varepsilon_q}(x)| = |\varepsilon_i w''_i(x_{q-1,q})| \leq C\sum_{q=1}^{m} \frac{B_{\varepsilon_{l_q}}(x_{q-1,q})}{\varepsilon_q} \leq C\frac{B_{\varepsilon_{l_q}}(x_{q-1,q})}{\varepsilon_q} \leq C\frac{B_{\varepsilon_{l_q}}(x)}{\varepsilon_q}.$$

For $x \in [x_{1,2}, d) \cup [d + x_{1,2}, 1]$, $w''_{i,\varepsilon_1}(x) = 0$.
For $x \in [0, x_{1,2}) \cup (d, d + x_{1,2})$.

$$|\varepsilon_i w''_{i,\varepsilon_1}(x)| = |\varepsilon_i w''_i(x) - \sum_{q=2}^{m} \varepsilon_i w''_{i,\varepsilon_q}(x)| \leq C\sum_{q=1}^{m} \frac{B_{\varepsilon_{l_q}}(x)}{\varepsilon_q} \leq C\frac{B_{\varepsilon_{l_1}}(x)}{\varepsilon_1}.$$

For the bounds on the first derivatives we have the relation

$$|w'_{i,\varepsilon_q}(x)| = \left| \int_x^{x_{q,q+1}} w''_{i,\varepsilon_q}(t)dt \right| \leq C \left| \int_x^{x_{q,q+1}} \frac{B_{\varepsilon_{l_q}}(t)}{\varepsilon_q} dt \right| \leq C \frac{B_{\varepsilon_{l_q}}(x)}{\varepsilon_q}.$$

3 Discretization of the Problem

In this section, we discretize the system of initial value problem (1) and (2) using a fitted mesh method composed of a backward difference scheme on a piecewise uniform variant of Shishkin mesh with points $\overline{\Omega}^N = \{x_i : i = 0, \ldots, N\}$. Let N = $2^k, k \geq 6$ be a positive integer. Define the transition parameter

$$\sigma_{\varepsilon_{lm}} := \min\left\{\frac{d}{2}, \frac{\varepsilon_m}{\alpha} \ln N\right\}, \qquad \sigma_{\varepsilon_{rm}} := \min\left\{\frac{(1-d)}{2}, \frac{\varepsilon_m}{\alpha} \ln N\right\},$$

$$\sigma_{\varepsilon_{l_k}} := \min\left\{\frac{\sigma_{\varepsilon_{l_{k+1}}}}{2}, \frac{\varepsilon_k}{\alpha} \ln N\right\}, \qquad \sigma_{\varepsilon_{r_k}} := \min\left\{\frac{\sigma_{\varepsilon_{r_{k+1}}}}{2}, \frac{\varepsilon_k}{\alpha} \ln N\right\},$$

$k = m-1, \ldots, 1.$

The interior points of the mesh are denoted by

$$\Omega^N = \left\{x_i : 1 \leq i \leq \frac{N}{2} - 1\right\} \cup \left\{x_i : \frac{N}{2} + 1 \leq i \leq N - 1\right\} = \Omega_1^N \cup \Omega_2^N.$$

Let $h_i = x_i - x_{i-1}$ be the ith mesh step and $\hbar_i = \frac{h_i + h_{i+1}}{2}$, clearly $x_{\frac{N}{2}} = d$. We divide the interval $[0, d]$ into $m+1$ subintervals $[0, \sigma_{\varepsilon_{l_1}}], [\sigma_{\varepsilon_{l_1}}, \sigma_{\varepsilon_{l_2}}], \ldots, [\sigma_{\varepsilon_{l_{m-1}}}, \sigma_{\varepsilon_{lm}}], [\sigma_{\varepsilon_{lm}}, d]$. On the subinterval $[0, \sigma_{\varepsilon_{l_1}}]$ a uniform mesh of $N/2^{m+1}$ mesh intervals, on each subinterval $[\sigma_{\varepsilon_{l_k}}, \sigma_{\varepsilon_{l_{k+1}}}], 1 \leq k \leq m-1$, a uniform mesh of $N/2^{m-k+2}$ mesh intervals, and on $[\sigma_{\varepsilon_{lm}}, d]$ a uniform mesh of N/4 mesh intervals are placed. Similarly, we divide the interval $[d, 1]$ into subintervals $[d, d+\sigma_{\varepsilon_{r_1}}], [d+\sigma_{\varepsilon_{r_1}}, d+\sigma_{\varepsilon_{r_2}}], \ldots, [d+\sigma_{\varepsilon_{r_{m-1}}}, d+\sigma_{\varepsilon_{rm}}], [d+\sigma_{\varepsilon_{rm}}, 1]$. On the subinterval $[d, d+\sigma_{\varepsilon_{r_1}}]$ a uniform mesh of $N/2^{m+1}$ mesh intervals, on each subinterval $[d+\sigma_{\varepsilon_{r_k}}, d+\sigma_{\varepsilon_{r_{k+1}}}], 1 \leq k \leq m-1$, a uniform mesh of $N/2^{m-k+2}$ mesh intervals, and on $[d+\sigma_{\varepsilon_{rm}}, 1]$ a uniform mesh of $N/4$ mesh intervals are placed. Let $h_{\varepsilon_{l_1}}$ and $h_{\varepsilon_{r_1}}$ be the mesh lengths on $[0, \sigma_{\varepsilon_{l_1}}]$ and on $[d, d+\sigma_{\varepsilon_{r_1}}]$ respectively. Let H_1 and H_2 be the mesh lengths on $[\sigma_{\varepsilon_{lm}}, d]$ and on $[d+\sigma_{\varepsilon_{rm}}, 1]$ respectively; $h_{\varepsilon_{l_k}}$ and $h_{\varepsilon_{r_k}}$ be the mesh lengths on $[\sigma_{\varepsilon_{l_k}}, \sigma_{\varepsilon_{l_{k+1}}}]$ and on $[d+\sigma_{\varepsilon_{r_k}}, d+\sigma_{\varepsilon_{r_{k+1}}}], k = 2, \ldots, m$ respectively.

Define the discrete finite difference operator $\boldsymbol{L}^N$ as follows:

$$\boldsymbol{L}^N \boldsymbol{U} = \boldsymbol{f}, \quad \text{for all } x_i \in \Omega^N, \quad \boldsymbol{U}(x_0) = \boldsymbol{p}, \tag{8}$$

where

$$\boldsymbol{L}^N := \boldsymbol{E}\boldsymbol{D}^- + \boldsymbol{A},$$

and at $x_{N/2} = d$ the scheme is given by

$$\boldsymbol{L}^N \boldsymbol{U}\left(x_{\frac{N}{2}}\right) = \boldsymbol{E}\,\boldsymbol{D}^- \boldsymbol{U}\left(x_{\frac{N}{2}}\right) + \boldsymbol{A}\left(x_{\frac{N}{2}}\right)\boldsymbol{U}\left(x_{\frac{N}{2}}\right) = \boldsymbol{f}\left(x_{\frac{N}{2}-1}\right),$$

where

$$D^- Z(x_i) = \frac{Z(x_i) - Z(x_{i-1})}{h_i}, \quad i = 1, \ldots, N.$$

Lemma 4 *Suppose that a mesh function* $\boldsymbol{W}$ *satisfies* $\boldsymbol{W}_0 \geq \boldsymbol{0}$ *and* $\boldsymbol{L}^N \boldsymbol{W} \geq \boldsymbol{0}$, *for all* $x_i \in \Omega^N$, *implies that* $\boldsymbol{W} \geq \boldsymbol{0}$ *for all* $x_i \in \overline{\Omega}^N$.

Lemma 5 *If* $\boldsymbol{U}$ *be the numerical solution of (1) and (2), then,*

$$\|\boldsymbol{U}\|_{\overline{\Omega}^N} \leq \max\left\{\|\boldsymbol{U}(0)\|, \frac{1}{\alpha}\|\boldsymbol{f}\|_{\Omega_1^N \cup \Omega_2^N}\right\}.$$

4 Error Analysis

By a Taylor expansion on regular and singular components, we have

$$|\varepsilon_k\left(\frac{d}{dx} - D^-\right)v_k(x_i)| \leq C\varepsilon_k \frac{(x_i - x_{i-1})}{2}|v_k|_2 \leq CN^{-1}, \tag{9}$$

and

$$|\varepsilon_k\left(\frac{d}{dx} - D^-\right)w_k(x_i)| \leq \begin{cases} C\varepsilon_k \frac{(x_i - x_{i-1})}{2}|w_k|_2 & (10)\\ C\varepsilon_k \max\limits_{[x_{i-1}, x_i]}|w_k'|, & (11)\end{cases}$$

where $k = 1, \ldots, m, \;\; i \neq \frac{N}{2}, \quad | z_k |_j := \max | \frac{d^j z}{dx^j} |, \;\; \forall j \in \mathbb{N}$. Now to evaluate the error estimates for the singular components on different subintervals considered as follows:

Case (i) For $x_i \in [\sigma_{\varepsilon_{lm}}, d) \cup [d + \sigma_{\varepsilon_{rm}}, 1]$.

Using (11) and bounds on singular components, we have, for $j = 1, \ldots, m$.

$$\begin{aligned} | ((\boldsymbol{L}^N - \boldsymbol{L})\boldsymbol{w})_j(x_i) | &\leq C\varepsilon_j \sum_{q=j}^{m} \frac{B_{\varepsilon_{lq}}(x)}{\varepsilon_q} \\ &\leq C \parallel B_{\varepsilon_{lm}} \parallel_{[x_{i-1}, x_i]} = B_{\varepsilon_{lm}}(x_{i-1}) \leq CN^{-1}. \end{aligned}$$

Similar arguments prove a similar result for the subinterval $[d+\sigma_{\varepsilon_{rm}}, 1]$. Hence, for $x_i \in [\sigma_{\varepsilon_{lm}}, d) \cup [d+\sigma_{\varepsilon_{rm}}, 1]$ we have,

$$| ((\boldsymbol{L}^N - \boldsymbol{L})\boldsymbol{w})_j(x_i) | \leq CN^{-1}.$$

Case (ii) For $x_i \in (0, \sigma_{\varepsilon_{l_1}}] \cup (d, d+\sigma_{\varepsilon_{r_1}}]$.
Using (10) and the bounds on the singular components yields

$$| ((\boldsymbol{L}^N - \boldsymbol{L})\boldsymbol{w})_j(x_i) | \leq C(x_i - x_{i-1})\|\varepsilon_j w_j''\| \leq h_i \sum_{q=1}^{m} \frac{B_{\varepsilon_{l_q}}(x)}{\varepsilon_q} \leq C(N^{-1}\ln N).$$

Case (iii) For $x_i \in (\sigma_{\varepsilon_{l_k}}, \sigma_{\varepsilon_{l_{k+1}}}) \cup (d+\sigma_{\varepsilon_{r_k}}, d+\sigma_{\varepsilon_{r_{k+1}}})$, where $1 \leq k \leq m-1$.
Using the decomposition in Theorem 4 of singular components and bounds on singular components gives

$$| ((\boldsymbol{L}^N - \boldsymbol{L})\boldsymbol{w})_j(x_i) | = | \sum_{q=1}^{m-1} \varepsilon_j \left(\frac{d}{dx} - D^-\right) w_{j,\varepsilon_q}(x_i) + \varepsilon_j \left(\frac{d}{dx} - D^-\right) w_{j,\varepsilon_m}(x_i)|. \tag{12}$$

Consider the first part of (12) and using the bounds on singular components, we obtain

$$|\sum_{q=1}^{m-1} \varepsilon_j \left(\frac{d}{dx} - D^-\right) w_{j,\varepsilon_q}(x_i)| \leq \| \sum_{q=1}^{m-1} \varepsilon_j w'_{j,\varepsilon_q} \|_{[x_{i-1},x_i]} \leq CB_{\varepsilon_{l_{m-1}}}(x_{i-1}) \leq CN^{-1}.$$

Using the bounds on singular components for the second part of (12), we have

$$|\varepsilon_j \left(\frac{d}{dx} - D^-\right) w_{j,\varepsilon_m}(x_i)| \leq \frac{h_i}{2} \| \varepsilon_j w''_{j,\varepsilon_m} \| \leq C(N^{-1}\ln N).$$

Case (iv) For $x_i \in \{\sigma_{\varepsilon_{l_k}}, d+\sigma_{\varepsilon_{r_k}}\}$, where $1 \leq k \leq m-1$.
Using the decomposition of the singular components and bounds on singular components defined in Theorem 4 gives

$$| ((\boldsymbol{L}^N - \boldsymbol{L})\boldsymbol{w})_j(x_i) | \leq | \sum_{q=1}^{m-1} \varepsilon_j \left(\frac{d}{dx} - D^-\right) w_{j,\varepsilon_q}(x_i) + \varepsilon_j \left(\frac{d}{dx} - D^-\right) w_{j,\varepsilon_m}(x_i)|. \tag{13}$$

Consider the first part of (13) for the case $j \leq k$, and using the definition of point $x_{i,j}$ we have

$$|\sum_{q=1}^{m-1} \varepsilon_j \left(\frac{d}{dx} - D^-\right) w_{j,\varepsilon_q}(x_i)| \le \| \sum_{q=1}^{m-1} \varepsilon_j w'_{j,\varepsilon_q} \|_{[x_{i-1},x_i]} \quad \le CN^{-1}$$

and if, $j > k$, using the bounds on singular component and the analysis in Case (i), we have

$$|\sum_{q=1}^{m-1} \varepsilon_j \left(\frac{d}{dx} - D^-\right) w_{j,\varepsilon_q}(x_i)| \le \| \sum_{q=1}^{m-1} \varepsilon_j w'_{j,\varepsilon_q} \|_{[x_{i-1},x_i]} \le CN^{-1}.$$

For the second part of (13), use bounds on singular components defined in Theorem 4, to obtain

$$|\varepsilon_j \left(\frac{d}{dx} - D^-\right) w_{j,\varepsilon_m}(x_i)| \le C\varepsilon_j h_{\varepsilon_k} \parallel w''_{j,\varepsilon_m} \parallel \le CN^{-1} \ln N.$$

Now at the point $x_{N/2} = d$,

$$\begin{aligned}
(\boldsymbol{L}^N(\boldsymbol{U} - \boldsymbol{u}))_1(d) &= f_1(d - H_1) - \varepsilon_1 D^- u_1(d) - \sum_{j=1}^{m} a_{1j}(d) u_j(d), \\
&= f_1(d - H_1) - \frac{\varepsilon_1}{H_1}(u_1(d) - u_1(d - H_1)) - \sum_{j=1}^{m} a_{1j}(d) u_j(d), \\
&= f_1(d - H_1) - \frac{\varepsilon_1}{H_1} \int_{t=d-H_1}^{d} u_1'(s)\, ds - \sum_{j=1}^{m} a_{1j}(d) u_j(d).
\end{aligned}$$

Using the bounds on derivatives of $\boldsymbol{u}$, we obtain

$$|(\boldsymbol{L}^N(\boldsymbol{U} - \boldsymbol{u}))_1(d))| \;\le\; C(N^{-1} \ln \mathrm{N}),$$

and similarly, we can prove for $2 \le j \le m$

$$|(\boldsymbol{L}^N(\boldsymbol{U} - \boldsymbol{u}))_j(d))| \;\le\; C(N^{-1} \ln \mathrm{N}).$$

We conclude this section with the following main result which follows by using the error analysis for the regular and singular components, and the discrete maximum principle.

Theorem 5 *Let* $\boldsymbol{u}$ *be the solution of given problem (1) and (2) and* $\boldsymbol{U}$ *be the solution of discrete problem, then*

$$\|\boldsymbol{U} - \boldsymbol{u}\|_{\overline{\Omega}^N} \le C(N^{-1} \ln N).$$

5 Numerical Results

Example 1 Consider the following singularly perturbed initial value problem with discontinuous source term (Table 1):

$$\begin{aligned}
\varepsilon_1 u_1'(x) + 4u_1(x) - u_2(x) - u_3(x) &= f_1(x), \quad x \in \Omega_1 \cup \Omega_2 \\
\varepsilon_2 u_2'(x) - u_1(x) + (4+x)u_2(x) - u_3(x) &= f_2(x), \quad x \in \Omega_1 \cup \Omega_2 \\
\varepsilon_3 u_3'(x) - 2u_1(x) - u_2(x) + (5+x)u_3(x) &= f_3(x), \quad x \in \Omega_1 \cup \Omega_2 \\
u_1(0) = 0, \quad u_2(0) = 0, \quad u_3(0) &= 0,
\end{aligned}$$

where

$$f_1(x) = \begin{cases} x & \text{for } 0 \le x \le 0.5 \\ 2 & \text{for } 0.5 < x \le 1, \end{cases} \qquad f_2(x) = \begin{cases} 1 & \text{for } 0 \le x \le 0.5 \\ 4 & \text{for } 0.5 < x \le 1, \end{cases} \quad \text{and}$$

$$f_3(x) = \begin{cases} 1 + x^2 & \text{for } 0 \le x \le 0.5 \\ 3 & \text{for } 0.5 < x \le 1. \end{cases}$$

The exact solution of the test example is not known. Therefore, we estimate the error for $\boldsymbol{U}$ by comparing it to the numerical solution $\widetilde{\boldsymbol{U}}$ obtained on the mesh $\widetilde{x}_j$ that contains the mesh points of the original mesh and their midpoints, that is, $\widetilde{x}_{2j} = x_j,\ j = 0, \ldots, N,\ \widetilde{x}_{2j+1} = (x_j + x_{j+1})/2, \quad j = 0, \ldots, N-1$.
For different values of N and $\varepsilon_1, \varepsilon_2, \varepsilon_3$, we compute

$$D^N_{\varepsilon_1,\varepsilon_2,\varepsilon_3} := \|(\boldsymbol{U} - \widetilde{\boldsymbol{U}})(x_i)\|_{\overline{\Omega}^N}.$$

Table 1 Maximum point-wise errors D^N, $D^N_{\varepsilon_1,\varepsilon_2,\varepsilon_3}$ with $\varepsilon_2 = 10^{-5}$ and $\varepsilon_3 = 10^{-3}$ for Example 1

$\varepsilon_1 = 10^{-j}$	$N = 64$	$N = 128$	$N = 256$	$N = 512$	$N = 1024$	$N = 2048$
$j = 5$	4.94E-02	3.34E-02	2.14E-02	1.29E-02	7.40E-03	4.13e-03
6	4.93E-02	3.34E-02	2.14E-02	1.29E-02	7.40E-03	4.13e-03
7	6.41E-02	4.89E-02	3.30E-02	2.12E-02	1.27E-02	7.29e-03
8	6.93E-02	5.37E-02	3.68E-02	2.40E-02	1.46E-02	8.49e-03
9	6.98E-02	5.42E-02	3.72E-02	2.43E-02	1.48E-02	8.62e-03
10	6.99E-02	5.43E-02	3.72E-02	2.43E-02	1.48E-02	8.64e-03
⋮	⋮	⋮	⋮	⋮	⋮	⋮
15	6.99E-02	5.43E-02	3.72E-02	2.43E-03	1.48E-02	8.64e-03
D^N	6.99E-02	5.43E-02	3.72E-02	2.43E-03	1.48E-02	8.64e-03
p^N	4.69E-01	6.74E-01	7.39E-01	8.42E-01	9.01E-01	

Then the parameter-uniform error is computed as

$$D^N := \max_{S_{\varepsilon_1,\varepsilon_2,\varepsilon_3}} \{D^N_{\varepsilon_1,\varepsilon_2,\varepsilon_3}\},$$

where the singular perturbation parameters take values in the set $S_{\varepsilon_1,\varepsilon_2,\varepsilon_3} = \{(\varepsilon_1, \varepsilon_2, \varepsilon_3)|, \varepsilon_1 = 10^{-j}, 0 \le j \le 15, \varepsilon_2 = 10^{-l}, 0 \le l \le j, \varepsilon_3 = 10^{-k}, 0 \le k \le l\}$ and the order of convergence is calculated using the formula

$$p^N = \frac{\ln(D^N) - \ln(D^{2N})}{\ln(2\ln N) - \ln(\ln(2N))}.$$

References

1. Cen, Z., Xu, A., Le, A.: A second-order hybrid finite difference scheme for a system of singularly perturbed initial value problems. J. Comput. Appl. Math. **234**, 3445–3457 (2010)
2. Dunne, R.K., Riordan, E.O'.: Interior layers arising in linear singularly perturbed differential equations with discontinuous coefficients. In: Proceedings of the Fourth International Conference on Finite Difference Methods: Theory and Applications. Lozenetz, Bulgaria, 26–29 August (2006). Rousse University, Bulgaria, 29–38 (2007)
3. Hemavathi, S., Bhuvaneswari, T., Valarmathi, S., Miller, J.J.H.: A parameter uniform numerical method for a system of singularly perturbed ordinary differential equations. Appl. Math. Comput. **191**, 1–11 (2007)
4. Meenakshi, P.M., Valarmathi, S., Miller, J.J.H.: Solving a partially singularly perturbed initial value problem on shishkin meshes. Appl. Math. Comput. **215**, 3170–3180 (2010)
5. Rao, S.C.S., Kumar, S.: Second order global uniformly convergent numerical method for a coupled system of singularly perturbed initial value problems. Appl. Math. Comput. **219**, 3740–3753 (2012)
6. Rao, S.C.S., Chawla, S.: Interior layers in coupled system of two singularly perturbed reaction-diffusion equations with discontinuous source term. NAA 2012, LNCS 8236, 445–453 (2012)
7. Roos, H.G., Stynes, M., Tobiska, L.: Robust numerical methods for singularly perturbed differential equations. Springer Series in Computational Mathematics, 2nd edn. Springer, Berlin (2008)
8. Shishkin, G.I.: Discrete approximation of singularly perturbed elliptic and parabolic equations. Russian Academy of Sciences, Ural Section, Ekaterinburg (in Russian)
9. Valarmathi, S., Miller, J.J.H.: A parameter-uniform finite difference method for singularly perturbed linear dynamical systems. Int. J. Numer. Anal. Model. **7**, 535–548 (2010)